MEMS: A Practical Guide to Design, Analysis, and Applications

Edited by

Jan G. Korvink
Institute for Microsystem Technology IMTEK
University of Freiburg
Freiburg, Germany

and

Oliver Paul
Institute for Microsystem Technology IMTEK
University of Freiburg
Freiburg, Germany

William Andrew Publishing

Norwich, NY, U.S.A.

Springer

Copyright © 2006 by William Andrew, Inc.
No part of this book may be reproduced or utilized in any form or by any means, electronic or mechanical, including photocopying, recording or by any information storage and retrieval system, without permission in writing from the Publisher.

Cover art © 2006 by Brent Beckley / William Andrew, Inc.

ISBN: 0-8155-1497-2 (William Andrew, Inc.)
ISBN: 3-540-21117-9 (Springer-Verlag GmbH & Co. KG)

Library or Congress Catalog Card Number: 2005023492
Library of Congress Cataloging-in-Publication Data

MEMS : a practical guide to design, analysis, and applications / edited by Jan G. Korvink and Oliver Paul.
  p. cm.
  Includes bibliographical references and index.
  ISBN 0-8155-1497-2 (alk. paper)
  1. Microelectromechanical systems.   I. Korvink, J. G. (Jan G.)
II. Paul, Oliver.
TK7875.M42 2005
621—dc22

2005023492

Printed in the United States of America
This book is printed on acid-free paper.

10 9 8 7 6 5 4 3 2 1

Co-published by:

William Andrew, Inc.
13 Eaton Avenue
Norwich, NY 13815
1-800-932-7045
www.williamandrew.com
www.knovel.com
(Orders from all locations in North and South America)

Springer-Verlag GmbH & Co. KG
Tiergartenstrasse 17
D-69129 Heidelberg, Germany
www.springeronline.com
(Orders from all locations outside North and South America)

NOTICE

To the best of our knowledge the information in this publication is accurate; however the Publisher does not assume any responsibility or liability for the accuracy or completeness of, or consequences arising from, such information. This book is intended for informational purposes only. Mention of trade names or commercial products does not constitute endorsement or recommendation for their use by the Publisher. Final determination of the suitability of any information or product for any use, and the manner of that use, is the sole responsibility of the user. Anyone intending to rely upon any recommendation of materials or procedures mentioned in this publication should be independently satisfied as to such suitability, and must meet all applicable safety and health standards.

Dedication

*To the memory of our fathers,
Gerrit Jörgen Korvink and Marius Paul.*

Contents

Foreword xvii
Preface xix

1 Microtransducer Operation 1
Oliver Paul

1.1 Introduction 1
1.2 Transduction 2
    1.2.1 Signal Domains 2
    1.2.2 Block Schematics of Transducers 6
    1.2.3 Transduction Effects 12
1.3 Microsystem Performance 13
    1.3.1 Figures of Merit 13
    1.3.2 Sensitivity, Selectivity, and Offset 18
    1.3.3 Noise 21
1.4 Transducer Operation Techniques 26
    1.4.1 Calibration 26
    1.4.2 Compensation 28
    1.4.3 Stabilization 36
    1.4.4 Multiple Measurements 38
    1.4.5 Circuitry 42
1.5 Powering Microsystems 44
    1.5.1 Local Energy Storage 44
    1.5.2 Miniaturized Fuel Cells 45
    1.5.3 Optical and Electromagnetic Energy Transmission 46
    1.5.4 Energy Harvesting 47
References 48

2 Material Properties: Measurement and Data 53
Osamu Tabata and Toshiyuki Tsuchiya

2.1 Introduction 53
2.2 Measurement Methods 55
    2.2.1 Internal Stress (σ) 55
    2.2.2 Young's Modulus (E) 58
    2.2.3 Poisson's Ratio 61
    2.2.4 Yield Strength and Fracture Strength 62
    2.2.5 Fracture Toughness 68

2.2.6 Fatigue 69
　　2.2.7 Thermal Conductivity and Specific Heat 70
　2.3 Data 74
　　2.3.1 Si 74
　　2.3.2 Poly-Si 75
　　2.3.3 Metal 82
　　2.3.4 Dielectrics 84
　References 87

3 MEMS and NEMS Simulation 93
Jan G. Korvink, Evgenii B. Rudnyi, Andreas Greiner, and Zhenyu Liu

　3.1 Introduction 93
　3.2 Simulation Scenario 93
　3.3 Generic Organization of a Computational Tool 101
　　3.3.1 Graphical User Interface or Front End 101
　　3.3.2 Input Files and Parsing 102
　　3.3.3 Preprocessing 102
　　3.3.4 Solving 103
　　3.3.5 Post-Processing and Program Interfacing 109
　3.4 Methods for Materials Simulation 111
　3.5 Computational Methods that Solve PDEs 122
　3.6 Design Automation Methods 138
　3.8 Case Studies 159
　3.9 Summary 180
　3.10 Acknowledgments 182
　References 182

4 System-Level Simulation of Microsystems 187
Gary K. Fedder

　4.1 Introduction 187
　4.2 Behavioral Modeling of MEMS Components 193
　　4.2.1 Micromechanical Plates 193
　　4.2.2 Micromechanical Flexures 195
　　4.2.3 Electrostatic Gaps 199
　　4.2.4 Reduced-Order Modeling 205
　4.3 Formulation of Equations of Motion 209
　4.4 Structured Design Tools 215
　　4.4.1 Signal-Flow Simulations 216
　　4.4.2 Conservative Network Simulations 216

	4.4.3 Analog Hardware Description Languages	218
	4.4.4 A Structured MEMS Methodology	220
4.5	Conclusions	224
References		224

5 Thermal-Based Microsensors ... 229
Friedemann Vöelklein

5.1	Introduction	229
5.2	Thermoresistors	230
	5.2.1 Metal Film Thermoresistors	230
	5.2.2 Semiconducting Thermoresistors	232
	5.2.3 Silicon Spreading Resistance Temperature Sensor	234
	5.2.4 Thermoresistors for the Detection of Thermal Radiation	235
	5.2.5 Pellistors	238
5.3	Silicon Diodes and Transistors as Thermal Microsensors	240
5.4	Thermoelectric Microsensors	243
	5.4.1 Microthermopiles as IR Radiation Detectors	250
	5.4.2 Thermopile Arrays	251
	5.4.3 Thermoelectric Vacuum Microsensors	255
	5.4.4 Gas Flow Microsensors	257
	5.4.5 AC/DC Thermoconverter	258
	5.4.6 Heat Flux Sensors	259
	5.4.7 Microelectromechanical Thermoelectric Cooler	262
5.5	CMOS-Compatible Thermal-Based Microsensors and Microactuators	266
5.6	Diagnostic Thermal-Based Microstructures	273
	5.6.1 Thermoelectric Microtips for AFM Temperature Sensors	273
	5.6.2 Diagnostic Microstructures for the Investigation of Thermal Properties of Thin Films	275
5.7	Conclusion	276
References		276

6 Photon Detectors ... 281
Arokia Nathan and Karim S. Karim

6.1	Introduction	281
6.2	Detectors	284
	6.2.1 Mo/a-Si:H Schottky Diode X-Ray Image Sensors	284
	6.2.2 ITO/a-Si:H Schottky Diode Optical Image Sensors	289
	6.2.3 ITO/p-i-n Optical Image Sensors	293

6.3 Thin-Film Transistors 296
    6.3.1 TFT Structures and Operation 296
    6.3.2 Threshold Voltage (VT) Metastability 299
6.4 Pixel Integration 308
6.5 Imaging Arrays 314
    6.5.1 Conventional Passive Pixel Sensor Arrays 314
    6.5.2 Amorphous Silicon Current-Mediated Active
          Pixel Sensor Arrays 320
    6.5.3 Amorphous Silicon Voltage-Mediated Active
          Pixel Sensor Arrays 324
    6.5.4 Integrated Amorphous Silicon Multiplexers for
          Imaging Arrays 331
6.6 New Challenges in Large-Area Digital Imaging 334
References 338

7 Free-Space Optical MEMS 345
Ming C. Wu and Pamela R. Patterson

7.1 Introduction 345
7.2 General Discussion of Micromirror Scanners 346
7.3 Electrostatic Scanners 349
    7.3.1 Scanners with Electrostatic Parallel-Plate Actuators 351
    7.3.2 Electrostatic Vertical Comb-Drive Scanners 356
7.4 Scanning Mirrors with Magnetic and
    Electromagnetic Actuators 361
7.5 Micromirror Arrays with Mirror Size ≤100 Micrometers 364
    7.5.1 Micromirrors for Dynamic Spectral Equalizers 370
7.6 2-D MEMS Optical Switches 370
    7.6.1 Switch Configuration, Requirements, and
          Expendability 370
    7.6.2 Vertical Chopper-Type Switch 376
    7.6.3 Switch with Pop-Up Mirrors 380
7.7 2 × 2 Switches 383
7.8 Optical Attenuator Array 386
7.9 Tunable WDM Devices 388
    7.9.1 Tunable Filters 388
    7.9.2 Tunable Lasers and Detectors 390
7.10 Diffractive Optical MEMS 390
7.11 Summary 394
Acknowledgment 394
References 394

8 Integrated Micro-Optics ... 403
Hans Zappe

- 8.1 Introduction ... 403
  - 8.1.1 Definitions ... 403
  - 8.1.2 Components ... 403
  - 8.1.3 Summary ... 404
- 8.2 Guided Waves ... 405
  - 8.2.1 Reflections at Boundaries ... 405
  - 8.2.2 Ray-Optic Model ... 409
  - 8.2.3 Modes and Propagation ... 413
  - 8.2.4 Electromagnetic Model ... 416
  - 8.2.5 Confinement Factor ... 418
  - 8.2.6 Solving a Waveguide ... 420
- 8.3 Stripe Waveguides ... 421
  - 8.3.1 Stripe Waveguide Structures ... 422
  - 8.3.2 Stripe Waveguide Modeling ... 426
- 8.4 Input/Output Coupling ... 429
  - 8.4.1 End-Fire Coupling ... 430
  - 8.4.2 Butt Coupling ... 431
  - 8.4.3 Numerical Aperture ... 432
  - 8.4.4 Tapers ... 433
- 8.5 Waveguide Characterization ... 434
  - 8.5.1 Modes ... 434
  - 8.5.2 Losses ... 435
- 8.6 Integrated Optical Devices ... 438
  - 8.6.1 Couplers ... 438
  - 8.6.2 Interferometers ... 441
  - 8.6.3 Active Optical Devices ... 444
- 8.7 Materials ... 445
  - 8.7.1 Silicon ... 446
  - 8.7.2 GaAs ... 446
  - 8.7.3 Glass ... 447
  - 8.7.4 Plastics ... 448
- 8.8 Applications ... 448
  - 8.8.1 Application Example: Monolithic Displacement Sensors ... 448
- References ... 450

9 Microsensors for Magnetic Fields ... 453
Chavdar Roumenin

- 9.1 Introduction ... 453

9.2 Magnetic Fields for Different Applications 453
    9.2.1 Methods for Sensing and Applications of
           Magnetic Fields 454
    9.2.2 (Micro) Sensors for a Magnetic Field 456
9.3 Main Figures of Merit of Magnetic Microsensors 457
    9.3.1 Classification of Magnetic Sensors: Figures of Merit ... 457
    9.3.2 Characteristics Related to OUT$(B)_C$ 457
    9.3.3 Characteristics Related to OUT$(C)_B$ 462
    9.3.4 Characteristics Related to the SD 463
9.4 Hall Microsensors 463
    9.4.1 The Lorentz Force 463
    9.4.2 Hall Effect 464
    9.4.3 Hall Effect as Sensor Action 464
       9.4.4 Hall Voltage Mode of Operation 466
       9.4.5 Hall Current Mode of Operation 470
       9.4.6 Diode Hall Effect 471
       9.4.7 Hall Effect Devices 471
9.5 Magnetoresistors 478
    9.5.1 Physical Magnetoresistance Effect 481
    9.5.2 Geometrical Magnetoresistance Effect 481
    9.5.3 Semiconductor Magnetoresistors 482
    9.5.4 Spin-Dependent Magnetoresistance 483
    9.5.5 GMR Sensors 484
9.6 Magnetodiodes 484
    9.6.1 Magnetoconcentration and Magnetodiode Effects 484
    9.6.2 Magnetodiode Microsensors 485
9.7 Magnetotransistors and Related Microsensors 488
    9.7.1 General Approach to Bipolar Magnetotransistor
           Design 488
    9.7.2 Principles of BMT Operation 490
    9.7.3 BMT Microsensors 491
    9.7.4 Sensors Related to the BMTs 494
9.8 Magnetic Field-based Functional Multisensors 495
    9.8.1 Functional Approach to Multisensors 495
    9.8.2 Linear Multisensors for Magnetic Field and
           Temperature 495
    9.8.3 Linear Multisensor for Magnetic Field,
           Temperature, and Light 498
    9.8.4 Functional Gradiometer Microsensors 499
    9.8.5 2-D and 3-D Vector Microsystems for
           Magnetic Fields 500
9.9 Interfaces and Improvement of Characteristics of
    Magnetic Microsensors 502

		9.9.1 Biasing Circuits and Signal Processing Electronics ...	502
		9.9.2 Improvement of Magnetosensor Characteristics	507
		9.9.3 Magnetic Systems for Contactless Measurements	512
	9.10	Conclusions and Outlook	514
References			516

10 Mechanical Microsensors **523**
Franz Laermer

	10.1	Introduction	523
		10.1.1 Automotive	524
		10.1.2 Computers and Peripherals	525
		10.1.3 Consumer Products	525
		10.1.4 Medical and Biological Applications	526
	10.2	Inertial Sensors	527
		10.2.1 Accelerometers	528
		10.2.2 Yaw-Rate Sensors	539
	10.3	Pressure Sensors	550
		10.3.1 Fundamentals	550
		10.3.2 Bulk-Micromachined Pressure Sensors	551
		10.3.3 Surface-Micromachined Pressure Sensors	553
		10.3.4 Signal Generation	554
	10.4	Force and Torque Sensors	560
		10.4.1 Linking the Macro World to the Micro World	561
		10.4.2 Fabrication, Protection, Test, and Calibration	561
		10.4.3 Conclusions	563
References			563

11 Semiconductor-Based Chemical Microsensors **567**
Andreas Hierlemann and Henry Baltes

	11.1	Introduction	567
	11.2	Thermodynamics of Chemical Sensing	574
	11.3	Chemomechanical Sensors	580
		11.3.1 Rayleigh SAW Devices	583
		11.3.2 Flexural-Plate-Wave or Lamb-Wave Devices	586
		11.3.3 Resonating Cantilevers	589
	11.4	Thermal Sensors	591
		11.4.1 Catalytic Thermal Sensors (Pellistors)	592
		11.4.2 Thermoelectric or Seebeck-effect Sensors	595
	11.5	Optical Sensors	598
		11.5.1 Integrated Optics	603
		11.5.2 Microspectrometers	608
		11.5.3 Bioluminescent Bioreporter Integrated Circuits	611

  11.5.4 Surface Plasmon Resonance Devices 613
11.6 Electrochemical Sensors 615
  11.6.1 Voltammetric Sensors 617
  11.6.2 Potentiometric Sensors 622
  11.6.3 Conductometric Sensors 636
  11.6.4 Combinations of Electrochemical Principles 646
Acknowledgments 648
References 648

12 Microfluidics 667
Jens Ducrée, Peter Koltay, and Roland Zengerle

12.1 Introduction 667
12.2 Properties of Fluids 670
  12.2.1 Volumes and Length Scales 670
  12.2.2 Mixtures 671
  12.2.3 Physical Properties 672
  12.2.4 Vapor Pressure 673
  12.2.5 Surface Tension 673
  12.2.6 Electrical Properties 674
  12.2.7 Optical Properties 674
  12.2.8 Transport Phenomena 675
12.3 Physics of Microfluidic Systems 678
  12.3.1 Navier-Stokes Equations 678
  12.3.2 Laminar Flow 679
  12.3.3 Dynamic Pressure 680
  12.3.4 Fluidic Networks 682
  12.3.5 Heat Transfer 683
  12.3.6 Interfacial Surface Tension 685
  12.3.7 Electrokinetics 686
12.4 Fabrication Technologies 689
  12.4.1 Silicon 690
  12.4.2 Plastics 692
  12.4.3 Quartz 694
  12.4.4 Glass 695
12.5 Flow Control 696
  12.5.1 Check Valves 697
  12.5.2 Capillary Breaks 698
  12.5.3 Active Microvalves 699
12.6 Micropumps 702
  12.6.1 Microdisplacement Pumps 702
  12.6.2 Charge-Induced Pumping Mechanisms 703
  12.6.3 Other Pumping Mechanisms 703

12.7 Sensors 703
    12.7.1 Flow Sensors 704
    12.7.2 Chemical Sensors 706
12.8 Pipettes and Dispensers 707
    12.8.1 Pipettes 707
    12.8.2 Dispensers 708
12.9 Microarrays 709
    12.9.1 Concept 709
    12.9.2 Fabrication 710
    12.9.3 Particle-Based Microarray Concepts 712
12.10 Microreactors 713
    12.10.1 Micromixers 713
    12.10.2 Heat Exchangers 714
    12.10.3 Chemical Reactors 715
12.11 Microanalytical Chips 715
    12.11.1 Lab-on-a-Chip Systems 715
    12.11.2 Chip-Based Capillary Electrophoresis 716
References 717

13 Biomedical Systems 729
Whye-Kei Lye and Michael Reed

13.1 Introduction and Overview 729
13.2 Materials and Fabrication Techniques 730
    13.2.1 Material Requirements 730
    13.2.2 Fabrication Techniques 733
13.3 Surgical Systems 735
    13.3.1 Sensors 738
    13.3.2 Motion Control 738
    13.3.3 Microinstruments 739
13.4 Tissue Repair 739
13.5 Therapeutic Systems 742
    13.5.1 Implantable Delivery Systems 743
    13.5.2 Mechanical Delivery Systems 744
13.6 Summary 745
References 746

14 Microactuators 751
Jack W. Judy

14.1 Introduction 751
14.2 Actuators: Transducers with Mechanical Output 752
    14.2.1 Transduction Mechanisms 752
    14.2.2 Scaling Advantages and Issues 753

	14.2.3 Electrical Microactuators	754
14.3	Electrostatic Forces	755
	14.3.1 Electrostatic Systems	755
	14.3.2 Forces in Electrostatic Systems	759
	14.3.3 Scaling Properties	760
14.4	Electrostatic Microactuator Configurations	765
	14.4.1 Gap-Closing Electrostatic Microactuators	766
	14.4.2 Examples of Gap-Closing Electrostatic Microactuators	770
	14.4.3 Constant-Gap Electrostatic Microactuators	778
	14.4.4 Examples of Constant-Gap Electrostatic Microactuators	780
	14.4.5 Hybrid Electrostatic Microactuators	785
	14.4.6 Electrostatic Induction	786
	14.4.7 Issues and Challenges	787
14.5	Piezoelectric Microactuators	787
	14.5.1 Piezoelectric Energy Density	789
	14.5.2 Piezoelectric Microactuator Configurations	790
	14.5.3 Piezoelectric Microactuator Design Issues	795
14.6	Electrostriction, Electrets, and Electrorheological Fluids	797
References		797

15 Micromachining Technology ... 805
Paddy J. French and Pasqualina M. Sarro

15.1	Introduction	805
15.2	Bulk Micromachining	805
	15.2.1 Wet Etching	806
	15.2.2 High-Aspect-Ratio Micromachining	816
15.3	Surface Micromachining	824
	15.3.1 Basic Process Sequence	824
	15.3.2 Materials and Etching	825
15.4	Epi-Micromachining	829
	15.4.1 SIMPLE	829
	15.4.2 SCREAM	830
	15.4.3 Black Silicon	831
	15.4.4 MELO	832
	15.4.5 Porous Silicon	833
	15.4.6 SIMOX	834
	15.4.7 Epi-Poly	835
	15.4.8 Release and Stiction	836
15.5	IC Compatibility Issues	837
	15.5.1 Compatible Bulk Micromachining	837

15.5.2 Compatible Surface Micromachining 840
15.5.3 Compatible Epi-Micromachining 844
15.6 Conclusions 844
References 845

16 LIGA Technology for R&D and Industrial Applications 853
Ulrike Wallrabe and Volker Saile

16.1 Introduction 853
16.2 The LIGA Process 854
    16.2.1 Mask Making 857
    16.2.2 Deep X-Ray Lithography 859
    16.2.3 Electroplating and Micromolding 862
    16.2.4 Sacrificial Layer Technique 865
    16.2.5 UV-LIGA Based on UV Lithography 867
16.3 Application in Modular Micro-Optical Systems 867
    16.3.1 Definition of a Modular Micro-Optical System 867
    16.3.2 Multifiber Connector from Polymer 869
    16.3.3 Heterodyne Receiver 871
    16.3.4 Spectrometer 874
    16.3.5 Distance Sensor 874
    16.3.6 Optical Cross-Connect with Rotating Mirrors 876
    16.3.7 Oscillating Modulator for Infrared Light 876
    16.3.8 Laser Scanner for Barcode Reading Actuated by Electromagnetics 879
    16.3.9 FTIR Spectrometer for Infrared Light 881
    16.3.10 Ultra-High X-Ray Lenses in SU8 884
16.4 Mechanical Applications 885
    16.4.1 Cycloid Gear System 885
    16.4.2 LIGA Gyroscope 889
    16.4.3 Microturbines for Cardiac Catheters 891
    16.4.4 Watch Pieces Made by UV-LIGA 891
16.5 Outlook 895
Acknowledgments 896
References 897

17 Interface Circuitry and Microsystems 901
Piero Malcovati and Franco Maloberti

17.1 Introduction 901
17.2 Microsensor Systems 902
17.3 Microsensor System Applications 905
    17.3.1 Automotive Sensors 907

17.3.2 Biomedical Sensors 908
17.3.3 Sensors for Household Appliances, Building
       Control, and Industrial Control 908
17.3.4 Environmental Sensors 909
17.4 Interface Circuit Architecture 909
17.4.1 Requirements and Specifications 910
17.5 Analog Front-End 912
17.5.1 Voltage Output 912
17.5.2 Current or Charge Output 916
17.5.3 Impedance Variation 918
17.6 A/D Converter 921
17.7 Digital Processing and Output Interface 931
17.7.1 Digital Signal Processing 931
17.7.2 Wired Output Interfaces 931
17.7.3 Wireless Output Interfaces 933
17.8 Conclusions 934
References 934

Contributors 943

Index 945

Foreword

MEMS are rapidly moving from the research laboratory to the marketplace. Many market studies indicate not only a tremendous market potential of MEMS devices; year by year we see the actual market grow as the technology matures. In fact, these days, many large silicon foundries have a MEMS group exploring this promising technology, including such giants as INTEL and Motorola.

Yet MEMS are fundamentally different from microelectronics. This means that companies with an established track record in these branches need to adapt their skills, whereas companies that want to enter the "miniaturization" market need to establish an entirely new set of capabilities. The same can be said of engineers with classical training, who will also need to be educated toward their future professional activity in the MEMS field.

Here are some questions that a company or technologist may ask:

I have an existing product with miniaturization market potential. Which technology should I adopt?

What are the manufacturing options available for miniaturization? What are the qualitative differences?

How do we maintain a market lead for products based on MEMS? Is there CAD support? Can we outsource manufacturing?

Which skills in our current capability need only adaptation? What skills need to be added?

Professors Jan Korvink and Oliver Paul have set out to answer these questions in a form that addresses the needs of companies, commercial practitioners, and technologists. They have collected together a set of world leaders in each of the areas that they have identified as significant for MEMS-based production. The experts have written chapters that lead the reader through the specialized knowledge of their field and guide them through the literature they may want to consult for further reading.

Microtechnology and, close on its heels, nanotechnology, are set to change many manufacturing and product paradigms. To stay ahead in this competitive world, we need to assess and reassess our options and establish products that have unique value that helps them stand apart from the rest. Microtechnology is one way to go about this, for it brings along small size, low power consumption, low per-unit manufacturing costs in a mass market, low environmental impact for discardable units, and high sophistication when combined with embedded systems. Professors Korvink and Paul have done an excellent job in providing a handbook that will help you to maintain your competitive advantage now and into the future.

Wolfgang Menz
Professor Emeritus
The Albert Ludwig University
Freiburg in Breisgau, Germany

Preface

MEMS, or microelectromechanical systems, claim to be the smallest functional machines that are currently engineered by humans. MEMS is an exciting field with rapidly growing commercial importance. When the field started, it was considered highly speculative, but early successes made researchers bold. Their enthusiasm spread to venture capitalists and eventually resulted in a range of commercially available and viable products. Certainly, MEMS technology is not as established as, for example, microelectronics, but every year shows growth in the commercial application of the technology. Consequently, many companies are under competitive pressure to evaluate whether or not MEMS technology has advantages for their own products. And as soon as this question is posed to the engineers of a company, it is our hope that this book will help them to formulate an informed opinion by filling the growing need for a practical collection of information that supports the product development engineer.

This book aims to provide workers in industry with access to comprehensive resources on MEMS devices, systems, manufacturing technologies, and design methodologies. It addresses the rapid evaluation of questions such as: *What is out there? What works? What is still speculative?* We believe that the decision to *design* implies having an application and a market potential, which is followed by selecting solutions (devices, systems, electronics, packaging), then selecting technologies, then selecting design support tools, and finally making business decisions. This book aims to help newcomers to the field ramp up their technology in the shortest time possible by making accessible the views of experts in the respective fields of application.

A central dilemma and at the same time one of the best features of MEMS is its incredibly wide range of applications. As this book has progressed through the stages of production, many MEMS paradigm shifts have occurred *out there*. These have had the effect of both proving the previous point of diversity of applications and of increasing the necessity of having a good starting resource. We think here of the importance of RF-MEMS for mobile telecommunications, micro-optics for internet hardware, and the rapidly growing importance of MEMS in the life sciences as miniaturized laboratories, catheter-based minimally invasive operating theater tools, and in applications for the *in vivo* monitoring of chemical levels paired by exact dosage of medication in ailing patients.

Whereas we cannot hope to foresee how MEMS applications will develop and diversify, we do believe that by looking at existing technologies and applications the experienced engineer can quickly extrapolate toward their own application area and speed up the evolutionary path to expertise. This book provides data for selecting solutions (devices, systems, electronics, packaging), selecting technologies, and selecting design support tools. It also provides rapid access to literature for additional details, as your design evaluation focuses and your information needs change. For this purpose, we aim to answer some central questions that a starting project may have:

Q. *What can I do with MEMS?*
A. MEMS allow you to build sensors and actuators, together with measurement, control, and signal conditioning circuitry, and equipped with power and communications, all in the tiniest space. This enables you to be more accurate, to manufacture more cost-effectively, to achieve autonomy of a larger system, and so on.

Chapter 1, Microtransducer Operation, by Oliver Paul, guides you through the various possibilities and helps to structure systems thinking about MEMS. Here you will find the big picture, with information on which effects are available, by which equations they are described, and how the individual chapters specialize the ideas.

Q. *How do I design for better MEMS products?*
A. As we progress through the evolution of a product, it becomes clear that the optimal operation of a device, or a system, will depend on too many parameters and that a more organized approach to design becomes necessary. Naturally we first think of CAD systems, and the questions to answer here are: *What is available? What are the capabilities? What is the best organization for tools and teams?* Three chapters deal with this very important area.

In Chapter 3, MEMS and NEMS Simulation, Jan G. Korvink, Evgenii B. Rudnyi, Andreas Greiner, and Zhenyu Liu, provide a comprehensive overview of simulation tools, techniques, and approaches, covering both microdevices and nanodevices. Many of the modeling tools are highly specific to MEMS because of the relative scaling of physical effects at decreasing dimensions, but also because of the special manufacturing processes. Here you will learn which tools are available for which effect and how to approach modeling, from the theory all the way to how a simulation team should be organized.

In Chapter 4, System-Level Simulation of Microsystems, Gary K. Fedder shows how the overall behavior of a microsystem can be predicted by relaxing the level of detail of a device simulation, in the guise of a compact model, without necessarily losing any predictive accuracy, thus gaining a kind of SPICE for MEMS.

These techniques are certainly important components of a design suite, but without accurate material property data, it is impossible to plan a design. For MEMS this often means creating special measurement devices or devising special measurement techniques, as can be found in Chapter 2, Material Properties: Measurement and Data, by Osamu Tabata and Toshiyuki Tsuchiya.

Q. *What devices and application areas are enabled by basing a design on MEMS?*

A. A vast variety of devices and applications are covered; the list is continuously growing. The currently most important areas are covered by individual chapters. Working through the chapters, you will see how experts have explored the possibilities offered by the technology and brought out the best of each new idea to achieve devices that not only emulate their macroscopic counterparts but greatly improve on performance and, in many cases, enable technical concepts that are otherwise unthinkable.

Take a look at Chapter 5, Thermal-Based Microsensors, by Friedemann Völklein for a comprehensive discussion on how thermal phenomena (such as temperature and radiation) are measured using MEMS technology. But more than this, thermal effects are also useful to use as intermediaries in a whole range of detector applications, for example in the measurement of electrical power or for gas flow velocity. Consequently, the chapter teaches us to reconsider frequently our notions of a measurement because, once a technology is established (say, making and measuring a very small thermopile), it may be used advantageously in a variety of *new* applications.

In Chapter 6, Arokia Nathan and Karim S. Karim discuss Large Area Digital Photon Imaging, an important technology for the detection of low-energy x-rays (replacing traditional film processing) and other electromagnetic sources. They clearly show that each new application area is a challenge that must be confronted afresh, with expertise built up as you expose the challenges set by a vision. The authors discuss new manufacturing materials, detector devices, electronics in a new technology, the challenge of large area integration, and the exiting area of working on flexible substrates.

For optical benches in miniature, Chapter 7, Free-Space Optical MEMS, by Ming C. Wu and Pamela R. Patterson, shows how MEMS technology not only captures this exciting application area but also how microdevices can leave the plane of the manufacturing substrate. Making truly 3-D structures has proven to be one of the toughest challenges of

MEMS, and I have personally extended the idea to an incredibly rich family of devices in an application area where precise 3-D positioning is one the most demanding requirements.

For many key applications, notably data communications, Integrated Micro-Optics (Chapter 8) is the preferred technology. Hans Zappe shows how this technology has been driven by himself and other workers to enable the design of entire optical microsystems, with exciting applications in accurate sensing, with enormous potential in miniaturization and speed, and ranging over communications, data storage, and sensors applications.

Chapter 9, Microsensors for Magnetic Fields, by Chavdar Roumenin, discusses the classical area of magnetic field measurement. Here we find no moving parts whatsoever; ingenious schemes are necessary to get a useful result from these complex silicon devices. The devices have captured numerous markets, including rotation sensors, proximity switches, compasses, and, lately, catheter devices for nuclear magnetoresonance sensing.

In Chapter 10, Mechanical Microsensors, by Franz Laermer, we encounter the exciting area of automotive sensing, which includes accelerometers for airbag applications and microgyroscopes. In sensing mechanical quantities, we usually need moving parts or are required to channel some of the application's mechanical deformation through our system; this remains a big challenge for any product.

Finally, Andreas Hierlemann and Henry Baltes discuss the very important area of Semiconductor-Based Chemical Microsensors in Chapter 11. Chemistry is rich in effects, in products, and in the vast range of measurement scales that are required for any device or system to become useful, so that this area offers both fantastic opportunities for products as well as huge challenges to implementation engineers. In trying to emulate some of the body's functions (smell or taste), we quickly discover the limits of engineering, and the authors guide us to techniques that try to overcome this limitation.

Q. *Where does the system aspect become important?*

A. Only the smallest numbers of microdevices are used as separate entities in engineering systems. For the vast majority, it is either necessary (due to the low level of signals) or advantageous to engineer entire systems at the microlevel. Because we are creating products for use in areas outside of microengineering, we have to address not only the needs but also the conventions of the application areas. The best new products feel familiar, cost less, and do a whole lot more. Systematic engineering of an application provides tremendous market advantages because products can evolve more rapidly once a technology is established.

In Microfluidics, Chapter 12, Jens Ducrée, Peter Koltay, and Roland Zengerle show us how systems microengineering of entire fluidic systems including pipes, pumps, and a vast range of sensors and manipulation tools results in exciting new applications such as a fountain pen that doesn't leak or a highly precise pipette.

Chapter 13, Biomedical Systems, by Whye-Kei Lye and Michael Reed, addresses the important area of new tools for the treatment of human ailments. With a small number of applications, this chapter only touches on the tip of an iceberg, for this topic could fill an entire book. We believe that this area may eventually dominate MEMS applications as medical techniques, biotechnology tools, and MEMS technology merge.

Q. *Sensors are fine, but can MEMS achieve significant actuation?*
A. MEMS became famous because of the first micromotor, which was certainly spectacular but without significant applications. In the ensuing years, MEMS actuators have grown to capture markets and application areas, with the best-known actuator probably being the digital micromirror device (DMD) array produced by Texas Instruments for video beamer applications.

This exciting area is discussed in Chapter 14, Microactuators, by Jack W. Judy. Here we discover how to get the most mechanical power out of a chip, whether we need force, large movement, linear behavior, or special kinematics. Many cases are worked through, including the Texas Instruments DMD.

Q. *How are MEMS made?*
A. More often than not, they are made in a cleanroom. But the cleanroom must be filled with life, and the naive days of just varying electronic manufacturing processes are certainly over. Choosing a technology will probably be the most critical cost factor for a company embarking on a MEMS adventure, so that the flexibility of the technology for use in other products, its availability as a service, and so forth, will be very important factors. From a design point of view, the most important concept in MEMS is to be able to mix physical effects (such as electrothermal or piezo-optical). This implies forming layers of different (perhaps incompatible) materials and structuring the materials, possibly even in 3-D.

In Chapter 15, Micromachining Technology, Paddy J. French and Pasqualina M. Sarro show how traditional silicon cleanroom equipment can be used and extended to enable a wide range of MEMS manufacturing capabilities. Many of the applications discussed in previous chapters are ideally suited to be manufactured using the techniques shown here.

The chapter also discusses the important issues related to compatibility of MEMS processed with traditional IC manufacturing.

In Chapter 16, LIGA Technology for R&D and Industrial Applications, Ulrike Wallrabe and Volker Saile present one of the first dedicated MEMS technologies capable of producing very high aspect ratio microcomponents, primarily out of metals and plastics. The many successful industrialization projects discussed also demonstrate how research facilities and industry can collaborate fruitfully.

Q. What role do electrical circuits play in MEMS?

A. Electronics represents the key means with which to get signals into and out of MEMS, as well as to provide signal conditioning, control, and a range of other functions that directly derive from traditional analog and digital circuit design.

How these techniques change when we are addressing MEMS applications is thoroughly discussed in Chapter 17, Interface Circuitry and Microsystems, by Piero Malcovati and Franco Maloberti. Future integrated microsystems will benefit significantly from progress in the VLSI field thanks to two key elements: the progress in batch-manufactured silicon sensors, and the introduction of new circuit techniques for designing interface circuits. These two factors will be essential in favoring the transition from *research-driven speculations* to *customer-driven activities*. Malcovati and Maloberti discuss the key issues in realizing integrated microsystems, describing the most suitable circuit techniques for interfacing and processing microsensor output signals. A number of examples of integrated devices illustrate the problems and suggest possible solutions, showing how essential interface circuits are in compensating for sensor shortages and increasing the functionality of microsensor systems.

Q. What is a good strategy to get started in the MEMS field?

A. Follow the steps outlined in this book:

Start reading the chapters on manufacturing techniques to see how typical MEMS manufacturing processes work.

Next, find an application that seems close to your own, and see how the authors solved design challenges and how they exploited the advantages of miniaturization.

Try to conceive of a complete system consisting of sensors, actuators, circuits, and packaging.

Next, detail the choice of effects used in the devices, the circuit techniques needed to extract and condition the signals, and the manufacturing processes needed at each stage.

Based on economics, partition the system into separate manufacturing entities and estimate the cost per unit. Remember that packaging will dominate cost and component complexity will work against yield.

Now that the first iteration is over, re-evaluate each decision you made and extend your data with the options that are available.

Now take the next steps, assuming that the answers to your questions were promising, yielding new possibilities for your application and your company:

Secure your IP by patenting and trademarking. Build up a literature base of what the competition is doing.

Build up a team that will take your concept further. If possible, take on new people who already know the basics of MEMS technology. This will save you lots of money.

Educate your team. Many courses are offered where workers can quickly gain hands-on experience. Make the designers join the lab people in practical courses, and send the entire team to design courses. This will ensure that the team bonds and that and team members talk the same technical language.

Jan G. Korvink
Karlsruhe, Germany
August 2005

1 Microtransducer Operation

Oliver Paul

Institute of Microsystem Technology IMTEK
University of Freiburg, Germany

1.1 Introduction

Rooted in mechanical, electrical, and chemical engineering and relying on physical insight, biological techniques, and materials science know-how, microelectromechanical systems (MEMS) engineering is a fundamentally interdisciplinary field. Its fascinating diversity often forces the research and development engineer to take into account a broader range of issues than in many classical, well-established technical disciplines. Simultaneously, the diversity creates the impression of a lack of unity, contrasting strongly with the classical disciplines of science and engineering. These are usually able to offer a core of thoughts stripped of unnecessary details, with well-established foundations and lines of thought, and representations accepted by the majority of researchers active in the field. More peripheral aspects of the disciplines can be built up from a solid basis of knowledge.

The situation is different in MEMS, and in particular in such a general domain as MEMS transducer operation. The bewildering diversity of materials, structures, and effects in MEMS translates into a wide range of operational concepts from which it appears at first sight difficult to extract unifying principles. Nevertheless, when considering the issue of transducer operation from a more generous distance, some unifying aspects may be peeled out of layers of technical details and individual preferences. It is the goal of this chapter to start from basic considerations and gradually build up the surrounding aspects useful for transducer operation.[1,2]

The purpose of microsystems is to collect physical and chemical information of various kinds about their environment and to make this information available in a form more suitable for the human senses and technical systems. It is clear that the task of gathering and transforming information is performed by many technical systems. However, the distinctive feature of microsystems is their ability to perform this task despite or even because of their small size. The definition of *microsystem* varies from researcher to researcher. Nevertheless, it will in general be accepted that a system has to be at most a few cubic centimeters in volume to qualify as a microsystem.

Transduction is performed by miniaturized elements with dimensions scaling from a few millimeters down to submicrometer lengths. Larger structures are usually considered as macroscopic. The description and analysis of such larger components and systems have been the subject of classical textbooks on transducers, sensors, and actuators.[3-5]

In the field of microsystems, the action of transforming information or signals is usually designated by the term *transductions* and *conversion*. Transduction is derived from the Latin verb "transducere," which means "to lead across."[6] In microsystems, information is indeed "led across" the boundaries between different signal domains, that is, it is transformed from one domain into another. In this sense, microsystems contain one or several microtransducers taking advantage of physical and chemical effects on the scale from centimeters down to atomic dimensions, and exploiting appropriate material properties to achieve the transduction goal.

To fill these rather general statements with a clearer meaning, the next section presents the signal domains in more detail and summarizes transduction principles implemented in microsystems to date. Section 1.3 describes important figures of merit of microtransducers, including their limits due to noise. Common techniques to improve transducer performance are then described in Section 1.4, while various methods to power microsystems are the object of Section 1.5.

1.2 Transduction

1.2.1 Signal Domains

In the context of microsystems, the term *information* has to be understood in a broad sense. Any signal with which the microsystem is able to interact is likely to carry a certain amount of information. As an example, the spectral energy density and propagation direction of electromagnetic radiation enable us to extract information from and draw conclusions about the thermodynamic state of the distant source of radiation. Analogously, the direction and amplitude of a magnetic field measured by a microsystem reveal some information about the orientation of the magnet in which the field originates. Similarly, the inertial forces experienced by a microsystem make it possible to determine the dynamics of the substrate carrying the microsystem.

What these three examples have in common is that the information extracted is associated with an energy field: the radiation field in the first example, the magnetic interaction energy in the second, and the mechanical potential in the third. This connection between information, signals, and energy is not surprising in view of thermodynamics, which teaches the intricate relationship between energy, states, entropy, and information.

Current microsystems operate mainly within six signal/energy domains:[6]

1. Mechanical signal/energy domain
2. Electrical signal/energy domain
3. Thermal signal/energy domain
4. Magnetic signal/energy domain
5. Radiant signal/energy domain
6. Chemical signal/energy domain

In addition to these, elementary particle interactions and gravitation in principle provide further signal/energy domains. However, the first is usually included in the radiant signal domain, since the dominant role of microsystems in elementary particle physical or nuclear research is to detect elementary processes via their decay products, which usually lead to particle fluxes "radiated" away from the location of the original process. The gravitational field of the earth provides a handy definition of the vertical direction and is used for example in tilt sensors. Since its main technically relevant effect is to exert a force on masses, it is natural to include it in the mechanical signal/energy domain. More subtle effects expected from the theory of gravitation, such as gravitational waves or black-hole evaporation, have not being relevant in MEMS devices, nor have microsystems been instrumental in elucidating such fundamental phenomena.

Finally, quantum mechanics provides a further, more abstract signal domain going beyond the rather intuitive domains mentioned so far. Quantum mechanics teaches that a considerable amount of information can be encoded in collections of bosonic or fermionic states by the technique of quantum mechanical superposition.[7,8] The resulting states have to comply with the symmetries requested by quantum mechanics from bosonic and fermionic states. States may show so-called entanglement. Preparing entangled states, performing operations on these, and reading out the result holds the promise of highly parallel, efficient computation and secure data transmission.[8,9] Microsystem technology has only started to play a role in providing microstructures useful for this purpose, i.e., the technical infrastructures necessary to hold the tiny quantum mechanical bits of information (qubits). Once breakthroughs have been made and techniques are established, it may well be that the quantum mechanical domain will have to be included with the others as a domain *inter pares* in future introductions to transducer operation.

The ensembles of signals constituting the individual signal/energy domains are listed here in more detail.[10]

The Mechanical Signal/Energy Domain

The mechanical signal/energy domain includes descriptors of the mechanical state of a system, such as:

- Position, orientation, tilt, velocity, angular velocity, acceleration, angular acceleration, relative position, displacement, level, proximity, topography, deformation, strain, stress, density, mass, and resonance frequency

Changes in the mechanical state descriptors often result from externally applied force configurations, such as:

- Localized forces, including amplitude and direction or equivalently perpendicular (normal) and in-plane (shear) components, and torques
- Inertial forces
- Distributed forces such as pressure, from shock waves to vacuum pressures
- Acoustic pressure, impedance, frequency, wavelength, and velocity
- Shear stress and mass flow

The measurement of several mechanical signals by mechanical microtransducers is described in Chapter 10. For the measurement of pressure, shear stress, and mass flow, the thermal techniques described in Chapter 5 are also used.

The Electrical Signal/Energy Domain

Electrical signals include:

- Voltage, electric field intensity and direction, current, power
- Charge, capacitance, dielectric constant, polarization, inductance, resistance, impedance
- Frequency, phase shift, dielectric loss tangent, decay time, duty cycle length
- Spectral distribution, e.g., noise spectral density and amplitude of a side band and its distance to the carrier band

This electrical signal domain benefits from the broad base in instrumentation and signal conditioning techniques contributed by the field of electrical engineering and in particular by microelectronics. Not surprisingly, transforming an initial signal into the electrical domain, conditioning it there, and transforming it into the digital domain, where it is immune to many external disturbances, has been found to be a sound way of proceeding in many microsystems. This approach also has the advantage of being compatible with modern information technology, where signals are usually stored, handled, combined, and distributed in the electrical domain.

For this reason, the electrical signal domain plays a central role in the thermal, magnetic, mechanical, and chemical microsystems described in Chapters 5, 9, 10, and 11, respectively, and most prominently in Chapter 17, which deals with microtransducer interface circuitry.

The Thermal Signal/Energy Domain

Thermal signals include:

- Temperature, entropy, free energy and free enthalpy, and changes thereof
- Heat capacity, thermal conductivity
- Heat quantity, thermal power, heat flux or flow
- Thermal resistance, conductance, impedance
- Thermal time constant and phase shift

The Magnetic Signal/Energy Domain

Magnetic signals include:

- Magnetic field and magnetic induction, both with amplitude and direction
- Magnetic moment, magnetization
- Magnetic permeability and susceptibility

The Radiant Signal/Energy Domain

Radiant signals include:

- Electromagnetic radiation energy density and flux density
- Polarization, coherence, phase shift
- Spectral density

- Reflectance, transmittance, absorptance
- Charged-particle passage, velocity, energy, and mass

As mentioned, signals due to charged particles pertain to nuclear or elementary particle physics. In view of the ionizing properties of corpuscular radiation, similar to energetic electromagnetic radiation, and its detection via radiative effects, such as scintillation, particle signals are generally merged with the radiant signal domain.

Chapters 7 and 8 show how radiant signals can be put to advantage to perform such interesting task as chemical sensing, communication network reconfiguring, and image projection.

The Chemical Signal/Energy Domain

Chemical signals include:

- Concentration, composition, pH
- Chemical potential, electrochemical potential, redox potential
- Reaction rate, equilibrium constants

The detection and measurement of chemical signals is described in Chapter 11.

1.2.2 Block Schematics of Transducers

Now that the questions of the relevant energy domains and the most important signals have been clarified to a first extent, common configurations of microtransducers and their arrangements into microsystems are described. At the same time we continue building up the general terminology used in the MEMS field.

As mentioned, the operation of signal transduction consists of transforming signals from one energy domain into another. For a miniaturized information processing system to be classifiable as a microsystem, a sufficient condition is that along its path from input to output, the signals be outside the electrical domain at some point. However, this definition does not draw a complete picture, since systems definitely classifiable as microsystems have been developed for the purpose of measuring electrical signals and translating them into a convenient electrical output, without the signal ever leaving the electrical domain. Miniaturized voltage probes, including miniaturized probe heads and neural probe arrays integrated with signal conditioning circuitry, are examples. In view of the fabrication technologies needed to achieve their miniaturization, there is no doubt that the structures in these examples can be viewed as microstructures. If they are extended by microelectronic components for further signal treatment, these systems will definitely be recognized as microsystems. Therefore this first attempt at providing a definition represents a sufficient but not a necessary condition.

It is proposed that this sufficient condition be completed to the following definition, drawing a narrower line around microsystems: A microsystem is a miniaturized system (1) involving signals that are often but not exclusively outside the electrical domain, and (2) fabricated using methods that differ from established fabrication techniques of electronics and microelectronics. Purely electrical or microelectronic devices and systems do not fall in this category. They have been treated extensively using the language of electrical engineering and microelectronics, from device fundamentals to fabrication and operation.[11-14]

Self-Generating and Modulating Transducer Elements

Microsystems are conveniently subdivided into schematic functional blocks that make it possible to follow the signal along its path from the input form to the output form. Each block represents an elementary transduction operation, where a signal is converted from one form into the next. In the figures of this chapter, transduction blocks are schematically drawn as black boxes, as illustrated in Fig. 1.1. In the simplest variant, a box has only two lines, for interaction with neighboring blocks, as shown in Fig. 1.1(a), representing the signal input and output paths. The method by which transduction is achieved is not specified at this point and is the subject of Section 1.2.3.[6,15]

A second box variant, shown in Fig. 1.1(b), has an additional line representing the need to drive the transducer element into its operation point. The first variant summarizes what is commonly called *self-generating transduction effects*; the second stands for *modulating transduction effects*. In this second category, the input signal exerts a modulating action on the transducer in the sense that the transduction of the bias signal into the output signal depends on the level of the input signal.

In practice, the output rarely depends exclusively on the desired input signal. Rather, the element is influenced in its operation by parasitic signals (shortened to *parasitics*) from the environment, such as ambient or operating temperature, magnetic fields from the earth or technical sources, water vapor, and other chemical species. A more complete picture is thus given by Fig. 1.1(c) and (d), where additional undesired input terminals

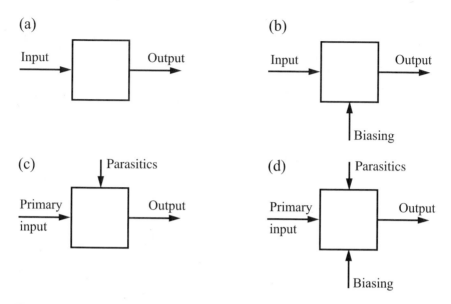

Figure 1.1 Schematic representation of self-generating (a,c) and modulating (b,d) transducer effects without (a,b) and with (c,d) parasitic inputs.

contribute to determine the value of the output. The transducer element is said to exhibit *cross-sensitivities* to the parasitic input signals.

Examples of self-generating effects include the Seebeck effect and the photovoltaic effect, where a temperature gradient and visible radiation, respectively, are directly transformed into a voltage. Modulating transduction is exemplified by the Hall effect, where an output signal (the Hall voltage) proportional to the magnetic field being measured is produced by the action of the magnetic field on a bias current. Absence of bias current implies the absence of Hall voltage. Similarly, some chemical sensors require their sensitive material to be heated to elevated operating temperatures. Without sufficiently high temperature, only a negligible chemical sensitivity is obtained.

Sensor and Actuator Elements and Microsystems

A second distinction is obtained by considering the particular combination of input and output signal domains connected by the microtransducer element. This is schematically shown in Fig. 1.2, where the input and output lines are labeled with the symbols φ, χ, and ε for the physi-

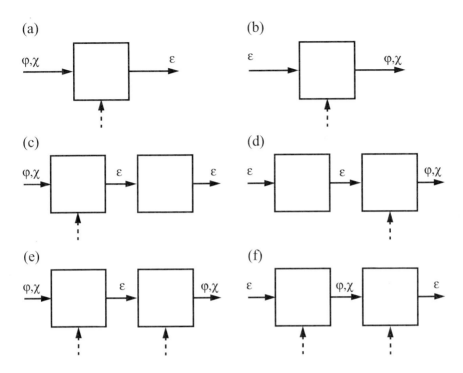

Figure 1.2 Schematic representation of (a) sensor and (b) actuator elements, (c) sensor and (d) actuator microsystems, and (e) sensactuator and (f) actuator-sensor systems. Signal domain labels are ε, electrical; χ, chemical; φ, all others.

cal, chemical, and electrical signal domains, respectively. The symbol φ is adopted to summarize compactly the mechanical, thermal, magnetic, and radiant energy domains. In the transducer of Fig. 1.2(a), a physical or chemical signal is transformed into the electrical domain; the opposite occurs in Fig. 1.2(b). The transducer element in Fig. 1.2(a) is conventionally termed a *sensor element* or *input transducer*, while that in Fig. 1.2(b) represents a so-called *actuator element* or *output transducer*. Both elements perform a conversion between nonelectrical and electrical domains and can be viewed as each other's reverses.

To amplify the output signal of a sensor element and turn it into a form suitable for further use, the sensor element is often followed by signal conditioning circuitry, as shown in Fig. 1.2(c), containing, e.g., preamplifiers, filters, amplifiers, and analog-to-digital (A/D) conversion units. This subsequent circuit block performs an electrical-to-electrical transduction. The combination of a sensor element and circuitry constitutes what is commonly termed a *sensor microsystem* or *smart sensor*. Well-known examples of smart sensors measure pressures, acceleration, rate of gyration, light intensity, magnetic fields, proximity, mass flow, shear rate, and concentration of chemicals. Techniques implemented in the electronic block are described in Chapter 17 and in the technical literature on the subject.[14,16]

Conversely, and as shown in Fig. 1.2(d), the initial electrical signal may have to be pretreated by an electrical-to-electrical transduction unit in order to be provided in a form and at a level suitable for an actuator element. This signal conditioning circuitry block may feature components for digital-to-analog (D/A) conversion, impedance matching, duty-cycle modulation, and bias ramping (see Chapter 17). The combination of a circuit block with an actuator element is usually termed an *actuator microsystem* or *smart actuator*. Note that schematically, sensor and actuator microsystems are opposites of each other. Examples of microactuators and actuator microsystems include microtweezers, adaptive optical elements, infrared radiation emission displays, ultrasound emitters, and micropositioners such as linear or rotational stages, scratch drives, and bio-inspired means of locomotion such as microcilia. Some of these are described in Chapter 14.

More complex systems for more demanding tasks are built by combining sensing and actuation within the same compact volume. Such combinations are schematically shown in Fig. 1.2(e) and (f) and are usually just termed *microsystem* for simplicity. To emphasize a particular signal evolution, the suggestive but clumsy terminology of *sensor-actuator*, *sensactuator* or *sensactor microsystem* [see Fig. 1.2(e)], and *actuator-sensor microsystem* [Fig. 1.2(f)] may be used. In the first case, a nonelectrical-to-electrical conversion is followed by electrical-to-nonelectrical conversion, while the initial and final signal forms in the second case are electrical, with a nonelectrical stage in between.

Examples of sensactuator microsystems include adaptive optical elements, where a wave-front signal drives an optical actuator; miniaturized flow controllers, where a mass flow signal drives a microvalve to maintain a stable flow; and position tracking systems, where a position-variation signal is used to lock a system into a desired position. An example of an actuator-sensor microsystem is provided by true root-mean-square (rms) AC-to-DC converters. In these devices, an electrical AC signal of unknown power content is first transduced into a temperature difference. The difference is used to generate an electrical DC signal with a power content identical to that of the input signal. The power content of the DC signal is easily determined. The process can be viewed as an electrical-to-thermal-to-electrical transduction.[17]

Tandem Transduction

Often, sensor and actuator elements can be further subdivided into a succession of transduction stages, as shown in Fig.1.3. The initial signal, φ_0, is transformed step by step into domains φ_1, φ_2, φ_3, etc., before finally being available in the desired form, ε. If more than two different energy domains are involved, the expression *tandem transduction* is appropriate. A nice example is the Golay cell, a highly sensitive radiation detector. In this device radiation is absorbed, thereby heating the gas in a closed volume, i.e.., by electromagnetic-to-thermal conversion. The expansion of the gas deflects a membrane, which amounts to a thermal-to-mechanical conversion. The displacement is sensed optically via mechanical-to-optical-to-electrical transduction. As an alternative, the tunnel effect, i.e., mechanical-to-electrical conversion, has been used for this last transduction step.[18]

Even if the signal remains in the signal space spanned by both input and output domains, the decomposition into elementary transduction steps can be convenient for analyzing the operation of a transducer. A closer look at many microsystems reveals that they can be decomposed in this way. Consider the example of the well-known silicon membrane pressure sensors. Overall, these devices perform a mechanical-to-electrical conversion by a modulating transduction effect. The measurement process can, however, be detailed in the following way, as illustrated in Fig. 1.4. First, the external pressure deforms the membrane (mechanical-to-mechanical transduction); the deformation strains the membrane material, which is sensed via integrated strain gauges using the piezoresistive effect (mechanical-to-electrical conversion). This raw output is then amplified and conditioned in a subsequent circuit block, turning the device into a smart microsensor. The first stage operates as a mechanical amplifier, making it possible to record sizable signals in the piezoresistive gauges.

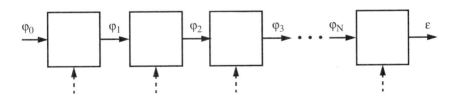

Figure 1.3 Chain of transduction stages representing a generalized tandem sensor, where the original signal φ_0 is successively converted into $\varphi_1,\ldots,\varphi_N$ before finally being converted into the electrical domain as signal ε.

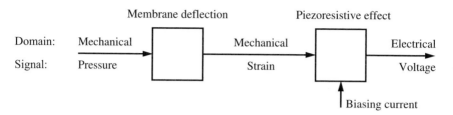

Figure 1.4 Decomposition of silicon-membrane-based pressure sensing into two transduction stages, with first a mechanical-to-mechanical transduction and finally a mechanical-to-electrical conversion.

In comparison, the direct effect of the applied pressure on a piezoresistor is small; as a standalone element, a piezoresistor would offer a very low pressure sensitivity (see Kloeck and de Rooij[19] and Chapter 10).

1.2.3 Transduction Effects

The transducer boxes in Figs. 1.1 through 1.4 do not specify any particular conversion effect. Also, in Section 1.2.2 a few common transducer effects were mentioned for illustration purposes. A more systematic base is now elaborated.

Transducer effects can be conveniently displayed as a matrix. The rows of the matrix list the input signal domains, while its columns represent the output signal domains. The field lying at the intersection of an input domain row and an output domain column stands for all physicochemical mechanisms able to convert signals from the row domain into the column domain. Since six main energy domains are considered here, the transduction matrix has six rows and six columns, with a total of 36 intersections.[6,15]

This does not offer the complete picture, however, since the two-dimensional structure is effective in categorizing two-terminal transduction boxes only, representing self-generating transduction effects with their two involved energy domains. Modulating transduction effects involve three domains, corresponding to the three terminals of the symbolic box in Fig. 1.1(b). This aspect has been taken into account by Middelhoek et al., who have added a third dimension to the transduction matrix and built up what they termed the sensor-effect cube.[15,20] The sensor cube can be viewed as a vertical stack of transduction effect matrices. The base layer corresponds to the self-generating transduction effects; the next layer contains transduction effects modulated by mechanical signals; the third layer tabulates transduction effects modulated by electrical signals; successively, the final four layers contain the transduction effects modulated by thermal, magnetic, radiant, and chemical signals. The resulting object has a total of $7 \times 36 = 252$ intersections able to receive all self-generating conversion mechanisms and all modulating transduction effects with at most two input lines. The structure covers all except less common and often elusive higher order effects. To identify the location of a particular transduction effect in the three-dimensional structures, the following procedure is proposed. First, determine the modulating signal and go into the appropriate sensor cube level. Next choose the energy domains of the driving/biasing and output signals, and select the corresponding matrix row and column, respectively. The location is uniquely determined by these three coordinates.

In view of its three-dimensional structure, the sensor cube is more useful in categorizing transduction effects than in tabulating them on paper. The following approach is therefore adopted to conform to the two-dimensional nature of conventional printing media. The various levels are projected onto the base level; thus self-generating and modulating effects are quoted in the same intersection. For the sake of clarity, the domain label of the modulating signal precedes the individual effects. Table 1.1 shows the resulting condensed transduction matrix.

The projection of the cube on a single level is facilitated by the fact that the cube levels corresponding to modulating transducer effects are by far less densely populated than the base level matrix of self-generating effects.

1.3 Microsystem Performance

1.3.1 Figures of Merit

Many parameters are used for specifying, characterizing, and comparing the performance of microsystems. This section briefly introduces some important figures of merit; subsequent sections discuss a few of them in more detail.[10]

- *Sensitivity* (often denoted S) or *responsivity* (often denoted R) denotes the proportionality constant between the output signal and the input signal. In the case of nonlinear devices, it is obtained in differential form by taking the derivative of the output with respect to changes of the input. In linear devices, sensitivity is a constant, independent of signal level. In nonlinear devices, sensitivity always has to be specified in

Table 1.1 Matrix of Transduction Effects Obtained by the Projection of the Sensor Cube[15] on a Single Plane. Self-generating effects are indicated using the label **sg**; the biasing inputs of modulating effects are indicated by the labels **mag** (magnetic), **mech** (mechanical), **rad** (radiative), **therm** (thermal), and **el** (electric).

Out \ In	Mechanical	Electrical	Thermal	Magnetic	Radiant	Chemical
Mechanical	**sg**: elastic, plastic deformation, fatigue, fracture, momentum conservation, density-driven buoyancy, forced convection	**sg**: piezoelectricity	**sg**: contact friction, thermoelastic friction	**sg**: magnetostriction	**sg**: sonoluminescence	**sg**: pressure-induced explosion, sonochemistry
Electrical	**sg**: electrostatic force, piezoelectric stress **mag**: Lorenz force	**sg**: Ohm´s law, pn junction effect, ferroelectric polarization loops **mech**: piezoresistance effect, tunnel effect **mag**: Hall effect, magnetotransistor effect, magnetoresistance **rad**: photoconduction, photodiode effect, phototransistor effect	**sg**: resistive heat dissipation, Peltier effect, phonon drag **therm**: Thomson effect **mag**: Ettingshausen effect	**sg**: Ampere´s law, Biot-Savart law	**sg**: electroluminescence, Cerenkov radiation **mech/el/mag**: synchrotron radiation	**sg**: electrolysis
Thermal	**sg**: thermal expansion, free convection	**sg**: Seebeck effect **mag**: Nernst effect	**sg**: heat conduction, thermal inertia (heat capacity), heat transport by free convection **mech**: heat transport by forced convection **el**: Thomson effect **mag**: Righi-Leduc effect	**sg**: Curie-Weiss law	**sg**: incandescence, grey/black-body radiation	**sg**: endothermal reaction

Magnetic	**sg**: magnetostriction, magnetostatic force	**sg**: law of induction		**sg**: magnetization loops, magnetic susceptibility	
Radiant	**sg**: radiation pressure	**sg**: photoelectric effect, photovoltaic effect	**sg**: radiation absorption, radiative heat transfer	**sg**: photomagnetic effect	**sg**: photoluminescence **mech**: photoelasticity **el**: liquid crystal effect **mag**: Faraday effect, Kerr effect	**sg**: photochemistry
Chemical	**sg**: explosion, osmotic pressure, concentration gradient forces	**sg**: Volta effect	**sg**: exothermal reaction		**sg**: chemiluminescence	**sg**: chemical reaction **el**: electroplating, electrochemical reactions

conjunction with operating conditions. In actuators, the terms *activity* or *effectivity* are more appropriate than *sensitivity*; they are not commonly used, however, and *performance* often has to play the same role instead.
- *Full-scale output* (*FSO*) quantifies the difference between the upper and lower end points of the output signals generated by the system.
- *Range* is the corresponding span of the input signal that can be handled by the system. In linear devices, the range is the full-scale output divided by the sensitivity.
- *Linearity/nonlinearity* describes the deviation of the measured response as a function of measurand (output vs. input) in comparison with the linear interpolation between the end points. Nonlinearity is often expressed as a percentage of the FSO. At a given level of nonlinearity, the experimental curve maximally deviates from the linear interpolation by the given percentage of the FSO.
- *Offset* denotes the residual output signal in the absence of the measurand. It can be translated into an equivalent input. For small offsets or with linear devices, the input-related offset value is obtained by dividing the output-related value by the sensitivity at small applied inputs.
- *Selectivity*. Devices are rarely sensitive to a signal input parameter. Most often other environmental parameters influence the operation of the device as well and translate into a contribution to the output signal. This contribution can erroneously be interpreted as the presence of an input signal of the desired kind. The goal of good microsystem design and operation is to guarantee high sensitivity for a single input only and to suppress the sensitivity to other input parameters as much as possible.
- *Repeatability* and *hysteresis* characterize the ability of the system to reproduce output values at repeated application of identical input signals. Constant operating conditions are assumed. Repeatability describes the statistical spread of the measurements, e.g., by their variance, while the hysteresis describes systematic differences between measurements where the input signal has been ramped up or down to the measured value.
- *Noise*. For thermodynamic reasons, every device, whether electrical, thermal, mechanical, or radiant, shows noise. Stochastic signal contributions are added at all stages of the conversion path. They appear as random output spreading the

value of the ideal output signal as a function of time. The input-related noise is the noise at the output terminal divided by the sensitivity, i.e., the equivalent noise at the input terminal that would produce the same noise at the output in an ideal, inherently noise-free microtransducer.
- *Signal-to-noise* (*S/N*) denotes the ratio of the output signal divided by the output noise.
- *Resolution* denotes the minimal difference in input signals that can be distinguished by a system. It is limited by the input-related noise.
- *Response time* is the time it takes for the system to relax from one signal level to another. Response time results from mechanical, thermal, and electrical inertia, i.e., from the limited effectivity of forces, heat exchange processes, and resistive electrical paths to change momentum, energy, and stored charge, respectively. In semiconductors, the relaxation of carriers to new equilibrium conditions is subject to transition probabilities that limit the effectiveness of recombinative processes.
- *Drift* and *stability*. A microsystem is expected to perform stably, i.e., identically over an extended period, possibly over its entire life span. If the output signal at a constant input changes with time, the output signal and the system are said to drift, and the device is recognized to show unstable performance. The phenomenon of drift can be caused by a changing offset and a changing sensitivity. If a system is prone to drift, it may require periodic recalibration. This is often a costly process that users prefer to avoid.
- *Overload characteristics* describe the ability of the microsystem to survive input signals far outside its specified range. A nice example is provided by automotive acceleration sensors. These devices are usually specified for acceleration levels up to 50 g, corresponding to a head-on automobile crash. However, the structures have to be certified to survive acceleration up to $10^4\,g$ during handling and mounting and to continue to operate reliably after such shocks.
- *Power consumption* is an issue with all battery-driven, autonomous microsystems such as hearing aids or devices called e-grains or smart motes that should remain functional without servicing for as long as possible. Power consumption directly determines the method for powering a device, e.g., by a fixed link connecting the microsystem to an external

macroscopic energy source or by a miniaturized energy storage unit internal to the system.

In addition to these parameters, each sensor field has developed its own figures of merit. Examples are the quantum efficiency η, the detectivity D^*, and the noise-equivalent power (NEP) of radiation detectors; the maximum conversion efficiency parameter $\alpha^2/\kappa\rho$ of thermoelectric materials, with the thermoelectric coefficient α, the thermal conductivity κ, and the resistivity ρ, convenient for comparing thermoelectric materials; and so on. Some of these additional parameters are defined and described elsewhere in this book.

1.3.2 Sensitivity, Selectivity, and Offset

The performance of a general transduction element, as represented by one of the boxes in Fig. 1.1, can be quantitatively described by its transfer function T. This is the function relating the vector of input signal A and parasitic inputs $P = \{P_1, ..., P_N\}$ to a response signal R at the output of the transducer element, subject to a number of biasing signals summarized by the biasing signal vector B maintaining a well-controlled operating point. In some cases it is advantageous to consider an output vector rather than a scalar output, as suggested by the single-output terminals in Figs. 1.1 through 1.4. Indeed, in many devices, the conversion leads to several output values that are fed into the subsequent stages of the conversion chain. An example is provided by silicon-based pressure sensors, where pressure induces strain in a micromachined membrane. The strain is sampled at several distinct locations of the membrane (typically four) by individual piezoresistors, corresponding to a four-dimensional output vector of the mechanical-to-mechanical transduction stage of integrated pressure sensors. In such cases, the transfer function is a vector-valued function rather than a scalar function of A, B, and P.

In the following, however, the discussion is restricted to scalar outputs, since the generalization to vector-valued output signals is straightforward. Similarly, only the case of a single, scalar-valued input A is considered. The generalization to vector-valued inputs is straightforward. The device response R of such a single-input, single-output transducer is therefore

$$R = T(A,B,P). \qquad (1)$$

A working point of the device is defined by a set of well-defined biasing conditions B_0, a primary input signal level A_0, and parasitic inputs P_0. The response of the transducer to small changes $\delta A = A - A_0$ in the primary input signal around A_0 is given by the Taylor expansion of T around the working point,

$$R = R_0 + \left(\frac{\partial T}{\partial A}\bigg|_{A_0, B_0, P_0}\right) \delta A, \qquad (2)$$

where the response in the working point $R_0 = T(A_0, B_0, P_0)$ and derivative of T with respect to the primary input signal, evaluated in the working point, constitute the offset and the primary sensitivity of the device. Generally, the derivatives of T with respect to primary and parasitic signal are termed *sensitivities* or *responsivities* and are here denoted by S_X, defined as

$$S_X(A_0, B_0, P_0) = \frac{\partial T}{\partial X}\bigg|_{A_0, B_0, P_0} \qquad (3)$$

with the subscript X denoting the particular measurand or parasitic input of interest. The response R_0 depends in particular on the level of the parasitic inputs. Therefore, when the parasitics are not well controlled but vary by small amounts $\delta P_1, ..., \delta P_M$, the output R is modulated by additional terms $\delta R_i = S_{P_i} \delta P_i$, and the resulting output signal is

$$R = R_0 + S_A \delta A + S_{P_1} \delta P_1 + ... + S_{P_M} \delta P_M \qquad (4)$$

The coefficients S_{P_i} to S_{P_M} are termed cross-sensitivities or parasitic sensitivities. The ratio of the primary sensitivity to a parasitic sensitivity defines a selectivity. Selectivity should be as high as possible. A transducer with small cross-sensitivities is said to be selective. As Eq. (4) shows, lack of control of the parasitic parameters translates directly into an uncertainty in the δA value inferred from the output values R_0 and R. Good transducer design therefore aims at minimizing parasitic sensitivities as much as possible, by selecting appropriate transduction mechanisms, optimizing materials and geometries, or taking the cross-sensitivities into account by calibration or suppressing them by compensation techniques, as described in Section 1.4.

The most important cross-sensitivity is that to temperature. Virtually every transducer shows some temperature dependence, because on the microscopic level, transport phenomena are subject to the laws of thermodynamics and statistical physics.[21,22]

Linear transducers have the convenient property that Eq. (4) takes the form

$$R = R_0 + S_A A + \Pi, \qquad (5)$$

where Π denotes parasitic terms. In other words, the response depends linearly on the applied input signal A over the entire range of the device. Consequently, the sensitivity is a single scalar value. Except for parasitics and the offset R_0, the level of A is extracted from the output value as

$$A = (R - R_0 - \Pi)/S_A. \qquad (6)$$

In transducer chains such as the one shown in Figs. 1.3 and 1.4, and especially in tandem transducers, the signal is handed down from one stage to the next. The transfer function of such a chain of length N is hierarchically defined as

$$R_N = T_{total}(A) = T_N(T_{N-1}...\{T_2[T_1(A, \boldsymbol{B}_1, \boldsymbol{P}_1), \boldsymbol{B}_2, \boldsymbol{P}_2]..., \boldsymbol{B}_{N-1}, \boldsymbol{P}_{N-1}\}, \boldsymbol{B}_N, \boldsymbol{P}_N). \qquad (7)$$

At each stage i, the input signal is provided by the output signal of the previous stage $i-1$ and is handed on via the transfer function T_i, while additional parasitic contributions $\boldsymbol{P}_i$ are added. As a consequence, the sensitivity of the entire chain with regard to the input A is the product of the individual sensitivities S_i ($i = 1,...,N$) of the individual stages,

$$S_{total} = S_N S_{N-1}...S_2 S_1. \qquad (8)$$

At the same time, parasitics added at a certain stage are handed onward by the sensitivity chain of all subsequent transduction stages. To suppress the cross-sensitivities of the entire chain, it is thus in the interest of good transducer design to make all stages as selective as possible. However, the earlier stages have to be given particular attention with regard to cross-sensitivity suppression, in view of the longer transmission path of early parasitics.

If the response R of a transducer is different from zero in the absence of an input signal, e.g., if $R = T(0, \boldsymbol{B}, \boldsymbol{P}) \neq 0$, this response value constitutes the offset of the devices. Any output comes on top of this residual offset. If the offset is not known, it may be mistaken for an input signal; this is the equivalent input offset. In a transduction chain, again, offsets are added at all possible stages. Earlier offsets pass through a longer portion of the chain and thus experience transmission by a larger number of sensitivities. Therefore, care has to be taken to design individual stages with minimal offset, in particular at the beginning of the chain, or to implement appropriate countermeasures, as described in Section 1.4.

1.3.3 Noise

Signals of both the desired and parasitic types contain deterministic and stochastic components. Among the signals with deterministic character are environmental conditions such as temperature, humidity, and radiation intensity, and technical influences such as stray magnetic fields, tilt, and accelerations, that all have the potential to contribute to the output signal in a well-defined way. Deterministic countermeasures can, in principle, be taken against these adverse influences. Some of them are discussed in Section 1.4.

Stochastic signals are perceived as noise; they are more difficult to handle. Noise is indeed ubiquitous and cannot be avoided in microsystems. The fundamental reason for noise is the particulate nature of matter and the tight connection between dissipative transport processes, on which microsystems rely, and the random processes occurring at the fundamental, microscopic level. This connection is known as the fluctuation-dissipation theorem due to Kubo.[22,23] In the transduction chains of microsystems, noise is added at every analog stage of the chain, ultimately limiting the resolution of the entire transducer. This is what is playing against MEMS: The smaller the microsystem, the more important noise becomes.

In analog microelectronic design and MEMS circuit development, noise issues attract considerable attention, and noise spectral density calculations are standard. Some of the approaches adopted in MEMS circuits to shape and filter noisy signals are described in Chapter 17. Using such techniques, instrumentation amplifiers with phenomenally low spectral noise power density have been developed, for example.[24]

However, MEMS design should not rely on the art of circuit design alone, in the hope of using it as a universal cure. Noise issues should be considered at all stages of transduction, because they are present in devices and transducers involving all energetic domains.

Resistors, Capacitors, and Other Electronic Noise Sources

Every dissipative element is a source of noise signals.[22] Across the contacts of a resistor R, a stochastically fluctuating voltage $v(t)$ is measured under open-circuit conditions, as illustrated in Fig. 1.5(a). Conversely, a low-impedance current-meter will record a randomly fluctuating current $i(t)$ under a short-circuit condition, as shown in Fig. 1.5(b). The resistor can be schematically represented as an ideal resistor and a noise voltage source labeled V_{nR}. The phenomenon of resistor noise has been given the name of Nyquist, Johnson, or just thermal noise. It originates in the random collision of carriers with the thermally vibrating lattice of the resistor material and among carriers themselves.

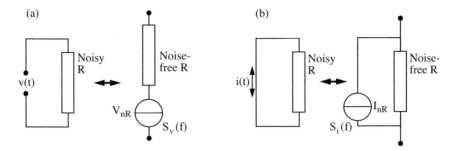

Figure 1.5 Real and equivalent circuits of an electrical resistor under (a) open-loop and (b) short-circuit conditions. In the equivalent circuits, auxiliary voltage and current sources with respective spectral noise power densities $S_v(f)$ and $S_i(f)$ are required.

The spectral noise power density $S_v(f)$ of the thermal voltage is the time average $<v^2(t)>$ of noise voltage contributions with frequencies in an interval of width $\Delta f = 1$ Hz around f, as a function of frequency f. For a resistor, $S_v(f) = 4kTR$, where $k = 1.38 \times 10^{-23}$ J/K and T denote the Boltzmann constant and the absolute temperature, respectively. S_v is independent of frequency, which is equivalent to saying that Nyquist noise is white noise. The corresponding spectral power density of the noise current, $S_i(f)$, is $4kT/R$. Consequently, the rms noise current $I_{nR} = (<i^2(t)>)^{1/2}$ under a short-circuit condition and the rms voltage $V_{nR} = (<v^2(t)>)^{1/2}$ under an open-loop condition in a frequency bandwidth Δf are $I_{nR} = \sqrt{4kT\Delta f / R}$ and $V_{nR} = \sqrt{4kTR\Delta f}$, respectively.[25]

As an example of the numbers involved, for a resistor with $R = 1$ kΩ, the current and noise voltage over a 1 Hz interval are 4.1 pA and 4.1 nV, respectively. These number are important for many electrical and thermal microsystems, and in transduction elements such as piezoresistors in mechanical transducers, heating resistors, thermal resistors, thermopiles, and semiconductor-based photodetectors. A semiconductor thermopile with 100 kΩ designed for detecting signals over a range of 1000 Hz contributes a noise voltage of 41 µV from thermal noise alone, which can easily exceed the thermoelectric voltages to be measured.

Based on the required bandwidth and the noise level contributed by the subsequent circuitry to the overall output, the spectral noise power densities of $S_i = (4.1)^2$ pA^2/(kΩ Hz) and $S_v = (4.1)^2$ nV^2/(kΩ Hz) enable the designer to select an appropriate value of the resistive element. Conversely, if a transducer resistance value is dictated by other constraints, these numbers are able to guide circuit design toward the specification of a tolerable circuit noise performance.

The rms noise voltage V_{nC} at the terminals of a capacitor with capacitance C is $\sqrt{kT/C}$. This is derived using the elementary circuit in Fig. 1.6:[25] A resistor noise voltage component $v_{nR}(t)$ with angular fre-

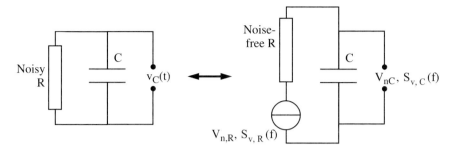

Figure 1.6 Real and equivalent circuit of a capacitor. The overall noise voltage V_{nC} measured at the output without further filtering does not depend on the value of R.

quencies concentrated around f is filtered by the capacitor and appears as corresponding noise voltage $v_{nC}(t) = v_{nR}(t)/(1 + 2i\pi fRC)$. Consequently, the noise voltage spectral densities $S_{vR}(f)$ and $S_{vC}(f)$ of the resistor noise source and at the capacitor terminals, respectively, are related by $S_{vC}(f) = S_{vR}(f)|1 + 2i\pi fRC|^2$. The mean-square noise voltage V_{nC}^2 of the capacitor is the integral of $S_{vC}(f)$ with respect to frequency. This leads to $V_{nC}^2 = kT/C$ and thus to the introductory statement of this paragraph.

This result does not depend on the resistor value. It is a property of the capacitor itself via its capacitance value C alone, and the resistor can be viewed merely as an auxiliary element, enabling the capacitor to go into thermal equilibrium with its environment. For a more physically founded picture, R can be thought of as the internal resistance of the capacitor.

In contrast to V_{nC}, a quantity that does depend on R in the RC circuit is the noise spectral density, which is constant at the level $4kTR$ roughly up to the corner frequency $1/(2\pi RC)$. A second consequence is that the average energy stored in the capacitor is $CV_{nC}^2/2 = kT/2$. In other words, the capacitor acts as a thermodynamic degree of freedom activated to the level predicted by the equipartition theorem of statistical physics.[26]

Smaller capacitors show larger noise. This is relevant for many capacitive elements, e.g., in capacitive acceleration sensors (Chapter 10) and electrostatic actuators (Chapter 14). With capacitance values in the pico-Farad range, noise voltage levels easily attain microvolt levels if no filtering is performed.

In addition to the white noise due to resistive elements, microelectronic circuits show so-called $1/f$ noise, also termed flicker noise.[25] This noise contribution is observed in field-effect transistors (FETs) and to a lesser extent in thin-film resistors, where carrier transport is strongly affected by surfaces at which carriers are captured and released. The spectral density of the noise voltage contributed by such elements, $S_v(f)$, is proportional to $1/f$. At low frequency, the overall spectral signal density $S_v(f)$ diverges and therefore has to be handled with care by appropriate filtering and noise-shaping techniques.

Furthermore, microelectronic components exhibit shot noise. This stochastic signal component originates in the particulate nature of current flow. An average DC current I_{DC} across a barrier consists of individual charge carriers that statistically cross the barrier. If the motion of the carriers is uncorrelated, the noise current associated with this random arrival is white and has the spectral power density $2qI_{DC}$, with the elementary charge q.

Thermal Resistors and Capacitances

The conclusions drawn for electrical resistors and capacitors can be directly translated into the thermal domain. The electrical resistor is translated into a thermal resistor R_{th} that establishes a thermal path between a thermal reservoir at temperature T_0 and a mass acting as a thermal capacitance C_{th}, as shown in Fig. 1.7; the electrical potential is replaced by temperature, and current becomes entropy flow. Random generation and conduction of phonons in R_{th} leads to a fluctuating thermal energy stored in the thermal capacitance. The fluctuations amount to $kT/2$ and are due to the random heat flow $i_{th}(t)$ [entropy flow $i_{th}(t)/T$] over the resistor. As a consequence, the temperature of the heat capacitance fluctuates by $\delta T_{nC} = \sqrt{kT^2/C_{th}}$, where $\delta T_{nC} = (<(T(t)-T_0)^2>)^{1/2}$ is defined in analogy with V_{nC} as the rms temperature excursion from the equilibrium value T_0. The spectral noise power density of the of the entropy flow across the thermal resistor is $S_{i_{th}/T} = 4kT/R_{th}$. As a consequence, the spectral noise power densities of the heat flow $i_{th}(t)$ and of the temperature excursion $\delta T(t)$ are $S_{i_{th}} = 4kT^2/R_{th}$ and $S_{\delta T} = 4kT^2 R_{th}$, respectively. The thermal corner frequency is $1/2\pi R_{th} C_{th}$, beyond which $S_{\delta T}$ decreases as $1/|1 + 2i\pi f R_{th} C_{th}|^2$.

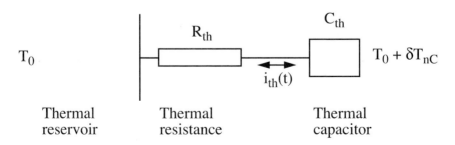

Figure 1.7 Schematic view of thermal microsystems with a small thermal mass (thermal capacitance) C_{th} connected to a thermal reservoir at the temperature T_0 by a thermal resistance R_{th}. The temperature of the thermal mass fluctuates by δT_{nC}.

This is relevant for thermal microsystems, in particular thermal infrared radiation detectors.[27] These detectors contain thermally insulated elementswhose temperature is measured in the process of determining the level of an external input. The temperature measurement cannot be more accurate than δT_{nC}. For example, the temperature fluctuations of the 50 × 50 µm² sized pixels of Honeywell's bolometer array amount to 36 µK.[28] This sets a theoretical lower limit to the detectable infrared radiation intensity and thus to the resolvable temperature change of thermal radiation.

Masses, Springs, and Damping Elements

Analogously, mass-spring systems, shown schematically in Fig. 1.8, are exposed to random motion around their equilibrium configuration. The ideal (noise-free) spring, with its spring constant K, stores an amount of potential energy fluctuating by $kT/2$. These fluctuations are due to the incessant Brownian motion of the mass M interacting via molecular collisions within the damping element. The rms displacement Δx_n of the mass from its equilibrium position is $\Delta x_n = \sqrt{kT/K}$. The spectral noise power density of the displacement has the bandwidth $\Delta f = K/(4D)$.

Delicate mechanical sensors, including accelerometers and gyroscopes, are subject to this type of mechanical fluctuation. As an example, accelerometers described in Chapter 10 consist of interdigitated electrodes spaced by a few micrometers. For these structures, Δx_n is calculated under reasonable assumptions to be a few Å. This may not seem like much, but a rough estimate of the mechanical situation shows that this random displacement corresponds to acceleration levels of 0.1 to 1 m/s².

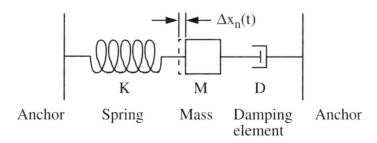

Figure 1.8 Schematic view of a mechanical microsystem by a mass connected elastically to an anchor structure. Excursions of the mass from its equilibrium position are damped. The position of the mass fluctuates around the theoretical equilibrium position.

Acceleration measurements subject to the entire displacement noise will show this level of uncertainty if no signal filtering limits the bandwidth of the signal handed over to the next stage of the system.

Radiation Noise

In its interaction with matter, radiation manifests its particle-like nature, as opposed to the continuous nature of electromagnetic radiation in Maxwell's formulation.[29] The absorption and emission of photons by matter is a stochastic process that adds its own noise contribution to optical microsystems.[30]

1.4 Transducer Operation Techniques

In transducer development, optimization has to be undertaken at all levels, from the initial concept, via fabrication issues and operating modes, to packaging techniques. The important roles played by modeling and numerical simulation in the initial stages of a microtransducer development are described in Chapters 3 and 4. One of the goals of simulation is to optimize such figures of merit as responsivity, range, and response time. Fabrication techniques for the realization of miniaturized systems are extensively treated in Chapters 15 and 16. This section reports common operating techniques that enable optimal performance to be extracted from imperfect microtransducer elements.

It has to be acknowledged that there are no miraculous tricks to save a flawed transducer concept or design. However, a range of methods are in use that enable microsystem engineers to tackle the three central issues of offset, cross-sensitivity, and noise.[31] Other performance parameters, including signal-to-noise, resolution, repeatability, and linearity, are secondary in the sense that they can often be derived from the more fundamental characteristics and thus can be optimized in conjunction with those.

1.4.1 Calibration

The technique of calibration consists of determining the transfer function T, as defined in Section 1.3.2, under controlled conditions and in using this knowledge to extract reliable input values from the output signals of the device. Calibration is a costly procedure, since in principle it requires well-controlled primary and parasitic signals to be applied to each device. A calibration procedure for a typical nonlinear sensor includes measuring an output curve in response to a primary input ramp at several temperature values. If the device shows negligible temperature dependence, calibration at a single temperature is sufficient. A requirement for calibration is therefore the availability of testing infrastructure able to apply the required input signal, e.g., pressure levels, gas flow velocities, humidity, and acceleration levels, to large numbers of devices, to measure their response, and to process and store the data. The cost intensity results from the fact that this is a serial process. The aim of smart calibration is therefore to keep the effort at a minimum.

The calibration effort directly reflects the severity of the device specifications. High-end products with more demanding specifications often have to be calibrated individually. In contrast, products intended for mass markets can be calibrated on a statistical basis using a sufficient number of sample devices, while the basic functionality is checked for all devices.

In the case of a linear sensor, a calibration curve is obtained from two measurement points, e.g., R_0 and R_{max} measured at zero primary input value ($A = 0$) and at the end of the range to be covered ($A = A_{max}$). The output at zero input is the offset $R_{off} = R_0$, while the slope $S_A = (R_{max} - R_0)/A_{max}$ is the sensitivity of the device. These two values determine the model response

$$R_{model}(A) = R_{off} + S_A A \tag{9}$$

of the device. They can be used to shift measured output data by R_{off} and compute the input signal corresponding to the shifted output. This data extraction can be done on-chip in integrated microsystems or off-chip in hybrid systems.

Determining nonlinear responses requires the acquisition of further data points at intermediate input levels. A more complicated model $R_{model}(A)$ can then be fit to the extended experimental values and used to extract the input levels from the measured output values. Two procedures can be applied. In the first, the system is provided with the computational power to calculate model values and solve the equation $R_{meas} = R_{model}(A)$ for A, e.g., by an iterative procedure. In the second procedure, model data can be stored in a look-up table on-chip or off-chip from which the value of A corresponding to the measured signal R_{meas} is retrieved. Alternatively, the look-up table may be filled with raw experimental data obtained by calibration.

Even in the simplest case of a linear transducer, a major cost position in calibration is due to the fact that the linear response depends on temperature.

To first order, therefore, the temperature dependence of the offset and the sensitivity of the device have to be determined. This is equivalent to repeating the calibration measurement at a second temperature level and extracting the temperature-dependent $R_{off}(T)$ and $S_{model}(T)$, in particular their temperature coefficients β_{off} and β_S, defined by $R_{off}(T) = R_{off,0} \times \{1 + \beta_{off}(T - T_0)\}$, and $S_{model}(T) = S_{model,0} \times \{1 + \beta_S(T - T_0)\}$, where $R_{off,0}$ and $S_{model,0}$ denote the respective values at the reference calibration temperature T_0. In nonlinear sensors, temperature coefficients of the model parameters can be defined and used in a similar way. Higher order, nonlinear temperature coefficients are extracted from measurements at more than two temperatures. Temperature-dependent calibration, even carried out at only two different temperatures, is time-consuming and requires dedicated equipment. When inferring input values from measurements with a microsystem, a local temperature sensor is required, either on-chip or off-chip. In the case of silicon-based systems, this increases the system cost only modestly, since silicon temperature sensors are relatively easy to implement.[32] The temperature characteristic of the temperature sensor also requires calibration.

Other parasitic influences such as pressure, humidity, and light intensity can be handled identically, by calibrating the microsystem at different levels of the parasitic signal and extracting the corresponding linear and nonlinear sensitivity coefficients. The information is then used to deduce an input level from a measured output.

In commercial products, the imperfect characteristics of microsystems are often adjusted to the desired values by trimming. This is a recursive process involving the measurement of system characteristics and trimming selected system parameters until a goal is reached. In the case of linear microtransducers, the goal of trimming is mainly the elimination of offset and the adjustment of the sensitivity to a desired value. A widespread method is to adjust resistive elements by laser trimming.

1.4.2 Compensation

Since calibration and trimming can be time-consuming and costly procedures, other means of reducing parasitic contributions at the source are needed. Compensation is one of these methods. Its aim is twofold:

1. Removal of parasitic influences
2. Indirect measurement of a primary signal

Methods for achieving the first objective make use of sensor/reference pairs and sensor bridges. These are illustrated in Fig. 1.9 and discussed below.

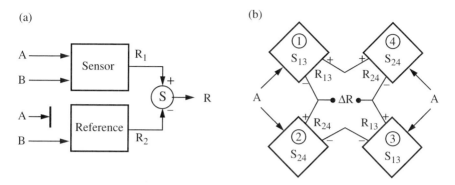

Figure 1.9 Schematic arrangements of sensor and auxiliary elements into (a) a sensor/reference pair and (b) a sensor bridge.

Sensor/Reference Pairs

A sensor/reference pair is designed to expose only one sensitive structure of a pair to the measurand of interest, A. Apart from this partial screening, both sensors are exposed to identical conditions. In particular, all parasitic influences are the same. The two output values of the two elements, $R_1 = T(A,B,P)$ and $R_2 = T(0,B,P)$, are subtracted from each other. The resulting difference $\Delta R = R_1 - R_2$ is

$$\Delta R = S_A A, \qquad (10)$$

which is valid for sufficiently small A values. The output is fully compensated for offset and for all parasitic influences, represented by the terms beyond $S_A \delta A$ in Eq. (4).

This compensation method is used in many sensors, including sensors for gas composition and flow, pressure, and radiation levels. Selected examples are shown in Fig. 1.10. Pellistors [Fig. 1.10(a)] are gas-sensitive pellets usually implemented as a pair.[33] One of the pellets is covered with a catalyst. The difference in thermal response is a measure of the presence of the gas reacting with the catalyzer. In recent arrays of thermal and resonant gas sensors, various elements have been covered with different gas-sensitive layers that are sensitive to a range of volatile organic compounds, as described in further detail in Chapter 11. Again, parasitic influences are eliminated, and gas composition is inferred from the different responses of the sensor array.

A thermal gas flow sensor is shown in Fig. 1.10(b). It consists of a central heating resistor on a micromachined thin-film membrane, and two thermopiles placed on either side of the heater. In the absence of a mass

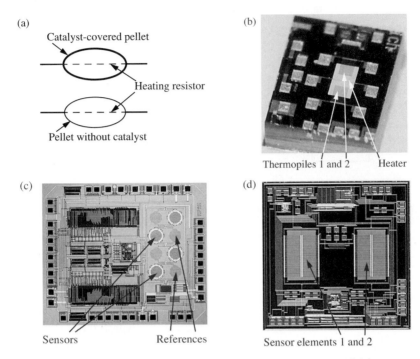

Figure 1.10 Four examples of sensor pairs for the measurement of (a) gas composition,[33] (b) mass flow,[35] (c) gas pressure,[37] and (d) infrared radiation.[38]

flow, both thermopiles measure the same temperature, thus providing the same output voltage V. Under mass flow, the downstream temperature sensor is cooled less than the upstream device. At small velocities, it is proportional to the average gas flow velocity v, as predicted by Eq. (10), although over a larger range, the difference signal depends nonlinearly on v. In comparison, the output voltages of the two individual thermopiles depend on v as $V_{\pm}(v) = V_0 - c_{\pm}v^{1/2} + \delta V_{\pm}(v)$, with a considerable offset V_0 and highly nonlinear v terms.[34-36] By the subtraction, the first two terms are eliminated rather efficiently and only the smoother term is left.

Similarly, thermal pressure sensors have been fabricated as pairs [Fig. 1.10(c)]. One element is surface-micromachined and measures the pressure-dependent thermal conductivity of the ambient gas, while the other has not been subjected to micromachining and thus is insensitive to the gas pressure.[37] Temperature variations of the supporting chip are effectively eliminated from the signal difference.

Finally, pyroelectric or thermoelectric infrared radiation detectors for presence detection also come in pairs, as illustrated by the integrated thermoelectric system shown in Fig. 1.10(d).[38] One detector element is exposed to the infrared radiation, while the second is obscured. The difference signal of the sensor pixels is insensitive to chip temperature and to constant temperature gradients.

Sensor Bridges

Sensor bridges are widely used in resistor-based sensors, such as piezoresistive pressure sensors[19] and gas flow sensors with resistive temperature monitors.[39] A Wheatstone bridge consists of four resistive elements arranged as two parallel voltage dividers, as shown in Fig. 1.9(b). The resistors are labeled 1 to 4. The bridge is driven by a constant voltage V_{in} or current I_{in}. Resistors 1 and 3 show the same sensitivity S_{13} to an input signal A, while resistors 2 and 4 have sensitivity S_{24}; the sensitivities of the two pairs are different and, if possible, of opposite sign. Alternatively, one branch can be designed with insensitive resistors, at the cost however of a halved overall bridge sensitivity. Upon application of an input signal A to the bridge in Fig. 1.9(b), an output voltage V_{out} equal to

$$V_{out} = \frac{(S_{13} - S_{24})A}{2R_0} V_{in} \qquad (11)$$

is measured between the contacts orthogonal to the driving voltage V_{in}. The output of voltage-biased Wheatstone bridges is temperature-compensated in the sense that the temperature sensitivities of the individual resistors do not appear in V_{in}. The remaining temperature dependence is that of the sensitivity only. In addition, as is evident from Eq. (11), the output of a symmetrical bridge with ideally matched resistors is offset-free.

In piezoresistive pressure sensors, resistors arranged in a Wheatstone bridge configuration are exposed to mechanical stress on a membrane deformed by differential pressures. This sensor principle is described in detail in Chapter 10. Resistive thermal gas flow sensors have used the same principle, with two temperature-sensing resistors placed upstream and downstream from a central heating element. This pair plays the role of resistors 1 and 4; the bridge is completed by two on-chip resistors insensitive to the gas flow. The output voltage starts at a residual offset and grows or decreases with the mass flow over the sensor, depending on flow direction.

Note that the bridge concept is not limited to resistors, as is often implicitly assumed. Any sensor type with voltage output that can be connected in the way sketched in Fig. 1.9(b) is a candidate for this procedure. A half-bridge constituted by the upper two devices, for example, is often able to perform the task of providing the required response ΔR. In fact, such half-bridges constitute sensor/reference pairs.

Sensactor Compensation

The second compensation method uses a different approach: It acts directly on the primary input channel of the device. The idea is to superimpose on the primary input A_1 a compensating input A_2 driven by the microsystem. The system is designed to show large sensitivities S_{A_1} and S_{A_2} to both inputs. The level of the compensating input is adjusted by the system so that the output of the device is zero. The response function of the system depends on the two inputs A_1 and A_2 as $R = T(A_1, A_2, B, P)$. In the linear case, including an offset R_{off}, the response is of the form $R = R_{off} + S_{A_1} A_1 + S_{A_2} A_2$, which is adjusted to 0 by an appropriate choice of A_2. By rewriting this as

$$A_2 = -\frac{R_{off}}{S_{A_2}} - \frac{S_{A_1}}{S_{A_2}} A_1 \qquad (12)$$

the compensating input A_2 is selected as the output signal of the transducer or system. It depends linearly on the input signal A_1 and can be compensated for the offset by calibration at $A_1 = 0$. Calibration also has to include the determination of the combined sensitivity S_{A_1}/S_{A_2}.

Usually the level of A_2 is adjusted by arranging the system in a control loop, as shown in Fig. 1.11. The generation of the secondary input is taken over by an actuator with effectivity E (defined by $A_2 = ER$), fed by the output of a controller with gain G. The sensor is assumed to exhibit an offset R_{off} and a sensitivity S to the total input signal resulting from the sum of the primary and secondary inputs. The overall response R of the system is then related to A_1 by

$$R = \frac{S_{A_1} G}{1 - S_{A_2} GE} A_1 + \frac{GR_{off}}{1 - S_{A_2} GE} \approx -\frac{S_1}{ES_{A_2}} A_1 - \frac{1}{ES_{A_2}} R_{off}, \qquad (13)$$

where the approximation is valid for sufficiently high gain G. Consequently, the effectivity E of the actuator has to be particularly well controlled, since it is the parameter determining the overall sensitivity of the system. Its sign has to be opposite to that of S_{A_1}/S_{A_2}. Also, as is evident from Eq. (13), sensor offset is not compensated. Neither is the offset of the

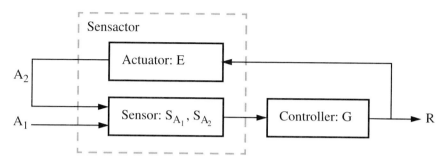

Figure 1.11 Schematic view of a sensactor system operated in a closed loop, enabling the sensor to be operated in a fixed working point. Even sensor nonlinearity can be eliminated if the actuator is linear.

controller, as a simple analysis of the system in Fig. 1.11 shows. If the actuator is linear, sensactor compensation provides an easy way for the compensation of a possibly nonlinear sensor characteristic, since the sensor is operated in a fixed working point (usually the null point). In fact, the overall system nonlinearity results from the actuator nonlinearity only.

The magnetic current sensor reported by Steiner et al.[40] and schematically shown in Fig. 1.12 provides a convincing implementation of this compensation scheme. An integrated Hall sensor is exposed to the primary magnetic field B_p generated by a loop of the primary current I_p to be determined. The Hall plate is surrounded by an on-chip secondary loop carrying the current I_c producing a field B_c required to compensate the

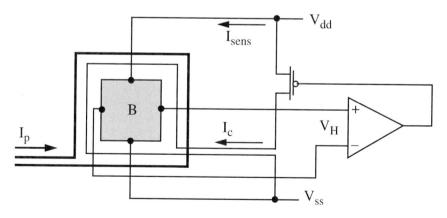

Figure 1.12 Magnetic current sensor with sensactor compensation.[40] The overall magnetic field B is the sum of primary and compensating fields B_p and B_c, respectively.

effect of B_p on the plate. Without compensation, the raw Hall voltage V_H would be the output signal. In the compensating system, it is used as the input for the circuit block stabilizing the system at $V_H \approx 0$ by tuning I_c, which becomes the output signal.

Similarly, acceleration sensors often implement force-balancing techniques, where an electrostatic force is applied to the seismic mass of the device, counteracting the inertial forces to which the seismic mass is exposed.

Instead of operating the system in the null point, other working points may be chosen that lend themselves similarly for the implementation of sensactor compensation. An example is the dew-point sensing for measuring relative humidity.[41] The sensor is a capacitive, conductive, or optical element experiencing a strong signal change upon condensation of humidity on the sensor surface. The actuator is often a Peltier element setting the temperature of the sensor. The fixed working point is given by the onset of dew condensation. The output is the temperature difference maintained by the Peltier element or, more precisely, the Peltier current needed to maintain this temperature difference.

A related but slightly trickier procedure was selected by Sunier *et al.* for the realization of a resonant magnetic sensor with frequency output.[42] The device relies on the excitation of a cantilever into resonance by an AC current flowing through a current lead on the cantilever. In an external magnetic field, the alternating current acts on the cantilever with a correspondingly alternating inductive force. The current level is set to be proportional to the out-of-plane deflection of the cantilever. This feedback elicits a linear dependence of the resonance frequency of the device on the magnetic field.

Chopping and Autozeroing

A simple method to eliminate cross-sensitivities and offset proceeds by periodically chopping the primary signal. Thus the desired signal is separated from undesired output contributions in the frequency domain. Subsequent filtering of the time-varying output in the frequency range of the modulation frequency suppresses all undesired influences outside the filter bandwidth. This method is schematically shown in Fig. 1.13. The various input and output signals have been supplemented with the frequency argument s, in view of the natural description of the system in the frequency (Laplace) domain.

Chopping is often applied in radiation-detection systems, where a mechanical shutter periodically blocks the sensor from the incoming radiation. It has been implemented in systems designed for optical signals,

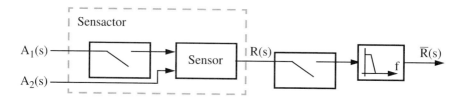

Figure 1.13 Cross-sensitivity suppression using the chopper-filter technique.

and also in thermoelectric and pyroelectric infrared detectors that require temperature changes of sub-milli-Kelvin amplitudes to be reliably detected.

Note that in circuit design, the chopping described here is also referred to as autozeroing. It is one of the preferred methods of eliminating amplifier offset, by periodically "zeroing" the amplifier, i.e., recalibrating its offset.

Cross-Sensitivity Cancellation in Saturating Sensors

Instead of turning the measurand on and off, as in the chopping technique, a highly nonlinear, saturating sensor can alternatively be exposed to a periodic signal that drives it into saturation. This approach is schematically shown in Fig. 1.14. The ideal sensor response, with its primary sensitivity S_1, goes through the origin and saturates at a value A_{sat}. Due to a cross-sensitivity to a signal P with sensitivity S_P, the sloped portion of the response curve is moved upward or, equivalently, shifted horizontally. A

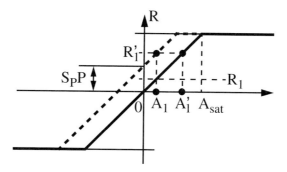

Figure 1.14 Cross-sensitivity cancellation in saturating devices. The vertical shift $S_P P$ due to the parasitic P is monitored by periodically driving the sensor into saturation.

primary input A_1 would therefore lead to the output R'_1, and thus would be erroneously interpreted as input A'_1. By driving the sensor into saturation under controlled conditions, the shift of the saturation point is recorded and R'_1 can be related to the correct value A_1. Variants of this scheme are applied in fluxgate magnetic sensors.[43]

1.4.3 Stabilization

A further method of compensating for a parasitic P consists of stabilizing the environment of the sensor with respect to a value of P. The most prominent example of stabilization is provided by the temperature stabilization implemented in many microsystems. Temperature is indeed an almost unavoidable influence affecting the sensitivity of virtually every sensor. Consequently, it is natural to try to keep this parameter fixed. Stabilization requires the implementation of auxiliary sensors and actuators arranged in feedback loops that can track a defined P value. This situation is schematically shown in Fig. 1.15. An external reference signal Y_{ref} sets the working point of the sensor with respect to the input $\bar{P}$. The external, varying value P is thus adjusted by the feedback loop to the stable value $\bar{P}$ that in turn affects the sensor in a stable way.

Stabilization is essential in many chemical sensors. Indeed, chemical detection processes (reaction rates, ionic conductivities, adsorption probability) often involve activation energies E_A that lead to strong temperature dependencies proportional to $\exp(-E_A/kT)$, where k and T denote Boltmann's constant and the absolute temperature, respectively. This temperature dependence is known as Arrhenius' law. The relevant process parameter can easily vary by several percent per degree. It is therefore of great importance to keep T constant within fractions of a degree. This has

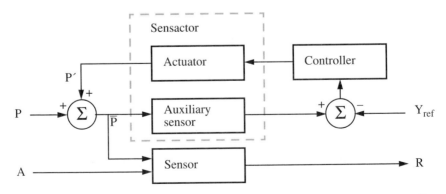

Figure 1.15 Sensor element operating at a stable level of the parasitic input $\bar{P}$.

been achieved in CMOS-based chemical sensors by locating the sensitive material on a thermally insulated membrane carrying a small silicon island with an integrated heating element and temperature monitor.[44]

Master-Slave Systems

Systems implementing the master-slave principle allow the simultaneous compensation of the nonlinearity and drift of the sensitivity, of cross-sensitivities, and of the offset. The general structure of such systems is schematically shown in Fig. 1.16. Ideally, the slave is a structure identical to the master sensor. It thus exhibits the same cross-sensitivities and offset as the master. While the primary signal incident on the master sensor is external to the system, the primary signal of the slave sensor is generated by the system in a control loop, ensuring the cancellation of the combined master and slave outputs.

The analysis of the system starts by noting the basic relationships between the outputs of the master and slave and their respective inputs:

$$R_m = S_A A + S_P \cdot P + R_{off} \quad (14)$$

and

$$R_s = S_{A'} A' + S_P \cdot P + R_{off}, \quad (15)$$

where it is assumed that master and slave show the same cross-sensitivities to parasitics and identical offsets. Ideally, the two structures are fabricated on the same substrate, i.e., on the same chip, in a compact arrangement, making sure that fabrication imperfections affect both structures as similarly as possible. The output of the controller is

$$R_c = G(R_m - R_s) = \frac{S_A G}{1 + S_{A'} GE} A \approx \frac{S_A}{ES_{A'}} A, \quad (16)$$

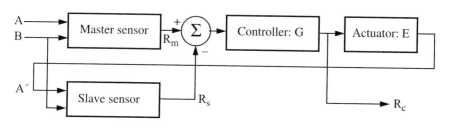

Figure 1.16 Schematic view of a master-slave system providing outputs R_s and R_c.

where E denotes the effectivity of the actuator. Obviously, the offsets and parasitics of both master and slave sensors are eliminated.

True rms thermal AC/DC converters have made use of this transduction principle. The purpose of thermal converters is to determine the integral power content of signals and to provide as the output signal the DC signal with the same power content. Several micromachined, true rms AC/DC converters have been reported.[17] Most consist of a pair of thermally insulated structures, each comprising a heating element and a temperature sensor, making it possible to determine the temperature increase caused by the heating. One of the structures is heated by the AC signal A of unknown power content. As a consequence of the heating, the temperature of the first structure is increased to the value R_m. The increase to R_m is constant if A' contains only components with frequencies significantly higher than the corner frequency $1/\tau$, where τ is the thermal time constant of the device. Simultaneously, the second structure is heated by a DC signal to the same temperature, $R_s = R_m$. In this particular case, A' rather than R_c is taken as the output of the system.

1.4.4 Multiple Measurements

Further compensation techniques build on the fundamental, quantitative understanding of the operation of a sensor and the origin of its cross-sensitivities and offset. The case is particularly simple to illustrate with a resistor used for temperature measurements, for example. As shown in Fig. 1.17, the device is contacted in a four-contact arrangement, with two contacts serving as current contacts and the two remaining contacts enabling an accurate measurement of the voltage drop over the device. Under ideal conditions, the voltage drop due to a current I_0 is $V_0 = I_0 \rho L/A$, where L and A denote the length and cross-sectional area of the device, respectively, and the resistivity corresponds to the device temperature T. If at the same time the resistor is exposed to a temperature gradient, the measured signal will be

$$V_+ = I_0 \rho_{av} L/A + \alpha \Delta T, \tag{17}$$

Figure 1.17 Schematic view of a resistive four-contact measurement. The voltage drop includes a parasitic thermoelectric voltage $\alpha \Delta T$ that can be canceled using a second measurement with reversed current.

with the relative Seebeck coefficient α between the resistor and contact materials, and the temperature difference ΔT between its two voltage contacts, and where ρ_{av} denotes the resistivity of the resistor averaged over its length. With the current flowing in the opposite direction, the voltage drop is

$$V_- = -I_0 \rho_{av} L/A + \alpha \Delta T. \tag{18}$$

By subtracting the two outputs, the voltage drop $V_0 = (V_+ - V_-)/2$, cleared of the thermoelectric cross-sensitivity, is determined and the desired average temperature of the sensor can be inferred. The disturbance $\alpha \Delta T$ may appear to be small in comparison with the voltage drop V_0. For example, with an assumed Seebeck coefficient of degenerately doped silicon of 100 μV/K against aluminum, a temperature difference of 1 K, a resistance value of 1 kΩ and a probe current of 1 mA leading to an ideal voltage drop of 1 V, the thermoelectric disturbance is only 100 ppm of V_0. Consequently, with a temperature coefficient of resistance of roughly 800 ppm/K, the temperature reading deduced from the sensor signal is offset by 0.125 K, which indeed may not be much, depending on the application.

In contrast, the situation is more critical in semiconductor-based Hall and stress sensors. An example of one such device is shown in Fig. 1.18. Integrated Hall plates usually consist of a diffusion with four peripheral contacts, a substrate contact, and possibly a further central contact of the diffusion for symmetric biasing. A current is passed across the device between opposite contacts, while an output signal is measured between the orthogonal contact pair. The output signal consists of several contributions, i.e., $V_{out} = V_G + V_H + V_{piezo} + V_{te}$, where V_G is caused by the imperfect

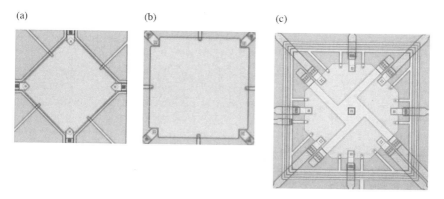

Figure 1.18 Optical micrographs of integrated stress sensors based on the pseudo-Hall effect. The devices (a) and (b) have different orientations with respect to the crystal axes, enabling different stress components to be extracted after elimination of magnetic and thermal parasitics by multiple measurements. Device (c) combines (a) and (b) in a single octagonal component.[47]

geometry of the device due, e.g., to imperfect photolithography and structuring, and inhomogeneous doping; V_H is the Hall signal proportional to the magnetic induction perpendicular to the plate; V_{piezo} is caused by the mechanical stress to which the chip with the sensor is exposed in its encapsulation and due to operating conditions; and V_{te} denotes a possible thermoelectric signal due to the component of the temperature gradient collinear with the Hall voltage contacts. If the direction of the primary current is successively switched in steps of 90°, the signs of V_G and V_{piezo} are inverted at each step while that of V_H remains the same; V_{te} is changed to a value $V_{te,orth}$ corresponding to the orthogonal component of the temperature gradient, then to $-V_{te}$ and $-V_{te,orth}$. The sign changes of V_G, V_{piezo}, and V_H are illustrated in Fig. 1.19(a) through (d). The sign changes are the consequence of the general properties of linearly conducting materials with four contacts.[45] These properties directly result from the symmetry properties of the conductivity tensor deduced from the theory of irreversible processes, as compactly stated in the "Onsager relation."[46]

In view of these sign changes, it is clear that

- Two-fold switching of the current by 180° and averaging of the resulting output signal eliminate the thermoelectric contribution
- Two-fold switching of the current between current direction separated by 90° eliminates geometrical imperfections and the stress influence
- Four-fold switching between all four current directions and averaging of the four outputs eliminate all contributions except the magnetic signal

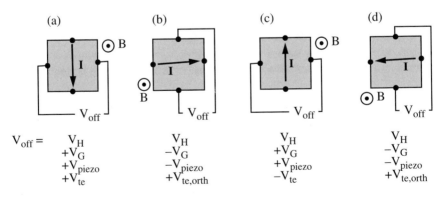

Figure 1.19 Sign change of various desirable and undesirable signal components in resistive four-point devices.

- Four-fold switching between all four current directions and averaging of the four outputs with the additional prefactors +1/−1/+1/−1 eliminate the thermoelectric and magnetic contributions; consequently, a device with ideal geometry can be operated as a stress sensor.[47]

In Hall sensors, reductions of the contributions V_{piezo} and V_G by four-fold switching from the mV range down to the lower μV range, i.e., by a factor of 10^3, are achievable. For this level of reduction, careful instrumentation is a necessity. Compensations similar to those listed can be achieved by placing two or four plates close to each other, letting currents flow through them in opposite or orthogonal directions, and connecting the output leads in parallel or oppositely, as shown in Fig. 1.20. However, possible gradients of the stress and temperature inhomogeneities over the area of the two or four devices make the compensation less accurate.

The idea of discrete switching has been generalized to the case of magnetic sensors and stress sensors with more than four contact pairs[47-50] and to the continuous rotation of the primary current.[51,52] The increase in reduction efficiency of these schemes beyond that of four-fold switching is modest, however.

Note that the techniques described in Sections 1.4.2 to 1.4.4 can be combined at leisure to build up more complex systems. As examples, the sensor in Fig. 1.15 may consist of a sensor pair or a bridge-type device, and in the auxiliary sensor, multiple measurement concepts may be applied.

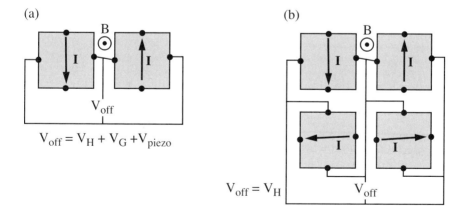

Figure 1.20 Schematic arrangement of multiple sensors for the inherent elimination of parasitic signal components from the overall output V_{off}.

1.4.5 Circuitry

Electronic and microelectronic circuitry, whether discrete or integrated, plays a large role in realizing the schemes described in Sections 1.4.1 through 1.4.4. Indeed, circuitry blocks may take over the functions of data storage units, amplifiers, filters, and actuators, among others. Techniques to implement these functions using the rich experience of circuit designers are described in detail in Chapter 17; this section summarizes a few central ideas.

In view of the small power required to drive and operate microsensors and of the small level of typical output signals, much effort in circuit development has been invested into techniques realizable in commercial IC technologies, in particular CMOS technologies. Despite the amazing capabilities offered by state-of-the-art CMOS technologies, circuit components, like sensors, have fundamentally remained imperfect technical objects. Their nonidealities limit the overall system performance, have to be taken into account in the overall system design, and possibly have to be eliminated or compensated for by appropriate techniques. The principal drawbacks of integrated circuits are noise, offset, frequency limitations, nonlinearity, and size.

Circuit noise contains thermal components with spectral density independent of frequency, $1/f$ noise, and discretization noise, among other contributions. Discretization noise dominates at small frequencies up to a characteristic corner frequency. Offset is due to imperfect matching of transistors, resistors/conductors, and capacitors. Both noise and offset have to be minimized under the additional constraint of minimal chip real estate and power consumption. Clearly, circuit designers are dealing with conflicting goals.

Amplifiers are essential components of microtransducer systems and form the core of gain stages and controllers, as implemented for example in Figs. 1.11, 1.15, and 1.16. The primary signals or signal changes offered by microtransducers are often small, in the range of mV and below, down to the µV level. The task of amplifiers is thus to boost the small output signals into the mV or V range, where they are robust against parasitic influences. It is best to perform this task as close to the microtransducer as possible, since any long lead connecting the transducer to the circuit contributes stray capacitance and acts as an antenna picking up the spectrum of electromagnetic radiation pervading the microsystem.

It is important to give the noise contributed by the circuitry sufficient attention in the context of the system. It is meaningless to amplify a low-noise transducer signal with a noisy amplifier stage. Neither does it make much sense to invest excessive effort in designing an amplifier with an input-related noise level far below the output noise of the transducer element.

A further important criterion for microelectronic components is power consumption. The art of sensor circuit design is to reconcile the conflicting goals of low power, low noise, and low offset.

Several techniques have been proposed to achieve low-offset, low-noise amplification. Among CMOS instrumentation amplifiers, a system with remarkable performance data shows a spectral noise-power density as low as 6.5 nV/√Hz (referred to the input), a residual offset of 200 nV, and a common mode rejection ratio larger than 155 dB.[24] This system is a switched-capacitor chopper amplifier realized in a double-differential architecture. Power consumption at 5 V is about 3 mW.

Chopper amplifiers hash the input signal and move the signal band to bands around odd harmonics of the chopper frequency. The sensor signal band is thus outside the range of the $1/f$ noise of the microelectronic components. Upon demodulation, the signal band is moved back to its original position around $f = 0$, while the $1/f$ noise distribution is moved away from the origin and only thermal noise remains in the signal band. After filtering with a low-pass filter matched in bandwidth to the signal bandwidth, the original signal is restored, distorted only by thermal noise within the signal bandwidth. Sigma-delta ($\Sigma\Delta$) amplifiers represent an alternative method to reshape the noise effectively in the signal bandwidth. Again, subsequent filtering of the conditioned signal is necessary to remove the noise power outside the signal bandwidth.[53]

Filters can be quite expensive to implement in terms of chip area. This is especially true for lower bandwidths, where use of external filters may be more advantageous. Resistor-capacitor filters are straightforward to realize in IC technologies. They have the largest area and power consumption of all methods. Switched capacitor filters are similar in structure. However, the resistors are replaced by switched capacitors. They occupy more modest area and benefit from low power dissipation. However, they show noise and aliasing, and they need clock circuits for the switching and anti-aliasing filters. Finally, transconductance-capacitor filters, where resistors are replaced by transconductance stages, can be realized as relatively compact designs with low power consumption. Unfortunately, transconductance stages are nonlinear and thus need to be linearized.

With the progressive shrinking of the critical dimensions of CMOS technologies and the capability of squeezing more and more computing power into the same small area, it is tempting for MEMS circuit developers to convert the sensor signals into the digital domain as quickly as possible and to have the system perform as much work as possible in the digital domain. Digital circuit design is attractive because it relies on sophisticated tools to translate digital logical procedures into microelectronic layouts in a highly automated way. It is therefore simpler than analog design, where intuition and time-consuming manual and numerical analyses are still part of the designer's job. Nevertheless, it is still advantageous to address and cure the imperfections of microtransducers at their root. Keeping the system in the analog domain allows much of the work that would otherwise have to be carried out digitally by the system to be eliminated by simple compensation tricks. Time-continuous analog techniques are limited by thermal, $1/f$, shot, and sensor noise. Digital signal handling also suffers from discretization noise, which requires high oversampling rates with high-resolution analog-digital converters for low-noise power densities. Oversampling and the processing of digitized high-resolution signals consume power and chip area.

1.5 Powering Microsystems

Active microsystems require some powering to acquire information, process it, and transmit it on. Most current microsystems receive their energy via electrical cables. Considering that the external power source may not be stable, it is advisable to provide the system with a voltage-stabilizing block such as a bandgap reference. However, for some applications, cables are inconvenient. This is the case with microrobots, where the tension in the cable affects the robot motion. Similarly, autonomous miniaturized sensor systems that can communicate, termed smart motes or e-grains, distributed over some observation space are expected to be able to carry out their tasks in a standalone mode. The requirement of minimal disturbance of the environment by such autonomous sensor networks is incompatible with cables connecting the components to a central unit or among each other.

A number of alternative powering schemes are therefore in use or are currently being developed.[54] They include:

- Local energy storage
- Miniaturized fuel cells
- Optical energy transmission
- Electromagnetic energy transmission
- Energy harvesting

1.5.1 Local Energy Storage

In the case of local energy storage, the energy source is made available within or close to the microsystem. Electrical or electrochemical energy is the most often used storage form. The energy is provided from a loaded battery or a charged condenser. Battery technology is well advanced and offers the advantage of durability and a relatively stable voltage. However, size is a concern, since batteries miniaturized to the dimensions typical of microsystems are not commercially available. The current situation leads to rather disproportionate demonstrations, where a mini-cell battery is the substrate for a much smaller actuator microsystem. All the system is able to do is to push the battery cell over a horizontal surface.[55] Undeniably, such systems demonstrate the idea of microsystem autonomy. Simultaneously, they illustrate the unsatisfactory state of battery development.

Attempts have been made to use thin-film technology to realize rechargeable micro lithium-ion batteries with a thickness of less than 1 µm. The element was rechargeable to potentials up to 3.6 V.[56] In the area of condenser research, miniaturized capacitors with a diameter of 6 mm, a thickness of 2 mm, and an electrical capacity of 0.33 F have been used, for example, by Seiko Epson to power a 1 cm^3 small microrobot for 5 minutes.

Alternative ways may be provided by storage of currents in superconducting loops. Unfortunately, as long as maximum transition temperatures of superconductors remain below room temperature, superconducting storage will remain an exotic idea.

1.5.2 Miniaturized Fuel Cells

Fuel cells consist of an electrode/electrolyte/electrode arrangement that allows electrical energy to be extracted from a chemical reaction. The process involves two reactants, e.g., H_2 and O_2, which are used here to sketch the general operating principle of fuel cells. First, the hydrogen is catalytically ionized by one of the electrodes. Electrons are collected by that electrode and reach the second electrode after crossing the load circuit connecting the two electrodes. The proton, on the other hand, diffuses through the electrolyte to the second electrode. There, two protons recombine with one oxygen atom and two electrons from the circuit into an H_2O molecule.

Fuel cell technologies are classified by operating temperature and their electrolyte. So-called low-temperature fuel cells typically operate at temperatures below 200°C. In view of their operating temperatures between 600 and 1000°C, high-temperature fuel cells are better suited for application at the industrial scale than in microsystems. Low-temperature fuel cell variants include the alkaline fuel cell (AFC), with the electrolyte KOH; the reacting species O_2 and H_2, and OH^- acting as the charge carrier in the electrolyte, at operating temperatures between 60 and 200°C; the polymer electrolyte membrane fuel cell (PEMFC), in which O_2 reacts with H_2 or methanol and where proton transport occurs in a polymer membrane at 60 to 100°C; the phosphoric acid fuel cell (PAFC), with liquid phosphoric acid as the electrolyte, ensuring proton transport, and the same reaction and charge carrier in the electrolyte as in the PEMFC.

Recently, significant effort has been invested in miniaturizing fuel cells.[57,58] Miniaturization implies several challenges, including the efficient transport of the "fuels" to the catalyst/electrode surfaces, the low-loss extraction of the electrical charge carriers, the elimination of the water that is a reaction product, and the thermal management of the compact cell with its increased power density. Müller et al.[59] have successfully addressed these issues. They have presented a stacked PEMFC with a volume of only a few cm^3 that is able to deliver 8 W of peak power and 6.5 W of continuous power. The power density generated by the device is larger than 1 W/cm^3. Each stack level consists of a carbon fiber-paper diffuser dispensing the hydrogen to the anodic side of a Gore Primea Select 5000 polymer sheet with its two electrode layers; the cathodic side features a thin carbon-fiber diffusion layer and a micro flow-channel field micromachined in brass, with a channel cross section of 300×300 μm^2. The flow-channel field ensures the efficient transport of water vapor out of the cell.

Miniaturization below 1 cm^3 will require further improvements in all components or the application of novel concepts. A first step in this direction was reported by Müller et al.[59] Their approach is based on the exclusive use of foils for all components of the cell. At a thickness of 570 µm, the device delivers a power density of 13 mW/cm^2.

1.5.3 Optical and Electromagnetic Energy Transmission

Energy is conveniently transmitted in the form of electromagnetic radiation. A beam of laser light can be focused on a miniaturized photovoltaic cell belonging to a microsystem. Photovoltaic elements fabricated in III-V technology can be optimally tailored to the laser wavelength. Conversion efficiencies up to 40% have thus been achieved.[60] In view of the high light intensities offered by lasers, energy transmission is an efficient method.

In environments where safety considerations preclude the use of lasers, telemetry offers an alternative. Electromagnetic radiation is dissipated by a primary coil at a frequency in one of the IMS (industrial, medical, and scientific) bands open to the public and is collected by a receiver coil integrated with the microsystem. It is then rectified, filtered, and made available to the system as a suitable voltage. The concept is now widespread in radio-frequency identification (RF-ID) systems, where the dimensions of the secondary coil are roughly those of a credit card. Smaller coils require higher electromagnetic radiation levels. These levels can be ensured by reducing the distance between the primary and secondary coils and by optimizing the design of the secondary coil with regard to its inductive, capacitive, and resistive characteristics.[61]

1.5.4 Energy Harvesting

Lately, the concept of energy harvesting has received increased attention, in view of the need for powering miniaturized autonomous systems. The idea consists of tapping into the "energy reservoir" surrounding and pervading the microsystem. Such harvesting involves the conversion of an energy present in any domain into a usable electrical form. Such concepts are subject to the fundamental laws of thermodynamics and can work only if the energy reservoir to be exploited is not in thermodynamic equilibrium with the temperature of the microsystem.

The vibrational energy present in many mechanical systems can be harvested using inverse electromechanical conversion. Systems for this kind of conversion usually comprise a micromechanical resonator, with a miniaturized magnetic element capable of generating a current in a coil by induction, or piezoelectric components translating the motion directly into the electrical domain. Alternatively, so-called electrets, i.e., materials presenting a fixed surface charge, mounted on the resonator are able to drive charge carriers capacitively into and out of nearby electrodes.[62,63]

Thermal energy can be converted using the thermoelectric effect. Unfortunately, even the thermoelectric materials with the highest figures of merit, belonging, for example, to the Bi-Sb-Te system, offer rather modest conversion efficiencies. Consequently, at best, ultra-low-power systems can be powered by miniaturized thermoelectric generators. A recently reported converter based on semiconductor thermoelectrics has been fabricated using standard CMOS technology and the combined field of n-polysilicon/p-polysilicon thermocouples with micromachining technology.

The thermophotovoltaic approach is based on the conversion of thermal radiation into electrical energy. Thermal photons emitted by bodies at elevated temperatures, e.g., components of combustion engines, are collected by photovoltaic cells with their bandgap matched to the energy of the thermal radiation. This approach benefits from expertise in bandgap and device engineering in the field of III-V and II-VI semiconductor technology.

Nuclear photovoltaics exploits the excitation of electron-hole pairs in photovoltaic elements by the decay products of nuclear processes and the subsequent separation in the photovoltaic element. In an interesting alternative, beta particles from nuclear decay processes charge the tip of a micromachined cantilever. This mechanical element is consequently attracted to the oppositely charged beta-active material. Upon contact, the charges are neutralized, and the beam returns to equilibrium by a damped vibration, the energy of which is harvested by a piezoelectric element. Both approaches are in the initial stages of their development and involve health issues that may restrict their usability.

Even more challenging concepts involve the harvesting of the electromagnetic radiation field, in particular the electrosmog in technical environments[65] and the consumption of organic matter in so-called biofuel cells.[66]

References

1. J. Fluitman, "Microsystems Technology: Objectives," *Sensors and Actuators*, A56(1–2):151–166 (1996)
2. W. Ko, "The Future of Sensor and Actuator Systems," *Sensors and Actuators*, A56(1–2):193–197 (1996)
3. *Sensors*, vols. 1–9 (W. Göpel *et al.*, eds.), Wiley VCH (1989–1995); *Sensor Update*, vols. 1–13 (H. Baltes *et al.*, eds.), Wiley VCH (1996–2004)
4. J. Fraden, *Handbook of Modern Sensors – Physics, Designs, and Applications*, 3rd ed., Springer (2003)
5. J. G. Webster, *The Measurement, Instrumentation and Sensors Handbook*, CRC Press (1998)
6. S. Middelhoek and S. A. Audet, *Silicon Sensors*, Delft University Press (1994)
7. A. Zeilinger, "Bell's Theorem, Information and Quantum Physics," in *Quantum [Un]Speakables. From Bell's Theorem to Quantum Information* (R. Bertlmann and A. Zeilinger, eds.), Springer (2002), pp. 241–254
8. A. Zeilinger, "Quantum Entangled Bits Step Closer to Information Technology," *Science*, 289:405–406 (2000)
9. G. Stix, "Best-Kept Secrets," *Sci. Am.*, 292(1):64–69 (2005)
10. S. M. Sze, "Classification and Terminology of Sensors," in *Semiconductor Sensors* (S. M. Sze, ed.), John Wiley & Sons, NY (1994)
11. C. Y. Chang and S. M. Sze, *ULSI Technology*, McGraw Hill (1996)
12. S. Sze, *Physics of Semiconductor Devices*, Wiley Interscience Publications (1981)
13. Y. Tsividis, *Operation and Modelling of the MOS Transistor*, 2nd ed., Oxford University Press (2003)
14. P. R. Gray, P. J. Hurst, S. H. Lewis, and R. G. Meyer, *Analysis and Design of Analog Integrated Circuits*, 4th ed., John Wiley & Sons (2001)
15. S. Middelhoek and D. J. W. Noorlag, "Three-Dimensional Representation of Input and Output Transducers," *Sensors and Actuators*, 2(1):29–41 (1981)

16. R. J. van de Plassche, J. H. Huijsing, and W. M. C. Sansen, *Analog Circuit Design: RF Analog-to-Digital Converters; Sensor and Actuator Interfaces; Low-Noise Oscillators, PLLs and Synthesizers*, Springer (1997)
17. D. Jaeggi, J. Funk, A. Haberli, and H. Baltes, "Overall System Analysis of a CMOS Thermal Converter," *Digest of Technical Papers, 8th International Conference on Solid-State Sensors and Actuators and Eurosensors IX, Stockholm, Sweden*, vol. 2 (1995), pp. 112–115
18. T. W. Kenny, W. J. Kaiser, S. B. Waltman, and J. K. Reynolds, "Novel Infrared Detector Based on a Tunneling Displacement Transducer," *Appl. Phys. Lett.*, 59(19):1820–1822 (1991)
19. B. Kloeck and N. F. de Rooij, "Mechanical Sensors," in *Semiconductor Sensors* (S. M. Sze, ed.), John Wiley & Sons, NY (1994)
20. S. Middelhoek and A. Hoogerwerft, "Classifying Solid-State Sensors – The Sensor Effect Cube," *Sensors and Actuators*, 10:1–8 (1986)
21. W. M. Deen, *Analysis of Transport Phenomena*, Oxford University Press (1998)
22. R. K. Pathria, *Statistical Mechanics*, Butterworth-Heinemann (1996), chap. 14
23. R. Kubo, "Fluctuation-dissipation theorem," *Rep. Prog. Phys.*, 29:255–284 (1966)
24. Q. Huang and C. Menolfi, "A 200 nV Offset 6.5 nV/square root Hz Noise PSD 5.6 kHz Chopper Instrumentation Amplifier in 1 mu m Digital CMOS," *Digest of Technical Papers, IEEE International Solid-State Circuits Conference ISSCC 2001*, IEEE (2001), pp. 362–363
25. S. D. Senturia, *Microsystem Design*, Kluwer Academic Publishers (2001), sect. 16.5
26. R. K. Pathria, *Statistical Mechanics*, Butterworth-Heinemann (1996), chap. 3
27. S. van Herwaarden and G. C. M. Meijer, "Thermal Sensors," in *Semiconductor Sensors* (S. M. Sze, ed.), John Wiley & Sons, NY (1994)
28. B. E. Cole, R. E. Higashi, and R. A. Wood, "Monolithic Two-Dimensional Arrays of Micromachined Microstructures for Infrared Applications," *Proc. IEEE*, 86(8):1679–1686 (1998)
29. J. D. Jackson, *Classical Electrodynamics*, 3rd ed., John Wiley & Sons (1998)
30. S. Sze, *Physics of Semiconductor Devices*, Wiley Interscience Publications (1981), chap. 13
31. A. Häberli, "Compensation and Calibration in IC Microsensors," Ph.D. dissertation, ETH Zurich, no. 12090 (1997)
32. G. C. M. Meijer and A. W. van Herwaarden, *Thermal Sensors*, Institute of Physics Publishing, Bristol (1994)
33. S. D. Kolev, M. Adam, I. Barsony, C. Cobianu, and A. van den Berg, "Thermal Properties of a Silicon Based Micro-Pellistor with Suspended Bridge Structure," *Proceedings of the 12th European Conference on Solid-State Transducers (Eurosensors XII)*, vol. 1 (1998), pp. 273–276
34. D. Moser and H. Baltes, "A High Sensitivity CMOS Gas Flow Sensor on a Thin Dielectric Membrane," *Sensors and Actuators*, A37–38:33–37 (1993)

35. O. Paul, J. Robadey, and H. Baltes, "Two-Dimensional Integrated Gas Flow Sensors by CMOS IC Technology," *J. Micromech. Microeng.*, 8:243–250 (1995)
36. F. Mayer, A. Häberli, G. Ofner, H. Jacobs, O. Paul, and H. Baltes, "Single-Chip CMOS Anemometer," *Tech. Digest Intl. Electron Devices Meeting IEDM '97, Washington DC*, IEEE (1997), pp. 895–898
37. A. Häberli, O. Paul, P. Malcovati, M. Faccio, F. Maloberti, and H. Baltes, "CMOS Integration of a Thermal Pressure Sensor System," *Proc. ISCAS 96, Atlanta*, vol. 1, IEEE (1996), pp. 377–380
38. R. Lenggenhager, D. Jaeggi, P. Malcovati, H. Duran, H. Baltes, and E. Doering, "CMOS Membrane Infrared Sensors and Improved TMAHW Etchant," *Tech. Digest Intl. Electron Devices Meeting IEDM 1994*, IEEE (1994), pp. 531–534
39. R. G. Johnson and R. E. Higashi, "A Highly Sensitive Chip Microtransducer for Air Flow and Differential Pressure Sensing Applications," *Sensors and Actuators*, 11:63–72 (1987)
40. R. Steiner, M. Schneider, F. Mayer, U. Munch, T. Mayer, and H. Baltes, "Fully Packaged CMOS Current Monitor Using Lead-On-Chip Technology," *Proc. MEMS 1998 Workshop* (1998), pp. 603–608
41. A. Steinke, H.-G. Ortlepp, G. Brokmann, and B. March, "Intelligent Hybrid Sensors," *Proc. NORTECH 2000, Bergen, Norway* (2000)
42. R. Sunier, Y. Li, K.-U. Kirstein, T. Vancura, H. Baltes, and O. Brand, "Resonant Magnetic Field Sensor with Frequency Output," *Tech. Digest IEEE MEMS 2005 Conference, Miami Beach, Jan. 2005*, IEEE (2005), pp. 339–342
43. S. Kawahito, A. Cerman, K. Aramaki, and Y. Tadokoro, "A Weak Magnetic Field Measurement System Using Micro-Fluxgate Sensors and Delta-Sigma Interface," *IEEE Trans. Instrum. Meas.*, 52(1):103–110 (2003)
44. M. Graf, S. Taschini, P. Käser, C. Hagleitner, A. Hierlemann, and H. Baltes, "Digital MOS-Transistor-Based Microhotplate Array for Simultaneous Detection of Environmentally Relevant Gases," *Tech. Digest IEEE MEMS 2004 Conference, Maastricht, Jan. 2004*, IEEE (2004), pp. 351–354
45. H. H. Sample, W. J. Bruno, S. B. Sample, and E. K. Sichel, "Reverse-Field Reciprocity for Conducting Specimens in Magnetic Fields," *J. Appl. Phys.*, 61(3):1079–1084 (1987)
46. L. Onsager, *Phys. Rev.*, 37:405 (1931); *Phys. Rev.*, 38:2265 (1931)
47. J. Bartholomeyczik, P. Ruther, and O. Paul, "Multidimensional CMOS In-Plane Stress Sensor," *IEEE Sensors J.* (2005), in press
48. A. A. Bellekom and P. J. A. Munter, "Offset Reduction in Spinning-Current Hall Plates," *Sensors and Materials*, 5(5):253–263 (1994)
49. R. Steiner, "Rotary Switch and Current Monitor by Hall-Based Microsystems," Ph.D. dissertation, ETH Zurich, no. 13135 (1999)
50. M. Doelle, P. Ruther, and O. Paul, "Novel Highly Miniaturized Multi-Stress Sensor Based on Eight-Terminal Field Effect Transistor," *Digest of Tech. Papers, 13th Intl. Conference on Solid-State Sensors and Actuators (Transducers 2005), Seoul, June 4–8* (2005), in press

51. R. Steiner, C. Maier, A. Haberli, F.-P. Steiner, and H. Baltes, "Offset Reduction in Hall Devices by Continuous Spinning Current Method," *Sensors and Actuators*, A66:167–172 (1998)
52. J. Bartholomeyczik, S. Kibbel, P. Ruther, and O. Paul, "Extraction of Compensated $\sigma_{xx}-\sigma_{yy}$ and σ_{xy} Stresses from a Single Four-Contact Sensor Using the Spinning Transverse Voltage Method," *Tech. Digest IEEE MEMS 2005 Conference, Miami Beach, Jan. 2005*, IEEE (2005), pp. 263–266
53. O. Bajdechi, "Systematic Design of Sigma-Delta Analog-To-Digital Converters," Ph.D. dissertation, Technical University of Delft (2003)
54. F. Arai and T. Fukuda, "Energy Source and Power Supply Method," in *Micromechanical Systems – Principles and Technology* (T. Fukuda and W. Menz, eds.), Elsevier (1998)
55. B. Warneke, B. Atwood, and K. S. J. Pister, "Smart Dust Mote Forerunners," *Tech. Digest IEEE MEMS 2001 Conference, Interlaken, Jan. 2001*, IEEE (2001), pp. 357–360
56. J. B. Bates, "Rechargeable Solid State Lithium Microbatteries," *Proc. Micro Electro Mechanical Systems Workshop, Fort Lauderdale, Feb. 1993* (1993), pp. 82–86
57. M. Müller, C. Müller, W. Menz, and C. Hebling, "Fuel Cell Using Micro-Structured Flow Fields," *Proc. Micro Syst. Technol. Conf., Düsseldorf* (2001)
58. E. Sakaue, "Micromachining/Nanotechnology in Direct Methanol Fuel Cells," *Tech. Digest, IEEE MEMS 2005 Conference, Miami Beach, Jan. 2005*, IEEE (2005), pp. 263–266
59. M. Müller, "Polymermembran-Brennstoffzellen mit mikrostrukturierten Strömungskanälen," Ph.D. dissertation, IMTEK, University of Freiburg (2002)
60. S. Glunz, Fraunhofer Institute for Solar Energy Systems, private communication
61. Klaus Finkenzeller, *RFID-Handbook: Fundamentals and Applications in Contactless Smart Cards and Identification*, 2nd ed., Wiley & Sons (2003)
62. T. Genda, S. Tanaka, and M. Esashi, "Micro-Patterned Electret for High-Power Electrostatic Motor," *Tech. Digest, IEEE MEMS 2004 Conference, Maastricht, Jan. 2004*, IEEE (2004), pp. 470–473
63. J. S. Boland, J. D. M. Messenger, H. W. Lo, and Y. C. Tai, "Arrayed Liquid Rotor Electret Power Generator System," *Tech. Digest IEEE MEMS 2005 Conference, Miami Beach, Jan. 2005*, IEEE (2005), pp. 618–621
64. M. Strasser, R. Aigner, M. Franosch, and G. Wachutka, "Miniaturized Thermoelectric Generators Based on Poly-Si and Poly-SiGe Surface Micromachining," *Digest of Tech. Papers, 11th Intl. Conference on Solid-State Sensors and Actuators (Transducers 2001), Munich, June 2001* (2001), pp. 26–29
65. L. Reindl, private communication
66. R. Zengerle and F. von Stetten, private communication

2 Material Properties: Measurement and Data

Osamu Tabata and Toshiyuki Tsuchiya
Kyoto University, Japan

2.1 Introduction

The performance of MEMS strongly depends on the properties of materials used not only as functional materials but also as structural materials. Therefore, the materials used for MEMS are required to possess many properties for applications with electrical, mechanical, thermal, magnetic, optical, and chemical requirements according to the specification of individual MEMS. Most of the MEMS use single-crystal silicon (SCS) as a substrate material because of its superior electrical and mechanical properties. The superiority of electrical properties of SCS has been widely accepted from the time they were first used as a substrate of large-scale integrations (LSIs). Furthermore, SCS has been used as a structural material for mechanical sensors such as piezoresistive pressure sensors and acceleration sensors since the early 1960s.[1] The great success of silicon-based mechanical sensors is attributable to the excellent mechanical properties of SCS as well as its well-established piezoresistive effect. A detailed account of mechanical sensors and their applications is given in Chapter 10.

For the silicon piezoresistive pressure sensors, an SCS wafer is micromachined using a silicon crystallographic anisotropic wet-etching technique to make a thin diaphragm structure that is a few mm square and a few tens of μm thick. This microfabrication process using SCS as the structural material is called bulk-micromachining. In the late 1980s, a surface-micromachining process using the sacrificial-layer-etching technique was proposed. With this technique, thin-film materials are used instead of SCS as structural materials. The importance of thin films as a structural material increased as the application of surface-micromachining technology increased. Micromachining technologies are described in detail in Chapter 15.

The difference between SCS and thin films from the viewpoint of MEMS is the dependence of material properties on the process conditions. Although the material properties of SCS are known, those of thin films are not known because they strongly depend on the deposition conditions, deposition apparatus, film thickness, and post-process, such as annealing. Furthermore, as the thickness of films shrinks, surface effects will no longer be negligible. Because of this complexity, it is difficult to know the exact material properties of a thin film without detailed information about the process condition, apparatus, film dimensions, and so on.

However, the material properties of thin films should be known and controllable for both design and process engineers. In the case of a MEMS design, a simulation should be carried out to estimate the MEMS performance based on accurate material property data used for the MEMS to be developed. The data values of material properties that appear in the literature could be different from those deposited in the process to be used. Therefore, accurate material property data used for the MEMS to be developed are required to carry out a precise simulation. In the case of process development, material properties should be measured and controlled to the values required by the design engineers by adjusting the process conditions. At the production stage, the material properties should be monitored to be stable in the developed process. Consequently, there is a strong demand from MEMS design and process engineers for a material properties database and measurement method for structural thin films used for MEMS.

According to the expanding application field of MEMS, the thin-film material properties related to MEMS performance have been expanded to electrical, mechanical, thermal, optical, and magnetic properties. The electrical, optical, and magnetic properties of thin films have already been extensively reported in the literature because these properties were required in LSI and optical devices. On the other hand, there is little information regarding the mechanical and thermal properties of thin films because the existing devices have been less demanding with respect to these properties. Therefore, the most urgent need of MEMS engineers is for information about the mechanical and thermal properties of thin films. This chapter focuses on these properties.

The measurement of mechanical and thermal properties of thin film has been difficult because the existing measurement apparatus for bulk material cannot be applied to a thin film. Since the thickness ratio of substrate and thin film is 100 to 1000 or more, it is difficult to measure the properties of thin films when they are attached to a substrate. Even if the thin film is peeled off a substrate, existing measurement setups cannot be used for this purpose due to the different measurement ranges and the difficulty of handling fragile thin films. In the late 1980s, the importance of mechanical property measurement of thin films for MEMS was recognized and several approaches were proposed to measure thin-film mechanical properties using microfabricated test structures comprising the thin film of interest as a structural material.[2] The advantages of using the microfabricated test structure itself to measure mechanical properties are (1) it can be used as an in-process monitoring technique, and (2) it can measure the local mechanical properties. For example, the most popular measurement technique for the internal stress uses the curvature of a wafer or a rectangular-shaped sample with the thin film of interest.[3,4] It is not applicable for measuring the internal stress distribution within a restricted sample area. For these reasons, microfabricated test structures have been widely recognized as a useful approach, and many improved methods have been published.

This chapter describes methods for measuring mechanical and thermal properties of thin-film materials important for MEMS and data acquired with these methods. Section 2.2 introduces methods for measuring thin-film properties using microfabricated test structures. First the methods for measuring internal stress, Young's modulus, and Poisson's ratio are explained; these properties govern the mechanical behavior such as deflection or deformation of the moving parts. Then methods of measuring yield strength, fracture strength, fracture toughness, and fatigue of the MEMS thin-film materials, including SCS, are shown. These are included because there has been strong and increasing interest in MEMS reliability related to these material properties as the MEMS field has matured. Even for SCS, there have been few reports about its strength and fatigue. Finally, thermal conductivity and specific heat measurement methods are introduced. These properties govern thermal behavior such as temperature distribution or change in MEMS.

Section 2.3 presents measured material data available from the literature for SCS and polycrystalline silicon (poly-Si), which are the most common structural materials of MEMS, and for other materials such as metal thin films and dielectric thin films.

2.2 Measurement Methods

2.2.1 Internal Stress (σ)

Internal stress (which equals residual stress, σ) causes the deformation of the microstructure and sometimes induces cracking of thin films. σ develops in a thin film because a deformation of the film is constrained by a substrate. If the thin film is constrained in a compressed state, σ is called compressive and expressed as a negative value. If a thin film is constrained in an expanded state, σ is called tensile and expressed as a positive value. The two major sources of σ are (1) thermal mismatch between a substrate and a thin film (σ_e, extrinsic stress), and (2) microscopic structural change of a thin film during deposition or post-processing due to film nucleation (σ_i, intrinsic stress). The former is related to the temperature difference between deposition and measurement. The stress σ is related to Young's modulus E and the strain ε with Hooke's law, i.e.

$$\sigma = \varepsilon E. \qquad (1)$$

If the E is known, σ can be deduced from the strain value. Eq. (1) applies to isotropic materials under uniaxial stress. In case of isotropic, biaxially stressed materials such as a thin film on a substrate, Eq. (1) reads $\sigma = \varepsilon E/(1-v)$, where v denotes Poisson's ratio of the thin film.

Figure 2.1 shows the layouts of three different micro test structures fabricated by surface micromachining to measure the ε of thin films: doubly supported beam,[5] ring and beam,[5] and rotation beam.[6] The typical dimensions of the structures are shown. The doubly supported beam, and the ring and beam patterns, are complementary techniques for measuring compressive and tensile stress, respectively. By preparing the pattern with incrementally increasing size, the critical length L_c of the straight beam,

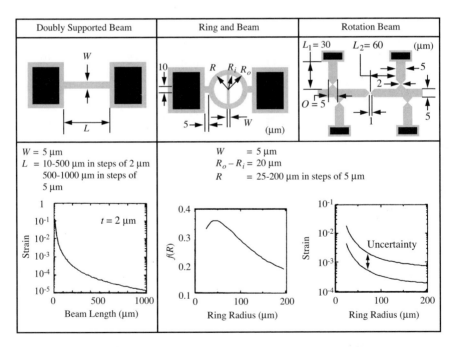

Figure 2.1 Three different micro test structure patterns for strain measurement.[5,6]

which causes the buckling, can be determined. For the doubly supported beam, the strain ε is deduced from the buckling length L_c of the beam using

$$\varepsilon = -\frac{\pi^2}{3}\left(\frac{t}{L_c}\right)^2, \tag{2}$$

where t the thickness of the thin film. For the ring and beam, the strain ε is deduced from the buckling radius R_c of the center beam using

$$\frac{\pi^2 t^2}{48 f(R) R_c^2} \leq \varepsilon \leq \frac{\pi^2 t^2}{12 f(R) R_c^2}, \tag{3}$$

where $f(R)$ is the function of ring radius R, as shown in Fig. 2.1. Note that the deduced ε has an uncertainty. For the rotation beam, the strain ε is deduced from the displacement of the tip Δy as

$$\varepsilon = \frac{O \Delta y}{2 L_1 (L_2 + \frac{O}{2})}. \tag{4}$$

Although the imperfections of the beam supports affect the buckling behavior,[7] the method using beam buckling to measure strain is convenient as a microfabricated test structure for process monitoring. The advantage of the rotation beam method is that it can be applied to both tensile and compressive strain. The drawback of this method is that a SEM observation is required to measure the tip displacement Δy with higher accuracy.

The ε value measured using the above technique is the value averaged over the thickness. As the film thickness decreases, ε varies because the interface between the thin film and the substrate affects the film growth. If this distribution of ε through thickness exists in a thin film, the released thin film from the substrate deforms upward or downward according to the distribution of ε through thickness.[8] The existence of ε variation through the thickness can be evaluated qualitatively from the direction and height of vertical deflection of the cantilever beam. For quantitative evaluation, sequential measurement of deflection with thinning of the thin film is required.

Another approach to measuring the compressive σ is the use of the ripple that appears in released wide cantilever beams or circular membranes with a large hole at their centers.[9,10]

2.2.2 Young's Modulus (*E*)

The deformation of a microstructure against the applied external force is dominated by Young's modulus (E) of the structural material and the shape of the microstructure. E is defined as the slope of the strain–stress (ε–σ) curve implicit in Eq. (1), which is a commonly used characterization method for the bulk material. Therefore, the most straightforward technique for measuring E is a tensile testing of a thin-film material, which is described in detail below. Another powerful technique, a load-deflection technique for measuring E together with σ, is explained here. This method was originally proposed in the MEMS field in 1983 using a circular membrane fabricated by isotropic etching of a substrate.[11] To improve the reproducibility of the thin-film membrane fabrication, a technique using a square membrane fabricated by anisotropic etching of the substrate SCS was proposed later;[12] it was improved by adopting a composite rectangular membrane.[13] The concept of the load-deflection technique is shown in Fig. 2.2. The thin film is deposited on an SCS substrate, followed by anisotropic etching of a part of the substrate to form a thin-film membrane. The deflection of the membrane center (d) is measured with changing pressure (p) across the membrane. An approximation to the pressure-deflection behavior of the membrane can be derived theoretically, using an energy minimization method, as

$$p = \frac{C_1(n)\sigma t d}{a^2} + \frac{C_2(n,v)E t d^3}{a^4}, \qquad (5)$$

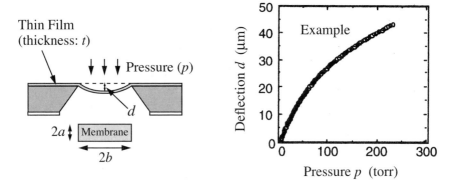

Figure 2.2 Concept of the load-deflection measurement technique for simultaneous σ and E measurement.

where t is the thickness of the thin film, a is the short side length of the membrane, n is the ratio between the short side of the membrane $2a$ and the long side of the membrane $2b$ ($n = a/b$), v is Poisson's ratio of the thin film, and $C_1(n)$ and $C_2(n,v)$ are dimensionless coefficients that depend on n and v. Equation (5) can be rewritten as

$$\frac{pa^2}{td} = C_1(n)\sigma + C_2(n,v)E\left(\frac{d}{a}\right)^2. \qquad (6)$$

The left-hand side of the equation is the normalized load, and $(d/a)^2$ is the normalized strain. Therefore, Eq. (6) corresponds to the strain-stress curve for a membrane with σ, its intercept with the y-axis is proportional to σ, and its slope is proportional to E. By fitting the measured load and deflection data sets to either Eq. (5) or (6), σ and E can be determined simultaneously. $C_1(n)$ and $C_2(n,v)$ are determined analytically to be

$$C_1(n) = \frac{\pi^4\left(1+n^2\right)}{64} \qquad (7)$$

$$C_2(n,v) = \frac{\pi^6}{32\left(1-v^2\right)}\left(\frac{9+2n^2+9n^4}{256}\right.$$
$$\left. - \frac{\left\{4+n+n^2+4n^3-3nv(1+n)^2\right\}}{2\left\{81\pi^2\left(1+n^2\right)+128n+v\left[128n-9\pi^2\left(1+n^2\right)\right]\right\}}\right) \qquad (8)$$

by assuming a cosine function for the shape of the membrane deflection. When the rectangular membrane shape becomes longer and longer, n decreases, $C_1(n)$ approaches the value 1.52, the dependency of $C_2(n,v)$ on n vanishes, and the dependency of $C_2(n,v)$ on v becomes weak. Since v is unknown for most thin films, v is assumed to take a value between 0 and 0.5, usually 0.25 for most cases. Therefore, a long, rectangular membrane is preferable for measuring E of a thin film with unknown v.

The load-deflection technique is easy to apply to thin films with tensile stress because the membrane is flat without load and it is easy to measure the load-deflection relationship. With a compressive-stress thin film, a composite membrane technique is used to keep this advantage. In this technique, a compressively stressed thin film to be measured is formed on a well-characterized tensile thin film to form a tensile composite thin-film membrane. As a well-characterized tensile membrane, silicon nitride (Si_3N_4) thin film deposited by a low-pressure chemical vapor deposition (LPCVD) apparatus can be used because it has a strongly tensile stress of 1 GPa and is chemically and mechanically stable. The following composite law of the material property between the individual thin films and composite thin film is used to calculate the material properties of interest:

$$\sigma_c(t_u + t_k) = \sigma_u t_u + \sigma_k t_k \qquad (9)$$

$$E_c(t_u + t_k) = E_u t_u + E_k t_k, \qquad (10)$$

where the subscripts c, u, and k are associated with the composite thin film, unknown thin film, and known thin film, respectively. The Poisson's ratios of these films are assumed to be identical. The error associated with different Poisson's ratios of thin films can be minimized using a rectangular membrane shape.

An analytical load-deflection equation applicable to both tensile and compressively stressed long, rectangular ($n<1/4$) membranes has recently been proposed.[14,15] Although numerical calculation is required to solve the analytical equation for the load-deflection curve, E and σ can be determined by measuring the initial buckled membrane height, and subsequent load-deflection behavior of the membrane even with compressive σ.

Another approach to measuring E is to use the resonant frequency of a cantilever beam made of the thin film of interest.[16] The resonant frequency (f_0) of a clamped-free cantilever beam is given as

$$f_0 = \frac{0.162 t}{L^2}\sqrt{\frac{E}{\rho}}, \qquad (11)$$

where t and L denote the thickness and length of the cantilever beam, respectively, and ρ is the density of the thin film. Excitation and detection of a cantilever beam vibration can be achieved by several methods, such as electrostatic excitation and optical detection. The resonant frequency ($f_{0d}(\sigma)$) of a doubly supported beam made of a thin film with internal stress σ is given as[17]

$$f_{0d}(\sigma) = f_{0d}(0)\sqrt{1 + 0.25\frac{\sigma L^2}{E t^2}} \qquad (12)$$

$$f_{0d}(0) = \frac{1.028t}{L^2}\sqrt{\frac{E}{\rho}}. \tag{13}$$

Therefore, E and σ can be deduced simultaneously for the measured results obtained from doubly supported beams of different lengths.

2.2.3 Poisson's Ratio

The general expression of strain-stress for anisotropic material is

$$\sigma_j = C_{jk}\varepsilon_k \ (j,k = 1,2,3\cdots,6), \tag{14}$$

where σ_j is a stress vector, ε_k is a strain vector, and C_{jk} is a compliance matrix. These quantities represent the engineering notation for the stress and strain tensors, both of rank 2, and the compliance tensor with rank 4. The numbers of independent elements of C_{jk} are 21 and 2 for asymmetric and isotropic materials, respectively. Most of the thin-film materials used in MEMS devices, such as silicon oxide, silicon nitride, poly-Si, and metals, can be handled as isotropic materials in an in-plane strain-stress field. Therefore two independent mechanical properties are necessary to completely define the mechanical properties of these materials. By measuring E and v, mechanical properties of thin films can be determined accurately.

For in-plane strain fields, v is defined as

$$v = -\frac{\varepsilon_j}{\varepsilon_i}, \tag{15}$$

where the subscripts i and j denote the direction parallel and perpendicular to the applied stress, respectively. The most straightforward measurement technique is the measurement of in-plane strains parallel and perpendicular to the applied stress during tensile testing.[18]

Another approach is the application of the load-deflection technique for noncomposite membranes. The coefficient $C_2(n,v)$ in the load-deflection relationship that is used to deduce E from measurement data is a strong function of v for a square membrane ($n = 1$) as shown in Eq. (8). If an exact value of E is known, the second term on the right-hand side of Eq. (5) or (6) can be used to deduce the v from the measured load-deflection data. In this case, a square membrane is preferable due to the strong dependency of its $C_2(n,v)$ on v. Even if E is unknown, by measuring the load-deflection data for square and rectangular membranes and comparing the deduced values from the second term on the right-hand side of Eq. (5) or (6), it is possible to eliminate the effect of E and deduce v. However, Eq. (8) is not accurate enough to determine v from this technique due to the assumed membrane deflection shape. FEM analysis should be used to formulate a more accurate load-deflection behavior of a membrane.[19]

It is also possible to determine v from the associated behavior of the ripple transition during pressurizing observed in a single compressively stressed long, rectangular membrane.[20]

2.2.4 Yield Strength and Fracture Strength

Strength properties defined by the yield stress σ_y and the fracture stress σ_f have to be evaluated to assure the reliability of MEMS. The σ_y is defined as a stress at which a material exhibits a specific limiting deviation from the proportionality of the stress-to-strain curve. If a stress larger than σ_y is applied to a structure of a device, the structure is plastically deformed and the device properties are changed irreversibly. The stress σ_f is the stress at which a structure of a device is fractured and thus fails. The σ_y of the brittle materials such as SCS, poly-Si, and dielectric films is equal to the fracture stress σ_f because these materials have no elastic limit. Consequently, the maximum stress applied to a microstructure should be designed much smaller than σ_y under any operational conditions.

The measurement methods of these properties are similar to those of the elastic properties such as a Young's modulus E or a Poisson's ratio v, because each method needs to determine the stress-strain curve of the material of interest. However, the following two points make strength measurements more difficult than an elastic measurement. (1) Stress applied to materials for strength measurements needs to be larger than that for elastic measurements. Therefore, the testing apparatus must deal with larger force or pressure. (2) To define an accurate strength value, a maximum stress in the specimen should be defined. However, this is very difficult because the maximum stress is easily and strongly affected by the specimen boundary configuration or specimen shape. For example, the thickness of chemically or physically deposited film is inhomogeneous over a wafer; a cross section of the fabricated specimen is often trapezoidal, not rectangular. These values and shapes must be checked carefully for accurate measurements.

Tensile test is the basic method among the strength-measurement techniques. The stress σ_s applied to a specimen is given by

$$\sigma_s = F / S, \qquad (16)$$

where F is the applied force along the displacement axis and S is the cross-sectional area of the specimen. Tensile testing for MEMS materials has some difficulties because a specimen is easily damaged or fractured when it is handled and immobilized on the chuck of a tensile tester. Therefore, several ideas and improvements have been proposed for accomplishing the tensile test.

Improvement of mechanical grip suitable for thin film is one of these approaches. Thick blocks are attached at the ends of the specimen to avoid damage by a mechanical grip of a conventional tensile tester.[21] A silicon wafer is often used as a substrate for this purpose. This support block makes specimen handling difficult because it has a large mass compared to the strength of the test materials. To overcome this drawback, both ends of the specimen are fixed to a support frame that is crafted from the same silicon substrate. This whole structure consists of the specimen, and the support frame is handled as one structure before testing. After the structure is immobilized on the chuck of the tester, the support frame is broken to start tensile testing.[22]

A pin-hooking system shown in Fig. 2.3 is also a commonly used technique. Specimens for this system are fabricated in a cantilever shape with a ring at the free end of the levers. The fixed end is directly connected to a substrate. The ring is hooked by a pin or a stylus positioned perpendicular to the substrate, and the ring is displaced along the tensile directions.[23] A nanoindenter with lateral force measurement function is suitable for hooking and load measurement. To reduce the bending moment at the fixed end, a link structure can be used.[24]

Adhesive glue is also a popular method for attaching the test structure to a gripping element with the shape of a sheet or a fiber for tensile testing. However, it is difficult to fix a specimen to a tester with a precisely aligned position. A special specimen holder that fixes 20 samples at the same time and tests these specimens simultaneously was developed by Koskinen.[25]

A useful new grip device based on an electrostatic force grip and dedicated to thin films was proposed by Tsuchiya[26] and is shown in Fig. 2.4.

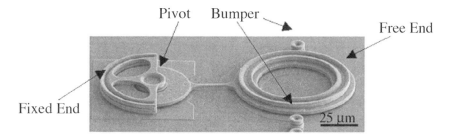

Figure 2.3 A pin-hooked type specimen with link structure.[24]

(a) **Conducting Film**

Top View

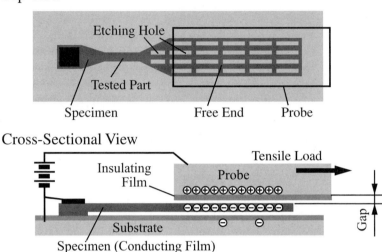

(b) **Insulating Film**

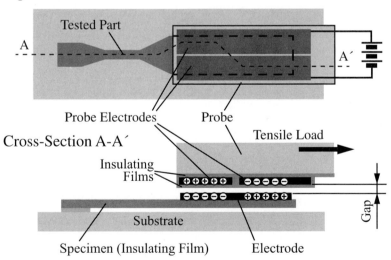

Figure 2.4 Electrostatic force grip for (a) a conductive film[26] and (b) an insulating film.[27]

A cantilever-beam-shaped specimen has large paddles on a free end; the other end is fixed to the substrate. The free end is attracted to the probe and fixed on it by electrostatic force. The specimen can be fixed to the tester without touching the thin film. The electrostatic force to hold the specimen free end is generated in one of two ways, depending on the material to be tested. For a conductive film, a probe made of a conductive material covered with an insulating layer is used. Voltage applied between the probe and the specimen generates electrostatic force. For an insulating film, a probe with two electrodes is used and an electrode is fabricated on the free end of the specimen. By applying voltage between the two electrodes on the probe, an electrostatic force is generated between the probe and the specimen.[27] To prevent slipping while the tensile force is loaded, enough force is needed to fix the specimen parallel to the probe. The friction force, F_F, caused by the surface normal force, F_V, is the dominant force to fix the specimen to the probe. In the case of the former grip system, the friction force, F_F, is described by

$$F_F = \mu F_V = -\mu \frac{\varepsilon_r \varepsilon_0}{2} \frac{S_c}{d^2} V^2, \tag{17}$$

where μ is the friction coefficient between the specimen and the probe; ε_r and ε_0 are the relative dielectric constant and the dielectric constant, respectively; S_c is the contacting area between the specimen and the probe; V is the applied voltage; and d is the thickness of the insulating layer. In the latter case, for insulating film, F_F is described by

$$F_F = \mu F_V = -\mu \frac{\varepsilon_r \varepsilon_0}{16} \frac{S_i}{d^2} V^2, \tag{18}$$

where S_i is the contacting area between the specimen and the probe. It is necessary for F_F to exceed the fracture tensile force, F_T,

$$F_F > F_T = \sigma_f wt, \tag{19}$$

where w and t are the width and thickness of the gauge part of the specimen, respectively. For specimens made of poly-Si whose tensile strength σ_f is in the order of several GPa, with gauge widths of several hundred μm and thicknesses of a few μm, the possible largest value of F_T is in the order of 0.1 N. On the other hand, the applied voltage, V, is about 100 V, taking into account an electrical breakdown strength of insulating films in the order of several MV/cm. Consequently, the required area of this electrode is more than 200×200 μm^2.

Another approach to exert a tensile force on thin-film specimens integrates the force conversion structure with the sample, producing a so-called "on-chip tensile testing device." One example is schematically shown in Fig. 2.5.[28] The device is made of micromachined SCS, and vertical force applied to the load lever is converted to horizontal force to stretch the specimen. The most sophisticated method integrates a loading mechanism into the specimen device; several approaches have been proposed. An electrostatic actuator cannot be used for this purpose because the force it generates is too small. Although a thermal actuator can generate large enough forces to perform tensile testing, the actuator displacement is very small and the specimen length needs to be less than 10 μm.[29]

The measurement of stress (force) and strain (displacement or elongation) is essential for the measurement mentioned above. If the cross-sectional area of a specimen is assumed to be in the order of 10^{-10} to 10^{-12} m^2, the fracture force of thin-film samples is in the order of 0.001 to 0.5 N. Therefore, the force sensor of a commercial load cell can be used for the force measurement. On the other hand, the strain measurement is difficult because the elongation is small, in the order of a few μm or less. Therefore, the effect of a finite stiffness of the whole testing system cannot be ignored. To eliminate this effect, a differential measurement method is often used. Using one specimen of length L_1, a stiffness S_1 is measured by dividing the tensile force by the displacement of the loading mechanisms, first. This stiffness includes the stiffnesses of the testing apparatus, of the specimen gauge part, and of the other parts of the specimen.[30] Then, by measuring the stiffness S_2 of a second specimen of different length L_2, Young's modulus E of the gauge part of the specimen can be derived as

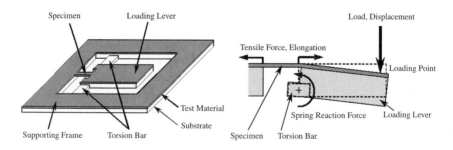

Figure 2.5 Schematic drawing of an on-chip tensile testing device.[28] (Courtesy of Prof. Sato.)

$$E = \frac{S_1 S_2 (L_1 - L_2)}{A(S_2 - S_1)}, \tag{20}$$

where A denotes the cross-sectional area of the gauge part of the specimen. The strain of the specimen can be calculated by dividing the applied force by EA. A direct elongation measurement can be realized using two small markers fabricated on a specimen. The displacements of these markers are measured using an image processing method with a CCD camera[31] or an interferometric method.[32]

The load-deflection technique, the so-called "bulge test," is another technique for measuring strength. In this case, higher pressure is applied to fracture a membrane than in the load-deflection technique to measure σ and E. The membrane stretched by the applied pressure generates biaxial tensile stress in the thin film. From Eq. (6), the nominal tensile stress in a pressurized membrane can be calculated as

$$\sigma_f = \frac{a^2}{C_1(n)td} p. \tag{21}$$

In case of internal stress $\sigma = 0$, the strain ε_f in the middle of the deflected film can be calculated using geometrical methods:[33]

$$\varepsilon_f = \frac{\pi^2 d^2}{16 a^2}. \tag{22}$$

These biaxial fracture stress and strain values have to be treated with care because the bending stress and stress concentration are neglected in Eqs. (21) and (22). The effect of stress concentration is clear from the fact that when pressure is applied to a membrane, the membrane often fractures at the edge. This is because high bending stress appears at the edge where a sharp edge corner is fabricated by the anisotropic etching of the silicon wafer.[34] This stress concentration value is difficult to estimate and is often underestimated even if finite element analysis is used.

A bending test is an easy and useful technique for strength measurement. A cantilever beam fabricated by bulk micromachining has been applied for this technique by pushing the end of the beam.[35] The maximum stress σ_m that appears on the upper surface of the fixed beam end is expressed as

$$\sigma_m = \frac{L}{wt^2}F = \frac{3}{2}\frac{Et}{L^2}\delta, \qquad (23)$$

where E is Young's modulus, t is the thickness, L is the distance between the loading point and the fixed end of the cantilever beam, F is the applied force, and δ is the vertical displacement at the loading point. This test is often conducted by a nanoindenter. For smaller specimens, with thicknesses of less than 0.2 μm, a probe of the atomic force microscope can be used.[36] A load is applied perpendicularly to the substrate in these tests. By using surface-micromachined structures integrated with a comb-drive electrostatic force actuator, in-plane bending tests can be performed.[37] These beam-bending tests have the same stress-concentration problem as the bulge test. Measured bending strengths often depend on the loading direction and the specimen shape. The fixed end of the beam especially tends to cause stress concentration and to affect the measured strength.

2.2.5 Fracture Toughness

The fracture strength of the brittle materials (Si, SiO_2, Si_3N_4, and ceramics) is drastically changed by the fabrication conditions because the fracture is dominated by the defects or the cracks in the structures. The strength property of these brittle materials is expressed as the fracture toughness, defined as follows:

$$K_{IC} = \sigma_f c \sqrt{a_c}, \qquad (24)$$

where σ_f is the fracture stress of a precracked specimen, a_c is the length of the crack, and c is a constant that depends on the shape of the crack. When the crack length a_c is much smaller than the specimen width, the shape constant c of a single-side-cracked specimen is 1.99.

The test procedure for measuring fracture toughness is identical to the tensile or bending test except for the shape of the specimen. The specimen must have cracks with known length. Cracks are formed to reduce stiffness of the beam and increase the stress applied to the beam. A crack can be fabricated easily using photolithography, and the size of the crack is well controlled and reproducible. However, the tip of the crack has a finite radius in the order of 0.5 μm, which is relatively large compared with the length of the crack. Therefore, a measured value obtained from the specimen prepared by photolithography cannot be treated as toughness; it should be treated as a stress-concentration problem. The crack fabricated by a microindenter or nanoindenter is appropriate for a toughness measurement. Note that when the specimen is small, the damage to the specimen might be large. To solve this problem, an indentation is made in the vicinity of a specimen before sacrificial etching so that most of the crack is not included in the specimen and only the crack tip reaches the specimen. Then the specimen is released from the substrate for the fracture test.[38]

2.2.6 Fatigue

Microstructures in the MEMS field, especially vibrating devices, are often exposed to cyclic stress during operation; this may induce fatigue. Fatigue properties include not only the fracture under cyclic loading or static loading, but also the time-dependent change in the elastic properties such as Young's modulus E and residual stress σ.

Fatigue tests are separated into two categories, by loading method: (1) static loading that provides a static fatigue test, and (2) cyclic loading test. The cyclic loading test using resonant vibration is often used for the fatigue testing of MEMS materials. This is because large vibration amplitudes can be obtained easily since MEMS devices have a large quality factor of vibration. A bending moment due to the large mass displacement is applied to the beam. The large numbers of the loading cycle are also easily achieved over a short period.

In resonant vibration test, a specimen consists of a mass and a suspended beam. The specimen can be oscillated by several methods such as a comb-drive actuator, a parallel-electrode electrostatic actuator, and an external piezoelectric actuator. One example is a single-crystal silicon resonator fabricated by bulk micromachining with gold upper electrodes for actuation and sensing.[39] Another example is a doubly supported beam resonator,[40] which has strain gauges on the support beam to measure the displacement. The piezoelectric actuator is used for actuating the resonator. The resonant frequency shift and the fracture by cyclic loading were reported. Surface-micromachined resonators were also used for fatigue testing. An in-plane resonant cantilever structure was fabricated using the poly-Si provided by the MuMPs process, as shown in Fig. 2.6.[41] The specimen has a large plate that serves as a resonant mass. The plate is connected to a short beam fixed to the substrate. Comb-drive actuators were used for both actuation and sensing. The beam has a large notch at the center to increase the stress. A nickel resonator was also applied for metal fatigue testing.[42]

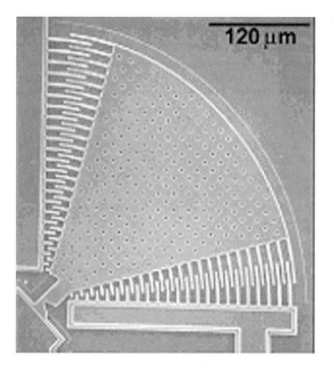

Figure 2.6 Poly-Si in-plane resonant fatigue specimen.[41]

2.2.7 Thermal Conductivity and Specific Heat

Thermal properties of materials include thermal conductivity, κ, and specific heat, c. Since MEMS integrate many components sensitive to temperature in a small area, thermal design is very important. Some MEMS use temperature change as a sensing principle or actuation principle.

The most commonly used technique for determining κ is the measurement of temperature change caused by known applied heat. Typical device structures used to measure κ based on this principle are shown in Fig. 2.7.[43] The upper-left SEM photograph shows a cantilever structure made of the thin film of interest that has a heater and a temperature sensor on its free edge. The SEM micrograph in the lower left part of Figure 2.7 shows a cantilever structure with a heater at a free end and two temperature sensors along the cantilever. The heater and the temperature sensors are electrically connected by thin suspension arms. Current applied to the heater generates Joule heat, which flows from the cantilever free edge to the substrate through the cantilever. This heat flow results in a temperature change along the cantilever. The κ of the can-

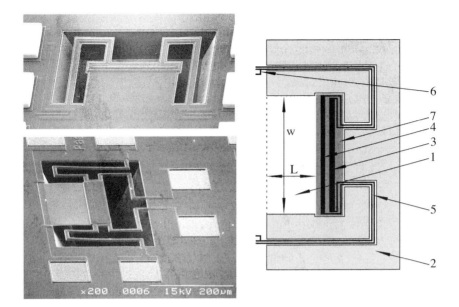

Figure 2.7 Typical device structures to measure κ. 1: cantilever; 2: windows to etch SCS under the cantilever; 3: poly-Si heater; 4: poly-Si temperature sensor; 5: metal connection to pads; 6: bifurcations into current and voltage lines for four-point measurements; 7: metal cover for temperature homogenization.[43]

tilever can be deduced from the amount of applied heat, the temperature change at the free end of the cantilever, the length of the cantilever, and the cross-sectional area of the cantilever. Once the thermal property of a cantilever is well characterized, this cantilever can be used as a reference to measure κ of any thin films deposited on it. When temperature is measured at two different positions along the cantilever, the measured temperature difference and the distance of the two temperature sensors are used to calculate κ.

Another approach to measure κ is the use of a van der Pauw Greek-cross pattern structure familiar to electrical conductivity measurement.[44] A SEM photograph of the test structure is shown in Fig. 2.8. The Greek-cross pattern is suspended by four thin arms. Each branch of the cross pattern contains a poly-Si resistor contacted in four-point configuration by four metal lines. This resistor serves both as a heater and as a temperature sensor. Extraction of κ is done by measuring the thermal response of the structure. Heat dissipation of powers P_j in the four resistors causes the temperature change of ΔT_i, where subscripts i and $j = 1, 2, 3, 4$ denote the four arms. By using the linear thermal response matrix M_{ij}, the relationship between P_j and ΔT_i can be written as

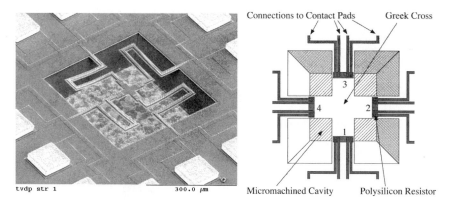

Figure 2.8 Use of van der Pauw Greek cross pattern structure to measure κ.[44]

$$\Delta T_i = M_{ij} P_j, \qquad (25)$$

where $i, j = 1, 2, 3, 4$ and M_{ij} denotes the temperature change of resistor i per unit heat power produced in resistor j. From the elements of M_{ij}, thermal conductances of the suspension arms are extracted. Then four thermal sheet resistance values of the Greek-cross material sandwich are extracted from the data. Finally, by using the thickness of the structure, an average κ value can be determined.

To determine the heat capacity (c), dynamic thermal response measurements need to be performed. Most often, $Jkg^{-1}K^{-1}$ is used as a unit of c for bulk materials. However, the volume is easier to obtain accurately than the density for thin films. It is therefore preferable to use a volumetric c in units of $Jm^{-3} K^{-1}$. The same structure shown in the lower left SEM photograph in Fig. 2.7 can be used to measure c. It is possible to measure the κ of the material by determining the static relationship between applied heat and temperature change of the membrane. When an AC current I_{AC} with an angular frequency of ω is applied to the heater, the resultant temperature change contains a constant DC term and dynamic temperature change with an angular frequency of 2ω. The temperature measured at the middle of the cantilever shows a phase shift with respect to the input signal. From the dependency of this phase shift on ω, thermal diffusivity c/k can be deduced.[45]

A structure used for c measurement based on the 3ω method is shown in Fig. 2.9.[46] A meandering, long poly-Si heater is formed on the membrane. An AC current I_{AC} with an angular frequency of ω is applied to the poly-Si heater, and the resultant temperature change contains a constant

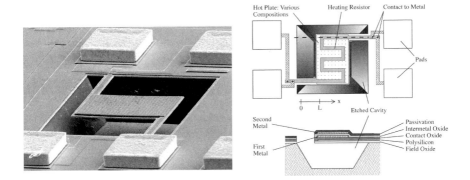

Figure 2.9 Typical device structure to measure c.[46]

DC term and a dynamic temperature change. Because the generated Joule heat is proportional to the square of the current and resistance, which varies linearly with the temperature change, the resultant temperature change term contains a third harmonic component phase-shifted with respect to the input signal. From the thermal modeling and analysis of the structure, the amplitude $A(\omega)$ and phase change ϕ of the third harmonic temperature change are extracted. The voltage change of the poly-Si heater due to the temperature change is expressed as

$$A(\omega)\cos(3\omega t - \phi(\omega)), \qquad (26)$$

$$A(\omega) = \frac{U_0}{\sqrt{1+(\tau\omega)^2}}, \qquad (27)$$

$$U_0 = \frac{(R_0)^2 L\beta}{4(2K+K_{rad})} I_{AC}^3, \qquad (28)$$

$$\phi(\omega) = \arctan(\tau\omega), \qquad (29)$$

$$\tau = \frac{2CL}{2K+K_{rad}}, \qquad (30)$$

where L is the length of arm, K is the thermal conductance of the supporting arms, K_{rad} is the thermal loss due to radiation, R_0 is the resistance of poly-Si at room temperature, and β is the temperature coefficient of resistance. Using these relationships, κ and c can be extracted. Finally, a pair of structures with heat capacitances c_1 and c_2 due to plate composition differing by a single thin film of volume V_{layer} is used to extract the volumetric c_{layer} of this thin-film material. It is given by

$$c_{layer} = \frac{c_2 - c_1}{V_{layer}}. \qquad (31)$$

2.3 Data

2.3.1 Si

Mechanical properties of bulk SCS are summarized in Table 2.1.[47] These properties are often referred to for comparison with measured data not only of SCS but also of poly-Si films. From Table 2.1, the E and the ν of some crystal directions can be obtained, as listed in Table 2.2.[48] Table 2.3 shows the measured values of E of SCS, which range within a small deviation from the reference values. This is because the structures used to measure these values are made of SCS, with thousands of atomic layers; they can be regarded as a bulk SCS.

SCS structures are free from the extrinsic stress caused by the thermal expansion coefficient difference from the substrate. However, doping SCS

Table 2.1 Bulk SCS Mechanical and Thermal Properties Data[47]

Density	at 25°C	2.329×10^3 kg^3
Stiffness	C_{11}	165.64 GPa
	C_{12}	63.94 GPa
	C_{44}	79.51 GPa
Thermal expansion coefficient	at 300 K	2.616×10^{-6}/K
Specific heat		0.713 J/g/K
Thermal conductivity		1.56 W/cm/K
Thermal diffusivity		0.86 cm^2/s

Table 2.2 Bulk SCS Elastic Moduli for Several Crystallographic Orientations[48]

Wafer Orientation	Direction	Young's Modulus	Direction	Poisson's Ratio
(100)	<011>	168.9 GPa	<011> <0-11>	0.064
(100)	<001>	130.2 GPa	<010> <001>	0.279
(110)	<111>	187.5 GPa	<1-11> <1-1-2>	0.182
(110)	<100>	130.2 GPa	<001> <1-10>	0.279
(111)	—	168.9 GPa	—	0.262

causes intrinsic stress by substituting dopant ions for Si atoms in the silicon crystal lattice. Boron doping causes tensile stress, and phosphorus and arsenic doping cause compressive stress. These stresses result from the mass and size of the substituting ions.

Both tensile and bending strength of the SCS are summarized in Table 2.3. These values depend on the process conditions and the specimen dimensions. Specimens where the defects on the surface have been removed by processes such as oxidization and HF etching have greater strength than the specimens with dry-etched or anisotropically etched surfaces. The smaller specimens show greater strength because they have a smaller probability of a large defect.

Fatigue fracture and resonant frequency shift caused by a cyclic loading are observed for SCS. The fatigue behavior by cyclic load induces the growth of small defects or cracks. Plastic deformation was also recently reported in the submicron-scale bending tests operated at relatively low temperatures between 100 and 300°C.[54]

2.3.2 Poly-Si

Mechanical properties of poly-Si are summarized in Table 2.4. The internal stress σ of poly-Si has been widely investigated in conjunction with LSI device process development.[60] As-deposited poly-Si and amorphous silicon by LPCVD using thermal decomposition of silane (SiH_4) gas possess a large compressive stress of -300 to -400 MPa. This compressive stress arises from the disordered grains and the oxygen content;[61] it is not related to a thermal stress because the thermal expansion coefficient of poly-Si is the same as that of the SCS substrate. The compressive stress is undesirable in MEMS applications because it induces buckling of the structures. Poly-Si under tensile stress can be obtained by

Table 2.3 SCS Data

Material	t (μm)	Deposition	Method	E (GPa)	ν	σ_f (GPa)	ε_f	m	References
Si <100>	5	SOI	Tensile	120 ± 19			0.5 - >2.3		28
Si <110>	5	SOI	Tensile	157 ± 26			0.8 - >1.5		
Si <111>	5	SOI	Tensile	180 ± 9			0.7		
Si <100>	10	SOI	Tensile	142 ± 9		1.73			49
Si <110>	3 - 5	SOI	Tensile	169.2 ± 3.5		0.6 - 1.2			50
Si <110>	15	Si (100) wafer	Bulge			6.8			34
Si <110>	0.255	SOI (100)	Bending			11.56 - 17.53		16.8 - 62.1	36
Si <110>	1.91	SOI (100)	Bending			7.68		7.2	
Si <110>	19	SOI (100)	Bending			3.70		4.2	
Si <110>	18 - 30	Si (100) wafer	Bending (lateral)			1.1 - 3.4			51
Si <110>	16 - 30	Si (100) wafer	Bending			3.6 ± 0.7 (front) 1.1 ± 0.2 (back)			52
Si <110>	15	Si (100) wafer B+ doped 5×10^{19}cm^{-3}	Bending	200 - 220		1.8			53
Si <110>	8 - 16	Si (100) wafer Etched pattern	Bending			3.9	2.0		35
Si <100>	8 - 16	Si (100) wafer Etched pattern	Bending			4.3	2.5		
Si <110>	8 - 16	Si (100) wafer Diffused pattern	Bending			3.9	2.0		
Si <100>	8 - 16	Si (100) wafer Diffused pattern	Bending			2.0	1.5		

Table 2.4 Poly-Si Mechanical Properties Data

Deposition	t (μm)	Method	E (GPa)	v	σ_0 (MPa)	Stress gradient (MPa/μm)	σ_f (GPa)	m	ε_f	K_{IC} MP$\sqrt{m}$	Refs.
LPCVD (MuMPs)	3.5	Tensile	170 ± 7	0.23			1.21 ± 0.16				18
MuMPs 13	3.5	Tensile								1.4 ± 0.65	32
MuMPs 19, 21	1.5	Tensile	153 ± 8 153 ± 20 146 ± 10							1.45 ± 0.19	30
LPCVD 620°C Grain Size (50,100, 500 nm)	1.0	Tensile	176 ± 25 164 ± 25 164 ± 25				2.86 ± 0.28 2.69 ± 0.30 3.37 ± 0.29				25
450 nm LPCVD +11.5μm Epi, polished, POC13 doped	10.5	Tensile	167 ± 4 164 ± 6 167 ± 5 160 ± 4		-10 -25 -115 -46	+0.9 -0.9 -9.9 +2.8	1.25 1.19 1.08 1.08	10.6 11.7 6.1 11.5			23
LPCVD 580°C P+ doped Anneal 950°C	2	Tensile					0.698	11			55
Epi In situ PH$_3$ doped	10	Tensile					1.07	7			
LPCVD 520°C(Si$_2$H$_6$) Anneal 1000°C	1.8	Tensile	163		76		2.0-2.8 ± 0.4-0.6	6.3			56
LPCVD 520°C(Si$_2$H$_6$) Anneal 1000°C POCl$_3$ 1000°C	2.4	Tensile	167		-7.0		2.0-2.5 ± 0.2-0.4	10.0			

(continued)

Table 2.4 Cont.

Deposition	t (μm)	Method	E (GPa)	v	σ_0 (MPa)	Stress Gradient (MPa/μm)	σ_f (GPa)	m	ε_f	K_{IC} MP$\sqrt{m}$	Refs.
Standard	4	Tensile (thermal)					3.1 ± 0.4				29
LPCVD 630°C	0.2	Bulge	160		-240						57
LPCVD 630°C N$_2$ anneal 1100°C	0.2	Bulge	194		-68						
LPCVD 630°C Anneal 1000°C	4	Bulge	41		189				0.10%		33
LPCVD 580°C Anneal 1100°C Nondoped, boron-doped	5.2	Bending			19 -32		5.0 4.2			3.5 (R = 1.0 μm)	37 38
LPCVD 580°C Anneal 1100°C Nondoped	5.2	Bending (resonance)			19*		3.3				
LPCVD 580°C Anneal 1000°C	3.5	Bending			12 ± 5					1.1 (crack)	58
LPCVD as deposited 1000°C annealed	1.27	Buckling							1.72% 0.93%		59

high-temperature annealing above 1050°C both in a furnace [62] and by a rapid thermal process,[63] due to stress relaxation.

Crystallization of an amorphous silicon film induces a large stress change, up to 1 GPa, from compressive to tensile due to film shrinkage.[64] This crystallization process occurs above 600°C, depending on the concentration of impurities in the film. In the crystallization process, larger grain sizes are realized by higher anneal temperatures. Using this crystallization-induced tensile stress, "as-deposited" tensile poly-Si film can be obtained by the deposition at a lower temperature near the amorphous-polycrystalline transition temperature of about 580°C.[65] This film is deposited as an amorphous material and is crystallized during the rest of the film deposition. The transition temperature depends on the deposition condition, such as silane pressure, silane flow rate, and deposition apparatus. These are summarized in Fig. 2.10.[65-69]

Doping poly-Si film is another way to control the film stress. However, the relationship between the doping condition and the stress is more complicated than with SCS, mainly due to grain growth. As with SCS, boron doping causes tensile stress, and phosphorus and arsenic cause compressive stress.[70] However, phosphorus ion implantation of high

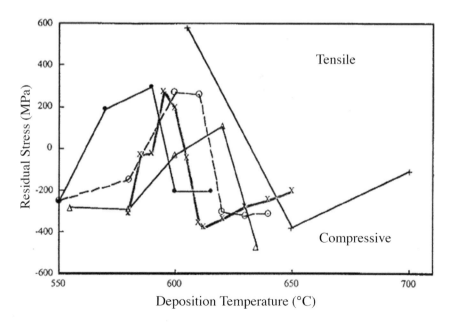

Figure 2.10 Residual stresses in poly-Si films as a function of deposition temperature.[66] Legend: + from Krulevitch et al.,[65] • from Yang et al.,[66] X from Oei and McCarthy,[67] o from Yu et al.,[68] D from O'Bannon.[69]

concentration (10^{19} cm^{-3}) results in tensile films. As-implanted films become amorphous by ion bombardment; the tensile stress is induced by subsequent crystallization.[71] Moreover, films heavily ion-implanted with both boron and phosphorus become more compressive because of dopant segregations to grain boundaries.[71]

Another important aspect of the internal stress is its depth profile, namely, the stress gradient. The stress gradient causes out-of-plane bending of fabricated structures. This stress gradient can also be controlled by many processes. Heavy ion implantation and rapid thermal annealing realize small stress gradients because the film become amorphous and homogeneously crystallized by this technique. High-temperature anneal to relax the mean stress is another way to reduce bending. Multistacked poly-Si processes are also low-temperature processes that can be used to obtain low-stress and bending-free films. By repeating the deposition of the thin poly-Si films, poly-Si/poly-Si interfaces with impurities of oxygen and phosphorus are formed in the film. These interfaces reduce the stress gradients.[72] Multi-stacks of tensile films (low-temperature-deposited fine-grained films) and compressive films (at standard deposition temperature, producing films with columnar grains) can have nearly zero stress and a small stress gradient.[66]

The *E values* of poly-Si films listed in Table 2.4 cluster around 160 GPa; most of the data coincide with this value within ±10%. These values agree well with the theoretically predicted values of randomly oriented and (110)- and (111)-oriented polycrystalline structures of SCS. The strength of poly-Si varies widely, ranging from 0.6 to 5 GPa. This variation is thought to be dominated by the surface roughness. The roughness of the top surfaces depends on the deposition conditions. Fine-grained and polished films have smooth surfaces; columnar-grained films have rough surfaces. The roughness of the side surface is formed by the RIE process. The difference in strength and E was compared among the multiuser processes in the United States;[73] it appeared that the measured strength of poly-Si is widely distributed, even for the same specimen shape.

A statistical analysis is required to evaluate the measured strength, especially for poly-Si. Weibull statistics, the most commonly used analysis for bulk ceramic materials, is also used for poly-Si films. The fracture probability F under the applied stress σ_s is expressed as

$$F = 1 - \exp\left\{-\frac{\sigma_s^m}{\alpha} \cdot V_E\right\}, \tag{32}$$

where m is the Weibull modulus, V_E is the effective volume (the volume of the tested part), and α is a constant. The reported Weibull modulus m is listed in Table 2.4.

Fracture toughness ranges from 1.1 to 3.5 MPa$\sqrt{m}$. The larger value must be caused by the inappropriate crack shape that has a relatively large radius of curvature on the top of the crack.

As an example, Fig. 2.11 shows the results of the fatigue fracture test plotted as stress-time or number-of-cycle diagrams. The lower stress results in the longer lifetime or the larger number of cyclic loading. As is known for bulk metal materials, a lower stress limit exists beyond which fatigue fracture is observed. This limit is termed the fatigue limit, which is defined as the fracture limit of 10^8 loading cycles in bulk metal materials. However, in MEMS structures, a test should be carried out for more than 10^{10} cycles of loading to quantify the fatigue because these structures have high resonant frequencies (up to 100 kHz) and these must have a lifetime as long as 10 years for automobile, aircraft, and satellite applications.

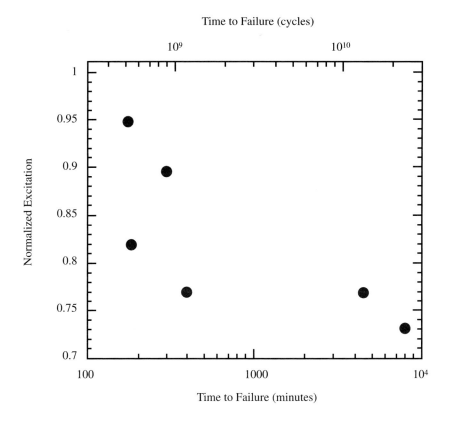

Figure 2.11 Time to failure for poly-Si in-plane resonant specimens. The vertical axis is scaled to the amplitude of motion associated with short-term failure.[41]

Thermal properties of poly-Si are shown in Table 2.5.[43,45,74] Comparing these to SCS, the thermal conductivity (κ) is much lower.

2.3.3 Metal

There are few reports regarding the mechanical and thermal properties of metal thin films. The measured σ and E of metal thin films commonly used in LSI processes for interconnection are shown in Figure 2.12. The films were deposited by sputtering, and measurements were carried out using a composite membrane load-deflection technique. The data show the effect of annealing at 450°C in H_2+N_2 atmosphere. Although research on fracture mechanisms using micro tensile tests was performed on similar metal thin films, the experimental conditions are very limited.[75]

For aluminum interconnection layers used in complementary metal-oxide semiconductor (CMOS) processes, systematic measurements have been done intensively regarding σ, E, ν, κ, and c. The results are summa-

Table 2.5 Gate Poly-Si Thermal Properties Data

κ [Wm^{-1}k^{-1}]	References
22.4 ± 0.7	
23.5 ± 0.7	
23.2 ± 0.7	
37.3 ± 1.0	
24.8 ± 2.4	43
34.1 ± 1.0	
28.6 ± 0.8	
28.4 ± 0.8	
28.6 ± 0.8	
18 ± 1	
17 ± 1	45
14 ± 1	
17	
19	74
22	
24	

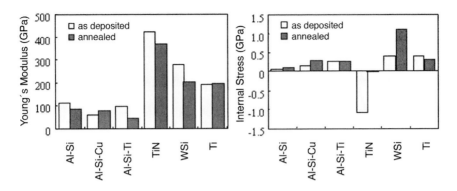

Figure 2.12 σ and E of the metal thin films deposited by sputtering. Measurements have been done by the composite rectangular membrane load-deflection technique.

rized in Table 2.6. The κ in metals is dominated by electrons and phonons in the material, and it was confirmed that, like in bulk Al, the electrical conductivity σ_e and κ obey the correlation predicted by the Wiedemann-Franz law:[76]

$$\frac{\kappa}{\sigma_e} = \left(\frac{\pi^2}{3}\right)\left(\frac{k_B}{e}\right)^2 T, \qquad (33)$$

Table 2.6 Metal Layer for CMOS Interconnection and Al Material Data

	σ [MPa]	E [GPa]	ν	κ [Wm^{-1}k^{-1}]	c [MJm^{-3}k^{-1}]	References
Metal 1	−73 ± 10	50 ± 10	0.3 ± 0.2	197 ± 5	—	45
Metal 2	30 ± 5	55 ± 10	0.2 ± 0.2	174 ± 2	—	
Metal 1	—	—	—	—	2.44 ± 0.12	
Metal 2	—	—	—	—	2.51 ± 0.24 2.77 ± 0.16 2.56 ± 0.24	46
Al	232 ± 24 (as deposited) 167 ± 17 (400°C annealed)	69	—	—	—	29
Bulk Al	—	70	0.34	237	2.43	—

where k_B is the Boltzmann constant, e is the electron charge, and T is the absolute temperature. From Table 2.6, it is clear that the material properties of metal thin films are different from the bulk.

Since electroplated Ni and Ni-Fe have been used for MEMS, their mechanical properties have been reported, as shown in Table 2.7. The measurements have been done by micro tensile test and load-deflection methods. The composition of Ni-Fe and the mechanical properties of electroplated Ni-Fe depend on the electroplating conditions.[49,77]

2.3.4 Dielectrics

The dielectric materials most commonly used in MEMS are silicon oxide and silicon nitride. They are deposited by various methods such as thermal growth, plasma chemical-vapor deposition (CVD), and low pressure CVD (LPCVD). Because of this variety of deposition methods and their conditions, the properties of these materials appearing in the literature cover broad ranges of values. The mechanical properties and thermal properties of silicon oxide and silicon nitride are summarized in Tables 2.8 and 2.9, respectively.

Table 2.10 summarizes data for ZnO. There are few data regarding strength and fatigue. Effects such as surface oxidation have been proposed as possible explanations for the instability of the properties.[82]

Table 2.7 Ni and Ni-Fe Data

	σ [MPa]	E [GPa]	$E/(1-v^2)$ [MPa]	Yield Strength [GPa]	Tensile Strength [GPa]	References
Ni	—	231 ± 12	—	1.55 ± 0.05	2.47 ± 0.07	49
Ni-Fe	—	155 ± 8	—	1.78 ± 0.02	2.26 ± 0.02	
Ni-Fe	63	—	139	—	—	
Ni-Fe	38	—	135	—	—	77
Ni-Fe	64	—	132	—	—	

Table 2.8 Silicon Oxide Data

	σ [MPa]	E [GPa]	ν	Tensile Strength [GPa]	ε_f	k [Wm^{-1}k^{-1}]	c [MJm^{-3}k^{-1}]	References
Thermal SiO$_2$	-290	70	0.15	—	—	—	—	7
Thermal SiO$_2$		74		—	2.5%			78
PECVD SiO$_2$				0.67 - 0.95 (in air) 1.58 - 1.84 (in a vacuum)				27 79
CMOS SiO				—	—	1.28 ± 0.11 (field) 1.32 ± 0.18 (contact) 1.16 ± 0.24 (intermetal)	1.05 ± 0.1	74
CMOS BPSG	-40 ± 10	20 ± 10						
CMOS BSG	-37 ± 6	65 ± 5	0.2 ± 0.2					45

Table 2.9 Silicon Nitride Data

Deposition	ε [strain]	σ [MPa]	E [GPa]	ν	Yield Strength [GPa]	Tensile Strength [GPa]	Fracture Toughness [MPa√m]	References
PECVD	-626 ± 17 (as deposited) -909 ± 25 (204°C anneal)	—	—	—	—	—	—	10
PECVD	-365 ± 7	-60.7 ± 1	130	—	—	—	—	80
PECVD	—	-67.2 ± 8.3	142.6 ± 1.4	0.253 ± 0.017	—	—	—	20
PECVD	—	3 ± 3.8 (tensile) -63 ± 12.4 (compressive)	134.4 ± 3.9 (tensile) 142 ± 2.6 (compressive)	—	—	—	—	14
PECVD	—	-1120	137	—	—	—	—	7
PECVD	—	82 ± 6	97 ± 6	0.13 ± 0.07	—	—	—	45
PECVD					4.15-4.64			79
PECVD LPCVD	— —	0.11×10^3 1.0×10^3	210 290	—	—	—	—	81
LPCVD	—	—	—	0.35	—	—	—	19
LPCVD	—	230 ± 11	—	—	—	—	—	82
LPCVD			370			3.8%		78
LPCVD							1.8 ± 0.3	83
LPCVD		1.2 ± 0.2	300 ± 20			7.0 ± 0.4 6.5 ± 0.4		84
PECVD		-1.1 ± 0.1	240 ± 30			5.7 ± 0.3 2.4 ± 0.2		

Table 2.10 ZnO Data

	σ [MPa]	E [GPa]	Reference
Sputtered ZnO	-16 to 135	114.6 ± 5.9 (on SiN/Si) 121.8 ± 2.7 (on Al/SiO$_2$/Si)	85

References

1. K. E. Petersen, "Silicon as a mechanical material," *Proc. IEEE*, 70(5):420–457 (1982)
2. S. D. Senturia, "Microfabricated structure for the measurement of mechanical properties and adhesion of thin films," *Proc. 4th Int. Conf. on Solid-State Sensors and Actuators* (1987), pp. 11–16
3. D. L. Smith, *Thin-Film Deposition, Principles & Practice*, McGraw-Hill, Inc. (1995), pp. 185–193
4. M. Madou, *Fundamentals of Microfabrication*, CRC Press (1997), pp. 220–230
5. H. Guckel, T. Randazzo, and D. W. Burns, "A simple technique for the determination of mechanical strain in thin films with applications to polysilicon," *J. Appl. Phys.*, 57(5):1671–1675 (1985)
6. B. P. van Drieenhuizen, J. F. L. Goosen, P. J. French, and R. F. Wolffenbuttel, "Comparison of techniques for measuring both compressive and tensile stress in thin films," *Sensors and Actuators*, A37–38:756–765 (1993)
7. W. Fang, "Determination of the elastic modulus of thin film materials using self-deformed micromachined cantilevers," *J. Micromech. Microeng.*, 9:230–235 (1999)
8. W. Fang and J. A. Wickert, "Determination mean and gradient residual stresses in thin films using micromachined cantilevers," *J. Micromech. Microeng.*, 6:301–309 (1996)
9. R. T. Howe and R. S. Muller, "Stress in polycrystalline and amorphous silicon thin films," *J. Appl. Phys.*, 54:4674–4675 (1983)
10. T. Kramer and O. Paul, "Surface micromachined ring test structures to determine mechanical properties of compressive CMOS IC thin films," *Proc. 14th European Conf. on Solid-State Transducers* (2000), pp. 25–28
11. E. I. Bromley, J. N. Randall, D. C. Flanders, and R. W. Mountain, "A technique for the determination of stress in thin films," *J. Vac. Sci. Technol. B*, 1(4):1364–1366 (1983)
12. M. G. Allen, M. Mehregany, R. T. Howe, and S. D. Senturia, "Microfabricated structures for the *in situ* measurement of residual stress, Young's modulus, and ultimate strain of thin films," *Appl. Phys. Lett.*, 51(4):241–243 (1987)
13. O. Tabata, K. Kawahara, S. Sugiyama, and I. Igarasi, "Mechanical property measurement of thin films using load-deflection of composite rectangular

membrane," *Proc. IEEE Int. Workshop on Micro Electro Mechanical Systems* (1989), pp. 152–156
14. V. Ziebart, O. Paul, U. Münch, J. Schwizer, and H. Baltes, "Mechanical properties of thin films from the load deflection of long clamped plates," *J. Micro Electro Mechanical Systems*, 7(3):320–328 (1998)
15. V. Ziebart, O. Paul, and H. Baltes, "Strongly buckled square micromachined membranes," *J. Micro Electro Mechanical Systems*, 8(4):423–432 (1999)
16. K. E. Petersen and C. R. Guarnieri, "Young's modulus measurements of thin films using micromechanics," *J. Appl. Phys.*, 50(11):6761–6766 (1979)
17. S. Bouwstra and B. Geijselaers, "On the resonance frequencies of microbridges," *Proc. 6th Int. Conf. on Solid-State Sensors and Actuators* (1991), pp. 538–542
18. W. N. Sharpe, B. Yuan, R. Vaidyanathan, and R. L. Edwards, "New test structures and techniques for measurement of MEMS materials," *SPIE Proc.*, 2880:78–91 (1996)
19. O. Tabata and T. Tsuchiya, "Poisson's ratio evaluation of LPCVD silicon nitride film," *IEEJ*, 116-E:34–35 (1996)
20. V. Ziebart, O. Paul, U. Münch, and H. Baltes, "A novel method to measure Poisson's ratio of thin films," *Proc. MRS*, 505:27–32 (1998)
21. C. T. RosenMayer, et al., *Mat. Res. Soc. Symp. Proc.*, 130:77–86 (1989)
22. F. Maseeh, M. A. Schmidt, M. G. Allen, and S. D. Senturia, "Calibrated measurement of elastic limit, modulus, and the residual stress of thin films using micromachined suspended structures," *Tech. Digest of Solid-State Sensors and Actuators Workshop* (1988), pp. 84–87
23. S. Greek, F. Ericson, S. Johansson, M. Fürtsch, and A. Rump, "Mechanical characterization of thick polysilicon films: Young's modulus and fracture strength evaluated with microstructures," *J. Micromech. Microeng.*, 9:245–251 (1999)
24. D. A. LaVan, T. Tsuchiya, G. Coles, W. G. Knauss, I. Chasiotis, and D. Read, "Cross comparison of direct strength testing techniques on polysilicon films," *Mechanical Properties of Structural Films, ASTM STP 1413* (2001). Available online at: www.astm.org/STP/1413/1413_06 (2001)
25. J. Koskinen, J. E. Steinwall, R. Soave, and H. H. Johnson, „Microtensile testing of free-standing polysilicon fibers of various grain sizes," *J. Micromech. Microeng.*, 3(1):13–17 (1993)
26. T. Tsuchiya, O. Tabata, J. Sakata, and Y. Taga, "Tensile testing of polycrystalline silicon thin films using electrostatic force grip," *Trans. IEE Japan*, 166-E(10):441–446 (1996)
27. T. Tsuchiya, A. Inoue, and J. Sakata, "Tensile testing of insulating thin films; humidity effect on tensile strength of SiO_2 films," *Sensors and Actuators*, A82:286–290 (2000).
28. K. Sato, T. Yoshioka, T. Ando, T. Shikida, and T. Kawabata, "Tensile testing of silicon film having different crystallographic orientations carried out on a silicon chip, *Sensors and Actuators*, A70:148–152 (1998)
29. H. Kapels, J. Urscher, R. Aigner, R. Sattler, G. Wachutka, and J. Binde, "Measuring fracture strength and long-term stability of polysilicon with a novel surface-micromachined thermal actuator by electrical wafer-level testing, *Proc. Eurosensors*, XIII:379–382 (1999)

30. W. N. Sharpe, K. T. Turner, and R. L. Edwards, "Tensile testing of polysilicon, *Experimental Mechanics*, 39(3):162–170 (1999)
31. H. Ogawa, K. Suzuki, S. Kaneko, Y. Nakano, Y. Ishikawa, and T. Kitahara, "Measurement of mechanical properties of microfabricated thin films," *Proc. IEEE Int. Workshop on Microelectromechanical Systems* (1997), pp. 430–435
32. W. N. Sharpe, B. Yuan, and R. L. Edwards, "Fracture tests of polysilicon," *MRS Proc.*, 505:51–56 (1998)
33. J. A. Walker, K. J. Gabriel, and M. Mehregany, "Mechanical integrity of polysilicon films exposed to hydrofluoric acid solutions," *J. Electronic Materials*, 20(9):665–670 (1991)
34. F. Pourahmadi, D. Gee, and K. Petersen, "The effect of corner radius of curvature on the mechanical strength of micromachined single-crystal silicon structures," *Proc. 6th Int. Conf. on Solid-State Sensors and Actuators* (1991), pp. 197–200
35. S. Johansson and J. A, Schweitz, "Fracture testing of silicon microelements *in situ* in a scanning electron microscope," *J. Appl. Phys.*, 63:4799–4803 (1988)
36. T. Namazu, Y. Isono, and T. Tanaka, "Nano-scale bending test of Si beam for MEMS," *Proc. Int. Conf. on MEMS* (2000), pp. 205–210
37. R. Ballarini, R. L. Mullen, H. Kahn, and A. H. Heuer, "The fracture toughness of polysilicon microdevices," *MRS*, 518:137–142 (1998)
38. H. Kahn, R. Ballarini, R. L. Mullen, and A. H. Heuer, "Electrostatically actuated failure of microfabricated polysilicon fracture mechanics specimens," *Proc. R. Soc. London*, A455:3807–3823 (1999)
39. S. B. Brown, G. Povirk, and J. Connally, "Measurement of slow crack growth in silicon and nickel mechanical devices," *Proc. IEEE Int. Workshop on Microelectromechanical Systems* (1993), pp. 99–104
40. T. Tsuchiya, A. Inoue, J. Sakata, M. Hashimoto, A. Yokoyama, and M. Sugimoto, "Fatigue test of single crystal silicon resonator," *Tech. Digest 16th Sensor Symposium* (1998), pp. 277–280
41. S. Brown, W. V. Arsdell, and C. L. Muhlstein, "Materials reliability in MEMS devices," *Tech. Digest Int. Conf. on Solid-State Sensors and Actuators* (1997), pp. 591–593
42. G. L. Povirk, J. Bernstein, and S. B. Brown, "Long-term stability of nickel in resonant micro-mechanical devices," *Proc. MRS Symp.*, 308:147–152 (1993)
43. M. Von Arx, O. Paul, and H. Baltes, "Process-dependent thin-film thermal conductivities for thermal CMOS MEMS," *J. MEMS*, 9(1):136–145 (2000)
44. O. Paul, P. Ruther, L. Plattner, and H. Baltes, "Thermal van der Pauw test structure," *IEEE Trans. on Semiconductor Manufacturing*, 13(2):159–166 (2000)
45. H. Baltes, O. Paul, and J. G. Korvink, "Simulation toolbox and material parameter data base for CMOS MEMS," *7th Int. Symp. on Micro Machine and Human Science* (1996), pp.1–8
46. M. Arx, L. Plattner, O. Paul, and H. Baltes, "Micromachined hot plate test structure to measure the heat capacity of CMOS IC thin films," *Sensors and Materials*, 10(8):503–517 (1998)

47. EMIS Data Reviews, Series No. 4, *Properties of Silcon*, INSPEC, London (1988)
48. W. A. Brantley, "Calculated elastic constants for stress problems associated with semiconductor devices," *J. Appl. Phys.*, 44(1):534–535 (1973)
49. J. A. Schweitz and F. Ericson, "Evaluation of mechanical materials properties by means of surface micromachined structures," *Sensors and Actuators*, 74:126–133 (1999)
50. T. Yi, L. Li, and C. J. Kim, "Microscale material testing of single crystalline silicon: process effects on surface morphology and tensile strength," *Sensors and Actuators*, 83:172–178 (2000)
51. C. J. Wilson and P. A. Beck, "Fracture testing of bulk silicon microcantilever beams subjected to a side load, *Proc. IEEE, J. MEMS*, 5(3):142–150 (1996)
52. C. J. Wilson, A. Ormeggi, and M. Narbutovskih, "Fracture testing of silicon microcantilever beams," *J. Appl. Phys.*, 79(5):2386–2393 (1996)
53. K. Najafi and K. Suzuki, "Measurement of fracture stress, Young's modulus, and intrinsic stress of heavily boron-doped silicon microstructures," *Thin Solid Films*, 181:251–258 (1989)
54. Y. Isono, T. Namazu, and T. Tanaka, "AFM bending testing of nanometric single crystal silicon wire at intermediate temperature for MEMS, *Proc. Int. Conf. on MEMS* (2001), pp. 135–138
55. S. Greek, F. Ericson, S. Johansson, and J. A. Schweitz, "In situ tensile strength measurement and Weibull analysis of thick film and thin film micromachined polysilicon structure," *Thin Solid Films*, 292:247–254 (1997)
56. T. Tsuchiya, O. Tabata, J. Sakata, and Y. Taga, "Specimen size effect on tensile strength of surface micromachined polycrystalline silicon thin films," *J. MEMS*, 7(1):106–113 (1998)
57. O. Tabata, S. Sugiyama, and M. Takigawa, "Control of internal stress and Young's modulus of Si_3N_4 and polycrystalline silicon thin films using the ion implantation technique," *Appl. Phys. Lett.*, 56(14):1314–1316 (1990)
58. H. Kahn, N. Tayebi, R. Ballarini, R. L. Mullen, and A. H. Heuer, "Fracture toughness of polysilicon MEMS devices," *Sensors and Actuators*, A82:274–280 (2000)
59. Y. C. Tai and R. S. Muller, "Fracture strain of LPCVD polysilicon," *Tech. Digest of Solid-State Sensors and Actuators Workshop* (1988), pp. 88–91
60. J. Adamczewska and T. Budzyński, "Stress in chemically vapor-deposited silicon films," *Thin Solid Films*, 113:271–283 (1984)
61. T. I. Kamins, "Design properties of polycrystalline silicon," *Sensors and Actuators*, A21–A23:817–824 (1990).
62. H. Guckel, J. J. Sniegowski, and T. R. Christenson, "Advances in processing techniques for silicon micromechanical devices with smooth surfaces," *Proc. IEEE Int. Workshop on MEMS* (1989), pp. 71–75
63. W. Huber, G. Borionetti, and C. Villani, "The behaviour of polysilicon thin film stress and structure under rapid thermal processing conditions," *Mat. Res. Soc. Symp. Proc.*, 130:389–393 (1989)
64. H. Miura, H. Ohta, N. Okamoto, and T. Kaga, "Crystallization-induced stress in silicon thin films," *Appl. Phys. Lett.*, 66(22):2746–2748 (1992)

65. P. Krulevitch, R. T. Howe, G. C. Johnson, and J. Huang, "Stress in undoped LPCVD polycrystalline silicon," *Proc. IEEE Int. Conf. on Solid-State Sensors, Actuators, and Transducers 91*, San Francisco, CA, June 24–27 (1991), pp. 949–952
66. J. Yang, H. Kahn, A.-Q He, S. M. Phillips, and A. H. Heuer, "A new technique for producing large-area as-deposited zero-stress LPCVD polysilicon films: the multipoly process, *J. MEMS*, 9(4):485–493 (2000)
67. D.-G. Oei and S. L. McCarthy, "The effect of temperature and pressure on residual stress in LPCVD polysilicon films," *Proc. MRS Symp,.* 276:85–90 (1992).
68. C.-L. Yu., P. A. Flinn, S.-H. Lee, and J. C. Bravman, "Stress and microstructural evolution of LPCVD ploysilicon thin films during high temperature annealing," *Proc. MRS Symp.*, 441:403–408 (1996)
69. L. S. O'Bannon, *Dictionary of Ceramic Science and Engineering*, New York, Plenum (1984)
70. M. Orpana and A. O. Korhonen, "Control of residual stress of polysilicon thin films by heavy doping in surface micromachining," *Proc. IEEE Int. Conf. on Solid-State Sensors, Actuators, and Transducers 91*, San Francisco, CA, June 24–27 (1991), pp. 957–960
71. D. Maier-Schneider, A. Köprülülü, S. Ballhausen Holm, and E. Obermeier, "Elastic properties and microstructure of LPCVD polysilicon films," *J. Micromech. Microeng.*, 6:436–446 (1996)
72. C. S. Lee, J. H. Lee, C. A. Choi, K. No, and D. M. Wee, "Effects of phosphorus on stress of multi-stacked polysilicon film and single crystalline silicon," *J. Micromech. Microeng.*, 9:252–263 (1999).
73. W. N. Sharpe, Jr., K. Jackson, and G. Coles, "Mechanical properties of different polyslilicons," *Proc. ASME MEMS*, 1:255–259 (2000)
74. O. Paul, M. von Arx, and H. Baltes, "CMOS IC layers," *Complete Set of Thermal Conductivities*, in *Semiconductor Characterization: Present and Future Needs* (W. M. Bullis, D. G. Seiler, A. C. Diebold, eds.), American Institute of Physics (1995), 197–201
75. M. Ignat, T. Marieb, H. Fujimoto, and P. A. Flinn, "Mechanical behaviour of submicron multilayers submitted to microtensile experiments," *Thin Solid Films*, 353:201–207 (1999)
76. N. Arokia and H. Baltes, *Microtransducer CAD, Physical and Computational Aspects*, Springer, Wien, New York (1999)
77. J. T. Ravnklide, V. Ziebart, O. Hansen, and H. Baltes, "Mechanical Characterisation of electroplated nickel-iron," *Proc. Eurosensors*, XIII:383–386 (1999)
78. T. Yoshioka, T. Ando, M. Shikida, and K. Sato, "Tensile testing of SiO_2 and Si_3N_4 films carried out on a silicon chip," *Sensors and Actuators*, 82:291–296 (2000)
79. T. Tsuchiya and J. Sakata, "Tensile testing of thin films using electrostatic force grip," *Mechanical Properties of Structural Films*, ASTM STP 1413 (2001). Available online at: www.astm.org/STP/1413/1413_01 (2001)
80. V. Ziebart, O. Paul, and H. Baltes, "Extraction of the coefficient of thermal expansion of thin films from buckled membranes," *Proc. MRS 546* (1999), pp. 103–108

81. O. Tabata, K. Kawahata, S. Sugiyama, and I. Igarashi, "Mechanical property measurements of thin films using load-deflection of composite rectangular membrane," *Sensors and Actuators*, 20:135–141 (1989)
82. R. Kazinczi, J. R. Mollinger, and A. Bossche, "New failure mechanism in silicon nitride resonators," *Proc. Int. Conf. on MEMS* (2000), pp. 229–234
83. L. S. Fan, R. T. Howe, and R. S. Muller, "Fracture toughness characterization of brittle thin films," *Sensors and Actuators*, A21–A23:872–874 (1990)
84. J. Koskinen and H. H. Johnson, „Silicon nitride fibers using microfabrication methods," *MRS Proc.*, 130:63–68 (1989)
85. S. Koller, V. Ziebart, O. Paul, O. Brand, H. Baltes, P.M. Sarro, and M. J. Vellekoop, "Determination of mechanical material properties of piezoelectric ZnO films, *SPIE Proc.*, 3328:102–109 (1998)

3 MEMS and NEMS Simulation

Jan G. Korvink, Evgenii B. Rudnyi, Andreas Greiner, and Zhenyu Liu

Institute of Microsystem Technology IMTEK
University of Freiburg, Germany

3.1 Introduction

Because MEMS and NEMS touch on so many application areas, the ideal simulation tool must follow suit and provide a vast range of coupled multidomain physical effects. In reality, no single tool caters to all the needs of the MEMS community. Hence, MEMS designers carry the burden to find the appropriate tools and strategy for their task. Fortunately, many alternative routes exist to achieve a given goal, but some insight is needed to get the most out of the simulators, especially if the target is to use simulation to achieve a design advantage. In this chapter we take a closer look at what is out there, and at the key features of each simulation method. We develop a simulation strategy to maximize the benefit from simulation, picking out a couple of key areas that are currently the focus of commercial applications. To round off the chapter, we illustrate the ideas with some concrete applications from our own work.

This chapter considers the available tools for simulation. It is organized according to what is being solved, and tabulates the commercial tools currently on the market. It also considers a simulation strategy, starting with a quick-and-dirty approach and ending with the integration within a successful product development cycle. It concludes with applications to some interesting MEMS simulation tasks, demonstrating the use of a variety of commercial and in-house simulation tools and illustrating the enormous diversity of applications in MEMS.[2-5]

3.2 Simulation Scenario

Numerical simulation has a special role in MEMS and NEMS engineering. (Hereafter, we mainly use the term MEMS but imply both.) Numerical simulation is useful primarily as tool to aid design processes, i.e., to test new designs, eliminate unfavorable alternatives, and search for best candidates; its usefulness is measured primarily by the competitive advantage it provides to the design team. This is in strong contrast to the nonengineering sciences, where simulation tools are primarily used to aid insight and test theories. MEMS simulation tools can be grouped into four broad usage categories:

1. Circuit and system simulators
2. Domain simulators (see the list below)
3. Simulation drivers: CAD integration and design automation
4. General numerical and symbolic tools

Circuit and system simulators can be viewed as providing a capability that targets design work based on mature manufacturing technologies, where most of the design effort takes place within well-defined design rules and where optimization of system performance is the primary design goal. System simulation tools are characterized by library elements with a predefined functionality (compact or lumped models, i.e., a movable mirror, a pump, a channel, a laser) that can be interconnected to form the whole simulation model [i.e., a micro total analysis system (μTAS)]. As usual, microelectronics provides the generic example: Circuit simulation tools such as SPICE enable circuit design evaluation but do not provide insight into the functioning of an individual transistor. System simulation for MEMS is extensively dealt with in Chapter 4.

In contrast, the *domain simulation* tools address deeper levels of detail, all the way from continuum to atomic resolution. These tools can be categorized according to usage as follows:

- Materials domain simulators
- Process domain simulators
- Equipment domain simulators
- Device domain simulators
- Environment domain simulators

Simulation drivers organize the workflow of circuit, system, and domain simulators, as well as providing higher order functionality, such as the automatic search for optimal designs. In contrast, general numerical and symbolic tools provide functionality at the most mathematical level of a simulation task.

We briefly look at the candidates of this classification. In Section 3.3, we classify the tools according to the way they perform a simulation, to gain insight into the analysis capabilities at our disposal to implement an efficient design strategy.

Circuit and System Simulators

System simulators are extensively treated in Chapter 4; they are not dealt with here. In the context of design automation, it is necessary to transport successful device models to the system environment and to put it in a form that makes it amenable to subsequent simulation. For example, device models usually are very large nonlinear systems of ordinary differential equations (ODEs). Thus the question here is one of interfacing, i.e., how to automatically convert large device models into smaller but equivalent system models (compact models). A further trend is shown by the following example: In some MEMS applications, many (identical) devices coexist side by side. (Think of the Texas Instruments DMD chip, with more that 1,000,000 micromirrors.) Including the DMD device's ordinary differential equation system in the system model is simply not practical. Here we look at numerical methods that perform the reduction automatically, with varying degrees of success. Chapter 4 addresses physically more direct methods for this process.

Usually, we are simply users of simulation tools. But often in MEMS work, we are confronted with needs that are not met by using commercial tools alone, in which case we have to assume the role of tool developers. These views are illustrated in Fig. 3.1.

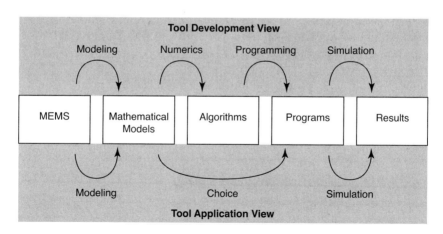

Figure 3.1 MEMS simulation entities and two interaction views with associated tasks.

Domain Simulators for Materials

Materials simulators are used to predict material behavior and material properties. These typically molecular-based simulation tools are becoming increasingly important to MEMS process developers as new materials with special behaviors are sought. For example, such tools can help predict microscopic behavior such as the onset of cracks and their propagation; they can be used to extract the anisotropic properties of grainy or layered materials, to predict the flow properties of molecular suspension fluids, and to determine the conduction mechanism in polymeric thin films. As device dimensions reduce toward truly nanometer sizes, molecular simulators gain in importance by becoming the preferred tools for device simulation as well. For MEMS, results from molecular simulators provide input for process simulators, for example, by predicting macroscopic silicon etch rates or the composition of deposited materials, and for device simulators, by providing the space-averaged constitutive properties of nonstochiometric solid-state materials or of molecular monolayers. Molecular simulators are already used extensively in chemistry and biochemistry, but are only now entering the MEMS field as technology moves away from using standard semiconductor materials for manufacturing.

Domain Simulators for Processes

Process simulators predict the geometric and compositional outcome of individual manufacturing steps. Very advanced tools are available for semiconductor processes, such as CMOS or bipolar processes. The idea is to emulate directly the clean-room process steps so that the programs merely need a process description and a mask layout. From these they produce a correct 3D topology of the resulting "chip" or portion thereof, as well as the state of the material in each of the resulting layers. By a state we mean the level of intrinsic stress or distribution of impurity atoms. These data are then used as the input to subsequent simulation programs. Process simulators use both particle-based and partial differential equation (PDE) type numerical methods.

Domain Simulators for Equipment

Equipment simulators form a relatively new field. Here the target is to simulate the behavior of a single process tool, such as a plasma reactor, based on its front panel controls. Such tools are meant to streamline and render more flexible the typical operations in a clean room, providing the machine operator with a method to predict accurately the outcome of new settings and ultimately to be able to run parallel processes in multifunctional clean rooms for better profitability. Process simulators use both particle-based and PDE-type numerical methods.

Domain Simulators for Devices

Device simulators form a very large class of tools in MEMS, as opposed to microelectronics, where electrothermal simulation covers the primary needs. In MEMS, the extraordinarily large number of application areas, ranging from optics and RF, through biology, to chemistry, require the consideration of many different energy domains, usually coupled with each other, to describe all phenomena. In general, simulation tools are expected to be predictive, and they should provide insight into the internal mechanisms and operation of a particular class of device. For CMOS MEMS, such tools must at least combine thermoelectrics with thermomechanics within the solid materials, and electrostatics overall. But CMOS MEMS is continuously being extended to a wider range of applications, such as by the addition of piezoelectric or chemically active layers, so simulation tools must be adapted to also include these operational effects. Device simulators obtain their computational geometries from standard CAD tools, as well as from process simulator output. The advantage of the latter is that, in addition to realistic geometries, physically correct material data is often provided, which otherwise need to be manually specified. Most device simulators solve PDEs, but increasingly, particle-like methods are gaining in importance for microfluidics, gas interactions, and molecular (polymeric) materials.

Domain Simulators for Environmental Effects

Environment simulators are needed increasingly as the MEMS field matures. One example here is packaging, which often cannot be separated from the behavior of its packaged devices when considering the behavior of a complete system. Whereas device simulators treat the MEMS device as an essentially isolated system, environment simulators couple the device with its engineering application and enable the analyst to consider the influence of electromagnetic interference, thermally induced stress, shocks, fatigue, corrosion, and so forth. More often than not, the environmental tools already exist for macroengineering; the main challenge is to couple these tools with device simulators and extend them with capabilities required by the MEMS community. The *de facto* standard tools here are the classical commercial finite element and finite difference packages, all of which solve PDEs. A notable exception is the ray tracers, which are used to predict radiation effects, for example in evaluating optical systems; they are not dealt with here.

Simulation Drivers: CAD Integration and Design Automation

CAD integration or technology CAD are terms for tools that combine all the above simulators and streamline their interactions. For standardized clean-room processes such as bipolar or CMOS, these tools are already in an advanced stage of development and can be thought of as a virtual manufacturing and testing environment. CAD tools can add more capabilities, though, mainly in the form of sophisticated simulation drivers and interfaces. Design automation requires more than mere streamlining of process, device, and system simulation, although streamlining is quite useful in itself. Design automation opens the possibility of a targeted search for a better design solution beyond the capabilities of the analyst or time constraints. For example, an optimal geometrical device layout that will guarantee correct operation, insensitivity to crosstalk and parasitics, and manufacturability can be sought, given the technology constraints of a process (set, for example, by the resolution of photolithography, the thermal budget of the process, the compatibility of materials, and the reliability of layer dimensions) and the system (amplifier noise, chip real estate, and number of masks). New tools are emerging that better interface domain solvers with system solvers, and that interact directly with experimentation. These higher level functions use the entire CAD suite of tools as black boxes and are based on sophisticated numerical methods that cannot be considered in the same context as the other phenomenon solvers.

The maturity of a manufacturing technology can be measured indirectly by the availability of design automation tools. When a technology is not mature, it is risky to develop design automation tools because the client base needed for a successful enterprise is missing. In turn, if a technology is established, market forces are huge, so design automation becomes imperative and the availability of the tools helps to spread the use of a technology quickly. Microelectronics (circuit design) and mechanical engineering (component stress analysis) provide often-cited examples, where universal access to design tools has ensured that technology and know-how have spread universally across engineering companies of all sizes. In MEMS, such tools are now also available, and their use is driven by the availability of multiproject wafer foundries with agreed-on design standards in Europe, the United States, and Japan.[1] To reiterate, design automation only makes sense if it rests on two essential prerequisites:

- Known, accessible, industry-standard manufacturing technology
- Widespread adopted design methodology

These requirements are fulfilled by, for example, the MUMPS process, which specializes in providing advanced polysilicon surface micromachining, and by many established multi-project-wafer (MPW) CMOS-MEMS processes, which provide sophisticated circuit capabilities together with bulk silicon micromachining.

General Numerical and Symbolic Tools

Commercial packages exist (e.g., Mathematica, Maple, Matlab) that implement general numerical and symbolic methods within a comfortable analysis environment. What makes these tools so powerful is that a new class of simulator can be rapidly prototyped efficiently and that the subsequent execution of the simulation can rely on robust and efficient implementations of core numerical methods. The creative modeler can use these tools in a variety of ways and needs only marginally to confront the details of the underlying numerical methods. In fact, we are observing the emergence in the marketplace of simulation programs that are built on these tools and that directly target the MEMS community. Often the implementers of such generalized tools rely on a large body of public domain numerical software, which they adopt and package in an environment that provides a comfortable high- to medium-level interpreted programming language, graphical output, and even specification capabilities for a graphical user interface (GUI). Before the general availability of such tools, implementers relied on the original public domain libraries, typically written in Fortran, C, and C++, which meant that the rapidness of the prototyping depended on the programming skills of the user.

Naturally, these tools do not target any of the preceding operational modes (system, domain, process, ...) alone, and applications are found across all disciplines, from materials to system simulation inclusive. Examples of general web resources are given in Table 3.1.

Table 3.1 General Web Resources

URL	Description
www.isgg.org	International Society of Grid Generation
www.andrew.cmu.edu/user/sowen/softsurv.html	Meshing Software Survey
www-users.informatik.rwth-aachen.de/~roberts/meshgeneration.html	Grid generation on the web
www.engr.usask.ca/~macphed/finite/fe_resources/fe_resources.html	Internet Finite Element Resources
www.netlib.org	NetLib
www.nag.com	Numerical Algorithm Group
www.lsc.nd.edu/research/mtl	Matrix Template Library
www.lsc.nd.edu/research/itl	Iterative Template Library
math.nist.gov/tnt/	TNT Home Page
www.oonumerics.org	O-O Numerics (Blitz++)
www-unix.mcs.anl.gov/otc/Guide/faq/linear-programming-faq.html#commercial	Linear Programming Resources
gams.nist.gov/	Guide to available mathematical software
www.biomems.net/OnLineResources/software.htm	MEMS software
www.memsnet.org/links/software.html	MEMS software
www.cfd-online.com	CFD resources
www.boundary-element-method.com	BEM resources
www-unix.mcs.anl.gov/otc/Guide/faq/nonlinear-programming-faq.html	Nonlinear programming resources
www.ari.net/ars/	Computational software for the Chemical Industry

3.3 Generic Organization of a Computational Tool

Most simulation programs (or tools) have the following organizational structure (see Fig. 3.2):

- Graphical user interface or front end
- Input files and parsing
- Preprocessing
- Solving
- Post-processing and input/output and interfacing

3.3.1 Graphical User Interface or Front End

Computational tools for complex tasks avoid excessive typing at the keyboard by implementing graphical user interfaces that aid data input,

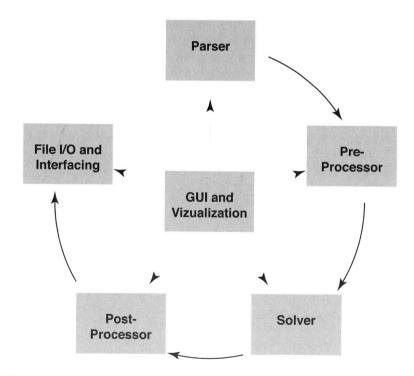

Figure 3.2 The program components of a generic simulation tool.

analysis control, and visualization. GUIs are also a convenient way for vendors to present the user with an analysis capability that relies on a whole suite of individual computer programs or computational kernels each of which may have no graphical capabilities and indeed may even execute their work on remotely connected computers. The roles of the GUI are:

- Graphical and menu-driven data input and manipulation
- Intuitive data consistency checking and reporting
- Graphical task scheduling
- Advanced graphical data processing
- Organization of tool suites
- Intuitive control over complex simulation procedures requiring user interaction

Whether the typical tasks of a simulation program are mapped into separate programs or within one executable code is frequently hidden from the user by the unified view provided by a GUI. The rest of Section 3.3 briefly describes each of these tasks.

3.3.2 Input Files and Parsing

The user defines the simulation assignment in a file using the application-specific input file language provided by the tool designer. The input file parser checks the grammatical correctness of these data and reads them into an internal data structure. Additional checks are performed, typically to ensure that a correctly stated simulation assignment is defined. For example, many tools will not proceed with analysis if certain minimal boundary or initial conditions are not specified or if material data are missing. Many tools provide a menu-driven application to generate such input files and ensure that these minimal conditions are met; for complex geometries, this has become a necessity. In an industrial setting, simulation experts are people who can get the most out of this phase by deeply understanding the mapping from a simulation assignment into the (possibly complex and large) input language of the tool.

3.3.3 Preprocessing

Preprocessing is a phase that prepares the internal data for subsequent solving and analysis. It is here that finite element solvers may provide sophisticated meshing capabilities and precompute data that are repeatedly used during the solution phase. For example, if only a certain region is nonlinear, the rest being linear, then the linear part of the equation coefficients can be computed *a priori*, once and for all, considerably reducing the time needed for the subsequent equation solving phase. Almost all domain solvers, including classical molecular dynamics programs, finite element solvers, boundary element solvers, and finite difference solvers, build a large system of ordinary differential equations from a combination of geometry, material data, partial differential equations, and boundary and initial conditions (BCs, ICs). Classically, this process is called the assembly phase, especially by the finite element community. Of course, it may be that stationary or quasi-stationary equations are formed, which determines which solver will subsequently be used. There is one important difference: So-called explicit solvers, including both molecular dynamics and rigid-body dynamics programs, completely avoid building matrix equation systems. Instead, they implement the inherent coupling within a single simulation time- step by sophisticated summation methods, after which the equations are decoupled and can be individually advanced one time-step. Section 3.3.4 discusses this mechanism.

3.3.4 Solving

A number of different solvers are usually provided by most of the (implicit) computational programs (see Fig. 3.3).

For stationary simulations, where all time derivatives are set to zero, linear and nonlinear solvers are applied.[6] Linear solvers use either a direct method (based on the Gauss elimination algorithm) or an iterative method (usually based on a so-called Krylov-subspace method).[7] In the latter, the method only requires the multiplication of the system matrix (assembled before) by a vector, which again results in a vector. An important point here is the way in which the equations are represented in the program: They are usually in the form of a matrix equation. PDE solvers such as the finite element method (FEM) and the finite volume method (FVM) produce sparse matrix equations, implying that only few of the matrix entries are actually nonzero (see Fig. 3.4). To allow program users to solve large simulation tasks, special sparse storage methods have been developed that minimize the storage of redundant information. This has consequences for the linear solver routines. The iterative solvers can optimally exploit the sparse data, for they only produce matrix-vector products. (In fact, they only produce dot products between matrix rows and a vector.) Direct solvers, however, implement some form of the Gauss elimination algorithm and need some "scratch" space in the matrix to do the operations. Reordering the unknowns of the matrix can have a profound

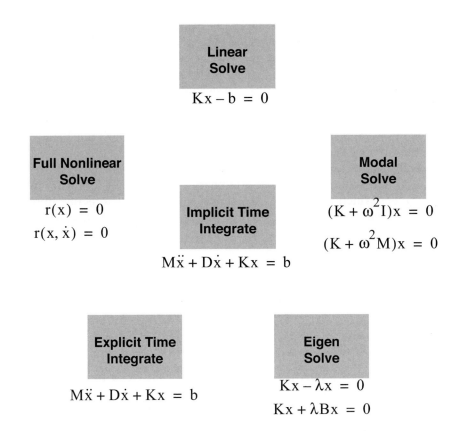

Figure 3.3 The most common solving kernels.

influence on the extra space required, which explains why many FEM solvers provide a variety of reordering algorithms as options for the solution (see Fig. 3.4). A popular reordering algorithm is that of Cuthill and McGee; one of the best is the Minimal Degree reordering algorithm.

Nonlinear solvers are usually found as Newton solvers or as path-following solvers. To start the nonlinear solution process, we require an initial guess for the first iteration. The choice of this initial guess can be simplified by solving a problem for a small value of the load (boundary conditions). The Newton solvers advance the solution from a stable convergence point to a solution point nearby by forming a generalized, truncated Taylor expansion of the equations about the converged point (see Fig. 3.5, right). This results in a linear system of equations to be solved. The matrix of this equation is called the Jacobian matrix; it is dependent on the value of the solution at the equilibrium point. Since forming the

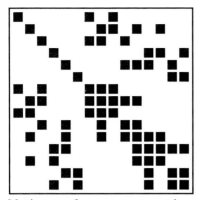
Matrices are often sparse; many entries are zero.

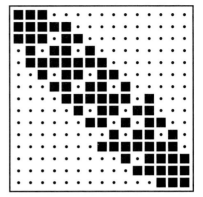
Mesh node renumbering can increase "bandedness" and hence efficiency of storage.

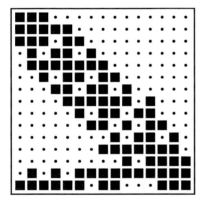
Constraints can destroy bandedness.

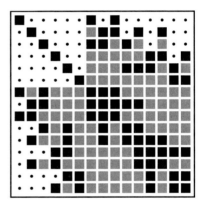
Factorization creates fill-in between the "skyline" and the diagonal.

Figure 3.4 Sparse matrix structures. Dots indicate zero-valued matrix elements. A dot indicates a zero value; a black square indicates a nonzero value; a gray square indicates a zero-valued number that needs to be stored.

Jacobian matrix is very expensive, many strategies exist to make this process more efficient. These strategies work especially well if the nonlinearity is mild and the solution curve does not contain instability points or bifurcations (see Fig. 3.5, left). Two popular techniques here are the Quasi-Newton methods, which keep an older version of the Jacobian matrix unaltered for as long as possible, and the so-called rank one update methods, which modify the Jacobian matrix as long as possible with computationally cheap updates that depend on the solution path. If the solu-

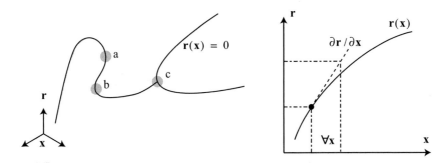

Figure 3.5 Nonlinear solver concepts. The gray dots (a) and (b) on the residual curve to the left indicate loss of stability, where the Jacobian is zero or infinite; the dot (c) marks a point of bifurcation where the Hessian vanishes. The right-hand plot illustrates a typical Newton step along the direction of the Jacobian. The black dot indicates the Taylor expansion point.

tion does contain instability and bifurcation points (which is often the case when solving the response of electrostatic actuators, or buckling membranes; see Fig. 3.5, left), then path-following methods (also called continuation methods) apply. A typical continuation method combines a predictor, usually a Euler time-step, with a corrector, usually a Newton or Quasi-Newton method.[8] The continuation method does not get into the same type of trouble as a purely Newton method because the predictor produces a guess that is already close to the solution curve. The Newton corrector method can exploit this proximity, usually with quadratic convergence behavior. The whole process is aided by an extra unknown (the homotopy parameter) that allows the applied boundary condition (for example, the applied potential to an electrostatic actuator, the buckling force) to be applied gradually. Clearly, the Newton step again requires the solution of a linear equation system, and all the known techniques can be applied here as well. Although the algebraic multigrid method[9] has achieved tremendous popularity over recent years as a very fast linear and nonlinear solver, it has not yet found its way into commercial solvers, although this situation could change quickly. Its less general form, the "normal" multigrid method, is mainly used in conjunction with discretization schemes where hierarchical meshes are available. In these cases, it is a highly robust and extremely fast method.

For transient simulations, the coupled equations are discretized in time, which again results in a linear or nonlinear equation system to be solved, once for each time-step. Therefore, this is again a very time-consuming problem. Some strategies are used to speed up the process. The concept of modal analysis for linear or small signal nonlinear analysis (Fig. 3.6[10]) is

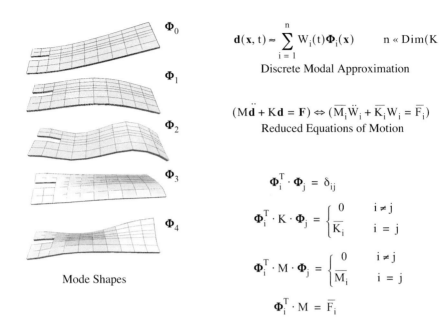

Figure 3.6 Modal analysis concepts. The key idea is that complex motion is reduced to the decoupled motion of very few orthogonal modal degrees of freedom Φ_j. The selected modes are used to project the global matrices (K, M, F) onto a smaller subspace, yielding small matrices ($\overline{K}_i, \overline{M}_i, \overline{F}_i$) and a fast solution algorithm. Adapted from Emmenegger et al.[10]

one way. Here we compute the dominant modes (the eigenvectors) of the system, which we use to produce a much smaller ODE system that is inexpensive to compute. Another related method is harmonic analysis: If the excitation is at a distinct frequency, we can eliminate the time derivative by assuming that the response is at the same frequency. This results in a linear system dependent on the excitation frequency. Sweeping over different values of the frequency and solving the equation system produces the frequency response of the system.

Explicit time integrators are best explained by considering each unknown in a vector to represent a particle with momentum traveling through space (see Fig. 3.7). We know each particle's current position and its momentum, and we wish to know where all particles will be after a small time-step. Instead of forming an equation system for the unknown positions, as an implicit solver would (Fig. 3.8), we instead compute the net force currently acting on each particle. By enforcing dynamic force

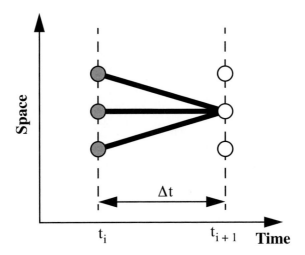

Figure 3.7 Explicit time integrator. The white dots indicate mesh nodes where unknowns are computed. Future solution values at a node depend on past solution values only. This can be done very efficiently, especially on vector and parallel processors.

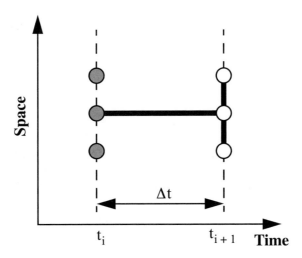

Figure 3.8 Implicit time integrator. The white dots indicate mesh nodes where unknowns are computed. Future solution values at a node depend on past solution values of the node as well as on the future values of neighboring nodes. Hence a simultaneous equation system has to be solved for each time step.

equilibrium for a particle, together with a known uniform time-step length for all particles, we can determine where each particle will move. At each new position we need to recompute the force on the particles. If the forces come from an FEM discretization, then the force calculation is particularly efficient because the connectivity among particles is very small. If, however, there is no underlying implicit mesh connectivity, as is the case for molecular dynamics or many-body simulators, then the force calculation can take up a major portion of the solution time. Techniques exist to speed up the force calculation. They run from fast summation methods (Ewald sums) to exploitation of the multipole expansion. (The multipole expansion treats a collection of charges by an efficient, single-multipole expression.) "Force" should not be taken too literally because many ODEs can be written in a form that mimics the structure of a Newtonian force computation.

Why do explicit solvers work? The trick lies first in the size of the time-step, which is usually extraordinarily small so as to limit the influence of cumulative errors. But more is required. Usually, the force is computed for particle positions between the computed time-steps, so-called leap-frog or symplectic methods; it can be shown that this technique further reduces the error of approximation.

To compute the modes (eigenvectors) of a system, many programs provide a built-in eigensolver. Computing all the modes of a large eigensystem is very time-consuming, growing with the third power of the number of equations (unknowns) in the system. Special strategies are available, such as subspace eigensolvers, which only compute the first few eigensolutions. This is often enough for engineering analysis because many systems have their energy stored in only a few of the modes or are excited only by a limited frequency band about a few modal frequencies.

3.3.5 Post-Processing and Program Interfacing

Post-processing is a generic term for all operations that are performed after a solution to the system of ODEs has been obtained, to bring the results into a form for evaluation and as input to the design process (Fig. 3.9). These operations are as follows:

- *Visualization.* Producing field plots, including contour plots, 3D plots, and (for transient simulations) animations of field plots, can provide additional insight.
- *Secondary Fields.* Many simulators compute the values of a scalar field quantity, for example, the temperature profile or the electric potential. If the heat flux or the electric field vec-

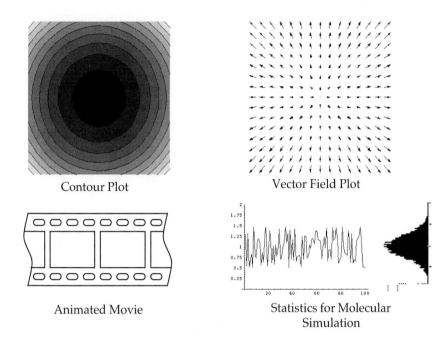

Figure 3.9 Typical post-processing actions: computing secondary fields from solutions, creating animations of solutions, analyzing time series data from molecular simulation.

tor is needed, this is computed from the scalar field by taking the gradient.
- *Reactions (lumping).* For a given field solution in the interior of an object, we may want to know some result at a boundary condition. For example, assume that we have computed the deformation of a 3D body. This implies a stress distribution throughout the body, i.e., the secondary field. Integrating the secondary field over a boundary will give the force acting at this boundary; if it happens to be a mechanical support, the force will be equal and opposite to the reaction force at the boundary.
- *Field plots.* In doing detailed design evaluation, we may want to inspect the variation of a field quantity over some surface or within a volume. Field plots provide us with this capability.
- *Interface conditions.* When using decoupled and hence iterative multiple field solvers, we often have to compute tem-

porary and approximate boundary conditions from the solution of one of the fields as input to the other field solvers.
- *Interfacing.* When many solvers interact, we often need to transfer data from one representation to another, for example, from a finite difference mesh in one solver to the finite element mesh of another, or between a field solver and an optimizer.

3.4 Methods for Materials Simulation

Modern materials (and, of course, NEMS) simulation is mostly based on molecular simulation that plays several roles in the simulation hierarchy shown in Fig. 3.10. First, it allows us to estimate macroproperties such as heat capacity, viscosity, elasticity, heat conductivity, equilibrium, and rate constants from microproperties. Second, it can be quite useful in the engineering of specific molecules with given optical, magnetic, or medical properties. Finally, it can be used in those cases where, because of small dimensions, continuum-based theory cannot be applied at all.

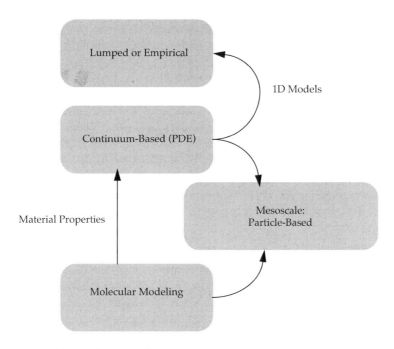

Figure 3.10 A hierarchy of models.

This section pays attention mainly to the first direction because it is the most important.[11-15]

There is an important exception in materials simulation related to modeling multiphase and multicomponent systems. In this case, the best approach available today is based on a phenomenological thermodynamics approach; this will also be covered.

Roughly speaking, at an atomic scale, we have nuclei in a sea of electrons that act as a glue to keep the nuclei altogether. An electron has a quantum nature and the laws of classical mechanics are not applicable. Nuclei are much heavier and often can be considered classical particles. As a result, according to the popular Born-Oppenheimer approximation, in many cases it is enough to say that there is some effective interatomic potential at the ground electronic state acting among point atomic masses and then to treat the movement of these masses classically.

Therefore, in the vast majority of cases, the problem of estimating macroproperties is divided into two steps:

- Obtaining the potential energy surface for a given molecular system
- Estimating the macroproperties for a given potential energy model

This allows us, in principle, to formulate a first-principles-based approach to solve any problem, that is, to estimate any desired property without any adjustable experimental parameters.[16,17] We consider this approach first.

Quantum Chemistry: *ab Initio* Potential Energy Surface

If nuclei positions are fixed, the solution of the electronic Schrödinger equation gives the energies of the different electronic states as well as the corresponding electron wave functions. Although we need only the energy of the ground state to build a potential energy surface, the ground-wave function allows us to estimate other useful molecular properties such as electron density and dipole moment. If we repeat this many times for different positions of nuclei, eventually the whole potential energy surface can be constructed.

Consequently, the solution of the electronic Schrödinger equations constitutes a first and basic computational problem in materials simulation. Unfortunately, this alone is a formidable computational task.

Figure 3.11[18] schematically shows the conventional procedure for solving the electronic Schrödinger equation. First, the electron wave func-

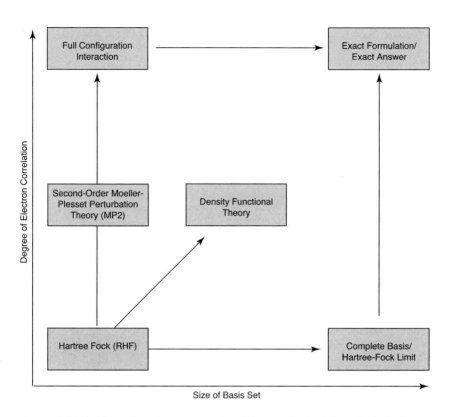

Figure 3.11 A hierarchy of quantum chemistry methods (after Mark S. Gordon, *www.msg.ameslab.gov/Group/GroupTutor.html*).

tion is factored into molecular orbitals; these, in turn, are approximated as a linear combination of atomic orbitals (the lower line in Fig. 3.11). This converts the partial differential electronic Schrödinger equation to a nonlinear eigenvalue problem. The main question at this level is which atomic basis to use. In other words, how can we choose the right basis among many cryptic acronyms made available by quantum chemistry software? Since the computational effort grows as the basis dimension to the fourth power (see Fig. 3.12[19]), it is important to find a reasonable compromise between affordable computational time and required accuracy.

In the ideal case, the atomic basis should be infinite, but even if it were, the "exact answer" would not be reached because the factorization to molecular orbitals does not take electronic correlation into account. To treat this case, it is necessary to move upward in Fig. 3.11, or in quantum chemistry slang, to apply a configuration interaction (CI). Even when the

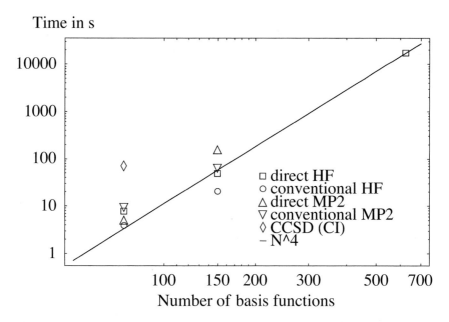

Figure 3.12 Quantum chemistry computational time as a function of a number of basis functions by means of Gaussian 98 on HP PA8600. Adapted from *www.emsl.pnl.gov:2080/docs/tms/abinitio-v4/cover.html*.

simplest ways to perform this move (such as using second-order Moeller-Plesset perturbation theory, often abbreviated MP2) are relatively cheap, this is usually even more expensive than the basis expansion.

Density function theory (DFT) allows us, for a single simulation run, to make a diagonal move in Fig. 3.11; this is what makes the technique very popular. However, the extent to which the electron correlation (a jump upward in Fig. 3.11) can be taken into account, for a given basis set, cannot be determined.

The choice as to which approximation level should be used depends on the particular target property in question. Some properties can be estimated quite well at the very basic level; others demand an extended basis as well as high-correlation methods. In general, the better the precision, the higher the computational price.

Statistical Mechanics: Molecular Dynamics or Monte Carlo

Strictly speaking, quantum chemistry results correspond to pure energetic considerations (in other words, to properties at a temperature of zero

K). In the general case, it is necessary to have a relationship between microproperties (positions and velocities of point atomic masses) and macroproperties. This is solved by statistical mechanics, which makes available a number of averaging operators in time or over the particle ensemble. However, in most cases these averages cannot be computed in closed form and it is necessary to run either molecular dynamics or Monte Carlo simulations.

A simulation starts with some random initial atomic configuration. The number of atoms to be included in the simulation is limited by hardware. This is why periodic boundary conditions (see Fig. 3.13) are necessary to obtain correct bulk macroproperties. In this case, the system is assumed to be extended periodically in all directions, yet the simulation is done for a master image only and other cells just copy its behavior. Without this technique, interface effects would be introduced into the system because the atoms at boundaries would have different interactions than bulk atoms. This could affect the results enormously because the number ratio of master image boundary atoms to bulk atoms is rather high for a typical simulation.

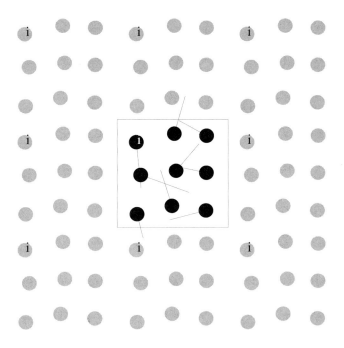

Figure 3.13 Periodic boundary conditions in two dimensions. Each particle represents many image particles. Particles leaving the simulation box at one edge reenter the domain at the opposite edge.

Molecular dynamics simulates the time development of a real system. The Newton equations of motion are solved numerically to advance from a current configuration to another. The idea is rather straightforward: The forces acting on each atom are computed for the current configuration, then positions and velocities of all atoms are updated according to the second Newton law to obtain a subsequent configuration for a chosen time-step. As a result, many configurations approximating the system trajectory in phase space are obtained and the results can be used to compute a time average to estimate the desired property.

The classical mechanics procedure just described can be used in most cases because the Born-Oppenheimer approximation introduced in the beginning of Section 3.4 works quite well during the nuclei motion when there is no change in the electronic state. If the latter is not the case, for example for photon absorption/emission phenomena, the quantum corrections must be applied. It is instructive to consider the computational effort required to perform a molecular dynamics simulation. Assume that the typical time required for the computation of a molecular force is 10^{-5} wall-clock second per particle. For solution accuracy, a typical simulation time-step will evolve the atomic movement for only a fraction of an atomic diameter, and typical time-step lengths are of the order of 10^{-14} "physical" seconds. So a simulation of 1,000 atoms for, say, 100 time-steps, requires 1 wall-clock second; this corresponds to 1 picosecond of physical time. One second of physical time therefore corresponds to a wall-clock time of around 30,000 years;[20] however, for most purposes, the simulation of nanosecond physical time is sufficient.

The goal of Monte Carlo simulation is to obtain atomic configurations that are distributed according to a given ensemble distribution. The computational algorithm is rather simple, especially when compared to the Markov chain theory it is based on. At the end of the simulation, a representative sample of the ensemble distribution is obtained that can be used to find an ensemble average. In a first principles approach, the electronic Schrödinger equation is solved "on the fly" to obtain the required energy for each configuration during molecular dynamics or Monte Carlo simulation. This is achieved by a so-called *quantum molecular dynamic* algorithm and is limited to highly parallel supercomputers (see Fig. 3.14).

Semi-Empirical Approaches

In order to make the steps described above computationally tractable, people have given up the idea of computing from first principles and have introduced some functional forms that possess the required theoretical properties with unknown parameters to be determined from experimen-

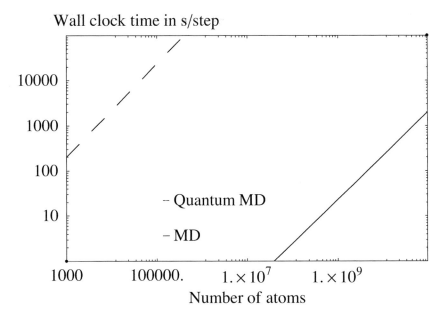

Figure 3.14 The computation time of molecular dynamics (solid line) and quantum molecular dynamics (dashed line) for linear-scaling schemes on 1,024 IBM SP3 processors according to Nakano et al.[17]

tally measured properties during a process called parameterization. The resulting approaches, in a way, looks like "data fitting"; the slight difference is that the functional forms are theoretically reasonable.

Within quantum chemistry, this idea leads to a variety of semi-empirical methods. Their success greatly depends on the parameterization set. If your molecular system is close to those experimental properties that have been used for the parameterization, a good outcome can be expected with a rather high probability. In the opposite case, one can only pray for the best. At the statistical mechanics level, this idea leads to the so-called molecular mechanics approach, or in other words, to empirical force fields. Here the potential energy surface is approximated with a linear combination of simple terms representing bond stretching, angle distortions, torsion movements, and so on. Its parameterization is obtained using available experimental values and ab initio calculations. Without doubt, this speeds up simulations by many orders of magnitude (see Fig. 3.14) compared to quantum molecular dynamics, and most molecular dynamics or Monte Carlo simulations are made on this level. Again, success greatly depends on the empirical potential used, the situation being rather close to that in the semi-empirical quantum methods: whether a molecular system in question is analogous to those that have been used during the parameterization.

In the general case, it may be necessary to first develop an appropriate empirical potential surface to obtain good simulation results.

Phenomenological Approach: Computational Thermodynamics

A phase diagram is of great importance for materials science.[21,22] In theory, it can be computed from first principles; in practice, this is almost unfeasible. An alternative approach is given by thermodynamics, which states that each phase can be described by a molar Gibbs energy function. Provided the Gibbs energies are known for each phase in the system in question, all system equilibrium properties can be computed, including phase equilibria and the complete phase diagram.

There are software packages that implement this idea of computational thermodynamics. The software contains a database with Gibbs energies and a computational engine that computes the properties desired, including phase diagrams, on the fly. Calculations are reasonably fast, especially if compared with molecular-level simulations. The Gibbs energy is described by some functional form and the unknown parameters within it are found during so-called thermodynamic assessment, by processing all available experimental results for a given system. This is quite similar to the semi-empirical approaches presented above, the difference being at another level of the parameterization.

The difference between the phenomenological approach and a pure empirical one should be stressed; the Bi-Se phase diagram shown in Fig. 3.15 is an example.[23] At first inspection, the question may arise whether we need thermodynamics at all to build the phase diagram from experimental points. In principle, we can just fit the liquidus lines according to some empirical polynomials. However, it is then impossible to use them for extrapolation. For example, how can the shape of the miscibility gap close to selenium be predicted from the available experimental points? An empirical approach does not allow us to predict it. In the phenomenological approach, there is a theory that supplies relationships among different properties, and a final model is developed based on this theory. In the example, there are many other thermodynamic experimental results for the Bi-Se system: the heat of mixing and the heat capacity of the melt, vapor pressures over the system, and emf values. The unknowns in the Gibbs energies are found from a joint treatment of all these values in addition to the phase equilibria results, then the extrapolation is done in terms of the Gibbs energies. This is much more reliable than the extrapolation of liquidus lines; it also allows us to estimate the

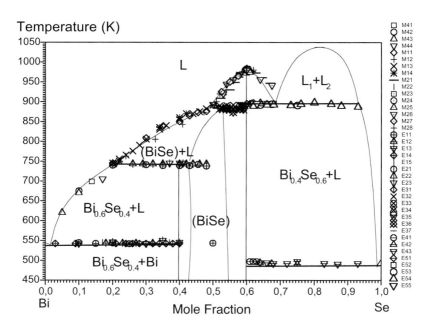

Figure 3.15 The Bi-Se phase diagram with experimental points according to Antipov et al.[23]

Gibbs energies in the multicomponent system and hence the phase diagrams from the low-order systems.

Empirical Correlations: Quantitative Structure-Property Relationship

In many cases, the methods presented above cannot be applied because of the high computational cost. Some properties, such as rate constants, hardness, and biological activity, are almost impossible to compute even for the case of empirical potentials arising from molecular mechanics, let alone first-principle approaches. At the same time, quite often the task is to predict properties for compounds in a homologous series when properties of some of them are known. This is a common task in organic chemistry where we search for the best functional group to increase the property required. In this case, it may be much more efficient to use pure empirical correlations, in which case the desired property is assumed to be an empirical polynomial function of molecular descriptors related to the molecular structure.

The main question for this approach is which molecular descriptors to use, especially for the case when the molecules have not yet been synthesized. A common strategy is to generate descriptors at the semi-empirical quantum chemistry level. The descriptor goal is to serve as an x-axis for the correlation. Its numerical value may not be accurate; rather, the descriptor values should give the right trend. Semi-empirical quantum chemistry usually satisfies this requirement rather well. The whole procedure then comprises three steps, as follows:

1. Compute several descriptors for the whole series of compounds.
2. Calibrate an empirical correlation equation by using the compounds with known properties by means of a modern multivariate technique such as principal component analysis or partial least squares.
3. Estimate the properties of other compounds from the empirical equation obtained above and choose a few potential candidates for experimental tests.

Discussion

We have reviewed material simulation methods that today can be considered routine. Table 3.2 lists the currently available material simulation programs. As usual, much more can be found at the frontiers of science. We conclude by mentioning two trends.

First, the methods described above concern bulk properties and are typically required for device simulation based on continuum theories. At present, the continuum-based theories can fail when device size diminishes sufficiently. In this case, the removal of periodic boundary conditions for molecular dynamics or Monte Carlo simulation presents a unique chance to simulate the system in question. Unfortunately, the current computational possibilities (see Fig. 3.14) allow us to treat only relatively small molecular systems. There is a range where the continuum-based methods may not be applied, yet it is impossible to run a full-scale atomistic model. New intermediate-level methods are emerging here; they will soon start playing an essential part in materials simulation. One of them, the so-called *dissipative particle dynamics* method, is already implemented within commercial software (e.g., MesoDyn in Accelrys' Materials Studio).

The second important area of research focuses on multiscale methods, that is, an attempt to use, within a single simulation, different levels of theory simultaneously to treat the huge spatial and temporary scales

Table 3.2 Software for Materials Simulation

Program Name	Application Area	Availability	Internet
CAChe	Quantum Chemistry/ Molecular Mechanics	Commercial	www.cachesoftware.com/cache/
Materials Explorer	Molecular Dynamics/ Monte Carlo	Commercial	www.cachesoftware.com/ materialsexplorer/index.shtml
Materials Studio	Quantum Chemistry/ Molecular Mechanics/ Molecular Dynamics/ Monte Carlo	Commercial	www.accelrys.com/mstudio/
GAMESS	Quantum Chemistry	Free for research	www.msg.ameslab.gov/GAMESS/
TURBOMOLE	Ab initio Quantum Chemistry	Free for research	www.turbomole.com
GROMACS	Molecular Dynamics	Free (GPL)	www.gromacs.org
CPMD	Quantum Molecular Dynamics	Free for research	www.cpmd.org
ThermoCalc	Computational Thermodynamics	Commercial	www.thermocalc.se
FactSage	Computational Thermodynamics	Commercial	www.factsage.com

needed for modern material simulation. The key problem here is how to perform the "handshaking" among different theory levels and keep the computational requirements needed under control.

3.5 Computational Methods that Solve PDEs

A partial differential equation describes the distribution of one or more fields over a portion of space. For example, the PDE of elastostatics describes the distribution of a vector-valued displacement field, and is for example used to calculate the deformation of a MEMS membrane subject to pressure. PDE methods spatially semi-discretize the governing equations to form a system of ordinary differential equations. They are almost always associated with a mesh that partitions the computational domain into smaller units. We will now discuss the most important of these methods (i.e., those that have found their way into commercial applications).

Finite Element Method

The finite element method (FEM)[24,25] has the following components:

- A *mesh*, made up of individual simple polytopes. (In 1D these are lines; in 2D, triangles, quadrilaterals, and infrequently other polygons; in 3D, tetrahedra, hexahedrons, triangular prisms, and infrequently other polytopes.) The polytopes are termed the elements. The elements completely cover the computational geometry and do not overlap (see Fig. 3.16[26]).
- *Nodes*, at which the fields of the PDE will be assigned discrete values. The nodes can be positioned anywhere within an element; they are typically found at the element vertices, sometimes along the edges, and, once in a while, in the interior of an element. A node is associated with one or more of the fields of the PDE. Therefore, they may carry more than one degree of freedom. Together, they determine the number of equations that will be in the ODE system.
- *Shape functions*, which determine how a field will be approximated over the domain of an element. The shape functions are typically simple polynomials, and their choice is governed by what level of continuity is needed by the analyst and what level of continuity is required by the PDE.

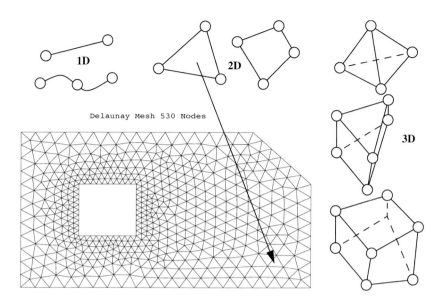

Figure 3.16 Finite element mesh concepts. Top and right: 1-D, 2-D, and 3-D element polytopes. Bottom left: a 2-D Delaunay mesh of triangular three-noded elements created using Easymesh (*http://www-dinma.univ.trieste.it/~nirftc/research/easymesh/*).

A given polytope can be associated with different collections of nodes and shape functions for a given PDE. Commercial finite element programs talk of "families" of elements. A mesh may mix different polytopes; indeed, it may mix 1D, 2D, and 3D elements. The mixing is only subject to the maintenance of required continuity conditions for the fields among the elements. At any rate, given an element, its material data, its nodal positions, and its shape functions, the finite element method discretizes the PDE into a relatively small element matrix system of ordinary differential equations (ODEs).

The finite element method uses a special technique to form the complete discretized equations (see Fig. 3.17). In essence, the FEM ensures that the error of approximation, measured over the whole element domain, is minimal in a special way. Mathematically, it is called the weighted residual method (which amounts to the principle of virtual work in mechanics); elucidating it goes beyond the scope of this text.

An *assembler* program passes over all the elements of the mesh, passing relevant information to an appropriate element subprogram, and receives back the small ODE coefficient matrices (see Fig. 3.18). Consider a simple example: Using tetrahedral elements, with four vertex

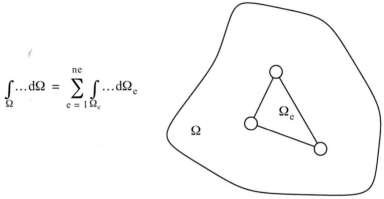

Figure 3.17 Finite element matrix equations typically arise from a bilinear and a linear form that are domain integrals (top). Here u is the unknown field (i.e., the temperature) and v is a known test function. Since the integrals are linear over their domain, all integrations can be performed element by element (bottom). Once the integrals are evaluated by the finite element software, a simultaneous equation system [A]{u} = {F} is obtained that is solved for the nodal unknowns (the values of field u at the finite element nodes).

nodes and linear shape functions, together with the PDE for heat flow in a solid, the element subprogram returns a small four-equation ODE. In contrast, the resulting ODE for the mesh could have any dimension, with the size dictated by the number of elements in the mesh.

The matrices contain only geometrical and material property data, and many clever schemes exist to make the preparation of this ODE system very efficient. The assembler manages the way element equation numbers map to the global numbers of the mesh; it quickly adds the element contributions into the large, sparse global system matrices. Once completed, the finite element program can proceed to solve the equation system, which presupposes the setting of initial or boundary conditions. This phase is similar for all PDE solvers.

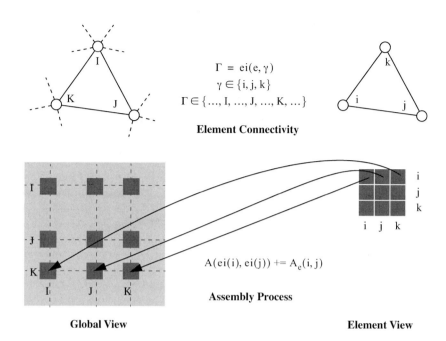

Figure 3.18 Finite element matrix assembly. Left: global equation system view of the element e and matrix A. Right: element-based view of the element e and matrix A_e. Mesh-generation software gives each node in the mesh a unique number, but the node numbers corresponding to one element are not contiguous. The element subroutines treat each element as having a small, identical set of node numbers, i.e., $\gamma \in \{1, 2, 3\}$. The array ei has an entry for each element of the mesh and relates the element node numbers γ to their global counterparts Γ. The array is used to automate the global matrix assembly process, i.e., to correctly insert the values of A_e into A.

The advantage of the FEM is immediately apparent: Once a program has its basic infrastructure, a finite element program can be extended to a variety of other PDEs, element shapes, and the like because the kernel discretization step takes place for a well-defined case and can then be handled by one of a set of little discretization subprograms. Table 3.3 lists currently available FEM programs.

Finite Volume Method

The finite volume method (FVM),[27] or control volume method, has meshing requirements similar to the finite element method, with the

Table 3.3 Software Based on the Finite Element Method

Program Name	MEMS Application	Availability	Internet
ANSYS	Multiphysics	Commercial	www.ansys.com
PATRAN			
MARC	Mechanical	Commercial	www.mscsoftware.com/products/
CFD-ACE+	Multiphysics	Commercial	www.cfdrc.com/datab/software/ aceplus/
FEMLAB	Multiphysics	Commercial; requires MATLAB	www.femlab.com
DIFFPACK	Multiphysics	Commercial	www.diffpack.com
ABAQUS	Multiphysics	Commercial	www.abaqus.com
FIDAP	CFD	Commercial	www.fluent.com/software/fidap/ index.htm
NM SESES	Multiphysics	Commercial	www.nmtec.ch

restriction that all interior element angles are smaller than 90 degrees (see Fig. 3.19). If the angle condition is not met, special integration techniques have to be added. There is a major difference in the way that the element matrix contributions are computed, however. In the FVM, a direct conservation equation is formed for each node (unknown) in the mesh (see Fig. 3.20). We usually associate a complementary "box" mesh with the primary mesh. This mesh (also called a Voronoi mesh) describes the conservation volumes about whose edges conservation is enforced. To benefit from the tremendous computational advantages of an element-based view, FVM programs also implement "element" subprograms. In this case, we again have to consider only a small variety of polytopes. (Note that the boxes can have any number of sides or faces.) The elements then contribute partially to the equations of a particular box. Apart from these

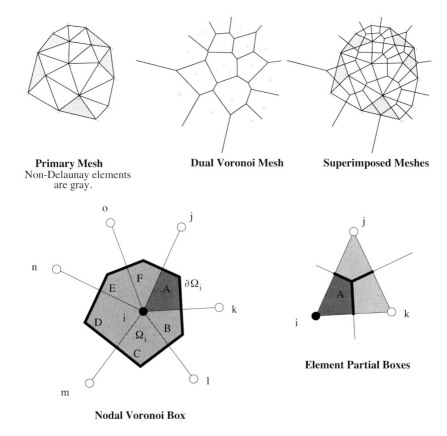

Figure 3.19 Finite volume mesh concepts. Above: general mesh constructions. Below: nodal and element view of the mesh. Adapted from Korvink and Baltes.[3]

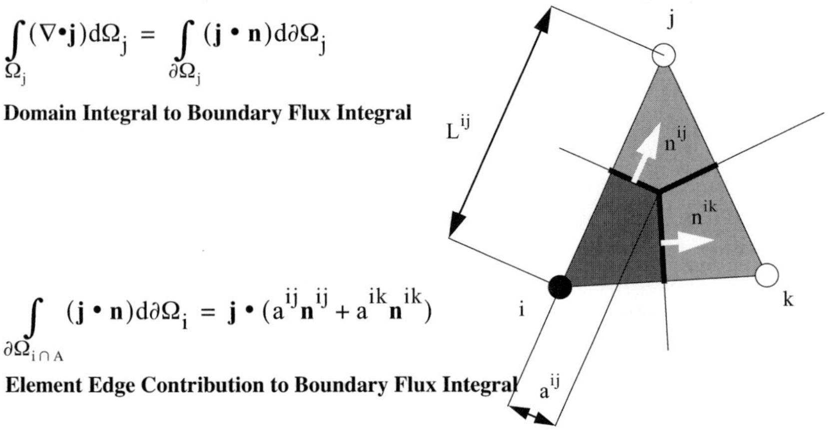

$$\int_{\Omega_j} (\nabla \cdot \mathbf{j}) d\Omega_j = \int_{\partial \Omega_j} (\mathbf{j} \cdot \mathbf{n}) d\partial \Omega_j$$

Domain Integral to Boundary Flux Integral

$$\int_{\partial \Omega_i \cap A} (\mathbf{j} \cdot \mathbf{n}) d\partial \Omega_i = \mathbf{j} \cdot (a^{ij} \mathbf{n}^{ij} + a^{ik} \mathbf{n}^{ik})$$

Element Edge Contribution to Boundary Flux Integral

Figure 3.20 Finite volume matrix equation structure. In converting a domain integral to a Voronoi box boundary integral, the matrix entries are enormously simplified.

two implementation details, the FVM fits perfectly into a FEM setting, and often FEM solvers also make available FVM discretizations within the same framework. Many mathematicians claim that the FVM is especially advantageous for the discretization of compressible flows with shock waves, although there are FEM approaches for this situation. Table 3.4 lists currently available FVM programs.

Finite Difference Method

The finite difference method (FDM)[27,28] uses an essentially Cartesian mesh, although the mesh may be defined in curvilinear coordinates (see Fig. 3.21). The nodes of the mesh are placed at the intersection points of the mesh lines. From a meshing viewpoint, the FDM is not nearly as general as the finite element method, which explains why it has not proliferated much in commercial settings. Of course a correction is due here. Intense research work over the years has devised schemes whereby the FDM can handle very general topologies, but there is no dispute that the implementation in a program of these generalities is very complex. One advantage is in the formulation of the ODEs; here the FDM is the undisputed champion. Discretization is relatively easy to do, and highly intuitive, so that quick-and-dirty implementations of PDE solvers are often FDM solvers. The focus here is typically on a single node and its immediate neighboring nodes along the mesh lines (see Fig. 3.22). The PDE is

Table 3.4 Software Based on the Finite Volume Method

Program Name	Application Area	Availability	Internet
FLUENT	CFD	Commercial	www.fluent.com/software/fluent/index.htm
CLAW PACK	CFD	Free for research	www.amath.washington.edu/~claw/
CFD-ACE+	Multiphysics	Commercial	www.cfdrc.com/datab/software/aceplus/
DESSIS	Semiconductor transport	Commercial	www.ise.ch/products/index.html
CFX	Multiphysics	Commercial	www.software.aeat.com/cfx
Flow3D	CFD	Commercial	www.flow3d.com

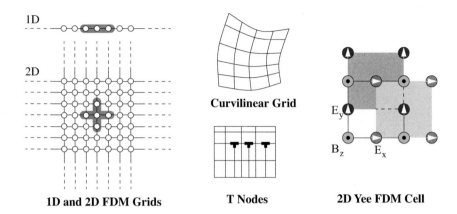

Figure 3.21 Finite difference mesh concepts. Left: Finite difference meshes are usually 1-D, 2-D (and 3-D) Cartesian grids; nodes for unknown values are indicated by circles at the grid line intersections; gray patches show nodes involved in a typical finite difference stencil formula. Middle, top: Grids may be formed in curvilinear coordinates. Middle, bottom: To aid mesh refinement, so-called T-nodes can be introduced. Right: Some PDEs require staggered grids; here, e.g., the 2-D Maxwell (Yee) grid.

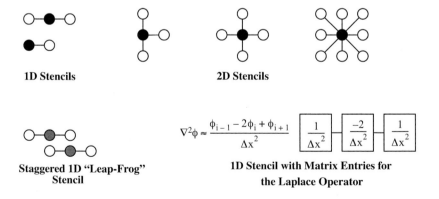

Figure 3.22 Finite difference nodal equation structure, showing typical stencil diagrams. Special stencils are required at the mesh boundaries, and multiple stencils are needed for some coupled PDEs (e.g., leap-frog schemes such as the 1-D Yee stencil). For a particular operator, a stencil may be drawn with its matrix entry values.

turned into a difference equation, and all interior nodes in the mesh have identical equations. The boundaries must be specially dealt with. The resulting ODE coefficients are collected into special banded matrices. Very fast methods exist to solve this special case (see Fig. 3.23).

The so-called finite-difference-time-domain method (FDTD) for electromagnetic propagation is a case where the FDM has dominated other numerical methods. It deserves special mention here because it demonstrates that there are situations where it is not always advantageous to consider the time and spatial discretization separately. Table 3.5 lists currently available FDM programs.

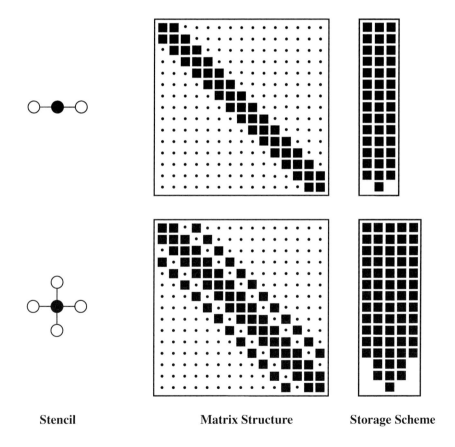

Stencil   **Matrix Structure**   **Storage Scheme**

Figure 3.23 Finite difference matrix structure depends strongly on the nodal numbering. The best numbering scheme leads to highly efficient matrix storage and solution techniques. Dots indicate zero-valued matrix elements.

Table 3.5 Software Based on the Finite Difference Method

Program Name	MEMS Application	Availability	Internet
FIDISOL	PDE	Commercial/ free for research	www.rz.uni-karlsruhe.de/ ~numerik/fidisol.cadsol/ index.html
XFDTD	RF devices, including entire mobile phones Optical wave guides Optical bandgap structures Solid state lasers (VCSEL, etc.)	Commercial	www.remcominc.com/html/ products.html
DIFFPACK	Multiphysics; requires programming	Commercial	www.diffpack.com

Boundary Element Method/Panel Method

The most distinguishing feature of the boundary element method (BEM)[29] is that the mesh is of one dimension lower than the computational geometry, i.e., only the boundaries between domains/materials are meshed. Since meshing is generally a tough job, this represents one of the advantages of the boundary element method. Typically, the same element polytopes are used as for the finite element method (see Fig. 3.24). The BEM can be used for the discretization of a PDE if the following conditions are met:

- The PDE can be rewritten as a *boundary integral* (see Fig. 3.25). If volume integrals remain (as is the case, for example, for the Poisson equation), these can also be handled. The BEM loses one of its important advantages, however, because in this case a volume-filling mesh is also needed to perform the numerical integration.

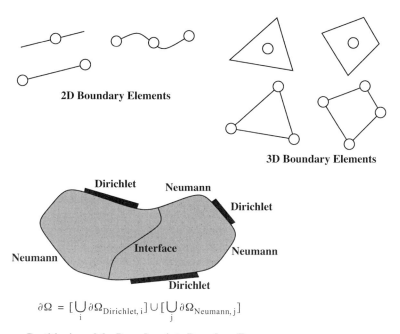

$$\partial\Omega = [\bigcup_i \partial\Omega_{\text{Dirichlet},i}] \cup [\bigcup_j \partial\Omega_{\text{Neumann},j}]$$

Partitioning of the Boundary into Boundary Types

Figure 3.24 Boundary element mesh concepts. The elements are one dimension lower than the simulation model, as they are used to discretize the interfaces between materials only. Elements are either of type Dirichlet, where the primary field is specified, or Neumann, where the flux is specified.

$$c\psi + \int_{\partial\Omega} \psi q^* \, d\partial\Omega - \int_{\partial\Omega} q\psi^* \, d\partial\Omega = 0$$

Boundary Element Method Integral Equation

$$\sum_{k=1}^{nn} h_{ik}\psi_k - \sum_{k=1}^{nn} g_{ik}q_k = 0$$

Boundary Element Method Matrix Equation

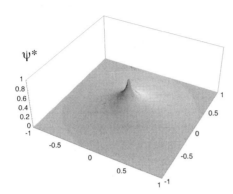

Laplace Equation Fundamental Solution

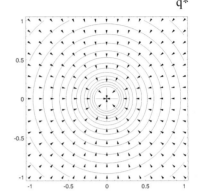

Laplace Equation Fundamental Flux

Figure 3.25 Boundary element matrix equations are composed mainly of boundary integrals if no interior source terms are present. Two special functions are needed to build the equations: the fundamental solution of the PDE ψ^* and its associated flux q^*, both depicted in the lower part of the figure for the Laplace equation. The symbols h and g represent the final matrices, and ψ_k and q_k are the unknown potential and surface flux density nodal values, respectively.

- There is a *Green function* for the PDE. In engineering terms, the Green function of a PDE is the analytical solution to a unit excitation. For example, in electrostatics, which is governed by a Poisson equation, the Green function is the potential field generated by a unit charge placed in infinite, empty space (see Fig. 3.25).

The boundary element method permits a straightforward interpretation. Since we know the analytical response to a unit excitation, we can use the linear superposition principle by expressing every excitation as a weighted sum of unit excitations. Then the response of the system is simply the weighted sum of the analytical Green functions.

Continuing with this engineering analogy, we remain with the electrostatic equation. On a boundary segment, we can either specify a voltage or we can place some charges, but we cannot do both. So either the charge or the voltage is known; the other quantity is unknown. The Green function has a complementary function, in this case, the response to a unit voltage at a point, and we also know this function: It is simply the gradient of the Green function. Now we can calculate the potential at a point in the domain. We simply take a weighted sum of the field of all unit charges about the charge boundaries, evaluated at the point of interest, and add to this the weighted sum of the unit voltage about all voltage boundaries, again evaluated at the point of interest.

The boundary element method automates this process, and adds a couple of refinements:

- Summation is replaced by integration, and the point-wise Green functions are applied as densities across a panel (i.e., charge densities, not charges, are approximated).
- By defining single-material zones, multiple materials are comfortably handled because interface boundaries get contributions from both sides of the element that separates the zones.
- The BEM naturally handles infinite domains, i.e., it can be used to embed a small domain in a much larger empty space without the need for large discretization meshes.

The BEM is not as popular as other domain solvers, mainly for the following reasons:

- *Speed.* Although the BEM mesh is potentially smaller than a FEM mesh for the same problem domain, the FEM unknowns are sparsely coupled, whereas the BEM unknowns are, in general, fully coupled for each subdomain. The BEM equations are also time-consuming to form, since the integrals are problem-dependent, whereas a FEM domain integral is over a predefined element shape and therefore allows tremendous speed-up possibilities. Even after the equations are formed, however, dense methods must be used for the BEM equations. Here the effort grows as the cubic power of the number of unknowns, whereas in the FEM, the effort grows somewhere between the first and second power of the number of unknowns. It should be mentioned, though, that tremendous speed-up has been achieved for Laplace solvers, mainly by making use of the multipole expansion.[30] The multipole allows us to evaluate the integrals up to speed n and to render sparse the equation system, so that in this special case, the BEM is the undisputed champion over the other general geometry methods.[31,32]

- *Source terms.* Since many PDEs have so-called source terms that must be evaluated using a domain integral, the BEM needs to set up a second volumetric mesh. The FEM has no difficulties here.
- *Nonlinearities.* If domain nonlinearities occur, the BEM is again hard to apply. First, a volumetric mesh is needed to carry the localized information. Second, the formulation becomes tricky and needs additional mathematical assumptions for the influence of the nonlinearity on the boundary integral terms.
- *Mathematics.* In general, it is hard to find the analytical Green functions for an arbitrary PDE. Many cases have been worked out, for example, for the Poisson equation (which covers linear heat conduction, electrostatics, and potential flow), for electrodynamics, and for elastostatics, but the list is quickly exhausted.

Table 3.6 lists currently available BEM programs.

Table 3.6 Software Based on the Boundary Element Method

Program Name	MEMS Application	Availability	Internet
FastCap, FastHenry	Multipole-accelerated capacitance and inductance analysis	Free for research	rleweb.mit.edu/vlsi/codes.htm www.fastfieldsolvers.com

Interface Tracking Methods

Interface tracking is of special interest to MEMS, in microfluidic applications, packaging, lithography, and deposition processing, and in general where an intermaterial interface moves in time. The modeling approach is either to model the interface explicitly or to use a representation of the phenomena that implicitly define the interface. Usually the latter approach is more successful when large-scale dynamics takes place and the interface topology varies dramatically during the simulation (e.g., when two separate droplets merge).

The volume-of-fluid (VOF) method models fluid dynamics by conventional means (for example, through the control volume method) but adds an equation for the level of fill of a computational cell to the mass conservation equation. It completes the description by adding surface tension forces.

Recently the level set method[33] has gained importance as an interface tracking technique, especially in the form of the fast level set method. The idea is rather clever because it avoids many of the topological difficulties of its predecessors. In the level set method we define a scalar field ϕ over the whole geometry, and we designate its $\phi = 0$ contour as the interface. The numerical method tracks the diffusion and convection of the whole field, although only the zero contour is of interest. To speed up the solution process, only a small band around the $\phi = 0$ contour is actually considered. Since ϕ is one dimension higher than the interface, i.e., the interface is implicitly defined, there is no problem in computing the merging or separation of interfaces. The method is straightforward to apply. The function ϕ is initialized to represent the initial state of the interface. It need only have a gradient in the immediate vicinity of the interface and can be constant elsewhere. Then we need a velocity function for the interface; this can be arbitrarily complex. The transient algorithm "transports" ϕ according to the velocity, thereby transporting the interface. At any time-step during the solution process we can extract $\phi(t) = 0$, which provides us with the current shape of the interface.

Table 3.7 lists currently available interface tracking programs.

Table 3.7 Software for Interface Tracking

Program Name	MEMS Application	Availability	Internet
Surface Evolver	Statics of liquid droplets; soldering; gluing;	Freeware	www.susqu.edu/ facstaff/b/ brakke/ evolver/evolver.html
CFD-ACE+	Volume of fluid	Commercial	www.cfdrc.com/ datab/software/ aceplus
FIDAP	Volume of fluid	Commercial	www.fluent.com/ software/fidap/ index.htm
Flotran	Volume of fluid	Commercial	www.ansys.com

3.6 Design Automation Methods

Design automation refers to tools that significantly reduce the burden of design, enabling the designer to focus attention on the design task rather than on the simulation task. These tools do not pretend to take over the designer's job, but they can significantly aid the discovery of good designs. We will discuss four important examples: optimization, mesh generation, mesh adaptivity, and model order reduction. Broadly speaking, optimization focuses on finding the best shape of an engineering component; mesh generation and adaptivity focus on providing the user with the best mesh for a computational task; and model order reduction targets the automatic generation of compact numerical models, originally from larger domain simulator models, for use in system simulators.

Optimization

There have been many definitions of optimization theory. Here is one definition: The objective of optimization theory is to select the best possible decision for a given set of circumstances without having to enumerate all the possibilities.[34] Generally, optimization problems are made up of three basic components: the *objective function* that we want to minimize or maximize; a set of *unknowns* or *variables* that affect the value of the objective function; and a set of *constraints* that allow the variables to take on certain values but exclude others. The characteristics of these three components just categorize the practical problems into different optimization strategies, even though these optimization strategies can often be transformed into each other. For example, some optimization problems have a single objective function, while others may have multiple objective functions. In most cases, problems with multiple objectives can be reformulated as a single objective either by forming a weighted combination of the different objectives or by replacing some of the objectives by constraints. Another example is that the optimization problem can be divided into an unconstrained method and a constrained method based on constraints. In practice, almost all problems really do have constraints. However, the equality constraints as part of the objectives by the *Lagrange multiplier method* and optimize the objectives can be transformed with unconstrained optimization algorithms. This is not possible for inequality constraints. Here we focus just on the structure of the optimization and on the material optimization problem.

Process of Optimization. The process of structural optimization can be summarized by the modules shown in Fig. 3.26. The *start* module represents a set of initial conditions, such as objective functions and initial values of variables that are set by human intervention before the optimization process. The *geometrical representation* module defines the design domain and design variables. The method used to represent the geometry domain will significantly limit the method used for the analysis and optimization module. After defining the geometrical domain, the *analysis* module converts the geometrical model to the analysis model if the geometrical model cannot be optimized directly.[35,36] An analytical or numerical method is formulated so that the value of the objective functions can be evaluated numerically. Later in this section we will select

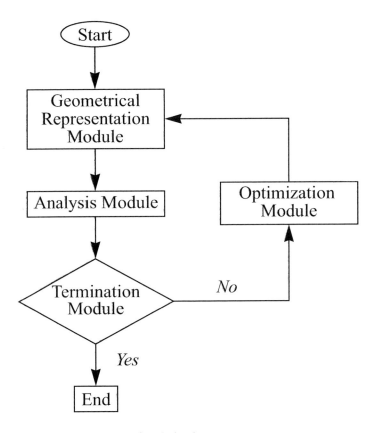

Figure 3.26 The general process of optimization.

one method for the analysis module; it depends on specific aspects of the objective function and the constraints. The *termination* module is checked to see if any of the termination conditions are met. Here the process branches, stopping if the termination module determines that the end module should be executed; otherwise, the branch is made to the optimization module. The *optimization* module makes modifications to variables that should result in an improved design. This is accomplished by interfacing with the analysis module to acquire information necessary to proceed with the optimization process. A mathematical programming algorithm is used to obtain the right strategy to update the objectives. Then the design variables are changed and sent to the geometrical module again. This loop is termed the optimization iteration. Because of the difficulty of interfacing the analysis module and the optimization module, many optimization techniques have been coupled into the analysis module. This is why most structural optimization software is developed based on sophisticated structural analysis software, for example, NASTRA, I-DEAS, and ANSYS. Tables 3.8 and 3.9 list currently available optimization programs.

Mathematical Programming Method. Generally, optimization algorithms are iterative. They begin with an initial guess of the variables and generate a sequence of improved estimates until they reach a satisfied solution. The strategy used to move from one iteration to the next distinguishes one method from another. Some algorithms accumulate information that is gathered from previous iterations, while others use only local information from the current solution. Regardless of these specifics, the final purpose is to generate a gradient-type direction vector and an optimal step size. Some categories of mathematical programming methods are shown in Fig. 3.27.[35] A good algorithm should maintain the following properties:[37]

- *Accuracy*. An algorithm should converge to the exact solution without being oversensitive to errors in the data or to arithmetic rounding errors caused by the computer.

Table 3.8 Software for Optimization

Program Name	MEMS Application	Availability	Internet
NEOS	Optimization over Web	Free	www-neos.mcs.anl.gov
GAMS	General purpose	Commercial	www.gams.com

Table 3.9 Software for Structure Optimization

Program Name	MEMS Application	Availability	Method/Type	Internet
OPTISTRUCT	Topology/Shape/Size	Commercial	FEM	www.altair.com
OPTISHAPE	Topology/Shape/Material	Commercial	FEM	www.quint.co.jp
GENESIS	Topology/Shape	Commercial	FEM	www.vrand.com
CATOPO	Topology/Shape	Commercial	FEM	www.ces-eckard.de
ANSYS	Topology	Commercial	FEM	www.ansys.com
TOSCA	Topology/Shape	Commercial	FEM	www.fe-design.com
PENOPT	Topology/Material	Commercial	FEM	www.inutech.de
TOPOPT	Topology	Free ware	FEM	www.topopt.dtu.dk

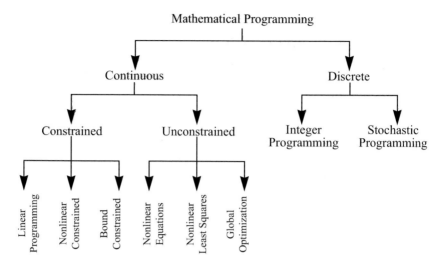

Figure 3.27 Mathematical programming methods. Adapted from *NEOS Guide Optimization Tree*.[36]

- *Robustness.* An algorithm is robust if it performs well for a wide variety of problems and copes with almost all reasonable choices of initial variables. Some optimization problems are very prone to converge to a local minimum. Global rather than local optimization methods are preferred to find the global optimum. However, global optimization is difficult to implement for most practical problems; generally, we just solve a sequence of local optimization problems instead.
- *Stability.* Although the physics of a given problem might imply a regular solution, optimization procedures based on discretized models may generate anomalies called numerical instabilities and often lead to ill-posed problems. Numerical instabilities often appear as numerical oscillations or converge to an unphysical solution. For the first case, numerical smoothness techniques are used to smooth the gradient vector or the objective functions in each iteration step. For the second case, regularization techniques are widely used to solve ill-posed inverse problems to avoid unphysical solutions.
- *Efficiency.* The algorithm should not require too much computer time or storage, and should converge in a finite number of iteration loops.

These requirements conflict. For example, a rapidly convergent method for nonlinear programming may require excessive amounts of computer storage for large-scale problems. On the other hand, a robust method may also be very slow. Finding a tradeoff of these properties is the task of optimization.

Structural Layout Optimization. The efficient use of material is an important aspect in many different areas, and it is also a key point in MEMS. The optimization of structural layout has a great effect on the performance of structures, through parameters such as weight, volume, stress distribution, and dynamic response. A number of successful algorithms are widely used at present for structural optimization, including some that have been incorporated into commercial software. Generally, we could classify these methods in three categories: size optimization, shape optimization, and topology optimization.

Size optimization is the oldest and simplest method with which to optimize the performance of a structure. Generally, only simple geometrical sizes, such as the thickness of a plate or the size of a truss, are used as the variables, and the shape of the structure is fixed throughout the optimization procedure. In *shape optimization*, additional variables (excluding size) are introduced to allow for boundary movement while keeping the structural topology unchanged. *Topology optimization*, generally called shape optimization, involves both size and shape modification, especially topology modifications that deal with the components of a structure. It is clear that topology optimization is the highest level of structural optimization with which to optimize the topology, shape, and size simultaneously. Some details of shape and topology optimization are given below.

Shape Optimization. One key point of shape optimization is to choose a method with which to represent the boundary of the design domain, because the optimal shape depends on design variables selected to represent modifiable boundaries. For example, the optimal shape of the cantilever beam shown in Fig. 3.28 depends on the boundary representation and the number of design variables. If the cantilever beam is modeled as shown in Fig. 3.28(a), with only one design variable (the height H of the free end), then the optimal shape will be different from the model shown in Fig. 3.28(b), where the boundary is represented by a cubic spline with control points along the length of the beam. The most commonly used methods to represent a boundary are based on polynomials and splines.

After the boundary representation is complete, the analysis module passes the geometrical model to the analysis model through an analytical or numerical method, because the geometrical model cannot be used directly in most cases. The FEM and BEM are generally applicable techniques. For example, FEM could generate a mesh inside the design domain automatically; adaptive remeshing or mesh refinement would fol-

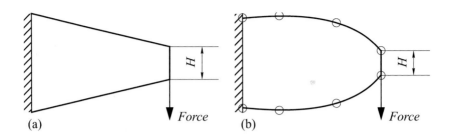

Figure 3.28 Shape optimization redistributes the boundary of a fixed topology to find an optimal shape. Adapted from Haftka and Gürdal.[34]

low the boundary movement. So it is straightforward to decouple the optimization module from the analysis module. We could use powerful FEM software as a black box that would provide the information used to evaluate the sensitivity of the objective function to design changes in the optimization module. In the optimization module, a mathematical programming method is used for most cases.

Topology Optimization. The initial purpose of topology optimization is to find the optimal layout of a structure. Based on prior development of optimization models and computational strategies, it has become a powerful optimization method that can find optimal topology, shape, and size simultaneously.[38] One commonly used method of topology optimization is the *ground structure* method, which discretizes one fixed design domain by FEM for the entire optimization process. Topology, shape, and size of the structure are depicted by a set of discretized distributed functions on each finite element, for example, the density of the constituent material. Then a mathematical programming method decides which elements should contain material and which should represent voids. Elements with intermediate density values are avoided by introducing some form of penalty that steers the value to become either void or dense. The number of elements used to discretize the design domain will influence the smoothness of the boundary (see Fig. 3.29).

So it is clear that the fundamental difference between shape and topology optimization is how the geometrical domain and the design variables are represented. The analysis module and optimization module remain essentially the same.

Material Optimization. Functional materials are also an important branch of MEMS. Materials with special behavior are often needed, yet they are difficult to include in designs when the designer is faced by typ-

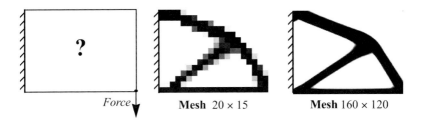

Figure 3.29 Topology optimization redistributes the material to find an optimal topology. Mesh resolution should not strongly influence the found topology, only the smoothness of the solution boundary. Also see Bendsoe and Sigmund.[38]

ical production constraints. When considering the material from a macroscopic point of view, topology optimization belongs to structural optimization. If we consider the material from the microscope point of view, topology optimization with homogenization can be used for composite material design and optimization.

The process of finding representative or effective properties of microstructure materials is called homogenization. The goal of material design is to synthesize a material with prescribed constitutive properties. For mechanical properties, the design variables are the material elastic tensor and the objectives are a specific Young's modulus or a specific thermal expansion coefficient. There are two ways to do this:

1. Changing the molecular composition. Here we need molecular simulation tools.
2. Changing the ratio of composition and distribution of homogenous "pure" materials (such as bimetallic strips) at a very small continuum length scale. This is identical to topology optimization.

Typical examples of material optimization are negative Poisson's ratio material, synthesized negative thermal expansion materials from the mixture of two positive thermal expansion phases, and a band gap material that allows wave propagation for certain frequency ranges only (Fig. 3.30).[39-41]

Mesh Generation and Mesh Adaptivity

For simulation tools based on meshes (e.g., FEM, CVM, FDM, BEM), it is well known that construction of a "good" mesh is a critical issue that

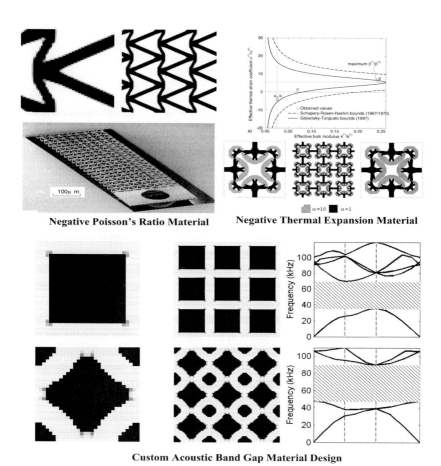

Figure 3.30 Topology optimization is highly effective for material microstructure optimization. Examples courtesy of Prof. O. Sigmund; also see Larsen *et al.*,[39] Sigmund and Torquato,[40] and Sigmund.[41]

has a huge influence on the subsequent solution accuracy and computational efficiency. Two techniques are generally in use (see Fig. 3.31):

1. Mesh generation[42] takes a computer representation of the simulation domain, for example, specified as a set of boundary descriptions, and creates a good mesh based on a variety of *a priori* quality criteria.
2. Mesh adaptivity[43] takes an existing mesh and, based on information about the discretization procedure and a computed solution, *a posteriori* improves the quality of the mesh and reduces the redundancy of the discretization.

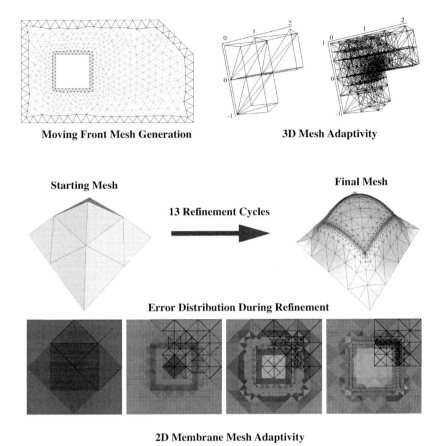

Figure 3.31 Mesh generation and mesh adaptivity concepts. Left: the advancing front of the mesh generator at an early stage of 2-D mesh generation. At this point, elements between the outer boundary and the front have been created. Right and below: Mesh adaptivity automatically constructs finer meshes based on mathematically formulated error estimates, until the error is both equilibrated (the same everywhere in the mesh) and below a user-defined threshold. Adapted from Müller.[39]

As the complexity of simulation domains increases, so mesh generation becomes absolutely essential in performing computational tasks efficiently. Fortunately, excellent tools already exist, and improvements continue to be made to them. Typically, the programs start with the definition of the computational geometry by constructive solid geometry (CSG) operations.[44] The operations include, for example, the instantiation of primitives (e.g., boxes, cylinders, cones), Boolean operations (e.g., union, intersection), sweeps (point to line, line to plane, plane to solid), and combinations of these operations grouped together as user-defined small program scripts. For MEMS, it is typical to define the operations in such a way as to emulate the manufacturing process; for example, deposition and etching can be represented conveniently and efficiently in this way and the costly process simulation operations otherwise needed are avoided.

Another important feature is the parameterization of the geometry, i.e., the definition of key dimensions as variables. In this case, it is then possible to create variants of the geometry automatically. (See also Shape Optimization, above.) Once the CSG model exists, it can be viewed in 3D, rotated, and checked for correctness. When the model is found to be error-free, a mesh can be generated. Here the user must choose an element geometry (or more than one). Often we can influence the size and grading of the element size over the geometry. Care must be taken to choose element shapes that lead to accurate discretization schemes (for example, tetrahedra are not necessarily good element shapes for all PDE discretizations) and to choose a grading that leads to reasonable solutions (not too coarse). Usually, mesh generators will report on the quality of the elements generated and will mark critical elements (for example, elements that are elongated, sliver-like).

Many mesh generation methods exist. Among those that claim to be fully automatic, two techniques appear to dominate:

1. Moving front methods
2. Tree-based decomposition methods

The moving front methods consider the discrete boundary as a front that is moved toward the interior of the CSG domain. In moving the front, new mesh nodes and elements are introduced between the new and old fronts. The method terminates when the front is empty, i.e., when it contains no more nodes (see Fig. 3.31).

The tree-based methods embed the geometry in a quadtree or octree that spatially localizes the boundary and provides a domain decomposition of the interior of the CSG domain. The tree is easily refined to obtain a finer decomposition. Each "leaf" of the tree is used to localize a mesh node. Because the neighborhood of a point is easy to find through tree traversal, subsequent mesh generation, i.e., connecting nodes to form the elements, is greatly facilitated.

Program users often have to choose between mesh generation available from within a domain solver and dedicated mesh generation tools that interface to domain solvers. The latter group offers the advantage that in-house know-how is retained regardless of the current choice of domain solver; the former can, under certain circumstances, result in a faster implementation and contain special features not found elsewhere.

Mesh adaptivity involves taking an existing mesh together with a computed solution and subdividing certain elements that are found to be too coarse. The basic steps in one adaptivity cycle are:

- Compute an error estimator for each element in the mesh. This computation is based on an already existing solution.
- Mark elements that do not satisfy a quality criterion. This step is often called the refinement strategy.
- Modify the marked elements (by subdivision or aggregation) to obtain a better mesh.
- Return the mesh for recomputation of the solution.

Mesh adaptivity is not nearly as widespread in computational tools, mainly because the theory of error estimation is fairly recent and still under development. Nevertheless, for those cases where capabilities exist, the technique is very powerful and removes a tremendous burden from the analyst, since the mesh becomes self-correcting (see Fig. 3.31).

Automatic Model Order Reduction

A designer often has to combine domain and system-level simulations. At an early design stage, a good solution is to start with quick-and-dirty models that perfectly fit within a system-level simulation. Such models are usually built by hand and work well where large-scale lumping into circuit equivalent models is intuitively apparent. However, quite often quick-and-dirty models describe the qualitative behavior well but fall short on the quantitative behavior; for the next phase of design, more accurate compact models are needed.

A typical way to produce compact models, which can be found in available software, is to make several domain-level simulations and then try to fit these results to some low-order lumped model. The functional form of the low-order models is not known *a priori*, however, so this process might take a lot of time by highly qualified personnel because success strongly depends on intuition.

An alternative is to use a mathematical theory that can perform the model order reduction completely formally or, in other words, automatically. For the linear case such a theory is already developed and known under the name of approximation of large-scale dynamic systems;[45,46] we briefly review this theory here (see Tables 3.10 and 3.11).

Table 3.10 Methods of Model Reduction Linear Dynamics Systems

Name	Advantages	Disadvantages
Control theory (Truncated Balanced Approximation, Singular Perturbation Approximation, Hankel-Norm Approximation)	Have a global error estimate, can be used in a fully automatic manner.	Computational complexity $O(n^3)$ can be used for systems with order less than a few thousand unknowns.
Padé approximants (moment matching) via Krylov subspaces by means of either the Arnoldi or Lancsoz process.	Very advantageous computationally, can be applied to very high-dimensional 1st order linear systems.	Does not have a global error estimate. It is necessary to select the order of the reduced system manually.
SVD-Krylov (low-rank Grammian approximants).	Have a global error estimate and the computational complexity is less than $O(n^2)$.	Just under development.

Table 3.11 Software for Automatic Model Reduction

Program Name	Application Area	Availability	Internet
Slicot	Control theory	Free for research	www.win.tue.nl/niconet/
Lyapack	Solution of the Lyapunov equations	Free, requires Matlab	www.netlib.org/lyapack/
mor4ansys	Implicit moment matching	Free	www.imtek.de/simulation/mor4ansys/

From an engineering viewpoint, any part of a system can be described as a black box with only a few inputs and outputs. The goal is to have a dynamic model to predict the behavior of the outputs as a function of the inputs. To this end, this part is modeled as a system of ODEs with some state vector. For the linear case, the relationship between the inputs, state vector, and outputs is written in terms of constant system matrices, schematically shown in the upper part of Fig. 3.32.

The problem is that during a discretization procedure of a domain simulator, the dimension of the state vector of the resulting system of ODEs can routinely reach a size of hundreds of thousands of entries. The goal of model-order reduction is to approximate such a system by a similar system with a low-dimensional state vector, as shown in the lower part of Fig. 3.32.

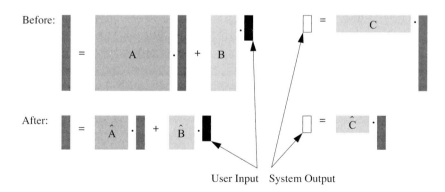

Figure 3.32 Sketch of the system equations before and after the model reduction step. The dimensions of the system matrices A, B, and C, and the internal state vectors are significantly smaller after model reduction. The input vector and output vector remain the same size. Adapted from Rudnyi and Korvink.[45]

The most advanced results are found in control theory, and we can safely state that the problem of model reduction of a linear dynamic system is solved in principle.[47,48] There are methods (truncated balanced approximation, singular perturbation approximation, Hankel norm approximation) with guaranteed error bounds for the difference between the transfer function of the original high-dimensional system and the reduced low-dimensional system. This means that model reduction based on these methods can be made fully automatic. A user just sets an error bound, then the algorithm finds the smallest possible dimension of the reduced system that satisfies that bound. Alternatively, a user specifies the dimension of the reduced system and the algorithm estimates the error bound for the reduced system. Unfortunately, the computational complexity of these algorithms is $O(n^3)$, where n represents the dimension of the large system. Hence, if the system order doubles, the time required to solve a new problem will increase about eight times. In other words, even though the theory is good for all linear dynamic systems, practically, we can use it for small-order systems only.

Most of the practical work in model reduction of large linear dynamic systems has been tied with Padé approximants (moment matching) of the transfer function via Krylov subspaces by means of either the Arnoldi or Lanczos process.[49] In the literature, there are some spectacular examples where the dimension of a system of ordinary differential equations was reduced using this technique by several orders of magnitude, almost without sacrificing precision. The disadvantage is that Padé approximants do not have global error estimates, and it is necessary to select the order of the reduced system manually.

Recently, there have been considerable efforts to find computationally effective strategies in order to apply methods based on Hankel singular values to large-scale systems, e.g., so-called SVD-Krylov methods based on low-rank Grammian approximants.[50,51] However, they are just under development, and engineers will have to wait as the experience of mathematicians grows in this area.

One of the most promising application areas for these methods is electrothermal analysis because its equations can be linearized relatively easily for many important cases. An example of the application of these theories to a microthruster is described in more detail in Section 3.8.

TCAD

A technology CAD (TCAD) tool is a computer program that presents the user with a virtual technology: Within one environment, the user can manufacture, measure and test, optimize, and generally qualify a new design using familiar modes of expression and a unified interface. This technique is most developed for the electronic semiconductor world, where a typical TCAD tool binds together mask design, process specification, equipment simulation, process simulation, device simulation, parameter extraction, circuit and system simulation, design of experiment, and optimization in such a way as to make data transfer among software tools completely transparent. The graphical interface emulates situations familiar to the engineers from laboratory and factory practice. Table 3.12 lists currently available TCAD systems.

Table 3.12 Software for TCAD

Program Name	Application	Availability	Internet
IntelliCAD	MEMS design	Commercial	www.intellisense.com
MEMSCAP	MEMS design	Commercial	www.memscap.com
Coventor	MEMS design	Commercial	www.coventor.com/coventorware/
ISE-TCAD	IC design	Commercial	www.ise.ch

3.7 A Simulation Strategy

Each MEMS application area has its special requirements from a simulation that requires a separate strategy to remain effective for design. We first consider some general ideas, then move on to specific areas:

- Material properties
- Microfluidics and life-science applications
- CMOS MEMS (solid-state MEMS)

There are a number of simulation tasks, however, that do not require a special strategy and for which currently available simulation tools are more than adequate. As a general rule, whenever microengineering tasks are down-scalable from routine macroengineering tasks, e.g., when no new theory is necessary to describe the phenomena, the chances are good that a well-established computer-aided engineering (CAE) tool exists and that it can be directly applied to the microengineering task at hand. Certainly the following tasks fall in this area:

- Linear and nonlinear elastostatics and elastodynamics (with the notable exception of thin-film damping effects); e.g., the elastic response of a microbeam or a dielectric membrane
- Linear and nonlinear heat transfer (with the exception of small cavities, rarified gases, and convective effects)
- Phenomena governed by the Poisson equation or the diffusion equation. These include electrostatics, heat transfer in solid materials, potential flow, and general diffusion phenomena. Since these tasks frequently occur and are covered by most general-purpose commercial finite element packages, it is certainly prudent to equip a design team with such a tool.

General Strategy

Figure 3.33 shows a simplified flow chart for a typical simulation job. It is necessary to make several interdependent decisions on different levels to reach success. Note that this is not a one-way process. The chart may appear strange at first. Why, for example, should the solution to a problem depend on design and modeling? The reason is that at the beginning, a typical engineering problem is expressed in an informal way. Engi-

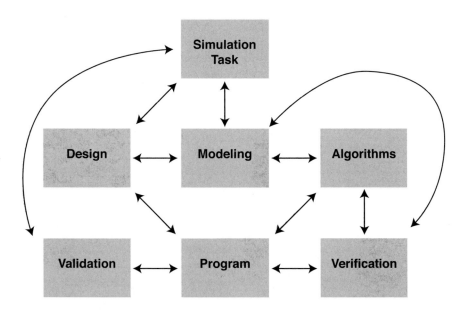

Figure 3.33 Simulation task flow.

neers would like to reach some goal by means of a simulation, but they do not yet know how this should be done. Then during discussions with experts, this problem is converted into another, more formal form. The latter depends on what is possible or feasible. As a result, the initial informal problem may be broken into several separate tasks, each of which can be treated by a simulation within a reasonable time.

The initial step in planning a simulation is to start with design and modeling. By design we mean a specification of the problem in terms of the inputs and outputs desired by the engineers. This by itself requires engineers to express their desire in some formal way and finally to write down the problem in the form of a compact chart. It is also crucial to try to divide the outputs into qualitative categories: essential, less important, good to have, and not critical.

Then it is necessary to make a mathematical model that can produce desired outputs from given inputs. It is possible to derive a model at different levels of theory with a conventional tradeoff of speed vs. accuracy, and it makes sense to start working on distinct models in parallel. Rough models can be used for a system-level simulation during the initial design; they can be replaced by more accurate models after further development.

Model development may require new inputs, such as material properties. Then it is necessary to go once more through the design step to understand where these inputs can be obtained (e.g., available handbooks, databases, additional experiments). This, in turn, could require consideration of the task at hand. The entire process is by nature iterative: Decisions made at lower levels may cause complete changes to the upper-level considerations.

In theory, modeling can currently be considered from first principles, in which case everything is determined computationally without any experimental values except for a few fundamental constants. However, we cannot recommend this approach. For the foreseeable future, it is much cheaper just to measure many important properties than to try to estimate them from first principles. Although the latter is of more academic interest, engineers should remain pragmatic. If some properties are already available or can easily be measured, why try to obtain them computationally?

During and after the mathematical model development, implementation issues and therefore available numerical algorithms have to be considered because almost all modern models cannot be solved in closed form. The final goal is not just to write down a model but rather to obtain results useful for design evaluation. Here the constraint is set by available computing power, and the final model should be computed within the required amount of time. Each year, according to Moore's law, computing power increases roughly by a factor of two, but for many important algorithms, this may not be fast enough.

Each algorithm is characterized by its computational complexity, that is, how many computer operations are needed for a given dimension of the computational task. If the computational complexity is proportional to the cubic power of the problem dimension, usually expressed as

$O(n^3)$, then an increase of 10 in the dimension increases the required time by a thousand. Here even Moore's law cannot help in the short term. As a result, many problems are solved in principle, i.e., there is a mathematical theory of how to obtain a necessary answer, but in practice, the required time is impractically long.

The development of fast algorithms with a computational complexity less than or equal to $O(n\log n)$ is a hot topic in modern computational science, and it is important to follow the current state of the art in this area.

The next step is the actual implementation, which depends both on the chosen algorithms and the design. During numerical solution, a vast amount of auxiliary data may be received. It is important to plan from the start what should be computed and saved to storage media. The main advice here is not to reinvent the wheel. If something is available in the commercial software domain, it is much cheaper to buy it rather than to try to program it from scratch. On the other hand, if programming is inevitable, it is better to start first with rapid prototyping in a development environment such as Mathematica or Matlab than to bet on low-level programming languages such as C++ and Fortran. The latter languages should be considered as last resorts, after all other means have failed.

In general, the implementation should be more at the software engineering level rather than resemble pure programming. This means that it should be possible to exchange data quickly among heterogeneous software packages to reach your goal.

After the implementation, it is time for verification and validation of the model. (Also see the Validation and Verification Mailing List web site.[52]) For a newly developed software package, it is necessary to check for bugs: Software without bugs does not exist. It is impossible to find all bugs because while the discovered bugs are being corrected, new errors are introduced. To this end, it is important to have or to develop test cases that should at least eliminate the most obvious errors.

Another verification task that applies to using commercial software packages is to make sure that numerical errors are within allowable tolerances. This is not a trivial task because in many cases error estimates are not available and it is impossible to choose the parameters such as mesh and time-step sizes automatically. A rule of thumb is to make several computations, varying mesh sizes and time-step lengths, and hope that the numerical error is comparable to the differences among these results. Again, good test cases are essential: An error can appear at any step, and it is important to be able to localize and characterize it. During validation, we need to check that the model is appropriate for the design task. The goal of the validation is to make sure that the level of theory chosen for a mathematical model is good enough. A mathematical model can be solved with high accuracy but be far from a real MEMS device because important effects have not been included in the model. To exclude this possibility, the developed software should be applied to a case with available experimental results to see that at least here a simulation can describe and reproduce the desired effects. The larger the range of verification, the greater the confidence of a designer in the predictive power of a tool. It is necessary to stress the iterative nature of the process once more. Verification and validation should be considered from the start of any project, test cases should be prepared and continuously extended, and the quality of the process should be systematically improved. Unfortunately, the straightforward path from a simulation task to a computer program and then to results is imaginable for routine problems only and does not apply to engineering practice in general. In addition, having access to the necessary hardware and software is not enough: It should be clear that for efficiency, it is imperative to have different experts manage the simulation effort to free designers to do what they do best.

Finally, for general implementation work, the resources of Table 3.1 are tremendously useful to the expert team. With these packages, it is possible to extend the power of a small set of commercial simulation packages greatly.

Material Properties

Any simulation depends crucially on knowing the correct values for material properties. In many cases they can be found in handbooks[53] or in specialized databases.[54,55] If they are not available there, the next step is to search literature databases such as Chemical Abstracts or the Web of Science. Conducting a successful literature search requires an investment in an expert who has the necessary training (usually a chemist). A high return on this investment can be expected because a successful literature search will provide access to the latest results, which may not be included in the material property database. The lag between publishing a result and having it included in the specialized database may be as long as 10 years. Unfortunately, there is another problem with a literature search: Results published by different research groups are often in disagreement with each other, and it takes an expert to choose the right one. Still, the importance of a literature search cannot be overestimated because the cost of experimental research to measure the properties required is much higher.

At present, we cannot exclude the situation where engineers are working closely with chemists using newly synthesized materials. The materials properties are unknown, and it is necessary to plan special research to determine them. Choosing between an experimental setup and molecular simulation highly depends on the property required and the material in question; in many cases, the best solution is a judicious combination of experiments and modeling.

Another special case that may arise is the use of a commercial semiconductor fabrication process, where material layers are grown according to proprietary recipes. In this case, property extraction experiments must be performed by designing special devices.[56] This technique is discussed extensively in Chapter 2.

Microfluidics and Life-Science Applications

A successful practical simulation of a microfluidic system can probably be done on a system level. (A simulation of the whole fluidic network at full physical level is very expensive.) Performing a simulation at the full physical level is unusual and cannot be recommended for the design of a microfluidic chip. A fluidic network is composed of separate elements (e.g., capillary tubes, valves, and pumps). In many cases there are compact models available and included in commercial software such as Coventor's.[57]

However, the challenge is that there are no good general behavioral models for many important parts, for example, a T-junction. Although an automatic model reduction can be useful for converting a device simulation to a compact model, most of the fluidic models are nonlinear. This is an active area of research.

- A complete tool suite should contain the following:
- A fluid dynamics solver, which is typically an FVM solver, preferably together with a volume of fluid (VOF) addition that can handle general free surface flows and surface tension effects is required.
- A tool such as the surface evolver may be necessary if electrocapillaries are considered. Here dynamics is not so important. Much freedom exists in modeling surface energy functions.
- A good molecular dynamics software is necessary if surface modifications or molecular reactions are targeted.
- A number of fluid solvers cover the appropriate effects, at least for continuum theory, if the fluid is electrochemically active.

- A system simulator is required to model complete microfluidic chips via component compact models.

CMOS MEMS (and Solid-State MEMS)

CMOS MEMS[58] are devices and systems manufactured using a commercially available CMOS manufacturing process together with specialized post-processing such as the selective removal or addition of materials. Typical CMOS MEMS exploit the circuit capabilities of CMOS to compensate for design rule limitations and material limitations and usually result in sophisticated system solutions.

To simulate these devices, much use can be made of existing microelectronic simulation tools, especially for CMOS process simulation. The post-processing operations must be handled by special tools that mimic the underlying technology. For example, most commercial process simulators do not yet include the capability to etch away material layers sacrificially or to deposit layers, e.g., by metal plating, spraying, screen printing, or anisotropically etching silicon. A strategy is therefore either to mimic these operations, where possible, by existing process operations (e.g., chemical vapor deposition) or to edit the resulting meshes (semi-) automatically to achieve etch-like processing.

At the end of a process simulation step, we obtain a mesh and a distribution of material properties. Very often the mesh obtained is too dense for subsequent device simulation since it was created to satisfy process simulation requirements. A good strategy here is to extract the material boundaries from the mesh, then to perform mesh generation on this geometry. This can be a formidable task because CMOS processing can result in very thin material layers that follow the underlying topography in complicated ways. It is often advisable to eliminate some material layers, such as the very thin gate oxides, just to get smaller mesh sizes.

In any case, the meshes so obtained are invariably volumetric meshes, and the question is whether these are adequate for a simulation. In general, the answer is affirmative. However, special situations exist where this is not so, for example, beams and membranes, where special lower dimensional discretization techniques (beam and plate elements) exist.

3.8 Case Studies

This section presents several case studies that illustrate a variety of concepts discussed in this chapter. In none of the cases did a ready-made tool exist for the job, and in each case, time and effort limitations excluded the development of a dedicated CAD package. Thus the emphasis is on strategy, i.e., getting the most from existing simulation tools tied together in useful ways.

Microthruster

A new class of microthrusters, for use in nanosatellites, is based on the integration of solid fuel with a silicon micromachined system.[59] In addition to the space application, a thruster can also be used as a gas generation device or a highly energetic actuator. The goal of the project was to fabricate a working array of independently addressed solid propellant microthrusters on a single chip, then to optimize the array's performance (see Fig. 3.34).

Beginning with the engineering problems faced by the developers, it was necessary to choose:

1. A wafer material (silicon, ceramics, or glass) and a technology to manufacture it

Figure 3.34 Firing a microthruster in a 4 x 4 array. Illustration courtesy of C. Rossi, LAAS-CNRS.

2. A technology for the low-temperature bonding of the wafers
3. A solid-fuel composition and a technology for filling it into millimeter-sized square cavities
4. A packaging technology
5. A flight qualification procedure

Then it was necessary to develop:

1. The optimal geometrical design for the microthruster array
2. An intelligent driving circuit to operate it

Pragmatically, the goal was to search for an optimal combination of different available technologies in order to have a microthruster array producing impulse-bits as required for minimum production cost.[60] In our view, therefore, the simulation should focus on the last two points, that is, to prepare the theory of operation of a microthruster and to choose the best approaches for its simulation from the perspective of the project goals. The main modeling goal is to help to choose the optimal geometry to fabricate the microthruster array as well as to develop an intelligent electrical circuit, schematically shown in Fig. 3.35.[61]

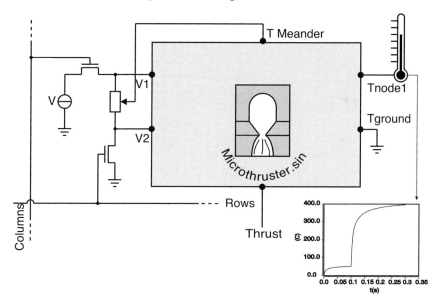

Figure 3.35 Compact model of the microthruster to allow for a joint simulation: The electrical current heats up the resistor, which causes fuel ignition, leads to sustained combustion, and finally produces the desired impulse. Adapted from Bechtold et al.[61]

The first iteration was a simulation design. We prepared a choice of possible models, summarized in Table 3.13. The time and resources necessary to implement models were compared with a would-be impact on each engineering problem listed above. These considerations are briefly described below.

The final output desired is clearly the same as for big rockets: a given impulse produced by a microthruster. The optimization of the nozzle shape plays an important role in conventional rockets. Nevertheless, this can hardly be applied to the microthruster array because, in the case of the geometry shown in Fig. 3.34, the technological degrees of freedom for the nozzle optimization are very limited. In the current design, it is only possible to change the diverging angle in the nozzle wafer slightly; it is not possible to consider designing a nozzle with optimal shape. As usual, an optimal design is possible in principle, but due to limiting production costs, it is beyond serious consideration because the nozzle lies in a direction perpendicular to the silicon wafer surface. From a technological point of view, the disadvantage of nonoptimal impulse production by an individual microthruster can be overcome by a higher integration density of microthrusters on a chip. The main challenge here deviates rather strongly from conventional rocket science, that is, how to predict the highest possible level of integration of microthrusters on a chip. A high level of integration can be achieved by reducing the size of a microthruster and by placing microthrusters closer to each other. The lower limit on microthruster size comes from the fact that, because of heat losses, conventional solid fuels do not display sustained combustion when the thruster diameter is too small. The limit on the distance between microthrusters is determined by thermal crosstalk. Hence, heat management becomes one of the main goals for microthruster simulation because it is necessary to isolate neighboring microthrusters thermally.

In the current project, the goal was to use available compositions of solid fuel. As a result, the modeling strategy was to rely on experimentally determined fuel properties such as ignition temperature, burning rate, and flame temperature for a given chamber with a qualitative understanding of how the burning rate scales to the microscale. The main quantitative emphasis was made on predicting and managing the heat flow.

There are two sources of heat that lead to thermal crosstalk: (1) the heat from the resistor to the solid fuel, which is controlled by the driving circuitry, and (2) the heat transfer from the hot gases to the wafer during the gas dynamics phase. There are many engineering degrees of freedom to influence this: a position and shape of the resistor, the introduction of special grooves between microthrusters, and so on.

Other degrees of freedom are tied to the driving circuitry. In the first phase, the circuitry has all means to control the resistor heating completely

Table 3.13 Models for Microthruster Operation

Task	Available Models	Modeling Strategy
Electrothermal ignition	• Coupled Poisson and heat transfer equations • Lumped resistor and heat transfer • Lumped heat transfer	The lumped resistor and 3D heat transfer model discretized with the finite element method and followed by a model order reduction.
Ignition and sustained combustion	• Detailed chemical kinetics • Quasi-state homogeneous solid one-dimensional model of the burning front • Adiabatic flame temperature	Experimental ignition temperature as input to the electro-thermal ignition problem, experimental burning rate and flame properties as inputs to gas dynamics. Qualitative discussion of the scaling of burning rate to the microscale.
Membrane rupture	• Molecular and multiscale simulations • Finite element method • Bulge test theory	Bulge test model and experiments
Gas dynamics	• Direct simulation Monte-Carlo • Navier-Stokes equations for the reactive flow • Local equilibria hypothesis to treat chemistry • Fixing chemical composition at the flame front • Quasi-1D approximation • Ideal rocket model	Ideal rocket model followed by empirical corrections

because it can directly measure the nonlinear resistor's temperature. After the onset of ignition, the circuit loses control (the current path is broken) and subsequent steps happen automatically. The circuit can influence sustained combustion indirectly only by means of preheating of the solid fuel before ignition, because the burning rate depends on the fuel temperature. This provides a natural way to decouple the simulation of the entire process: a first part models the electrothermal process, during which the fuel will reach ignition temperature, and a second part models the gas dynamics, wherein the fuel characteristics will be used as input parameters.

At the current level of development, the model for a microthruster array must include an involved electrothermal simulation, and the model should be able to make a joint simulation of the array simultaneously with the driving circuitry. All other aspects of microthruster operation can be modeled just on a basic level. Oddly enough, the development challenges of the microthruster array happen to be closer to microsystem packaging than to rocket science. This fact finally allowed us to find the best compromise in terms of resources/yield and allowed us to suggest the recommended modeling strategy presented in Table 3.13. The work was complemented by different experimental measurements to validate the models and to determine empirical correction factors if necessary.

The next phase of the project was designing the software implementation. The software to be developed was divided into the five separate low-level modules listed in Table 3.14; the user-specified inputs and outputs for each software package were also defined.

The commercial software (ANSYS[62] and Mathematica) was chosen for conventional tasks. Programming in C++ was planned only for model-order reduction because commercial software was not available for this case. However, programming ANSYS scripts is by itself a large task, which can be performed only by a professional. The crucial step with ANSYS was to verify that numerical errors remained below acceptable limits. To this end, results using low- and high-order elements available in ANSYS were compared.

The handling of the simulation software was made user-friendly. For this purpose, a graphic user interface was developed in webMathematica, which allows a user to operate all software with mouse clicks from a standard Internet browser to switch to each subsystem, to modify default values for each input and output parameter, and to run it (see Table 3.14). The results are displayed as output pages. Through the coupling of model-order reduction with ANSYS simulation, an automatic path was set up to convert a physical model of the device into its behavioral (HDL) model, ready for system simulation.

Finally, we will demonstrate how model reduction allows us to make a compact thermal model for an electrothermal simulation of the

Table 3.14. Software Implementation for a Microthruster (http://www.imtek.de/simulation/pyros)

Name	Short Description	Type	User Input	Program Output
EleThermo	Simulating electro-thermal ignition	ANSYS script	Specifications of the microthruster device	System matrices or transient simulation results
FilmCoef	Estimating film coefficient	Mathematica notebook	Properties of the flow	Value of film coefficient
HeatTran	Simulating heat transfer during sustained combustion	ANSYS script	Specifications of the microthruster device	System matrices or transient simulation results
Thrust	Estimating impulse produced by the microthruster	Mathematica notebook	Properties of the solid fuel and the nozzle	Thrust and impulse produced by a microthruster device
CoGen	Performing model order reduction compiler	Binary (executable) produced by C++	System matrices produced by EleThermo or HeatTran	Compact model for system-type simulations in HDL format

microthruster.[61] The results have been obtained for a simplified 2D axisymmetrical model of the microthruster that, after the FEM discretization, produced ODEs of order 1071. First, Fig. 3.36 shows the decay of the Hankel singular values for a microthruster model, computed by the balanced truncation approximation (BTA) method. The steep decay of the Hankel singular values shows that the finite element basis obtained after mesh generation is highly redundant. In this case, a low-dimensional model should accurately capture the behavior of the original high-dimensional finite element model. We believe that this holds for many electrothermal models.

Figure 3.37 confirms this conclusion, comparing a solution of the full 1071-order system for a node close to the resistor with those for a seven-order system obtained after model reduction by the BTA and Arnoldi methods. The reduction of the order of the original system by a factor of more than 100 leads to a difference in results that lies within the line thickness of the plot. On the other hand, a conventional Guyan approach implemented in ANSYS leads to a considerable difference in this case.

The Arnoldi method permits us to make a good approximation not to a single node but, rather, to all of them. In Fig. 3.38, the mean-square relative difference for all the nodes between the original and an Arnoldi-based reduced model is shown during the initial 0.05 sec. The error is already negligible for a reduced model of the order of 20.

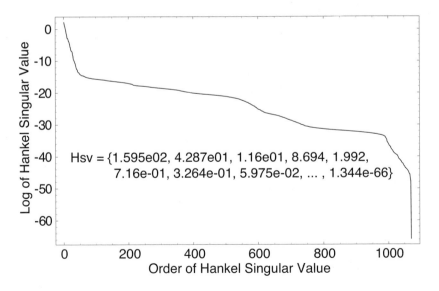

Figure 3.36 Decay of the Hankel singular values for a microthruster model of order 1071. Adapted from Bechtold *et al*.[61]

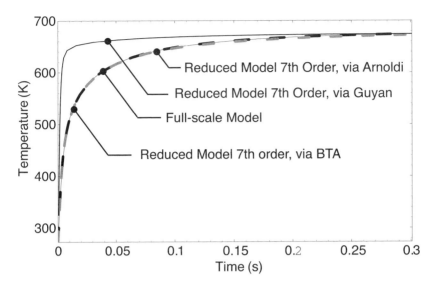

Figure 3.37 Solution of the full system (order 1071) and of the reduced systems for a single fuel node for the microthruster. Adapted from Bechtold et al.[61]

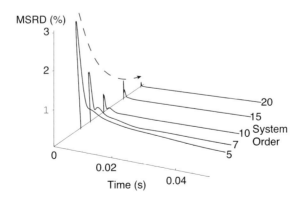

Figure 3.38 Mean-square relative difference (MSRD) for all the nodes during the initial 0.05 s, for an Arnoldi-based reduction from order 1071 to 20, 15, 10, 7, and 5. Adapted from Bechtold et al.[61]

Microfluidic Assembly

Microengineering provides a wide range of methods that allow for massive downscaling of components and single devices with functionalities in different energy domains, e.g., optics, electronics, and fluidics. If material incompatibilities arise in monolithic integration, an efficient microassembly method[63] for hybrid systems is required. One possibility is to do a pick-and-place operation. Exploiting microfluidic capillary effects for self-assembly seems to be a feasible way to meet the need for massive parallel processing,[64] where microparts are packaged on binding sites (see Fig. 3.39[65]). The key issue with this assembly method is control of the hydrophobic and hydrophilic surface properties of both binding sites and microparts. When implementing this method for an industrial assembling process, several questions arise regarding accuracy goals for self-assembly in the fluidic phase. One specific problem to be addressed is whether the binding site and micropart shape are of crucial significance for the uniqueness of assembling. This has been investigated in detail.[65]

Another key issue is to what extent the strengths of the capillary forces and the potential free surface energy shape with respect to the degrees of freedom are crucial for eventual accurate alignment. To optimize fluidic systems parameters for self-assembly, e.g., micropart size, and properties of liquids and surfaces, it is of enormous advantage to apply simulations to different configurations in order to reduce extensive experimental effort.

Assuming that the final state of alignment goes along with the system moving to an energy minimum, the simulation of the fluidic model system

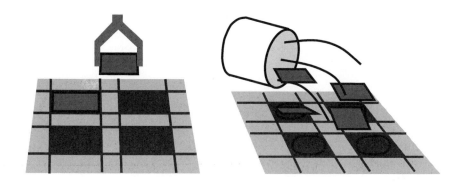

Figure 3.39 Pick and place versus parallel microfluidic assembly. Adapted from Rudnyi et al.[60]

tells us that the existing potential energy minimum will always result in force or torque softening. This means that accurate alignment is improved by influencing the shape of the potential energy minimum. We will see that this is achieved by careful control of the liquid volume between the micropart and the binding site.

The simulated system consists of square-shaped binding sites and microparts with exactly the same dimensions. For now, we do not address the problem of unique assembly configuration. In an experimental setup, the binding sites are coated with a hydrophobic alkanethiol layer (a self-assembled monolayer, or SAM), and a lubricant liquid is applied to the binding sites while the entire system is immersed in water. Lubricant droplets of well-controlled volume form on the binding sites. The microparts are coated with the same SAM and are attracted by the lubricant sitting on the binding part due to capillary forces. The surface energies of the water-lubricant interface (46 mJ/m^2) and the coating-water interface (52 mJ/m^2) show similar values, while the lubricant-coated interface (<1 mJ/m^2) is of two orders of magnitude less.[65] This difference of nearly two orders of magnitude in the surface energies is the driving force that assures self-alignment.

In our simulations, we calculated the total surface energies for different configurations using the surface evolver software by Brakke.[66] The potential energy for various displacements of the micropart relative to the binding site (see Fig. 3.40) and for different lubricant volumes was calculated. These displacements are schematically reproduced in Fig. 3.41.[68] They include a shift of the micropart against the binding site, a lift of the

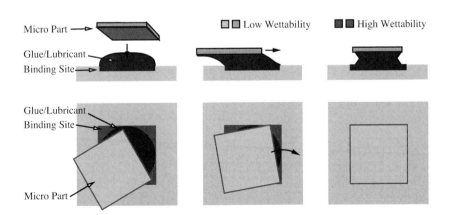

Figure 3.40 Self-assembly process due to capillary forces for shift displacement and twist displacement. Adapted from Lieuemann.[67]

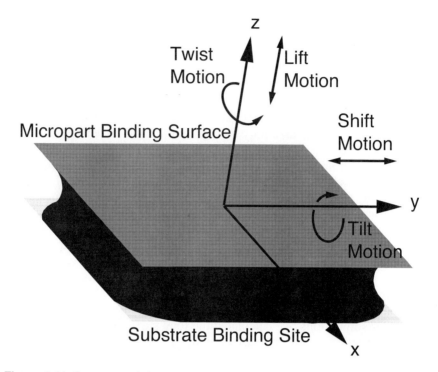

Figure 3.41 Geometry of the system and degrees of freedom for the displacements of the micropart, with respect to the substrate pad. The shape of the liquid meniscus was computed numerically. Adapted from Greiner et al.[68]

micropart in the direction perpendicular to the binding site, a twist rotation with the rotation axes taken as the z-axes, and a tilt motion, i.e., a rotation, with respect to an axis in the plane of the micropart. Different lubricant volumes were investigated. There is a clear dependence of the final alignment precision on the variations in the alignment forces and corresponding potential field.

The quadratic increase in surface energy for shift and twist displacements results in a decrease of the restoring force and the restoring torque near the point of perfect alignment. To reduce this restoring force, softening the lubricant volume must be minimized. The reduction by a factor of three for an already low lubricant amount of 150 nl increases the stiffness of the system remarkably, i.e., the increase of restoring forces and torques is higher the less lubricant is present. Slight imperfections may nevertheless cause local minima and prevent a final parallel alignment of the binding site edges and the edges of the micropart. This results in a tradeoff between technologically feasible lubricant volume control and acceptable design variations and is achieved by simulation and modeling of the fluidic self-assembly process.[67, 68]

Magnetic Field Sensors

This example shows the use of a symbolic computational tool (Mathematica) to prototype two solvers rapidly: a finite-element field solver for the anisotropic Hall effect and a topology optimization tool that uses FEM solvers as black-box components.

Magnetic field sensors based on the Hall effect make use of the Lorenz force acting on charge carriers moving in the plane of the device in a direction perpendicular to the carrier velocity vector. Since the charge carriers are driven to one side of the device, they build up a potential difference between those two otherwise electrically symmetric surfaces, which is a measure for the magnetic-field intensity.

The equipotential lines therefore twist by an angle of $\theta_B = -\mathrm{atan}(\mu_B B)$, with μ_B denoting the Hall mobility of the electrons (see Fig. 3.42[69]).

In the production process of an integrated circuit (IC) device, two different lithographic masks are used to define the outlines of the doped well and the metal contacts on top. Since the Hall displacement of the equipotential lines is rather small for common magnetic fields, a notable variation in the offset appears if the two masks are misaligned.[70] The equivalent magnetic field is $B = \Delta l/(w\mu)$ for a misalignment of Δl and may assume values of several mT. This drawback is compensated for by applying sophisticated electronic circuits that interchange the operating current and measurement contacts.[71,72]

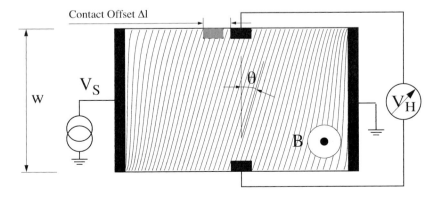

Figure 3.42 Hall device biasing conditions and layout; see Lieumann et al.[69]

With the topology optimization method, so far applied mainly to structural mechanics,[73-75] it is possible to find a shape that maximizes the sensitivity for a given maximum offset. The complexity of the driving circuitry is thereby reduced. The aim is to find a dopant distribution that optimizes these objectives. Since it is easier to manufacture a uniform distribution, a design with a conductivity distribution of either 0 or 1 (which means no intermediate concentrations) is preferred. Although it is difficult to find an optimal design without the intermediate concentrations, the use of a penalty function can lead to a clearer design.

The simulated sensor is symmetric with respect to the operating current contacts (see Figure 3.43). This ensures that both polarities of the magnetic field are measured the same way. Another reason is provided by our objective function for the optimizer: Maximization of the voltage between the measurement contacts would cause a short-circuit of each of the measurement contacts to one of the operating current contacts, resulting in a Hall voltage to input voltage ratio of 1. The symmetric design of the sensor excludes this trivial solution, and only the Hall effect can contribute to the asymmetric potential distribution, thus meeting our objective. Therefore, only one-half of the element densities are used as design

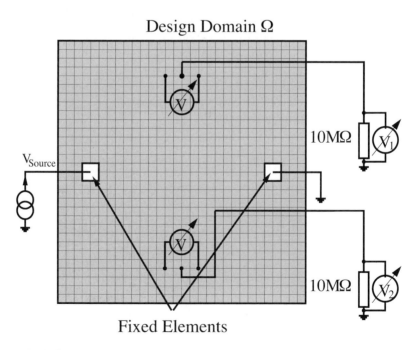

Figure 3.43 Geometrical configuration for the Hall sensor optimization process; see Lieuemann et al.[69]

variables. The elements at the contact pads are fixed to a very high value of ρ_e to model metal contact pads. They do not belong to the set of design variables. A flow chart of the optimization process is shown in Fig. 3.44. For every iteration, a finite element method (FEM) analysis is performed for a given magnetic field perpendicular to the sensor plane. Based on the result, the sensitivity of the Hall voltage and the offset with respect to the design variables are calculated and fed into the optimization algorithm along with the offset and volume constraints. The optimization algorithm now changes the values of the design variables according to the given sensitivities. The iteration with FEM analysis and optimization is repeated until the maximum change in the design variables meets the convergence criterion. Convergence of optimization is achieved if the maximum change in the design variables falls below a value δ.

Since the optimum is very flat, considering only the change of the objective function would cause the optimization to end prematurely. For a maximum offset of 0.01 V_{hall} and a penalty of 2, we got the final shape of Fig. 3.45 after 304 iterations. It is difficult to estimate the number of iterations due to the complexity of the problem, but high demands on the offset usually result in a slower convergence.

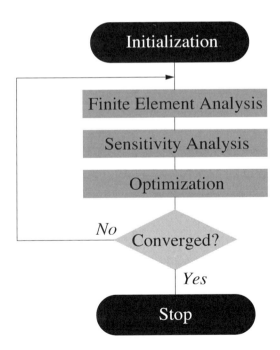

Figure 3.44 Topology optimization flowchart; see Lieuemann et al.[62]

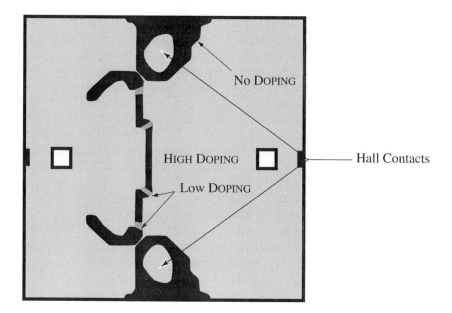

Figure 3.45 Final extracted profile for the optimized Hall sensor; see Lieuemann et al.[69]

Sensitivities achieved were up to four times as high as for a simple Hall plate while still meeting strict offset constraints. The generality of the applied method allows for various applications in optimizing Hall sensors and therefore should be considered a very promising tool.

Wire Bond Sensor

Bond pad sensors, integrated as piezoresistive n^+-doped channels underneath the bonding pads of a CMOS chip, are used to monitor and characterize ball-bonding processes.[77] Such devices are sensitive to ultrasound stresses and insensitive to bonding force stresses and temperature. Improvements of the sensor signal can be achieved by adequate selection of design parameters related to the sensor geometry. A fabrication series with variations of geometry parameters implies high costs and time consumption, so that simulation techniques are required with which to optimize the sensor. The layout of the sensor is depicted in Fig. 3.46. In discussions with the engineers, the following requirements for the simulation were issued:

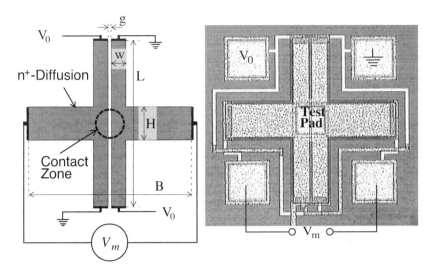

Figure 3.46 Device layout and SEM of the ball pad sensor. Adapted from Osorio et al.[77]

1. Calculate the sensitivity S of the sensor, which is defined as signal voltage V_m divided by applied operating voltage V_0: $S = V_m/V_0$.
2. Vary the geometry parameters to extract the optimum sensor geometry in terms of maximum S.
3. Investigate the effect of bond-wire misalignment on the performance of the sensor.

Since time and budget were at a premium, simulation results for the ultrasonic excitation of the pad were available (they were performed using ANSYS), yet no convenient piezoresistive simulation tool was available, it was decided to achieve the results as follows:

1. Import into the Mathematica program the inhomogeneous stress field from ANSYS simulation results for the coupling.
2. Transform the strain field to an anisotropic piezoresistance field based on measured piezoresistive coefficient data for the CMOS process.
3. Implement a piezoresistively coupled anisotropic current continuity solver in Mathematica, based on the finite difference method, because the sensor geometry is Cartesian and

the FDM is fast to implement. The anisotropic resistivity is a superposition of an intrinsic value and the strain-induced resistivity.
4. Parameterize the sensor geometry.
5. Create a simulation driver that can seek optimal sensor layouts subject to CMOS technology layout constraints. Optimality is measured as specified in Section 3.6 under the discussion of optimization.
6. Perform the simulations.

By taking components of the conductivity tensor along the coordinate axes, being careful to relate these correctly to the crystal axes, we obtained the plots shown in Fig. 3.47. Equipped with this information, we were able to use the FDM solver to compute the potential field in the sensor (see Fig. 3.48) and extract the sensitivity. The simulation procedure

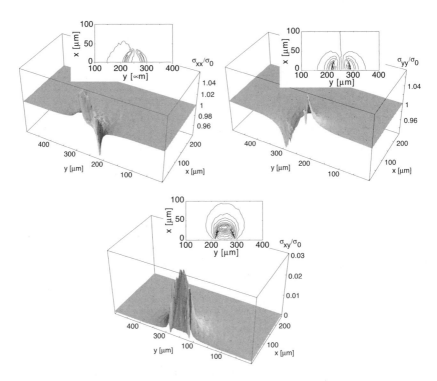

Figure 3.47 2-D and 3-D plots of the calculated, stress-induced conductivity deviations underneath a bond pad during ball bonding. Each figure represents a different component of the conductivity tensor. Adapted from Osario et al.[77]

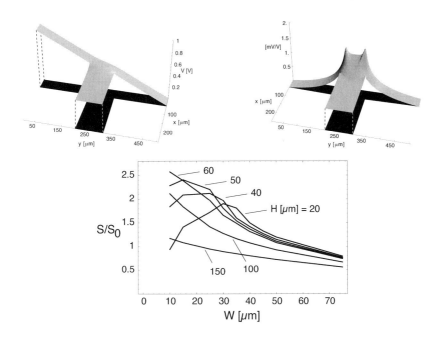

Figure 3.48 The first two figures show the total and differential voltage signal across a typical sensor domain. The right-hand graph shows the variation of the sensor sensitivity for varying parameters from which optimal sensor designs were extracted. Adapted from Osario et al.[77]

offered us the capability of extracting optimal values for geometry parameters. From the simulation series in which parameters W, H, and L were varied, it was demonstrated that reasonable criteria for the optimum value of these design parameters can be extracted in terms of maximum sensor sensitivity S. Using the program, the sensor dimensions were optimized, resulting in a predicted sensitivity that was 4.5 times higher than that of the fabricated prototype. The suggested geometry also reduced the required chip area. In addition, for the optimized sensor, an analysis of bond wire misalignment with respect to the sensor's center was also performed, rounding out the information that was needed for further design improvements.

The key message here is that the right mix of a commercial FEM simulator (ANSYS) and a rapid prototyping simulator (Mathematica) enabled us to obtain the desired design parameters in a short time.

VCSEL Laser

This example demonstrates the combination of different modeling methods and existing commercial solvers to yield a dedicated computational tool for an important effect that would otherwise require considerable development effort.

Vertical cavity lasers (VCLs) are diode lasers with the cavity defined normal to the semiconductor chip surface. There is considerable academic and industrial interest in the further development and application of VCLs because of their low threshold current, single longitudinal mode operation, good near- and far-field characteristics, and two-dimensional array capabilities.[78]

The VCL structure, shown in Fig. 3.49, has a circularly symmetric cavity such that the polarization direction of the laser emission is not *a priori* defined. As a result, the polarization may assume any one of a number of directions during laser operation and is also known to "flip" direction as the laser operation parameters are varied. Such polarization uncertainty and variability are drawbacks to a number of VCL applications. Considerable research effort is directed at gaining better control over VCL polarization direction.

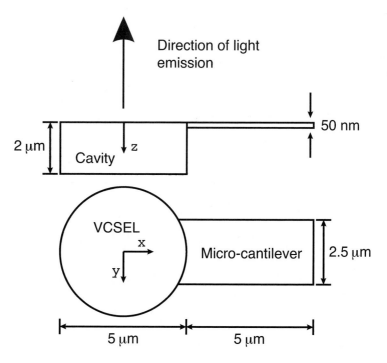

Figure 3.49 Vertical cavity laser schematics.

It is known that application of strain to a VCL can affect the modal distribution. Strain can be applied by mechanical forces. This approach is particularly useful for etched "air post" VCLs. To understand the influence of strain on the laser behavior, it is necessary to (1) calculate the refractive index profile due to applied strain, and (2) calculate the modal and polarization behavior of the laser that results from operation with a cavity in which the refractive index is no longer homogeneous. In the present case, strain is induced by the weight of a microcantilever attached to the top of the VCL cavity. The resultant inhomogeneous distribution of refractive index has the effect of "pinning" the polarization direction. A novel means for inducing controlled strain and thus a desired refractive index profile is the use of an asymmetric VCL structure, such as the beam attached to one side of the VCL post (see Fig. 3.50).

The beam, typically fabricated from a single layer of the materials forming the upper Bragg mirror of the laser, bends under its weight and induces an inhomogeneous refractive index profile inside the semicon-

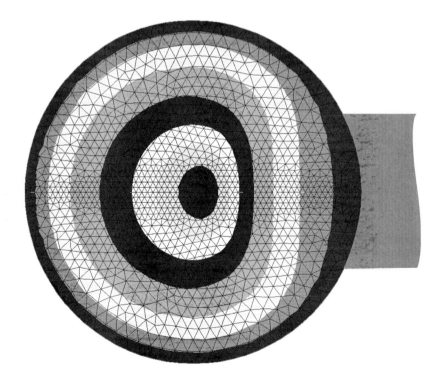

Figure 3.50 Strain profile of the active layer. The highest strain is on the border; the lowest strain is in the middle. The microcantilever is indicated on the right in grey. Adapted from Sigmund.[75]

ductor post due to the inhomogeneous mechanical strain. The simulation was performed under the assumption that the material was ideally homogeneous.

The induced strain was computed using the finite element software ANSYS 5.6, assuming isotropic material properties. This is a simplification. In fact, the orientation of stress relative to the crystal axes is important. The total strain distribution that results from the isotropic simulation of a beam structure with dimensions as indicated in Fig. 3.49 is shown in Fig. 3.50. The resulting strain profile allows for the calculation of the refractive index change due to elasto-optic coupling. The first-order elasto-optic effect is described by a linear material equation. The effect is present in any material that shows a change in dielectric properties when it is exposed to mechanical stress. Once its value is quantified and the refractive index change is calculated, the resulting changes in laser modes and polarization can be determined by solving the respective equations of motion for the laser field distribution, where the refractive index enters as a distributed material property. We performed this second computation with Mathematica, using a singularly perturbed mode expansion approach.

3.9 Summary

We now briefly summarize this chapter with some observations:

- No MEMS simulation tool will ever cover all possible design situations. As designs evolve, so will the importance of varying aspects of the design; as the MEMS industry establishes itself, so will the focus of applications of the technology. This is in strong contrast to microelectronics.
- No simulation technique will be the best for all possible tasks. MEMS simulation will always require a variety of techniques, often in tandem, often provided by different tools.
- It is unrealistic to expect computer-aided engineering tools to be fully automatic and guaranteed correct.

The availability of computing power has made it possible for an engineer to have a variety of software packages accessible from the corporate network or even on his or her own computer to model and simulate many important engineering problems. (See Tables 3.2 through 3.9, 3.11, and 3.12. Our software lists are not exhaustive, and we do not endorse the programs as the best ones for the jobs.) This allows engineers to shorten a development cycle and in turn drives the software industry to develop even more tools. Nevertheless, this process does not always proceed smoothly; there are many cases where companies have not received any return on their investment in simulation software.

Software packages available out of the box make ideas accessible that were developed by previous generations of scientists from different application areas. A problem whose solution could bring a researcher worldwide fame 20 years ago is now considered routine. It is especially noteworthy that hardware progress as expressed by Moore's law has often lowered the computational cost for the solution of such a problem to next to nothing. A modern notebook with a price of about $2K is much more powerful than a mainframe that cost millions of dollars 20 years ago.

Yet the current status of software development is such that an engineer can hardly be expected to achieve simulation success without any knowledge of what is going on within a program under consideration. Almost anyone can complete a typical software tutorial included with the package and run examples (and, in this way, routinely solve problems that were at the frontiers of science 20 years ago). Unfortunately, this is not enough, because a real engineering problem involves nonroutine work. To solve a real problem, an engineer must have additional knowledge and expertise beyond mastering the software as an operator.

First, the engineer must understand some basics of computer science. A realistic engineering assignment will drive the software to work at the hardware limit, however powerful the software is. To estimate how the limit for the problem in question is dictated by available hardware, it is necessary to know how computational complexity grows with problem size. This is especially important at the planning stage, when decisions are taken to purchase software and hardware. A program may have different options, and they may need to be tuned to run the program effectively on given hardware.

Second, it is very important to comprehend the different steps made by a program during a simulation and to predict possible performance bottlenecks. For a simulation of PDEs, this limit is typically the mesh size. If a mesh element is too big, the results obtained may be wrong. On the other hand, a fine mesh made for the whole device may not be solved at all. Unfortunately, at the present time, this problem is not automated; we believe it will take another generation of research before meshing can be considered routine. In quantum chemistry, an analogous problem is to choose a correct basis set. A further problem is to choose the best method to simulate, for example, a particular discretization method. This may involve a choice between different options within a single software package or a choice among different programs. In some cases, different discretization schemes produce almost the same result; in some (for example, compressible flow with shock waves), the choice of discretization can crucially affect the quality of results.

Finally, and probably most important, an engineer should have basic knowledge of the physical models used in a particular software product. Without it, the engineer cannot know to what extent to believe in the simulation results. Remember, if a physical model is inappropriate, the simulation may not be valid. In short, it is not enough to purchase software and hardware: A company that wants to be successful requires experts.

3.10 Acknowledgments

The authors thank their numerous Ph.D. students and collaborators for their contributions. In particular, we thank Mr. Jan Lienemann for his contributions to microfluidic assembly and magnetic field sensor simulations; Prof. Dr. Ole Sigmund for introducing us to the topology optimization method; Mr. Ricardo Osorio and Dr. Michael Mayer for their work on the bond sensor model; Drs. Markus Emmenegger and Martin Bächtold for their work on modal analysis of MEMS; Mr. David Kallweit and Prof. Hans Zappe for their contributions to the VCSEL model; Ms. Tamara Bechtold for her contributions to model order reduction; Dr. Jens Müller for numerous contributions to finite element mesh adaptivity; Dr. Carole Rossi for discussions on all aspects of the microthruster; Profs. Peter Benner and Boris Lohmann for numerous discussions on model order reduction; and Prof. Henry Baltes for many years of fruitful collaboration. The authors gratefully acknowledge the University of Freiburg (operating and guest scientist grants); the German Foundation for Research (grants KO-1883/3-1 and KO- 1883/6); the Autonomous Province of Trento (PAT) and the Italian Research Foundation (CNR), Italy (Ph.D. grant); and the European Union (grant EU IST-1999-29047, Micropyros) for their generous financial support.

References

1. For Europe, see *www.europractice.com* (Europractice); for the United States, see *http://www.memsrus.com/CIMSsvcs.html* (MUMPS); for Japan, see *http://www.wtec.org/mems1/views/berlin/sld009.htm*
2. A. Nathan and H. Baltes, *Microtransducer CAD: Physical and Computational Aspects*, Springler-Verlag, Wien (1999)
3. A. Nathan (ed.), "Special issues on modelling," *Sens. Mater.*, 6(2-4) (1994)
4. J. G. Korvink and H. Baltes, invited review, "Sensors Update," chap. 6 in *Microsystem Modeling*, vol. 2, VCH Verlag, Weinheim (1996), pp. 181-209

5. J. G. Korvink, A. Nathan, and H. Baltes, invited review, "Optical Microsystem Modeling and Simulation," chap. 4 in *MEMS and MOEMS Technology and Applications*, vol. PM85 (P. Rai-Choudhury, ed.), SPIE Press, Bellingham (2000), pp. 109-168
6. G. H. Golub and C. F. Van Loan, *Matrix Computations*, 3rd ed., Johns Hopkins University Press, Baltimore, MD (1996)
7. Y. Saad, *Iterative Methods for Sparse Linear Systems*, PWS Publishing Co., Boston, MA (1996)
8. E. L. Allgower and K. Georg, "Numerical Path Following," in *Handbook of Numerical Analysis*, vol. 5 (P. G. Ciarlet and J. L. Lions, eds.), North-Holland, Amsterdam (1997), pp. 3-207
9. W. Hackbusch, *Multi-Grid Methods and Applications*, Springer-Verlag, Berlin and New York (1985)
10. M. Emmenegger, J. G. Korvink, M. Bächtold, M. von Arx, O. Paul, and H. Baltes, "Application of harmonic finite element analysis to a CMOS heat capacity measurement structure," *Sens. Mater.*, 10:405-412 (1998)
11. M. P. Allen and D. J. Tildesley, *Computer Simulation of Liquids*, Clarendon Press, New York, and Oxford University Press, Oxford, England (2001)
12. D. Frenkel and B. Smit, *Understanding Molecular Simulation: From Algorithms to Applications*, Academic Press, San Diego, CA (1996)
13. J. M. Haile, *Molecular Dynamics Simulation: Elementary Methods*, Wiley, New York, NY (1997)
14. A. R. Leach, *Molecular Modelling: Principles and Applications*, Longman Publishing Group, White Plains, NY (1996)
15. D. C. Rapaport, *The Art of Molecular Dynamics Simulation*, Cambridge University Press, Cambridge, UK (1995)
16. M. Parrinello, "Simulating complex systems without adjustable parameters," *Computing Sci. Eng.*, 2:22-27 (2000)
17. A. Nakano, M. E. Bachlechner, R. K. Kalia, E. Lidorikis, P. Vashishta, G. Z. Voyiadjis, T. J. Campbell, S. Ogata, and F. Shimojo, "Multiscale simulation of nanosystems," *Computing Sci. Eng.*, 3:56-66 (2001)
18. M. S. Gordon, www.msg.ameslab.gov/Group/GroupTutor.html
19. www.emsl.pnl.gov:2080/docs/tms/abinitio-v4/cover.html
20. J. N. Israelachvili, *Intermolecular and Surface Forces*, 2nd ed., Academic Press, San Diego, CA (1992)
21. N. Saunders and A. P. Miodownik, *CALPHAD: Calculation of Phase Diagrams: A Comprehensive Guide*, Pergamon, Oxford and New York (1998)
22. M. Hillert, *Phase Equilibria, Phase Diagrams, and Phase Transformations: Their Thermodynamic Basis*, Cambridge University Press, Cambridge and New York (1998)
23. A. V. Antipov, E. B. Rudnyi, and Z. V. Dobrokhotova, "Thermodynamic evaluation of the Bi-Se system," *Inorg. Mater.*, 37:126-132 (2001)
24. P. P. Silvester and R. L. Ferrari, *Finite Elements for Electrical Engineers*, 3rd ed., Cambridge University Press, New York, NY (1996)
25. O. C. Zienkiewicz and R. L. Taylor, *The Element Method*, 5th ed., Butterworth-Heinemann, Oxford and Boston (2000).

26. http://www-dinma.univ.trieste.it/~nirftc/research/easymesh/
27. S. Selberherr, *Analysis and Simulation of Semiconductor Devices*, Springer-Verlag, Wien, Austria, and New York (1984)
28. A. Iserles, *First Course in Numerical Solution of Differential Equations*, Cambridge University Press, Cambridge, UK (1996)
29. C. A. Brebbia, J. C. F. Telles, and L. C. Wrobel, *Boundary Element Techniques: Theory and Applications in Engineering*, Springer-Verlag, Berlin and New York (1984)
30. L. Greengard and S. Wandzura, "Fast multipole methods," *IEEE Comp. Sci. Eng.*, 5(3):16-18 (1998)
31. M. Kamon, M.J. Ttsuk, and J.K. White, "FASTHENRY: A multipole-accelerated 3-D inductance extraction program," *IEEE Trans Microwave Theory and Techniques*, 42(9):1750-1758 (1994)
32. M. Bächtold, J. G. Korvink, and H. Baltes, "Enhanced multipole acceleration technique for the solution of large Poisson computations," *IEEE Trans. CAD Integrated Circuits and Systems*, 12:1541-1546 (1996)
33. J. A. Sethian, *Level Set Methods and Fast Marching Methods: Evolving Interfaces in Computational Geometry, Fluid Mechanics, Computer Vision, and Materials Science*, 2nd ed., Cambridge University Press, Cambridge and New York (1999)
34. R. T. Haftka and Z. Gürdal, *Elements of Structural Optimization*, Kluwer Academic Publishers Group, Dordrecht (1992)
35. E. J. Haug, K. K. Choi, and V. Komkov, *Design Sensitivity Analysis of Structure System*, Academic Press, New York (1986)
36. *NEOS Guide Optimization Tree*; available at http://www.ece.northwestern.edu/OTC/
37. J. Nocedal and S. J. Wright, *Numerical Optimization*, Springer, Berlin (1999)
38. M. P. Bendsoe and O. Sigmund, *Topology Optimization - Theory, Methods and Applications*, Springer, Berlin and Heidelberg (2002)
39. U. D. Larsen, O. Sigmund, and S. Bouwstra, "Design and fabrication of compliant micromechanisms and structures with negative poisson's ratio," *IEEE J. MEMS*, 6:99-106 (1997)
40. O. Sigmund and S. Torquato, "Design of materials with extreme thermal expansion using a three-phase topology optimization method," *J. Mech. Phys. Solids*, 45:1037-1067 (1997)
41. O. Sigmund, "Microstructural design of elastic band gap structures," *Proceedings of the Fourth World Congress of Structural and Multidisciplinary Optimization*, Liaoning Electronic Press, Dalian, China (2001)
42. P. L. George, *Automatic Mesh Generation: Application to Finite Element Methods*, J. Wiley, Chichester, New York, and Masson Publishers, Paris (1991)
43. Jens Müller, *Accurate FE-Simulation of Three-Dimensional Microstructures*, Fakultät für Angewandte Wissenschaften, Albert-Ludwigs Universität Freiburg im Breisgau, 2001; available at http://www.freidok.uni-freiburg.de/volltexte/251/pdf/thesis.pdf

44. H. Chiyokura, *Solid Modelling with Designbase: Theory and Implementation*, Addison-Wesley, Singapore and Reading, MA (1988)
45. E. B. Rudnyi and J. G. Korvink, "Review: automatic model reduction for transient simulation of MEMS-based devices, *Sensors Update*, 11:3-33 (2002)
46. A. C. Antoulas and D. C. Sorensen, "Approximation of large-scale dynamical systems: an overview," Rice University Technical Report, Houston, TX (2001); *http://www-ece.rice.edu/ ~aca/mtns00.pdf*
47. G. Obinata and B. D. O. Anderson, *Model Reduction for Control System Design*, Springer Verlag, London, Berlin, and Heidelberg (2000)
48. A. Varga, "Model reduction software in the SLICOT library," in *Applied and Computational Control, Signals, and Circuits*, vol. 2 (B. N. Datta, ed.), Kluwer Academic Publishers, Boston, MA (2001), pp. 239-282
49. R. W. Freund, "Reduced-order modeling techniques based on Krylov subspaces and their use in circuit simulation," in *Applied and Computational Control, Signals, and Circuits*, (B. N. Datta, ed.), Birkhauser, Boston, MA (1999), pp. 435-498
50. T. Penzl, "A cyclic low-rank Smith method for large sparse Lyapunov equations," *SIAM J. Sci. Comput.* 21:1401-1418 (2000)
51. J. Li and J. White, "Low rank solution of Lyapunov equations," *SIAM J. Matrix Anal. Appl.*, 24:260-280 (2002)
52. Online website: *http://www.cs.clemson.edu/~steve/ ivandv/*
53. *CRC Handbook of Chemistry and Physics: A Ready-Reference Book of Chemical and Physical Data*, Chemical Rubber Company, CRC Press, Boca Raton, FL (1999/2000)
54. *NIST WebBook, http://webbook.nist.gov/*
55. *Evaluated Kinetic Data*, IUPAC: Subcommittee for Gas Kinetic Data Evaluation, *http:// www.iupac-kinetic.ch.cam.ac.uk/*
56. O. Paul, J. G. Korvink, and H. Baltes, "Determination of thermophysical properties of CMOS IC polysilicon, *Sensors and Actuators*, A41-42:161-164 (1993)
57. Coventor: *www.coventor.com/coventorware/architect/behavioral_models.html*
58. H. Baltes, "CMOS as sensor technology," *Sensors and Actuators*, A37-A38:51- 56 (1993)
59. C. Rossi, "Micropropulsion for space A survey of MEMS-based micro thrusters and their solid propellant technology," *Sensors Update*, 10:257-292 (2002)
60. E. B. Rudnyi, T. Bechtold, J. G. Korvink, and C. Rossi, "Solid propellant microthruster: theory of operation and modelling strategy," *Nanotech 2002, At the Edge of Revolution*, September 9-12, 2002, Houston, TX, AIAA paper 2002-5755
61. T. Bechtold, E. B. Rudnyi, and J. G. Korvink, "Automatic generation of compact electrothermal models for semiconductor devices," *IEICE Trans. Electronics*, E86C:459-465 (2003)
62. ANSYS: *www.ansys.com*

63. K. F. Böhringer, R. S. Fearing, and K. Y. Goldberg, "Microassembly," in *The Handbook of Industrial Robotics*, 2nd ed., John Wiley & Sons, NY (1999), pp. 1045-1066
64. U. Srinivasan, R. T. Howe, and D. Liepmann, "Microstructure to substrate self-assembly using capillary forces," *J. MEMS*, 10:17-24 (2001)
65. K. F. Böhringer, U. Srinivasan, and R. T. Howe, "Modeling of capillary forces and binding sites for fluidic self-assembly, *IEEE Workshop on Micro Electro Mechanical Systems (MEMS)*, Interlaken, Switzerland, January 21-25, 2001
66. K. A. Brakke, *Surface Evolver Manual*, Mathematics Department, Susquehanna University, Selingsgrove, PA (1999)
67. J. Lienemann, "Modelling and simulation of the microfluidic self-assembly of microparts," diploma thesis, Albert Ludwig University, Freiburg (2002)
68. A. Greiner, J. Lienemann, J. G. Korvink, X. Xiong, Y. Hanein, and K. F. Boehringer, "Capillary forces in micro-fluidic self-assembly," *Technical Proceedings of the 2002 International Conference on Modeling and Simulation of Micorsystems NanoTech 2002-MSM2002*, Puerto Rico, pp. 198-201 (2002)
69. J. Lienemann, A. Greiner, J. G. Korvink, and O. Sigmund, "Optimization of integrated magnetic field sensors," *Technical Proceedings of the MSM 2001 4th International Conference on Modeling and Simulation of Microsystems*, Hilton Head Island, USA (2001), pp. 120-123
70. S. Middelhoek and S. A. Audet, *Silicon Sensors*, Department of Electrical Engineering, Laboratory of Electronic Instrumentation, Delft University of Technology (1994)
71. P. J. A. Munter, "A low-offset spinning current Hall plate," *Sensors and Actuators*, A21-A23:743-746 (1990)
72. R. Steiner, A. Haeberli, F.-P. Steiner, C. Maier, and H. Baltes, "Offset reduction in Hall devices by continuous spinning current," *Sensors and Actuators*, A66:167 (1998)
73. M. P. Bendsøe and N. Kikuchi, "Generating optimal topologies in optimal design using a homogenization method," *Comput. Methods Appl. Mech. Eng.*, 71:197-224 (1988)
74. M. P. Bendsøe: *Optimization of Structural Topology, Shape and Material*, Springer, Berlin and Heidelberg (1995)
75. O. Sigmund, "On the design of compliant mechanisms using topology optimization," *Mech. Structures and Machines*, 25(4):495-526 (1997)
76. M. P. Bendsøe and O. Sigmund, "Material interpolations in topology optimization," *Archive Appl. Mech.*, 69:635-654 (1999)
77. R. Osorio, M. Mayer, J. Schwitzer, H. Baltes, and J. G. Korvink, "Simulation procedure to improve piezoresistive microsensors used for monitoring ball bonding," *Sensors and Actuators*, 92(1-3):299-304 (2001)
78. C. W. Wilmsen, *Vertical Cavity Surface Emitting Lasers*, Cambridge University Press, Cambridge, England (1999)

4 System-Level Simulation of Microsystems

Gary K. Fedder

*Department of Electrical and Computer Engineering and The Robotics Institute
Carnegie Mellon University, Pittsburgh, Pennsylvania*

4.1 Introduction

Microelectromechanical systems require a multidisciplinary approach to design, including knowledge of fabrication technology, mechanics, electromechanics, and electronics. In the majority of cases, good MEMS design requires the evaluation of tradeoffs among the fabrication process, micromechanical topology, and sizing. Sensor interface circuits, signal conditioning, and feedback provide additional interactions that affect design choices. This chapter provides an introductory overview to the system-level simulation of microsystems in support of the design effort. Discussion is devoted to microelectromechanical structures with electronics. However, the general methods outlined here have equal applicability in other microsystems, such as microfluidic and micro-optical systems.

Batch fabrication prevalent in MEMS provides the manufacturing capability to make very complex systems with multiple interconnected MEM devices. System complexity for MEMS is measured by both the number of interacting devices and the number of interacting physical domains. For example, a typical surface-micromachined capacitive microaccelerometer incorporates about 5 to 100 devices, depending on the specified granularity of the system partitioning. These elements have mechanical, electrical, thermal, electrostatic, and fluidic interactions. This combined complexity presents a possible design bottleneck that is best handled using structured design techniques, where models are built hierarchically by interconnecting smaller components. At their lowest level, the components are described by behavioral models, equations that directly describe physical behavior. Structured design was successfully introduced to the digital IC design world in the 1980s.[1] Analog electronic circuit design has borrowed these concepts to handle complex analog designs. In 1995, the U.S. National Science Foundation sponsored a seminal workshop to discuss the application of structured design to MEMS.[2] Most of this chapter addresses issues identified at that workshop.

Design is an iterative process where no exact sequence of steps is always followed. Nevertheless, some structure can be imposed on the design process. Any design procedure must start with identifying the function and metrics required or desired by the system. The performance must then be quantified into a set of specifications. How the specifications are met is the purpose of design. For the case of MEMS design, the effort can be partitioned into three general areas: process design, device design, and system design. Successful design efforts may require a combination of innovation in all three areas, or may simply require merging of existing technology from the three areas. Some brief comments about process and device design follow; details on these topics are found in other chapters of this book.

Process design is the combining of individual process steps to form a realizable process flow. The choice of process flow is critical to MEMS performance. A working knowledge is required of material properties and interactions with other materials, fabrication steps and equipment, and photolithographic limitations. New process flows are designed for a variety of reasons, which include:

- Fabricating micromechanical devices of the required dimensions (nm, µm, or mm size)
- Forming devices with materials that have special sensing or actuation properties
- Providing isolated electrical interconnections for microstructures
- Merging micromechanics with electronic circuitry
- Increasing manufacturing yield and reducing cost

These goals are often at odds with each other. Each step in the process flow must be compatible with succeeding steps, accounting for limits to microfabrication capabilities. Microstructural patterning is limited by photolithography over widely varying surface topographies. Deposited thin-film materials must adhere to the underlying surface and not diffuse with time. The temperatures of process steps must not affect materials from prior steps. Etching steps have limits of critical dimensions in both the lateral and vertical dimensions. Material combinations must be chosen with appropriate etch selectivities. The etch steps must avoid undesired side effects such as sidewall deposition. New materials and processes continue to push these limitations.

Device design is the positioning and sizing of functional materials to achieve a given performance. If the fabrication processes can be varied, then device and process design for MEMS are usually closely linked. However, at some point in development, the process flow becomes fixed and constrains the device design space. The necessary base knowledge for MEMS device design includes mechanics of materials, circuit theory, electrostatic field theory, electromechanics, and fluid mechanics. Other domain-specific knowledge of particular device physics is often required. For example, a knowledge of semiconductor physics is desired for design of piezoresistive and magnetoresistive transducers, while an understanding of optics is important for design of micromirror surfaces. The wide multidisciplinary range of topics underscores why MEMS design is often difficult.

System design is the composition, placement, and sizing of components to meet application specifications. One designer's system is often considered to be another designer's component, which makes it difficult to find the dividing line between device and system design. The distinction is made by the approach to design, not the object being designed. In system design, a structured hierarchical approach to design is enforced; devices are represented with interoperable behavioral models and then interconnected together to form a system. An example of hierarchy is the dual-resonator bandpass filter shown in Fig. 4.1. Such microelectro-

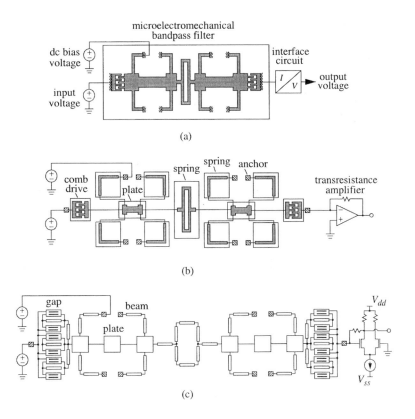

Figure 4.1 Hierarchical partitioning for design of a microelectromechanical bandpass filter. (a) Top level of abstraction; (b) intermediate level; (c) low level.

mechanical filters, scaled to resonate at Mhz to GHz frequencies, are proposed for miniature low-power communications applications.[3] At the top level of abstraction in the hierarchy, the filter is partitioned into a microelectromechanical structure that shapes the resonant pass band and an interface circuit that converts motional displacement current into an output voltage. At an intermediate level, the micromechanical filter block is broken down into springs, plates, and comb-drive actuators. At a low level, the intermediate components may be further partitioned into finite beam, plate, and gap components. Behavioral models may be formed for components at any level. Alternatively, components in the higher levels may be structured by interconnecting lower level components. Tradeoffs in simulation accuracy and time to completion, along with verification, usually create a need for behavioral models at all levels when evaluating complex systems. Therefore, as microsystems are becoming ever more complex, there is an increasing interest in rapid reduced-order modeling and in subsequent model reuse to aid in more rapid system design.

A system design flow for MEMS, analogous to that for integrated electronic circuits, is illustrated in Fig. 4.2. Independent of the design hierarchy, a variety of MEMS representations, also called views, are available to support system design. The layout view describes the masks necessary for photolithographic microfabrication. It provides the two-dimensional (2-D) physical dimensions of the material layers. Additional process technology information supplies the process flow, materials, and layer thicknesses. Armed with this information, process cross sections can be automatically generated to help visualize designs. A second MEMS view is the three-dimensional (3-D) solid model. Solid models are partitioned into smaller elements to form a meshed model that is suitable for

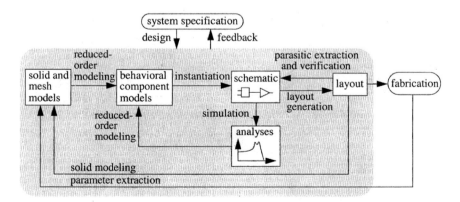

Figure 4.2 A design flow for MEMS, paralleling that of electronic design.

finite-difference, finite-element, or boundary-element analysis. A third view, the MEMS schematic, is critical to structured design because it describes components and their interconnections. Components associated with schematics have a symbol for placement, terminal pins for interconnection, and a model describing the element's behavior in terms of its terminal variables. Ideally, the designer can move seamlessly among schematic, layout, and solid model views. Several MEMS computer-aided design vendors support interaction among these views.

An idealized, top-down design approach traverses the following path: The design process begins with user input, leading to identification of the system specifications. An appropriate system topology is then proposed and partitioned by function into components. This step forces the designer to create input-output relations for each component and to confront crude tradeoffs while meeting the overall system performance goals. Detailed component behavior is then modeled, either through further partitioning into a more fine-grained structure or by creating a behavioral model. Once the system has a basis of behavioral models, individual blocks and eventually the full system are debugged and verified though simulation. At any step, the designer can opt to refine selected component models to include more degrees of freedom (DOFs) or to create reduced-order behavioral models of larger components initially modeled through composition of smaller components. Mechanical and electrical parasitics are extracted from the layout and added to the overall system simulation as a final verification of system performance.

Typically, the desired absolute accuracy in simulation of a microfabricated device is around 1%; fabrication tolerances are usually much worse at about 10 to 20%. Simulation accuracy of component sensitivities to geometric variation should also be about 1% in order to assess the impact of manufacturing variation on system performance. Model accuracy is coupled intimately with assumptions about the necessary physics. Inclusion of all physics requires low-level numerical simulation of the general physical equations. Quasistatic Maxwell's equations govern electrostatics, the theory of elasticity governs mechanical behavior, and Navier-Stokes equations govern air damping. These are all partial differential equations, which finite-element and boundary-element methods handle.

System-level simulation requires formulation of the system dynamics as ordinary differential equations, coupled with efficient solvers for transient simulation. Analog hardware definition languages (AHDLs) are a flexible way to encode either signal flow or Kirchhoffian network formulations, although other representations are also used. In general, there are three key aspects to the components that constitute the system. First, each component model must faithfully predict behavior. Second, each model must be chosen at a particular granularity to include important effects while excluding insignificant effects. Third, each component must interoperate efficiently with the components connected to it. Interoperability is primarily a language support and standardization issue.

Component physical behavior, which continuously varies over space and is typically modeled with thousands of DOFs in finite-element solvers, must be lumped into a much smaller number of variables to be compatible with system simulation. A fine partitioning is technically possible and will provide simulation data on high-order modes of displacements and forces at many locations. However, this degree of detail comes at the cost of more computational resources, reflected in simulation time and computer processing capability. As systems become more complex, a tradeoff of simulation detail with speed is necessary. In some cases, such as bending of beams and plates, accurate reduced-order behavioral models may be crafted directly from the underlying partial differential equations by choosing a set of basis functions that relate lumped variables to the distributed variables. In more complex cases, numerical model-order reduction strategies may be used to generate a set of linear basis functions and coefficients directly from numerical simulation results. This latter approach is general and can be automated through commercial finite-element and boundary-element simulators.

Automated MEMS modeling tools allow relatively rapid custom model generation within a system design flow. However, if only fixed-parameter models are available, new models must be generated when layout or process parameters change. Behavioral models with geometric parameters and material property parameters allow modifications to the layout and process, and they avert the need for repeated model development within design iterations. Such parameterization of models is necessary, even in single-use situations, assuming that simulation in the face of manufacturing variations is important. However, the number of possible geometric parameters defining a given design topology grows linearly with the number of elements. The number of coefficients relating these parameters inside a model can grow exponentially. Therefore, making behavioral models that span the design parameter space becomes increasingly difficult and time consuming. General parameterized models are usually restricted to simple structures with ten or fewer parameters. As discussed earlier, the composition of systems by interconnecting such simple components at the system level allows greater design reuse of models. The system designer must ultimately make the decision either to generate behavioral models or to build structural models in the design loop. The best design procedure usually involves some hybrid of both techniques.

The remainder of this chapter outlines, in a bottom-up way, the three main techniques in structured design for MEMS:

1. Behavioral modeling of components
2. Formulation of ODEs from these models
3. The use of structured design tools

4.2 Behavioral Modeling of MEMS Components

System simulation requires the physics underlying all components to be embedded in behavioral models. Extended discussions of physical modeling in microsystem design are given by Senturia[4] and by Nathan and Baltes.[5] The present discussion of modeling is restricted to suspended electrostatic MEMS, which span a wide range of applications, including pressure sensors, accelerometers, gyroscopes, resonators, relays, RF filters, variable capacitors, micromirrors, microactuators, and acoustic and ultrasonic transducers. Suspended electrostatic MEMS are made from beam flexures, plate masses, and electrostatic gap sensors and actuators. Microfluidic systems have been partitioned into a similar hierarchy.[6,7]

Components of suspended MEMS are modeled with local mass, stiffness, damping, and force coefficients. Next, selected aspects of plate, flexure, and electrostatic gap components are considered to introduce the basic physical principles and to exemplify common macromodeling techniques.

4.2.1 Micromechanical Plates

In applications such as low-frequency resonators, accelerometers, and gyroscopes, a suspended micromechanical plate can be considered a rigid body. Theory and models for flexible plates, for example for pressure sensors, are given in mechanics texts;[8-10] they will not be considered further here. Primarily in-plane forces will be considered throughout this chapter. The inertial forces acting on a rigid plate are

$$\begin{bmatrix} F_{x,m} \\ F_{y,m} \\ F_{z,m} \end{bmatrix} = \begin{bmatrix} m & 0 & 0 \\ 0 & m & 0 \\ 0 & 0 & I_z \end{bmatrix} \begin{bmatrix} \ddot{x}_m \\ \ddot{y}_m \\ \ddot{\theta}_m \end{bmatrix}, \qquad (1)$$

where $m = \rho h A$ is the mass, ρ is the density, h is the thickness, and A is the area of the plate. Coordinates are chosen at the center of mass of the plate. For a rectangular plate, the mass moment of inertia, I_z, is

$$I_z = \int_{-L_{y/2}}^{L_{y/2}} \int_{-L_{x/2}}^{L_{x/2}} \rho h(x^2 + y^2) dx dy = \frac{m}{12}(L_x^2 + L_y^2), \qquad (2)$$

where L_x and L_y are the plate dimensions.

Viscous air damping is the dominant dissipation mechanism for microstructures that operate at atmospheric pressure. Structural damping induced by strain within the structure becomes dominant at sufficiently low pressure. Squeeze-film damping arises from relative motion compressing or expanding the air volume between two parallel plates. The motion creates an opposing pressure in the thin film of air that is proportional to the velocity of the plates. The Navier-Stokes equations can be solved for the squeeze damping force as an infinite series,

$$F_{squeeze} = b_{squeeze} \dot{z} = \frac{\eta L_x^3 L_y}{g^3} \dot{z} \sum_{i=0}^{\infty} \sum_{j=0}^{\infty} \frac{786/\pi^6}{(2i+1)^2(2j+1)^2 \left[(2i+1)^2 + \frac{L_x}{L_y}(2j+1)^2\right]}, \qquad (3)$$

where b is the damping coefficient, g is the air gap, and η is the viscosity of the gas in the gap.[11] The infinite series in Eq. (3) converges rapidly. Only the terms $i = 0, j = 0..4$ and $j = 0, i = 1..3$ are required for 1% accuracy over all geometries. At low pressures, an effective viscosity is substituted.[12] Squeeze-film damping occurs in vertical gaps between plates and substrate, as well as in lateral gaps between microstructures. Equation (3) must be modified for squeeze damping in rotational modes, which is dependent on the direction of rotation, and for plates with holes. At higher frequencies, the squeeze forces are dispersive and have both a damping component and a spring component (proportional to displacement). Only damping force needs to be considered if mechanical dynamics in the system have frequency components such that

$$f \ll \frac{\pi g^2 P}{24\eta}\left(\frac{1}{L_x^2} + \frac{1}{L_y^2}\right). \qquad (4)$$

Shear flow of gas between laterally moving surfaces creates a Couette damping force, given by

$$F_{Couette} = b_{Couette}\dot{x} = \frac{\eta L_x L_y}{g}\dot{x}, \tag{5}$$

where g can be either a vertical or lateral gap between surfaces. The plate width parameter in Eq. (5) that is orthogonal to the direction of motion is sometimes increased by an amount on the order of the gap dimension. This bloated effective width accounts for edge effects of the shear flow. Viscous drag on lateral plate motion with open surfaces (no gap) may be modeled with an effective gap.[13]

4.2.2 Micromechanical Flexures

Micromechanical flexures are used in MEMS versions of accelerometers, gyroscopes, resonant sensors, RF switches, and mirrors. In most cases, it is desirable to have a flexure compliant in one DOF and very stiff in other DOFs. For example, the proof mass for a lateral microaccelerometer is designed to move easily in one direction in-plane with the substrate. Compliance in other directions usually increases sensitivity to cross-axis acceleration. Most micromechanical flexures constrain motion to a rectilinear direction and are created from straight beams. Alternative topologies incorporating spiral springs and other curved-beam flexures are possible but are rarely necessary.

A simple beam element is shown in Fig. 4.3. Common flexure examples are the cantilever, guided-end, and fixed-fixed beams, which are defined by the boundary conditions at the ends of the beam. A cantilever has zero moment at its end; a guided-end beam has its angle constrained to zero at its ends; and a fixed-fixed beam has its displacement and angle constrained to zero at both ends.

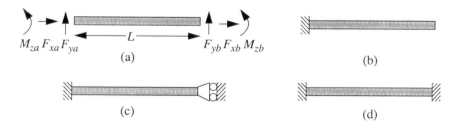

Figure 4.3 In-plane simple beams. (a) General free-body diagram; (b) cantilever; (c) guided-end beam; (d) fixed-fixed beam.

Axial displacement, u_x, along a beam of length L, is linear with position and is given by

$$u_x = \mathbf{a}_x \begin{bmatrix} x_a \\ x_b \end{bmatrix} = \begin{bmatrix} 1-\xi & \xi \end{bmatrix} \begin{bmatrix} x_a \\ x_b \end{bmatrix} \qquad (6)$$

where $\xi = x/L$ is the normalized distance along the beam, x_a and x_b are the axial displacements at the ends of the beams, and $\mathbf{a}_x$ is a matrix of shape functions. Concentrated force, F_x, acting axially on the beam, is related to the end displacements by

$$F_x = k_{xx}(x_b - x_a) = \frac{EA}{L}(x_b - x_a), \qquad (7)$$

where A is the beam cross-sectional area, E is the Young's modulus of elasticity, and the axial spring constant, also called the stiffness coefficient, is $k_{xx} = EA/L$.

Beam bending for small deflections and neglecting shear stress is governed by the Euler-Bernoulli equation

$$\frac{d^4 u_y}{dx^4} - N \frac{d^2 u_y}{dx^2} = f(x), \qquad (8)$$

where u_y is the transverse displacement of the beam, N is axial tensile force in the beam, and $f(x)$ is the distributed force along the beam length. Beams with large deflections require an exact nonlinear expression of the beam curvature, which modifies Eq. (8).[14] Effects of shear stress in the beam on displacement are handled by "Timoshenko beams" and become increasingly important for beams with length-to-width ratios of less than 10.[15] Matrix approximations that handle shear and axial stress in three dimensions are found in Przemieniecki.[8]

In Eq. (8), $f(x) = 0$ in cases where only concentrated forces are applied at the ends of the beam. Additionally, if the effect of axial force on bending is assumed to be negligible, then Eq. (8) is solved using a cubic polynomial in x. Given arbitrary displacements $y_a = u_y(0)$ and $y_b = u_y(L)$ and rotation θ_a and θ_b at the ends of the beam, the resulting displacement along the beam length is

$$u_y = \mathbf{a}_y^T \begin{bmatrix} y_a \\ y_b \\ \theta_a \\ \theta_b \end{bmatrix} = \begin{bmatrix} 1 - 3\xi^2 + 2\xi^3 \\ 3\xi^2 - 2\xi^3 \\ L(\xi - 2\xi^2 + \xi^3) \\ L(-\xi^2 + \xi^3) \end{bmatrix}^T \begin{bmatrix} y_a \\ y_b \\ \theta_a \\ \theta_b \end{bmatrix}, \qquad (9)$$

where $\mathbf{a}_y$ is a shape function matrix for bending. Bending moment in the beams is

$$M_z = EI_z \frac{d^2 u_y}{dx^2}, \qquad (10)$$

where I_z is the beam bending moment of inertia and the shear force is

$$F_y = -\frac{dM_z}{dx}. \qquad (11)$$

General concentrated force-displacement relations for simple beams can be derived using Eq. (7) and Eqs. (9) through (11). For a simple cantilever beam with its left end fixed ($y_a = \theta_a = 0$), these relations are

$$\begin{bmatrix} F_{xb} \\ F_{yb} \\ M_{zb} \end{bmatrix} = \begin{bmatrix} k_{xx} & k_{xy} & k_{x\theta} \\ k_{xy} & k_{yy} & k_{y\theta} \\ k_{x\theta} & k_{y\theta} & k_{\theta\theta} \end{bmatrix} \begin{bmatrix} x_b \\ y_b \\ \theta_b \end{bmatrix} = \begin{bmatrix} \dfrac{EA}{L} & 0 & 0 \\ 0 & \dfrac{12EI_z}{L^3} & -\dfrac{6EI_z}{L^2} \\ 0 & -\dfrac{6EI_z}{L^2} & \dfrac{4EI_z}{L} \end{bmatrix} \begin{bmatrix} x_b \\ y_b \\ \theta_b \end{bmatrix}, \qquad (12)$$

where the matrix of k_{ij} coefficients in Eq. (12) is the linear stiffness, or spring constant, matrix of the beam. A similar relation exists between F_{zb}, M_{yb} and z_b, ϕ_b (ϕ_b is rotation around the y axis), with I_z replaced by I_y. Spring constants k_{xx}, k_{yy}, and k_{zz} for the examples of the cantilever beam, the guided-end beam, and the fixed-fixed beam are given in Table 4.1, assuming a rectangular beam cross section, with width w and thickness h. The fixed-fixed beam is treated as two guided-end beams of length $L/2$ connected together in the middle. Axial stress in fixed-fixed beams arises from residual stress, and additional tensile stress is generated when the beam is displaced. A full solution of Eq. (8) is then required to achieve sufficient simulation accuracy.

The stiffness ratio of the axial to lateral in-plane motion, k_x/k_y, is proportional to $(L/w)^2$. For a cantilever with $L/w = 100$, the stiffness ratio is 40,000:1. The stiffness ratio of the vertical to lateral in-plane motion, k_z/k_y, is equal to $(h/w)^2$. If restricted vertical motion is desired, the beam thickness must be much larger than the width. Many polysilicon surface-

Table 4.1 Spring Constants for Concentrated Load

	Cantilever	Guided-End Beam	Fixed-Fixed Beam
k_{xx}	$\dfrac{Ehw}{L}$	$\dfrac{Ehw}{L}$	$4\dfrac{Ehw}{L}$
k_{yy}	$\dfrac{1}{4}\dfrac{Ehw^3}{L^3}$	$\dfrac{Ehw^3}{L^3}$	$16\dfrac{Ehw^3}{L^3}$
k_{zz}	$\dfrac{1}{4}\dfrac{Eh^3w}{L^3}$	$\dfrac{Eh^3w}{L^3}$	$16\dfrac{Eh^3w}{L^3}$

micromachining processes can produce 1-μm-wide by 2-μm-thick beams, corresponding to $k_z/k_y = 4$.

The distributed mass along the beam's length may be modeled as effective inertial forces lumped at the ends of the beam. A general effective inertial force is found by assuming that the static mode shapes of the beam, given by Eqs. (6) and (9), are sufficiently accurate approximations for the mode shapes of velocity and acceleration under dynamic conditions. With this assumption, the resulting effective inertial forces are

$$\begin{bmatrix} F_{xa,m} & F_{ya,m} & M_{za,m} & F_{xb,m} & F_{yb,m} & M_{zb,m} \end{bmatrix}^T$$
$$= \rho ALS \begin{bmatrix} \ddot{x}_a & \ddot{y}_a & \ddot{\theta}_a & \ddot{x}_b & \ddot{y}_b & \ddot{\theta}_b \end{bmatrix}^T. \tag{13}$$

The weighting matrix, **S**, is calculated directly from the mode shapes:

$$\mathbf{S} = \int_0^1 \mathbf{a}^T \mathbf{a} \, d\xi, \tag{14}$$

where **a** is the matrix combining shapes from Eqs. (6) and (9).[8] Integrating Eq. (14), the mass matrix weights are

$$\mathbf{S} = \begin{bmatrix} \frac{1}{3} & 0 & 0 & \frac{1}{6} & 0 & 0 \\ 0 & \frac{13}{35} & \frac{11}{210}L & 0 & \frac{9}{70} & -\frac{13}{420}L \\ 0 & \frac{11}{210}L & \frac{1}{105}L^2 & 0 & \frac{13}{420}L & -\frac{1}{140}L^2 \\ \frac{1}{6} & 0 & 0 & \frac{1}{3} & 0 & 0 \\ 0 & \frac{9}{70} & \frac{13}{420}L & 0 & \frac{13}{35} & -\frac{11}{210}L \\ 0 & -\frac{13}{420}L & -\frac{1}{140}L^2 & 0 & -\frac{11}{210}L & \frac{1}{105}L^2 \end{bmatrix}. \quad (15)$$

Viscous damping is also distributed along the beam's length and is modeled as effective damping forces in an identical way to the effective inertial force. If Couette damping is present in a gap, g, between the beam and substrate, then lateral damping forces are

$$\begin{bmatrix} F_{xa,b} & F_{ya,b} & M_{za,b} & F_{xb,b} & F_{yb,b} & M_{zb,b} \end{bmatrix}^T$$
$$= \frac{\eta L w}{g} \mathbf{S} \begin{bmatrix} \ddot{x}_a & \ddot{y}_a & \ddot{\theta}_a & \ddot{x}_b & \ddot{y}_b & \ddot{\theta}_b \end{bmatrix}^T. \quad (16)$$

4.2.3 Electrostatic Gaps

Micromechanical devices are at such a small scale that the electromechanical effects are quasistatic; that is, the behavior at any instant in time is accurately found from static electric field solutions. Electrostatic elements within a microsystem always assume a dual role as actuators and sensors, although usually only one role is intended. Electrostatic elements are referred to as actuators in most of this discussion. Actuators may be made from two or more electrodes; however, only two electrodes are intentionally used in most designs. The charge on an electrode in a system with n electrodes is related to the voltages on the electrodes by

$$q_k = \sum_{l=1}^{n-1} C_{kl} V_l, \qquad (17)$$

where C_{kl} are coefficients of the capacitance matrix, and the voltage reference is taken as the nth electrode. Off-diagonal terms have negative values because a positive voltage on one electrode creates negative charge on another electrode. A time-varying charge on an actuator electrode gives rise to a displacement current,

$$i_k = \frac{dq_k}{dt} = \sum_{l=1}^{n-1} \left(C_{kl} \frac{dV_l}{dt} + V_l \frac{dC_{kl}}{dt} \right). \qquad (18)$$

Electronic detection of this current is the basis for capacitive motion sensing. The first term in the summand is the electrical displacement current arising from time-varying voltage across electrodes, and is proportional to displacement for small motions. The second term is the mechanical displacement current arising from time-varying capacitance of movable micromechanical electrodes and is proportional to velocity for small motions. The predominant term depends on the nature of the applied voltages and electrodes. The electrical displacement current is zero for dc applied voltage and increases with the frequency of the voltage. The mechanical displacement current is zero for static electrodes and increases with the frequency of the mechanical displacement.

Electrostatic force, F_e, acting in an arbitrary direction x, is found from conservation of energy:

$$\frac{dU_e}{dt} = \sum_{k=1}^{n-1} V_k \frac{dq_k}{dt} - F_e \frac{dx}{dt}, \qquad (19)$$

where the total change in stored energy, U_e, in the system is equal to the electrical power into each electrode pair less the rate of mechanical work done by the system. The stored energy is expressed in terms of electrode charge and displacement in the system; however, electrode charge is rarely used to control electrostatic force in practical microsystems. To obtain an expression for force in terms of applied voltages, a co-energy, $\tilde{U}_e$, is defined such that

$$\tilde{U}_e + U_e = \sum_{k=1}^{n-1} V_{kn} q_k. \qquad (20)$$

Substituting Eq. (20) into the differential from of Eq. (19), the expression for differential co-energy is

$$d\tilde{U}_e = \sum_{k=1}^{n-1} q_k dV_k + F_e dx = \sum_{k=1}^{n-1} \frac{\partial \tilde{U}_e}{\partial V_k} dV_k + \frac{\partial \tilde{U}_e}{\partial x} dx. \qquad (21)$$

Comparing coefficients in Eq. (21), the electrostatic force is

$$F_e = \left.\frac{\partial \tilde{U}_e}{\partial x}\right|_{V_{1\ldots n-1}}. \qquad (22)$$

Note that from Eq. (22), the electrostatic force will always act in a direction to maximize the co-energy. The co-energy is determined by integrating Eq. (21). Since the system is assumed to be conservative, the integrals can be performed in any order and will produce the same result. A convenient choice is to integrate over the x displacement first. Then all voltages are zero, so F_e and the integral of $F_e dx$ are zero. Substituting Eq. (17) into Eq. (21) and integrating, the co-energy expression is

$$\tilde{U}_e = \sum_{k=1}^{n-1} \sum_{l=1}^{k} \int_0^{V_k} C_{kl} V_l \, dV_k'. \qquad (23)$$

and subsequent substitution into Eq. (22), produces the relation for electrostatic force,

$$F_e = \sum_{k=1}^{n-1} \sum_{l=1}^{k} \frac{dC_{kl}}{dx} \int_0^{V_k} V_l dV_k' = \sum_{k=1}^{n-1} \sum_{l=1}^{n-1} \frac{1}{2} \frac{dC_{kl}}{dx} V_l V_k. \qquad (24)$$

Equation (24) holds in three dimensions for any generalized electrostatic force (force or moment) acting along a generalized coordinate (translation or rotation). Equations (18) and (2
4) show that the electrical and electromechanical behavior of a microactuator with rigid electrodes is completely modeled by determining the capacitance matrix and its sensitivities to all six degrees of freedom of each electrode.

The primary canonical actuator topology in MEMS is the parallel plate, shown in Fig. 4.4 in a 2-D configuration where the plates are offset in x. A large number of electrostatic microactuator designs can be approximately modeled as a parallel-plate capacitance. Damping for a parallel-

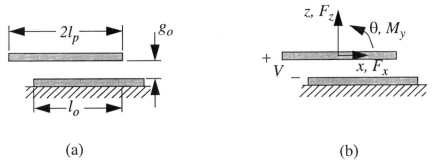

Figure 4.4 Parallel-plate capacitor, with the bottom plate fixed. (a) Cross-section view at the local origin; (b) definition of local origin, electrostatic forces, and applied voltage.

plate air gap was discussed earlier in connection with plate models. Capacitance for an ideal parallel plate is

$$C = \frac{\varepsilon_0 A}{g}, \qquad (25)$$

where $\varepsilon_o = 8.854 \times 10^{-12}$ F/m is the permittivity of space in the gap, g is the gap between the plates, and $A = l_y(l_o + x)$ is the overlapping area of the plates. Fringe fields at the edges of the plates add capacitance; however, that will be neglected in this analysis. The capacitance as a function of generalized local displacement of the top plate is determined by integrating differential capacitance across the overlap area. Assuming partial plate overlap ($-l_o < x < 2l_p - l_o$) and approximating for small angle, θ, the capacitance is

$$C(x,z,\theta) = \frac{\varepsilon_0 l_y}{\theta} \ln \frac{g_0 + z + l_p \theta}{g_0 + z + (l_p - l_0 - x)\theta}, \qquad (26)$$

where g_0 is the gap of rest. Using Eq. (24), the electrostatic generalized forces are

$$\begin{bmatrix} F_x \\ F_z \\ M_y \end{bmatrix} = \varepsilon_0 l_y \frac{V^2}{2} \begin{bmatrix} \dfrac{1}{g_0 + z + (l_p - l_0 - x)\theta} \\[6pt] \dfrac{-(l_0 + x)}{(g_0 + z)^2 + (g_0 + z)(2l_p - l_0 - x)\theta + l_p(l_p - l_0 - x)\theta^2} \\[6pt] \dfrac{1}{\theta}\left(-\dfrac{1}{\theta}\ln\dfrac{g_0 + z + l_p \theta}{g_0 + z + (l_p - l_0 - x)\theta} \right. \\[6pt] \left. + \dfrac{(l_0 + x)(g_0 + z)}{(g_0 + z)^2 + (g_0 + z)(2l_p - l_0 - x)\theta + l_p(l_p - l_0 - x)\theta^2} \right) \end{bmatrix}. \qquad (27)$$

As with most nonlinear relations, linearization of Eq. (27) about the actuator operating point ($x = 0$, $z = 0$, and $\theta = 0$) helps in understanding the physical significance of the forces. Taking the first two terms of a Taylor's series expansion of Eq. (27), the linear approximation of forces for the parallel plate is

$$\begin{bmatrix} F_x \\ F_z \\ M_y \end{bmatrix} \approx \varepsilon_0 l_y \frac{V^2}{2} \left(\begin{bmatrix} \frac{1}{g_0} \\ -\frac{l_0}{g_0^2} \\ -\frac{l_0(2l_p - l_0)}{2g_0^2} \end{bmatrix} + \begin{bmatrix} 0 & -\frac{1}{g_0^2} & \frac{l_0 - l_p}{g_0^2} \\ -\frac{1}{g_0^2} & \frac{2l_0}{g_0^3} & \frac{l_0(2l_p - l_0)}{g_0^3} \\ \frac{l_0 - l_p}{g_0^2} & \frac{l_0(2l_p - l_0)}{g_0^3} & \frac{2l_{x0}}{g_0^3}\left(l_p^2 - l_p l_0 + \frac{l_0^2}{3}\right) \end{bmatrix} \begin{bmatrix} x \\ z \\ \theta \end{bmatrix} \right).$$

(28)

From Eq. (28), F_x is positive, and F_z and M_y are negative when the plate is at its local origin. The forces act on the top plate to increase capacitance, and therefore increase overlap area and decrease the gap. The matrix in Eq. (28) relating forces to displacements is an "electrical stiffness" matrix. The diagonal terms are positive, act to lower the overall stiffness of a system, and lead to unstable behavior when the voltage is sufficiently large. The actuator plates snap together when any of the diagonal electrical stiffness terms equal the corresponding mechanical spring constant constraining the plate. Note that this snap-in behavior can occur in either the z or θ mode. The off-diagonal terms couple motion in the plate into motion in the other coordinates.

A very common set of actuator designs uses interdigitated comb fingers, as shown in Fig. 4.5. An array of N rotor fingers can be roughly approximated as a rigid body connected to an ensemble of $2N$ parallel plates, where the gaps are between the fingers. This approximate model is very useful for quickly evaluating sizing tradeoffs in design; however, it is not accurate for gaps of small width. Conformal mapping techniques

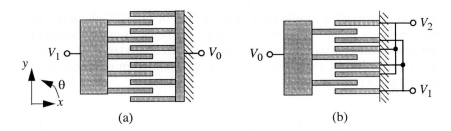

Figure 4.5 Examples of comb finger micromechanical capacitors. (a) Five-finger lateral comb drive; (b) three-finger differential comb sensor.

may be used to form more detailed, physics-based parameterized models.[16,17] Some comb finger designs, such as in Fig. 4.5(b), have more than two electrodes, so multiple capacitances must be determined to calculate the displacement current from Eq. (18) and the electrostatic force from Eq. (24).

Comb finger actuators in surface micromachined processes are often designed with a planar electrode underneath to inhibit a vertical actuation force pulling down to the substrate. The cross section of this configuration is shown in Fig. 4.6. The correct calculation of x-axis force is

$$F_x = \frac{1}{2}\frac{dC_{11}}{dx}(V_1 - V_0)^2 = \frac{1}{2}(C_m + C_1 - C_1')(V_1 - V_0), \quad (29)$$

where C_m, C_1, and C_1' are capacitances per unit length in the x direction, as defined in Fig. 4.6. The capacitance C_{11} of the two-conductor actuator is defined by Eq. (17). The total change in capacitance with x must account for the decrease $(C_1' - C_1)x$ in the capacitance from the stator to

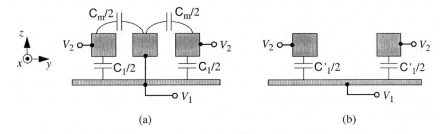

Figure 4.6 Cross-section of a single comb finger actuator with an underlying conducting plane. (a) Capacitances per unit length in the section where the central rotor finger is engaged; (b) capacitances per unit length in the section where the rotor finger is unengaged.

the underlying conducting plane when the rotor finger becomes engaged. For a beam where comb width, thickness, and gap are equal, calculation of comb force from only C_m gives a 44% overestimate. The correct identification of all capacitance terms must be carried out whether the calculations are carried out analytically or through numerical simulation.

4.2.4 Reduced-Order Modeling

Reducing the spatial granularity of the constituent models in a system simulation is important in order to complete simulations in a reasonable time. Model-order reduction techniques are also often called macromodeling. The most common approaches to macromodeling for MEMS are semianalytical modeling and numerical model-order reduction from finite-element and boundary-element formulations.[18] System identification techniques are a third form of model-order reduction; they perform fitting of observed input-output data to a specified linear transfer function, i.e., a set of poles and zeroes. System identification is popular in applied control engineering,[19] and a toolbox covering many techniques is available in Matlab™.[20]

Semianalytical macromodeling begins with an *ad hoc* specification of a set of equations, chosen from basic ordinary differential equations that physically describe the device behavior. The choice of equations demands significant physical insight into the operation of the device being modeled. In most cases, coefficients in the equations are introduced and fit to numerical simulations to provide increased accuracy. One advantage of this approach is that it is possible to embed physically meaningful parameters into the model and maintain acceptable accuracy. Another advantage is the ability to model nonlinear behavior by simply specifying nonlinear equations.

Elastic behavior of flexures made from multiple beams is one example where semianalytic modeling is very accurate. If the higher order modes of a flexure are deemed negligible in the system dynamics, then force-displacement characteristics of flexures made from multiple beams may be lumped into a single stiffness matrix, similar to Eq. (12). Instead of solving Eq. (8) for each beam in the flexure and matching boundary conditions, however, Castigliano's theorem is readily used to determine the stiffness matrix. Castigliano's theorem states that, for in-plane force-displacement relations,

$$x = \frac{\partial U^*}{\partial F_x}\bigg|_{F_y,M_z} ; \; y = \frac{\partial U^*}{\partial F_y}\bigg|_{F_x,M_z} ; \; \theta = \frac{\partial U^*}{\partial M_z}\bigg|_{F_x,F_y}, \tag{30}$$

where U^* is the complementary strain energy (co-energy) of the flexure. If the strain energy is assumed to be due only to linear effects, then the total strain energy, U, in the flexure is equal to the co-energy. For in-plane linear bending, the co-energy is given by

$$U^* = U = \sum_{i=1}^{n} \int_0^1 \frac{M_{z,i}^2}{2EI_{z,i}} d\xi, \tag{31}$$

where ξ is the normalized displacement along each beam segment in the flexure. This method can provide linear stiffness coefficients for a given spring topology as analytic equations of geometric parameters, which simplifies the evaluation of system performance for different spring sizing.

As an example application of Castigliano's theorem, the stiffness coefficients for the crab-leg spring in Fig. 4.7(a) will be solved. Arbitrary forces F_{xo} and F_{yo} and moment M_{zo} are applied at the load end of the spring. First, the moments along each beam segment in the spring are determined by evaluating the free-body diagram of the spring in Fig. 4.7(b). For static displacement, the sum of forces as well as the sum of moments acting at the ends of each beam segment must equal zero. The moments are linear functions of the local normalized coordinates, ξ_i, along each beam segment and are found from inspection of the free-body diagram.

$$M_{z,1} = M_{zo} + F_{xo} L_a \xi_1 \tag{32}$$

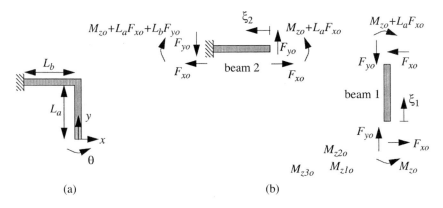

Figure 4.7 A single crab-leg flexure. (a) Layout; (b) free-body diagram with the flexure partitioned into simple beams.

$$M_{z,2} = M_{zo} + L_o F_{xo} + F_{yo} L_b \xi_2 \tag{33}$$

The spring flexibility matrix is found by substituting Eqs. (32) and (33) into Eq. (31), then solving Eq. (30). Inverting the flexibility matrix yields the stiffness matrix relating forces to displacements:

$$\begin{bmatrix} F_{xo} \\ F_{yo} \\ M_{zo} \end{bmatrix} = \frac{EI_z}{L_a^3 L_b^3 (L_a + L_b)} \times \begin{bmatrix} 3L_b^3(4L_a + L_b) & -9L_a^2 L_b^2 & -3L_a L_b^3(2L_a + L_b) \\ -9L_a^2 L_b^2 & 3L_a^3(L_a + 4L_b) & 3L_a^3 L_b^2 \\ -3L_a L_b^3(2L_a + L_b) & 3L_a^3 L_b^2 & L_a^2 L_b^2(4L_a + 3L_b) \end{bmatrix} \begin{bmatrix} x \\ y \\ \theta \end{bmatrix}, \tag{34}$$

where the moments of inertia for all beam segments are assumed to be equal. Section 4.3 describes how the stiffness matrix of the crab-leg spring is incorporated into the system's equations of motion.

Effects of local and surrounding electrode geometry on capacitance make macromodeling of electrostatic actuators more difficult than modeling of elastostatic beams. A set of basis functions for the electrode shapes must be assumed to build the capacitance model. Most microactuators are designed with stiff electrodes, relative to beam flexures elsewhere in the system. In these cases, electrodes can be considered rigid, and the appropriate basis functions for local macromodeling of the actuator are the six-DOF relative orientations of the electrode surfaces. An example of this is the parallel-plate actuator analysis described earlier. A prominent counterexample is a micromechanical RF relay using a fixed-fixed beam as an actuator pull-down electrode. The relay electrode is not rigid; it bends. Selected static bending modes of the electrode structure (i.e., the fixed-fixed beam) may be chosen as the basis functions.[21,22] As is the case with the rigid-body shape functions, the lumped generalized displacement variables are the amplitudes of the modal basis functions.

Energy contributions in the electric fields usually extend far away from an electrostatic actuator. Therefore, when an actuator is incorporated in a system, the surrounding structures will act as additional electrodes that may need to be included in the model. More electrodes add terms to the current and force calculations, and may alter the capacitance coefficients associated with the original actuator electrodes. In composable models, the assumption is made that all possible electrodes are accounted in the model, and that the designer is not placing other conducting structures nearby. However, even small perturbations in current or force can be important to system performance, so guidelines depend on the application. In MEMS made from a single planar mechanical layer, this design constraint translates into a minimum spacing between actuators before they can be modeled as independent elements. Capacitances decay as $1/r$ and electrostatic forces decay as $1/r^2$ in the surrounding volume. As a guideline, it is prudent to combine the modeling of actuators spaced less than ten times the minimum actuator gap if 1% modeling accuracy in force is desired. Attention to capacitive coupling is of primary importance on high-impedance sensing nodes and may require modeling of electrode effects from longer distances.

Exact analytic expressions for capacitance as a function of displacement exist only for a handful of canonical problems. For many MEMS processes and designs, the electrode geometry is very complex, and numerical simulation must be used to model the electric field distribution. Curve fitting to the numerical simulation data is the only recourse for forming accurate capacitance functions of displacement variables. The function may be based on physics, such as an ideal parallel-plate capacitor formula with fitting parameters, or a more complex analytic formula found with conformal mapping. A more general fitting technique uses a rational fraction of multivariate polynomials of displacement variables:[21]

$$C_{kl}(\mathbf{q}) = \frac{\sum_{i_1=0}^{R_1}\sum_{i_2=0}^{R_2}\cdots\sum_{i_m=0}^{R_m} a_{i_1 i_2 \ldots i_m} q_1^{i_1} q_2^{i_2} \cdots q_m^{i_m}}{\sum_{i_1=0}^{S_1}\sum_{i_2=0}^{S_2}\cdots\sum_{i_m=0}^{S_m} b_{i_1 i_2 \ldots i_m} q_1^{i_1} q_2^{i_2} \cdots q_m^{i_m}}, \qquad (35)$$

where the terms $q_1, \ldots, q_m$ are the generalized displacements, a and b are the fitting coefficients, and $R_1, \ldots, R_m$ and $S_1, \ldots, S_m$ are the polynomial orders chosen for the fitting function. The Levenberg-Marquardt algorithm[23] is used to fit the coefficients in Eq. (35) to numerical simulation data. Higher order polynomials with hundreds of fitting coefficients may be necessary to achieve adequate accuracy over the displacement space. In Gabbay et al.,[21] an asymmetrically suspended plate with an electrostatic pull-down actuator was modeled with five modal basis functions and 863 fitting coefficients, using 250 boundary-element simulations to determine capacitance data points for the fit. Although the resulting equations may be very large, the explicit function of capacitance and force in terms of displacements is ideal for time-domain, system-level simulation. Automated macromodeling algorithms have been combined with some commercial numerical analysis tools.[24,25]

In principle, general parameterized capacitance models for design purposes can be determined by expressing the fitting coefficients, a and b, as rational fractions of multivariate polynomials of electrode geometric parameters. However, construction of this form of parameterized model for design will require tens of thousands of numerical simulations. Difficulty also arises in nonlinear modeling where no general method has been found to determine basis functions automatically. A method for finding basis functions that account specifically for beam stiffening nonlinearity in the modeling of distributed electrostatic actuators is given in Mehner *et al.*[22] For other cases, an appropriate basis function can be determined by reverting to physics-based approaches.

It is not possible, at least at present, to create completely general actuator models that can be composed hierarchically into larger systems. Nevertheless, the parallel-plate and comb-finger actuators are specific geometries commonly used in microsystem design, and are therefore worth modeling parametrically for design reuse. Less common geometries are best modeled directly with specific geometry values to determine capacitance and sensitivities as functions of displacement.

4.3 Formulation of Equations of Motion

The intent in design of most microelectromechanical devices is to allow motion of certain parts of a structure in a specific direction, while inhibiting other motions. In actuator applications, such as RF relays and scanning mirrors, forces are applied by microactuators designed into the system. In sensor applications, such as accelerometers and microphones, forces arise from transduction of physical events external to the system. One of the crucial steps in MEMS design is to identify the relationship between dynamic forces and displacements, known as the equations of motion.

Micromechanical components can be partitioned by function into multiple discrete spring, mass, and damper elements. Electrostatic actuators and capacitive sensors add active gap elements, which produce external force. In general, each lumped mass has six degrees of freedom: translational DOFs x, y, z and rotational DOFs θ_x, θ_y, and θ_z. A mechanical system with n degrees of freedom can be described in terms of n generalized coordinates, $q_1, q_2, ..., q_n$, and time, t. The generalized coordinates are usually expressed directly as the DOFs of the masses, but may be expressed as some other linear basis of these DOFs. A general method of determining the equation of motion for each DOF involves use of Lagrange's equation:

$$\frac{d}{dt}\left(\frac{\partial L}{\partial \dot{q}_i}\right) - \frac{\partial L}{\partial q_i} = Q_{ext,i} - \frac{\partial D}{\partial \dot{q}_i}; i = 1,\ldots,n, \qquad (36)$$

where $L = T - V$ is the Lagrangian operator, T is the total kinetic energy of the system, and V is the potential energy of the system, arising from conservative forces. A derivation of Lagrange's equations from the principle of conservation of energy is given in Meirovitch.[26] The two terms on the right-hand side of Eq. (36) are associated with nonconservative forces. External generalized force (i.e., force or moment) along the ith DOF is given by $Q_{ext,i}$. Viscous damping terms (damping proportional to velocity) are represented by the Rayleigh dissipation function, D. In lumped systems, the kinetic energy, potential energy, and dissipation function have the forms:

$$T = \frac{1}{2}\sum_{i=1}^{n}\sum_{j=1}^{n} m_{ij}\dot{q}_i\dot{q}_j, \qquad (37)$$

$$V = \frac{1}{2}\sum_{i=1}^{n}\sum_{j=1}^{n} k_{ij}q_i q_j, \qquad (38)$$

$$D = \frac{1}{2}\sum_{i=1}^{n}\sum_{j=1}^{n} b_{ij}\dot{q}_i\dot{q}_j, \qquad (39)$$

where m_{ij} are inertia coefficients, k_{ij} are stiffness coefficients, and b_{ij} are damping coefficients of the system. The inertia, stiffness, and damping coefficients can be alternatively represented by matrices **m**, **k**, and **b**, respectively. The generalized coordinates and external forces can be represented by vectors **q** and $\mathbf{Q}_{ext}$, respectively. Substituting Eqs. (37) through (39) into Eq. (36) yields the equations of motion:

$$\mathbf{m}\ddot{\mathbf{q}} + \mathbf{b}\dot{\mathbf{q}} + \mathbf{k}\mathbf{q} = \mathbf{Q}_{ext}. \qquad (40)$$

The conservative energy functions in Eqs. (37) and (38) are invariant with coordinate transformations, and the dissipation function is assumed invariant. Therefore, it is usually easiest to find these functions as a sum of the energy of individual elements in their local coordinates. The functions can then be expressed in terms of the generalized variables through coordinate transformations.

Every beam or plate in the micromechanical system has a nonzero value of mass, stiffness, and damping. However, specific properties of each element are often assumed dominant in a first cut at analytical system analysis to simplify derivations. If analytic analysis is undertaken, a comparison of results to numerical analysis should be performed to verify that all important attributes of the system's elements are included.

The micromechanical resonator shown in Fig. 4.8 is a canonical problem to illustrate formation of the equations of motion. The resonator is a plate mass suspended by beam springs and with electrostatic parallel-plate actuators on the left and right sides of the plate. The microresonator may be partitioned into lumped elements as shown in Fig. 4.8(c). For brevity, only in-plane motion is considered, so the generalized coordinates are $q_1 = x$, $q_2 = y$, and $q_3 = \theta_z$ of the central resonator plate.

The central resonator plate has kinetic energy

$$T = \frac{1}{2} m\dot{x}^2 + \frac{1}{2} m\dot{y}^2 + \frac{1}{2} I_z \dot{\theta}_z^2, \tag{41}$$

The coordinates in the chip (i.e., the on-chip system) frame of reference are used in Eq. (41), since the chip is assumed not to be moving in an inertial frame. If the chip is moving, for example in accelerometer or vibratory-rate gyroscope applications, the kinetic energy is initially expressed in the coordinates of the inertial frame. In this first analysis, the comb elements attached to the central plate are assumed to be part of the rigid body mass and can be lumped together. If other lumped mass elements in the system are considered, additional generalized coordinates are defined, adding extra DOFs and extra kinetic energy terms. In the present case, the inertial coefficients are written as the mass matrix given in

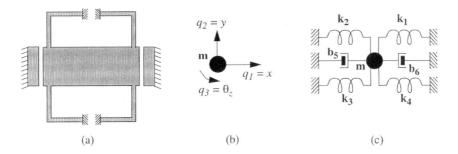

Figure 4.8 A lateral microresonator with two parallel-plate electrostatic actuators. (a) Layout; (b) definition of in-plane generalized coordinates; (c) functional system schematic of dominant effects.

Eq. (1). For any lumped system, the mass matrix will always be diagonal if generalized coordinates are chosen as the translational and rotational displacements of the rigid body masses.

The beam flexures connected to the central plate have symmetric stiffness matrices. The matrix for the rth flexure, $\mathbf{k}_r$, relates generalized forces to displacements in the local coordinate system defined in Fig. 4.9(a). The local stiffness matrix for the crab-leg flexure is given by Eq. (34). In the microresonator problem, all local stiffness matrices have identical coefficient values by design; however, the actual values may differ between flexures due to manufacturing variation of geometry and of material properties.

The spring forces in Eq. (34) are not expressed in terms of the system's generalized coordinates. Although it is possible to calculate the spring force-displacement relations for the system directly, Lagrange's equation provides a relatively simple method for finding the stiffness coefficients that is usually less error prone. The mirror symmetry of the flexures about the x axis and y axis provides the geometric relations between generalized and local flexure coordinates,

$$\begin{bmatrix} x_1 \\ y_1 \\ \theta_1 \end{bmatrix} = \begin{bmatrix} x - l_y\theta \\ y + l_x\theta \\ \theta \end{bmatrix}; \begin{bmatrix} x_2 \\ y_2 \\ \theta_2 \end{bmatrix} = \begin{bmatrix} -x + l_y\theta \\ y - l_x\theta \\ -\theta \end{bmatrix}; \begin{bmatrix} x_3 \\ y_3 \\ \theta_3 \end{bmatrix} = \begin{bmatrix} -x - l_y\theta \\ -y + l_x\theta \\ \theta \end{bmatrix}; \begin{bmatrix} x_4 \\ y_4 \\ \theta_4 \end{bmatrix} = \begin{bmatrix} x + l_y\theta \\ -y - l_x\theta \\ -\theta \end{bmatrix},$$

(42)

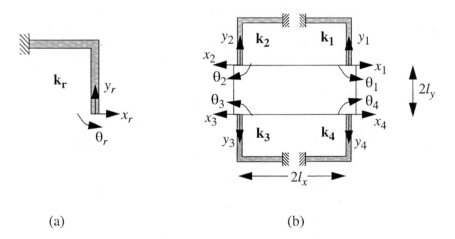

(a)   (b)

Figure 4.9 Layout and local coordinates of the crab-leg flexure, where r = 1,2,3,4. (a) Layout of a flexure element; (b) mirror symmetric placement of the four flexures in the microresonator system.

where the local coordinates are defined in Fig. 4.9(b). The local x and y displacements are related through trigonometric relations to the rotation. The displacements in Eq. (42) are linearized around $\theta = 0$, assuming small rotation. The total mechanical potential energy is

$$V = \frac{1}{2} \sum_{r=1}^{4} \begin{bmatrix} x_r \\ y_r \\ \theta_r \end{bmatrix}^T \begin{bmatrix} k_{xxr} & k_{xyr} & k_{x\theta r} \\ k_{xyr} & k_{yyr} & k_{y\theta r} \\ k_{x\theta r} & k_{y\theta r} & k_{\theta\theta r} \end{bmatrix} \begin{bmatrix} x_r \\ y_r \\ \theta_r \end{bmatrix}. \tag{43}$$

Note that it is straightforward to describe the potential energy first in terms of the local coordinates of the flexures. Substituting Eq. (42) into (43), the system stiffness coefficients, k_{ij}, in Eq. (38), are

$$k_{11} = k_{xx1} + k_{xx2} + k_{xx3} + k_{xx4} \tag{44}$$

$$k_{12} = k_{xy1} - k_{xy2} + k_{xy3} - k_{xy4} \tag{45}$$

$$k_{13} = k_{x\theta 1} + k_{x\theta 2} - k_{x\theta 3} - k_{x\theta 4} + l_y(-k_{xx1} - k_{xx2} + k_{xx3} + k_{xx4}) \\ + l_x(k_{xy1} + k_{xy2} - k_{xy3} - k_{xy4}) \tag{46}$$

$$k_{22} = k_{yy1} + k_{yy2} + k_{yy3} + k_{yy4}$$
$$k_{23} = k_{y\theta 1} + k_{y\theta 2} - k_{y\theta 3} - k_{y\theta 4} + l_x(k_{yy1} - k_{yy2} - k_{yy3} + k_{yy4}) \tag{47}$$
$$+ l_y(-k_{xy1} + k_{xy2} + k_{xy3} - k_{xy4})$$

$$k_{33} = k_{\theta\theta 1} + k_{\theta\theta 2} + k_{\theta\theta 3} + k_{\theta\theta 4} + l_y^2 k_{11} + l_x^2 k_{22} - 2l_y(k_{x\theta 1} + k_{x\theta 2} + k_{x\theta 3} + k_{x\theta 4})$$
$$+ 2l_x(k_{y\theta 1} + k_{y\theta 2} + k_{y\theta 3} + k_{y\theta 4}) - 2l_x l_y(k_{xy1} + k_{xy2} + k_{xy3} + k_{xy4}). \tag{48}$$

When the four flexures are perfectly matched, all the off-diagonal system stiffness coefficients are zero. Any flexure topology connected to a plate mass with mirror symmetry about the center of mass will exhibit this property. Sensitivities to the system stiffness coefficients to mismatch in the flexure stiffnesses are found directly from Eqs. (44) through (48). Nonzero off-diagonal stiffness coefficients will arise from gradients across the device in manufacturing variations, for example from beam width variations.

In the microresonator example, the narrow gap between the parallel plates creates significant squeeze-film damping in the x direction, given by Eq. (3). Couette damping in Eq. (5) accounts for the dissipation in the y direction. Rotational damping arises from both squeeze and Couette contributions. The resulting simplified dissipation function is

$$D = \frac{1}{2} \sum_{r=5}^{6} \begin{bmatrix} x_r \\ y_r \\ \theta_r \end{bmatrix}^T \begin{bmatrix} b_{xxr} & 0 & 0 \\ 0 & b_{yyr} & 0 \\ 0 & 0 & b_{\theta\theta r} \end{bmatrix} \begin{bmatrix} x_r \\ y_r \\ \theta_r \end{bmatrix}. \quad (49)$$

Damping for each comb element is lumped, and from the layout is seen to be symmetric about its local origin, so off-diagonal terms are zero. However, the details of off-axis terms must be included if effects of manufacturing variations across a device are to be analyzed. The local coordinates of the comb elements have indices numbered 5 and 6, as indicated in Fig. 4.8(b), and are related to the generalized coordinates by

$$\begin{bmatrix} x_5 \\ y_5 \\ \theta_5 \end{bmatrix} = \begin{bmatrix} x \\ y - l_{xc}\theta \\ \theta \end{bmatrix} \text{ and } \begin{bmatrix} x_6 \\ y_6 \\ \theta_6 \end{bmatrix} = \begin{bmatrix} x \\ y + l_{xc}\theta \\ \theta \end{bmatrix}, \quad (50)$$

where the center of each comb element is positioned a distance l_{xc} along the x axis. The resulting system dissipation coefficients, b_{ij}, in Eq. (39) are

$$\mathbf{b} = \begin{bmatrix} b_{xx5} + b_{xx6} & 0 & 0 \\ 0 & b_{yy5} + b_{yy6} & l_{xc}(-b_{yy5} + b_{yy6}) \\ 0 & l_{xc}(-b_{yy5} + b_{yy6}) & b_{\theta\theta 5} + b_{\theta\theta 6} + l_{xc}^2(b_{yy5} + b_{yy6}) \end{bmatrix}, \quad (51)$$

where small angular displacements are assumed. Off-diagonal terms in the damping matrix are zero, since the actuators are symmetrically placed about the system center of mass, and assuming the gaps are identical.

Time-varying capacitance [Eq. (26)] of the actuators gives rise to displacement currents, given by Eq. (18). The voltages across the gap are specified by external electrical interconnections. The external generalized forces generated by the actuators are

$$\mathbf{Q}_{nc} = \begin{bmatrix} F_{x5} & F_{x6} & 0 & 0 \end{bmatrix}^T, \qquad (52)$$

where only an x-directed electrostatic force is generated by each actuator due to symmetry. The equations for the parallel-plate electrostatic force are given in Eq. (27), where the z local actuator force is transformed into $-x$ and x in the system for the left and right actuators, respectively.

The system equations of motion for the resonator are found by substituting Eqs. (41), (44) through (48), (51), and (52) into Eq. (40). Extending the equations to include out-of-plane motion is straightforward and adds a considerable number of terms to the equations. These equations of motion provide a first-order analysis of the microresonator system. This level of analysis may be sufficient for design; however, additional effects that are important in many designs include distributed mass and damping in the beams and bending of the plate. Critical design specifications should be verified through numerical simulation. Verification of an entire system may be extremely difficult, especially for systems with several functional elements that span multiple energy domains. Numerical simulation of smaller parts of the system for partial verification is often a practical substitute.

4.4 Structured Design Tools

Mixed-energy-domain system simulation is the process of solving the ordinary differential equations (ODEs) that describe the microsystem. At each time step, the process involves formulation of the ODEs, formation of a set of nonlinear algebraic equations using implicit integration methods, generation of a sparse set of linear algebraic equations using iterative methods, and application of matrix solution techniques. Simulators provide absolute and relative tolerance parameters that specify when the iterative methods for each time step have converged to sufficient accuracy and can be terminated. Integrated design environments that support simulation include schematic capture for defining system structure; one or more simulators for dc, ac, and transient analysis; and tools for plotting and post-processing of the simulated data.

As described earlier, the system ODEs may be formed directly as part of the behavioral modeling process. Further formulation of the ODEs is performed by defining structure through interconnected behavioral models.

These structural interconnections have two primary formulations: signal flow graphs and conservative networks. Bondgraphs are a third formulation that graphically represent flow of energy through systems;[27] they will be not considered further here.

4.4.1 Signal-Flow Simulations

Signal-flow simulations are represented by partitioning the system into blocks with input and output ports, related by a behavioral model. A single signal value is associated with each interconnection node in the system. An input port connected to a node reads the value of that node; an output port connected to a node sets the value of that node. The lumped models and equations of motion derived in Sections 4.2 and 4.3 can be readily implemented in signal-flow form. A popular tool for creating and simulating signal-flow diagrams is Matlab™ and its graphical user interface toolbox, Simulink™.[28]

4.4.2 Conservative Network Simulations

Conservative networks have two values associated with each port connected to components in the system. The value of a node connected to a port is called the *across* variable or potential. All potentials connected to a given node are equal, so the potential around any interconnected loop in the system equals zero. The second value associated with a port is called the *through* variable, or flow. In the simulator, flow is conserved such that the sum of all flows into ports connected to the same node equals zero. In addition to iteration tolerances, absolute and relative tolerances are adjustable for each type of across and through variable in most simulators.

Electrical circuit simulators formulate ODEs as conservative networks that enforce Kirchhoff's current and voltage laws. Current is the flow variable and voltage is the potential variable. Spice, a general-purpose circuit simulation program originally developed at the University of California at Berkeley in 1971, is the *de facto* standard for simulation of analog electronic circuits. Spice has built-in behavioral models for electronic components, such as resistors, inductors, capacitors, transistors, voltage sources, and current sources. These models, especially transistor models, have been refined over the years and achieve high accuracy in dc, ac, and transient simulation. Most versions do not provide the ability for the user to add new behavioral models, as these are compiled into the code. Structural descriptions are supported through subcircuit modules that can be instantiated multiple times within a circuit. Although there are many extensions to the capabilities within Spice, its ability as a tool to design arbitrary physical systems is limited, and at best awkward. A primary drawback is the availability of only electrical components. However, microsystems can be simulated in Spice through translation of nonelectrical physical behavior to an equivalent electrical model, then simulated voltages and currents can be interpreted in the original physical domain. Examples of MEMS modeling and simulation in Spice include electromechanical microsystems[29,30] and electrothermal microsystems.[31,32]

In the electromechanical case, lumped equations of motion are modeled in Spice as analogous L-C-R electrical circuits. Either the series-connected or parallel-connected LCR circuit can model each mode of the second-order mechanical system; the series-connected LCR equivalent circuit for a single mass-spring-damper is shown in Fig. 4.10. Voltages represent forces, currents represent velocities, and charges represent displacements. Each mechanical parameter is represented by an equivalent electrical parameter.

$$L\{m\} = Am \tag{53}$$

$$R\{b\} = Ab \tag{54}$$

$$C\{k\} = \frac{1}{Ak} \tag{55}$$

$$V_F\{F_z\} = AF_z \tag{56}$$

$$V_c = Akz \tag{57}$$

$$V_z\{z\} = Az \tag{58}$$

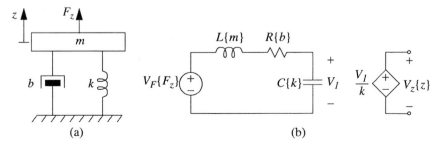

Figure 4.10 Electrical analogy to a 1-DOF second-order mechanical system. (a) Mechanical mass-spring-damper; (b) equivalent LCR circuit.

The scaling factor, A, keeps the electrical impedance values of the LCR circuit from being very small, thereby avoiding convergence problems in Spice. A useful value for A is 10^6 for micromechanical components, which translates volts to micronewtons, and amps to microns/second.

Electrostatic actuator models can be made using nonlinear dependent sources, available in most versions of Spice, as shown in Fig. 4.11. Dependent voltage sources define the capacitance and force, for example, using relations given in Eqs. (26) and (27) for a parallel-plate actuator. The current through the capacitor, C_s, provides the derivative of the charge on the modeled time-varying capacitance. The value of C_s is a normalizing factor, chosen to be 1 pF to increase the impedance of the equivalent circuit. Additional dependent voltage sources are used to calculate forces from off-diagonal coupling coefficients in the system equations of motion.

4.4.3 Analog Hardware Description Languages

Analog hardware description languages (AHDLs) combine the ability to build larger components hierarchically using smaller components with the ability to model component behavior directly in any physical domain. These languages support both signal flow and conservative network formulations. The creation of AHDLs was inspired by the impressive success of digital hardware description languages in linking architectural system descriptions in software through multiple levels of abstraction to the physical transistor-level descriptions. The intent of AHDLs is to manage design complexity of mixed-domain systems by enabling both top-down and bottom-up system design, and to encourage reuse of design intellectual property through standardized interoperable model descriptions.

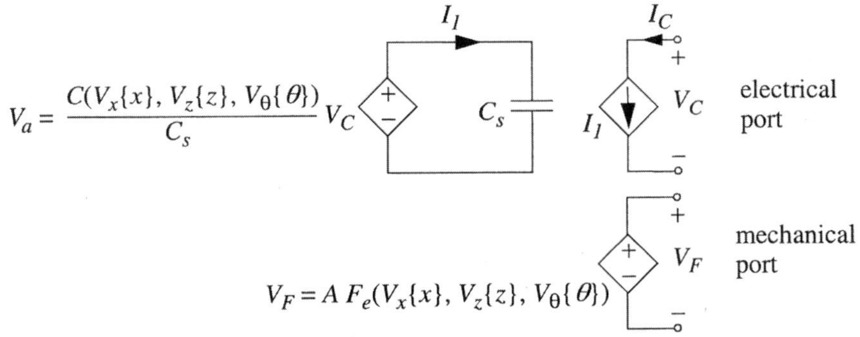

$$V_a = \frac{C(V_x\{x\}, V_z\{z\}, V_\theta\{\theta\})}{C_s}$$

$$V_F = A\, F_e(V_x\{x\}, V_z\{z\}, V_\theta\{\theta\})$$

Figure 4.11 Electrostatic actuator model in Spice.

Several commercial implementations of AHDLs exist, including MAST, supported by the Avant! Saber™ simulator;[33,34] HDL-A, supported by the Mentor Graphics Eldo™ simulator;[35] and Verilog-A,[36] supported by Cadence Design Systems Spectre™ simulator.[37] Groups are refining specifications of single-language solutions for mixed digital and analog applications (i.e., all physical energy domains). The two competing single-language standards are Verilog-AMS (analog mixed-signal)[38,39] and IEEE standard 1076.1-1999, also known as VHDL-AMS.[40] Syntaxes of the various AHDLs are different, but the architectures of the languages are very similar. As MEMS becomes more integrated with electronics, the role of these analog mixed-signal languages is expected to become increasingly important and to provide greater interoperability among vendors.

Examples of general MEMS behavioral modeling using AHDLs is given by Romanowicz.[41] In AHDLs, types of across and through variables are described in a "nature" definition, which includes the physical units, the mnemonic of the access function used within the code, and the absolute tolerance for the variable within the simulation. A grouping of two natures, one through variable and one across variable, is defined as a *discipline*. Standard disciplines within Verilog-A are encoded within an include file, and are summarized in Table 4.2. The information in parentheses is added to better indicate the physical intent of the variables within a simulation. Users are free to add or change the disciplines or natures; however, models and simulations developed with custom natures and disciplines are likely to be incompatible with those developed elsewhere.

Table 4.2 Verilog-A Standard Definitions for Disciplines and Natures

Discipline	Across (Flow) Nature	Through (Potential) Nature	Across Access Function	Through Access Function
Electrical	Voltage	Current	V	I
Kinematic (translational)	Position (or translational displacement)	Force	Pos	F
Rotational	Angle (or rotational displacement)	Angular_Force (moment or torque)	Theta	Tau
Magnetic	Magneto_Motive_Force	Flux (magnetic flux)	MMF	Phi
Thermal	Temperature	Power (heat flux)	Temp	Pwr

Certain choices of through and across variables can lead to convergence problems.[42] There are no comprehensive standards for disciplines and natures of MEMS at the present time.

4.4.4 A Structured MEMS Methodology

Several academic groups have independently developed a MEMS design methodology where a reusable set of components is used to construct ODEs from structure. Sugar, developed at UC Berkeley, incorporates models in a Spice-like syntax with implementation in Matlab.[43-45] Nodas,[46,47] developed at Carnegie Mellon, and Lorenz's work at the University of Bremen[17] implement models in AHDLs. Increasing attention is being given to commercial support of this methodology.[48-51]

The methodology builds on a reusable component library with geometrically parameterized behavioral models, akin to the methodology supported by Spice for circuit design and to the basic lumped element models available in mechanical finite-element analysis packages. At the lowest level of the hierarchy for suspended MEMS are plates, beams, electrostatic gaps, and anchors, which are at a similar structural level to resistors, capacitors, and inductors in electronic design. Additional behavioral macromodels may be defined at any level of the hierarchy. Microsystems lying in this very large and useful design space can be formulated by structure if this finite set of low-level elements is predefined with sufficiently general geometric parameters and material properties. The microresonator filter in Fig. 4.1 is an example of a design that can be partitioned into low-level beam, plate, anchor, and gap components.

The conservative network representation of MEMS with AHDLs given above has three key attributes:

1. One-to-one correspondence with micromechanical layout
2. Ability to construct multilevel behavioral models with interlevel compatibility
3. Interoperability with electronic components

MEMS design is inherently driven by shape, because shape determines function. A one-to-one correspondence of components to layout provides a geometrically intuitive user interface for micromechanical system design. This correspondence to layout enables generation of a computer-assisted layout from a schematic view. Conversely, extraction tools enable automated generation of netlists (MEMS components and interconnect) from a layout definition of the system.[52] A paradigm shift is occurring where MEMS design starts with schematic entry, instead of layout entry. As is the case with analog circuit design, layout extraction then creates an executable simulation netlist that may be compared with the design schematic for verification.

Adoption of consistent terminal conventions allows designers to partition their system into blocks that can be replaced with models that have fewer or more degrees of freedom. At the lowest level, it is possible to include the equivalent of finite-element models directly as ADHL components. New behavioral models may be generated, or new structured cells may be generated by reusing pre-existing components. In this way, a design hierarchy may be defined and traversed in a top-down or bottom-up way, supported by system simulation at every iteration.

The ability to simulate complete microsystem performance with electronics is important to many design efforts. Full system simulation of any on-chip electronic system is essential to verify proper connectivity. Simulation of the interaction of capacitive sensors with on-chip or external electronic interfaces is especially critical.

A simple design of two beams is shown in Fig. 4.12(a) to exemplify the representation used in Nodas. The MEMS schematic in Fig. 4.12(b) is a structural representation of the design incorporating two instantiations of a reusable beam component. The beam component has parameters for width, w, length, L, thickness, t, and layout rotation, $angle$. In-plane displacements (x, y, θ) and voltage, v, are across variables, and forces, torques (F_x, F_y, M_θ), and current, i, are through variables. Fig. 4.12(c) through (e) illustrates the sign convention of the mechanical variables. The x and y across variables are chosen positive along the positive-axis direction, and θ is positive in a counterclockwise rotation (right-hand rule) around the z axis. Positive-valued through variables going *into* a node are interpreted as providing force in the positive-axis direction and providing torque in a counterclockwise rotation around the z axis. Mechanical variables on the a side of the beam symbol are associated with the left side of the physical beam relative to its local origin. Similarly, variables on the b side of the symbol are associated with the right side of the physical beam. Since these associations are tied to the beam's local origin, they are invariant with the layout angle parameter. For example, a through force flowing into the x_b pin and out of the x_a pin represents an axial tensile force acting on the beam. A through force flowing into the x_a pin and out of the x_b pin represents an axial compressive force acting on the beam. A net through force flowing into the x_a and x_b pins represents a net force in the positive x direction, resulting in acceleration of the beam.

A Verilog-A model of a micromechanical beam is shown in Fig. 4.13 as an example of a general interoperable component in the structured design methodology. The terminal variables are declared conservative by using the keyword *inout* and correspond to the variables defined in Fig. 4.12(c). User-

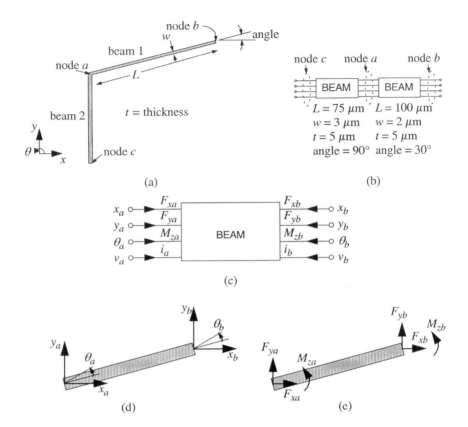

Figure 4.12 Example of a structured representation. (a) Layout of two connected micromechanical beams; (b) structured representation with two lumped-model beam components; (c) expanded view of the beam-1 component, showing across and through variables for in-plane displacement; (d) physical view illustrating the sign convention for the generalized displacements; (e) physical view illustrating the sign convention for the generalized forces and current.

defined parameters for beam length, width, thickness, layout angle, gap from substrate, and bloat dimensions for Couette damping are declared and given default values. Starting at the "analog begin" statement, the behavioral equations implement Eqs. (12), (13), and (16), representing the beam spring, inertial, and damping forces, respectively. This beam model can be instantiated multiple times and interconnected in a general way to formulate the ODEs. A simulation of the system in Fig. 4.1 using this beam model has been verified experimentally.[53]

Electrostatic gap models are best formed through fitting, as in Eq. (35), or through physics-based functions for capacitance and force, as

```
real cos_angle, sin_angle, resistance, mass, Iz, dampx, dampy, dx, dy,
    Fm_xn, Fm_xp, Fm_yn, Fm_yp, Mm_zn, Mm_zp, Fb_xn, Fb_xp, Fb_yn, Fb_yp, Mb_zn,
    Mb_zp, Fk_xn, Fk_xp, Fk_yn, Fk_yp, Mk_zn, Mk_zp;

//Begin behavioral model description
analog begin
    sin_angle = sin(angle*`M_PI/180);
    cos_angle = cos(angle*`M_PI/180);
    resistance = `resistivity*l/w;
    mass = `density*w*l*thickness;
    Iz = thickness*w*w*w/12;
    dampx = `visc_air*l*(w+bloat)/z_air_gap;
    dampy = `visc_air*w*(l+bloat)/z_air_gap;

    //Calculate beam's electrical relation, I=V/R
    I(vp, vn) <+ V(vp, vn)/resistance;

    //Transform displacements into beam's local frame of reference
    dx = cos_angle*Pos(xp, xn) + sin_angle*Pos(yp, yn);
    dy = cos_angle*Pos(yp, yn) - sin_angle*Pos(xp, xn);

    //Differentiate in beam's local frame to obtain velocity and acceleration
    Pos(Vxn) <+ ddt(cos_angle*Pos(xn) + sin_angle*Pos(yn));
    Pos(Vyn) <+ ddt(cos_angle*Pos(yn) - sin_angle*Pos(xn));
    Omega(Vphim) <+ ddt(Theta(phim));
    Pos(Vxp) <+ ddt(cos_angle*Pos(xp) + sin_angle*Pos(yp));
    Pos(Vyp) <+ ddt(cos_angle*Pos(yp) - sin_angle*Pos(xp));
    Omega(Vphip) <+ ddt(Theta(phip));

    //Calculate inertial forces and moments
    Fm_xn = mass/6 * ddt(2*Pos(Vxn)+Pos(Vxp));
    Fm_xp = mass/6 * ddt(Pos(Vxn)+2*Pos(Vxp));
    Fm_yn = mass/420 * ddt(156*Pos(Vyn)+54*Pos(Vyp)+
        l*(22*Omega(Vphim)-13*Omega(Vphip)));
    Fm_yp = mass/420 * ddt(54*Pos(Vyn)+156*Pos(Vyp)+
        l*(13*Omega(Vphim)-22*Omega(Vphip)));
    Mm_zn = mass*l/420 * ddt(22*Pos(Vyn)+13*Pos(Vyp) +
        l*(4*Omega(Vphim)-3*Omega(Vphip)));
    Mm_zp = mass*l/420 * ddt(-13*Pos(Vyn)-22*Pos(Vyp) +
        l*(-3*Omega(Vphim)+4*Omega(Vphip)));

    //Calculate damping forces and moments
    Fb_xn = dampx/6 * (2*Pos(Vxn)+Pos(Vxp));
    Fb_xp = dampx/6 * (Pos(Vxn)+2*Pos(Vxp));
    Fb_yn = dampy/420 * (156*Pos(Vyn)+54*Pos(Vyp)+
        l*(22*Omega(Vphim)-13*Omega(Vphip)));
    Fb_yp = dampy/420 * (54*Pos(Vyn)+156*Pos(Vyp)+
        l*(13*Omega(Vphim)-22*Omega(Vphip)));
    Mb_zn = dampy*l/420 * (22*Pos(Vyn)+13*Pos(Vyp) +
        l*(4*Omega(Vphim)-3*Omega(Vphip)));
    Mb_zp = dampy*l/420 * (-13*Pos(Vyn)-22*Pos(Vyp) +
        l*(-3*Omega(Vphim)+4*Omega(Vphip)));

    //Calculate spring forces and moments
    Fk_xp = `E*thickness*w/l*dx;
    Fk_xn = -Fk_xp;
    Fk_yp = 12*`E*Iz/(l*l*l)*dy - 6*`E*Iz/(l*l)*(Theta(phim)+Theta(phip));
    Fk_yn = -Fk_yp;
    Mk_zp = -6*`E*Iz/(l*l)*dy + `E*Iz/l*(2*Theta(phim) + 4*Theta(phip));
    Mk_zn = -6*`E*Iz/(l*l)*dy + `E*Iz/l*(4*Theta(phim) + 2*Theta(phip));

    //Transform forces back from beam's local frame to chip frame
    //Syntax: F(x) <+ 1; specifies 1N flow out of pin
    F(xn) <+ -((Fm_xn+Fb_xn+Fk_xn)*cos_angle - (Fm_yn+Fb_yn+Fk_yn)*sin_angle);
    F(yn) <+ -((Fm_xn+Fb_xn+Fk_xn)*sin_angle + (Fm_yn+Fb_yn+Fk_yn)*cos_angle);
    F(xp) <+ -((Fm_xp+Fb_xp+Fk_xp)*cos_angle - (Fm_yp+Fb_yp+Fk_yp)*sin_angle);
    F(yp) <+ -((Fm_xp+Fb_xp+Fk_xp)*sin_angle + (Fm_yp+Fb_yp+Fk_yp)*cos_angle);
    Tau(phim) <+ -Mm_zn-Mb_zn-Mk_zn;
    Tau(phip) <+ -Mm_zp-Mb_zp-Mk_zp;
end
endmodule
```

Figure 4.13 Verilog-A model of an in-plane linear beam.

exemplified in Eqs. (26) and (27). However, special attention is required for handling the blow up in capacitance and force in electrostatic gaps during snap-in. One practical technique is to assume the electrodes are covered with an insulating material that is of negligible thickness compared to the gap at rest. This assumption is actually true for polysilicon and silicon MEMS, where about 20 nm of native oxide covers the electrodes. When the gap closes to the point where the insulating layers contact, the first-order contact model generates a reaction force from compression of the electrodes to constrain the beam to close no further. The electrostatic force is essentially held at a constant value while in contact.

4.5 Conclusions

System modeling and simulation of MEMS requires a background knowledge of the underlying physics. For suspended electrostatic MEMS, this physics is primarily elastostatics, kinematics, air damping, and electrostatics. Modeling is the key to useful simulation, and the current state of the art still requires the system designer to act also as the modeler. Automated verification from architectural definition to physical layout is still a long-term goal. However, application of structured design methodologies to MEMS is maturing rapidly. Automated modeling tools simplify the task of reduced-order modeling. The advent of standard analog hardware description languages enables construction of reusable design libraries for MEMS. A key factor in seeing MEMS design mature further is the adoption of standard model interface definitions, which may be *de facto*, as was the case with Spice.

General, accurate Spice models for transistors were developed over two decades; general, accurate MEMS models may require a similarly dedicated effort. The behavior of virtually any additional effects to suspended microsystems can be incorporated in AHDL models in an incremental way over time. Important effects that are of interest include nonlinear mechanics, residual stress gradients, piezoelectric sensing and actuation, piezoresistive sensing, thermal flow, thermal stress, resistive heating, noise, and optical flow. Although not discussed explicitly in this chapter, microfluidics is an application area that will eventually leverage the same kind of MEMS system-level design methodology.

References

1. C. A. Mead and L. A. Conway, *Introduction to VLSI Systems*, Addison-Wesley, Reading, MA (1980)

2. E. K. Antonsson, "Structured design methods for MEMS," *Final Report on NSF Sponsored Workshop on Structured Design Methods for MEMS, Nov. 12–15, 1995*; on-line at *http://design.caltech.edu/NSF_MEMS_Workshop*
3. C. T.-C. Nguyen, "Frequency-selective MEMS for miniaturized low-power communication devices," *IEEE Trans. Microwave Theory and Technique*, 47 (8):1486–1503 (Aug. 1999)
4. S. D. Senturia, *Microsystem Design*, Kluwer Academic Publishers, Boston (2000)
5. A. Nathan and H. Baltes, *Microtransducer CAD: Physical and Computational Aspects*, Springer-Verlag, Wien (1999)
6. CoventorWare MicroFlume webpage, *http://www.coventor.com/software/coventorware/*, Coventor, Inc. 4001 Weston Parkway, Cary, NC 27513 (Nov. 2001)
7. P. Voigt, G. Schrag, and G. Wachutka, "Electrofluidic full-system modelling of a flap valve micropump based on Kirchhoffian network theory," *Sensors and Actuators*, A66(1–3):9–14 (April 1998)
8. J. S. Przemieniecki, *Theory of Matrix Structural Analysis*, McGraw-Hill, New York (1968)
9. H. Reismann and P. S. Pawlik, *Elasticity, Theory and Applications*, John Wiley & Sons, New York (1980)
10. S. P. Timoshenko and J. N. Goodier, *Theory of Elasticity*, 3rd ed., McGraw-Hill, New York (1987)
11. J. J. Blech, "On isothermal squeeze films," *J. Lubrication Technol.*, 105:615–620 (1983)
12. T. Veijola, H. Kuisma, J. Lahdenpera, and T. Ryhanen, "Equivalent-circuit model of the squeezed gas film in a silicon accelerometer," *Sensors and Actuators*, A48:239–248 (1995)
13. Y.-H. Cho, A. P. Pisano, and R. T. Howe, "Viscous damping model for laterally oscillating microstructures," *J. MEMS*, 3(2):81–87 (June 1994)
14. J. M. Gere and S. P. Timoshenko, *Mechanics of Materials*, 4th ed., PWS Publishing Co., Boston (1990)
15. S. Timoshenko, *Strength of Materials, Part I: Elementary Theory and Problems*, 3rd ed., R. E. Krieger Publishing Co., Huntington, NY (1976)
16. W. A. Johnson and L. K. Warne, "Electrophysics of micromechanical comb actuators," *J. MEMS*, 4(1):49–59 (March 1995)
17. G. Lorenz, "Network Simulation of Micromechanical Systems," Ph.D. thesis, University of Bremen, Germany (1999)
18. T. Mukherjee, G. K. Fedder, D. Ramaswamy, and J. White, "Emerging simulation approaches for micromachined devices," *IEEE Transactions on Computer-Aided Design of Integrated Circuits and Systems*, 19(12):1572–1589 (Dec. 2000)
19. L. Ljung, *System Identification—Theory for the User*, 2nd ed., Prentice-Hall, Upper Saddle River, NJ (1999)
20. Matlab System Identification Toolbox, The MathWorks; on-line at *http://www.mathworks.com*
21. L. D. Gabbay, J. E. Mehner, and S. D. Senturia, "Computer-aided generation of nonlinear reduced-order dynamic macromodels—I: Non-stress-stiffened case," *J. MEMS*, 9:262–269 (June 2000)

22. J. E. Mehner, L. D. Gabbay, and S. D. Senturia, "Computer-aided generation of nonlinear reduced-order dynamic macromodels—II: Stress-stiffened motion," *J. MEMS*, 9:270–278 (June 2000)
23. W. H. Press, *Numerical Recipes in C: The Art of Scientific Computing*, 2nd ed., Cambridge University Press, Cambridge, UK (2000)
24. N. R. Swart, S. F. Bart, M. H. Zaman, M. Mariappan, J. R. Gilbert, and D. Murphy, "AutoMM: Automatic generation of dynamic macromodels for MEMS devices," *Proc. IEEE MEMS Conference, 1998*, pp. 178–183
25. B. Affour, P. Nachtergaele, S. Spirkovitch, D. Ostergaard, and M. Gyimesi, "Efficient reduced order modeling for system simulation of micro-electromechanical systems (MEMS) from FEM models," *Proc. SPIE*, 4019:50–54 (2000)
26. L. Meirovitch, *Analytical Methods in Vibrations*, Macmillan Publishing Co., Inc., New York (1967)
27. J. U. Thoma, *Simulation by Bondgraphs*, Springer-Verlag, New York (1990)
28. Matlab Simulink Toolbox, The MathWorks, on-line at *http://www.mathworks.com*
29. H. Tilmans, "Equivalent circuit representation of electromechanical transducers: I. Lumped-parameter systems," *J. Micromech. Microeng.*, 6:157-176 (1996); erratum, 6:359 (1996)
30. H. Tilmans, "Equivalent circuit representation of electromechanical transducers: II. Distributed-parameter systems," *J. Micromech. Microeng.*, 7:285–309 (1997)
31. N. R. Swart and A. Nathan, "Mixed-mode device-circuit simulation of thermal-based microsensors," *Sensors and Materials*, 6:179–192 (1994)
32. H. H. Pham and A. Nathan, "Compact MEMS-SPICE modeling," *Sensors and Materials*, 10:63–75 (1998)
33. I. Getreu, "Behavioral modelling of analog blocks using the SABER simulator," *IEEE Proc. 32nd Midwest Symposium on Circuits and Systems,* Aug. 14–16, 1989, vol. 2, pp. 977–980
34. Saber simulator and MAST modeling language, Avant! Corporation, 46871 Bayside Parkway, Fremont, CA 94538, on-line at *http://www.avanticorp.com* or *http://www.analogy.com*
35. Mentor Graphics Corp., 8005 SW Boeckman Road, Wilsonville, OR 97070, on-line at *http://www.mentorgraphics.com*
36. D. Fitzpatrick and I. Miller, *Analog Behavioral Modeling with the Verilog-A Modeling Language*, Kluwer Academic Publishers, Boston (1998)
37. Cadence Design Systems, Inc., 555 River Oaks Parkway, San Jose, CA 95134, on-line at *http://www.cadence.com*
38. Verilog Analog Mixed-Signal Working Group, on-line at *http://www.eda.org/verilog-ams*
39. K. S. Kundert and O. Zinke, *The Designer's Guide to Verilog-AMS*, Kluwer Academic Publishers, Boston, MA (2004)
40. VHDL Analog Mixed-Signal Working Group, on-line at *http://www.eda.org/vhdl-ams*

41. B. Romanowicz, *Methodology for the Modeling and Simulation of Microsystems*, Kluwer Academic Press, Boston (1998)
42. S. Iyer, Q. Jing, G. K. Fedder, and T. Mukherjee, "Convergence and speed issues in analog HDL model formulation for MEMS," *Proc. Fourth Int. Conf. on Modeling and Simulation of Microsystems Semiconductors, Sensors and Actuators (MSM '01), Hilton Head Island, SC, March 19–21, 2001*, pp. 590–593
43. Z. Bai, D. Bindel, J. Clark, J. Demmel, K. S. J. Pister, and N. Zhou, "New numerical techniques and tools in SUGAR for 3D MEMS simulation," *Proc. Fourth Int. Conf. on Modeling and Simulation of Microsystems, March 27–29, 2000*, pp. 31–34
44. SUGAR: A simulation program for MEMS, UC Berkeley, on-line at *http://www-bsac.EECS.Berkeley.EDU/~cfm/index.html, 2001*
45. N. R. Lo, E. C. Berg, S. R. Quakkelaar, J. N. Simon, M. Tachiki, H. J. Lee and K. S. J. Pister, "Parameterized layout synthesis, extraction, and SPICE simulation for MEMS," *Proc. ISCAS, Atlanta, GA, 1996*, pp. 481–484
46. T. Mukherjee and G. K. Fedder, "Hierarchical mixed-domain circuit simulation, synthesis and extraction methodology for MEMS," *Journal of VLSI Signal Processing Systems for Signal Image and Video Technology*, vol. 21, Kluwer Academic Publishers (July 1999), pp. 233–249
47. G. K. Fedder and Q. Jing, "A hierarchical circuit-level design methodology for microelectromechanical systems," *IEEE Trans. Circuits and Systems II*, 46(10):1309–1315 (Oct. 1999)
48. D. Teegarden, G. Lorenz, and R. Neul, "How to model and simulate microgyroscope systems," *IEEE Spectrum*, 35(7):66–75 (July 1998)
49. M. A. Maher and H. Lee, "MEMS systems design and verification tools," *Proc. SPIE 5th Annual Int. Symp. on Smart Structures and Materials, San Diego, CA, March 1–5, 1998*
50. CoventorWare Architect webpage, *http://www.coventor.com/software/coventorware/architect.html*, Coventor, Inc., 4001 Weston Parkway, Cary, NC 27513 (Nov. 2001)
51. D. Moulinier, P. Metsu, M.-P. Brutails, S. Bergeon, and P. Nachtergaele, "MEMSMaster: A new approach to prototype MEMS," *Design, Test, Integration, and Packaging of MEMS/MOEMS 2001, Apr. 25–27, 2001, Cannes, France*, pp.165–174
52. B. Baidya and T. Mukherjee, "Extraction for integrated electronics and MEMS devices," *Proc. Int. Conf. on Solid-State Sensors and Actuators (Transducers '01/Eurosensors XV), 2001*, pp. 280–283
53. Q. Jing, H. Luo, T. Mukherjee, L. R. Carley, and G. K. Fedder, "CMOS micromechanical filter design using a hierarchical MEMS circuit library," *Proc. IEEE Int. Conf. on MEMS, Miyazaki, Japan, January 23–27, 2000*, pp. 187–192

5 Thermal-Based Microsensors

Friedemann Vöelklein

*FH Wiesbaden, University of Applied Sciences
Rüsselsheim, Germany*

5.1 Introduction

Users require inexpensive, reliable sensors and actuators compatible with modern signal processing circuitry. This demand can be satisfied by microsensors and microactuators (microelectromechanical systems, MEMS), notably based on silicon with on-chip circuitry fabricated by using integrated circuit (IC) technology. A large number of such MEMS are based on thermal and thermoelectric principles. They use thermoresistive and thermoelectric thin films for sensor or actuator operation and the concepts of micromachining for device optimization. Indeed, a variety of thermal-based microsensors and microactuators fabricated by standard semiconductor technologies have been demonstrated.[1-4]

In the field of measurement and control, temperature (or, rather, temperature difference) is one of the most important physical quantities. Temperature sensors can be divided into self-generating and modulating sensors. Self-generating sensors, on the one hand, require no source of power other than the quantity they are intended to measure. Thermocouples are one example of self-generating sensors. Modulating sensors, on the other hand, are temperature sensors based on the thermal modulation of electric currents or voltages supplied by auxiliary energy sources. Thermoresistors, bolometers, diodes, and transistors belong to this category. Thermal-based sensors are also often used to measure nonthermal measurands. For example, it is possible to design an airflow sensor that is based on the measurement, on a sensor chip, of small temperature differences that are caused by the flow velocity. Radiation intensities or the heat generated by chemical reactions can be transferred into temperature variations of the sensitive sensor components. These variations are finally converted into electrical signals. It appears that multifunctional transducers for radiation, pressure, position, level, flow, or biological and chemical reactions can all be constructed on the basis of temperature or temperature-difference sensors. Hence, complex microsystems, e.g., for applications in molecular biology and biochemistry,[5-6] all contain thermal-based microdevices.

Thermal-based microsensors and actuators are obtained by combining established IC technologies with a few additional processing steps, which are specific to the pertinent device function and compatible with the preceding IC process.[7-8] On-chip circuitry, high reliability, and low-cost batch-process fabrication are the outstanding advantages of this approach. The most important standard semiconductor (IC) technologies applied to microsensor and microactuator fabrication are:

- Thermal oxidation of silicon
- PVD and CVD deposition for metallizations, sensor/actuator films (e.g., polysilicon), and passivations (e.g., SiO_2, Si_3N_4)
- Doping and implantation
- Planar and deep optical or X-ray lithography for lateral and three-dimensional micropatterning
- Wet chemical and plasma dry etching of films and substrates

A detailed account of micromachining techniques useful for the fabrication of thermal microstructures is given in Chapter 15.

5.2 Thermoresistors

Thermoresistors are devices for temperature measurement that use the temperature sensitivity of electrical conductive materials like metals or semiconductors. The dependence of the resistivity of these materials on temperature has been intensively investigated, so that by measuring the resistance, the temperature can be deduced directly from tables or curves.

5.2.1 Metal Film Thermoresistors

The resistivity ρ of metal film thermoresistors varies almost linearly with temperature between –200 and 1000°C. The electrical conductivity σ (the inverse resistivity ρ) is a function of the concentration n and the mobility μ of the charge carriers.

$$\sigma(T) = 1/\rho(T) = en\mu(T), \qquad (1)$$

where e is the electronic charge. Since the concentration of electrons in metallic materials is nearly constant, the dependence of the conductivity on temperature is determined by the mobility μ, with

$$\mu(T) = e\tau(T)/m^*, \qquad (2)$$

where m^* denotes the effective mass of the electrons and $\tau(T)$ their relaxation time. Combining Eqs. (1) and (2), we obtain for the resistivity

$$\rho(T) = m^* / (e^2 \tau(T) n). \tag{3}$$

The relaxation time represents the mean time between collisions of the electrons with lattice atoms and imperfections (such as impurities). The total inverse relaxation time can be represented by the sum of the inverse relaxation times of the various scattering processes:

$$\tau(T)^{-1} = \tau_{lattice}^{-1} + \tau_{imp}^{-1}, \tag{4}$$

where $\tau_{lattice}$ denotes the relaxation time associated with scattering on lattice vibrations and τ_{imp} describes the scattering due to imperfections. The inverse relaxation time is proportional to the scattering probability. For scattering on lattice vibrations, this probability is proportional to the amplitude of the lattice vibrations, which in turn is proportional to the absolute temperature. Hence, $\tau_{lattice}^{-1} \propto T$ and $\rho(T)$ can be represented as

$$\rho(T) = \rho_{lattice}(T) + \rho_{imp}, \tag{5}$$

where $\rho_{lattice}(T) \propto T$ is the resistivity due to lattice vibration scattering. For pure metals with temperatures above $-200°C$, the lattice vibration scattering is the dominating scattering process; therefore, the resistivity is roughly proportional to the absolute temperature, in agreement with experiments. At very low temperatures, however, the lattice vibrations are frozen out, and metals show a resistivity that is determined by impurities and is almost independent on temperature. For this temperature range, germanium or carbon thin films are usually applied as thermoresistors.

Appropriate materials are platinum, nickel, copper, tungsten, or copper-nickel. For high-accuracy thermometers, platinum is the thermoresistor most often applied. The resistance of a platinum film as a function of temperature between 0 and 600°C is shown in Fig. 5.1. For temperatures between 73 K and 1123 K, the resistance of a standard platinum temperature sensor can be represented by

$$R(T) = R_0 \{1 + AT + BT^2 + CT^3(T - 100°C)\}, \tag{6}$$

with $A = 3.908 \times 10^{-3}$ K^{-1}, $B = -5.802 \times 10^{-7}$ K^{-2}, and $C = -4.273 \times 10^{-12}$ K^{-4}, and with T in °C.[9] Metal thin-film resistors are prepared by thin-film deposition (sputtering or thermal evaporation) of about 1-µm-thick metallic films onto oxidized silicon wafers or onto Al_2O_3 ceramic substrates. After deposition, the films are patterned by photolithography and dry

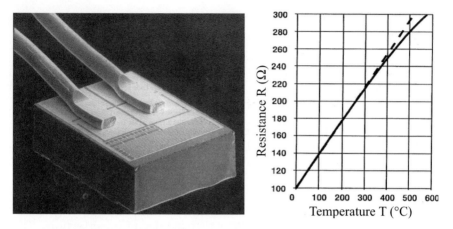

Figure 5.1 Pt-100 Ω thermoresistor with meander-shaped platinum film and its resistance R as a function of temperature between 0 and 600°C.

plasma-etching as meander-shaped thin-film resistors. The resistance R_0 at 0°C is adjusted by laser-trimming to values of, for example, 100, 500, or 1000 Ω. Figure 5.1 shows a Pt-100 Ω thermoresistor with meander-shaped platinum film on an Al_2O_3 ceramic substrate.

5.2.2 Semiconducting Thermoresistors

The conductivity and resistivity for intrinsic semiconductors can be expressed by

$$\sigma(T) = 1/\rho(T) = e(n\mu_n + p\mu_p), \tag{7}$$

where $n = p$ is the intrinsic concentration of electrons and holes, respectively, and μ_n, μ_p are the corresponding mobilities. The temperature dependence of the intrinsic carrier concentration can be expressed by

$$n(T) = p(T) = cT^{3/2} \exp(-E_g/2kT), \tag{8}$$

where c is a constant, k is the Boltzmann constant, and E_g represents the band gap between valence and conduction band. The dependence of the charge carrier mobility on temperature for intrinsic semiconductors is given by

$$\mu(T) \propto T^{-5/2}. \tag{9}$$

Thus we obtain, for the temperature dependence of the conductivity and resistivity,

$$\sigma(T) = 1/\rho(T) \propto T^{-1} \exp(-E_g/2kT). \qquad (10)$$

Because of the dominant exponential term, the temperature coefficient is negative and very large. Therefore, intrinsic silicon should be a very useful material for thermoresistors. However, in the technically interesting temperature range around room temperature, the resistivity of silicon is governed by the presence of impurities and does not exhibit intrinsic behavior. The intrinsic carrier density of silicon at room temperature is 1.5×10^{10} cm^{-3}, whereas the best high-resistivity silicon that can be made today has an impurity concentration of 10^{12} cm^{-3}, which is still much too high for intrinsic behavior.

The resistivity of extrinsic semiconductors as a function of temperature displays a rather complex pattern, as illustrated in Fig. 5.2.[10] In the extrinsic range, the electron concentration of an n-type silicon [as shown

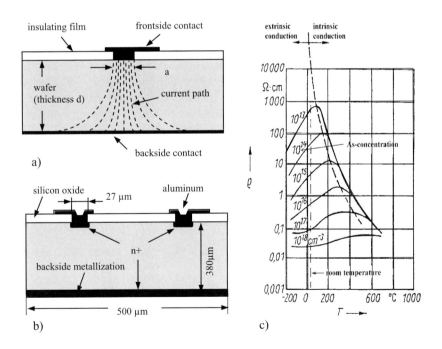

Figure 5.2 (a) Schematic cross section of a spreading resistance thermoresistor; (b) device with two spreading contacts and typical chip dimensions;[10] (c) resistivity ρ of silicon as a function of temperature for various As doping concentrations.

in Fig. 5.2(c)] or the hole concentration of p-type material is determined solely by impurity concentration. All donor (acceptor) atoms are ionized in this temperature range and the carrier concentration is almost constant. Therefore, the temperature dependence of the resistivity is governed by the temperature dependence of the mobility only. When temperature is increased, the mobility decreases and, consequently, the resistivity decreases, as is indicated in Fig. 5.2(c). In the high-temperature (intrinsic) region, electrons are thermally activated to move from the valence band to the conduction band. With increasing temperature, the concentration of additional electrons in the conduction band and of holes in the valence band increases and exceeds the number of electrons due to the donor concentration. The doped silicon starts to exhibit intrinsic behavior. The increasing number of carriers with increasing temperature overcompensates the decreasing mobility; therefore, resistivity decreases.

Because a unique relationship exists between resistivity and temperature, the extrinsic conduction range is the technically interesting region for semiconducting thermoresistors. The extension of the extrinsic range, the resistivity, its temperature coefficient, and thus the sensitivity of the thermoresistor can be influenced by the donor concentration, as shown in Fig. 5.2(c). At high impurity concentrations, the relaxation time depends primarily on impurity scattering and not on lattice vibration scattering; thus mobility does not depend on temperature. Since electron concentration in the extrinsic range is also independent of temperature, the resistivity of heavily doped silicon is almost insensitive to temperature variations,[11] This is indicated in Fig. 5.2(c) for the highly doped material with an As concentration of 10^{18} cm^{-3}. For lower impurity concentrations, lattice vibration scattering is the dominating scattering process. Thus with increasing temperature, mobility decreases and the extrinsic semiconductor shows a metal-like increase of the resistivity with increasing temperature.

5.2.3 Silicon Spreading Resistance Temperature Sensor

As shown in Fig. 5.2(c), lightly doped silicon exhibits a metal-like increase of resistivity in the region between −50 and 150°C. Based on this effect, silicon thermoresistors were made by diffusing meander-shaped n-type resistors in p-type substrates, where the resistor is isolated by the pn-depletion zone. However, these thermoresistors suffered from a large spread of nominal value and temperature coefficient. This was caused by the fact that both the pn-junction and the leakage current depend on temperature. The so-called spreading resistance temperature sensor overcomes these problems. To measure the resistivity of a semiconductor wafer, a common technique is to press a needle on the surface of the wafer.[12] When the diameter a of the needle is much smaller than the thickness of the wafer, the resistance between needle contact and a large-area contact on the back of the wafer appears to be

$$R(T) = \rho(T)/2a. \tag{11}$$

Thus the needle-silicon wafer combination provides a resistor that depends only on the resistivity of the silicon and the diameter of the needle, not on the other dimensions of the resistor. It does not require pn-junctions for insulation.

Thermoresistors based on this structure can be fabricated using standard semiconductor technology. Of course, the needle contact should be replaced by something more reliable. A small contact on the front can easily be made by the usual silicon planar technology. As shown in Fig. 5.2, such a device consists of a silicon chip ($0.5 \times 0.5 \times 0.38$ mm^3) with one large flat ohmic contact at the back and a point contact with a diameter of 20 to 30 µm at the front of the wafer. A silicon oxide film serves as electrical insulation between the silicon wafer and point contact metal lines. Moreover, n$^+$ layers are diffused both at the back and at the point contact to obtain a resistor with ohmic behavior. Figure 5.2(b) demonstrates a thermoresistor with two spreading contacts, which are used in commercial thermoresistors to increase the symmetry of the device. The temperature dependence of such thermoresistors can be approximated by

$$R(T) = R_0 + AT + BT^2, \tag{12}$$

where R_0 is the resistance at 0°C. For a resistivity of 5 Ωcm at 0°C and a point contact diameter a = 27 µm (as shown in Fig. 5.2), we obtain with Eq. (11) the resistance R_0 = 925 Ω. Thermoresistors based on spreading contacts have been made for temperatures up to 350°C.

5.2.4 Thermoresistors for the Detection of Thermal Radiation

When applying thermoresistors for temperature measurement, the resistor is usually in excellent thermal contact with the solid, liquid, or gaseous medium to prevent inaccuracies due to thermal resistances. However, there are many situations where a direct thermal contact of sensor and measuring object cannot be realized, e.g., for moving objects or for media with extremely high temperatures. Then contactless temperature measurements can be performed by measuring the infrared radiation emitted by the object. According to the Stefan-Boltzmann law, the total hemispherical irradiance M emitted from a surface of temperature T can be expressed by

$$M \propto \varepsilon \sigma_B T^4 \quad (\text{W/m}^2), \tag{13}$$

where $\sigma_B = 5.87 \times 10^{-8}$ Wm^{-2}K^{-4} denotes the Stefan-Boltzmann constant and ε the emissivity. When the emissivity of the surface is known, its temperature can be calculated from the measured total irradiance. Thermoresistors are appropriate sensors for such thermal radiation measurements because their sensitivity is independent of the radiation wavelength, in contrast to the sensitivity of photon detectors (like photodiodes). In general, various thermal effects (e.g., thermoresistive, thermoelectric, or pyroelectric effects) that generate changes in the properties of sensor materials resulting from temperature changes can be applied for thermal radiation detectors. These effects do not depend on the photon nature nor on the spectral content of the incident radiation; the heating and temperature change of sensor materials depend only on the radiant power. This assumes that the mechanism responsible for absorption of radiation is itself independent of the wavelength. Because heating or cooling of a sensor material is a relatively slow process, thermal radiation detectors as a class are slower in their response time than photon detectors. However, by using micromachining, radiation microsensors with response times between milliseconds and microseconds have been achieved because of their extremely small thermal capacity.

Bolometers are thermoresistors designed for the detection of thermal radiation. Because of the often very small incident radiation powers between μW and nW, thermoresistors are prepared as thin films with excellent thermal insulation from bulk substrates to increase the bolometer sensitivity. Figure 5.3 shows a schematic view of a microbolometer fabricated by thin-film technology and micromachining. A 0.8-μm thin silicon oxide/silicon nitride film is deposited by CVD processes onto (100)-oriented silicon wafers. A meander-shaped 200-nm thin platinum film is deposited and patterned by photolithography onto this electrically insulating film. After photolithographic patterning, the insulating film together with the platinum resistor can be prepared as a suspended membrane by anisotropic etching of the silicon substrate. The etching process can be performed from the front or the back of the wafer. Excellent thermal insulation of the thermoresistor can be achieved by this technique, resulting in a thermal resistance of such devices between 10^5 and 10^6 K/W. Consequently, an absorbed radiation power of only 1 μW causes a temperature increase of the thin-film resistor between 0.1 and 1 K. Thus even very small radiation intensities generate a resistance change ΔR, which can be detected in an electrical

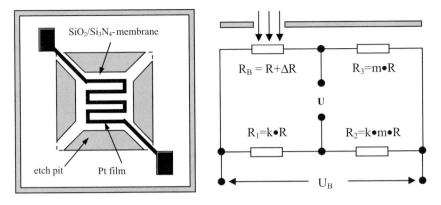

Figure 5.3 Schematic setup of a microbolometer and Wheatstone bridge for detection of the resistance change ΔR caused by irradiation of a bolometer R_B.

circuit. Absorption of the radiation in a large wavelength area can be obtained by an appropriate absorbing film (e.g., gold black or silver black), which is deposited above the thin-film resistor (not shown in Fig. 5.3). The absorbing film determines the radiation-sensitive area of the bolometer and enables a radiation detection independent of the spectral content of incoming radiation intensity.

Bolometers are usually applied in a Wheatstone bridge circuit, as indicated in Fig. 5.3. The signal voltage U can be calculated with the relative resistance change $\Delta R/R$ of the thin-film thermoresistor, the power supply voltage U_B and the imbalance factor m (with $m = 1$ for a symmetrical and $m \neq 1$ for a nonsymmetrical bridge):

$$U = \frac{m}{(m+1)^2} \frac{\Delta R}{R} U_B. \qquad (14)$$

The relative resistance change is determined by the temperature change ΔT of the thermoresistor and the temperature coefficient of resistivity β. For an almost linear dependence of resistivity on temperature, the resistance change can be expressed by

$$\Delta R / R = \beta \Delta T. \qquad (15)$$

The temperature increase caused by radiation depends on the absorbed radiation power N, and the thermal resistance R_T of the bolometer:

$$\Delta T = R_T N. \tag{16}$$

Combining Eqs. (14), (15), and (16), we obtain the signal voltage of the sensor as a function of the absorbed radiation:

$$U = \frac{m}{(m+1)^2} \beta \Delta T U_B = \frac{m}{(m+1)^2} \beta R_T N U_B, \tag{17}$$

and the sensitivity S of the bolometer, defined as the ratio of signal voltage and absorbed radiation power:

$$S = \frac{U}{N} = \frac{m}{(m+1)^2} \beta R_T U_B. \tag{18}$$

We learn from Eq. (18) that a high sensitivity can be achieved applying thin-film materials with high temperature coefficients of resistivity β, a symmetrical Wheatstone bridge, and a sensor design with a large thermal resistance R_T. U_B cannot be increased arbitrarily, since high power supply voltages generate an undesirable electrical heating of the bolometer. Typical values for U_B are in the order of 1 V. Preferred materials with appropriate β values and close to linear temperature dependence of resistivity in the temperature range of the sensor operation are nickel, platinum, or doped polysilicon films with β between $(0.8...10) \times 10^{-3}$ K^{-1}. By geometrical optimization of the design, microbolometers with extremely high thermal resistances as mentioned above can attain sensitivities of about 100 V/W at $U_B = 1$ V. Figure 5.4 shows a microbolometer[13] with a meander-shaped platinum resistor and an absorbing area of 0.1×0.1 mm^2 (left) and a silicon wafer that demonstrates the batch-process fabrication of such radiation sensors (right). Bolometer arrays with high sensitivity and excellent resolution are commercially available[14] for infrared cameras.

5.2.5 Pellistors

Pellistors are thermoresistors that detect the heat of chemical reactions caused by catalytic combustion. The heat of reaction leads to a temperature increase of the resistor, which causes a resistance change that can be detected electrically. Pellistors are applied for the measurement of concentrations of gases, since the amount of heat of reaction at the catalytic surface is proportional to the gas concentration. Conventionally, pellistors consist of platinum wires embedded into a sintered ceramic pill. The surface of the ceramic pill is coated with a catalytic film. The platinum wires

Figure 5.4 Microbolometer with meander-shaped 50-nm-thick platinum film and part of a silicon wafer with batch-process-fabricated bolometers.[13]

serve as heaters that increase the temperature of the catalytic surface to amounts between 100 and 600°C, which are necessary for the catalytic combustion. Simultaneously, the heating resistor can be applied as a temperature sensor to detect the temperature increase ΔT due to the heat of chemical reaction N_{chem}:

$$\Delta T = R_T N_{chem}, \qquad (19)$$

where R_T denotes the thermal resistance of the pellistor. Thus the principle of pellistors corresponds to that of bolometers. Only the types of detected heating powers are different: Bolometers are made sensitive to radiation heat, whereas pellistors are made sensitive for heat of chemical reactions N_{chem}. The catalytic combustion of methane to CO_2, described by $CH_4 + 2O_2 \Rightarrow CO_2 + 2H_2O$, is a representative reaction that can be detected by pellistors. The oxygen necessary for this combustion is supplied by air. The heat of combustion N_{chem} is proportional to the concentration of methane; thus the methane concentration can be detected by the temperature increase ΔT and the corresponding resistance change ΔR of the pellistor. The resistance change is converted into an electrical voltage by a Wheatstone bridge. Usually an identical pellistor without catalytic film, and therefore without chemical reaction, is used as a compensation resistor in the Wheatstone circuit.

Micropellistors have been fabricated by applying the benefits of semiconductor technology and micromachining. Figure 5.5 shows a schematic view of such a silicon planar-pellistor.[15-16] The sensitivity of

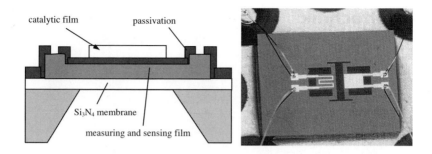

Figure 5.5 Schematic view of a silicon planar-pellistor (left) and pellistor chip for the detection of combustible gases like H_2, butane, and methane; chip size: 2.84 × 2.46 mm².[17]

the device is proportional to its thermal resistance. Therefore an excellent thermal insulation of the heating and sensing metallic film has to be realized. This can be achieved by fabrication of a thin suspended membrane of silicon nitride (with low thermal conductivity) and by deposition both of the heating/sensing film and the catalytic film onto this membrane. Figure 5.5 represents a pellistor chip[17] with an operating temperature of 400 to 500°C, which can be achieved with an electrical heating power of only 100 mW. The meander-shaped resistor on the left-hand side is a compensation resistor without a catalytic layer, whereas the resistor on the right-hand side is covered by a platinum catalytic film. Besides metal films like platinum, integrated bipolar transistors are also applied for heating and temperature sensing.[18] Metallic heating films are designed as meander-shaped resistors to optimize the homogeneity of the catalyst temperature.

5.3 Silicon Diodes and Transistors as Thermal Microsensors

In 1962, diodes had already been proposed as simple and low-cost devices for temperature measurement in the range between −50 and 150°C.[19] The forward voltages as a function of the temperature for a constant small forward current are usually measured. For many types of diodes between 20 and 300 K, a linear behavior of the forward voltage U can be found. The diffusion current I flowing through a forward-biased diode is given by[20]

$$I = (KT^s/\eta)\exp\left[e(U-U_g)/kT\right], \quad (20)$$

where K is a constant that depends on the junction design and fabrication process; s is a constant that depends on impurity concentration and is related to the temperature dependence of the mobility of minority carriers; and η is the ionization factor that characterizes the ionization of the impurity atoms, which is close to 1 for temperatures higher than about 20 K. The term eU_g denotes the band gap energy at 0 K. Using Eq. (20), the forward voltage can be expressed by

$$U = U_g + (kT/e)\left[\ln I - \ln(KT^s/\eta)\right]. \quad (21)$$

Since the second term inside the square brackets is much smaller than the first one, the voltage U varies linearly with temperature at constant current I. For the theoretical temperature coefficient β of the device, we find

$$\beta = dU/dT \approx (k/e)\ln(I/I_s), \quad (22)$$

where $I_s = KT^s/\eta$ denotes the very small saturation current. The temperature coefficient at $I = 10$ µA is between 1 and 3 mV/K for many different diodes. The individual characteristics of diodes depend strongly on their structure and the processing of the junction. Therefore, their interchangeability is poor; this impedes their broad use in industrial temperature measurement. Transistors and dual-transistor structures have much better characteristics.

The use of a transistor as a temperature sensor is based on the characteristic of the base-emitter voltage U_{BE} as a function of the emitter current I_C. In a transistor, I_C consists mainly of the diffusion current (because of the thin base), whereas the surface leakage and the recombination components are small. The expression for the dependence of U_{BE} on the collector current I_C and the temperature T corresponds to Eq. (21):

$$U_{BE} = U_g + (kT/e)\left[\ln I_C - \ln(KT^s/\eta)\right]. \quad (23)$$

The value U_g varies between 1.12 and 1.19 V, and s between 3 and 5 for many commercially available devices. Based on Eq. (23), a transistor can be used as an electrical thermometer similar to the application of diodes. For short-circuit of base and collector [as shown in Fig. 5.6(a)], the voltage U_{BE} at the pn-junction of the base-emitter contact can be expressed by

$$U_{BE} = U_g + (kT/e)\ln\left(\frac{I_C}{I_s}\right), \quad (24)$$

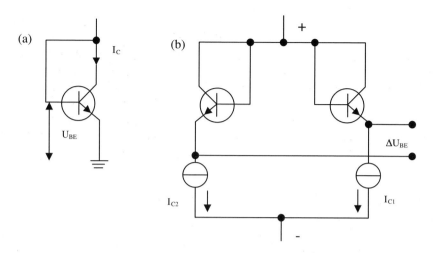

Figure 5.6 Application of bipolar transistors as temperature sensors.

where $I_s = KT^{\nu}/\eta$ is the very small collector saturation current, which can be measured for the base-emitter junction under reverse bias. When applying two different collector currents, I_{C1} and I_{C2}, two different base-emitter voltages, U_{BE1} and U_{BE2}, respectively, will be measured. We learn from Eq. (23) that the difference of these two voltages ΔU_{BE} is exactly proportional to the absolute temperature T:

$$\Delta U_{BE} = (kT/e)\ln(I_{C1}/I_{C2}). \tag{25}$$

The temperature coefficient β of this voltage difference is constant and can be expressed by

$$\beta = d(\Delta U_{BE})/dT = (k/e)\ln(I_{C1}/I_{C2}). \tag{26}$$

Further, both ΔU_{BE} and β are independent of material properties or transistor parameters. The temperature coefficient can be adjusted by appropriate choice of the collector currents. Such devices are known as PTAT (proportional-to-absolute-temperature) sensors. Technical realizations of these sensors make use of two matched transistors operating at two different collector current levels [Fig. 5.6(b)].

By using integrated-circuit technology, the matching of transistor pairs is performed without problems; therefore, most silicon temperature sensors used today are based on such pairs of identical transistors. Furthermore, devices have been developed where the currents through both transistors are equal but the emitter areas are chosen to be different, to achieve different I_s-values.[21] For different transistors, ΔU_{BE} can be expressed by

$$\Delta U_{BE} = (kT/e)\ln(v\frac{I_{C1}}{I_{C2}}), \tag{27}$$

where v is the ratio of the saturation currents. For identical collector currents, we obtain a voltage difference that depends only on the ratio v. This ratio can be adjusted by an appropriate choice of emitter areas. To obtain a well-defined ratio between the emitter surfaces, it is advisable to design the large transistor in such a way that it consists of v small transistors in parallel, if v is the desired ratio. The monolithic integration of the two transistors in a small silicon chip ensures that both junctions are at the same temperature. The sensitive transistors are usually integrated with additional microelectronic components like current sources and resistors for the adjustment of the collector currents and the output voltage.

Commercially available sensors have a sensitivity $\Delta U_{BE}/\Delta T$ in the order of 10 mV/K. Their nonlinearity in the temperature range between –50 and 150°C can be smaller than 0.5°C. Sensors with adjusted output voltage show a voltage signal (in V), which is equal to the absolute temperature (e.g., U_{BE} = 2.982 V for T = 298.2 K).[22]

5.4 Thermoelectric Microsensors

Thermoelectric microsensors are based on the thermoelectric effects discovered and analyzed by Seebeck in 1826,[23] Peltier in 1834,[24] and Thomson (Lord Kelvin) in 1857.[25] These effects are widely used for sensors and actuators. One of the first applications of thermoelectricity was the infrared detector, developed by Melloni in 1833.[26] Meanwhile, thermoelectric generators have become important constituents of power supply systems in space, and thermoelectric refrigerators are used for cooling. Seebeck's discovery has become an important sensor effect: Thermoelements are "active" or "self-generating" transducers converting temperature differences directly into electrical voltages without an external power supply.

The Seebeck Effect

A thermoelement or thermocouple is a junction of two electrically conducting materials (metal or semiconductor) electrically connected at a "hot" point of temperature T_1 (see Fig. 5.7). The nonconnected ends of both legs are kept at another temperature T_0 ("cold" point). Under open

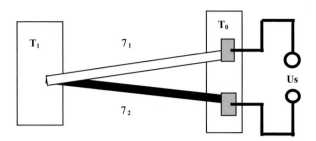

Figure 5.7 Schematic view of a thermocouple: generation of a thermoelectric voltage U_s due to a temperature difference $(T_1 - T_0)$.

circuit conditions, the net current flow through the thermoelement is zero and the thermoelectric or Seebeck voltage

$$U_s = \int_{T_0}^{T_1} \left[\alpha_1(T) - \alpha_2(T)\right] dT \qquad (28)$$

can be observed between the thermocouple leads at the cold point. For small temperature differences, $(T_1-T_0)/T_0 \ll 1$, the Seebeck voltage can be expressed by

$$U_s = (\alpha_1 - \alpha_2)(T_1 - T_0) = \alpha_{1/2}(T_1 - T_0), \qquad (29)$$

where α_1 and α_2 denote the Seebeck coefficients (or thermopower or thermoelectric power) of both materials at $(T_0 + T_1)/2$. They are specific transport properties, determined by band structure and carrier transport mechanisms of the materials. Seebeck coefficients are negative for charge transport mainly by electrons (n-type materials) and positive for conduction dominated by holes (p-type materials). Legs of p-type and n-type should be combined to achieve large voltages, since then the absolute value of the resulting Seebeck coefficient $\alpha_{1/2} = \alpha_1 - \alpha_2$ is the sum of the absolute values of α_1 and α_2.

The Seebeck voltage arises from the carrier drift caused by the applied temperature gradient. This drift induces a carrier concentration gradient and hence an electrical potential difference. The Seebeck coefficient relates the applied temperature gradient to the induced carrier gradient or to the gradient of the Fermi level E_F, according to

$$\alpha \nabla T = (1/e) \nabla E_F. \qquad (30)$$

The thermoelement can be connected with a load resistance for thermoelectric energy conversion, which can be used in thermoelectric generators with up to 10% conversion efficiency.

Modern microsensors based on the Seebeck effect have been fabricated by using silicon micromachining, thin-film technology, and photolithographic patterning. The crucial feature of such devices is the high thermal insulation of "hot" contacts, which can be performed by etching thin suspended membranes or preparing cantilever beams. The sensitivity can be increased with so-called thermopiles, where a plurality of thermoelements are connected in series. Such thermopiles can be realized by photolithographic patterning of the thermoelectric thin films. Figure 5.8

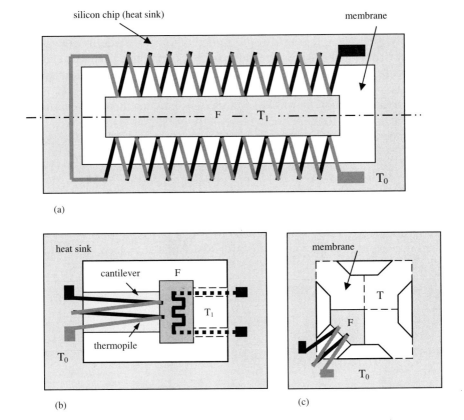

Figure 5.8 Schematic view of some types of thermoelectric microsensors: (a) membrane-supported thermopile with sensitive area F at temperature T_1 in the center of the membrane; (b) beam-supported thermopile with electrically heated sensitive area F; (c) part of a beam-supported thermopile after anisotropic etching of silicon.

shows examples of such thermopiles. Figure 5.8(a) demonstrates a membrane-supported thermopile with the sensitive area F in the center of the membrane. The sensitive area can be an absorbing layer for thermal radiation, a catalytic film for the detection of heat of chemical reactions, or an electrically heated area. This area has a good thermal connection to the hot contacts of the thermopile, whereas the cold contacts are arranged on the bulk silicon rim, which acts as a heat sink with constant temperature T_0. The thin membrane of silicon oxide/silicon nitride is prepared by anisotropic etching of a silicon chip.

Figure 5.8(b) shows a beam-like thermopile with a free-standing area, F, which can be radiation-sensitive or electrically heated. For this case, electrical connections (illustrated by dashed lines) are needed, which should have a negligible thermal conductance in comparison with the thermopile beam. Figure 5.8(c) shows a beam-supported thermopile produced by, for example, anisotropic underetching of oxide/nitride layers from the front of the wafer. Because of the symmetry, it is sufficient to model only one beam with a partitioned sensitive area F as indicated in Fig. 5.8(c).

Modeling of such thermoelectric devices enables the optimization of crucial sensor parameters such as sensitivity, signal-to-noise ratio, and specific detectivity. The sensitivity (or responsivity)

$$S = \frac{U}{N} \quad \text{(V/W)} \tag{31}$$

is the ratio of the signal voltage U and the input power N. The noise equivalent power NEP is the value of N, for which the signal voltage U of the sensor is equal to its noise voltage U_N.

$$NEP = \frac{U_N}{S} = \frac{\sqrt{4kTR\Delta f}}{S} \quad \text{(W)}. \tag{32}$$

For thermoelectric sensors, the Johnson noise with

$$U_N = \sqrt{4kTR\Delta f} \quad \text{(V)} \tag{33}$$

is usually the crucial noise voltage, where R denotes the resistance, T the absolute temperature and Δf the bandwidth of the detector. For such sensors, the *NEP* is related to a bandwidth $\Delta f = 1$ Hz. The detectivity represents the signal-to-noise ratio of the sensor in response to incident power *N*. To compare various sensors with different sensitive areas *F*, the specific detectivity D^* ("dee-star") is used as an area-independent figure of merit. D^* is defined as the signal-to-noise ratio in a 1-Hz-bandwidth per unit incident power times the square root of detector area.

$$D^* = \frac{\sqrt{F}}{N}\frac{U}{\sqrt{4kTR}} = \frac{S\sqrt{F}}{\sqrt{4kTR}} \quad (\text{cm}\sqrt{\text{Hz}}/\text{W}). \tag{34}$$

Simulation tools[27] have been developed for the optimization of the sensor parameters, and both analytical and numerical simulations have been performed. The modeling includes the analysis of heat balance and the geometrical optimization together with an appropriate choice of materials. With respect to material properties, the so-called thermoelectric efficiency $z = \alpha^2\sigma/\lambda$ is the crucial figure of merit, where α is the Seebeck coefficient, σ is the electrical conductivity, and λ is the thermal conductivity. Table 5.1 summarizes these transport properties for both some useful thermoelectric thin-film materials and for monocrystalline silicon. In view of the thermoelectric efficiency, monocrystalline silicon and polysilicon are not the most powerful materials, mainly because of their large thermal

Table 5.1 Thermoelectric Properties of Thin-Film Materials Used for Thermoelectric Microsensors and of Bulk Monocrystalline Silicon

Film Material	Seebeck Coefficient α (μV/K)	Electrical Conductivity σ ($10^4/\Omega$m)	Thermal Conductivity λ (W/mK)	Thermoelectric Efficiency $z=\alpha^2\sigma/\lambda$ (10^{-3}K^{-1})
p-Bi$_{0.5}$Sb$_{1.5}$Te$_3$	230	5.8	1.05	2.9
n-Bi$_{0.87}$Sb$_{0.13}$	-100	14.0	3.10	0.45
Sb	35	100.0	13.0	0.09
Bi	-65	28.5	5.2	0.23
n-Si* monocrystalline	-450	2.86	150	0.039
n-polysilicon*	-420	0.2	29	0.012
n-SiGe*	-136	9.9	4.45	0.41
p-SiGe*	144	7.6	4.8	0.33
n-PbTe	-170	5.0	2.5	0.58

* Dependent on doping level

conductivity. Nevertheless, these materials are interesting for integrated thermoelectric sensors and actuators because of their tremendous technological potential. Their Seebeck coefficients depend on the kind of doping (p- or n-type), and the absolute value of α decreases strongly with increasing doping concentration. Figure 5.9 shows the Seebeck coefficient of silicon and polysilicon.[28,29]

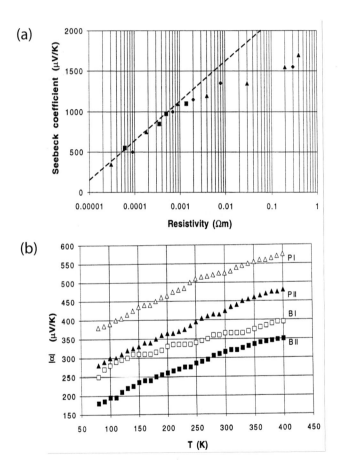

Figure 5.9 (a) Seebeck coefficient of bulk silicon as a function of resistivity at 300 K. The dashed line represents a curve according to $\alpha = 2.6(k/e)\ln(\rho/\rho_0)$, with $\rho_0 = 5 \times 10^{-6}$ Ωm.[28] (b) Absolute value of the Seebeck coefficient α of boron-doped (BI and BII) and phosphorus-doped (PI and PII) polysilicon as a function of temperature.[29]

The Peltier Effect

Peltier discovered that heat is transferred from, or to, the ambient when an electrical current I flows through the junction of two materials (see Fig. 5.10). Absorption or emission of heat depends on the direction of the current flow. The transported heat per unit charge is determined by band structure and scattering mechanisms of the materials. The absorbed or emitted "Peltier heat" per unit time Q_P (in watts) is proportional to the number of charge carriers through the junction, i.e., the current flow I:

$$Q_P = \Pi_{1/2} I, \qquad (35)$$

where $\Pi_{1/2}$ is the Peltier coefficient, which is related to the Seebeck coefficients by $\Pi = \alpha T$ and $\Pi_{1/2} = \alpha_{1/2} T$, with T as the absolute temperature of the junction.

The Thomson Effect

When the junctions of a Peltier element are at different temperatures T_1 and T_0 (Fig. 5.10), heat must be absorbed or released by the legs because of energy conservation.[25] Heat absorption or emission depends on the directions of current and of temperature gradient. The "Thomson heat" per unit volume and time Q_{TV} can be expressed by

$$Q_{TV} = j\tau \nabla T, \qquad (36)$$

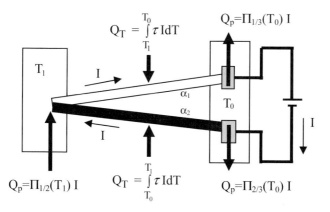

Figure 5.10 Schematic setup of a Peltier element.

where j is the current density and τ is the Thomson coefficient. For a thermoelectric leg with constant cross section, the total Thomson heat Q_T is

$$Q_T = \int_{T_0}^{T_1} \tau I dT. \qquad (37)$$

The Thomson coefficient is related to the Seebeck coefficient by $\tau = (d\alpha/dT)T$.

5.4.1 Microthermopiles as IR Radiation Detectors

Single-element infrared radiation thermopile sensors with thermocouples made of n-doped polysilicon and aluminum or with BiSbTe films in an IC compatible technology are well established.[30-31] A membrane is prepared on a silicon wafer by deposition of a thin SiO_2/Si_3N_4 layer and by anisotropic etching on the back of the silicon substrate (see Fig. 5.11). The hot junctions of the thermocouples are formed on this thermally insulating membrane, and the cold junctions are placed on the rim of the silicon chip, which acts as a heat sink. The absorbing film, which converts the IR radiation into a temperature rise, defines the sensitive area.[30] Miniaturized sensor modules, including a single-element thermopile sensor, mirror optics, and an ambient temperature compensation, are

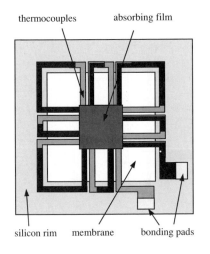

Figure 5.11 Schematic view of a microthermopile and SEM picture of a device with 50 thermocouples on a SiO_2 membrane, prepared by micromachining.

described by Schieferdecker et al.[31] These single-element thermopile sensors were transferred to mass production in the past 10 years. They have found a satisfying diversity of applications. Figure 5.11 shows the schematic of such a sensor chip and a SEM picture of a thermopile with 50 thermocouples arranged with their hot contacts under an absorbing film on a thin SiO_2/Si_3N_4 membrane.

Microthermopiles based on thin silicon cantilever beams were also realized by a standard IC technology.[32-33] The p-type cantilever shown in Fig. 5.12 is underetched by an anisotropic electrochemical etching process. Because of the high boron doping, the cantilever is resistant against the etching bath. The thermopile consists of thermocouples formed by n-type silicon and aluminum, where the n-type silicon legs are integrated into the p-type cantilever by diffusion. The pn-depletion zone between n- and p-type regions serves as electrical insulation of the n-type legs. The aluminum legs are structured by photolithographic patterning after PVD (sputtering).

5.4.2 Thermopile Arrays

The detection of position, presence, and direction of movement or counting of slowly moving human beings or objects is impossible using only one sensor element. Linear and two-dimensional IR sensor arrays can be used, for example, for out-of-position surveillance of the passenger seat or as a comfort sensor for contactless sensing of the temperature of different spots within a chosen area. Large sensor arrays are too expen-

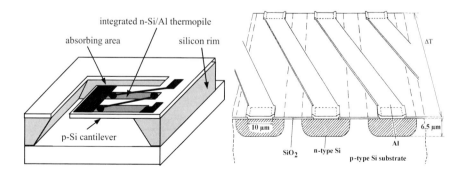

Figure 5.12 Schematic setup of an IR radiation thermopile based on a p-type silicon cantilever (left) and representation of integrated n-type silicon/aluminum thermocouples in detail (right).[32-33]

sive for mass-production applications. The solution can be a thermopile linear or two-dimensional array with only few sensor elements.[34-35] Whereas high sensitivity, high detectivity, and good linearity are important for single-element sensors, the additional optimization parameters for array sensors are homogeneous sensitivity, a high fill factor, and small thermal and electrical crosstalk. Efficient signal processing close to the sensor chip becomes very important. Table 5.2 gives some design parameters for the linear sensor array and the matrix array, which are represented in Fig. 5.13.

For the linear array, eight elements with a sensitive area of 400×400 μm^2 are arranged with a center-to-center spacing of 500 µm on a membrane formed by micromachining. The overall size of the chip is kept as small as possible so that it can fit into a small housing and its price can be kept low. The area of the suspended membrane also has to be kept small to achieve good mechanical stability. Each thermopile consists of 26 thermocouples made of n-doped polysilicon and aluminum connected in series. The pixel size is defined by the IR-radiation-absorbing layer, which covers and heats the hot junctions.

The two-dimensional 3×5 matrix array was fabricated with a similar technology. The individual elements have an absorber area of 375×425 μm^2 with a vertical and horizontal element pitch of 500 and 1100 µm, respectively. The overall size of the sensor chip is kept small to allow its integration with an ASIC chip in a small TO-5 housing. The readout chip integrates a multiplexer for selection of individual elements and an amplifier. A temperature reference based on silicon diodes can also be integrated on the chip.

Table 5.2 Design Parameters for Thermopile Arrays[34-35]

Parameter	Linear Array	Array
Number of thermopiles	1×8	3×5
Substrate	(100)-silicon	(100)-silicon
Chip size	5.2×1.8 mm^2	3.5×3.7 mm^2
Element pitch	500 µm	500 µm (vertical) × 1100 µm (horizontal)
Element size	400×400 µm^2	375 µm (vertical) × 425 µm (horizontal)
Number of thermocouples	26	26
Thermoelectric materials	n-polysilicon/Al	n-polysilicon/Al

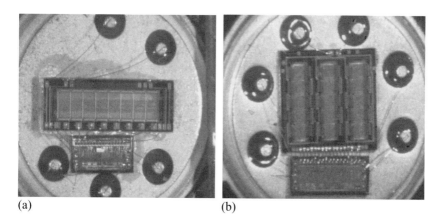

Figure 5.13 (a) TO-5 baseplate with thermoelectric linear array chip and readout chip[34]; (b) TO-5 baseplate with thermoelectric matrix array and ASIC.[35]

To meet security and automotive requirements, the sensor contains a self-test circuit. Externally applied digital pulses at the self-test pins generate a thermal excitation for each individual element of the array. As a result, the sensor provides an electrical output sequence to allow the required self-test function during the measurement.

A transistor cap with an infrared window is hermetically sealed in a dry nitrogen atmosphere. The size of the window is 5.2 × 4.2 mm². The transmission range is typically 6 to 14 µm, but other filter ranges can be chosen in accordance with the application. The most important sensor parameters are represented in Table 5.3.

Table 5.3 Sensor Parameters of Thermopile Arrays[34-35]

Parameter	Values Linear Array	Matrix Array	Units
Sensitivity, S	40	23	V/W
Resistance, R	40	25	kΩ
Noise voltage, U_N	89	87	nV/√Hz
NEP	2.2	3.8	nW/√Hz
Detectivity, D^*	0.2×10^8	0.1×10^8	cm√Hz/W
Time constant	24	15	ms

The sensor application for simple imaging purposes is very easy, since neither cooling nor mechanical choppering is necessary. The total size of the sensor system (containing filter, optics, and sensor package) can be less than 1 cm^3. The small size combined with low cost allows many new applications in security, automotive (passenger occupation detection for smart airbags), industrial, and medical market areas. As an example, Fig. 5.14 shows the sensor response of the matrix array when an object with a surface temperature of 34°C moves through the aperture of the sensor. The height of the columns corresponds to the signal voltages of the individual elements.

A linear array for higher resolution of thermal images[36] is shown in Fig. 5.15. Here the microthermocouples are formed with Sb and BiSb thin films, whereas the silicon substrate with the dielectric membrane is realized by standard techniques of micromachining.

A technology for the industrial fabrication of IR thermopile arrays, which is fully CMOS compatible, is reported by Münch et al.[37] and by Schneeberger.[38] The individual sensor pixels are thermally separated by gold lines, which are prepared by an industrial gold-bumping technology. The gold lines act as a heat sink and thus represent the cold junctions of the thermopiles. Furthermore, the gold lines stiffen the membrane mechanically. Hundreds of individual sensor pixels can be placed on a single membrane that consists of the CMOS dielectrics. A pixel yield over 99% can be achieved with this technology after wafer dicing for imagers with up to 32 by 40 pixels.

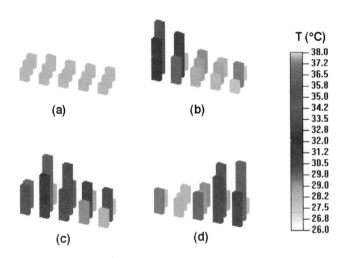

Figure 5.14 Sensor response of the thermoelectric 3 x 5 matrix array.[35]

Figure 5.15 Thermoelectric linear array with high resolution due to 256 thermoelectric pixels formed by Sb/BiSb thin-film microthermocouples.[36]

5.4.3 Thermoelectric Vacuum Microsensors

Thermal vacuum sensors (Pirani gauges) are based on the principle that heat transfer between two surfaces is proportional to the number of molecules (and hence to the pressure) transferring the heat, when the mean free path in the gas is larger than the distance between the surfaces.[39-42] As an example, a thermopile vacuum sensor is presented for the pressure range typically between 10 mPa and 10 kPa. Figure 5.16 shows a thin-film thermopile on a stress-compensated SiO_2/Si_3N_4 membrane (with a thickness of 0.8 μm) prepared by anisotropic etching of the silicon substrate. The thermopile contains 50 thermocouples with their hot contacts in the center of the membrane; the cold contacts are placed on bulk silicon. The thermopile is covered with a thin electrically insulating film (SiO_2). A meander-shaped thin-film heater is deposited in the area enclosed by the hot junctions. By using a constant heating power, the temperature of hot junctions and, consequently, the thermopile output voltage are functions of the surrounding gas pressure.

To increase the influence of pressure changes on output voltage and, consequently, to raise the sensitivity of the sensor, the thermal conductance of the gas was enhanced by using a bulk silicon "bridge" (lid). This component was also produced by anisotropic etching and is mounted on the sensor chip by anodic bonding. This measure leads to a heat exchange between membrane and lid across the small gap between them, instead of the free convection heat exchange with the surrounding gas. Figure 5.17 demonstrates the effect of various gap dimensions and shows the increasing sensitivity of the sensor in the high-pressure range with decreasing gap width.

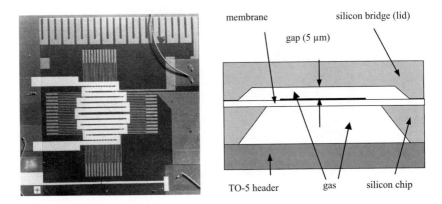

Figure 5.16 Schematic view and scanning electron micrograph (left) of a thermoelectric vacuum sensor.[39]

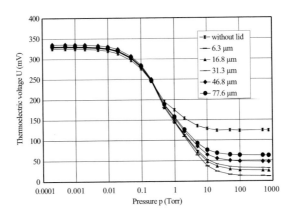

Figure 5.17 Output voltage U (mV) as a function of pressure p (torr) of the thermoelectric vacuum sensor shown in Fig. 5.16 at a heating power of 84 µW and for various gap dimensions.

5.4.4 Gas Flow Microsensors

Gas flow sensors[43-45] based on the Seebeck effect can be made by using the aluminum/polysilicon contacts provided by a standard IC process. The thermopile gas flow sensor shown in Fig. 5.18 contains a polysilicon/aluminum thermopile together with a polysilicon heater. The hot contacts are placed with good thermal connection to the heater in the center of a thin silicon oxide/silicon nitride membrane; the cold contacts are arranged on bulk silicon. The excellent thermal insulation provided by the 0.8-µm-thick dielectric membrane leads to a high sensitivity. The variation of the Seebeck coefficient of the thermocouples with temperature compensates the corresponding variation of the thermal resistance. Hence, the sensor shows a sensitivity that is almost independent of the ambient temperature. Two modes of operation can be applied: operation with constant heating power, or operation with constant temperature of the heater. In the second mode, the electrical power N necessary to keep the heater temperature constant is measured. This power is a function of the flow velocity, as indicated in Fig. 5.18.

The thermopile flow sensor shows a sensitivity with respect to the heating power of 30 V/W. With laminar nitrogen flow, an output voltage of 9 mV per m/sec for 5 mW heating power and a sensitivity of 1.8 (V/W)/(m/sec) were measured.

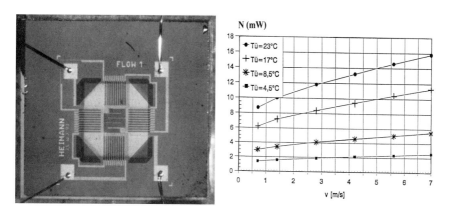

Figure 5.18 Scanning electron micrograph of a thermopile gas flow sensor with a polysilicon heater in the center of the silicon oxide/silicon nitride membrane (left) and the heating power N (mW) necessary for a constant heater temperature $T_{ü}$ as a function of flow velocity v.

For applications in a rough ambient, flow sensors with very thin membranes cannot be used. However, a good thermal insulation of the substrate is required for thermoelectric sensor principles. The use of porous silicon can be a solution.[46] Figure 5.19 shows a thermoelectric flow sensor, where the hot junctions are located on an area of porous silicon that is integrated in a silicon wafer. Porous silicon has a very low thermal conductivity of only about 1.5 W/mK.[46] The porous silicon area is formed by electrochemical etching.[47] On porous silicon area A, a thermopile is patterned that consists of polysilicon legs (layer B) and aluminum legs (layer C). The aluminum film also forms the bonding pads for the thermopile. A thin heating stripe is structured in the center of the porous silicon area simultaneously with the polysilicon thermoelectric legs. The heater generates a thermoelectric voltage, which depends on the heater temperature and, consequently, on the flow rate of a surrounding gas or fluid. The sensitivity related to the heating power is 9.8 (mV/W)/(m/sec).

5.4.5 AC/DC Thermoconverter

In a thermoconverter, electrical power is converted into thermal power by an ohmic resistor, and the resulting heat is measured with a temperature sensor. Thermoconverters have been used for AC power measurement in microwave instrumentation, true root-mean-square voltage measurement, and AC/DC calibration. The thermoconverter chip[48-49] illustrated in Fig. 5.20 is based on a thin-film NiCr heating

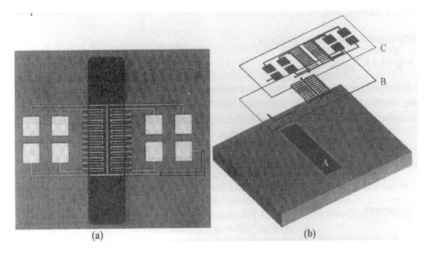

Figure 5.19 Thermoelectric flow sensor on silicon substrate with an area of porous silicon (A), polysilicon heater and thermoelectric legs (layer B), and aluminum thermoelectric legs and bonding pads (layer C).[46]

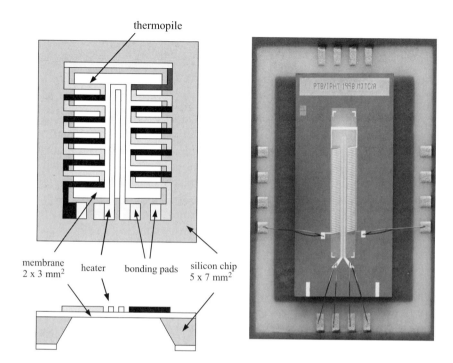

Figure 5.20 Schematic setup of a thin-film thermoconverter and optical micrograph of a thermoconverter chip used for AC current and voltage calibration at PTB (Germany) and NIST (USA).[49]

resistor and a thermopile integrated on an oxide/nitride membrane. The AC power source to be calibrated is connected to the heating resistor. The dissipated heat is transformed into the thermopile output voltage. The output voltage caused by the unknown AC power is compared with the thermopile output voltage caused by a calibrated DC power source (true rms measurement). The thermoconverter presented has a sensitivity of 16 V/W in air and about 120 V/W in vacuum. The high-frequency AC/DC voltage transfer difference is a function of the frequency and the heater resistance.

5.4.6 Heat Flux Sensors

Measurements of heat flux densities, heat transfer coefficients, and heat transmission coefficients are important for applications in building physics, for the determination of thermophysical properties, and for the optimization of thermal insulating devices. For example, the heat flux density q (in W/m²) that passes through a wall is given by

$$q = k(T_1 - T_2), \qquad (38)$$

where T_1, T_2 are the temperatures on both sides of the wall and k is the heat transmission coefficient. At the present time, heat flux sensors are designed by using a bulk auxiliary plate with well-known thermal resistance R_T and temperature sensors on both sides of the plate. Such heat flux sensors are attached with thorough thermal contact on a wall or surface, which is penetrated by a heat flux density q. This heat flux density generates a temperature difference ΔT_z across the auxiliary plate, which is measured by the temperature sensors. Using ΔT_z and the thermal resistance of the auxiliary plate, the heat flux density can be calculated according to

$$q = \Delta T_z / R_T. \qquad (39)$$

Applying a differential thermopile, the temperature difference ΔT_z is proportional to its thermoelectric voltage U, and q is given by

$$q = U / (n\alpha_{1/2} R_T), \qquad (40)$$

where $\alpha_{1/2}$ is the Seebeck coefficient of the thermoelectric junction and n is the number of thermocouples connected in series. The application of this measuring technique requires auxiliary plates with small thermal resistance, since heat flux density must not be changed by the plate itself. This demand leads to small ΔT_z and, consequently, to small thermoelectric voltage U, low sensitivity, and inferior signal-to-noise ratio of the sensor. The heat flux sensors presented by Völklein and Kessler[50] are designed for the measurement of flux density q with high sensitivity and are realized as microelectromechanical systems. To achieve high sensitivity and small temperature coefficient of responsivity, $Bi_{0.87}Sb_{0.13}$ films (n-type) and pure Sb films (p-type) are used as thermopile materials and patterned by photolithography. Figure 5.21 shows the schematic cross section of one membrane area and a top view of part of the thermopile. The thermoelectric films are deposited on 800-nm-thick membranes of silicon oxide/silicon nitride. The sensor consists of 10 membranes with 889 thermocouples on each membrane.

The high density of thermocouples on a small membrane size of 22 mm² enables the high sensitivity of 5.87 mV/(W/m²), whereas the sensitivity of conventional heat flux sensors is only 0.07 mV/(W/m²). This increase of sensitivity is achieved by the microelectromechanical design. The sensor can be calibrated absolutely by using an integrated thin-film

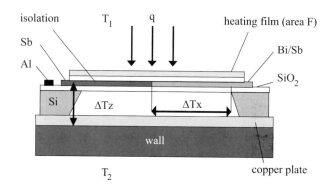

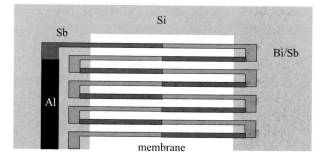

Figure 5.21 Schematic cross-sectional view of one membrane area and top view of part of the BiSb/Sb thermopile.

heater, which generates a well-defined heat flux (self-calibration). For good thermal contact, the sensor chip is mounted on a copper plate with high thermal conductivity.

When the sensor is directly attached to the wall, the heat flux q will generate a temperature difference $\Delta T_z \propto q$ across the bulk silicon rim, analogous to the temperature difference caused in bulk auxiliary plates. This temperature difference for the sensor presented by Völklein and Kessler[50] is given by

$$\Delta T_z = R_{T1} q, \qquad (41)$$

with $R_{T1} = 5.1 \times 10^{-6}$ (K m^2)/W. Because of the high thermal conductivity of copper and silicon and the small thickness of the membrane, the thermal resistance of the bulk region and ΔT_z is very small. On the other hand, the heat flux will also generate a temperature difference $\Delta T_x \propto q$ across the membrane, which is several orders of magnitude higher than ΔT_z due to the low thermal conductance of the membrane system. The temperature difference ΔT_x between the center of the membrane, where the sensing junctions are located, and the silicon heat sink is given by

$$\Delta T_x = R_{T2} q, \qquad (42)$$

with $R_{T2} = 7.5 \times 10^{-3}$ (K m^2)/W, which is several orders of magnitude higher than ΔT_z, calculated according to Eq. (41). The temperature difference ΔT_x causes a signal voltage

$$U = n\alpha_{1/2} \Delta T_x \propto q \qquad (43)$$

proportional to the heat flux q to be determined. n is the number of thermocouples connected in series, and $\alpha_{1/2}$ is the Seebeck coefficient of a thermocouple. With $n = 8890$ and $\alpha_{1/2} = 130$ µV/K, the theoretical sensitivity is

$$S = U/q = 8.6 \, \text{mV/(W/m}^2). \qquad (44)$$

The thermal response time (time constant) is 40 to 50 msec. Therefore, the dynamical behavior of heat fluxes can also be investigated with these sensors.

The sensor operation is now described with respect to the measurement of heat flux density across a wall due to a temperature difference $T_1 - T_2$ ($T_1 > T_2$). The sensor is attached with its copper baseplate on the hot side (temperature T_1) of the wall. Hence, the heat flux q will generate a signal voltage U. Then the heating film is switched on and the heating power is controlled to double the signal voltage: $U \Rightarrow 2U$. The heating power per unit area can be calculated, since the area F of the thin-film heater is known. The heating power per unit area q_h doubling the signal voltage U is equal to the heat flux density q across the wall. Obviously, by use of integrated heating film, no additional calibration of the sensor is necessary. When the sensor is attached on the cold side of the wall (temperature T_2), the heat flux will generate a signal voltage $-U$. Now the electrical heating power that eliminates the signal voltage must be determined: $-U \Rightarrow 0$. Then this heating power per unit area is equal to the heat flux density. Measurements on the hot and cold sides usually lead to almost identical results, with inaccuracies less than 10%.

5.4.7 Microelectromechanical Thermoelectric Cooler

Thermoelectric coolers (Peltier devices) are widely used in microelectronics to stabilize the temperature of laser diodes, to cool infrared detectors and charge-coupled devices (CCDs), and to reduce unwanted noise of integrated circuits. A conventional thermoelectric cooler usually consists of a number of n- and p-type bulk semiconductor thermoelements connected electrically in series by metallizations and sandwiched between two electrically insulating but thermally conducting ceramic plates.[51] The dimensions of commercially available coolers vary from about $4 \times 4 \times 3$ mm^3 to around $50 \times 50 \times 5$ mm^3. Although in principle, the dimensions can be reduced further, the fabrication of conventional thermoelectric coolers is a bulk technology and is incompatible with microelectronic fabrication processes.

Thermoelectric microcoolers that are designed using micromachining have recently been proposed.[52-53] The cooler fabrication is compatible with standard semiconductor technology. Therefore, such coolers can be integrated with microelectronic circuits. The small thermal bypass of thin membranes enables devices with sufficient cooling power and temperature difference.

Figure 5.22 shows the top view and cross-sectional view of the thermal model. The cooler consists of a number n of thermocouples connected in series and arranged with their cold junctions (temperature T_C) around

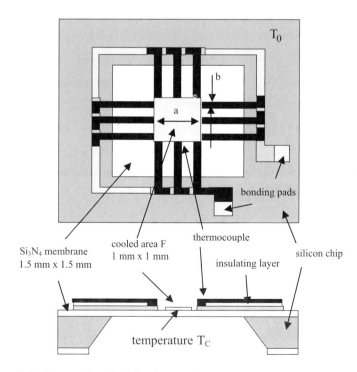

Figure 5.22 Thermal model of a thermoelectric microcooler.

the cooled area F; their hot junctions are in good thermal contact with the heat sink. The thermopile and the cooled area are supported by a very thin membrane with low thermal conductivity. Heat is pumped laterally from the central region to the silicon substrate rim and then dissipated vertically to an external heat sink.

Each thermocouple is a stack of two thermoelectric films with an insulating layer between them. The width of a thermocouple is b; its length (the distance between the heat sink and cooled area F) is l. For operation in high vacuum (pressure $<10^{-4}$ mbar), heat flux by convection can be neglected. The modeling is performed for identical leg thicknesses $d_1 = d_2 = d$. Then the maximum temperature difference between cooled area (T_C) and ambient (T_0) can be expressed as

$$(T_c - T_0)_{max} = -zT_c^2 / \left[2 + (\lambda_s d_s / \lambda d) + (\gamma_F Fl / nb\lambda d) \right], \quad (45)$$

where z is the thermoelectric efficiency and λ, λ_s, and d_s denote the thermal conductivity of the legs, the thermal conductivity of the membrane plus the insulating layer, and the thickness of the membrane plus the insulating layer, respectively. The parameter γ_F is defined as $\gamma_F = 8\varepsilon_F \sigma_B T_0^3$, where ε_F and σ_B denote the emissivity of the cooled area and the Stefan-Boltzmann constant, respectively. To calculate the maximum temperature difference, thermoelectric properties of thin-film materials have to be compared. At room temperature, p-type $(Bi_{0.5}Sb_{1.5})Te_3$-films show the best thermoelectric efficiency. These films have been prepared by flash evaporation[54] or MBE[55], with subsequent annealing. The properties of these optimized films are close to the bulk material and are represented in Table 5.4.

The preparation of n-type films with high figures of merit is much more complicated. The modeling is based on the (somewhat optimistic)

Table 5.4 Material Properties and Geometrical Dimensions Used for Calculations of Microcooler Parameters

Materials or Geometrical Parameters	T_C = 250 K	T_0 = 293 K
Seebeck coefficient $\alpha_1 = -\alpha_2$ (µV/K)	185	230
Resistivity $\rho_1 = \rho_2$ ($10^{-5}\Omega$m)	1.33	1.70
Thermal conductivity $\lambda_1 = \lambda_2 = \lambda$ (W/mK)	1.28	1.08
Efficiency $z_1 = z_2 = z = \alpha^2/(\rho\lambda)$ (10^{-3}/K)	2.00	2.88
Emissivity ε, ε_F (worst case!)	1	1
Substrate's thermal conductivity λ_S (W/mK)	2.0	2.2
Length of the legs l (mm)	0.3	0.3
Lateral length of cooled area a (mm)	1	1

assumption of a highly efficient n-type film with properties like those of the best p-type films. By further improvement of the preparation techniques, certainly it will be possible to realize such highly efficient n-type films.

Figure 5.23 shows the maximum temperature difference as a function of leg thickness d for substrate thickness $d_s = 1$ µm, calculated according to Eq. (45), with $T_0 = 293$ K and the film data at $T_C = 250$ K. Note that Eq. (45) neglects the contact resistance of the thermocouples. Therefore, the temperature difference that can be obtained in practice will be smaller than the calculated one. On the other hand, the contact resistance of such photolithographically patterned thin-film devices is only a small part of the total resistance. Therefore, it will not drastically reduce the calculated curve.

By using the microcooler in a normal atmosphere, heat flux by convection has to be involved. The calculated temperature difference for this case, using the parameters of Table 5.4, is also represented in Fig. 5.23. The results demonstrate that a microcooler with thermoelectric films of $z = 2 \times 10^{-3}/$ K can achieve a maximum temperature difference of about 20 K at normal atmosphere. Therefore, dewpoint sensors based on such microelectromechanical coolers can be realized by batch-process fabrication. The calculated maximum heat load is about 10 to 30 mW. These parameters can be achieved with an optimum current density $j_{opt} = 1.3 \times 10^3$ A/cm^2.

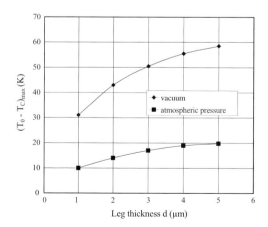

Figure 5.23 Maximum temperature difference $(T_0-T_C)_{max}$ of a thermoelectric microcooler for operation in a vacuum and at atmospheric pressure.

5.5 CMOS-Compatible Thermal-Based Microsensors and Microactuators

Inherent features of CMOS technology and the etching properties of its (100)-silicon substrate allow the fabrication of micromechanical structures based on silicon oxide or silicon nitride by maskless anisotropic silicon etching performed after the completion of the CMOS process. This is achieved in three steps:

1. Design of a suitable layout for a CMOS process
2. Processing by CMOS standard technology
3. Post-processing by final anisotropic etching

As an example, the layout of an oxide cantilever beam realized with a standard CMOS process is shown in Fig. 5.24. The sandwiched oxide layer consisting of thermal oxide and CVD oxides of the CMOS process

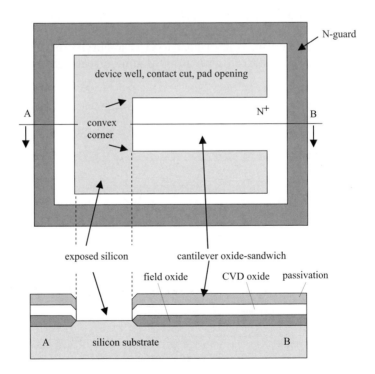

Figure 5.24 CMOS oxide microbeam; top view of layout and cross-section resulting after completion of CMOS process.[8]

serves as a "natural" etching mask. Windows in that layer (areas where the silicon surface is to be exposed to the final etchant) are obtained by superimposing a device well, a contact cut, and a pad opening in the IC layout. This unusual layout design prevents the formation of thermal oxide, CVD oxide, and passivation at the desired locations, thus allowing the underetching of free oxide structures, such as cantilever beams, bridges, or suspended membranes, for the purpose of thermal insulation or mechanical degrees of freedom. N^+ and N guard exclusion masks prevent doping and hence any related undesirable increase in silicon etching time in the area surrounding the microstructure.

Figure 5.24 also shows the cross section to be expected after completion of the IC process but before the post-processing etching step. The silicon is exposed to the final etchant in such a way that the sandwiched oxide layer, which eventually forms the beam, shows a convex corner to be undercut by the final etchant. The post-processing etching can be performed using EDP or KOH. No additional mask is required, since the oxide layers are not significantly attacked by the anisotropic etchant. The release of mechanical stress inherent in underetched layers can result in buckling or sagging. This problem is eased by modern CMOS technology achieving minimal stress on the metal lines and on other sensor films by stress-compensated dielectric layers that include silicon oxide and nitride, with oxide and nitride showing stresses of the opposite sign (compressive vs. tensile). Frequency mixing during the PECVD process for SiO_2 or Si_3N_4 film deposition is also useful for stress-compensated preparation.

Typical CMOS cross sections are shown in Fig. 5.25. By appropriate design, polysilicon and aluminum films can be included between the oxide or oxide/nitride layers that form a micromechanical cantilever beam. Metal lines or areas can be used for electrical connections or optical mirrors. Polysilicon structures can be used for resistive heating or thermometry. Metal/polysilicon or polysilicon/polysilicon contacts can serve as thermocouples. While the designer is free to choose the lateral dimensions of such structures within the design rule limits, the thickness of each layer is imposed by the CMOS process technology.

As an example, Fig. 5.25 shows the cross sections of a p-polysilicon/n-polysilicon thermopile on an oxide beam before and after the final etching step that removes bulk silicon for the purpose of thermal insulation. The fabrication of closed membranes rather than of open cantilever beams or bridges requires more, and different, post-processing steps. In contrast to "open" structures, the etching has to be performed from the back of the wafer. A PECVD silicon nitride film has to be deposited on the back. By photolithographic patterning and reactive ion etching (RIE), etch windows are opened in the Si_3N_4. Then the anisotropic etching of the

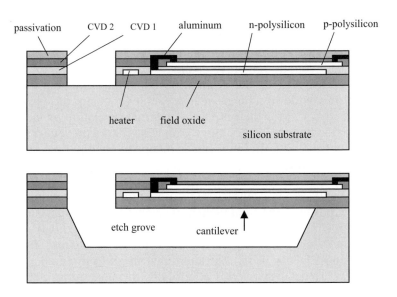

Figure 5.25 Cross section of p-polysilicon/n-polysilicon thermopile on a cantilever beam, prepared by standard CMOS process, before and after the post-processing.[8]

wafer can be performed with KOH. The front of the wafer has to be protected from the etchant. This is provided by a mechanical setup.

In the described way, thermal and mechanical microtransducers such as thermoelectric and resistive mass flow sensors,[56] infrared radiation sensors,[57] vacuum sensors,[58] AC power sensors (AC/DC thermoconverter),[59] thermal heat flux sensors,[60] and acoustic resonators for proximity sensors[61] have been obtained. As an example, Fig. 5.26 shows a CMOS thermopile IR radiation sensor. Radiation-absorbing films are arranged at the tip of the cantilever beams, where the hot junctions of n-polysilicon/p-polysilicon thermocouples are located. The cold junctions rest on the bulk silicon. A small beam thickness of about 2 µm enables excellent thermal insulation and high thermoelectric output voltage even for very small IR radiation intensity. The patterning of the cantilever tips is performed to achieve increased radiation absorption.

By using this approach, very complex sensor systems have been realized together with CMOS signal conditioning circuits. Figure 5.27 shows a multisensor chip for air conditioning systems of a building, which integrates single thermoelectric microsensors, resistors, and capacitances for humidity, air flow, and temperature sensing and the signal conditioning.[62] The output of the humidity sensor is a capacitive variation, while the flow

Figure 5.26 IR radiation thermopiles realized by a standard CMOS process with post-process underetching of cantilever beams. n-poly-silicon/p-polysilicon thermocouples are included in the sandwich of various CVD-oxide and passivation films given by the standard CMOS process. The microstructures on the tip of the beams increase the IR absorption.[57]

sensor is a thermoelectric device that provides a voltage signal like the temperature sensor. Taking advantage of the flexibility of the cointegrated $\Sigma\Delta$ modulator, the same A/D converter for the three sensors is used, without any additional conditioning circuit.

CMOS-compatible thermoresistors have also been realized.[58,63] Figure 5.28 shows a vacuum sensor that was fabricated by a standard CMOS process and sacrificial-layer technology. In the framework of sacrificial-layer micromachining, first sacrificial layers with a thickness of only a few micrometers are deposited on the substrate. Then top layers that cover the sacrificial films are deposited. These top layers may be dielectric materials, metallizations, or semiconducting films like polysilicon. After their deposition, the sacrificial layer is removed by a selective etching process using wet chemical etchants or dry plasma-etching techniques. To perform this removal, the top layers are patterned with openings, which permit the etching solution or etching gas to attack the sacrificial layer. Useful sacrificial materials are dielectric SiO_2 films, prepared by CVD processes or thermal oxidation, as well as metallizations like aluminum or copper. Besides membranes, cantilever beams are also useful structures realized by surface micromachining. Such (movable) beams are used not only for thermoelectric sensors but also for accelerometers and AFM devices. For the vacuum sensor of Fig. 5.28, a meander-shaped heater/resistor is deposited on a top layer of silicon oxide, which is thermally insulated from the substrate after removal of an aluminum

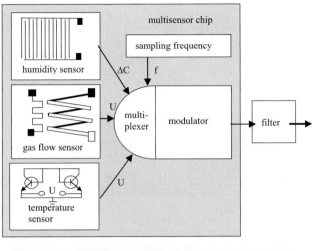

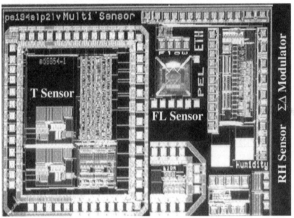

Figure 5.27 Schematic of a multisensor chip for air conditioning systems and SEM picture of the single sensor devices together with CMOS curcuits.[62]

sacrificial layer. The eight openings for the wet chemical etching can be seen. For a given heating power, the temperature of the heater/resistor is a function of the vacuum pressure of the surrounding gas. The temperature of the thermoresistor can be measured with the methods mentioned above. The resistor without thermal insulation shown on the right-hand side of Fig. 5.28 can be used for compensation of ambient temperature drift.

The integration of a low power thermoelectric generator (LPTG) into a CMOS environment is represented by Strasser *et al.*[64] This is of great

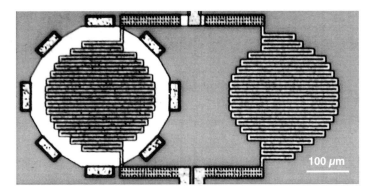

Figure 5.28 Meander-shaped thin-film resistor on thermal insulating membrane prepared by surface micromachining.[58,63]

interest for powering miniaturized sensor systems and for replacement of batteries in applications where a few μW are required for operating the electronic circuitry. The design of such generators with fabrication based on CMOS process technology is shown in Fig. 5.29.

The actual generator layer is sandwiched between the silicon substrate below and a convection-cooled heat sink mounted on top. In this way, a vertical heat flux is applied to the device. The thermoelements consist of couples of p-polysilicon and n-polysilicon, which are standard materials of CMOS technology. The device differs from conventional thermoelectric generators by the large number of thermocouples that can be arranged on a small area. Using the CMOS process available, 56,400 thermocouples have been integrated on an area of only 6 mm^2. To achieve a temperature gradient between the two ends of the polysilicon legs, oxide barriers can be generated below their hot side (see Fig. 5.29). As a consequence of such local thermal barriers, the heat flux is delivered, leading to a rather lateral heat flow direction. Using the local oxidation of silicon (LOCOS) process, thermal oxides up to 1.6-μm thickness have been produced. The 400-nm-thick polysilicon layer is partly phosphorous-implanted to generate the n-legs and partly boron-implanted in other regions to form the p-legs. Aluminum bridges are used to avoid pn-junctions that otherwise would occur between adjacent thermoelectric legs. A second metal bridge is added to improve the thermal coupling to the (cold) surface of the device. Common surface micromachining techniques are used to increase the temperature difference between hot and cold junctions. The LOCOS oxide barriers below the hot polysilicon contacts are removed by buffered HF wet chemical etching. Alternatively, cavities have been etched into the silicon substrate isotropically with CF_4 (Fig. 5.29). Furthermore, devices

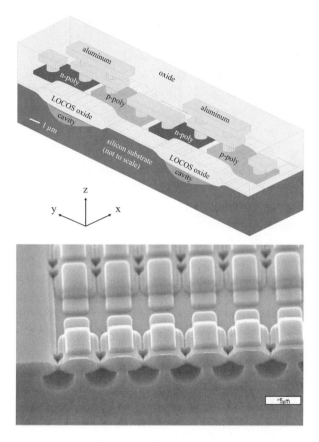

Figure 5.29 Design of a low-power thermoelectric generator based on CMOS process technology (top) and SEM picture of a device with etched cavities in the silicon substrate to increase the temperature difference between "hot" and "cold" contacts (bottom).[64]

with polycrystalline SiGe thermoelectric layers have been investigated. SiGe, which is usually not available in a silicon foundry, is more promising because of its low thermal conductivity of 4.5 W/mK compared with polysilicon, which has thermal conductivities between 25 and 30 W/mK. Measured values of the thermoelectric properties of polysilicon and polycrystalline SiGe films with various doping levels are also reported.[64] The expected parameters for the optimized generators are about 3 V output voltage and 1 µW output power for a given temperature difference of 1 K across the chip.

5.6 Diagnostic Thermal-Based Microstructures

Currently microsystems are applied increasingly as diagnostic microstructures in the submicron or nanoscale range. A thermoelectric microtip for temperature detection in the submicron range and diagnostic microstructures for the investigation of thermal properties of thin films are presented as examples.

5.6.1 Thermoelectric Microtips for AFM Temperature Sensors

These devices can be applied for the localization of hot spots in integrated circuits or for the detection of reaction heat in microcavities, since the chemical reaction technique using very small quantities and performing in microreactors is a very promising new field of research (chemical laboratory on a chip). Hereby, thermoelectric microtips for temperature measurement in very small volumes can be a very important tool for investigations. The microtip represented by Oesterschulze and Kassing[65] is realized on a cantilever beam similar to the fabrication of AFM tips on cantilevers. The principle of tip fabrication is based on the anisotropic etching of (100)-silicon wafers by using the underetching of convex corners. Figure 5.30 demonstrates schematically the etch process. A quadratic passivation layer is patterned on a (100)-silicon surface. Using KOH solution, the convex corners of the passivation are underetched, whereby the (411)-plane and the (100)-plane are attacked with nearly the same etch rate.

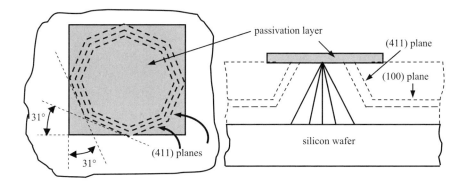

Figure 5.30 Preparation of microtips by anisotropic etching of (100)-silicon wafers using passivation films with convex corners.

Figure 5.31 shows the fabrication process of a complete tip on a cantilever beam. After formation of the silicon tip on the front according to Fig. 5.30, the wafer is deposited with an insulating film of silicon oxide. This insulation has to be removed at the top of the tip by a wet chemical-etching process. Then a metallization is deposited and patterned to realize a silicon/metal thermocouple. The electrical contacts of both legs of this thermocouple are performed by bonding pads. To achieve a movable cantilever, after the opening of the insulating film and after the metallization, the wafer is etched through from the front by a dry plasma-etching process. Figure 5.32 shows SEM pictures of two steps of the fabrication sequence. The photograph (left) represents a silicon tip that, in contrast to Fig. 5.30, is prepared by an isotropic plasma-etching process. However, the passivation layer on the (100)-silicon wafer has been used as shown in Fig. 5.30. The SEM picture (right) shows the microtip after deposition of an insulating layer and after removal of the insulation at the top of the tip. The tip radius is only 20 to 30 nm.

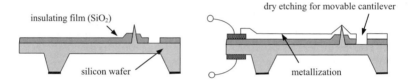

Figure 5.31 Fabrication steps for realization of a thermoelectric microtip.

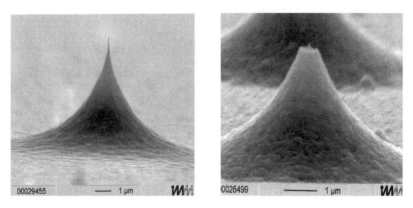

Figure 5.32 SEM pictures of fabrication steps of thermoelectric microtips: (left) after isotropic etching of silicon; (right), after deposition and partial removal of the insulating layer.

5.6.2 Diagnostic Microstructures for the Investigation of Thermal Properties of Thin Films

Many new experimental tools for the investigation of properties of thin films have been developed by using microsystems. One advantage of this approach is that such microsystems can be integrated into silicon wafers that are processed for device fabrication. Consequently, process-specific material properties can be characterized with these diagnostic structures; sometimes even *in situ* measurements are possible. As an example, Fig. 5.33 shows two microstructures for the investigation of (a) thermal conductivity and (b) thermoelectric properties of thin films used in a standard CMOS process.[66-67] The microstructures are part of the CMOS fabrication masks and can be used after completion of the CMOS process. Only one additional post-processing step is necessary to prepare the diagnostic device. Similar devices have been developed for the investigation of mechanical stress,[68] specific-heat capacity,[69] and optical properties of thin films.[70] A more detailed account of such efforts is given in Chapter 2. To illustrate, two test structures are briefly described in the following paragraphs.

The test structure of Fig. 5.33(a) shows a cantilever beam[66] that is composed of CMOS dielectrics. It contains the polysilicon film to be investigated. An aluminum heater/resistor is integrated at the tip of the beam. When this heater/resistor is applied, the temperature of the cantilever beam is increased and the temperature rise ΔT_1 is measured. This

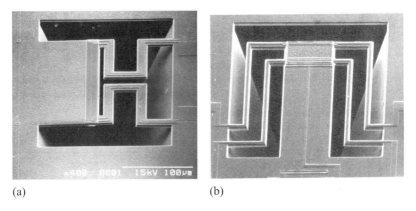

(a) (b)

Figure 5.33 Diagnostic microstructure for the investigation of (a) thermal conductivity[66] and (b) thermoelectric properties of polysilicon,[67] fabricated in a standard CMOS process.

temperature rise is compared with the temperature rise ΔT_2 caused in a cantilever beam [not shown in Fig. 5.33(a)] with identical structure, but without the polysilicon film. The thermal conductivity of polysilicon film can be calculated with the measured ΔT_1 and ΔT_2.

The micromachined test structure[67] of Fig. 5.33(b) consists of a 100-µm-wide and 260-µm-long beam with four narrow arms suspended over a cavity. The beam is composed of all CMOS dielectrics and contains the contacted polysilicon film to be characterized. The total thickness of the dielectric sandwich, including the polysilicon layer is 3.8 µm. The polysilicon film is 246 µm long and 92 µm wide. Using a polysilicon heater, a temperature gradient is created along the cantilever and with a polysilicon temperature sensor the temperature difference ΔT between cantilever tip and heat sink (bulk silicon wafer) is measured. At the hot and cold contacts, where aluminum films are connected with the polysilicon layer, the thermoelectric voltage U across the polysilicon is detected. Using U and ΔT the Seebeck coefficient of the investigated film can be calculated.

5.7 Conclusion

Examples for thermal-based microsensors and microactuators realized by combination of IC process technology with silicon micromachining have been presented. In favorable cases the approach to microsensors and microactuators outlined here will lead to cost-effective sensor/actuator-based products with co-integrated driving and signal processing circuitry. The improvement of such microsensors and microactuators needs further investigations, notably the systematic measurement of process-dependent properties of materials and the optimization of design by using device simulation. Furthermore, systematic investigations of long term stability and reliability will become important for wide industrial application and commercial success.

References

1. P. Krummacher and H. Oguey, *Sensors and Actuators*, A21–23: 636–638 (1990)
2. T. Nakamura and K. Maenaka, *Sensors and Actuators*, A21–23:762–769 (1990)
3. H. Baltes and A. Nathan, in *Sensors—a Comprehensive Survey*, vol. 1 (T. Grantke and W. H. Ko, eds.), VCH, Weinheim, Germany (1991), pp. 195–215
4. J. Kramer, P. Seitz, and H. Baltes, *Digest of Technical Papers Transducers '91, San Francisco* (1991), pp. 727–729

5. A. van den Berg and P. Bergveld, *Micro Total Analysis Systems*, Kluwer Academic Publishers, Dordrecht, The Netherlands (1994)
6. J. Köhler, U. Dillner, A. Mokansky, S. Poser, and T. Schulz, "Micro channel reactors for fast thermocycling," *Proc. 2nd Int. Conf. Microreaction Technol., New Orleans* (1998), pp. 241–247
7. H. Baltes, *Technical Digest of the 10th Sensor Symposium, Tokyo* (T. Nakamura, ed.), Institute of Electrical Engineers of Japan, Tokyo (1991), pp. 17–23
8. H. Baltes, D. Moser, and F. Völklein, in *Sensors—a comprehensive survey*, vol. 7 (W. Göpel, J. Hesse, and J. N. Zemel, eds.), VCH, Weinheim, Germany (1993), pp. 13–55
9. T. Ricolfi and J. Scholz, *Thermal Sensors*, VCH, Weinheim, Germany (1990)
10. Halbleiter-Sensoren, product information, Siemens AG, München (1991)
11. H. F. Wolf, *Silicon Semiconductor Data*, Pergamon Press, Oxford (1969)
12. R. Holm, *Electric Contacts, Theory and Applications*, Springer, Berlin (1967)
13. J.-S. Shie and P. K. Weng, "Fabrication of micro-bolometer on silicon substrate by anisotropic etching technique," *Digest of Technical Papers Transducers '91, San Francisco* (1991), pp. 627–630
14. B. E. Cole, R. E. Higashi, and R. A. Wood, "Monolithic two-dimensional arrays of micromachined microstructures for infrared applications," *Proc. IEEE*, 86(8):1679–1686 (1998)
15. G. Scheller, *Proc. Sensor 93, Nürnberg*, vol. 1 (1993), pp. 87–93
16. R. Aigner et al., *Digest of Technical Papers, Transducers '95, Stockholm* (1995), pp. 839–842
17. Microsens SA, Neuchatel/Switzerland, product information (1991)
18. F. Nuscheler, *Archiv f. Elektronik u. Übertragungstechnik*, vol. 42 (1988), pp. 80–84
19. A. G. McNamara, "Semiconductor diodes and transistors as electrical thermometers," *Rev. Sci. Instrum.*, 33:330–333 (1962)
20. S. Middelhoek and S. A. Audet, *Silicon Sensors*, Academic Press, London (1989)
21. M. P. Timko, "A two-terminal IC temperature transducer," *IEEE J. Solid State Circuits*, 11:784–788 (1976)
22. R. A. Pease, "A new Fahrenheit temperature sensor," *IEEE J. Solid-State Circuits, SC-19(6):971–977 (1984)*
23. T. J. Seebeck, *Pogg. Ann.*, VI:133 (1826)
24. J. C. A. Peltier, *Ann. Chem. Phys.*, 56:371 (1834)
25. W. Thomson (Lord Kelvin), *Proc. R. Soc. Edinburgh Trans. 21 Part I* (1857), p. 123
26. M. Melloni, *Ann. Phys.*, 28:371 (1833)
27. F. Völklein and H. Baltes, "Optimization tool for the performance parameters of thermoelectric microsensors," *Sensors and Actuators A*, 36:65–71 (1993)
28. A. W. van Herwaarden and P. M. Sarro, "Thermal sensors based on the Seebeck effect," *Sensors and Actuators*, 10:321–346 (1986)
29. F. Völklein and H. Baltes, "Thermoelectric properties of polysilicon films doped with phosphorus and boron," *Sensors and Materials*, 3:325–334 (1992)

30. F. Völklein, A. Wiegand, and V. Baier, *Sensors and Actuators A*, 29:87–91 (1991)
31. J. Schieferdecker, R. Quad, E. Holzenkämpfer, and M. Schulze, *Sensors and Actuators A*, 46–47:422–427 (1995)
32. A. W. van Herwaarden and P. M. Sarro, *Sensors and Actuators*, 10:321–346 (1986)
33. A. W. van Herwaarden, D. C. van Duyn, and B. W. van Oudheusden, *Sensors and Actuators A*, 21–23:621–630 (1989)
34. J. Schieferdecker, M. Schulze, R. Quad, and A. Beudt, *Proc. Sensor 93, Nürnberg*, vol. 1 (1993), pp. 613–618
35. M. Simon, J. Schieferdecker, M. Schulze, R. Gottfried-Gottfried, M. Müller, and R. Jähne, *Proc. Sensor 97, Nürnberg*, vol. 2 (1997), pp. 83–88
36. IPHT Jena, Germany, with permission
37. U. Münch, D. Jaeggi, N. Schneeberger, A. Schaufelbühl, O. Paul, H. Baltes, and J. Jasper, "Industrial fabrication technology for CMOS infrared sensor arrays," *Digest of Technical Papers, Transducers '97, Chicago* (1997), pp. 205–208
38. N. Schneeberger, CMOS microsystems for thermal presence detection, Ph.D. thesis, ETH Zurich, No. 12675 (1998)
39. F. Völklein and W. Schnelle, *Sensors and Materials*, 3:41–48 (1991)
40. A. W. van Herwaarden, D. C. van Duyn, and J. Groeneweg, "Small-size vacuum sensors based on silicon thermopiles," *Sensors and Actuators A*, 25–27:565–569 (1991)
41. A. W. van Herwaarden and P. M. Sarro, *J. Vac. Sci. Technol.*, A5:2454–2457 (1987)
42. P. K. Weng and J.-S. Shie, *Rev. Sci. Instr.*, 65:492–499 (1994)
43. D. Moser, R. Lenggenhager, and H. Baltes, "Silicon gas flow sensors using industrial CMOS and bipolar IC technology," *Sensors and Actuators A*, 25–27:577–581 (1991)
44. B. W. van Oudheusden and A. W. van Herwaarden, "High-sensitivity 2-D flow sensor with an etched thermal isolation structure," *Sensors and Actuators A*, 21–23:423–430 (1990)
45. R. G. Johnson and R. E. Higashi, "A highly sensitive silicon chip microtransducer for air flow and differential pressure sensing applications," *Sensors and Actuators*, 11:63–72 (1987)
46. G. Kaltsas, *Proc. Eurosensors XII, Southampton*, vol. 2 (1998), pp. 757–760
47. O. Tabata, *IEEE Trans. Electron. Devices*, ED-33:361–365 (1986)
48. M. Klonz and T. Weimann, *IEEE Trans. Instr. Meas.*, 38:335–336 (1988)
49. H. Dintner, M. Klonz, A. Lerm, F. Völklein, and T. Weimann, *IEEE Trans. Instr. Meas.*, 42:612–614 (1993)
50. F. Völklein and E. Kessler, *Proc. XVII. Int. Conf. on Thermoelectrics, Nagoya* (1998), pp. 214–217
51. R. Marlow and E. Burke, "Module design and fabrication," in *CRC Handbook of Thermoelectrics* (D. M. Rowe, ed.), CRC Press, Boca Raton, FL (1995), pp. 597–607

52. D. M. Rowe, Gao Min, and F. Völklein, "A high performance thin film thermoelectric cooler," *Proc. 33rd Int. Engineering Conf. on Energy Conversion, Colorado Springs* (1998), pp. 24–28
53. F. Völklein, Gao Min, and D. M. Rowe, "Modelling of a microelectromechanical thermoelectric cooler," *Sensors and Actuators A*, 75:95–101 (1999)
54. F. Völklein, V. Baier, U. Dillner, and E. Kessler, *Thin Solid Films*, 187:253–262 (1990)
55. A. Boyer and E. Cisse, "Properties of thin film thermoelectric materials: application to sensors using the Seebeck effect," *Mater. Sci. Eng.*, B13:103–111 (1992)
56. D. Moser, CMOS flow sensors, Ph.D. thesis, ETH Zurich, No. 10059 (1993)
57. R. Lenggenhager, CMOS thermoelectric infrared sensors, Ph.D. thesis, ETH Zurich, No. 10744 (1994)
58. O. Paul and H. Baltes, "Thermal vacuum sensor by CMOS IC technology and sacrificial metal etching," *Sensors and Materials*, 6:245–249 (1994)
59. D. Jaeggi, Thermal converters by CMOS technology, Ph.D. thesis, ETH Zurich, No. 11567 (1996)
60. V. M. Meyer, N. Schneeberger, and B. Keller, "Epoxy-protected thermopile as high sensitive heat flux sensor, *Digest of Technical Papers Transducers '97, Chicago* (1997), pp. 1267–1270
61. O. Brand, M. Hornung, D. Lange, and H. Baltes, "CMOS resonant microsensors," *Proc. SPIE*, 3514:238–250 (1998)
62. P. Malcovati, CMOS thermoelectric sensor interfaces, Ph.D. thesis, ETH Zurich, No. 11424 (1996)
63. A. Häberli, Compensation and calibration of IC microsensors, Ph.D. thesis, ETH Zurich, No. 12090 (1997)
64. M. Strasser, R. Aigner, and G. Wachutka, "Analysis of a CMOS low power thermoelectric generator, *Proc. Eurosensors XIV, Copenhagen* (2000), pp. 17–20
65. E. Oesterschulze and R. Kassing, "Thermal and electrical imaging of surface properties with high lateral resolution, *Proc. XVI. Int. Conf. on Thermoelectrics, Dresden* (1997), pp. 719–725
66. F. Völklein and H. Baltes, *J. Electromech. Syst.*, 1:193–196 (1992)
67. M. von Arx, Thermal properties of CMOS thin films, Ph.D. thesis, ETH Zurich, No. 12743 (1998)
68. H. Kapels, "Material and device characterization in silicon micromachining," *Proc. Symposium on Microtechnology in Metrology and Metrology in Microsystems, Delft* 2000, pp. 109–119
69. M. von Arx, L. Plattner, O. Paul, and H. Baltes, "Micromachined hot plate test structures to measure the heat capacity of CMOS IC thin films," *Sensors and Materials*, 10:503–517 (1998)
70. R. F. Wolffenbuttel, "Optical characterization of silicon-compatible materials," *Proc. Symposium on Microtechnology in Metrology and Metrology in Microsystems, Delft* 2000, pp. 79–90

6 Photon Detectors

Arokia Nathan
University of Waterloo, Ontario, Canada

Karim S. Karim
Simon Fraser University, Vancouver, Canada

6.1 Introduction

Current interest in hydrogenated amorphous silicon (a-Si:H) technology extends well beyond active-matrix liquid-crystal displays and solar cells; it stems from the variety of desired material and technological attributes.[1-3] The high optical absorption; low-temperature deposition (<300°C); high uniformity over large areas; few constraints on substrate size, material, or topology; standard integrated circuit lithography processes; and low capital equipment cost associated with the a-Si:H material offer a viable technological alternative for improved imaging of optical signals and high-energy radiation. Notable application areas include contact imaging for document scanning, digital copiers, and fax machines; color sensors/imaging; position/motion detection; and radiation detection/imaging of high-energy X-rays in biomedical applications, gamma-ray space telescopes, airport security systems, and nondestructive testing of the mechanical integrity of materials or structures.

There are two architectures currently used in large-area imaging arrays (or flat panel imagers): (1) the linear architecture, used in photocopiers, fax machines, and scanners, and (2) the two-dimensional or area array architecture, used in digital (including video) lensless cameras as well as X-ray imaging systems. In both linear and 2-D architectures, the basic imaging unit is the pixel, which consists of an image sensor and a switch. The pixel is accessed by a matrix of gate and data lines, and operated in storage (or integration) mode. Here, during the off-period of the switch, the sensor charge is integrated. When the switch is turned on, the charge in the sensor is transferred to the data line, where it is detected by a charge-sensitive amplifier. Currently, either a diode[4] or a thin-film transistor (TFT)[5] is used for the switching element. Although the diode has fewer masking steps and simpler connectivity, it comes with high capacitance and nonlinear current-voltage characteristics. In contrast, the TFT has a lower capacitance, linear switching characteristics, and low leakage current, and is thus more widely used despite its potentially large shift in threshold voltage (V_T) after prolonged gate bias. The reduction of both leakage and instability remains a key challenge from the standpoint of materials, processing, and device design.[6]

In large-area optical imaging applications, the sensing elements most commonly used are based on the Schottky or p-i-n a-Si:H photodiode structures. The latter typically has p-type amorphous silicon carbide to yield a large barrier height and hence lower dark current. The lowest dark current reported for p-i-n photodiodes is below 10 pA/cm^2,[7] which is at least two orders lower than the best reported dark current of 1 nA/cm^2 for the Schottky photodiode.[8,9] The other device characteristics, such as I_{photo}/I_{dark}, ideality factor, quantum efficiency, dark-current instability, and response time, are similar between both structures; typical values are 10^3 to 10^4, 1.2 to 1.5, 0.7 at 550 nm, <10% at -2 V, and <1 µs, respectively. There are many applications, however, where a dark current of 1 to 10 nA/cm^2 is low enough to achieve a frame rate of 30 fps. Therefore the Schottky structure can still be a candidate if its leakage current can be further reduced. For example, if an insulating nitride layer is incorporated at the metal/semiconductor interface to form a metal insulator semiconductor (MIS) structure, a reduction in the dark current and an increase in photocurrent has been observed to yield a hundred-fold improvement in dynamic range.[10] However, the quality of the dielectric and its thickness are critical because these characteristics have a strong bearing on the tunneling transmission of carriers through the dielectric.

For large-area imaging of X-rays, different detection schemes can be used. One configuration uses a phosphor layer, which converts X-rays into visible photons; these photons are detected by an a-Si:H array of optical image sensors and thin-film transistors.[11] The efficiency of detection is determined largely by the X-ray absorption efficiency of the layer, the efficiency of conversion of X-ray photons to visible light, the modulation transfer function (MTF), and the (optical) quantum efficiency of the photodetector (typically 0.7). The first three of these factors depend on the type of phosphor material. With a cesium iodide (CsI) phosphor layer co-integrated with the a-Si:H electronics, good success has been demonstrated at relatively high X-ray energies for digital radiology applications. An alternative arrangement uses a photoconductor such as bulk amorphous selenium (a-Se) for direct photoelectric conversion[12] of X-rays to electron-hole pairs, which are subsequently read out using a-Si:H electronics. Here a thick a-Se layer is required for high X-ray absorption. This detection arrangement offers high resolution and conversion efficiency, including at lower X-ray energies. A sample image from an a-Se flat panel imager is depicted in Fig. 6.1.[13] However, the a-Se material requires a high electric field (10 V/µm) for efficient electron-hole separation and collection, which is achieved by applying a large bias

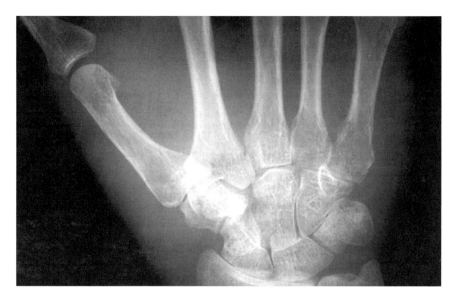

Figure 6.1 Radiograph of a hand phantom made with a prototype amorphous selenium flat panel imager with active matrix readout.

(4 to 10 kV) across the a-Se layer. Other more recent direct detection sensors include Mo/a-Si:H Schottky diodes[14] and Mo/a-Si:H sandwiched structures.[15] Here, interaction of X-ray photons with Mo leads to ejection of high-energy electrons, by virtue of the photoelectric effect, into a reverse-biased a-Si:H depletion layer, where electron (avalanche) multiplication yields a gain.

Extensive reviews on applications of the a-Si:H flat panel imager technology to biomedical X-ray imaging are contained in a recently published *Handbook of Medical Imaging*,[16] which describes the physics of the different imaging modalities. This chapter reviews recent developments in large-area a-Si:H electronics associated with photon detectors. The focus is on image sensors and this family of detectors in terms of their operating principles, materials-related and processing issues, electrical and optoelectronic characteristics, and stability, along with the new challenges that lie ahead for reduction of dark current. Issues pertinent to sensor-TFT integration are also discussed, along with new pixel architectures for high-fill-factor imaging arrays with reduced parasitic capacitance, processing conditions for reduced V_T shift (ΔV_T) and leakage current, and enhanced mechanical integrity. Selected results are shown for X-ray and optical detectors and integrated X-ray pixel structures. Extension of the current fabrication processes to low temperature (~120°C), enabling fabrication of flexible (curved) imagers (on polymeric substrates) for high light collection efficiency, are also discussed.

6.2 Detectors

6.2.1 Mo/a-Si:H Schottky Diode X-Ray Image Sensors

X-ray photon detection in Mo/a-Si:H Schottky diodes is based on the photoelectric interaction of X-ray photons in the Mo layer and ejection of energetic electrons into the a-Si:H layer.[14] The device schematic and operating principle are depicted in Fig. 6.2. Inside the a-Si:H layer, the energetic electrons undergo various scattering events by which they transfer their energies, leading to generation of e-h pairs. The e-h pairs are then separated by the electric field in the depletion region of the reverse-biased Schottky diode. The detector quantum yield, i.e., the ratio of the measured number of electrons to the number of incident X-ray photons, can be studied by arranging the underlying transduction processes into various stages of conversion as indicated in Fig. 6.3. The overall transduction efficiency of the detector (G) for a mono-energetic X-ray beam can be shown in terms of the efficiencies of different stages as[17]

$$G = \eta (\zeta \gamma \chi + \kappa), \tag{1}$$

where η is the photoelectron generation efficiency, ζ is the ejection efficiency of the photoelectrons, γ is the e-h generation efficiency in the a-Si:H by the ejected photoelectrons, χ is the e-h separation efficiency in a-Si:H, and κ is the contribution to thermionic emission caused by photo-

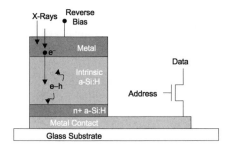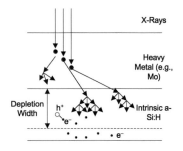

Figure 6.2 Device schematic and operating principle of direct X-ray sensor based on a Mo/a-Si:H Schottky diode for large-area imaging applications.

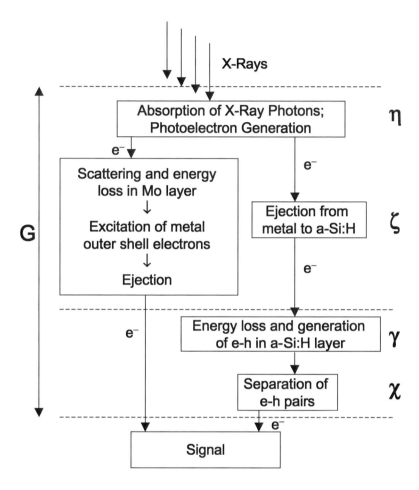

Figure 6.3 Schematic representation of the transduction process in the fabricated detectors. Greek symbols refer to the efficiencies of each process.

electrons absorbed in the Mo layer. Equation (1) is of particular importance in order to examine the impact of different parameters on the detector performance. A quantitative analysis of the various efficiencies in Eq. (1) can be found in Nathan et al.[17]

The fabrication process of the Schottky diode is described as follows. First, a thin (100 nm) chromium (Cr) layer is sputter-deposited on a glass wafer and patterned with mask 1. This is followed by plasma-enhanced chemical vapor deposition (PECVD) of highly doped amorphous silicon (n^+ a-Si:H) and intrinsic amorphous silicon (i-a-Si:H) layers in one pump-down to preserve the interface quality. The thicknesses of n^+ a-Si:H and i-a-Si:H are 100 nm and 1 µm, respectively. The n$^+$ a-Si:H layer is deposited to provide an ohmic contact with the bottom electrode metal. Then a heavy metal (in this case, Mo) is deposited to form the Schottky barrier interface and patterned with mask 2. Mask 3 patterns the top amorphous silicon nitride (a-SiN$_x$) layer used as the etch stop against KOH solution, which etches the amorphous silicon layers. Finally, mask 4 is used to open the contact windows on the top nitride layer where needed. The deposition parameters of the various thin films are listed in Table 6.1.

The choice of the top electrode material needs to consider X-ray absorption, electron ejection, and the intrinsic mechanical stress. The stress in thin films is discussed in more detail in association with pixel configurations. The work function of the heavy metal determines the Schottky barrier height, which is crucial because it determines the detector leakage. The leakage current of a Mo/a-Si:H Schottky diode is shown in Fig. 6.4. The fabricated diodes show rectifying characteristics with an on/off ratio of 10^6 at ±1 V. The barrier height was measured to be 0.77 eV, and the ideality factor was 1.2. The diodes have a low reverse current density (<10 nA/cm^2 at −1 V); this is crucial for charge retention during imaging. The thickness of the (top electrode) heavy metal is based on a compromise between the absorption of X-rays and the ejection of the energetic electrons from the metal into the depletion layer.

Typical X-ray measurement results of a (200 µm)^2 Mo/a-Si:H Schottky diode operated at 2 V reverse bias are illustrated in Fig. 6.5. These confirm the linear response of the detector to the number of absorbed photons obtained from simulations;[17] the linearity is consistent with data measured for other values of X-ray currents at constant X-ray source voltage. Figure 6.6 illustrates the impact of the Mo thickness on the performance of the detector.[17] At low thicknesses, the performance is limited by the absorption of X-ray photons inside the Mo layer. An increase in the thickness of Mo leads to an increase in the number of absorbed photons, hence the output signal. However, beyond an optimum thickness, the

Table 6.1 Deposition Conditions of Thin Films

	System	Temp. (°C)	Pressure (mTorr)	Power (W)	Deposition rate (Å/min.)
Cr	DC sputtering	25	5	210	112
Mo	DC sputtering	25	5	80	45
a-SiN$_x$:H	PECVD	260	400	110	95
i-a-Si:H	PECVD	260	160	10	110
n$^+$ a-Si:H	PECVD	260	160	10	50

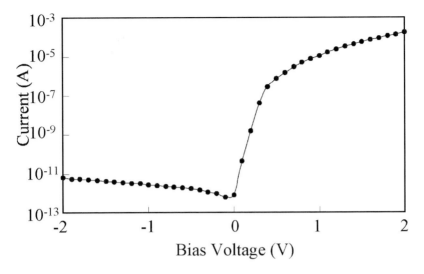

Figure 6.4 Forward and reverse current characteristics of a Mo/a-Si:H Schottky diode X-ray image sensor.

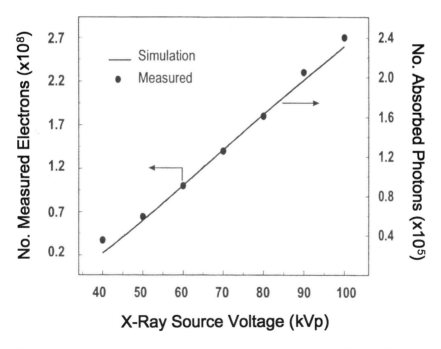

Figure 6.5 Response of a sensor for various X-ray source voltages (kVp), collected over a period of 500 ms.

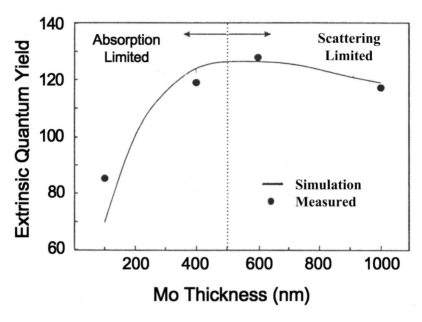

Figure 6.6 Variation of sensor response with the thickness of the Mo layer, taken at 60 kVp and 25 mAs.

number of electrons reaching the a-Si:H depletion region is limited by the scattering and absorption inside the Mo film. For the X-ray source voltage of 60 kVp, a thickness of Mo ~ 500 nm yields the optimum performance. At higher energies, this optimum thickness moves to thicker films.

The thickness of the a-Si:H layer is another important parameter; it mainly determines the efficiency of stopping the ejected photoelectrons and absorbing their kinetic energies inside a-Si:H. The number of measured electrons increases as the thickness of the a-Si:H film increases. However, at large thicknesses, the charge collection efficiency at constant bias reduces, since the depletion width remains constant but the neutral region increases. Therefore, the optimum thickness is a compromise between the range of electrons and carrier diffusion length in the a-Si:H. Measurement results give a thickness of ≈1 μm as the optimum thickness for the a-Si:H layer. With respect to bias conditions, the choice is based on a compromise between signal level, noise, reverse current, and reverse current stability.[18-20] Although a low bias results in stable characteristics, low noise, and low leakage current, it yields lower charge collection efficiency. At higher biases, the charge collection improves, but the stability, noise, and leakage current degrade. Typical values for the noise, stability, detector saturation, and dynamic range, in terms of numbers of collected electrons over a 500 ms integration period, are 8×10^5, 10^7, 5×10^8, and 50:1, respectively. The use of other metals (e.g., W) would yield a larger barrier height with a-Si:H, thus allowing higher voltage operation to enhance charge collection. A larger barrier height would also result in a wider depletion region and a reduction in the density of trapped charges, yielding an improvement in detector stability.

6.2.2 ITO/a-Si:H Schottky Diode Optical Image Sensors

Indium tin oxide/amorphous silicon (ITO/a-Si:H) Schottky diodes have been used in solar cells and optical imaging arrays. When coupled with a scintillating (phosphor) layer (for conversion of X-ray photons into visible light to be subsequently detected by a-Si:H photosensors), they enable imaging of X-rays. In large-area imaging applications, an improvement in the electrical characteristics of this family of sensors is needed particularly with the dark current and its stability (i.e., its variation in time). The dark current is a source of noise, and its accumulation limits the integration time, especially when imaging low signal levels.

Contributions to the dark current stem from contact injection, bulk thermal generation, and edge effects.[21] Bulk thermal generation is a feature of the active layer and is generally negligible; edge effects are important only when the sensor area is small. Contact injection depends on interface structure and is the main source of the dark current. It is determined not only by the work functions of the optically transparent ITO and a-Si:H, but also by the ITO/a-Si:H interface integrity, which includes interface states, interface charges, and interface nonuniformity. A high density of interface states increases tunneling, which in turn increases the dark current. This is particularly true in ITO/a-Si:H Schottky diodes, where there is significant diffusion of oxygen and indium from the ITO into a-Si:H if the ITO is deposited at elevated temperatures.[22] The presence of these impurities also gives rise to instability in the dark current. Therefore, there is a clear need for good-quality ITO that can be deposited at relatively low temperature and yet maintain good optical transmission.

Following the considerations outlined above, a deposition process for polycrystalline ITO at room temperature has been developed[23] for realization of stable and low dark-current Schottky photosensors.[9] The device schematic and operating principle are illustrated in Fig. 6.7. Here, the ITO resistivity is below 6×10^{-4} Ωcm, and the optical transmittance of the film, within the range of visible wavelengths, is in excess of 80%.[23] The fabrication process starts with a 120-nm Cr layer that acts as the bottom electrode, followed by deposition of a highly doped n^+ a-Si:H layer

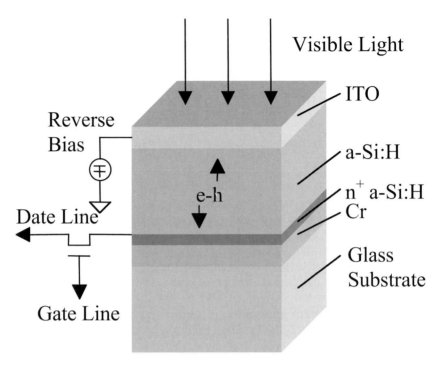

Figure 6.7 Schematic of an ITO/a-Si:H Schottky diode photosensor.

(with a thickness of 50 nm) for ohmic contact. Thick (1μm), intrinsic a-Si:H is then deposited to serve as the active layer. For all devices, the a-Si:H is etched in buffered HF to remove any surface oxide before the samples are loaded into the sputtering system. Polycrystalline ITO (with a thickness of 80 nm) is then deposited in pure argon at room temperature for the Schottky contact and light window.

Figure 6.8 shows the I-V characteristics of the ITO/a-Si:H Schottky photodiode. The dark current at a reverse bias of -2 V is observed to be low (~7 × 10^{-10} A/cm^2). Secondary ion mass spectroscopy (SIMS) measurements indicate that this notable improvement in device characteristics stems from reduced diffusion of oxygen, rather than indium, from the ITO into the a-Si:H layer, thus preserving the integrity of the Schottky interface.[9] The diffusion profiles of the indium and oxygen into the a-Si:H layer at different annealing temperatures are shown in Fig. 6.9. Here the key variation observed is with oxygen; it diffuses more deeply by 20 nm in sensors annealed at 260°C as opposed to sensors with the as-deposited ITO. No big difference is observed, however, with sensors annealed at 150°C, which shows that sensors can be annealed at about 150°C, if

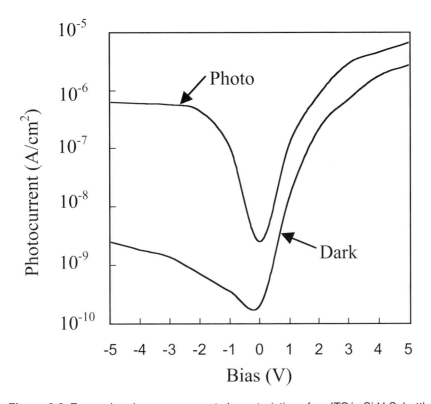

Figure 6.8 Forward and reverse current characteristics of an ITO/a-Si:H Schottky diode photosensor.

needed, to recover from light-soaked conditions. A comparison of the dark-current characteristics of a sensor with an area of 300 x 300 µm and an a-Si:H thickness of 0.75 µm for annealed and unannealed conditions is depicted in Fig. 6.10. The difference in dark current is pronounced at high reverse voltages. This dramatic behavior of dark current can be explained in terms of the increased oxygen diffusion at 260°C (see Fig. 6.9) and the underlying transport processes.[24] The oxygen creates positively charged defects in the bulk a-Si:H layer in the vicinity of the interface; these in turn actively participate in tunneling. In particular, at high bias voltages, the increased band bending results in increased interface states that contribute to tunneling, leading to a rapid increase in leakage current.

The measured shift in the dark current of the photodiode is insignificant at low reverse voltages. For example, the shift stabilizes to a value less than 9% when biased at -2 V (see Fig. 6.11). Even at high reverse voltages ($V_{rev} = -10V$), the dark current increases by no more than a factor of 3; its value is smaller than that observed in the Mo/a-Si:H Schottky

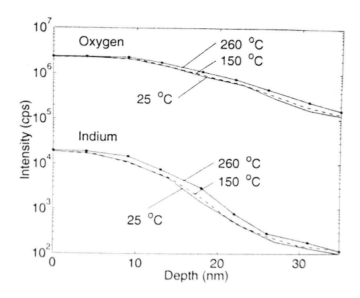

Figure 6.9 Oxygen and indium distributions in the a-Si:H layer for ITO deposited at room temperature (no annealing) and when subjected to annealing at 150 and 260°C.

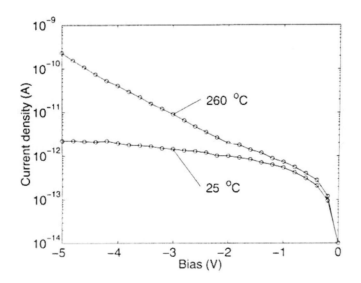

Figure 6.10 The dark-current characteristics of the ITO deposited at room temperature and when annealed at 260°C.

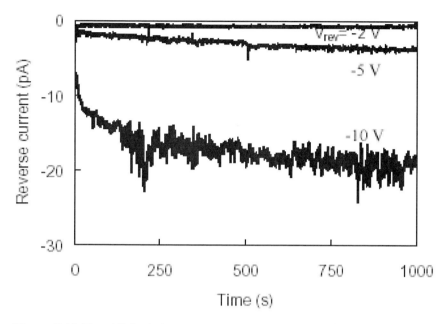

Figure 6.11 The shift in dark current at −2, −5, and −10 V.

diode.[18] A further improvement of both dark current and photocurrent has been shown possible by introducing a thin a-SiN insulator film between the a-Si:H and ITO layers.[10] The spectral response of the photodiode is shown in Fig. 6.12 for wavelengths in the range of 400 to 800 nm. As expected, the behavior of photocurrent is governed by the absorption characteristics of a-Si:H. The maximum photocurrent is at 570 nm, and the spectral response is well matched to the light emitted from phosphors when subjected to X-rays.[2]

6.2.3 ITO/p-i-n Optical Image Sensors

An a-Si:H photodiode can be configured in a number of different ways, depending on the type of blocking contact to the 1 to 2 μm-thick intrinsic a-Si:H layer that absorbs the incident visible wavelength photons. The most common approach is to use homo- or hetero-junctions (i.e., junctions made from the same or different types of a-Si:H, respectively) incorporating p- and n-type doped layers of amorphous or microcrystalline (μc-Si:H) silicon. These doped layers may also be alloyed with other elements such as carbon or germanium to adjust their optical prop-

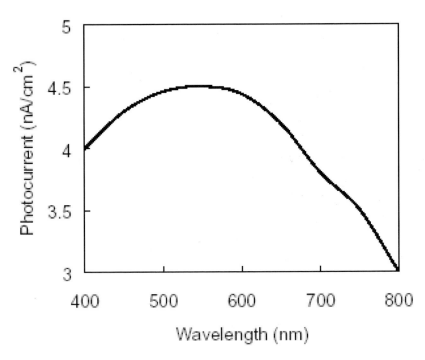

Figure 6.12 Spectral response of an ITO/a-Si:H Schottky diode photosensor.

erties.[25,26] Figure 6.13 shows a cross section of an n-i-p photodiode made using microcrystalline silicon, μc-Si. Although this is only one example of the many possible designs, it will illustrate the pertinent features. The bottom metallic contact is chromium. This is followed by a 10 to 50-nm-thick n^+ blocking layer, a ~1.5-μm-thick intrinsic a-Si:H layer, a ~10 to 20-nm-thick p^+ μc-$Si_{1-x}C_x$:H blocking layer, a ~50-nm layer of transparent ITO, and a surface passivation layer of oxy-nitride, which is very important in keeping the properties of the array stable. Sensors are usually oriented such that the p^+ layer is toward the incoming light. This allows the movement of electrons, which have the higher mobility in a-Si:H, to be the main contributor of signal charge. Light enters the intrinsic layer through the passivation layer, the ITO, and the p^+ layer. Each of these layers absorbs some light, thereby reducing the absorption in the intrinsic layer and hence the signal measured in the photodiode.

To lessen this detrimental effect, the passivation and ITO layers are made as thin as possible, and the μc-Si:H in the p^+ layer is alloyed with carbon to increase its optical band gap, and hence transparency, and improve the diode reverse leakage current properties.[27] Photocharge generated in the p^+ and n^+ layers does not contribute to the signal charge

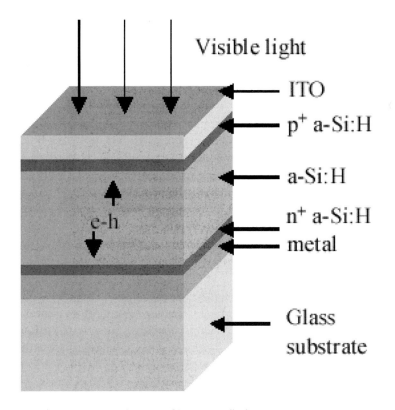

Figure 6.13 Cross section of n-i-p a-Si:H photodiode structures.

because of the very short drift lengths of carriers in these heavily doped layers. An externally applied reverse bias voltage of ~–5 V applied to the ITO creates a depletion layer[28] across the thickness of the intrinsic a-Si:H. Each light photon absorbed in the depletion layer generates a single e-h pair. The electron moves toward the n^+ layer and the hole toward the p^+ layer under the action of the applied electric field. Because of the thin nature of the intrinsic a-Si:H layer and the relatively low intensities of incident photons from the overlying phosphor, geminate and nongeminate recombination[29] of the charge carriers in the bulk of the sensor is minimal. The majority of the photogenerated charge in the intrinsic layer is transferred to the n^+ and p^+ contacts. Some loss of charge may result from back diffusion of carriers at the p^+/i-layer interface, but this is minimal at the low intensities typical of medical imaging applications. The quantum efficiency drops at higher wavelengths because of increased penetration of the intrinsic layer and at lower wavelengths due to increased absorption in the p^+ layer. The photodiode also acts as a capacitor to store the photogenerated charge. The intrinsic capacitance of the n-I-p structure is described by the parallel-plate capacitor formula

$$C_{pd} = \varepsilon_r . \varepsilon_o . A/d, \qquad (2)$$

where ε_r is the relative dielectric constant of the intrinsic semiconductor (~12 for a-Si:H), ε_o is the permittivity of free space (8.85 × 10^{-14} F/cm), A is the geometric area, and d is the thickness of the photodiode. Thus a 100 x 100 μm, 1-μm-thick photodiode has a capacitance of ~1 pF. The total charge stored on the photodiode is its capacitance multiplied by the applied reverse bias voltage (e.g., ~5 pC at –5 Vbias).

6.3 Thin-Film Transistors

6.3.1 TFT Structures and Operation

The lateral thin-film transistor can be made in one of four basic structures, defined by the order in which the source and drain contacts, semiconductor layer, gate insulator, and gate electrode are deposited onto a glass substrate. Figure 6.14(a) and (b) shows staggered-electrode structures where the source and drain contacts are on one side of the semiconductor and the gate electrode is on the opposite side. In the staggered structure of Fig. 6.14(a), the metal source and drain electrodes are first deposited on the insulating substrate, separated by a narrow space for the channel. The semiconductor, insulating layer, and gate are then deposited in alignment with the channel. In the inverted-staggered structure of

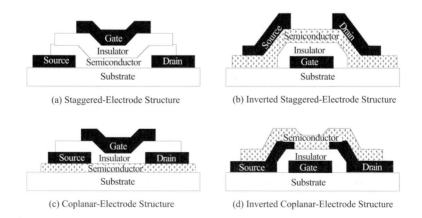

Figure 6.14 Types of thin-film transistors.

Fig. 6.14(b), the gate layer is the first layer deposited on the glass. Figure 6.14(c) and (d) shows coplanar-electrode structures where the source and drain electrodes and the gate electrode are on the same side of the semiconductor.

The most popular structure for a-Si:H TFTs is the inverted-staggered structure with SiN_x used as the insulator layer in state-of-the-art applications.[30] Chromium and molybdenum are commonly used gate materials, and aluminum is commonly used for source and drain contacts. There are many different technologies used in fabricating this particular structure, with two dominating.[30] In the first type, there is a consecutive deposition of gate SiN_x, intrinsic a-Si:H, and n^+ a-Si:H or n^+ μc-Si:H (n^+ henceforth) in a single growth step without breaking the deposition chamber vacuum. The n^+ is then etched from the channel region of the transistor, followed by the contact metallization. Finally, a top passivating layer, which could be SiN_x or a polyimide, is deposited. This type of TFT is shown in Fig. 6.15(a).

In the second common type of inverted-staggered TFT, gate SiN_x, intrinsic a-Si:H, and a second SiN_x layer are deposited consecutively in a single chamber pumpdown. The top nitride is then etched from the contact regions and the n^+ is deposited and patterned. Finally, the contact metallization is deposited and patterned. This type of TFT is shown in Fig. 6.15(b). Typically, the inverted-staggered, top-nitride type TFTs show superior leakage current performance because the consecutive deposition gives rise to an improved interface between the two SiN_x films and the intrinsic a-Si:H film.[31]

The basic operation of a-Si:H TFTs is similar to that of crystalline silicon MOSFETs, with differences resulting from their individual energy band structures. In MOSFETs, the gate-induced charge in the semiconductor is primarily made of free carriers. In TFTs, the localized states trap most of the induced charge. The density of free carriers in the channel of an a-Si:H TFT rarely exceeds 10% of the total gate-induced charge.[32] Therefore, the distribution of the localized states is an important factor in understanding TFT behavior.

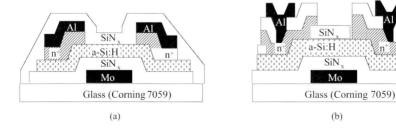

Figure 6.15 Two main types of inverted staggered-electrode TFTs.

Like the MOSFET, the a-Si:H TFT has two basic regions of operation. Figure 6.16 schematically illustrates the band bending and presents a simple DOS diagram at equilibrium and with positive applied gate voltages, V_g. Neglecting the flat-band voltage, at equilibrium ($V_g = 0$), no band bending occurs and E_F is at the intrinsic Fermi-level position [Fig. 6.16(a)]. When a low positive gate voltage is applied ($V_g > 0$), the bands begin to bend downward [Fig. 6.16(b)]. E_F moves through the deep states toward the conduction band mobility edge, E_C, and most of the induced electrons are trapped within the deep states. There are, however, a few electrons trapped in the tail states; some of them go into the extended states, leading to an exponential increase in the source-drain current. This regime of operation is referred to as the *subthreshold* or *below threshold* regime.[30,32]

As the gate voltage is increased further ($V_g >> 0$), the majority of the gate-induced electrons become trapped in the tail states rather than the deep states. This leads to a dramatic change in the I-V characteristics of the TFT. This regime of operation, where the device behavior is dominated by the tail states, is known as the *above-threshold* regime. As the gate voltage is increased, the Fermi level is pushed closer to the conduc-

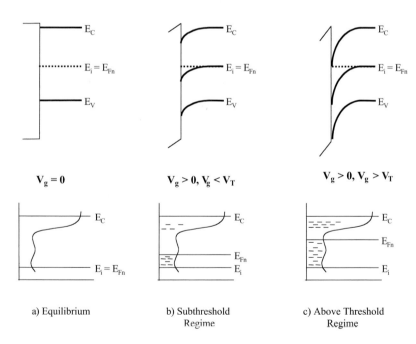

Figure 6.16 Basic operation of an a-Si:H TFT showing band bending and density of states.

tion band edge, causing an increase in the number of electrons that are trapped in the tail states. The gate voltage at which the transition from the subthreshold regime to the above threshold regime occurs is called the *threshold* voltage (V_T). More detail on TFT operation and modeling is given by Powell,[30] Shur and Hack,[32] and Leroux.[33]

6.3.2 Threshold Voltage (V_T) Metastability

Studies of a-Si TFTs have concluded that two mechanisms can explain the electrical instability.[34,35] The first is carrier trapping in the gate insulator, which occurs primarily in PECVD a-Si:N gate insulator TFTs where the high density of defects in the insulator can trap charges when the TFT gate undergoes bias stress. Injection of electrons into the insulator layer of a-Si TFTs has long been known to cause ΔV_T. Electrons are first trapped within the localized interfacial states at the a-Si/a-SiN boundary and they then thermalize to deeper energy states inside the a-SiN layer, either by variable-range hopping[34] or through a multiple-trapping and emission process.[36-38] The broad distribution of energies for the trapping levels with the a-SiN layer leads to two kinds of electron trapping behavior.[38]

The first is due to interfacial traps with fast re-emission times (or fast states). These traps were found to cause a hysteresis in the I_{DS}-V_{GS} behavior of TFTs.[36,39,40] The density of these fast states decreases when the optical band gap of the nitride layer is increased. Moreover, the optical band gap of the nitride was found to increase when the nitrogen content of the layer was increased.[39,40] Nitrogen-rich nitrides are generally obtained by using a higher ratio of NH_3/SiH_4 (~10) during PECVD deposition.

The second nitride-related degradation effect is due to interfacial and bulk traps with slow re-emission times (slow states). These traps cause a metastable increase in the threshold voltage.[34,37,38,40,41] However, the number of these deep traps was found to decrease for wider bandgap, nitrogen-rich a-SiN layers.[39] As such, the use of nitrogen-rich a-SiN has become prevalent for TFTs because it improves the reliability and reduces degradation associated with both slow and fast states in the gate insulator. The in-house TFT fabrication process uses a ratio of NH_3/SiH_4 ~10 for the a-SiN films to achieve good metastability characteristics.

The second mechanism related to metastability is point-defect creation in the a-Si layer or at the a-Si/SiN interface that increases the density of deep-gap states.[42] When a positive gate bias causes electrons to accumulate and form a channel at the a-SiN/a-Si interface, a large number of these induced electrons reside in conduction-band tail states, as noted previously. These tail states have been identified as weak silicon-silicon bonds that, when occupied by electrons, can break to form silicon dangling bonds (deep-state defects).[43,44] Deep-state defect creation forms the basis of the defect-pool model,[42] which has some similarities to the Staebler-Wronski effect,[45] in which photo-generated carriers result in the generation of dangling bonds.[46] In the defect-pool model, the rate of creation of dangling bonds is a function of ϕ, the barrier to defect formation, the number of electrons in the tail states, and the density of the weak bond sites.[42] The metastable increase in the density of deep defect states is shown in Fig. 6.17. The location of these newly created defect states in the gap depends on the polarity of the applied stress bias and, in some cases, can cause the V_T of the TFT to shift.[42,47]

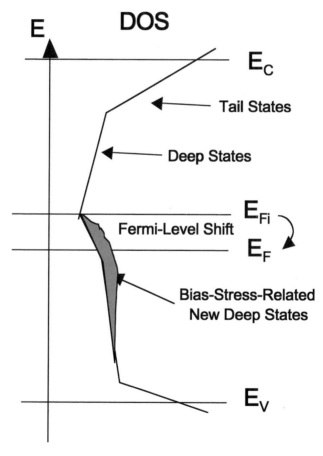

Figure 6.17 Bias-induced metastable defect states in a-Si TFTs. Most of the states are below E_F near the valence band.

The charge trapping and defect-state creation mechanisms distinguish themselves from each other (for TFTs with an a-Si:N gate insulator) in that charge trapping in the nitride has been observed to occur at high gate-bias voltages and long stress times. Here, the shift in the threshold voltage can be positive or negative depending on the polarity of the trapped charge (electron or hole). In contrast, defect-state creation dominates at lower stress voltages and at smaller stress times. At small stress voltages, Powell showed that while state creation in the lower part of the gap dominated at positive biases, state removal from the lower part of the gap occurred for negative bias voltages.[42]

Powell distinguished between defect-state creation and charge trapping in 1987 by experimenting with ambipolar TFTs.[48] If charge is trapped in the nitride under positive bias stress, the V_T for both electron and hole conduction will shift to more positive values. In contrast, defect-state creation in the a-Si layer manifests itself by the V_T becoming more negative for hole conduction and more positive for electron conduction since the Fermi level has to move through the newly created states to establish the channel. Powell showed that for his TFTs, state creation was the dominant mechanism up to 50 V;[42] charge trapping became significant thereafter. The critical voltage at which charge trapping overtakes state creation depends on the bandgap of the a-Si:N gate insulator, which also supports the theory that the charge-trapping process is nitride-related.[49] The defect-creation process appeared to be independent of the nitride characteristics, implying that state creation takes place in the a-Si film.[34]

Powell determined that state creation is characterized by a power law time dependence and is strongly affected by temperature. Charge trapping, on the other hand, has a logarithmic time dependence and is weakly dependent on temperature. A logarithmic time dependence and temperature independence is generally observed for charge trapping in an insulator where the injection current depends exponentially on the density of previously injected charge and the charge is trapped near the gate insulator interface.[35,50] As noted before, the process of state creation observed in a-Si film was reported to have several features in common with observations of state creation through illumination or thermal equilibration. One proposed model is that Si dangling bonds are created by breaking weak Si-Si bonds, which are then stabilized by the dispersive diffusion of hydrogen.[30] The power-law time dependence is found to be consistent with a defect-creation model.[51]

The subthreshold slope in the transfer characteristics of n-channel transistors does not change after positive bias stress, confirming the location of the newly created states below the flat-band Fermi level (see Fig. 6.17). This is consistent with the idea that the created states are Si dangling bonds or deep-state defects. The process of creating states below the flat-band Fermi level shifts the Fermi level nearer to midgap, which accounts for the increase in threshold voltage.

For negative biases, there are a few newly created states located above the flat-band Fermi level in addition to state removal in the lower part of the gap. This explains why there is some subthreshold slope degradation in n-channel TFTs during negative bias stress. For imaging applications using n-channel a-Si TFTs, TFT OFF voltages are chosen primarily by the requirement for low leakage current, as explained later in this chapter. The magnitude of typical TFT OFF voltages ranges from 0 to –5 V, implying that subthreshold slope degradation will be minimal. Therefore, although a-Si TFT metastability comprises both ΔV_T and subthreshold slope degradation, the focus in this chapter is on ΔV_T management.

Modeling of the two ΔV_T mechanisms is summarized as follows. Defect-state creation is characterized by a power-law dependence of ΔV_T over time, i.e., $\Delta V_T \propto t^\beta$. Jackson and Moyer[35] and Berkel and Powell[48] claimed that their data mapped a relationship,

$$\Delta V_T(t) = A(V_{ST} - V_{Ti})^\alpha t^\beta, \quad (3)$$

for the defect-state creation mechanism where α takes on a value of unity and V_{Ti} is the V_T before stress is applied. In contrast, charge trapping is associated with a logarithmic time dependence,[51] which can be represented as

$$\Delta V_T(t) = r_d \, log(1 + t/t_0), \quad (4)$$

where r_d is a constant and t_0 is some characteristic value for time.

On the other hand, Libsch and Kanicki[36] reported that an empirical stretched-exponential function could describe the stress-time and stress-voltage dependence of V_T in a-Si TFTs irrespective of the ΔV_T mechanism:

$$V_T(t) = (V_{ST} - V_{Ti}) [1 - exp(-(t/\tau)^\beta)], \quad (5)$$

where τ is some characteristic extractable time constant and β is the stretched-exponential exponent, which is temperature-dependent. For a short effective stress time ($t \ll \tau$), Eq. (5) can be simplified to a form similar to Eq. (3),

$$\Delta V_T(t) = (V_{ST} - V_{Ti}) \tau^{-\beta} t^\beta, \quad (6)$$

where

$$A = \tau^{-\beta}. \quad (7)$$

The main difference here is that Kanicki extracts nonunity values for α, making the function empirical. Our results, discussed in the following sections, appear to be in accord with Powell's theory: We noticed a unity value for α for the range of positive stress voltages applied as well as a power-law time dependence for the ΔV_T.

DC Bias Stress

Figure 6.18(a) shows the overall ΔV_T behavior plotted against the various gate-stress voltages applied during the experiment. Here, the ΔV_T increases for positive values and decreases for negative voltages. For

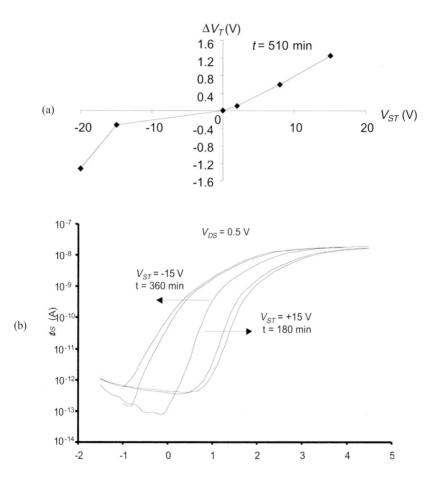

Figure 6.18 (a) ΔV_T as a function of TFT stress voltage, V_{ST}. V_{ST} was applied to the gate of the TFT for a stress time, T_{ST}, of 510 minutes. (b) Effect of gate-bias stress on the I_{DS}–V_{GS} characteristics of an a-Si:H TFT.

identical stress times, an equivalent decrease in the magnitude of ΔV_T appears at much larger negative stress voltages compared to the increase in ΔV_T at positive gate-stress voltages. For positive stress voltages, some authors have claimed that ΔV_T is solely due to charge trapping; others have suggested that defect-state creation and charge trapping coexist. For the range of negative voltages in Fig. 6.18(a), Powell described the small decrease in ΔV_T as a property of the a-SiN/a-Si interface, where state creation and reduction mechanisms are occurring simultaneously.[52-54] The mechanism that dominates (reduction or creation) is a consequence of the type of nitride used as well as the interface. Figure 6.18(b) shows the change in TFT I_{DS}-V_G characteristics for positive and negative gate-bias-stress voltages. Here, although V_T changes significantly for both positive and negative gate-bias-stress, the subthreshold slopes stay relatively constant. However, there is a slight change in subthreshold slope for the negative stress that is not present in the positive bias-stress measurements.

Positive DC Bias Stress

As shown in Fig. 6.19(a), if ΔV_T is plotted against $V_{ST} - V_{Ti}$, ΔV_T is directly proportional to $V_{ST} - V_{Ti}$ for positive voltages. The unity value for α is in agreement with the defect-creation model. The relationship of the ΔV_T as a function of time for positive stress voltages is shown in Fig. 6.19(b). Here, the value of the time exponent, β_1 (shown as β in the figure), is extracted to be around 0.3 for the range of measured stress voltages. To quantify, the $\Delta V_T = A(V_{ST} - V_{Ti})t^{\beta_1}$ relation can be fit to the positive stress data with $\alpha = 1$, $\beta_1 = 0.3$, and $A = 0.0115$, yielding an error of less than 5%.

Negative DC Bias Stress

In contrast to positive stress voltage measurements, the relationship between the measured data and theory is not evident for negative voltages, and much of the work done in this area relies on empirical models.[55,56] If the ΔV_T data for negative stress voltages are plotted with respect to gate-stress voltage, V_{ST}, we notice a value of around 2.49 for the term α, as shown in Fig. 6.20(a). Similarly, plotting against time yields a 0.28 value for the time-power dependence, β_2 [shown as β in Fig. 6.20(b)], where ΔV_T is increasing in the negative direction.

Powell claimed that defect-state reduction is responsible for the ΔV_T at low negative voltages, but there might be some charge-trapping effects as the stress voltage becomes more negative. With respect to the results in Fig. 6.20,

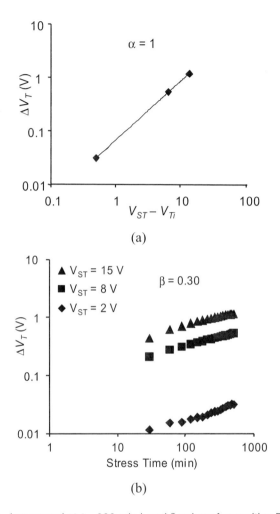

Figure 6.19 α (measured at $t = 600$ min.) and β values for positive DC bias values.

it is challenging to confirm the dominant ΔV_T mechanism for negative bias stress on the in-house TFTs. However, ΔV_T appears to have a power law dependence on both gate bias and stress time. To quantify empirically, a $\Delta V_T = -B \, |V_{ST}|^\alpha \, t^{\beta_2}$ relation is fit to the negative stress data with $\alpha = 2.49$, $\beta_2 = 0.28$, and $B = 0.12 \times 10^{-3}$, which yields an error of less than 5% for the range of voltages tested. Table 6.2 shows values for α and β published by various international researchers for a-Si TFTs over the past decade. The values extracted from in-house TFTs for both positive and negative gate-bias stress are evidently in agreement with some of the previously reported results.

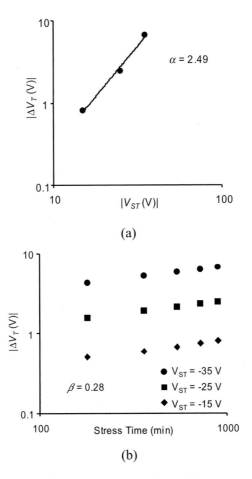

Figure 6.20 (a) α and (b) β values for negative DC bias-stress voltages.

Managing ΔV_T

Similar to DC bias stress, positive and negative unipolar gate pulses tend to shift V_T in the positive and negative directions, respectively, for n-channel TFTs. Therefore, using bipolar pulses offers an attractive means for reducing the overall ΔV_T. In practice, when TFT ON and OFF periods are comparable, the a-Si TFTs used in the AMLCD industry[55,56] require bipolar gate pulses for long-term ΔV_T management. The effect of these bipolar pulses is discussed in[55] where it was shown that the positive and negative ΔV_T resulting from positive (during the TFT ON cycle) and negative (OFF cycle) pulse bias cancel each other in a linear fashion. Experimental results from in-house TFTs shown in Fig. 6.21 appear to

Table 6.2 α and β Values Extracted by Various Researchers for a-Si TFT ΔV_T

Positive Bias Stress	α	β
In-house	1	0.30
Powell[51]	1	0.45 (at 40°C)
Libsch[36]	1	0.25
Kanicki[55]	1.9	0.50
Negative Bias Stress		
In-house	2.5	0.28
Kanicki[55]	2.4	0.32
Tsukada[56]	3.8	0.25

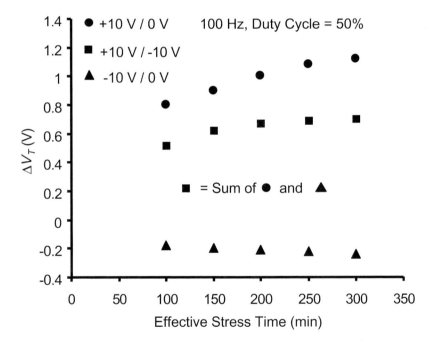

Figure 6.21 ΔV_T for a bipolar pulse and its two unipolar components.

corroborate the findings by Chiang et al.,[55] who attribute the maximum error of about 15% to measurement inaccuracy in the unipolar −10 V pulse measurement. As before, measurement interruptions to obtain TFT transfer characteristics for V_T extraction may mitigate the full effect of the negative bias pulse on the TFT's ΔV_T.

6.4 Pixel Integration

A high fill factor is a crucial design requirement in large-area imaging. In an attempt to achieve high fill factor, the Schottky diode was stacked on top of the TFT switch (Figs. 6.22 and 6.23), and its performance was compared with the conventional pixel structure commonly used in imaging applications (Fig. 6.24). In all pixels, the detector is 200×200 μm.

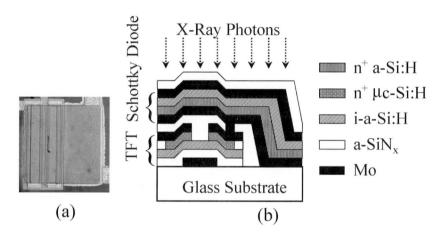

Figure 6.22 Fully overlapped pixel; (a) top and (b) cross-sectional views.

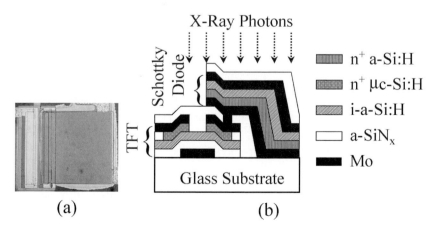

Figure 6.23 Partially overlapped pixel; (a) top and (b) cross-sectional views.

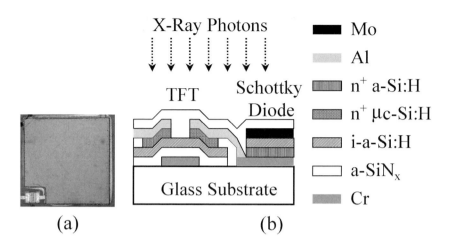

Figure 6.24 Nonoverlapping pixel; (a) top and (b) cross-sectional views.

When devices are stacked on top of each other as shown in Figs. 6.22 and 6.23, the intrinsic stress of films becomes an important fabrication issue. When a thick Mo layer is deposited on the stacked pixel structure, the layers underneath are unable to withstand the shear stress, causing the films to crack and peel off the substrate (see Fig. 6.25).[57] Therefore, the characterization of stress in each thin-film layer and subsequent optimization of process parameters are necessary. The measured intrinsic stresses of thin films on glass substrates are listed in Table 6.3.[58] Characterization methods to determine thin-film stresses and strengths are described in further detail in Chapter 2 of this book.

When deposited on glass substrates as the gate dielectric layer of the TFT, the magnitude of the tensile stress in a-SiN$_x$ is 2.22×10^8 N/m^2. When the same silicon nitride film is deposited on i-a-Si:H as the top dielectric, its resulting stress is tensile, with a measured value of 5.69×10^7 N/m^2. After the tri-layer deposition needed for the TFT, the compressive stress of very thin i-a-Si:H, sandwiched by the tensile gate and top dielectric layers, diminishes, and the total stress of all three layers is tensile, with a measured value of 6.20×10^7 N/m^2.

Unfortunately, the canceling out of film stresses is not guaranteed when fabricating Schottky diodes. The Mo layer (500 nm), deposited on a thick (1 μm) i-a-Si:H layer to form the Schottky barrier, can be in either compressive or tensile stress depending on the deposition pressure and the RF power. The measured intrinsic stress of a 500-nm-thick Mo layer deposited on i-a-Si:H is shown in Fig. 6.26. This graph shows that the Mo layer deposited at 200 W has stresses smaller in magnitude (either

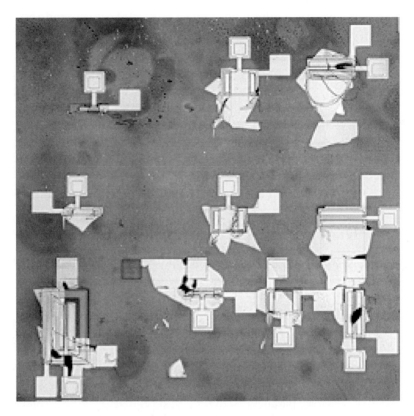

Figure 6.25 Damaged films due to high intrinsic film stress.

Table 6.3 Measured Intrinsic Stress of Thin Films Deposited on Corning 7059 Glass Substrate

	Thickness (nm)	Stress (10^8 N/m^2)	State
i-a-Si:H	1000	1.16	Compressive
n$^+$ a-Si:H	750	3.05	Compressive
n$^+$ µc-Si:H	200	14.7	Compressive
a-SiN$_x$:H	250	2.22	Tensile
Al	820	0.16	Tensile
Cr	600	5.75	Tensile

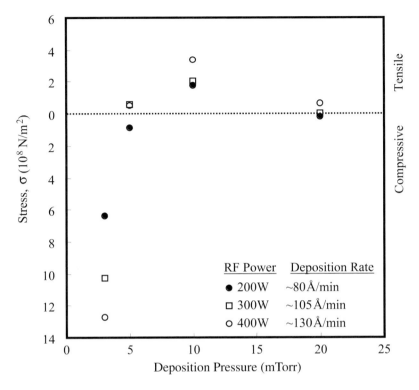

Figure 6.26 Measured stress of 500 nm Mo film deposited at 25°C on i-a-Si:H.

compressive or tensile) than the layer deposited at 400 W. However, Fig. 6.27 shows that the film density of the Mo layer deposited at 200 W is quite low compared to layers deposited at higher RF power. Since the X-ray detection principle of the Schottky diode is based on the photoelectric interaction of the X-ray photons with Mo atoms, a lower density leads to a lower X-ray absorption efficiency. Therefore, it is necessary for the Mo layer for the Schottky interface to be deposited at high RF power and low process pressure. However, Mo films deposited at 3 mTorr peel off the i-a-Si:H due to high compressive stress, regardless of the RF power. Without significantly undermining the film density (only 2 to 3% loss), Mo can be deposited at 5 mTorr, which drastically reduces the stress, as shown in Fig. 6.26.

Another important requirement in terms of pixel performance is a low leakage current. The dark current of the sensor (see Fig. 6.4) in the different pixels is the same regardless of configuration. However, the same is not true for the TFTs. Figure 6.28 shows that the leakage current

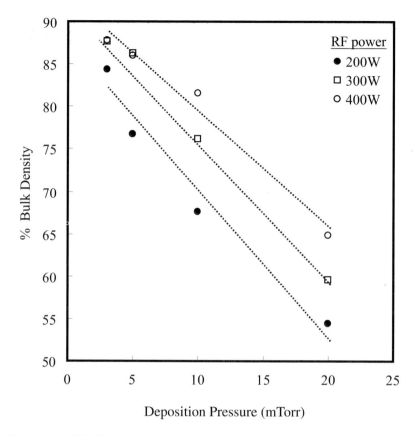

Figure 6.27 Mo film density (deposited at 25°C) relative to bulk.

in the fully overlapped structure is very high compared to the other designs. In the fully overlapped pixel structure, when the TFT gate voltage is negative, the electrons in the a-Si:H experience an electric field, which forms a steady parasitic back channel at the a-Si:H/a-SiN$_x$ interface and therefore provides a high conducting path from drain to source. This leakage can be reduced by using thicker dielectric layers or by suitably modifying the detector structure to reduce the vertical electric field at the top interface.[59]

The partially overlapped and nonoverlapping pixels are exposed to X-rays for 50 ms at various X-ray doses in the range of 20 to 200 mR. The number of measured electrons ranges from 3 to 20 million for the partially overlapped pixel (Fig. 6.29). However, in the nonoverlapping pixel structure, the charge transfer for the same bias and measurement conditions is reduced because of the different TFT geometry.

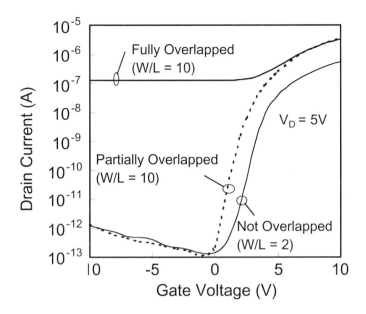

Figure 6.28 Leakage currents in the different pixel structures.

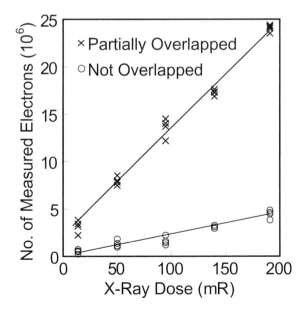

Figure 6.29 Measured electrons vs. X-ray dose.

6.5 Imaging Arrays

6.5.1 Conventional Passive Pixel Sensor Arrays

The passive pixel, first introduced by Weckler in 1967, is the workhorse of the digital X-ray imaging industry today. A passive pixel sensor (PPS) consists of a sensor element (e.g., a photodiode) and an integrated TFT connected to signal lines (see Fig. 6.30). The data signal line is connected to a charge amplifier to convert the charge content into an appropriate voltage level for image processing in external electronics. The photosensor is biased to a voltage optimal for integration of photo-induced charge on the diode capacitance. For this particular architecture, the TFT gate is connected to a common row gate signal line and a scanning clock generator addresses all the rows. During read-out, a row of TFTs is activated via a gate line by the clock generator sequentially. The signal charge from each photodiode is transferred through the common column data lines into the column charge amplifiers. The amplified signals are then digitized by an A/D converter and sent for image processing.[16] A physical storage capacitor, C_{st}, may be added between the node V_X and ground. Also, the sensor element (assumed to be either a photodiode or a photoconductor operated in reverse bias) is connected between a positive bias voltage and the V_X node. The complete model for analysis is shown in Fig. 6.31.

The integrated capacitor C_{st} in this PPS architecture allows for design and interface flexibility with different sensing devices as well as improved performance. With a PPS architecture, either charge or voltage can be

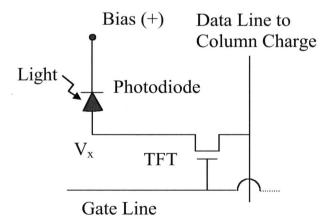

Figure 6.30 Passive pixel sensor architecture.

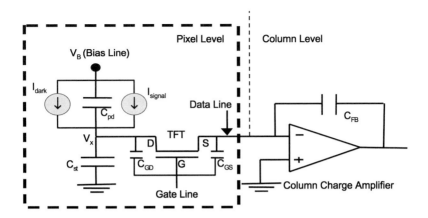

Figure 6.31 Analysis model for PPS architecture.

sensed. For a voltage to be properly sensed, sufficient voltage must accumulate on the V_X node to avoid degradation of the signal-to-noise ratio (SNR). For applications that have low input signals, such as diagnostic medical fluoroscopic imaging, the PPS architecture faces fundamental challenges. During image readout, the pixel charge is discharged exponentially with a time constant τ_{ON} given by

$$\tau_{ON} = R_{ON} C_{pixel}, \quad (8)$$

where R_{ON} is the ON resistance of the switching TFT and C_{pixel} is the pixel capacitance.

A sample calculation of this for a TFT switch of W/L = 80/20 and a $(200 \ \mu m)^2$ a-Si:H photodiode would be as follows:

$$R_{ON\text{-}TFT} = \frac{1}{\left[\left(\dfrac{W}{L}\right)\mu_n \Gamma_G \left(V_{ON} - V_T\right)\right]} = 1.4 \ M\Omega, \quad (9)$$

where W/L = 80/20, μ_n = 1.0 cm²/Vs, V_{ON} = 12 V, V_T = 3 V, and Γ_G = 2×10^{-8} F/cm².

Here μ_n is the electron mobility, V_{ON} is the TFT ON voltage, V_T is the threshold voltage, and Γ_G is the gate capacitance per unit area of the TFT. Next, assuming an intrinsic amorphous silicon layer thickness ($d_{a\text{-}Si:H}$) of 1 μm for the photodiode, the photodiode capacitance C_{pd} can be approximated as:

$$C_{pd} = \frac{\varepsilon_0 \varepsilon_{\text{a-Si:H}} A_{PD}}{d_{\text{a-Si:H}}} = 3.9 \text{ pF}, \quad (10)$$

where $\varepsilon_0 = 8.85 \times 10^{-12}$ F/m, $\varepsilon_{\text{a-Si:H}} = 11$, $A_{PD} = 200 \times 200$ μm², and $d_{\text{a-Si:H}} = 1$ μm.

Assuming the TFT parasitic capacitances, C_{GS} and C_{GD}, are at least ten times smaller than C_{pd}, the pixel capacitance, C_{pixel}, can be approximated by C_{pd}. Using values from Eqs. (9) and (10), Eq. (8) gives an *RC* time constant of 5.45 μs. For complete charge readout (95 to 99.3%), about 3 to 5 time constants are usually sufficient, which gives a total pixel readout time of 16.3 μs (for 3 time constants). This implies the charge amplifier has 16.3 μs to integrate the charge and produce an output voltage. For a direct detection scheme such as an a-Se photoconductor, the capacitance can be estimated to be on the order of 10 fF since the a-Se layer is very thick ($t_{a\text{-}Se} \approx 500$ μm[16]). This capacitance is very small, and a small pixel capacitance allows a large voltage to accumulate on the TFT drain node during charge integration. Voltages in excess of even 40 V may cause breakdown of some TFTs. For this reason, a storage capacitor of about 1 pF is often used for a-Se passive pixel sensors[16] (see Fig. 6.31). In both cases, the results are the same for readout time.

For the PPS architecture, the main sources of electronic noise are due to the readout TFT (thermal and flicker) and the charge amplifier. During readout, one TFT is turned ON and its associated photodiode is connected to the data line with a TFT ON resistance of about 1 MΩ. This resistive component in the circuit generates thermal noise. The 1/f noise in a-Si:H TFTs and the shot noise from a-Si:H photodiodes increase with current under the conditions of high drain current and high sensor leakage current, respectively. However, because even the photocurrent level is very low in these diagnostic X-ray imaging systems (~1 fA to 10 pA), 1/f and shot noise sources may be neglected as a first-order approximation.[60] From Fujieda[60] and referring to Fig. 6.32, the final expression for equivalent input TFT ON resistance thermal noise in electrons is as given below. (C_{pixel} and C_{amp} are the pixel and amplifier input capacitances, respectively.)

$$N_{th} = \sqrt{\frac{C_t}{q^2} kT \left[1 - \frac{2}{\pi} \tan^{-1}\left(\frac{2\pi RC_t}{\tau_{int}}\right)\right]}, \quad C_t = \frac{C_{pixel} C_{amp}}{C_{pixel} + C_{amp}}. \quad (11)$$

For a $C_{pixel} = 6$ pF, $C_{amp} = 100$ pF, $R_{ON\text{-}TFT} = 1.4$ MΩ, and an integration time $\tau_{int} = 33$ μs, N_{th} is estimated to be 620 electrons using the derived for-

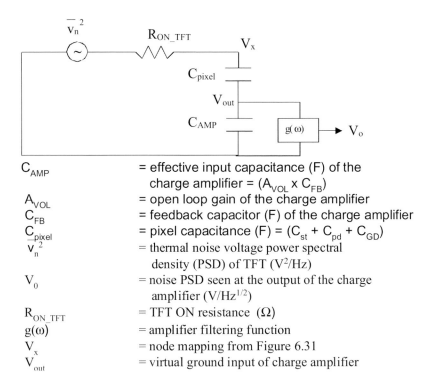

Figure 6.32 Equivalent PPS block diagram for noise extraction.

mula. A typical value for TFT thermal noise for the PPS architecture in diagnostic medical imaging applications is about 700 electrons.[60]

The amplifier noise can be modeled as having a fixed noise component N_{amp0} in addition to an input-capacitance-dependent component:[12]

$$N_{amp} = N_{amp0} + \delta C_d. \quad (12)$$

Here δ is a constant determined by the design properties of the charge amplifier (e.g., input FET noise) and C_d is the external capacitance loading the amplifier input node. This includes the parasitic capacitances on the data line such as C_{GS} of readout TFT as well as the overlap capacitance of data and gate lines. A typical value for amplifier noise is ~1700 electrons.[12]

There are two challenges to address for the PPS. The first concerns the accumulated charge and hence the question whether the equivalent photocurrent has a sufficient magnitude to overcome the dark current of the a-Se photoconductor. State-of-the-art a-Se photoconductors as well as a-Si p-i-n photodiodes[16] can achieve the requirements for diagnostic medical imaging.

The second requirement is that the voltage accumulating at the drain of the TFT be large enough to be sensed by the readout amplifier. For the fluoroscopy exposure range of 0.1 to 10 µR (assuming a pixel capacitance of 0.5 pF), a voltage swing of 0.32 mV (here, 1000 electrons is assumed as the minimum signal[16]) to 16 mV (49100 electrons) at the V_x node is expected, which is quite small. Past studies show[60] that voltage-sensitive amplifiers are not suited for this type of low-noise operation. In the voltage amplifier configuration, the signal charge from C_{pixel} is transferred to the data line capacitance, C_{ext}; this creates a tradeoff between the charge sensitivity of the amplifier and the charge transfer efficiency (CTE) from C_{pixel} to C_{ext}. In other words, if the C_{ext} is small, then insufficient charge, Q_{ext}, will be transferred from C_{pixel} to C_{ext} and the CTE will be degraded,

$$Q_{ext} = Q_{sig} \times \left[\frac{C_{ext}}{(C_{ext} + C_{pixel})} \right]. \tag{13}$$

From Eq. (13), to get half of Q_{sig} at the Q_{ext} output, C_{ext} must be equal to C_{pixel}. This implies that a large C_{ext} is needed for almost complete charge transfer. On the other hand, if C_{ext} is large, most of Q_{sig} will be transferred. However, the voltage, V_{ext}, developed at the input of the amplifier will be smaller than V_x since C_{ext} is larger than C_{pixel}. Hence, charge sensitivity is degraded.

$$V_{ext} = \frac{Q_{ext}}{C_{ext}} \tag{14}$$

So the current approach is to connect a charge amplifier directly to the pixel output to collect the entire signal charge, as shown in Fig. 6.31.

From the point of view of noise, the readout TFT and amplifier noise are two independent noise sources that add in quadrature,

$$N_{total}^2 = N_{th}^2 + N_{amp}^2. \tag{15}$$

The total noise, N_{total}, at the TFT drain is about 1830 electrons, compared to the 620 noise electrons from the TFT. It can therefore be concluded that the amplifier is the dominant noise source. The signal charge accumulating on C_{pixel} at the TFT drain node, V_x (for the a-Se photoconductor), ranges from 1000 electrons (i.e., minimum 1 X-ray detected) to 49100 electrons. This range is due to the exposure range of fluoroscopy (0.1 to

10 uR). The lower end of the range implies that part of the lower signal (1000 electrons) is drowned out by the noise (1830 electrons). Table 6.4 summarizes the performance of an a-Si p-i-n diode sensor detection scheme for fluoroscopy, mammography, and radiography. The fluoroscopic integration time of 33 ms comes from the real-time operation requirement on this modality.

For detectors such as a-Se where the detector capacitance is small, adding a storage capacitance can prevent a very large voltage from appearing at the V_x node and causing the switch TFT to break down. Similarly, for the a-Si:H p-i-n or Schottky photodiodes, adding the storage capacitor prevents the V_x node voltage from rising excessively and altering the photodiode bias. Therefore the reverse bias of the diode is maintained as well as its charge-collection ability. Controlling large V_x changes is especially important for applications such as mammography where the input signal is relatively high but resolution requirements demand a small pixel size and C_{pixel} therefore becomes small. Here, the integrated C_{st} provides an easy method of increasing the pixel capacitance. There is a tradeoff created in that increasing C_{pixel} increases the RC time constant and hence the readout time. In addition, a large pixel capacitance increases the amplifier noise, as seen in Eq. (11).

Fluoroscopy is the most challenging modality because the input signal can be so low that it is susceptible to degradation by the photodiode

Table 6.4 PPS Performance for Indirect Detection Using a-Si:H p-i-n Photodiodes Coupled with a CsI Phosphor

	Radiography	Mammography	Fluoroscopy
Pixel area (µm^2)	200 x 200	50 x 50	250 x 250
Charge/pixel (e$^-$/pixel/mR)	1.2×10^6	3.0×10^4	3.75×10^6
Exposure range (mR)	0.03 to 3	0.6 to 240	0.0001 to 0.01
Signal charge (Q_{sig}) on C_{pixel} (e$^-$)	3.6×10^4 to 3.6×10^6	1.8×10^4 to 7.2×10^6	1.0×10^3 to 3.75×10^4
Noise (e$^-$) from readout electronics	1.83×10^3	1.83×10^3	1.83×10^3
Signal integration time (s)	0.33	0.33	0.033
I_{photo} (A) = Q_{sig} / T_{int}	17.5 fA to 1.75 pA	8.73 fA to 3.4 pA	5.2 fA to 0.182 pA
J_{photo} (A/cm^2) = I_{photo}/Area	43.7 pA/cm^2 to 4.37 nA/cm^2	350 pA/cm^2 to 136 nA/cm^2	8.0 pA/cm^2 to 0.29 nA/cm^2
J_{dark} (A/cm^2)	10 pA/cm^2	10 pA/cm^2	10 pA/cm^2
C_{pixel} (F) = C_{pd}	3.89 pF	0.34 pF	6.1 pF

dark current as well as the leakage current of the readout TFT. In addition, noise from the sensor and external amplifier is very prohibitive. To build a fluoroscopic detector array, two requirements on the pixel are a low sensor dark-current density (~1 pA/cm^2) and a low TFT leakage current (~1 fA). More significantly, noise from the readout electronics should not exceed the minimum signal level. However, for current PPS technology, the external electronics noise drowns out the lower range of input signal, making this technology unsuitable for diagnostic medical fluoroscopy.

6.5.2 Amorphous Silicon Current-Mediated Active Pixel Sensor Arrays

While the PPS architecture has the advantage of being compact and thus amenable to high-resolution imaging, reading the small output signal of the PPS for low input signal, large-area applications (e.g., fluoroscopy[61]) is extremely challenging and requires costly, high-performance, and sometimes custom-made charge amplifiers.[62] More importantly, if external noise sources (e.g., charge amplifier noise[61] and array data line thermal noise[62]) are comparable to the input, there is a significant reduction in pixel dynamic range. The current-mediated amplified pixel is an integrated pixel amplifier circuit using a-Si TFTs based on CMOS active pixel sensors (APS).[63] The APS performs *in situ* signal amplification, providing higher immunity to external noise, thereby preserving the dynamic range. In addition, the performance and cost constraints on external charge amplifiers are relaxed.

Unlike a conventional PPS, which has one TFT switch, there are three TFTs in the APS pixel architecture. This could undermine fill factor if conventional methods of placing the sensor and TFTs are used.[2] Therefore, in an effort to optimize fill factor, the TFTs may be embedded under the sensor to provide high-fill-factor imaging systems.[61] Central to the APS illustrated in Fig. 6.33 is a source follower circuit, which produces a current output (C-APS) to drive an external charge amplifier. The C-APS operates in three modes:

1. Reset mode: The RESET TFT switch is pulsed ON and C_{PIX} charges up to Q_P through the TFT's ON resistance. C_{PIX} is usually dominated by the detector (e.g., a-Se photoconductor[61] or a-Si photodiode[62] detection layer) capacitance.
2. Integration mode: After reset, the RESET and READ TFT switches are turned OFF. During the integration period, T_{INT}, the input signal, $h\nu$, generates photocarriers discharging C_{PIX} by ΔQ_P and decreases the potential on C_{PIX} by ΔV_G.

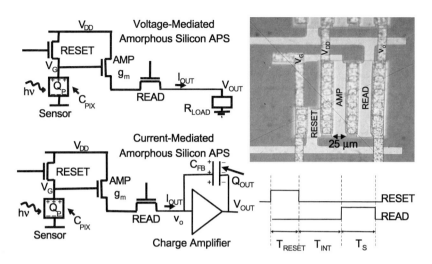

Figure 6.33 Amorphous silicon voltage (V-APS) and current (C-APS) mediated active pixel schematic, readout timing diagram and circuit micrograph The R_{LOAD} and charge amplifier appear at the end of each column.

3. Readout mode: After integration, the READ TFT switch is turned ON for a sampling time T_S, which connects the APS pixel to the charge amplifier, and an output voltage, V_{OUT}, is developed across C_{FB} proportional to T_S.

In the C-APS circuit, the characteristic threshold voltage shift of a-Si TFTs is manageable[64] since the TFTs have a duty cycle of ~0.1% in typical large-area applications. Therefore, appropriate biasing voltages in the TFT ON and OFF states can minimize the threshold voltage shift. Operating the READ and RESET TFTs in the linear region reduces the effect of interpixel threshold voltage (V_T) nonuniformities. Although the saturated AMP TFT causes the C-APS to suffer from FPN, using CMOS-like off-chip double sampling techniques[63] can alleviate this problem.

The linearity of the C-APS architecture is obtained from a sensitivity analysis[65] of the change in output current, ΔI_{OUT}, with respect to the input illumination, $h\nu$ in Fig. 6.33,

$$\gamma = \frac{d\left[\log |\Delta I_{OUT}|\right]}{d\left[\log |h\nu|\right]} = \frac{d\left[\log |\Delta Q_P|\right]}{d\left[\log |h\nu|\right]} \cdot \frac{d\left[\log |\Delta V_G|\right]}{d\left[\log |\Delta Q_P|\right]} \cdot \frac{d\left[\log |\Delta I_{OUT}|\right]}{d\left[\log |\Delta V_G|\right]}, \quad (16)$$

where $\gamma = 1$ for an ideal linear sensor and ΔV_G is the change in the gate voltage of the AMP TFT due to ΔQ_P. The first term is linear if the detector gives a linear change in the charge on C_{PIX}, ΔQ_P, with changing $h\nu$. The second term depends on the voltage change ΔV_G across C_{PIX} with changing ΔQ_P where

$$\Delta Q_P = \Delta V_G C_{PIX}. \qquad (17)$$

The second term is linear, provided C_{PIX} stays constant under the changing bias conditions. The last term imposes a linear small-signal condition on the AMP TFT gate input ΔV_G,

$$\Delta V_G \ll 2(V_G - V_T), \qquad (18)$$

where V_G is the DC bias voltage at the AMP TFT gate and V_T is its threshold voltage.

When photons are incident on the detector, electron-hole pairs are created, leading to a change in the charge given by Eq. (17). In small-signal operation, the change in the amplifier's output current with respect to a small change in gate voltage, ΔV_G, is

$$\Delta I_{OUT} = g_m \Delta V_G = g_m v_{in}, \qquad (19)$$

where g_m is the transconductance of the AMP and READ TFT composite circuit,[66] and v_{in} represents the small-signal voltage at the gate of the AMP TFT. Using Eqs. (17) and (19), the charge gain,[65] G_i, stemming from the drain current modulation is

$$G_i = |\Delta Q_{OUT}/\Delta Q_P| = (\Delta I_{OUT} T_S)/\Delta Q_P = (g_m T_S)/C_{PIX}. \qquad (20)$$

The charge gain amplifies the input signal, making it resilient to external noise sources.

The C-APS test pixel, consisting of an integrated a-Si amplifier circuit in a 250 × 250 μm² pixel area, was fabricated in-house and is shown in Fig. 6.33. Based on the test setup and data in Fig. 6.34, the small-signal linearity is within 5% of the theoretical value. Gain measurements were performed on a C-APS test circuit using the charge amplifier of Fig. 6.33, the test circuit shown in Fig. 6.34, and $C_{FB} = 10$ pF, and varying the READ pulse width, T_S. A commercially available charge amplifier, Burr-Brown IVC102P, was used with a DC bias of $V_G = 16$ V at the gate of the AMP TFT. Using Eq. (19) and assuming constant ΔI_{OUT}, ΔV_{OUT} for the charge-integrating circuit in Fig. 6.33 can be written as

$$\Delta V_{OUT} = -\frac{1}{C_{FB}} \int_0^{T_S} \Delta I_{OUT} dt = \frac{\Delta I_{OUT} T_S}{C_{FB}} = \frac{(g_m \Delta V_G) T_S}{C_{FB}} = \frac{(g_m v_{in}) T_S}{C_{FB}}. \qquad (21)$$

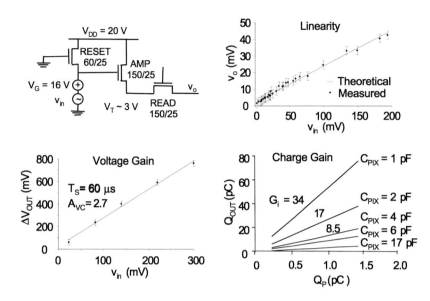

Figure 6.34 C-APS measurement results of small signal linearity and gain.

Theoretical voltage gain, A_V [based on Eq. (21), where $A_V = \Delta V_{OUT}/v_{in}$], and experimental results agree reasonably well with a maximum discrepancy of about 20%. Using Eqs. (20) and (21), it can be shown that

$$G_i = |\Delta Q_{OUT}/\Delta Q_P| = A_V(C_{FB}/C_{PIX}). \tag{22}$$

The verified theoretical model of Eq. (21) was extended to predict charge gain using Eq. (22) for different values of C_{PIX} and $T_S = 60$ μs in Fig. 6.34. Theoretically, using a low-capacitance sensor (i.e., small C_{PIX}) provides a higher charge gain, which minimizes the effect of external noise. However, a tradeoff between pixel gain and amplifier saturation places an upper limit on the achievable charge gain. In addition to external noise, investigations of noise added by the APS architecture to the input signal[67] indicate that intrinsic APS noise is minimized for small C_{PIX}, implying the feasibility of low-capacitance detectors (see, for example, Zhao and Rowlands[61] and Karim et al.[67]). Minimizing C_{PIX} will also reduce the reset time constant (which comprises mainly the RESET TFT on resistance and C_{PIX}), thereby reducing image lag.[68] For example, assuming column-parallel readout, a typical array[61,67] comprising 1000 × 1000 pixels operating in real time at 30 frames/s allows 33 μs for each pixel's readout and reset. Typical values for a-Si RESET TFT on resistance (~1 MΩ) and C_{PIX} (~1 pF for a-Se[61]) yield an RC time constant of 1 μs, implying that 5 μs resets would eliminate image lag and still allow sufficient time for readout with double sampling. Like other current mode circuits,[65] the C-APS, operating at 30 kHz, is susceptible to sampling clock jitter. However, off-chip, low-jitter clocks using crystal oscillators can alleviate this problem.

6.5.3 Amorphous Silicon Voltage-Mediated Active Pixel Sensor Arrays

The previously discussed current-mediated a-Si TFT APS readout circuit incorporated on-pixel signal amplification for improved SNR in digital fluoroscopy. This section examines a voltage-mediated a-Si TFT APS (V-APS) readout circuit that eliminates the need for external amplifiers in digital mammography or radiography applications. There are three TFTs in the V-APS pixel, similar to the C-APS pixel. Central to the V-APS is a source-follower circuit, which produces an output current that is converted to a voltage by a resistive load. The V-APS circuit is popular in CMOS imaging[63] and is illustrated in Fig. 6.35. The V-APS operates in three modes:

1. Reset: The RESET TFT is switched ON and the pixel capacitance, C_{PIX}, charges up to Q_P through this TFT's ON resistance, R_{ON_RESET}.
2. Integration: After reset, the RESET TFT is switched OFF for an integration period, T_{INT}. During T_{INT}, the X-ray input signal, hv, generates photocarriers that discharge C_{PIX} by ΔQ_P, decreasing V_G.
3. Readout: After integration, the READ TFT is switched ON for a sampling time T_S, where a load resistance converts the output current into a voltage.

In the APS pixel circuit, the characteristic threshold voltage (V_T) shift of the a-Si TFTs is not an issue[64] since again, the TFTs have a duty cycle of less than 0.1%, which is typical of most large-area applications. Referring to Fig. 6.35, the large-signal expression in Eq. (23) relates the V-APS readout circuit intermediate node voltage, V_B, to the input, V_G. Here the drain current of the saturated AMP TFT during V-APS readout has been set equal to the current through R_D, the sum of the READ switch ON resistance, R_{ON_READ} (approximated as a constant, and R_{LOAD},

$$V_B = \frac{K_{AMP}}{2}\left(V_G - V_B - V_T\right)^2 R_D, \tag{23}$$

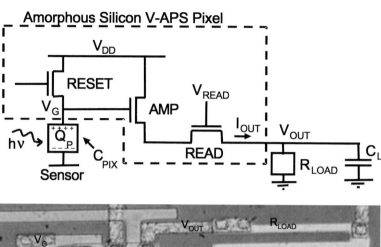
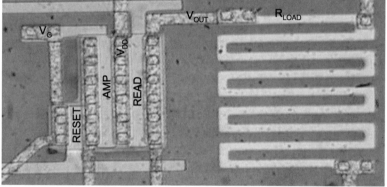

Figure 6.35 Amorphous silicon V-APS circuit schematic and fabricated pixel.

where K_{AMP} denotes the usual product of parameters, i.e., $\mu_{FET}C_G W/L$. Solving Eq. (23) and differentiating V_B with respect to V_G yields

$$\frac{\Delta V_{OUT}}{\Delta V_B} = 1 - \frac{1}{(1 + 2K_{AMP}R_D(V_G - V_T))^{1/2}}. \qquad (24)$$

From Eq. (24), the large-signal voltage gain of the AMP TFT source follower approaches unity for a large ($K_{AMP}R_D$) product and $V_G > V_T$. Since a constant value for Eq. (24) implies linearity, a large ($K_{AMP}R_D$) product also minimizes gain variations for different values of V_G. Lastly, the voltage divider formed by R_{ON_READ} and R_{LOAD} causes an additional linear drop in gain, so the overall V-APS voltage gain, A_{VV}, becomes

$$A_{VV} = \frac{\Delta V_{OUT}}{\Delta V_B} \cdot \frac{\Delta V_B}{\Delta V_G} = \frac{\Delta V_{OUT}}{\Delta V_G}$$

$$\approx \frac{R_{LOAD}}{R_{LOAD} + R_{ON\_READ}} \left(1 - \frac{1}{\left(1 + 2K_{AMP} R_D (V_G - V_T)\right)^{1/2}} \right). \quad (25)$$

In general, R_{LOAD} is designed to be much larger than R_{ON_READ}, so the drop in gain is minimal.

Large-area medical imaging applications such as mammography have been reported to require imaging arrays with 3600 × 4800 pixels and a maximum readout time of 2 to 5 seconds per frame.[16] The readout time requirement arises from the sensor's dark current degrading the integrated signal charge stored on each pixel's C_{PIX} over time. The lower currents and larger resistances (in MΩ) of a-Si TFTs as well as the large column bus capacitances in large-area arrays[2] make V-APS readout relatively slow. The large readout time for a frame required for an a-Si V-APS pixel can make it comparable to the dark-current limited maximum frame readout time. Therefore the V-APS must be designed to meet the frame readout requirement. The column bus capacitance (C_L) comprises primarily the sum of the gate-to-source capacitances (C_{GS}) of the READ TFTS in all of the APS pixels connected to a particular column. C_L usually ranges from 10 to 100 pF for typical large-area imaging arrays.

Referring to Fig. 6.35, the V-APS circuit can have a load that is either an integrated n$^+$ a-Si film resistor (R_{LOAD}) or a TFT.[69] The load TFT (LD TFT), with a gate bias of V_{LOAD}, is operated in the saturation regime (i.e., $V_{OUT} \geq V_{LOAD} - V_T$), since its channel resistance, $R_{LD(sat)}$, stays relatively constant and is insensitive to any variations in V_{OUT}. A V_{OUT} insensitive load is necessary for linearity, as shown by the constant R_D in Eq. (24). The channel resistance of a TFT operated in the linear region $R_{LD(lin)}$ is dependent on V_{OUT}, which, using Eq. (24), implies nonlinear V-APS operation.

The V-APS readout time consists of a rise and fall time component. To estimate the V-APS rise time, t_{rise}, KCL equations at node B and V_{OUT} yield a second-order differential equation that is easier to solve numerically via a circuit simulator. However, some insight into the circuit may be gained if the READ TFT is neglected in the analysis altogether. Neglecting the READ TFT and using an LD TFT as the load, the sum of currents at the V_{OUT} node yields

$$K'_{AMP} (V_G - V_{OUT} - V_T)^2 = I_{LD} + C_L \frac{dV_{OUT}}{dt}, \quad (26)$$

where $K'_{AMP} = \mu_{FET} C_G W/2L$ and I_{LD} is the current in the LD TFT. Equation (26) is a first-order ordinary differential equation that can be rewritten as

$$\int dt = \int \frac{C_L}{\left[K'_{AMP}\left(V_G - V_{OUT} - V_T\right)^2 - I_{LD}\right]} dV_{OUT}. \tag{27}$$

The solution to Eq. (27) is of the form

$$t_{rise} = \frac{C_L}{\sqrt{K'_{AMP} I_{LD}}} \tanh^{-1}\left[\frac{\left(\sqrt{K'_{AMP}}\left(V_G - V_{OUT} - V_T\right)\right)}{\sqrt{I_{LD}}}\right] + K, \tag{28}$$

where K is an integration constant. While circuit simulation software (e.g., Spectre, Hspice) gives results to a desired accuracy, Eq. (28) illustrates how C_L, K_{AMP}, and I_{LD} relate to t_{rise}. For a fast t_{rise}, C_L should be minimized, I_{LD} should be maximized, and there exists some optimum K'_{AMP} that can be determined from simulation. Reducing the size of the READ TFT in each APS pixel decreases C_{GS} and minimizes C_L. I_{LD} is made larger by choosing either a larger V_{LOAD} or a larger W/L for the LD TFT. As described by Karim et al.[69], increasing V_{LOAD} decreases dynamic range by increasing the minimum valid V_{OUT} voltage level of the V-APS since the LD TFT now enters the linear region at a larger value of V_{OUT}. Thus, increasing the size (W/L) of the LD TFT W is a preferable alternative in achieving a lower t_{rise}.

While increasing the current I_{LD} through the LD TFT reduces t_{rise}, it also has the adverse effect of reducing the largest achievable V_{OUT} voltage level and hence the dynamic range. The maximum achievable V_{OUT} due to increased I_{LD} can be estimated by noting that when $V_{OUT(MAX)}$ settles at equilibrium, the current I_{AMP} through the AMP TFT must be equal to I_{LD}. Equating $I_{AMP(sat)}$ to $I_{LD(sat)}$, and assuming both TFTs are in saturation,

$$K_{AMP}\left(V_G - V_{B(MAX)} - V_T\right)^2 = K_{LD}\left(V_{LOAD} - V_T\right)^2. \tag{29}$$

Solving Eq. (29) for the intermediate node voltage, $V_{B(MAX)}$, gives

$$V_{B(MAX)} = V_G - V_T - \sqrt{\frac{K_{LD}}{K_{AMP}}}\left(V_{LOAD} - V_T\right). \tag{30}$$

$V_{OUT(MAX)}$ is $V_{B(MAX)}$ reduced by the voltage divider formed with R_{ON_READ} and $R_{LD(sat)}$,

$$V_{OUT(MAX)} = V_{B(MAX)} \left[\frac{R_{LD(sat)}}{R_{LD(sat)} + R_{ON\_READ}} \right]. \quad (31)$$

In addition to reducing the dynamic range, a larger $I_{LD(sat)}$ gives a smaller $R_{LD(sat)}$, where $R_{LD(sat)}$ is the usual $(\lambda I_{LD(sat)})^{-1}$ and λ is the channel length modulation parameter. Using Eq. (25) with $R_{LOAD} = R_{LD(sat)}$, it can be seen that a large I_{LD} reduces the linearity and gain of the V-APS. So it can be concluded that in designing for a small t_{rise}, a tradeoff between readout speed, dynamic range, linearity, and gain exists. The t_{rise} for an n⁺ a-Si film resistor with R_{LOAD} in place of the LD TFT is similar to the t_{rise} with an LD TFT. A smaller R_{LOAD} gives a faster rise time but trades off dynamic range, linearity, and gain.

For the V-APS fall time, t_{fall}, the situation is considerably simpler since the READ TFT is OFF. If an LD TFT is used, it initially behaves as a constant current source discharging V_{OUT} until it enters the linear region of operation, where it approximates the discharging of a single RC time-constant circuit. Thus, t_{fall} of a V-APS with a LOAD TFT can be written as

$$t_{fall} = \frac{(V_{OUT} - V_{LOAD} - V_T)C_L}{I_{LD(sat)}} + mR_{LD(lin)}C_L, \quad (32)$$

where m is the number of time constants (typically five) required for complete readout, $R_{LD(lin)}$ is the average ON resistance of the LD TFT in the linear region, and $I_{LD(sat)}$ is the LD TFT saturation current. For the n⁺ a-Si film resistor, R_{LOAD}, t_{fall} is given by a single RC time constant,

$$t_{fall} = mR_{LOAD}C_L. \quad (33)$$

Depending on the value of R_{LOAD} or $R_{LD(lin)}$, t_{fall} can be several times larger than t_{rise}. However, dynamic range may be traded for an increase in the frequency of V-APS operation by preventing the circuit from discharging completely.

The V-APS test pixel, consisting of an integrated a-Si amplifier circuit in a 250 × 250 μm² pixel area with a 1.3 GΩ n⁺ a-Si film load resistor was fabricated in-house and is shown in Fig. 6.35. Large-signal linearity measurements of the V-APS are shown in Fig. 6.36, where the large R_{LOAD}

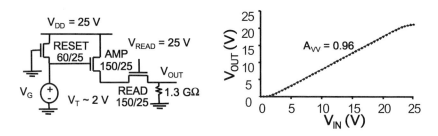

Figure 6.36 Linearity and gain of V-APS pixel.

gives an almost ideal gain, $A_{VV} = 0.96$, in the linear region of the curve. Here, Keithley Model 236 SMUs are used to supply and measure the input and output voltages, respectively. The measured data match simulations using a previously developed in-house a-Si TFT model[6,70,71] to within 5%. In Fig. 6.36, V_{OUT} levels off at lower voltages (around V_T) due to the AMP TFT not turning on. At higher voltages, V_{OUT} levels off as given in Eq. (31).

Tests performed on the V-APS readout time are shown in Fig. 6.37. Here, a Burr-Brown instrumentation amplifier INA116 was connected in a unity-gain configuration at the V-APS output node to facilitate

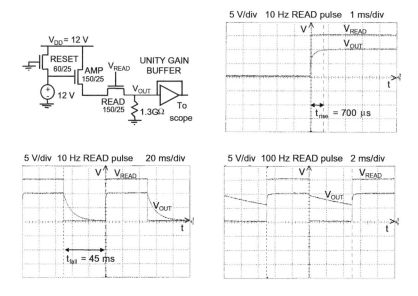

Figure 6.37 Transient performance of V-APS pixel.

measurement of V_{OUT}. The unity-gain op-amp buffer minimized the loading effect of the 1.3 GΩ resistor on the oscilloscope input probe, which was essential for accurate measurement of V_{OUT}. The load capacitance, C_L, was supplied by the INA116's input capacitance of 7 pF. Initially, the READ pulse was operated at 10 Hz (100 ms period) to allow for complete rise and fall times. From Fig. 6.37, it can be seen that t_{rise} ~ 700 μs while t_{fall} ~ 50 ms. Both measurements agree with simulations to within 5%. In addition, t_{fall} can be estimated from Eq. (33) as being 45.5 ms. Lastly, a readout time measurement for a READ pulse of 100 Hz is presented in Fig. 6.37. As illustrated, speeding up the readout operation causes a reduction in dynamic range since V_{OUT} does not discharge completely between READ cycles.

The main benefit of using an n^+ a-Si film resistor over an a-Si LD TFT is its relative immunity to metastability. For the TFT, a continuous V_{LOAD} bias at the gate causes a time-dependent shift in the threshold voltage, which serves to reduce the dynamic range (by increasing the minimum valid V_{OUT} level) as well as increasing readout time (by decreasing I_{LD}). In contrast, initial stress tests conducted on a 500 MΩ n^+ a-Si film resistor for different biases revealed a maximum change in resistance of 6% over 15 hours. These results indicate better stability than a-Si TFTs, where the threshold voltage change was measured to be as much as 20% over a similar time period. The primary advantage of using an a-Si LD TFT in place of a resistor is attaining fast readout times without sacrificing linearity or gain. During readout, the AMP TFT behaves as a voltage-dependent current source charging up C_L, while the LD TFT provides an opposing (and smaller) constant current source discharging C_L. The worst case for the charging process, i.e., to $V_{OUT(MAX)}$, can by symbolically represented as

$$t_{rise} = \frac{V_{OUT(MAX)} C_L}{(I_{AMP}(V_{OUT}(t)) - I_{LD})}, \quad (34)$$

where I_{AMP} is the V_{OUT}-dependent charging current through the AMP TFT. From Eq. (34), reducing $V_{OUT(MAX)}$ and hence the dynamic range reduces the rise time, t_{rise}. Referring to Eq. (30), choosing the appropriate LD TFT to AMP TFT aspect ratio gives a corresponding decrease in $V_{OUT(MAX)}$, while reducing V_{LOAD} maintains a low value for I_{LD}. Since $R_{LD(sat)} = (\lambda I_{LD(sat)})^{-1}$, keeping a low $I_{LD(sat)}$ provides a large $R_{LD(sat)}$, thereby preserving gain and linearity [see Eq. (25)]. In contrast, decreasing the resistance of the n^+ a-Si film decreases t_{rise} due to a reduced dynamic range, but gain and linearity are not preserved as they are with a saturated LD TFT.

The 100 Hz READ pulse frequency measurement in Figure 6.37 illustrates another way in which the dynamic range can be traded off for a reduction in readout time. For a typical mammographic digital imaging array, there are 3600 × 4800 pixels and an X-ray induced signal charge of 1.8×10^4 to 7.2×10^6 electrons.[16] Using the input charge and a nominal 1 pF pixel capacitance, the input signal voltage ranges from 3.84 mV to 1.536 V, which implies that a dynamic range of 2 V is sufficient. Similarly, the acceptable dynamic range for radiography is less than 1 V. In calculations of readout time, a column-parallel readout architecture (i.e., a row of pixels read out simultaneously) and V-APS loads (TFTs or resistors) on both ends of the large-area array are assumed. Hence 1800 pixel rows need to be read out simultaneously on each side in less than 2 seconds, which allows 1.1 ms of readout time per row of pixels (1 kHz V-APS operation). Following the design procedure highlighted in Section 6.5.2, a 1 ms readout time per pixel row is achievable with current state-of-the-art a-Si technology.

The primary advantage of the V-APS circuit over traditional PPS circuits is that external amplifiers are not required to obtain a readable output voltage, which reduces the array component count and cost. However, the reduction in gain in achieving real-time readout (30 kHz pixel operation) of the V-APS circuit makes it suitable for higher input signals such as static chest radiography or mammography applications. Lastly, the V-APS architecture is suitable for direct connection to an integrated multiplexer,[72] which has the potential to reduce external bond-pad connections.

6.5.4 Integrated Amorphous Silicon Multiplexers for Imaging Arrays

In typical large-area applications, depending on the array size and pixel resolution, the number of gate and data lines can be large; all require external multiplexer chips. Moreover, the high-threshold voltage (from 2 to 4 V) and the poor current drives of a-Si:H TFTs require high operating voltages (around 15 to 25 V). Thus, the input pins of the large-area applications require a large number of level-shifting buffers that add a significant cost to the system and make it bulky. These issues can be resolved by implementing on-chip integrated multiplexer circuits compatible with the existing a-Si:H technology.

The large number of gate lines can be addressed by using either a shift register or a pass-transistor logic (PTL) based multiplexer. The primary disadvantage of the shift register is the large number of transistors and inverters required.[73] Due to low mobility of holes in a-Si:H, complementary logic is not feasible for circuits with acceptable performance. In addition, due to the large defect density in a-Si:H, substitutional doping is very difficult and depletion mode TFTs are not popular in a-Si:H technology. Thus, the inverter circuits can be implemented using enhancement-mode TFTs, where the gate of the load TFT is connected to the positive power supply (V_{DD}). This inverter configuration exhibits stability problems since the load TFT is always "ON" and thus it suffers large ΔV_T. The stability of the inverter circuit can be improved by replacing the load TFT by a passive resistor, which is feasible in an n⁺ a-Si:H layer compatible with the standard TFT process.[31] Inverter simulations were performed using the physically based model developed at the University of Waterloo.[70] Simulation results show that an inverter circuit with a minimum-sized driver TFT (W/L = 20 μm/20 μm) and a resistive load of 100 MΩ exhibits excellent switching characteristics. The "ON" resistance of the driver TFT (R_{ON_TFT}) was found to be approximately 1/40 times smaller than the load resistance, i.e., $R_{ON\_TFT} \approx 2.5$ MΩ. However, the driver TFT of the inverter shows a significant ΔV_T that directly affects the output of the shift register.

On the other hand, the PTL-based multiplexer is more general in application than the shift register since it can also be used to multiplex the output data lines of an imaging array.[67] The multiplexer offers the additional flexibility of selecting the gate lines in an arbitrary order. A PTL-based multiplexer requires a fewer inverter circuits, which are prone to ΔV_T. Moreover, the ΔV_T in the inverter circuit affects the inverter output only when it is logic "LOW," i.e., when the pass transistor is "OFF." Thus the PTL-based multiplexer circuit shows a better ΔV_T robustness than the shift register. A circuit schematic of a PTL-based 1:8 demultiplexer driving the gate lines of an active matrix array is shown in Fig. 6.38. In this circuit, pull-down resistors are provided for each branch so that the output voltage of a particular branch can be forced to logic "LOW" when the select lines S_1-S_3 do not address the branch. If V_{DD} is the voltage at the input pin, D_0, and V_T is the threshold voltage of the pass transistors, the output of the selected branch reaches a voltage of approximately ($V_{DD} - V_T$), provided the logic "HIGH" value of select lines is also V_{DD}. Increasing the logic "HIGH" voltage of the select pins S_1-S_3 can reduce this voltage drop of V_T. Typically, each output pin of the PTL-based demultiplexer drives gate capacitances of all the TFTs of that gate line, which can be as large as 100 to 200 pF.[67] This makes the discharging time of the gate line too large to be useful for the typical values for the pull-down resistors (~100MΩ to 1GΩ). The discharging time can be made equal to the charging time by removing the pull-down resistances and replacing the input constant voltage source (V_{DD}) by an external pulse voltage source that switches between V_{DD} and V_{SS}. The gate capacitance is estimated to be ≈100 fF for an a-Si:H TFT with W/L = 20/20, nitride thickness t_{a-SiN} = 250 nm, nitride dielectric constant $\varepsilon_{SiN} \approx 7$, and gate area A_{TFT} = 20 × 20 μm = 400 μm². Thus the gate-line capacitance with "N" TFTs can be calculated

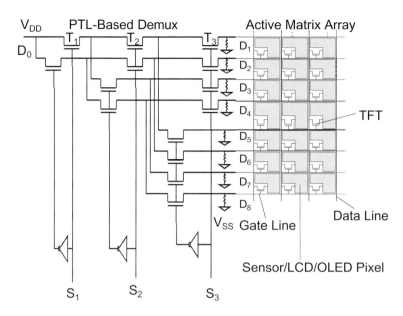

Figure 6.38 Circuit schematic of PTL-based demultiplexer driving the gate lines of active matrix arrays.

as $C_{GATE\_LINE} \approx N \times 100$ fF. If a 1:2^M PTL-based demultiplexer drives this gate line, the charging and discharging times can be approximated as $\approx 5 M R_{ON\_TFT} C_{GATE\_LINE}$, where R_{ON_TFT} is the approximate "ON" resistance of an a-Si:H TFT. From simulation results, R_{ON_TFT} for an a-Si:H TFT with $W/L = 20/20$ is found to be 2.5 MΩ. Having two demultiplexers on both sides of the array can reduce the C_{GATE_LINE}, and hence the charging and discharging times of the gate lines, to half. The charging and discharging times can be further improved by increasing the size (W/L) of the pass transistors. However, the maximum size of the pass transistors is limited by the pixel size. The pixel capacitance in an amorphous selenium photoconductor scheme is usually given by the nominal storage capacitance value of 1 pF,[67] which gives an approximate read-out time $t_{READ} \approx 5 R_{ON\_TFT} C_{PIX} = 12.5$ μs.

The PTL-based multiplexer circuit is symmetrical with respect to the input and output pins; therefore, it can also be used to multiplex the output pins of the imaging arrays in a-Si:H technology, as shown in Fig. 6.39. In this configuration, the output of the active pixel sensor is a current, and a current-to-voltage conversion is achieved by placing a resistor at the output D_0. This resistor should be relatively large compared to the "ON" resistance of a multiplexer branch to minimize the voltage drop across the

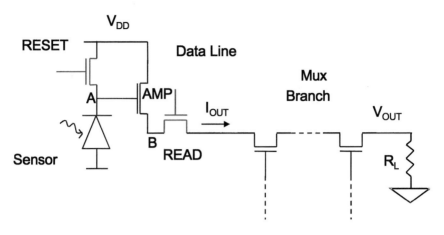

Figure 6.39 Application of PTL-based multiplexer to multiplex the output pins of an imaging array.

branch. However, a very large resistor causes an increase in the fall time of the output node. Although increasing the size of the pass transistors can reduce the "ON" resistance of the multiplexer branch, it increases the feed-through due to larger gate-to-source/drain overlap capacitance. A larger output resistor retains the effect of feed-through for a longer duration due to a larger RC constant. Hence the selection of output resistance is a tradeoff between a smaller voltage drop across the multiplexer branch and a smaller output fall-time.

The lifetime and reliability of a-Si:H TFT based circuits are reduced due to bias-stress-induced metastability issues. In addition, the variations in device parameters due to prolonged gate-bias stress lead to unsteady characteristics. An on-chip integrated PTL multiplexer using a-Si:H TFTs can lower the component count, and hence the cost of a-Si:H imagers or displays. The lifetime of the multiplexer is improved (at low operational frequencies) by the application of a negative voltage on the pass-transistor gates during the "OFF" cycle. The controllable ΔV_T of the PTL multiplexer is promising with respect to the creation of long-lifetime, ΔV_T-insensitive integrated multiplexers for large-area a-Si:H imaging or display arrays.

6.6 New Challenges in Large-Area Digital Imaging

Recently there has been growing interest in the fabrication of large-area electronics, including imaging arrays, on thin flexible plastic substrates. This stems from the need for lightweight, unbreakable, foldable computer screens or large-area sensor arrays for imaging of nonplanar (curved) surfaces. Although there are cost-related advantages to using plastic substrates, several technological issues need to be addressed. First, the upper working temperature for most plastics does not exceed 160 to 200°C. This requires the deposition temperatures for the a-Si:H, a-SiN$_x$, and microcrystalline (μc-Si:H) films to be reduced from the conventional 250 to 300°C. Secondly, due to the reduced thickness and Young's modulus, the plastic substrate cannot serve as a mechanical support for the multilayer device structures, in which bending-induced mechanical stresses, in addition to the internal stress, can cause the layers to break and peel off. Thus mechanical stress in single layers as well as in the composite device structure needs to be optimized.

Reducing the deposition temperature in PECVD a-Si:H technology using an SiH$_4$ source gas causes a decrease of the hydrogen surface diffusion coefficient. This leads to increased polyhydride bonding in the film, which results in poor electronic properties.[74] However, a-Si:H films with predominantly monohydride bonding have been deposited at low temperature by use of hydrogen or helium dilution followed by post-deposition annealing at 160°C.[75] More recently, there have been reports of transistor fabrication at 160°C on Kapton® E[76] and at 110°C on PET,[77] showing performance characteristics close to those fabricated at 250 to 300°C.

We developed a low-temperature fabrication process for which we have optimized the deposition process for a-Si:H and a-SiN$_x$:H films at 120°C with reduced mechanical stress in the structure.[78] Kapton® HN films 2 mil (51 μm) thick and 3 inches in diameter were used as substrates. Before fabrication, 0.5-μm-thick silicon nitride coatings were deposited on both sides of the plastic substrate. This coating offers protection against humidity and liquid chemical agents and serves as a good mechanical support for the devices. Also, since the thermal expansion coefficient of Kapton® HN is about one order of magnitude higher than that of silicon nitride, this coating reduces the thermally induced strain.

Unlike the processes used for film deposition on glass, dry etching was used for the silicon nitride, a-Si:H, and n$^+$ layers, and wet etching was used for the metal layers. The maximum deposition temperature was determined by the photoresist postbake. The a-Si:H films were deposited from a 10% SiH$_4$ + 90% H$_2$ gas mixture. For the n$^+$ a-Si:H contact layer, a hydrogen-diluted 1% PH$_3$ + 99% SiH$_4$ gas mixture was used. Amorphous silicon nitride films were deposited from a SiH$_4$ + (50% NH$_3$ + 50% N$_2$) + He gas mixture.

The current-voltage and transfer characteristics of the transistor switch (Fig. 6.40) in an active matrix on plastic are shown in Figs. 6.41 and 6.42, respectively. The OFF current is less than 10^{-12} A, and the ON

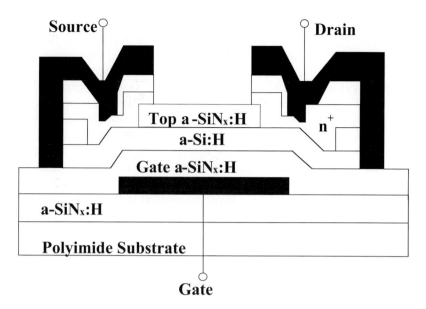

Figure 6.40 Cross section of an active matrix array switch based on the inverted staggered configuration.

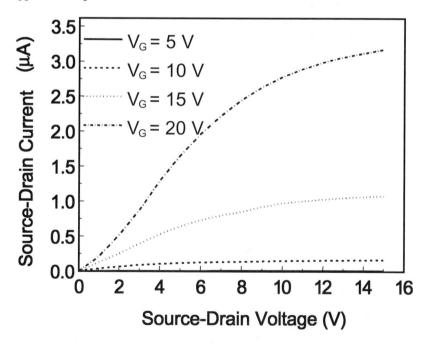

Figure 6.41 Current-voltage characteristics of the TFT at different gate voltages.

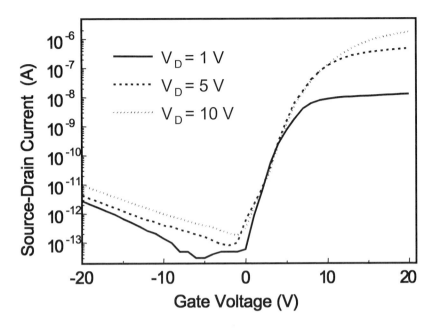

Figure 6.42 Transfer characteristics of the TFT at different drain-source voltages.

current is more than 10^{-7} A, yielding an ON/OFF current ratio of more than 10^6. Using a linear approximation, we obtain an effective device mobility, $\mu_{eff} \sim 0.4$ cm^2/Vs, using a dielectric constant value $\varepsilon_{SiN} = 6$.

Considering the longer term, there are still several issues with this technology that need to be resolved. The fabrication temperature needs to be further decreased to permit use of transparent substrates such as Mylar®. Further work is also needed in low-temperature n$^+$ μc-Si:H to increase the conductivity because these layers are needed at the source and drain regions for decreased contact resistance (see Charania et al.[79]). Finally, we believe that further significant improvements in device performance and, in particular, in the field-effect mobility may still be possible with use of substrates with reduced surface roughness. Work along these lines is currently in progress at our laboratory and others.

One of the key issues lies in the reduction of pixel size. We believe this can be achieved using the approaches outlined in Section 6.4, where we discussed stacked pixel architectures. Such architectures provide a 100% fill factor. However, the challenges here lie in reducing the pixel's parasitic capacitance, leakage current, and mechanical stress associated with the multiple thin-film stack. Alternatively, the pixel size in the nonoverlapping pixel architecture can also be reduced, without compromising the fill factor, through use of a vertical TFT structure (VTFT)[80,81] for the active matrix switch. In contrast to the conventional lateral structure, the channel length in the VTFT is defined by the a-SiN$_x$:H layer thickness in a drain-source island; thus the channel length can be precisely controlled by film thickness. Crucial benefits of this device are not only in high fill-factor imaging but also for other high-resolution applications, including displays. With flexible imagers on plastic substrates, the fabrication temperature needs to be further decreased and optimized for materials of electronic grade (low defect density and low stress). As mentioned, further significant improvements in device performance and, in particular, in the field-effect mobility may still be possible with use of substrates with reduced surface roughness.

References

1. *Amorphous and Microcrystalline Semiconductor Devices, Material and Device Physics* (J. Kanicki, ed.), Artech House, Norwood, MA (1992)
2. R. A. Street and L. E. Antonuk, "Amorphous silicon arrays develop a medical image," *IEEE Circuits and Devices*, 9(4):38–42 (1993)
3. *Amorphous and Microcrystalline Semiconductor Devices, Optoelectronic Devices* (J. Kanicki, ed.), Artech House, Norwood, MA (1992)
4. C. van Berkel, M. J. Powell, and S. C. Deane, "Physics of a-Si:H switching diodes," *J. Non-Crys. Solids*, 164–166:653–658 (1993)
5. P. G. LeComber, W. E. Spear, and A. Ghaith, "Amorphous silicon field effect device and possible application," *Electron. Lett.*, 15:179–181 (1979)
6. R. V. R. Murthy, Q. Ma, A. Nathan, and S. G. Chamberlain, "Effect of NH3/SiH4 gas ratios of top nitride layer on stability and leakage in a-Si:H thin film transistors," *MRS Symp. Proc.*, 507:73–78 (1998)
7. R. A. Street, X. D. Wu, R. Weisfield, S. Ready, R. Apte, M. Nguyen, and P. Nylen, "Two dimensional amorphous silicon image sensor arrays," *MRS Symp. Proc.*, 377:757–765 (1995)
8. M. J. Powell, I. D. French, J. R. Hughes, N. C. Bird, O. S. Davies, C. Glasse, and J. E. Curran, "Amorphous silicon image sensor arrays," *MRS Symp. Proc.*, 258:1127–1137 (1992)
9. Q. Ma, A. Nathan, and R. V. R. Murthy, "ITO/a-Si:H Schottky photodiode with low leakage current and high stability," *MRS Symp. Proc.*, 558:231–236 (1999)
10. S. Tao, Q. Ma, D. Striakhilev, and A. Nathan, "ITO/a-SiN$_x$/a-Si:H photodiode with enhanced photosensitivity and reduced leakage current using polycrystalline ITO deposited at room temperature," *MRS Symp. Proc.*, 609:A12.2.1–A12.2.6 (2000)
11. L. E. Antonuk, Y. El-Mohri, J. H. Siewerdsen, J. Yorkston, W. Huang, V. E. Scarpine, and R. A. Street, "Empirical investigation of the signal performance of a high resolution indirect detection active matrix flat-panel imager

(AMFPI) for fluoroscopic and radiographic operation," *Med. Phys.*, 24:51–70 (1997)
12. W. Zhao and J. A. Rowlands, "Large-area solid state detector for radiology using amorphous selenium," *Proc. SPIE*, 1651:134–143 (1992)
13. J. A Rowlands, W. Zhao, I. M. Blevis, D. F. Waechter, and Z. Huang, "Flat panel digital radiology with amorphous selenium and active matrix readout," *RadioGraphics*, 17:753–760 (1997)
14. K. Aflatooni, A. Nathan, R. Hornsey, I. A. Cunningham, and S. G. Chamberlain, "a-Si:H Schottky diode direct detection pixel for large area x-ray imaging," *Tech. Digest IEEE IEDM 1997* (1997), pp. 197–200
15. Y. Naruse and T. Hatayama, "Metal/amorphous silicon multilayer radiation detectors," *IEEE Trans. Nucl. Sci.*, 36:1347–1352 (1989)
16. *Handbook of Medical Imaging, Vol. 1, Physics and Psychophysics* (J. Beutel, H.L. Kundel, and R.L. Van Metter, eds.), SPIE (2000), www.spie.org/bookstore/
17. A. Nathan, R. Hornsey, and K. Aflatooni, "Transduction principles of a-Si:H Schottky diode X-Ray image sensors," *IEEE Trans. Electron Devices*, 7(11):2093–2100 (Nov. 2000)
18. R.I. Hornsey, K. Aflatooni, and A. Nathan, "Reverse current transient behaviour in amorphous silicon Schottky diodes at low biases," *Appl. Phys. Lett.*, 70:3260–3262 (1997)
19. K. Aflatooni, R. Hornsey, and A. Nathan, "Reverse current instabilities in amorphous silicon Schottky diodes: modeling and experiments," *IEEE Trans. Electron Devices*, 46:1417–1422 (1999)
20. K. Aflatooni, A. Nathan, and R. I. Hornsey, "Low frequency noise behaviour in a-Si:H Schottky barrier devices," *MRS Symp. Proc.*, 420:747–752 (1996)
21. R. A. Street, "Long-time transient conduction in a-Si:H pin devices," *Philos. Mag. B*, 63(6):1343–1363 (1991)
22. M. Hoheisel, N. Brutscher, and H. Wieczorek, "Ambient-induced defect states at a-Si:H/ITO interfaces," *J. Non-Crys. Solids*, 115:114–116 (1989)
23. Q. Ma and A. Nathan, "Room temperature sputter deposition of polycrystalline ITO for photodetectors," *Proc. ECS*, 98–22:408–420 (1999)
24. J. K. Arch and S. J. Fonash, "Using reverse bias currents to differentiate between bulk degradation and interfacial degradation in hydrogenated amorphous silicon pin structures," *J. Appl. Phys.*, 72:4483 (1992)
25. R. A. Street and D. K. Biegelsen, "The spectroscopy of localized states," in *Topics in Applied Physics 56*, Springer Verlag, New York (1984), p. 242
26. R. S. Sussman and R. Ogden, "Photoluminescence and optical properties of plasma deposited amorphous SixC1-x alloys," *Phil. Mag.*, B44:137–158 (1981)
27. R. L. Weisfield and C. C. Tsai, "The role of carbon in amorphous silicon nip photodiode sensors," *MRS Symp. Proc.*, 192:423–428 (1990)
28. R. A. Street, "Measurement of depletion layers in hydrogenated amorphous silicon," *Phys. Rev. B*, 27:4924–4932 (1983)
29. F. Carasco and W. E. Spear, "Photogeneration and geminate recombination in amorphous silicon," *Phil. Mag. B*, 47:495–507 (1983)

30. M. J. Powell, "The physics of amorphous silicon thin film transistors," *IEEE Trans. Electron. Devices*, 36:2753–2762 (1989)
31. A. M. Miri, "Development of a novel wet etch fabrication technology for amorphous silicon thin-film transistors, Ph.D. thesis, University of Waterloo (1995)
32. M. Shur and M. Hack, "Physics of amorphous silicon based alloy field-effect transistor," *J. Appl. Phys.*, 55:3831–3842 (1984)
33. T. Leroux, "Static and dynamic analysis of amorphous silicon thin film transistors," *Solid-State Electron.*, 29:47–58 (1986)
34. M. J. Powell, "Charge trapping instabilities in amorphous silicon-silicon nitride thin-film transistors," *Appl. Phys. Lett.*, 43(6):15 (1983)
35. W. B. Jackson and M. D. Moyer, "Creation of near interface defects in hydrogenated amorphous silicon-silicon nitride heterojunctions: the role of hydrogen," *Phys. Rev. B*, 36:6217 (1987)
36. F. R. Libsch and J. Kanicki, "Bias-stress-induced stretched-exponential time dependence of charge injection and trapping in amorphous silicon thin-film transistors," *Appl. Phys. Lett.*, 62:1286–1288 (1993)
37. F. R. Libsch, "Steady state and active-matrix liquid-crystal displays," *Tech. Digest IEEE, IEDM 1992* (1992), pp. 681–684
38. F. R. Libsch and J. Kanicki, "Bias-stress-induced stretched-exponential time dependence of charge injection and trapping in amorphous silicon thin-film transistors," *Extended Abstracts of the 1992 International Conference on Solid State Devices and Materials, Tsukuba, Japan*, Business Center for Acad. Soc. Japan, Tokyo, Japan (1992), pp. 155–157
39. N. Lustig and J. Kanicki, "Gate dielectric and contact effects in hydrogenated amorphous silicon-silicon nitride thin-film transistors," *J. Appl. Phys.*, 65:3951–3957 (1988)
40. R. A. Street and C. C. Tsai, "Fast and slow states at the interface of amorphous silicon and silicon nitride," *Appl. Phys. Lett.*, 48:1672–1674 (1986)
41. N. Nickel, R. Saleh, W. Fuhs, and H. Mell, "Study of instabilities in amorphous silicon thin-film transistors by transient current spectroscopy," *IEEE Trans. Electron Devices*, 37: 280–283 (1990)
42. M. J. Powell, C. van Berkel, A. R. Franklin, S. C. Deane, and W. I. Milne, "Defect pool model in amorphous silicon thin-film transistors," *Phys. Rev. B*, 45:4160–4170 (1992)
43. M. Stutzmann, "Weak bond-dangling bond conversion in defects in amorphous silicon," *Philos. Mag.*, 56:63–70 (1987)
44. Z. E. Smith and S. Wagner, "Band tails, entropy, and equilibrium defects in hydrogenated amorphous silicon," *Phys. Rev. Lett.*, 59:688–691 (1987)
45. D. L. Staebler and C. R. Wronski, "Reversible conductivity changes in discharge-produced amorphous Si," *Appl.Phys. Lett.*, 31:292–294 (1977)
46. M. Stutzmann, W. B. Jackson, and C. C. Tsai, "Light-induced metastable defects in hydrogenated amorphous silicon," *Phys. Rev. B*, 32:23–47 (1985)
47. M. J. Powell *et al.*, "A defect-pool for near-interface states in amorphous silicon thin-film transistors," *Philos. Mag.*, 63:325–336 (1991)

48. C. van Berkel and M. J. Powell, "The resolution of amorphous silicon thin film transistor instability mechanisms using ambipolar transistors," *Appl. Phys. Lett.*, 51:1094 (1987)
49. M. J. Powell, C. van Berkel, I. D. French, and D. H. Nicholls, "Bias dependence of instability mechanisms in amorphous silicon thin film transistors," *Appl. Phys. Lett.*, 51: 1242 (1987)
50. R. H. Walden, "A method for the determination of high field conduction laws in insulating films in the presence of charge trapping," *J. Appl. Phys.*, 43:1178 (1972)
51. M. J. Powell, C. van Berkel, and J. R. Hughes, "Time and temperature dependence of instability mechanisms in amorphous silicon thin-film transistors," *Appl. Phys. Lett.*, 54(14):1323 (1989)
52. M. J. Powell, S. C. Deane, and W. I. Milne, "Bias-stress-induced creation and removal of dangling-bond states in amorphous silicon thin-film transistors," *Appl. Phys. Lett.*, 60(2): 207 (1991)
53. M. J. Powell, I. D. French, and J. R. Hughes, "Evidence for the defect pool concept for Si dangling bond states in a-Si from experiments with thin-film transistors," *J. Non-Crys. Solids*, 114:642 (1989)
54. M. J. Powell, C. van Berkel, A. R. Franklin, S. C. Deane, and W. I. Milne, "Defect pool in amorphous-silicon thin-film transistors," *Phys. Rev. B*, 45(8):4160–4170 (1992)
55. C. Chiang, J. Kanicki, and K. Takechi, "Electrical instability of hydrogenated amorphous silicon thin-film transistors for active-matrix liquid-crystal displays," *Jpn. J. Appl. Phys., part 1*, 37(9A):4704–4710 (1998)
56. R. Oritsuki, T. Horii, A. Sasano, K. Tsutsui, T. Koizumi, Y. Kaneko, and T. Tsukada, "Threshold voltage shift of amorphous silicon thin-film fransistors during pulse operation," *Jpn. J. Appl. Phys.*, part 1, 30(12B):3719–3723 (1991)
57. B. Park and A. Nathan, "Fully-vs non-overlapping pixel configurations for direct x-ray detection: process and performance considerations," *Proc. ECS*, 98–22:381–390 (1999)
58. B. Park, K.S. Karim, and A. Nathan, "Intrinsic film stresses in multi-layered imaging pixels," *J. Vac. Sci. Technol. A*, 18(2):688–692 (2000)
59. R.L. Weisfield, "Amorphous silicon TFT X-ray imager sensors," *Tech. Digest IEEE IEDM 1998* (1998), pp. 21–24
60. I. Fujieda, R.A. Street, R.L. Weisfield, S. Nelson, P. Nylen, V. Perez-Mendez, and G. Cho, "High sensitivity readout of 2D a-Si:H image sensors," *Jpn. J. Appl. Phys.*, 32:198–204 (1993)
61. W. Zhao and J. A. Rowlands, "X-ray imaging using amorphous selenium: feasibility of a flat panel self-scanned detector for digital radiology," *Med. Phys.*, 22(10):1595–1604 (1995)
62. M. Maolinbay, Y. El-Mohri, L.E. Antonuk, K.-W. Jee, S. Nassif, X. Rong, and Q. Zhao, "Additive noise properties of active matrix flat-panel imagers", *Med. Phys.*, 27(8):1841–1854 (2000)
63. S.K. Mendis, S.E. Kemeny, and E.R. Fossum, "CMOS active pixel image sensor," *IEEE Trans. Electron Devices*, 41(3):452–453 (1994)

64. K. S. Karim, A. Nathan, and J. A. Rowlands, "Amorphous silicon active pixel sensor readout circuit architectures for medical imaging," in *Amorphous and Heterogeneous Silicon-Based Films* (J. R. Abelson, J. B. Boyce, J. D. Cohen, H. Matsumura, and J. Robertson, eds.), *MRS Symp. Proc.*, 715:661–666 (2002)
65. Z. Huang and T. Ando, "A novel amplified image sensor with a-Si:H photoconductor and MOS transistors," *IEEE Trans. Electron Devices*, 37(6):1432–1438 (1990)
66. N. Matsuura, W. Zhao, Z. Huang, and J.A. Rowlands, "Digital radiology using active matrix readout: amplified pixel detector array for fluoroscopy," *Med. Phys.*, 26(5):672–681 (1999)
67. K.S. Karim, A. Nathan, and J.A. Rowlands, "Alternate pixel architectures for large area medical imaging", in *Medical Imaging 2001: Physics of Medical Imaging* (L. Antonuk and M. Yaffe, eds.), *Proc. SPIE*, 4320:35–46 (2001)
68. H. Tian, B. Fowler, and A. El Gamal, "Analysis of temporal noise in CMOS photodiode active pixel sensor," *IEEE J. Solid-State Circuits*, 36(1):92–101 (2001)
69. K.S. Karim, A. Nathan, and J.A. Rowlands, "Active pixel sensor architectures in a-Si for medical imaging," *J. Vac. Sci. Technol. A*, 20(3):1095–1099 (2002)
70. P. Servati, "Modeling the static and dynamic characteristics of a-Si:H TFTs," M.A.Sc. thesis, University of Waterloo (2000)
71. K. S. Karim, P. Servati, N. Mohan, A. Nathan, and J. A. Rowlands, "VHDL-AMS modeling and simulation of a passive pixel sensor in a-Si technology for medical imaging", in *Proc. IEEE International Symposium on Circuits and Systems 2001, Sydney, Australia*, 5:479–482 (2001)
72. N. Mohan, K.S. Karim, and A. Nathan, "Design of multiplexer in amorphous silicon technology," *J. Vac. Sci. Technol. A*, 20(3):1043–1047 (2002)
73. K. Suzuki, T. Aoki, M. Ikeda, T. Higuchi, M. Akiyama, M. Dohjo, T. Niiyama, and Y. Oana, *J. Inst. Television Eng. Jpn.*, 40(10):974–979 (1986)
74. R.A. Street, *Hydrogenated Amorphous Silicon*, Cambridge University Press (1991)
75. E. Srinivasan, D.A. Lloyd, and G.N. Parsons, "Dominant monohydride bonding in hydrogenated amorphous silicon thin films formed by plasma enhanced chemical vapor deposition at room temperature", *J. Vac. Sci. Technol. A*, 15:77–80 (1997)
76. H. Gleskova, S. Wagner, and Z. Suo, "a-Si:H TFTs on Kapton," *MRS Symp. Proc.*, 508:73–78 (1998)
77. G.N. Parsons, C.-S. Yang, T.M. Klein, and L. Smith, "Surface reactions for low temperature (110°C) amorphous silicon TFT formation on transparent plastic substrates," *MRS Symp. Proc.*, 507:19–24 (1998)
78. A. Sazonov, A. Nathan, R.V.R. Murthy, and S.G. Chamberlain, "Fabrication of a-Si:H TFTs at 120°C on flexible polyimide substrates," *MRS Symp. Proc.*, 558:375–380 (1999)
79. T. Charania, B. Park, A. Sazonov, D. Striakhilev, and A. Nathan, "Characterization of n^+ μc-Si: H for TFTs fabricated at 120°C on plastic substrates,"

Proceedings of the 198th Meeting, The Electrochemical Society, M2 - Thin Film Transistor Technologies V, Phoenix, Arizona, October 22–27, 2000, 2000-31:54–62 (2001)
80. Y. Uchida, Y. Nara, and M. Matsumura, *IEEE Electron Device Lett.*, EDL-5:105–107 (1984)
81. I. Chan, B. Park, A. Sazonov, and A. Nathan, "Process considerations for small area a-Si:H vertical thin film transistors," *Proceedings of the 198th Meeting, The Electrochemical Society, M2 - Thin Film Transistor Technologies V, Phoenix, Arizona, October 22–27, 2000,* 31:63–69 (2001)

7 Free-Space Optical MEMS

Ming C. Wu
University of California, Berkeley, California

Pamela R. Patterson
HRL Laboratories, Malibu, California

7.1 Introduction

Microelectromechanical systems (MEMS) technology enables the creation of micro-optical elements that are inherently suited to cost-effective manufacturability and scalability because the processes are derived from the very mature semiconductor microfabrication industry. The inherent advantages of applying microelectronics technology to silicon micromechanical devices, including optical MEMS, were presented in 1982 by Petersen in the now-classic paper, "Silicon as a mechanical material."[1] The ability to steer or direct light is a key requirement for free-space optical systems. In the 20 years since Petersen's silicon scanner,[2] the field of optical MEMS has seen explosive growth.[3,4] In the 1980s and early 1990s, displays were the main driving force for the development of micromirror arrays. Portable digital displays are now commonplace, and head-mount displays are also commercially available. In the past decade, telecommunications have become the market driver for optical MEMS. The demand for routing ever-increasing Internet traffic through fiber-optic networks pushes the development of both digital and scanning micromirror systems for large port-count, all-optical switches. In the health care arena, scanning optical devices promise low-cost, optical cross-sectioning endoscopic microscopy for *in vivo* diagnostics.

This chapter summarizes the state of the art of optical MEMS technologies and applications. There has been rapid commercialization of optical MEMS in the past several years. Many of these developments are posted on companies' websites and are presented in various business publications. However, because most of these technologies are proprietary, the technical details are not given. The discussions in this chapter are therefore based on archived papers from technical journals and conference proceedings.

This chapter is organized as follows: Section 7.2 discusses the general considerations for micromirrors and the design parameters related to their optical performance. Section 7.3 discusses electrostatic scanning micromirrors with medium-size mirrors (a few hundred micrometers to one millimeter). Section 7.4 deals with magnetic and electromagnetic scanners with large mirror size (greater than one millimeter). Section 7.5 describes micromirror arrays for various telecommunication applications. The two-dimensional (2-D) micromirror arrays for switching applications (2-D MEMS switches) are discussed in detail in Section 7.6, while non-scalable 2 × 2 optical switches are described in Section 7.7. Section 7.8 presents variable optical attenuator arrays. Sections 7.9 and 7.10 present short summaries of tunable wavelength-division-multiplexed (WDM) devices and diffractive optical MEMS, respectively. Section 7.11 summarizes the discussions in the chapter.

7.2 General Discussion of Micromirror Scanners

MEMS micromirrors are candidates for virtually any application where miniaturization, light weight, low-energy consumption, and/or reduced cost provide the incentive to replace existing macromirrors. The design space for MEMS scanners encompasses optical, electrical, and mechanical parameters. MEMS still represent new microfabrication technology as compared to, e.g., standard CMOS, and considerable flexibility in both design and fabrication may be exercised to achieve the desired objectives. Generally speaking, performance of a scanning micromirror is determined by the maximum scan angle, resonant frequency, resolution, and surface quality with respect to flatness and smoothness. The chosen microfabrication technology as well as the performance is closely linked to the size of the micromirror; therefore length scale is a convenient way to classify MEMS scanners. Figure 7.1 summarizes the applications and main fabrication technologies of various micromirror devices, with length scale ranges from micrometers to several millimeters.

The mirror size plays a critical role in the performance of the optical systems. High-resolution imaging applications require large mirrors and large scan angles. Resolution as defined by the number of resolvable spots is a function of the beam divergence and the scan range (Fig. 7.2), as shown in the following equalities:

$$N = \frac{\theta_{max}}{\delta\theta} = \frac{\theta_{max} D}{a\lambda}, \tag{1}$$

where $\delta\theta$ is beam divergence, θ_{max} is the total optical scan range, D is the mirror diameter, λ is the wavelength of incident light, and a is the aperture shape factor ($a = 1$ for a square aperture and 1.22 for a circular aperture).

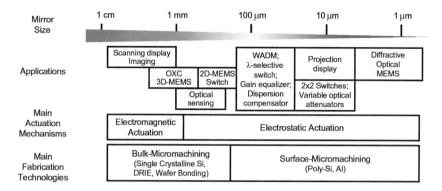

Figure 7.1 Applications and main fabrication technologies of micromirror devices with length scales from micrometers to centimeters.

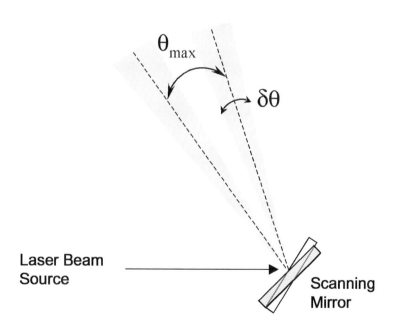

Figure 7.2 Schematic diagram illustrating the maximum scan angle and beam divergence of a scanning micromirror.

For imaging or display applications involving raster scanning, a dual-axis system with fast scanning capability is required (Fig 7.3).[5] This may be accomplished by combining two one-dimensional (1-D) scanners or by a single biaxial scanner, typically designed with a gimbal structure (Fig 7.4). The maximum scan speed is determined by mechanical properties

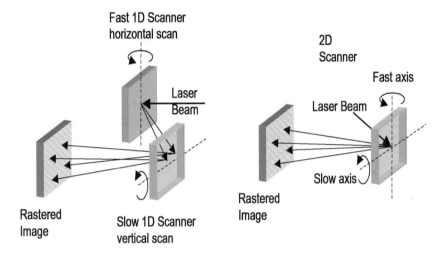

Figure 7.3 Optical setup for raster scanning with cascaded 1-D scanners and with a single 2-D scanner (after Hagelin and Solgaard[5]).

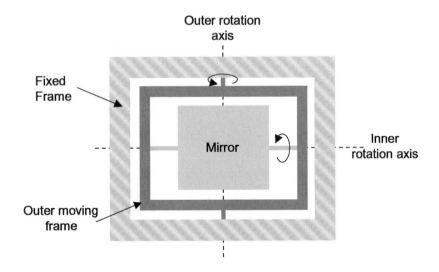

Figure 7.4 Double gimbal structure for a 2-D scanning torsion mirror.

(mass and spring constant) that dictate the resonant frequency of the device.

The natural frequency for torsional displacement is given by

$$\omega_0 = 2\pi f_0 = \sqrt{\frac{k}{I}}. \qquad (2)$$

For a square mirror,

$$I = \frac{1}{12} mL^2 \qquad (3)$$

where I is the mass moment of inertia, k is the spring constant, L is the mirror diameter, ω_0 is the natural frequency, and m is the mass of the mirror.

Very large area mirrors with a length scale >2 mm benefit from electromagnetic actuation, since the torque produced by the Lorentz force is proportional to the mirror area. To maintain the flatness of large mirrors, such devices are typically fabricated in single-crystal silicon (SCS) derived from a bulk substrate or silicon on insulator (SOI). In the 250 µm to 1 mm range, electrostatic actuation is more advantageous, and current trends favor higher force electrostatic comb drives over parallel-plate-type actuators and SOI MEMS over surface micromachining. For micromirror devices below 250 µm, surface micromachining offers higher integration and multilevel interconnect, features that are ideal for large arrays of micromirrors. The following sections first discuss electrostatic scanners, followed by electromagnetic scanners, then micromirror arrays.

7.3 Electrostatic Scanners

The first single-crystal silicon scanning micromirror was reported by Petersen.[2] The mirror was fabricated by wet, anisotropic etching and subsequently glued to a glass substrate where a central offset mesa-maintained recessed cavity defines the small gap between the top and bottom electrodes (Fig. 7.5). Life testing through ~10^{11} cycles in resonant mode was accomplished by electrically driving the device at the frequency required to sustain mechanical resonance. The results confirmed that silicon was a robust mechanical material, and the first scanning micromirror was born.

Electrostatic micromirrors with torsional rotation can be described as follows. When voltage is applied between the movable and the fixed elec-

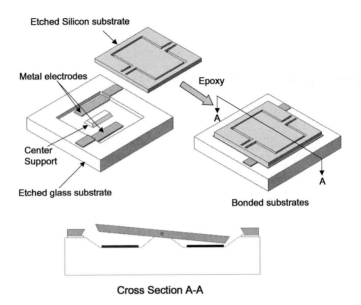

Figure 7.5 Early single-crystal silicon electrostatic micromirror (from Petersen [1]).

trodes, the mirror rotates about the torsion axis until the restoring torque and the electrostatic torque are equal. The torques can be expressed as

$$T_e(\theta) = \frac{V^2}{2}\frac{\partial C}{\partial \theta} \quad (4)$$

$$T_r(\theta) = k\theta, \quad (5)$$

where V is the applied voltage, C is the capacitance of the actuator, θ is the rotation angle, and k is the spring constant. The capacitance is determined by the area of the electrode overlap and the gap between the electrodes. For simple parallel-plate geometry, the capacitance can be expressed by

$$C = \frac{\varepsilon_0 A}{g}, \quad (7)$$

where ε_0 is the permittivity of free space, A is the area of electrode overlap, and g is the gap between fixed and moving electrodes.

There are two main types of electrostatic scanners. The first is based on parallel-plate actuators; the other is based on comb-drive actuators. For the parallel-plate type mirror, the area of the electrode overlap is essentially the area of the fixed electrode. The gap for the parallel-plate actuator (mirror) is a function of the rotation angle, and there is a tradeoff because the initial gap spacing needs to be large enough to accommodate the scan angle but small enough for reasonable actuation voltage. This is further exacerbated by a pull-in phenomenon, which renders 60 to 66% of the gap spacing unstable.[6,7] The vertical comb drive offers several advantages: The mirror and the actuator are decoupled, and the gap between the interdigitated fingers of the comb drive is typically quite small, on the order of several microns. In the comb drive, the gap is constant, and the area of the electrode overlap is a function of the rotation angle.

7.3.1 Scanners with Electrostatic Parallel-Plate Actuators

A schematic of the parallel-plate scanner is show in Fig. 7.6. The device can be realized by using surface and/or bulk micromachining technologies. Surface-micromachined scanners can be fabricated using a standard foundry process, such as the Multi-User MEMS Process (MUMPs).[8] However, assembly is required to raise the mirror above the substrate. Fan and Wu[9] have reported a 400 × 400 μm^2 parallel-plate scanner. The polysilicon mirror is lifted vertically to 200 μm above the substrate using a microactuated 3-D structure called a microelevator by self-assembly (MESA).[10] Lucent Technologies has used a self-assembly technique that is driven by the residual stress in deposited thin films (Cr/Au on polysilicon) to raise two-axis polysilicon scanners (500 μm mirror diameter) to a fixed position 50 μm above the substrate.[11,12] The

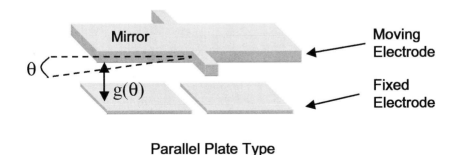

Figure 7.6 Schematic of the parallel-plate scanner.

scanning electron micrographs (SEMs) of both scanners are shown in Fig 7.7.[13] Two-axis scanning is achieved by electrostatic force between the mirror and the quadrant electrodes on the substrate. Fan's scanner has a static optical scan range of 28° and a drive voltage of 70 V. Resonant frequency for the mirror was measured at 1.5 kHz.

Telecommunication switches with large port count have been the main driver for the two-axis scanner in the past few years. With an increasing number of wavelengths and bandwidth in dense wavelength-

(a)

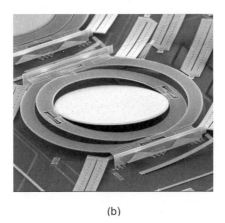

(b)

Figure 7.7 SEMs of surface-micromachined two-axis scanners from (a) UCLA,[9] and (b) Lucent Technologies.[13] [Part (b) Courtesy of Lucent Technologies Inc. ©2003 Lucent Technologies Inc. All rights reserved.]

division-multiplexed (DWDM) networks, there is a need for optical cross-connect (OXC) with a large port count.[14-16] The dual-axis analog scanning capability is key for these applications since each mirror associated with the input fiber array can point to any mirror associated with the output fiber array. Implementation of N × N OXC using two arrays of N analog scanners is illustrated in Fig. 7.8. This switch is often called a 3-D MEMS switch because the optical beams propagate in three-dimensional space. Since the optical path length is independent of the switch configuration, uniform optical insertion loss (2 to 3 dB) can be achieved. Square mirror arrays with up to 256 scanning micromirrors have been demonstrated with this surface-micromachined process for OXC applications.[14,17,18]

The port count of the 3-D MEMS switch is limited by the size and flatness of the micromirrors, as well as their scan angle and fill factor. Syms provides a complete discussion of scaling laws for MEMS free-space optical switches.[16] For a large port count (approaching 1000 × 1000), single-crystal micromirror scanners are necessary to achieve a large mirror size with the required flatness. Micromirrors fabricated by standard polysilicon surface-micromachining processes can have significant curvature due to residual stress of the deposited thin films. Even for flat mirrors with balanced stress, the difference in thermal expansion coefficients between the thin mirror and the metal coating causes the mirror curvature to change with temperature. Techniques to improve the

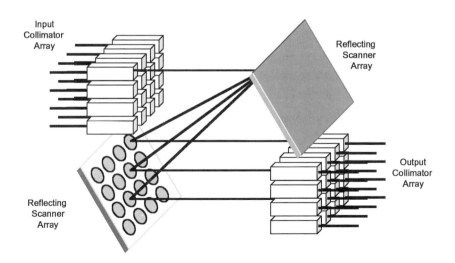

Figure 7.8 Configuration for a 3-D optical switch for N input fibers and N output fibers (N × N) that requires 2N analog scanning mirrors.

curvature of polysilicon micromirrors have been reported: by incorporating an outer-edge folded frame in the mirror film formed by depositing the polysilicon in an etched trench;[19] by exploiting the tensile stress of the polysilicon to create a drum-like mirror structure on a rigid solid frame;[20] and by a combination of high-temperature annealing and thick polysilicon cross-bar support structures.[21]

Hybrid structures that use single-crystal silicon for the mirror structure and polysilicon for the actuators offer an alternative to polysilicon mirrors. Su *et al.* showed that the mirror curvature can be improved by 150 times (from 1.8 to 265 cm) by bonding a 22.5-µm-thick single-crystal silicon (SCS) mirror to the surface-micromachined actuator.[22,23] Bonding may be performed at wafer scale. Scanners with 1-mm SCS honeycomb micromirrors created by deep reactive ion etching (DRIE) and silicon fusion bonding have also been reported to reduce the mirror weight and increase resonant frequency.[24] Fabrication of hybrid devices has also been demonstrated using a fluidic self-assembly technique using low-temperature adhesives and selective surface treatments of the polysilicon receptor and a thin (15 µm) SCS mirror.[25] Gold compression bonding of two different surface-micromachined MUMPS chips has been used to increase the number of layers and the design flexibility for mirror arrays.[26,27] Similar bonding can be applied to SCS mirrors to improve the mirror quality.

A clear trend for electrostatic scanning micromirrors with mirror size in the 0.5 to 1 mm range is toward the use of SOI-MEMS and simplified fabrication methods.[28] The schematic diagram SEM of the two-axis scanner reported by Analog Devices is shown in Fig. 7.9.[29] The scanner is realized using a three-layer approach. The top two layers are from an SOI wafer with a 10-µm-thick device layer. The top layer is used for mirrors and mechanical springs, as well as integrated control electronics. After backside etching, the wafer is bonded to a bottom wafer with polysilicon electrodes for electrostatic actuation and capacitive sensing of mirror position. The electrodes are contacted through polysilicon-filled via posts. A similar electrostatic parallel-plate scanner has been recently reported from Corning Intellisense.[30] Corning's mirrors were derived from a single 100-µm SOI that was patterned and etched down to 3 to 4 µm in areas intended for the torsion flexures, and to 14 µm in areas intended for the mirrors. The SOI was anodically bonded to a glass substrate with patterned metal electrodes. Arrays containing 320 scanners each at 750×800 µm^2 were demonstrated with component mirror mechanical scan range of 10° at 60 V. The resonant frequencies for both mirror axes were reported in the 200 to 400 Hz range. Using bulk micromachining, NTT has developed a "terraced bottom electrode" to reduced the operating voltage.[31] Note that the reso-

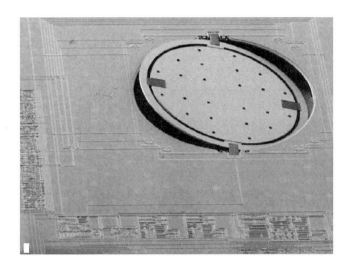

Figure 7.9 Scanning electron micrograph of the Analog Devices SOI MEMS scanner.[31]

nant frequency requirement is relaxed in optical switching applications (compared to imaging applications) since switching times of several milliseconds are acceptable.

Using two chips of two-axis SCS micromirror arrays, each with 36 × 36 = 1296 elements, Ryf *et al.* recently reported a 1296-port OXC.[32] Light was coupled to the micromirror arrays through a matched 2-D fiber collimator array. The maximum operating voltage of the scanning mirror was 200 V. Connections measured between 60 input ports and 60 output ports exhibited a mean insertion loss of 5.1 ± 1.1 dB. Crosstalk during beam scanning was less than −38 dB, and after the connection was established, the crosstalk to adjacent channels was −58 dB. The switching time was 5 msec. For OXCs with smaller port count, low loss can be obtained even with surface-micromachined scanning micromirrors. Aksyuk *et al.* reported a mean insertion loss of 1.33 dB and maximum loss of 2.0 dB for all 56,644 connections in a 238 × 238 switch fabric.[33] The improved performance is attributed to tight control of the MEMS and optical components, including a large radius of curvature of the mirrors (loss due to mirror curvature <0.3 dB), low aberration (~λ/10) and focal length variation (~1%) of microlens arrays, and low position and point error of fibers in the array. The optical path length in the switch fabric has also been minimized. Kim *et al.* have tested the mechanical stability by switching the mirror over 18 billion cycles in 38 weeks in dry ambient at room temperature.[34] No measurable change in device performance was found. When the switch was operated without invoking feedback control, the loss variation was less than 0.1 dB over 120 hours of testing for a connection with fixed bias on the mirrors.

7.3.2 Electrostatic Vertical Comb-Drive Scanners

Electrostatic comb drives are efficient at using the driving voltage due to the narrow gap between the fixed and moving electrodes. The first MEMS electrostatic comb drives, demonstrated by Tang et al.[35] in 1989, were lateral comb drives formed in polysilicon. For lateral comb drives, the moving comb travels in-plane relative to the fixed comb, parallel to the substrate. Lateral comb drives have been used for scanning micromirrors,[36] but vertical comb drives are much more prevalent. In the vertical comb drive, the motion of the moving comb is out of the fixed comb plane and perpendicular to the substrate. The first vertical comb drive was introduced by Selvakumar et al.[37] and featured a fixed comb etched in the silicon substrate and a moving comb formed in polysilicon, with the entire structure spanning an etch pit in the substrate.

Vertical comb-drive actuation offers some key advantages over parallel-plate type actuation for electrostatic scanning micromirrors. Decoupling of the mirror and actuator removes the restriction on the maximum deflection (scan angle) imposed by the geometry of the parallel plate. In addition, the pull-in associated with the parallel plate can be avoided in the vertical comb drive. The large electrostatic torque allows for the design of a relatively stiff torsion bar; this increases the resonant frequency and therefore the scan speed and bandwidth of the device. Higher resonant frequencies also provide insurance against unwanted excitation from environmental vibrations. The vertical comb drive requires a disruption of the electrostatic field between the combs for actuation. Devices designed strictly for resonant mode operation require only a slight asymmetry between the fixed and moving areas, as demonstrated by Schenk et al.[38] However, for quasistatic or DC operation, the distance from the top of the moving comb to the top of the fixed comb (in the unactuated state) determines the range of motion, so the fixed and moving combs are designed in separate planes.

There are several ways to combine vertical comb-drive actuators with micromirrors. As shown in Fig. 7.10, the vertical combs can be integrated (a) underneath the mirror, (b) on a shaft parallel to the torsion bar, or (c) on the edge of the mirror. Alignment of the top and the bottom combs plays a critical role in the performance of the scanner. Misalignment of the combs could lead to lateral instability and limit the scan range. Recent efforts have focused on the development of self-aligned processes or

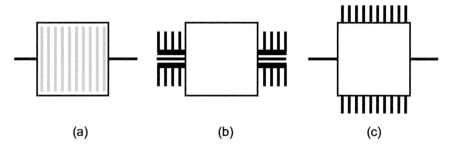

Figure 7.10 Top views of various types of vertical comb-drive-based scanners: (a) scanner with combs underneath the mirror, (b) scanner with combs integrated on a shaft parallel to the torsion bar, and (c) scanner with combs integrated at the edge of the mirror.

device structures, as discussed below. Another important parameter is the gap spacing between the movable and the fixed combs. The electrostatic torque is inversely proportional to the gap spacing. The minimum spacing is determined by the comb alignment and the aspect ratio of the etched comb fingers.

Figure 7.11 shows a large-area (1.5×1.2 mm^2) scanning mirror with a movable comb directly integrated underneath the mirror.[39] The scanners were fabricated with DRIE, and fixed and moving combs were fabricated in two separate SCS layers that were bonded together with a gold-tin layer using a flip-chip aligner. The bottom silicon layer was anodically bonded to glass before the fixed combs were etched. The device

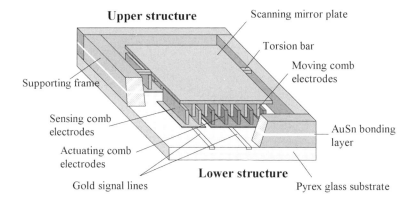

Figure 7.11 Schematic of a bulk-micromachined SCS scanner with bonded vertical combs underneath the mirror. (Picture courtesy of Jin-Ho Lee. Reprinted from Lee et al.[39] with permission from Elsevier.)

achieved a 12° optical scan range with a 60 Hz sinusoidal signal at 28 V and a DC bias voltage of 35 V. The resonant frequency for the device was reported at 1.353 kHz. When fixed and moving combs are fabricated in separate layers, the alignment step is critical since it controls the electrode gap (5 μm in the reported device).

A scanning mirror with staggered vertical comb-drive actuation was demonstrated by Conant et al.[40] by patterning the fixed and moving teeth of the comb drive in two SCS layers separated by oxide (Fig. 7.12). The fixed teeth were patterned in the silicon substrate to a depth of 100 μm. The moving teeth were patterned in a 50-μm SCS layer created by fusion bonding after the substrate etch. Opened holes in the top silicon layer provided 0.2-μm alignment between the fixed and moving teeth. The device reported had very stiff torsion bars (50 μm thick × 15 μm wide × 150 μm), which resulted in an extremely high resonant frequency (32 kHz) for the 550-μm-diameter mirror. The total optical scan angle in resonant mode was reported as 24.9° at 171 V_{rms}. For such a stiff device, the resonant mode is required to achieve usable actuation at reasonable voltage, but the staggered vertical comb-drive design is suitable for quasistatic actuation of more compliant torsion springs. Self-aligned vertical comb drives have been described by Krisnamoorthy and Solgaard,[41] and are achieved by

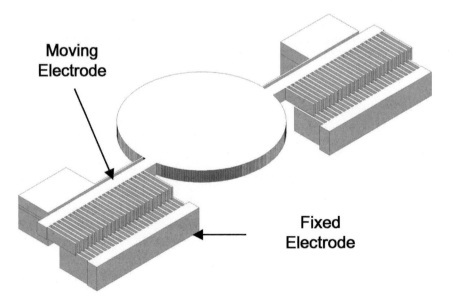

Figure 7.12 Schematic of the staggered vertical comb-drive-actuated SCS scanner (after Conant et al.[40]).

using sacrificial comb teeth such that the final etch defines both the fixed and moving comb.

At UCLA we developed a truly self-aligned process with the angular vertical comb (AVC) drive.[42] The self-aligned comb fingers are patterned in a single etch of the SOI layer. A polymer hinge is then applied to fasten the moving combs to the fixed combs and to provide the force for the surface tension assembly.[43] After release, heating of the polymer causes reflow, which rotates the moving comb out of the wafer plane. Upon cooling, the thermoplastic polymer fastens the combs together at a fixed rotation angle (see Fig. 7.13). This initial angle determines the maximum rotation angle for the device. An AVC device (1 mm diameter) was demonstrated with a photoresist hinge derived from a 25-μm SOI layer. However, yield was low due to photoresist residue in the high-aspect-ratio features, and the best static scan angle measured was 3.2° (mechanical range) at 108 V for a scanner with a resonant frequency of 1.4 kHz.[44] We have recently reported a process using benzocyclobutene (BCB), a negative photoactive material for the AVC.[45] The BCB process has yielded more AVC scanning

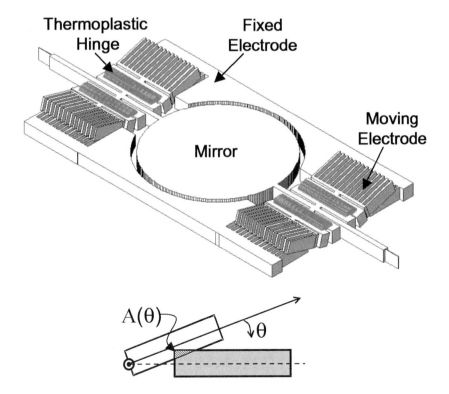

Figure 7.13 Schematic of an angular vertical comb scanner.

devices covering a broader design space, and improved device performance has been measured on devices with more compliant spring designs. We have achieved a static scan angle of 14° (mechanical range) at 65 V for a device with a resonant frequency of 598 Hz.

All of the vertical comb scanners covered so far, with the exception of the offset electrode device,[38] have been 1-D scanners. Two-dimensional (2-D) vertical comb-drive scanners are more difficult to implement than parallel-plate-type scanners since four isolated electric leads are needed for the fixed combs. This requires isolation schemes that allow the inner-axis electrical path to cross over the outer axis to reach the terminal pads on top or connections from the bottom. A 2-D vertical comb-drive scanner recently reported by Kwon *et al.*[46] uses SOI and backside islands (insulated from the substrate by buried oxide) to achieve isolation of the inner and outer frames while maintaining mechanical continuity. The device was also designed for control of up and down, i.e., piston, motion.

Fujitsu has reported a 2-D vertical comb-drive scanner with the comb fingers extended from the edge of the mirror.[47] Since the fingers are located at the edge, where linear displacement is largest, much taller comb fingers are necessary to achieve the same scan range (100 μm in the reported device). The scanner was derived from a symmetric SOI wafer with both top and bottom silicon layers 100 μm thick separated by a 1-μm buried oxide (Fig. 7.14).[48] The top and bottom layers were differentially etched with DRIE and an oxide hard mask, so that thinned torsion bars

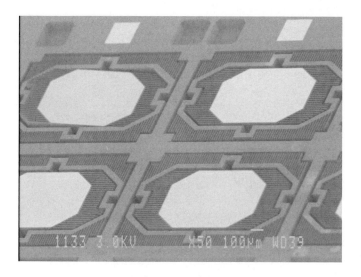

Figure 7.14 SEM of the Fujitsu scanner.[47] (Photograph courtesy of Yoshihiro Mizuno. Reprinted with permission from Fujitsu.)

and thinned electrical connection bars were possible in the top and bottom layers, respectively. The device achieves five electric potentials by electrically connecting to the four bottom isolated combs through gold-bump bonds to another substrate with patterned electrodes. The gold bumps also create a 120-μm gap to accommodate the mirror motion. The process was used to fabricate an 80 × 80 switch. The scan angle range was reported as 10° at 60 V under static conditions.

7.4 Scanning Mirrors with Magnetic and Electromagnetic Actuators

As stated previously, magnetic actuation is practical when the mirror dimensions are on the millimeter scale since the magnetic torque (generated by the magnetic device interacting with an external magnetic field) scales with volume for permanent magnetic materials and with total coil area for electromagnets. (For an analysis of magnetic torque see Judy and Muller.[49]) In addition, the overall system size must accommodate the magnets (permanent or electric coils) used to generate the external magnetic field. Therefore, motivation for this type of scanner may be the price reduction afforded with batch fabrication techniques or lower power consumption rather than miniaturization *per se*. Miyajima *et al.*[50] pointed out that conventional optics use optical beams with spot sizes in the millimeter range and that relatively large area scanners are required if MEMS are to be integrated into such systems.

Magnetically actuated pop-up type scanning mirrors (Fig. 7.15) were demonstrated by Miller and Tai[51] using bulk-silicon micromachining and by

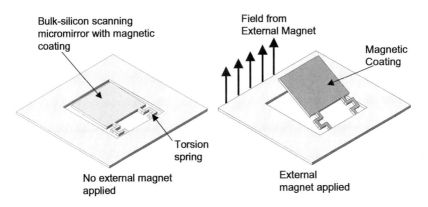

Figure 7.15 Pop-up type 1-D scanner with magnetic actuation (after Hagelin and Solgaard[5]).

Judy and Muller[49] using surface micromachining. In Miller and Tai's system, the scanning mirror was used to create and retrieve hundreds of stored holograms created by angle multiplexing.[51] The mirror angle was primarily controlled by the coil current of an external electromagnet. Micromirror devices with permalloy coatings only and with permalloy coatings plus patterned coils were demonstrated. The on-mirror copper-patterned coils provided better angular resolution, which improved the hologram spacing from 0.143° to 0.03° when compared to the mirrors with magnetic coating only. Both types of mirrors were approximately 4×4 mm^2 and achieved deflection angles of greater than 60° with response times ~30 msec.

Miyajima et al. at Olympus[52] have developed MEMS 1-D scanners (mirror size ~4.2×3.0 mm^2) for insertion into a commercial laser-scanning confocal microscope. Their MEMS scanners (Fig. 7.16) were designed for fast or horizontal scanning in resonant mode at ~4 kHz. The scanners were fabricated from SOI wafers, essentially providing two SCS structural layers: the top silicon layer, used to define thinner torsion bars, and the entire substrate thickness of 300 μm used to stiffen the mirror against deformation caused by residual stress and dynamic operation. (For analysis and discussion of dynamic deformation in MEMS scanners, see Conant et al.[53] and Urey et al.[54]) A combination of DRIE and wet anisotropic etching was used for the top layer and the substrate, respectively. The mirror coils were formed from electroplated copper, and sensing coils of sputtered aluminum were added to provide closed-loop feedback for accurate mirror positioning. The authors reported that the device met or exceeded the requirements for the laser-scanning confocal microscope application, including stability and repeatability. In this application, where the overall system is large (tabletop size), Olympus uses the MEMS scanner in conjunction with a macroscale scanner to achieve 2-D raster scanning. For smaller systems, two electromagnetic MEMS mirrors may be cascaded[55] or, with the tradeoff of design complexity, a single 2-D scanner may be used.

An early 2-D electromagnetic scanner demonstrated by Asada et al.[56] was fabricated by bulk micromachining from a 200-μm-thick silicon substrate with 16-μm-thick electroplated copper coils. The inner plate (y-axis) was 4×4 mm^2 and the outer frame (x-axis) was 7×7 mm^2, with measured resonant frequencies of 1450 and 380 Hz, respectively. Reported scan angles were low; even in resonant mode, the scanner achieved less than 3° for the x-axis and less than 1° for the y-axis at 35 mA coil current.

The current state of the art (as reported in the literature) shows improved performance for MEMS electromagnetic 2-D scanners. Ahn and Kim[57] have reported an SCS (70-μm-thick substrate) scanner with layered thin films of BCB (benzocyclobutene), silicon nitride, and aluminum forming the torsion suspensions and serving as a multilevel interconnect

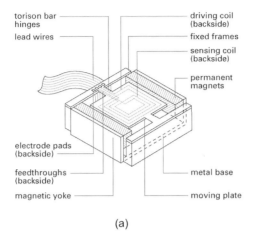

Figure 7.16 (a) Schematic and (b) photograph of a packaged electromagnetic 1-D scanner. (Courtesy of Dr. Hiroshi Miyajima, Olympus.)

for independent control of the electromagnetic coils for the inner-mirror and outer-frame rotation axes. The mirror was 3.5×3.5 mm^2 and the outer frame was 5.7×5.7 mm^2, with measured resonant frequencies of 380 and 150 Hz, respectively. The mirror inner coil currents must flow on the outer frame to reach the electrical terminals. To avoid crosstalk between the inner and outer axes, the mirror current was looped back through a parallel yet isolated path along the outer frame. Thus a compensation current with equal magnitude yet opposite direction effectively cancelled the Lorenz force in these overlapping areas. Reported maximum scan angles (total optical scan range) were 5.44° at 30 mA in resonance for the inner axis and 51.34° at 130 mA in resonance for the outer axis.

An interesting approach for achieving remotely actuated 2-D scanning by magnetostrictive effect has been reported by Bourouina *et al.*[58,59] In this work, 2-D scanning was achieved by simultaneously exciting two modes of motion, bending and twisting, in a cantilevered mirror. The scanner was fabricated from 20-μm SOI with a differential DRIE etch technique to obtain a thin, compliant section through which the more rigid 100-μm mirror section attaches to a fixed frame. The device includes silicon piezoresistive sensors for feedback control, consisting of resistor bridges formed on the thin section with boron ion implantation into the n-type SOI. The backside of the device was sputter-coated with a 4.5-μm magnetostrictive thin film of layered TbFe and CoFe developed for this application. The large devices (20×10 mm^2) were driven to bimodal resonance by an external electromagnet with superimposed coil currents at 189 and 1890 Hz, corresponding to the resonant frequencies for bending and torsion, respectively. Optical scan ranges for the modes were reported as 8.7° for the horizontal axis (torsion) and 7.4° for the vertical axis (bending).

A combined electrostatic/electromagnetic 2-D scanner has been developed by Microvision for retinal scanning displays.[54,60] These two-axis scanners use electromagnetic actuation for the outer frame that provides the vertical or slow axis, and electrostatic actuation for the inner mirror axis that provides the horizontal or fast axis. The devices were bulk-micromachined using both wet and dry anisotropic etching, and electroplating was used to form electromagnetic coils on the outer frame. These scanners must be stiff to remain flat and withstand the forces developed in resonant scanning mode. The scanners also incorporated piezoresistive strain sensors on the torsion flexures for closed-loop control. Scanner performance included 13.4° and 9.6° total mechanical scan range for the horizontal and vertical axes, respectively. The vertical axis drive signal (nonresonant mode) was at 60 Hz, and the horizontal axis resonant mode drive signal was at 19 kHz. The scanners are designed to meet SVGA video standards that require 800×600 resolution.

Besides the magnetic, electromagnetic, and electrostatic actuation discussed here, there have been scanners demonstrated with piezoelectric actuation[61-63] and thermal bimorph actuation.[64,65]

7.5 Micromirror Arrays with Mirror Size ≤100 Micrometers

Texas Instruments' Digital Micromirror Device (DMD) is probably the most well-known micromirror device.[66] It is now widely used in digital projection displays.[67] Since the operating principle, design, fabrication, and testing of DMD have been extensively discussed in textbooks,[68] they are not discussed here. Though DMD was developed primarily for projection display applications, there are some interesting nondisplay applications. For example, a variable optical attenuator with 11-bit digital control was implemented by using a DMD as a large aperture mirror with variable reflecting area.[69]

Micromirror arrays have many applications in WDM photonic networks. By placing a 1-D digital micromirror array in the focal plane of a grating spectrometer, a dynamically reconfigurable wavelength add-drop multiplexer (WADM) has been realized.[70] The optical layout of the WADM is shown in Fig. 7.17. Light from an input optical fiber is collimated and then dispersed by a diffraction grating. The diffracted signal is focused by a lens onto the micromirror array. Each mirror corresponds to a WDM channel. An SEM of the digital micromirror is shown in Fig. 7.18. It is fabricated using the three-polysilicon-layer surface-micromachining process through the MUMPs service offered by Cronos.[8] The mirror area is 30×50 μm^2, with a pitch of 57 μm, which matches the 200-GHz WDM channel spacing. Like the DMD, the micromirror here operates in a bistable mode. It has a pull-in angle of $\pm 5°$ at 20 V bias. The mirror is driven by an AC voltage to avoid electrostatic charging. The fiber-to-fiber optical insertion loss is from 5 to 8 dB, and the switching contrast ranges from 32 to 47 dB. A quarter-wave plate that rotates the polarization of the reflected beam by 90° minimizes the polarization dependent loss (0.2 dB). Switching time is 20 μsec.

The WADM concept can be extended to switches with more than one output port, as illustrated in Fig. 7.19. This is generally called wavelength-

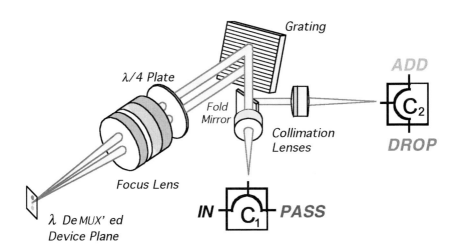

Figure 7.17 Schematic optical layout of a free-space WADM. (Picture courtesy of Joe Ford. Reprinted from Ford *et al.*[70] with permission.)

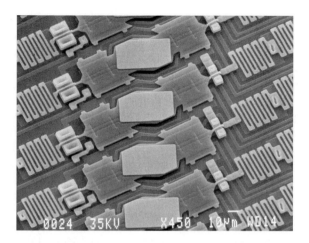

Figure 7.18 SEM of the digital micromirror array used in a WADM. (Photograph courtesy of Joe Ford. Reprinted from Ford et al.[70] with permission.)

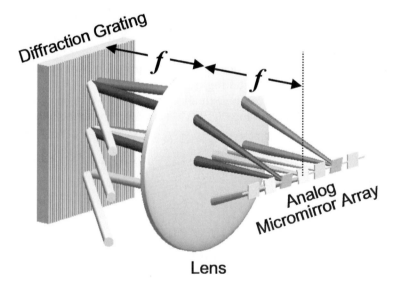

Figure 7.19 Schematic of a wavelength-selective 1 × N switch.

selective 1 × N switch,[71,72] or multiport WADM, or WDM routers.[73] To address more than two input/output ports, the micromirrors need to have more than two discrete angles. Analog micromirrors with large continuous scan range enable individual wavelengths to be directed to any of the N output fibers. In addition, high fill factor is desired to minimize the gaps between WDM channels. The mirror size is usually several times larger than the focused spot size of the Gaussian beam to attain a flat spectral shape for the passband ("flat-top" spectral response). Electrostatically actuated parallel-plate micromirror scanners have been used for multiport WADM; however, they suffer from high actuation voltage, limited scanning angle due to unstable pull-in effect, and high crosstalk between adjacent mirrors. A double-hinged micromirror has been used to amplify the scan angle (14° at 57 V); however, the scan range is asymmetric with the rotation axis offset to one edge of the mirror, which results in out-of-plane displacement of the mirror when it rotates.[74]

UCLA has developed a low-voltage analog micromirror array for wavelength-selective 1 × N switches. The schematic of the micromirror is shown in Fig. 7.20(a). The actuator and the suspension spring are completely covered by the mirror, so that a high fill factor is achieved along the array direction. As in larger scanning micromirrors, the vertical comb-drive actuator provides much larger torque than the parallel-plate actuator. Furthermore, the maximum scan angle will not be limited by the pull-in instability inherent in parallel-plate actuators. The analog micromirror arrays are fabricated using the Sandia Ultra-Planar, Multi-Level MEMS Technology-5 (SUMMiT-V)[75] foundry process, which is a five-polysilicon-layer surface-micromachining process with two chemical-mechanical planarization (CMP) steps. The first CMP process separates the lower and upper vertical combs so that narrow gap spacing (0.5 μm) between the upper and lower fingers can be achieved. Significantly higher force density is attained (240 times larger than that of comparable parallel-plate actuators) since it is inversely proportional to the square of the gap spacing. This enables the design to achieve low operating voltage and large scan angle, while still maintaining a high resonant frequency. The second CMP process produces a smooth, flat micromirror surface without reflecting the mechanical structures underneath the mirror. An SEM of the fabricated analog micromirror array is shown in Fig. 7.20(b).

The DC scan angles of the analog micromirror arrays with 1-μm comb-finger spacing are shown in Fig. 7.21. The maximum mechanical scan angle of ±6° is achieved at 9 V actuation voltage. For 0.5-μm gap spacing, the actuation voltage is as low as 6 V. In comparison, micromirrors with parallel-plate actuators that are fabricated on the same chip have a scan range of ±4° at a much higher voltage of 22 V. A prototype system constructed using the micromirror array exhibited a flat-top spectral response with a switching ratio of 35 dB. The switching time is less than 400 μsec. One of the most critical issues for MEMS-based optical systems is the stability of the micromirror. It is known that many electrostatically driven micromirrors suffer from slow drifting of mirror angle due to dielectric charging. In the vertical comb structure, we have carefully designed the

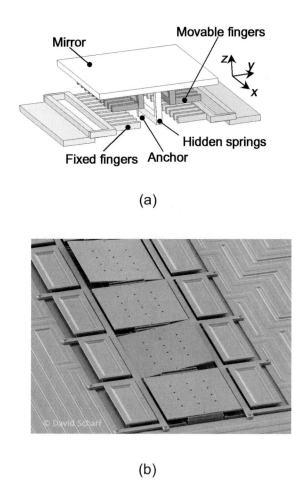

Figure 7.20 (a) Schematic and (b) SEM of the analog micromirror with hidden vertical-comb-drive actuators. (SEM taken by David Scharf ©.)

structure to eliminate most exposed dielectric areas and grounded all the conducting structures around the mirror. Extremely low drift has been achieved for both the micromirror (±0.00085°) and the switch (variation of fiber-to-fiber loss <±0.0035 dB) over three hours of continuous operation without any feedback control. This suggests the switch can operate in an open-loop condition, which has enormous cost benefits.

Lucent Technologies and JDS Uniphase have reported system performance of wavelength-selective 1 x 4 switches with 128 WDM channels at 50 GHz spacing. The optical setup for the Lucent switch is shown in Fig. 7.22.

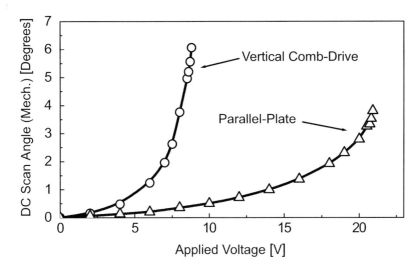

Figure 7.21 DC scan characteristics of an analog micromirror array with hidden vertical comb drives. The scan characteristics of a parallel-plate-actuated micromirror with the same mirror dimensions are also shown for comparison.

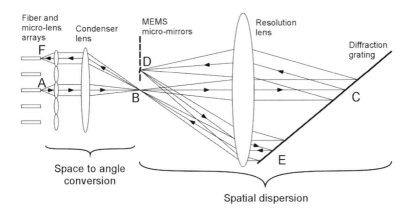

Figure 7.22 Schematic of the optical system for a wavelength-selective 1 × 4 switch. (Picture courtesy of Dan M. Marom. Reprinted from Marom et al.[71] with permission.)

The fibers are first imaged onto a common spot by a magnification of 3x. A resolution lens and a 1100-gr/mm grating provide the necessary spatial dispersion. Using a micromirror with ±8° at 115 V made on an SOI wafer, a flat-top spectral response and an optical insertion loss of 3 to 5 dB were attained.

7.5.1 Micromirrors for Dynamic Spectral Equalizers

Dynamic spectral equalizations are important components for advanced lightwave communication systems. There are two types of dynamic spectral equalizers (DSE): (1) continuous DSE for flattening the gain of optical amplifiers;[76] and (2) channelized DSE that attenuate each optical channel independently to an extinction ratio greater than 30 dB.[77] The latter is used in optical systems with broadcast-and-select topologies and add-drop capability.

The basic optical setup for DSE is similar to that for WADM (Fig. 7.17) except the digital micromirror array is replaced by a membrane mirror with variable reflectivity. A folded telecentric 4f imaging system consisting of a lens and a grating in a Littrow configuration images individual wavelengths onto micromirrors. A continuous mirror surface with individual electrodes for modulating the reflectivity offers a continuous operating wavelength spectrum with no passbands. This is useful for equalizing gain profiles of optical amplifiers. It was implemented by modifying the mechanical antireflection switch (MARS).[76] The MARS consists of a ¼-λ-thick silicon nitride suspended above the silicon substrate by a ¾-λ gap (Fig. 7.23), where λ is the optical wavelength. At zero bias, MARS acts as a mirror with 70% reflectivity. When the nitride is pulled down to ½ λ above the substrate, the layer becomes an antireflection coating and the reflectivity becomes nearly zero. By patterning an array of 32 electrodes spaced at a 28-μm pitch, a 32-channel equalizer is realized. The packaged component exhibits 9-dB excess loss, 20-dB dynamic range, and 10-μsec response time.

The micromirror array used in channelized DSE is very similar to arrays used in wavelength-selective 1 × N switches. By fine-tuning the angle of the analog micromirror array, each wavelength channel can be attenuated independently. Unlike continuous DSE, large variation between adjacent channels is possible. Therefore, this device can be used as an add-drop multiplexer as well. The mirror tilting direction plays an important role in the response of interchannel regions. It has been shown that micromirrors tilting in the direction orthogonal to the array can minimize the undesirable interchannel response.[78]

7.6 2-D MEMS Optical Switches

7.6.1 Switch Configuration, Requirements, and Expendability

The 2-D MEMS optical switch is basically an optical crossbar switch with N^2 micromirrors that can selectively reflect the optical beams to

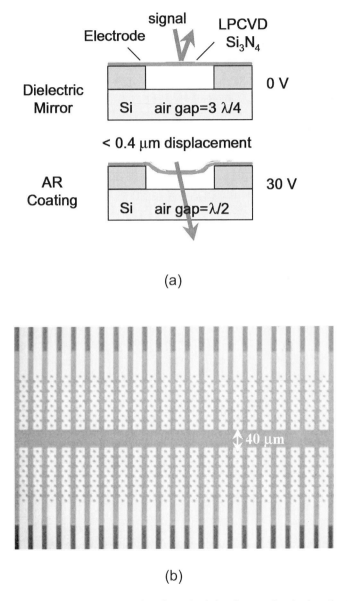

Figure 7.23 (a) Schematic illustrating the principle of a mechanical anti-reflection switch (MARS). (b) Photograph of a MARS variable reflectivity strip mirror. (Picture and photograph courtesy of Joe Ford. Reprinted from Ford and Walker[76] with permission.)

orthogonal output ports or pass them to the following mirrors. They are often referred to as 2-D switches because the optical beams are switched in a two-dimensional plane. This is in contrast to the 3-D switch in which the optical beams are steered in three-dimensional space. A generic configuration of the 2-D switch is shown in Fig. 7.24. The core of the switch is an $N \times N$ array of micromirrors for a switch with N input fibers and N output fibers. The optical beams are collimated to reduce the diffraction loss. The micromirrors intersect the optical beams at 45° and can be switched in and out of the optical beam path. The micromirrors are "digital," that is, they are either in the optical beam path (ON) or completely out of the beam path. When the mirror in the ith row and j^{th} column (M_{ij}) is ON, the ith input beam is switched to the jth output fiber. Generally, only one micromirror in a column or a row is ON. Thus during operation of an $N \times N$ switch, only N micromirrors are in the ON position while the rest of the micromirrors are in the OFF position. MEMS 2-D switches were first reported by Toshiyoshi and Fujita.[79] Several different ways of switching micromirrors have been reported, including rotating, sliding, chopping, and flipping motions. The switches are usually actuated by electrostatic, electromagnetic, or piezoelectric mechanisms. Before discussing different types of 2-D switches in more detail, we need to under-

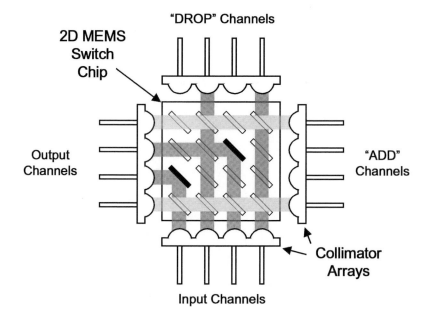

Figure 7.24 Schematic of 2-D MEMS optical switches.

stand the performance requirements and design tradeoffs as well as the expandability of the switch.

The most critical parameters for the optical performance of 2-D switches are the accuracy and uniformity of mirror angles, and the fill factor (mirror size divided by switching cell size), size, and curvature of the mirrors. The performance and scaling limit of the 2-D switches can be derived using Gaussian beam analysis.[80] As shown in Fig. 7.25, the optical beam is expanded between the input and output collimators. To minimize the optical loss, confocal geometry is often used. The fibers are placed slightly behind the focal point of the collimators so that the optical beam focuses slightly before it diverges again. The optical beam waist is placed halfway between the input and output optical collimators. Gaussian beam analysis of 2-D switches has been reported in detail by Lin *et al.*[81]

The Gaussian beam width is a function of the axial distance from the beam waist. If we define the origin of the axial coordinate to be zero at the beam waist (see Fig. 7.25), the $1/e$ half-width of the Gaussian beam is described by

$$w(z) = w_0 \sqrt{1 + \left(\frac{z}{z_0}\right)^2},$$

where w_0 is the $1/e$ half-width at the beam waist, z_0 is the Rayleigh length given by $z_0 = \pi w_0^2 / \lambda$, and λ is the wavelength of the optical beam. When $z = z_0$, the optical beam width expands by a factor of $\sqrt{2}$: $w(z_0) = w_0 \sqrt{2}$. Thus z_0 is a good measure of how far a Gaussian beam propagates before it diverges significantly. The optical loss can be calculated by the over-

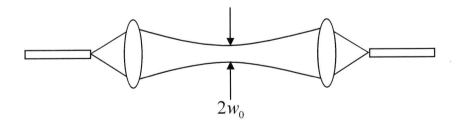

Figure 7.25 Schematic illustrating the propagation of a Gaussian beam.

lap integral of the input and output optical fields evaluated at the same plane.

The optical path length depends on the switch configuration. The longest optical beam path for an N × N switch is $L_{max} = 2Np\Delta + 2\Delta$, where p is the pitch of the micromirror switching cell and Δ is the distance between the collimator lens and the nearest micromirror. On the other hand, the shortest optical path is simply $L_{min} = 2\Delta$. Invariably, there is a distribution of optical insertion loss due to different optical path lengths. It is desirable to minimize the distribution. To first order, we can set the Rayleigh length to be equal to half of the median optical beam path, $z_0 = Np/2 + \Delta$. Large optical beams have slower divergence and less variation in loss, but they also require larger mirror size and switching cell pitch, as well as longer propagation length.

The expandability of the 2-D switch is easily understood when the following assumption is made: The fill factor of the mirror is $\eta = 2R/p$, where R is the radius of the mirror in the vertical direction. Since the mirrors intersect the optical beam at 45°, the mirror shape is actually elliptical, with the horizontal width equal to $\sqrt{2}$ times the vertical width. The mirror size needs to be larger than the optical beam to minimize the clipping loss, $R = aw_0$, where a is a constant whose value is approximately 1.5 to 2. Combining the above equations, we find that the number of ports, N, is proportional to the beam waist ω_0:

$$z_0 = \frac{\pi w_0^2}{\lambda} = \frac{Np + \Delta}{2} \approx \frac{Np}{2} = \frac{N}{2}\frac{2R}{\eta} = N\frac{aw_0}{\eta}$$

$$\Rightarrow N \approx \frac{\pi \eta}{a\lambda} w_0.$$

A larger beam waist extends the propagation distance and accommodates more fiber ports. The number of ports also increases with the fill factor of the micromirror. This is illustrated in Fig. 7.26. However, the maximum beam waist will be limited by several practical constraints. First, the chip size ($\approx z_0 = \pi w_0^2/\lambda$) increases with the square of the beam waist. The yield drops with larger chip areas. Second, as the chip size increases, the optical insertion loss due to mirror tilt increases very rapidly. Using the above assumptions, the coupling efficiency to the output collimator reduces as $exp[-(\theta \pi w_0/\lambda)^2]$ for mirror tilt angle θ. Depending on the accuracy and repeatability of the fabrication process, eventually the accuracy and uniformity of mirror angles limit the maximum port count of 2-D switches.

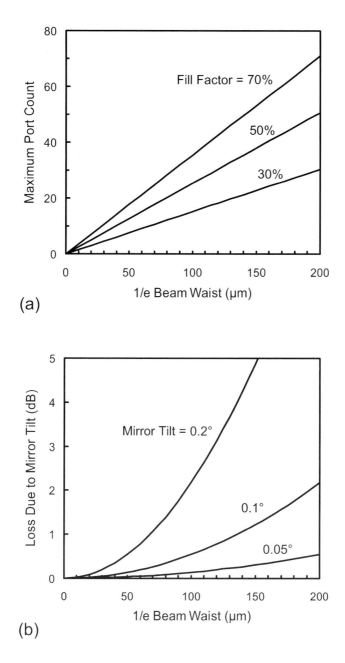

Figure 7.26 (a) Number of fiber ports versus the optical beam waist. (b) Optical loss due to the mirror tilt as a function of the beam waist.

7.6.2 Vertical Chopper-Type Switch

As discussed in the previous section, the mirror angles and their uniformity play a critical role in the performance and the scalability of 2-D switches. The reproducibility of mirror angles over switching cycles determines the repeatability of insertion loss. The "chopper-type" 2-D switch uses a vertical mirror whose height can be changed by MEMS actuators.[82,83] The mirror angle is fixed during switching, and excellent repeatability of insertion loss has been reported.

Figure 7.27 shows the schematic diagram of OMM's 2-D switch.[82] The mirror is assembled vertically at the tip of a long actuator plate. The plate is tilted upward and fixed by microlatches to raise the mirror height. A large traveling distance is achieved by extending the actuator arm. A displacement of several hundred microns has been achieved with this configuration. The switch is actuated electrostatically by applying a voltage between the actuator plate and a bottom electrode on the substrate. The mirror moves in the vertical direction, and the mirror angle is maintained at 90° during the entire switching cycle. The actuator is basically an angular gap-closing actuator. A mechanical stopper defines the lower position of the mirror. OMM uses a curved landing bar with a single-point contact to minimize stiction and increase reliability (see Fig. 7.27). More than 100 million cycles have been demonstrated with repeatable mirror angle and

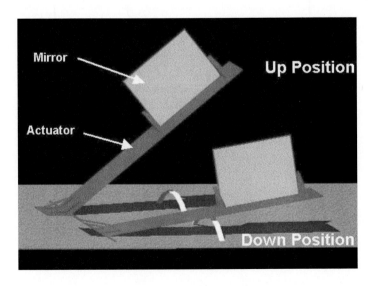

Figure 7.27 Schematic of a switching element in the OMM 2-D switch. (Picture courtesy of Li Fan. Reprinted from Fan et al.[82] with permission.)

performance. The landing bar also provides a cushion that helps reduces mirror ringing and improves switching time. OMM has demonstrated a switching time of 12 msec using a square-wave driving voltage without preshaping the waveform.

OMM's switch is fabricated with polysilicon surface-micromachining technology. The mirrors and the actuators are batch-assembled into the 3-D structures. Figure 7.28(a) shows the SEM of a 16 x 16 switch. The distribution of mirror angles is shown in Fig. 7.28(b). The uniformity is better than ±0.1° for 256 mirrors. The switch is hermetically packaged with optical collimator arrays. Extensive testing has been performed for the packaged switches. The maximum insertion loss is less than 3.1 dB, and the crosstalk is less than –50 dB. Loss variation over the wavelength range of 1280 to 1650 nm is less than 1 dB. Return loss is greater than 50 dB, and maximum temperature variation is <1 dB over a temperature range of 0 to 60°C. Polarization-dependent loss (PDL) is <0.4 dB, and polarization mode dispersion (PMD) is <0.08 ps. Vibration tests showed <0.2 dB change under operation, and three-axis shock tests confirm no change of operational characteristics with accelerations under 200 g (1 g = 9.81 ms^{-2}).

To reduce the operating voltage, UCLA has developed an optical switch using a curled cantilever beam.[83] The schematic drawing of the switching element is shown in Fig. 7.29. The mirror is attached to the end of the curled cantilever beam. The initial gap spacing between the curled cantilever beam and the substrate electrode is very small, which enables pull-in at low voltage. Once the pull-in occurs, it "zips" through the entire beam.[84] Therefore, large mirror displacement and low actuation voltage are achieved simultaneously. The curled cantilever switch can be realized using the MUMP process.[8] By depositing a stressed layer of Cr/Au on a polysilicon cantilever beam, the stress bimorph causes the cantilever to curve upward. Tip displacement larger than 300 μm is achieved. Operating voltage below 20 V has been demonstrated for switches with 300 × 175 μm mirrors. Using commercially available discrete collimators, low optical insertion loss (0.7 dB) and negligible crosstalk (<–80 dB) were attained.

Bulk-micromachined vertical mirrors eliminate the need for the assembly that is required for surface-micromachined micromirrors. By clever use of KOH anisotropic etching on (100) silicon wafers, both vertical mirrors and self-aligned V-grooves are etched at the same time.[85] As shown in Fig. 7.30, the vertical mirrors are parallel to the <100> directions, while the V-grooves are parallel to <110> directions. The angle between the mirrors and the V-grooves is exactly 45°. To prevent excessive lateral etching, the vertical mirror is etched by DRIE first, followed by KOH etching to reveal crystallographic mirror surfaces. The dimen-

(a)

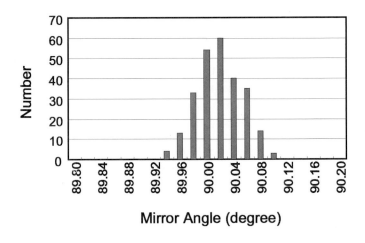

(b)

Figure 7.28 (a) SEM of the OMM 16 × 16 switch. (b) Measured distribution of mirror angles for the 16 × 16 switch. (Picture courtesy of Li Fan. Reprinted from Fan et al.[82] with permission.)

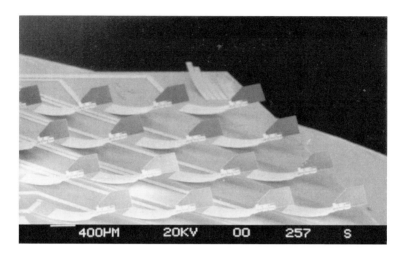

Figure 7.29 SEM of a 4 × 4 switch with curled cantilever beam actuators. (Photograph courtesy of Rich Chen and Hung Nguyen.)

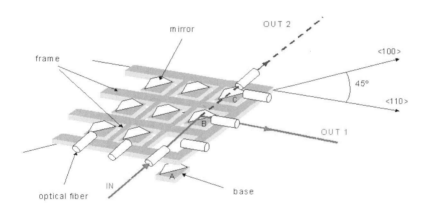

Figure 7.30 Schematic of bulk-micromachined vertical mirror arrays and self-aligned V-grooves for 2-D matrix switches. (Picture courtesy of Hiroyuki Fujita and Lionel Houlet. Reprinted from Houlet *et al*.[85] with permission.)

sions of the mirrors are 185 μm in height, 450 μm in length, and 10 μm in thickness. Permalloy is deposited under the base, and the mirror is actuated electromagnetically by a matching microcoil matrix fabricated on a separate wafer. Graded index (GRIN) lenses with matching diameter to single-mode fibers (125 μm) are used to collimate the optical beams. The beam spot size is 55 μm.

7.6.3 Switch with Pop-Up Mirrors

2-D switches using various types of flip-up (or pop-up) mirrors have been reported.[79,86,87] They are realized by either bulk[79] or surface[86] micromachining, or by a combination of both.[87] Unlike the chopper-type switch with vertical mirror, the mirror lies originally in the plane of the substrate, and no assembly or special fabrication of the vertical mirror is needed. However, since the mirror angle is changing continuously during switching, a common challenge for this type of switch is the reproducibility of the mirror angle. Three different switches are discussed here: bulk-micromachined torsion mirrors,[79] surface-micromachined pop-up mirrors,[86] and electromagnetically actuated flip-up mirrors.[87]

The first 2-D matrix optical switch was reported by Toshiyoshi and Fujita.[79] Their switch structure, which consists of two bonded wafers, is shown in Fig. 7.31. The mirrors are suspended by a pair of torsion beams in the plane of the top wafer. The biasing electrodes are fabricated on the bottom wafer. When a voltage is applied between the mirror and the bottom electrode, electrostatic actuation rotates the mirror downward by 90°. The mirror angle in the ON (down) state is controlled by a stopper on the bottom wafer. Since the mirror angle is defined by the relative positions of two wafers, precise alignment is necessary to achieve accurate and uniform mirror angles. A single-chip electrostatic pop-up mirror has recently been reported.[88] The actuation and mechanical stopper are realized between a back-flap and a vertical trench etched in the silicon substrate. The angular accuracy and uniformity of the mirrors depend on the etched sidewall profile and the lithographic alignment accuracy.

AT&T Labs has reported surface-micromachined 2-D switches with free-rotating hinged mirrors.[86] The schematic drawing and the SEM of the switch are shown in Fig. 7.32. The switch is fabricated using the MUMP process.[8] The mirror is pivoted on the substrate by microhinges. A pair of pushrods is used to convert in-plane translations into out-of-plane rotations of the mirrors. The switch is powered by scratch-drive actuators.[89] Though scratch drive actuators do not move at high speed, fast switching time is achieved (700 μsec) because only a short traveling distance (22 μm) is needed for the mirror to reach 90°. The free-rotating microhinges have an inherent 0.75-μm clearance between the hinge pins and the staples, which could result in large variations in mirror angles. Using an improved design and mechanical stoppers that are insensitive to lithographic misalignment during fabrication, an angular repeatability of the mirrors better than 0.1° was experimentally demonstrated.[81] The mirror flatness was improved by using a multilayer structure with phosphosilicate glass (PSG) sandwiched between two polysilicon layers. The largest switch size demonstrated was 8 × 8 due to the foundry-imposed

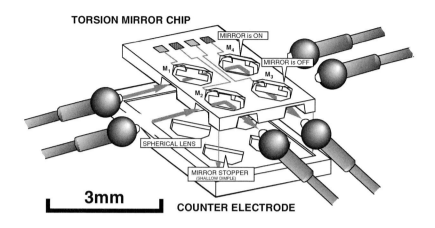

Figure 7.31 Schematic and SEM of a bulk-micromachined 2-D switch with free-rotating torsion mirrors. (Pictures courtesy of Hiroshi Toshiyoshi. Reprinted from Toshiyoshi and Fujita[79] with permission.)

chip-size limits of 1×1 cm. One of the potential issues is the constant tear and wear of the free-rotating hinges and actuators. This might affect the reliability of the switch and the accuracy and uniformity of mirror angles over many switching cycles.

Magnetic actuation offers some unique advantages, including large force, large deflections in directions both parallel and perpendicular to the plane of the wafer, and actuation using magnetic fields generated by off-

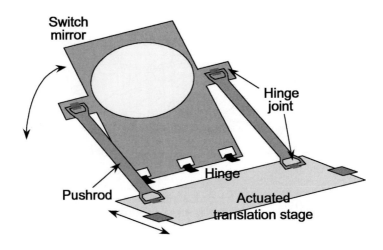

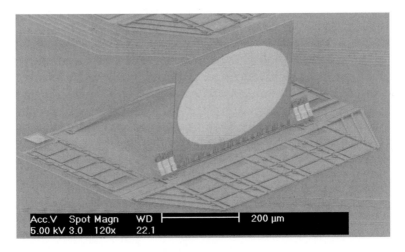

Figure 7.32 Schematic and SEM of the surface-micromachined free-rotating hinged mirrors reported by AT&T. (Picture courtesy of Lih Y. Lin. Reprinted from Lin and Goldstein[15] and Lin et al.[86] with permission.)

chip sources.[90] Actuation of torsion-suspended mirrors using electroplated magnetic thin films such as NiFe have been demonstrated. To define precisely the mirror angles in the ON position, researchers at UC Berkeley have proposed to electrostatically clamp the mirrors to vertically etched sidewalls formed on a top-mounted (110)-silicon chip.[87] The cross-section of the mirror is shown in Fig. 7.33. The mirror can be clamped horizontally or vertically, depending on the position of the mir-

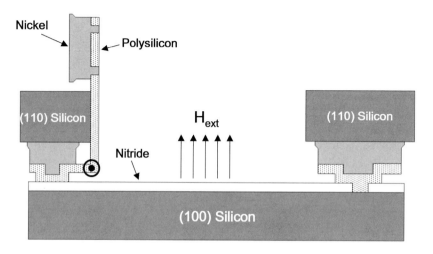

Figure 7.33 Cross section of a magnetically actuated mirror with electrostatic clamping at dual positions (after Hehin et al.[87]).

ror when the clamping voltage is applied. To switch a mirror to the vertical (ON) position, the clamping voltage is first turned off and the magnetic field is turned on, lifting the mirror up close to the vertical position. The mirror voltage is then turned on to clamp the mirror to the vertical sidewalls. Mirrors staying in the OFF state can be clamped horizontally. The clamped mirrors will not be affected by the magnetic field since the electrostatic force is larger than the magnetic force. After switching, the magnetic field can be turned off. To improve the mirror uniformity in the ON state, anisotropic etching using KOH along the <110> crystal axis of (110)-silicon is used to define the precise orientation of the vertical clamping electrodes. The vertical angle of a single mirror in their research prototype device was reproducible to within 0.04°. Stiction is a potential issue for using clamping to define mirror angle. Precise alignment of the top and bottom chips is also needed for accurate and uniform mirror angles. Alternative magnetic micromirror arrays with single-crystalline silicon mirrors and aluminum spring have also been reported.[91]

7.7 2 × 2 Switches

The simplest type of 2 × 2 switch requires only one micromirror between two pairs of orthogonal fibers. Since the propagation distance is short (~ a few tens of μm), the optical beams can be coupled directly into output fibers without using any focusing lenses. The theoretical optical insertion loss is only 0.2 dB when the fiber ends are separated by 20 μm. Therefore, the micromirror only needs to be a couple of times larger than the fiber core (8 to 9 μm for typical single-mode fibers). The displacement required is comparable to the mirror size. The mirror pitch that is critical for larger 2-D switches is not important here. The actuator area could be much larger than the active mirror. Therefore, there are more technology choices for nonscalable 2 × 2 switches.

A simple, elegant solution for 2 × 2 switches is using SOI-based optical MEMS.[92,93] Electrostatic comb-drive actuators and vertical micromirrors can be fabricated on SOI wafers. The fabrication process is shown in Fig. 7.34. The micromirror and the U-grooves for holding optical fibers are patterned and etched simultaneously in one single DRIE step on an SOI wafer. The diameter and the core size of single-mode fibers are 125 and 9 μm, respectively. SOI wafers with a 75-μm-thick silicon device layer were chosen so that the optical spot is 10 to 20 μm below the wafer surface. The structure is released in HF by timed etching. The mirrors can be coated with metal by angle evaporation to increase their reflectivity.

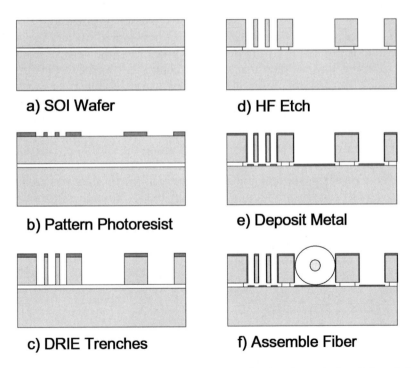

Figure 7.34 A SOI MEMS fabrication process for 2 × 2 optical switches (after Noell et al.[93]).

The schematic structure and the SEM micrograph of the 2 × 2 switch are shown in Fig. 7.35. The most critical part of the process is the etching of thin (<2 μm) vertical mirrors with smooth sidewalls. A thin vertical mirror is required for such 2 × 2 switches because the offset of the reflected optical beams from the opposite sides of the mirror caused by the finite thickness of the mirror will introduce additional optical loss. The Institute of Microtechnology (IMT) at Neuchatel has perfected the mirror-etching technology by DRIE.[94] A surface roughness of 36 nm has been achieved. The switch has excellent optical performance: 0.3 to 0.5 dB optical insertion loss and 500 μsec switching time, and very low polarization dependence.[93]

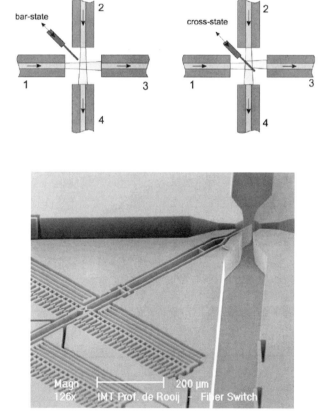

Figure 7.35 Schematic and SEM of a 2 × 2 switch fabricated by SOI MEMS. (Picture courtesy of Nico de Rooij. Reprinted from Marxer and Rooij[92] with permission.)

Another 2 × 2 switch configuration using four micromirrors has also been proposed for multichannel add-drop applications. The mirrors are oriented at ±45° between two pairs of parallel fibers. The parallel arrangement lends itself to monolithic integration of 1-D switch arrays for multichannel add-drop switches. This has been demonstrated using both surface-micromachining[95] and bulk-micromachining technologies.[96]

7.8 Optical Attenuator Array

The variable optical attenuator (VOA) is a simple but important component in advanced optical networks and subsystems. The main applications include power management of optical cross-connects or reconfigurable add-drop multiplexers, gain control of optical amplifiers, and power tuning of lasers and detectors. The advantages of MEMS-based VOAs compared with other technologies are better optical performance (large dynamic range, low optical insertion loss, low polarization- and wavelength-dependent loss), lower power consumption, compact size, and low-cost batch manufacturing. These are particularly attractive in WDM networks in which a large number of VOAs in array form is required.

There are two main types of MEMS-based VOAs: the transmission type with moving shutters, and the reflection type with moveable mirrors. Both can be fabricated by either bulk- or surface-micromachining processes. In many applications, the VOAs are actually integrated with other components such as a 2-D switch or a wavelength add-drop multiplexer. The technology choice would be clear in those cases. For discrete VOA arrays, consideration will be based on manufacturing cost, especially the packaging cost. This section discusses two examples of VOAs, one based on bulk-micromachining, the other based on surface-micromachining.

Bulk-micromachined VOAs using a one-step DRIE process have been reported.[97] The fabrication process is the same as that for 2 × 2 switches reported by the same group. A mirror or shutter is fabricated at the same time as the comb-drive actuator on a 75-μm-thick SOI wafer. U-grooves for input and output fiber are fabricated at the same time. Both transmission and reflection VOAs have been reported.[98] The schematics and the SEM of the fabricated device are shown in Fig. 7.36. In the transmission-type VOA, a shutter is moved back and forth by an electrostatic comb-drive actuator. This design is more compact and easier to package since the input and output fibers are collinear. When the fiber ends are separated by 20 μm, the theoretical optical insertion loss is only 0.2 dB. No imaging lens or other components are needed. Index-matching liquid

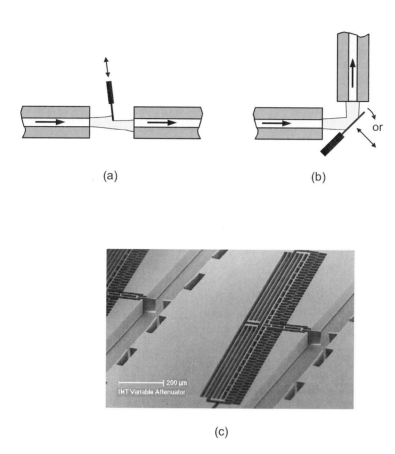

Figure 7.36 (a) Shutter-type and (b) reflection-type VOAs. (c) SEM of a shutter-type VOA. (Reprinted from Marxer et al.[97] with permission.)

is used to reduce the optical insertion loss from 2.5 dB to 1.5 dB.[97] To reduce the back reflection, the shutters and the fiber end faces are tilted by 8°. A dynamic range of 50 dB, maximum voltage of 32 V, back-reflection less than −37 dB, and switching time of 5 msec have been achieved.

The reflection-type device uses a mirror that can either be tilted or translated to vary the coupling loss of reflected light. The reflective design shows better polarization-dependent loss (PDL). The mean PDL for the reflective design is 0.085 dB, while the shutter-based VOA has a mean PDL of 0.3 dB, both measured at 10 dB attenuation. On the other hand, the shutter-based VOA has less wavelength-dependent loss. Similar VOAs use electromagnetic actuation.[99]

VOAs with a similar structure have also been fabricated by the surface-micromachining technique.[100] A gold-coated 55-μm-high micromirror is vertically assembled at the tip of a long (500 to 575 μm) cantilever. The other end of the cantilever is connected to a gap-closing electrostatic actuator through a pair of short beams. Due to the lever effect, a small displacement of the actuator produces 20 μm vertical motion for the micromirror. A 1-dB excess insertion loss, 50-dB dynamic range, 100-μsec response time, and 0.12-dB PDL have been demonstrated. Some variation in PDL was observed at intermediate shutter position, for example, at 10-dB attenuation. This was attributed to vectorial diffraction effects from light grazing exposed silicon on the shutter.

7.9 Tunable WDM Devices

7.9.1 Tunable Filters

MEMS-actuated tunable Fabry-Perot (FP) filters offer advantages such as wide continuous tuning range (~100 at 1550 nm wavelength), polarization-insensitive operation, and monolithic integration of 2-D arrays. Their applications include WDM networks, spectroscopy, and optical sensing systems. The filter consists of two distributed Bragg reflectors (DBR) separated by a variable air gap. Usually the top DBR is attached to a cantilever or membrane that is actuated electrostatically or thermally. The large tuning range stems from the wide free spectral range of the short optical cavities. Figure 7.37 shows various designs of MEMS-actuated FP tunable filters. Two original designs are based on electrostatically actuated cantilevers [Fig. 7.37(a)][101] and a deformable membrane [Fig. 7.37(d)].[102] In gap-closing actuators, the displacement of the top DBR is restricted to one-third of the total gap spacing due to pull-in instability. This limits the maximum tuning range that can be achieved. Two approaches have been proposed to overcome this limit. The first is to use an asymmetric seesaw structure suspended by torsion springs [Fig. 7.37(b)].[103] The top DBR is attached to the long arm of the seesaw while the actuator is implemented on the short, wide arm. Using the lever effect, a small displacement on the short arm can move the top DBR over one free-spectral range without running into the pull-in limit. A tuning range of over 100 nm has been achieved.

An alternative approach is to use thermal bimorph actuation [Fig. 7.37(b)]. A thermally tunable filter has been demonstrated using the difference of thermal expansion coefficients between GaAs and GaAlAs.[104,105] A tuning range of 23.2 nm has been demonstrated. To

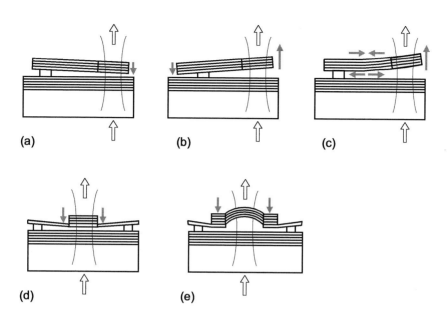

Figure 7.37 Various designs of MEMS-based FP tunable filters: (a) electrostatically actuated cantilever, (b) electrostatically actuated torsion structure, (c) thermally actuated cantilever, (d) electrostatically actuated membrane, and (e) electrostatically actuated membrane with a spherical refector.

increase the finesse of the FP cavity, a concave-top DBR has been proposed to form a stable cavity. As shown in Fig. 7.37(b), the stress gradient in the top DBR layers is controlled such that it forms a concave spherical shape when released.[106,107] The linewidth of the passband was measured to be between 0.25 to 0.27 nm for wavelengths around 1.55 μm, which is narrower than filters with flat DBRs. Fiber-to-fiber coupling loss can also be reduced by mode matching. (1.1 dB was reported by Tayebati et al.[106])

There are several possible choices of materials for the DBRs in tunable filters. Epitaxially grown III-V multilayers in GaAs-based or InP-based material systems have been developed for vertical-cavity lasers.[101,102,104] Because of the small difference in refractive indices, a large number of quarter-wave layers (~20 or more) are needed to achieve high reflectivity. Dielectric mirrors with larger index difference (e.g., SiO_2, SiN_x, or TiO_2) require only five to ten pairs.[105-107] An even larger refractive index difference can be obtained using DBRs with air gaps between semiconductor layers.[108,109] Only three pairs of reflectors are needed. High-index-contrast DBRs also have broader reflection bandwidth. Using InP-air gap DBRs, a record tuning range of 112 nm has been reported.[109]

7.9.2 Tunable Lasers and Detectors

Broadly tunable lasers and detectors are useful components for WDM networks.[110] They can be used as spare lasers for multiple WDM channels or next-generation, wavelength-agile WDM networks. There are two approaches for broadly tunable lasers. The first is based on in-plane photonic integrated circuits, such as distributed Bragg reflector lasers with sampled gratings.[110] These types of lasers are beyond the scope of this chapter since no MEMS components are involved. The other type of monolithic tunable laser is micromechanical vertical-cavity surface-emitting lasers (VCSEL).[111-113] The tunable VCSEL and photodetector structures are very similar to those of tunable FP filters except that active layers are embedded inside the FP cavity. In fact, most research groups publishing on MEMS tunable filters also publish on tunable lasers and detectors. Tunable VCSELs and detectors are discussed in detail in review articles by Chang-Hasnain[114] and Harris,[115] and in the references therein.

The tunable external-cavity diode laser is another approach to achieve wide tunability.[116] As shown in Fig. 7.38, it is a miniaturized version of the conventional bulk external-cavity tunable laser using lead zirconate titanate (PZT) actuators. In this hybrid device, an anti-reflection-coated diode laser is placed inside a Littman/Metcalf tunable resonator. A micromachined mirror is mounted on a rotary comb-drive actuator bulk-micromachined on an SOI wafer. The angle of the mirror selects the lasing wavelength. With proper layout of the optical cavity, the grating wavelength and cavity mode are tuned synchronously to give continuous tuning over a 40-nm range without mode hops. Though this hybrid device is bulkier than the monolithic devices discussed earlier, it has higher output power (10 dBm).

7.10 Diffractive Optical MEMS

Diffractive optical MEMS are based on the wavelength nature of light – diffraction and interference. Among the first diffractive optical MEMS elements demonstrated is the grating light modulator.[117] An improved version (called grating light valve, or GLV) of the grating light modulator is shown in Fig. 7.39.[118] It consists of parallel rows of reflective ribbons. Alternative rows of ribbons can be pulled down approximately one-fourth wavelength to create diffraction effects on incident light. When all the rib-

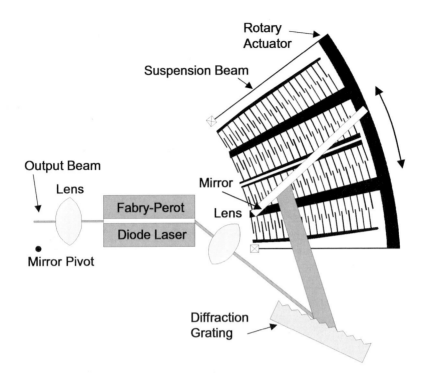

Figure 7.38 Schematic of a MEMS-actuated tunable external cavity diode laser (after Jerman and Grade[116]).

bons are in the same plane, incident light is reflected from their surfaces. When the alternate movable ribbons are pulled down, diffraction produces light at an angle determined by the grating pitch. The ribbons are made of suspended silicon nitride films with aluminum coating to increase its reflectivity. The silicon nitride film is under tensile stress to make the ribbons optically flat. The tension also reduces the risk of stiction and increases the frequency response of the device. The GLV materials are compatible with standard CMOS foundry processes. GLV can be made into one-dimensional or two-dimensional arrays for projection-display applications.[118] Recently, there has also been intense interest in using GLV or similar diffractive optical MEMS devices for dynamic spectral equalization, wavelength add-drop multiplexers, and variable optical attenuator applications.[119]

The MIT/Honeywell/Sandia National Laboratories team recently developed a different MEMS-based programmable diffraction grating using polysilicon surface-micromachining technologies.[120,121] The

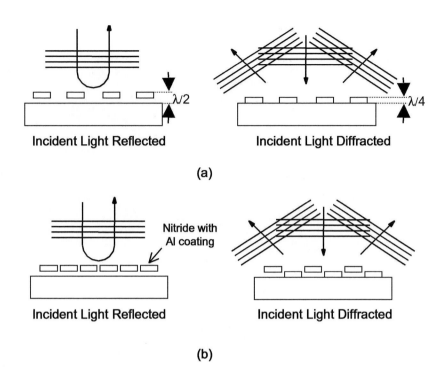

Figure 7.39 Schematic cross section of (a) the original grating light modulators (after Solgaard et al.[117]), and (b) an improved version of grating light valves (after Bloom et al.[118]).

device, called a polychrometer, consists of a parallel array of 20-μm-wide mirror elements, each of which can be individually pulled down with an analog signal. An aperiodic grating is realized with a fully programmable optical transfer function. When illuminated with white light, the spectral content at a fixed viewing angle can be controlled by adjustment of the various mirror element positions, as illustrated in Fig. 7.40(a). The polychrometer enables a new form of correlation spectroscopy. It can synthesize the spectrum of a molecule and replace the reference cell containing that molecule in the correlation radiometers. It can also be used as a programmable spectral filter to enhance the selectivity of the detection process. The polychrometer is actuated by a two-level structure that separates the top reflective optical layer from the lower actuation layer. The structure also permits a very long (1 cm) grating beam without bending. This enables large-aperture programmable gratings, which are needed in correlation spectroscopy. The leverage-bending mechanism enables the beam to move across the entire gap.[122] The device was designed for operation in the 3 to 5 μm spectral range for remote sensing.

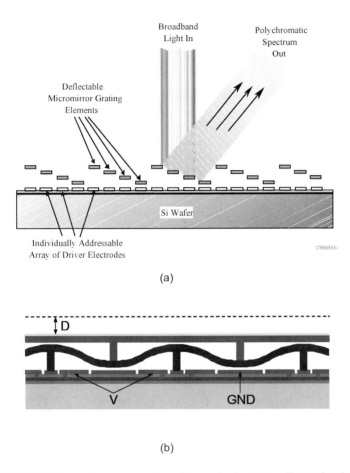

Figure 7.40 (a) Schematic cross section of the polychrometer illustrating its operating principle. The spectral content of the diffracted beam can be varied in real time (~ msec) for correlation spectroscopy. (b) Cross section of the polychrometer. (Pictures courtesy of Steve D. Senturia and Mike Butler. Reprinted from Butler *et al.*[121] with permission.)

Compared with "reflective" or micromirror-based optical MEMS, diffractive optical MEMS offer some advantages. First, small displacement (on the order of one-quarter wavelength) is required since it uses the interference effect. Second, stiffer springs can be used because of the small displacement. Together with the small mass of the ribbon, very high resonant frequencies can be achieved. A response time of 20 nsec was reported by Bloom.[118]

There are also some fundamental limitations. First, diffractive optical MEMS are wavelength-sensitive, so in principle they are narrow-band devices. Second, the diffraction efficiency is low compared with the high reflectivity of micromirrors. To increase the efficiency, both positive and negative as well as higher order (third-order) diffraction beams should be collected. Finally, diffraction is polarization-sensitive. This issue can be mitigated by spatially multiplexing gratings with different orientations.

7.11 Summary

We have described the recent advances in optical MEMS technologies, devices, and applications. Micromirror devices with mirror size ranges from a few tens of micromirrors to several millimeters have been discussed in detail. Applications presented include optical scanning display and imaging systems; telecommunication switches, including 2-D and 3-D MEMS switches, and wavelength-selective switches; and dynamic wavelength-division-multiplexing (WDM) devices such as dynamic spectral/channel equalizers, tunable filters, detectors, and lasers.

Acknowledgment

The authors would like to thank Dr. Lih Y. Lin and Dr. Li Fan for valuable discussions.

References

1. K. E. Petersen, "Silicon as a mechanical material," *Proc. IEEE*, 70(5):420–457 (1982).
2. K. E. Petersen, "Silicon torsional scanning mirror," *IBM J. R&D*, 24:631–637 (1980).
3. M. C. Wu, "Micromachining for optical and optoelectronic systems," *Proc. IEEE*, 85:1833–1856 (1997).
4. R. S. Muller and K. Y. Lau, "Surface-micromachined microoptical elements and systems," *Proc. IEEE*, 86:1705–1720 (1998).
5. P. M. Hagelin and O. Solgaard, "Optical raster-scanning displays based on surface micromachined polysilicon mirrors," *IEEE J. Sel. Topics Quantum Electron.*, 5:67 (1999).
6. O. Degani, E. Socher, A. Lipson, T. Leitner, D. J. Setter, S. Kaldor, and Y. Nemirovsky, "Pull-in study of an electrostatic torsion microactuator," *IEEE J. MEMS*, 7(4):373–379 (1998).
7. E. Hah, H. Toshiyoshi, and M. C. Wu, "Design of electrostatic actuators for MOEMS," *Proc. SPIE, Design, Test, Integration and Packaging of MEMS/MOEMS 2002*, May 2002, Cannes, France.

8. Multi-User MEMS Processes is offered by Cronos Microsystems; for details see *http://www.memsrus.com/nc-pmumps.refs.html*
9. L. Fan and M. C. Wu, "Two-dimensional optical scanner with large angular rotation realized by self-assembled micro-elevator," *Proc. IEEE LEOS Summer Topical Meeting on Optical MEMS, Paper WB4, August 20–22, 1998, Monterey, CA*
10. L. Fan, M. Wu, K. Choquette, and M. Crawford, "Self-assembled microactuated XYZ stages for optical scanning and alignment," *International Conference on Solid-State Sensors and Actuators (Transducers '97), June 1997, Chicago, IL*, pp. 319–322
11. V. A. Aksyuk, F. Pardo, C. A. Bolle, S. Arney, C. R. Giles, and D. J. Bishop, "Lucent Microstar micromirror array technology for large optical crossconnects," *Proc. SPIE, MOEMS and Miniaturized Systems, Sept. 2000, Santa Clara, CA*, pp. 320–324
12. V. A. Aksyuk, M. E. Simon, F. Pardo, S. Arney, D. Lopez, and A. Villanueva, "Optical MEMS design for telecommunications applications," *2002 Solid-State Sensor and Actuator Workshop Tech. Digest, June 2–6, Hilton Head, SC*, pp.1–6
13. *http://www.lucent.com/pressroom/images/Lambda3.jpg*
14. D. T. Neilson et al., "Fully provisioned 112x112 micro-mechanical optical crossconnect with 35.8 Tb/s demonstrated capacity," *Optical Fiber Communication Conference, OFC 2000, March 7–10, Baltimore, MD*, vol. 4, pp. 202–204
15. L. Y. Lin and E. L. Goldstein, "Opportunities and challenges for MEMS in lightwave communications," *IEEE J. Sel. Topics Quantum Electron.*, 8(1):163 (2002)
16. R. R. Syms, "Scaling laws for MEMS mirror-rotation optical cross connect switches," *IEEE J. Lightwave Technol.*, 20(7):1084 (2002)
17. J. Kim, A. Papazian, R. E. Frahm, and J. V. Gates, "Performance of large scale MEMS-based optical crossconnect switches," *IEEE/LEOS Conference 2002, Nov. 11–14, Glasgow, Scotland, UK*, pp. 411–412
18. D. R. Neilson and R. Ryf, "Scalable micro mechanical optical crossconnects" *IEEE/LEOS Conference 2000, Nov. 13–16, Rio Grande, Puerto Rico*, vol. 1, pp. 48–49
19. H.-Y. Lin and W. Fang, "Torsional mirror with an electrostatically driven lever mechanism," *Proc. IEEE/LEOS International Conference on Optical MEMS, Kauai, Hawaii, August 21–24, 2000*, pp. 113–114
20. J. Nee, R. Conat, M. Hart, R. Muller, and K. Lau, "Stretched-film micromirrors for improved optical flatness," *Proc. IEEE MEMS 2000, Miyazaki, Japan, Jan. 23–27, 2000*, pp. 704–709
21. J. Drake and H. Jerman, "A micromachined torsional mirror for track following in magneto-optical disk drives," *2000 Solid-State Sensor and Actuator Workshop, Hilton Head, SC*, pp. 10–13
22. G.-D. Su, H. Toshiyoshi, and M. C. Wu, "Surface-micromachined 2-D optical scanners with high-performance single-cystalline silicon micromirrors," *IEEE Photonics Technol. Lett.*, 13:606 (2001)

23. G.-D. J. Su, P. R. Patterson, and M. C. Wu, "Surface-micromachined 2D optical scanners with optically flat single-crystalline silicon micromirrors," Silicon-Based and Hybrid Optoelectronics III, San Jose, CA, 23–24 Jan. 2001, *SPIE Symp. Proc.*, 4293:46–53 (2001)
24. P. Patterson, D. Hah, G.-D. Su, H. Toshiyoshi, and M.Wu, "MOEMS electrostatic scanning micromirrors design and fabrication," in *Integrated Optoelectronics, Proceedings of the First International Symposium* (M.J. Deen, D. Misra, and J. Ruzyllo, eds.), Electrochemical Society, Inc., Pennington, NJ (2002), p. 369
25. U. Srinivasan, M. Helmbrecht, C. Rembe, R. S. Muller, and R. T. Howe, "Fluidic self-assembly of micromirrors onto surface micromachined actuators," *2000 IEEE/LEOS International Conference on Optical MEMS, Kauai, HI*, pp. 59–60
26. M. A. Michalicek and V. Bright, "Flip-chip fabrication of advanced micromirror arrays," *14th IEEE International Conference on Micro Electro Mechanical Systems (MEMS 2001), Interlaken, Switzerland, 21–25 Jan. 2001*, pp. 152–167
27. Y.-A. Peter, E. Carter, and O. Solgaard, "Segmented deformable micro-mirror for free-space optical communication," *2002 IEEE/LEOS International Conference on Optical MEMS, Lugano, Switzerland*, pp.197–198
28. S. Blackstone and T. Brosnihan, "SOI MEMS technologies for optical switching," *IEEE/LEOS International Conference on Optical MEMS, Sept. 25–28, 2001, Okinawa, Japan*
29. T. Roessig, "Mirrors with integrated position sense electronics for opticalswitching applications," *Analog Dialogue*, 36(4) (2002) www.analog.com/library/analogDialogue/archives/36–04/mirrors/index.html
30. M. R. Dokmeci, S. Bakshi, M. Waelti, A. Pareek, C. Fung, and C. H. Mastrangelo, "Bulk micromachined electrostatic beam steering micromirror array," *2002 IEEE/LEOS International Conference on Optical MEMS, Lugano, Switzerland*, pp.15–16
31. R. Sawada, J. Yamaguchi, E. Higurashi, A. Shimizu, T. Yamamoto, N. Takeuchi, and Y. Uenishi, "Single Si crystal 1024ch MEMS mirror based on terraced electrodes and a high-aspect ratio torsion spring for 3-D cross-connect switch," *2002 IEEE/LEOS International Conference on Optical MEMs*, pp. 11–12
32. R. Ryf et al., "1296-port MEMS transport optical crossconnect with 2.07 petabit/s switch capacity," *Optical Fiber Communication Conference, OFC 2001, March 17–22, Anaheim, CA*, vol. 4, p. PD28-1
33. V. A. Aksyuk et al., "238x238 surface micromachined optical crossconnect with 2 dB maximum loss," *Optical Fiber Communications (OFC) Conference, March 2002, Anaheim, CA* (postdeadline paper)
34. J. Kim, A. R. Papazian, R. E. Frahm, and J. V. Gates, "Performance of large scale MEMS-based optical crossconnect switches," *IEEE/LEOS Annual Meeting, Glasgow, Scotland, November 10–14, 2002*, paper WG-1
35. W. C. Tang, T.-C. H. Nguyen, M. W. Judy, and R. T. Howe, "Electrostatic-comb drive of lateral polysilicon resonators," *Sensors and Actuators*, A21(1–3):328–331 (1990)

36. V. Milanovic, M. Last, and K. S. J.Pister, "Torsional micromirrors with lateral actuators," *Transducers 01, Munich, Germany*, pp. 1290–1301
37. Selvakumar, K. Najafi, W. H. Juan, and S. Pang, "Vertical comb array microactuators," *8th IEEE International Conference on Micro Electro Mechanical Systems (MEMS 1995), Amsterdam, Netherlands*, pp. 43–48
38. H. Schenk, P. Dürr, T. Hasse, D. Kunze, U. Sobe, and H. Kück, "Large deflection micromechanical scanning mirrors for linear scans and pattern generation," *IEEE J. Selected Topics in Quantum Electron.*, 6(5):715–722 (2000)
39. J.-H. Lee, Y.-C. Ko, D.-H. Kong, J.-M. Kim, K. B. Lee, and D.-Y. Jeon, "Design and fabrication of scanning mirror for laser display," *Sensors and Actuators A*, A96(2–3):223–230 (2002)
40. R. A. Conant, J. T. Nee, K. Lau, and R. S. Mueller "A flat high-frequency scanning micromirror," *2000 Solid-State Sensor and Actuator Workshop, Hilton Head, SC*, pp. 6–9
41. U. Krisnamoorthy and O. Solgaard, "Self-aligned vertical combdrive actuators for optical scanning micromirrors," *IEEE/LEOS International Conference on Optical MEMS, Sept. 25–28, 2001, Okinawa, Japan*, p. 41
42. P. R. Patterson, D. Hah, H. Chang, H. Toshiyoshi, and M. C. Wu, "An angular vertical comb drive for scanning micromirrors," *IEEE/LEOS International Conference on Optical MEMS, Sept. 25–28, 2001, Okinawa, Japan*, p. 25
43. R. R. A. Syms, C. Gormley, and S. Blackstone, "Improving yield, accuracy and complexity in surface tension self-assembled MOEMS," *Sensors and Actuators A*, 88:273–283 (2001)
44. P. R. Patterson, D. Hah, H. Nguyen, R.-M. Chao, H. Toshiyoshi, and M. C. Wu, "A scanning micromirror with angular comb drive actuation," in *15th IEEE International Micro Electro Mechanical Systems Conference (MEMS 2002), Jan. 20– 24, 2002, Las Vegas, NV*
45. H. Nguyen, D. Hah, P. R. Patterson, W. Piywattanametha, and M. C.Wu, "A novel MEMS tunable capacitor based on angular vertical comb drive actuators," in *Solid-State Sensor and Actuator Workshop (Hilton Head 2002), June 2–6, 2002, Hilton Head Island, SC*
46. S. Kwon, V. Milanovic, and L. P. Lee, "A high aspect ratio 2D gimbaled microscanner with large static rotation," *2002 IEEE/LEOS International Conference on Optical MEMS, Lugano, Switzerland*, pp. 149–150
47. Y. Mizuno, O. Tsuboi, N. Kouma, H. Soneda, H. Okuda, Y. Nakamura, S. Ueda, I. Sawaki, and F. Yamagishi, "A 2-axis comb-driven micromirror array for 3D mems switches" in *2002 IEEE/LEOS International Conference on Optical MEMS, Lugano, Switzerland*, pp. 17–18
48. M. Yano, F. Yamagishi, and T. Tsuda, "Optical MEMS for photonic switching—compact and stable optical crossconnect switches for simple, fast, and flexible wavelength applications in recent photonic networks," *IEEE J. Selected Topics in Quantum Electron.*, 11:383 (2005)
49. J. W. Judy and R. S. Muller, "Magnetically actuated, addressable microstructures," *IEEE J. MEMS*, 6(3):249–256 (1997)

50. H. Miyajima, N. Asoaka, M. Arima, Y. Minamoto, K. Murakami, K. Tokuda, and K. Matsumoto, "A durable, shock-resistant electromagnetic optical scanner with polyimide-based hinges," *IEEE J. MEMS*, 10(3):418–424 (2001)
51. R. Miller and Y.-C. Tai, "Electromagnetic MEMS scanning mirrors," *Optical Eng.*, 36(5):1399–1407 (1997)
52. H. Miyajima, N. Asoaka, T. Isokawa, M. Ogata, Y. Aoki, M. Imai, O. Fujimori, M. Katashiro, and K. Matsumoto, "Product development of a MEMS optical scanner for a laser scanning microscope," *15th IEEE International Micro Electro Mechanical Systems Conference (MEMS 2002), Jan. 20–24, 2002, Las Vegas, NV*, pp. 552–555
53. R. A. Conant, J. T. Nee, K. Y. Lau, and R. S. Muller, "Dynamic deformation of scanning mirrors," *2000 IEEE/LEOS International Conference on Optical MEMS, Kauai, HI*, pp. 49–50
54. H. Urey, D. W. Wine, and T. D. Osborn, "Optical performance requirements for MEMS-scanner based microdisplays," *Proc. SPIE, MOEMS and Miniaturized Systems, Sept. 2000, Santa Clara, CA*, pp. 176–185
55. N. Asada, M. Takeuchi, V. Vaganov, N. Belov, S. Hout, and I. Sluchak, "Silicon micro-optical scanner," *Sensors and Actuators A*, 83:284–290 (2000)
56. N. Asada, H. Matsuki, K. Minami, and M. Esashi, "Silicon micromachined two-dimensional galvano optical scanner," *IEEE Trans. Magnetics*, 30(6):4647–4649 (1994)
57. S.-H. Ahn and Y.-Y. Kim, "Galvanometric silicon scanning mirror with 2 DOF," *2002 IEEE/LEOS International Conference on Optical MEMS, Lugano, Switzerland*, pp. 87–88
58. Garnier, T. Bourouinna, H. Fujita, E. Orsier, T. Masuzawa, T. Hiramoto, and J.-C. Peuzin, "A fast, robust and simple 2-D micro-optical scanner based on contactless magnetostrictive actuation," *13th IEEE International Micro Electro Mechanical Systems Conference (MEMS 2000), Jan. 2000, Miyazaki, Japan*, pp. 715–720
59. T. Bourouina, E. Lebrasseur, G. Reyne, A. Debray, H. Fujita, A. Ludwig, E. Quandt, H. Muro, T. Oki, and A. Asaoka, "Integration of two degree-of-freedom magnetostrictive actuation and piezoresistive detection: application to a two-dimensional optical scanner," *IEEE J. MEMS*, 11(4):355–361 (2002)
60. D. W. Wine, M. P. Helsel, L. Jenkins, H. Urey, and T. D. Osborn, "Performance of a biaxial MEMS-based scanner for microdisplay applications," *Proc. SPIE*, 4178:186–196 (2000)
61. K. Yamada and T. Kuriyama, "A novel asymmetric silicon micro-mirror for optical beam scanning display," *IEEE Eleventh Annual International Workshop on Micro Electro Mechanical (MEMS 98), 25–29 Jan. 1998, Heidelberg, Germany*, pp. 110–115
62. H.-J. Nam, Y.-S. Kim, S.-M. Cho, Y. Yee, and J.-U. Bu, "Low voltage PZT actuated tilting micromirror with hinge structure," *2002 IEEE/LEOS International Conference on Optical MEMS, Lugano, Switzerland*, pp. 89–90
63. J. Tsuar, L. Zhang, R. Maeda, and S. Matsumoto, "2D micro scanner actuated by sol-gel derived double layered PZT," *15th IEEE International Micro Electro Mechanical Systems Conference (MEMS 2002), Jan. 20–24, 2002, Las Vegas, NV*, pp. 548–551

64. M. A. Sinclair, "High frequency resonant scanner using thermal actuation," *15th IEEE International Micro Electro Mechanical Systems Conference (MEMS 2002), Jan. 20–24, 2002, Las Vegas, NV*, pp. 698–701
65. H. Xie, Y. Pan, and G. Fedder, "A SCS CMOS micromirror for optical coherence tomographic imaging," *15th IEEE International Micro Electro Mechanical Systems Conference (MEMS 2002), Jan. 20–24, 2002, Las Vegas, NV*, pp. 495–498
66. L. J. Hornbeck, "Digital light processing for high-brightness, high-resolution applications," *Proc. SPIE Projection Displays III*, 3013:27–40 (1997)
67. P. F. Van Kessel, L. J. Hornbeck, R. E. Meier, and M. R. Douglass, "A MEMS-based projection display," *Proc. IEEE*, 86:1687–1704 (1998)
68. See, for example, S. Senturia, *Microsystem Design*, Kluwer Academic Publishers (2001), chapter 20
69. N. A. Riza and S. Sumriddetchkajorn, *Optics Lett.*, 24(5):282 (1999)
70. J. E. Ford, V. A. Aksyuk, D. J. Bishop, and J. A. Walker, "Wavelength add-drop switching using tilting micromirrors," *IEEE J. Lightwave Technol.*, 17(5):904–911 (1999)
71. D. M. Marom *et al.*, "Wavelength-selective 1x4 switch for 128 WDM channels at 50GHz spacing," *2002 Optical Fiber Communication (OFC) Conference, Anaheim, CA*, postdeadline paper, FB7
72. T. Ducellier *et al.*, "The MWS 1 ? 4: a high performance wavelength switching building block," *Proceedings of the European Conference on Optical Communication 2002*, session 2.3.1
73. D. Hah, S. Huang, H. Nguyen, H. Chang, H. Toshiyoshi, and M. C. Wu, "A low voltage, large scan angle MEMS micromirror array with hidden vertical comb-drive actuators for WDM routers," *2002 Optical Fiber Communication (OFC) Conference, Anaheim, CA*, paper TuO-3
74. D. Lopez *et al.*, "Monolithic MEMS optical switch with amplified out-of-plane angular motion," *Proc. 2002 IEEE/LEOS International Conf. Optical MEMS*, Piscataway, NJ, 2002, pp. 165–166
75. M. S. Rodgers and J. J. Sniegowski, "Designing microelectromechanical systems-on-a-chip in a 5-level surface micromachine technology," 2nd International Conference on Engineering Design and Automation, Maui, Hawaii, August 9–12, 1998
76. J. E. Ford and J. A. Walker, "Dynamic spectral power equalization using micro-opto-mechanics," *IEEE Photonics Technol. Lett.*, 10(10):1440–1442 (1998)
77. D. T. Neilson *et al.*, "High-dynamic range channelized MEMS equalizing filters," *2002 Optical Fiber Communication (OFC) Conference, Anaheim, CA*, paper ThCC3
78. D. M. Marom and S. H. Oh, "Filter-shape dependence on attenuation mechanism in channelized dynamic spectral equalizers," *2002 IEEE/LEOS Annual Meeting, Glasgow, Scotland*, paper WG3
79. H. Toshiyoshi and H. Fujita, "Electrostataic micro torsion mirrors for an optical switch matrix," *IEEE J. MEMS*, 5:231 (1996)
80. B. E. A. Salah and M. C. Teisch, *Fundamentals of Photonics*, Wiley-Interscience (1991)

81. L.-Y. Lin, E. L. Goldstein, R. W. Tkach, "On the expandability of free-space micromachined optical cross connects," *IEEE J. Lightwave Technol.*, 18:482–489 (2000)
82. L. Fan, S. Gloeckner, P. D. Dobblelaere, S. Patra, D. Reiley, C. King, T. Yeh, J. Gritters, S. Gutierrez, Y. Loke, M. Harburn, R. Chen, E. Kruglick, M. Wu, and A. Husain, "Digital MEMS switch for planar photonic crossconnects," *2002 Optical Fiber Communication (OFC) Conference, Anaheim, CA*, paper TuO4
83. R. T. Chen, H. Nguyen, and M. C. Wu, "A high-speed low-voltage stress-induced micromachined 2×2 optical switch," *IEEE Photonics Technol. Lett.*, 11:1396–1398 (1999)
84. R. Legtenberg, E. Berenschot, M. Elwenspoek, and J. Fluitman, "Electrostatic curved electrode actuators," *Proc. IEEE MEMS 1995*, pp. 37–42
85. L. Houlet, P. Helin, T. Bourouina, G. Reyne, E. Diffour-Gergam, and H. Fujita, "Movable vertical mirror arrays for optical microswitch matrixes and their electromagnetic actuation," *IEEE J. Selected Topics in Quantum Electron.*, 8(1):58–63 (2002)
86. L. Y. Lin, E. L. Goldstein, and R. W. Tkach, "Free-space micromachined optical switches with submillisecond switching time for large-scale optical crossconnects," *IEEE Photonics Technol. Lett.*, 10:525–527 (1998)
87. B. Hehin, K. Y. Lau, and R. S. Muller, "Magnetically actuated micromirrors for fiber-optic switching," in *Solid-State Sensors and Actuators Workshop, Hilton Head Island, SC, 1998*
88. Y. Yoon, K. Bae, and H. Choi, "An optical switch with newly designed electrostatic actuators for optical cross connects," in *2002 IEEE/LEOS International Conference on Optical MEMS, Lugano, Switzerland*
89. R. Akiyama and H. Fujita, "A quantitative analysis of scratch drive actuator using buckling motion," *Proc. 8th IEEE International MEMS Workshop, 1995*, pp. 310–315
90. J. W. Judy and R. S. Muller, "Magnetically actuated, addressable microstructures," *IEEE J. MEMS*, 6(3):249–256 (1997)
91. C.-H. Ji and Y. K. Kim, "Addressable electromagnetic micromirror array with single crystal silicon mirror plate and aluminum spring," *2001 IEEE/LEOS International Conference on Optical MEMS, Okinawa, Japan, 2001*
92. Marxer and N. F. de Rooij, "Micro-opto-mechanical 2×2 switch for single-mode fibers based on plasma-etched silicon mirror and electrostatic actuation," *IEEE J. Lightwave Technol.*, 17(1):2–6 (1999)
93. W. Noell, P.-A. Clerc, L. Dellmann, B. Guldimann, H.-P. Herzig, O. Manzardo, C. R. Marxer, K. J. Weible, R. Dandliker, and N. de Rooij, "Applications of SOI-based optical MEMS," *IEEE J. Selected Topics in Quantum Electron.*, 8(1):148–154 (2002)
94. Marxer, C. Thio, M.-A. Gretillat, N. F. de Rooij, R. Battig, O. Anthamatten, B. Valk, and P. Vogel, "Vertical mirrors fabricated by deep reactive ion etching for fiber-optic switching applications," *J. MEMS*, 6(3):277–285 (1997)
95. S.-S. Lee, L.-S. Huang, C.-J. Kim, and M. C. Wu, "Free-space fiber-optic switches based on MEMS vertical torsion mirrors," *IEEE J. Lightwave Technol.*, 17(1):7–13 (1999)

96. Y. Kato, T. Norimatsu, O. Imaki, T. Sasaki, K. Kondo, and K. Mori, "Development of a multi-channel 2×2 optical switches," *2002 IEEE/LEOS International Conference on Optical MEMS, Lugano, Switzerland*, paper ThB3
97. C. Marxer, P. Griss, and N. F. de Rooij, "A variable optical attenuator based on silicon micromechanics," *IEEE Photonics Technol. Lett.*, 11(2):233–235 (1999)
98. C. Marxer, B de Jong, and N. F. de Rooij, "Comparison of MEMS variable optical attenuator designs," *2002 IEEE/LEOS International Conference on Optical MEMS, Lugano, Switzerland*, paper FA1
99. C. H. Ji, Y. Yee, J. Choi, and J. U. Bu, "Electromagnetic variable optical attenuator," *2002 IEEE/LEOS International Conference on Optical MEMS, Lugano, Switzerland*, paper WB2
100. C. R. Giles, V. Aksyuk, B. Barber, R. Ruel, L. Stulz, and D. Bishop, "A silicon MEMS optical switch attenuator and its use in lightwave subsystems," *IEEE J. Selected Topics in Quantum Electron.*, 5(1):18–25 (1999)
101. C. Vail, M. S. Wu, G. S. Li, L. Eng, and C. J. Chang-Hasnain, "GaAs micromachined widely tunable Fabry-Perot filters," *Electron. Lett.*, 31(3):228–229 (1995)
102. M. C. Larson and J. S. Harris, Jr., "Broadly-tunable resonant-cavity light-emitting diode," *IEEE Photonics Technol. Lett.*, 7(11):1267–1269 (1995)
103. C. F. R. Mateus, C. Chih-Hao, L. Chrostowski, S. Yang, D. Sun, R. Pathak, and C. J. Chang-Hasnain, "Widely tunable torsional optical filter," *IEEE Photonics Technol. Lett.*, 14(6):819–821 (2002)
104. R. Amano, F. Koyama, and M. Arai, "GaAlAs/GaAs micromachined thermally tunable vertical cavity filter with low tuning voltage," *Electron. Lett.*, 38:(14):738–740 (2002)
105. J. Peerlings, A. Dehe, A. Vogt, M. Tilsch, C. Hebeler, F. Langenhan, P. Meissner, and H. L. Hartnagel, "Long resonator micromachined tunable GaAs-AlAs Fabry-Perot filter," *IEEE Photonics Technol. Lett.*, 9(9):1235–1237 (1997)
106. P. Tayebati, P. Wang, M. Azimi, L. Maflah, and D. Vakhshoori, "Microelectromechanical tunable filter with stable half symmetric cavity," *Electron. Lett.*, 34(20):1967–1968 (1998)
107. H. Halbritter, M. Aziz, F. Riemenschneider, and P. Meissner, "Electrothermally tunable two-chip optical filter with very low-cost and simple concept," *Electron. Lett.*, 38(20):1201–1202 (2002)
108. Spisser, R. Ledantec, C. Seassal, J. L. Leclercq, T. Benyattou, D. Rondi, R. Blondeau, G. Guillot, and P. Viktorovitch, "Highly selective 1.55 mu m InP/air gap micromachined Fabry-Perot filter for optical communications," *Electron. Lett.*, 34(5):453–455(1998)
109. J. Daleiden, V. Rangelov, S. Irmer, E. Romer, M. Strassner, C. Prott, A. Tarraf, and H. Hillmer, "Record tuning range of InP-based multiple air-gap MOEMS filter," *Electron. Lett.*, 38(21):1270–1271 (2002)
110. L. A. Coldren, "Monolithic tunable diode lasers," *IEEE J. Selected Topics in Quantum Electron.*, 6(6):988–999 (2000)
111. M. S. Wu, E. C. Vail, G. S. Li, W. Yuen, and C. J. Chang-Hasnain, "Tunable micromachined vertical cavity surface emitting laser," *Electron. Lett.*, 31:1671–1672 (1995)

112. M. C. Larson, A. R. Massengale, and J. S. Harris, Jr., "Continuously tunable micromachined vertical cavity surface emitting laser with 18 nm wavelength range," *Electron. Lett.*, 32:330–332 (1996)
113. D. Vakhashoori, P. D. Wang, M. Azimi, K. J. Knopp, and M. Jiang, "MEMs-tunable vertical-cavity surface-emitting lasers," *Proc. Optical Fiber Communication Conference (OFC), Post-Conference Edition,* 2:TuJ1-3 (2001), IEEE Cat. no. 01CH37171
114. C. J. Chang-Hasnain, "Tunable VCSEL," *IEEE J. Selected Topics in Quantum Electron.*, 6(6):978–987 (2000)
115. J. S. Harris, Jr., "Tunable long-wavelength vertical-cavity lasers: the engine of next generation optical networks?" *IEEE J. Selected Topics in Quantum Electron.*, 6(6):1145–1160 (2000)
116. H. Jerman and J. D. Grade, "A mechanically-balanced, DRIE rotary actuator for a high-power tunable laser," in *2002 Solid-State Sensor and Actuator Workshop Tech. Digest, June 2–6, 2002, Hilton Head, SC*
117. O. Solgaard, F. S. A. Sandejas, and D. M. Bloom, "Deformable grating optical modulator," *Optics Lett.*, 17(9):688–690 (1992)
118. D. M. Bloom, "The grating light valve: revolutionizing display technology," *SPIE Symp. Proc., Projection Displays III,* 3013:165–171 (1997)
119. O. Solgaard, "Dynamic diffractive optical elements based on MESM technology," *Proc. 2002 Optics-Photonics Design and Fabrication (ODF) Conference, Tokyo, Japan*, paper WP01
120. S. D. Senturia, "Diffractive MEMS: the polychromator and related devices," *Digest 2002 IEEE/LEOS International Conference on Optical MEMS*, pp. 5–6
121. M. A. Butler, E. R. Deutsch, S. D. Senturia, M. B. Sinclair, W. C. Sweatt, D. W. Youngner, and G. B. Hocker, "A MEMS-based programmable diffraction grating for optical holography in the spectral domain," *Technical Digest 2001 International Electron Devices Meeting (IEDM)*, pp. 41.1.1–41.1.4
122. E. S. Hung and S. D. Senturia, "Extending the travel range of analog-tuned electrostatic actuators," *J. MEMS*, 8:497 (1999)

8 Integrated Micro-Optics

Hans Zappe

Institute of Microsystem Technology IMTEK
University of Freiburg, Germany

8.1 Introduction

Microelectromechanical systems (MEMS), as the name suggests, are predisposed to the use of electrons and mechanical movement. By adding optics to the palette of MEMS capabilities, the resultant micro-optoelectromechanical systems (MOEMS) or micro-opto-mechanical systems (MOMS) provide increased functionality while retaining the attractive features of MEMS technology. As the spectrum of potential applications can thus be substantially increased, research and development work on optical MEMS has recently seen considerable activity.[1,2]

8.1.1 Definitions

We can usefully divide optical MEMS into two classes: free-space systems and integrated systems. The former represents "classical" microoptical systems where the light is directed through free space using lenses, gratings, or mirrors;[3] two-dimensional arrays of micromirrors,[4,5] silicon-optical-bench systems,[6,7] and three-dimensional silicon micro-optics[8] represent established examples of this category.

Integrated optical systems, in contrast, use optical waveguides to guide light through the system.[9,10] Such waveguide-based systems represent "optical circuits" that are closely analogous to electrical integrated circuits and allow a broad spectrum of optical functions to be integrated with established MEMS technology.[11] This chapter focuses on integrated micro-optics for MEMS; the free-space variant is considered in Chapter 7.

8.1.2 Components

The basic building block of an integrated micro-optical system is the waveguide. As will be thoroughly discussed in Section 8.2, the waveguide represents in essence an "optical wire" with cross-sectional dimensions on the order of the wavelength in the material, several µm.[12] Fabricated using glass, semiconductors, or plastics, the waveguide is used to transmit the optical signal, where curves, couplers, and splitters are used to combine, divide, and transfer the light. Optical fibers represent a macroscopic analog to integrated optical waveguides,[13] as the light-guiding principle is identical; the size of the fibers generally limits their use in MEMS to external input/output coupling of an optical signal.

Using waveguides, optical circuits with a wide variety of optical functions can be fabricated. Integrated optical interferometers are useful for a wide variety of applications; Fabry-Perot, Mach-Zehnder, Michelson, and Sagnac interferometers are representative examples. In addition, $1 \times N$ couplers (one optical input divided into N optical outputs) and optical crossbars are popular components in optical communications systems.

Using physical structures based on waveguides, a number of additional optical and opto-electronic components can be used in integrated micro-optical systems. The semiconductor laser diode is the most prominent, whose active region is typically configured as a waveguide structure forming the Fabry-Perot cavity.[14] Photodetectors are quite similar in structure, as are waveguide-based modulators. The former take advantage of material absorption in a particular waveguide material ,and the latter frequently make use of electro-optical effects for the generation of phase or intensity modulation.[15]

8.1.3 Summary

This chapter undertakes a basic introduction to waveguides and waveguide-based integrated micro-optics as they are relevant for optical MEMS. We concentrate on an analysis of guided waves in one and two dimensions, deriving the relationships necessary to predict the number and characteristics of modes and optical fields. Basic input and output coupling techniques as well as the approaches for measurement required to determine waveguide parameters experimentally are discussed. This analysis will allow us to understand the operating principles and experimental behavior of the various waveguide-based devices useful for optical MEMS.

A brief look at passive waveguide optics, including couplers, splitters, and various types of interferometers subsequently provide an overview of the types of integrated micro-optical structures likely to be encountered in an optical MEMS. Following a discussion of the materials commonly used in integrated optics, including the semiconductor materials most useful for integration with classic MEMS technology, the chapter concludes with a summary of the most important application areas.

8.2 Guided Waves

The waveguide is the fundamental optical component in integrated optical microsystems. Using multiple reflections of an optical wave on the sidewalls of a transparent medium, the waveguide represents an "optical wire" used to transmit light on a well-defined path. Conceptually similar to a light guide, as used for example in endoscopes and shown schematically in Fig. 8.1, the waveguide is distinguished from its macroscopic precursor by its dimensions. A light guide can have a diameter on the order of mm or cm, while the waveguide has dimensions on the order of the light wavelength in the material; typical optical waveguide cross sections are thus in the µm range.

The dimensions of the waveguide result in an altogether different mode of light propagation in the waveguide as compared to a light guide. Whereas the optical field can propagate quasi-continuously in the macroscopic light guide, the interference effects that result from light propagating in a waveguide lead to discrete modes of transport. Thus light in a waveguide is transmitted in discrete modes, whose number and form are a strong function of wavelength and waveguide dimensions. This section analyzes propagation in waveguides and will lead to an understanding of the propagation behavior of light in these microscopic structures.

8.2.1 Reflections at Boundaries

Propagation of light down a waveguide is based on multiple reflections of the optical wave with the sides of the waveguide. Waveguide structures are generally fabricated from dielectric materials, so that propagation is based on reflection at the boundaries between materials with differing refractive indices.

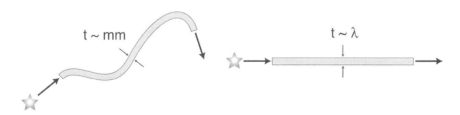

Figure 8.1 Comparison between a light-guide (left) and a waveguide (right). A waveguide has dimensions on the order of the wavelength in the material, typically µm.

Reflection of light at a boundary between two dielectrics is characterized simply by the angle of reflection, θ_r, which is equal to the angle of incidence of the light beam, θ_i, so:

$$\theta_i = \theta_r, \qquad (1)$$

independent of the material refractive indices. Note that the angles are measured with respect to the normal of the interface. Transmission is characterized by Snell's law, which relates the angles of incidence and transmission to the refractive indices of the materials. Defining n_i and θ_i as the refractive index and incidence angle of the incident beam, and n_t and θ_t those of the transmitted beam, as shown in Fig. 8.2, Snell's law states:

$$n_i \sin\theta_i = n_t \sin\theta_t. \qquad (2)$$

For the situation where $n_i > n_t$, the so-called internal incidence case, the angle of transmission can reach 90 degrees for an incidence angle greater than the critical angle, given by:

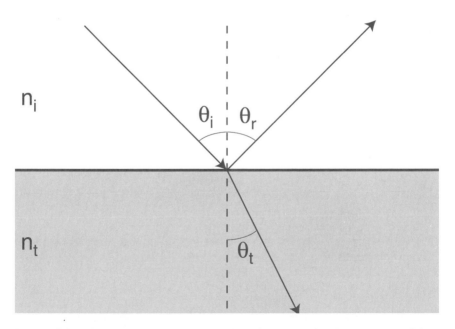

Figure 8.2 Reflection and transmission of light at a boundary between two dielectric materials. Light is incident from material with refractive index n_i.

$$\theta_c = \sin^{-1}\left(\frac{n_t}{n_i}\right). \tag{3}$$

For this case, all light is reflected at the boundary and none is transmitted across the interface, a situation known as total internal reflection (TIR). For TIR to take place, incidence must be from the material with higher refractive index and must be at an angle greater than the critical angle, θ_c.

This simple picture of reflection and transmission can be refined somewhat by consideration of the electromagnetic fields at the boundary. For an incident transverse electromagnetic (TEM) wave, where the electric and magnetic fields are normal to one another and are both normal to the propagation direction, we can consider two cases for the polarization of the fields with respect to the interface. As shown in Fig. 8.3, for the transverse electric (TE) case, the electric field is always parallel to the dielectric boundary. The transverse magnetic (TM) case results when the magnetic field remains parallel to the interface. The general case for arbitrary incidence of an optical field onto an interface can always be reduced to a sum of the TE and TM cases.

For reflection of an optical beam from a perfect mirror, the light is phase-shifted by π. In contrast, when a light beam is reflected by TIR, it undergoes a phase shift that is a function of the incidence angle as well as the refractive indices of the dielectric materials on both sides of the interface.

These phase shifts differ for TE and TM polarizations, and are given by:[16]

$$\Delta\phi_{TE} = 2\tan^{-1}\sqrt{\frac{\sin^2\theta_i - \left(\frac{n_t}{n_i}\right)^2}{\cos^2\theta_i}}, \tag{4}$$

and

$$\Delta\phi_{TM} = 2\tan^{-1}\sqrt{\frac{\left(\frac{n_i}{n_t}\right)^2 \sin^2\theta_i - 1}{\left(\frac{n_t}{n_i}\right)^2 \cos^2\theta_i}}. \tag{5}$$

We use these expressions in Section 8.2.2 when we calculate the phase condition for transmission of light through a waveguide.

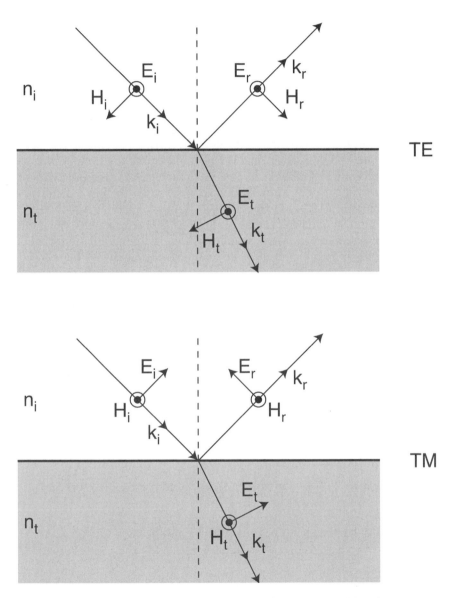

Figure 8.3 Electric and magnetic field vectors for TE and TM incidence onto a boundary. Propagation is given by the vector k.

8.2.2 Ray-Optic Model

A waveguide takes advantage of TIR to guide light. The ray-optic model treats the light as a "beam" whose propagation vector bounces from surface to surface, as shown schematically for a one-dimensional cross-section in Fig. 8.4. An alternative waveguide model, using an electromagnetic representation derived from a solution of Maxwell's equations, is presented in Section 8.2.4 and yields the same predictions concerning mode propagation.

Figure 8.4 shows the basic structure of a slab waveguide, which can be envisaged as a sandwich of three planar dielectrics; the center core region is the layer through which light propagates by TIR and thus must have the highest refractive index of the three regions. The core is surrounded on either side by a cladding layer, where the top and bottom cladding layers may be the same or differing materials. The only requirement is that $n_{core} > n_{cladding}$ for both top and bottom layers.

From the discussion of Section 8.2.1, we see that light rays with an incidence angle $\theta > \theta_c$ will propagate by TIR; steeper angles of incidence will result in transmission of a portion of the light into the cladding, leading to a complete loss of intensity after a few reflections. However, even for angles $\theta > \theta_c$, only certain discrete values of θ are allowed. This phenomenon represents the essential difference between light guides and waveguides and can be understood by examining the phase conditions of a light ray in the waveguide.

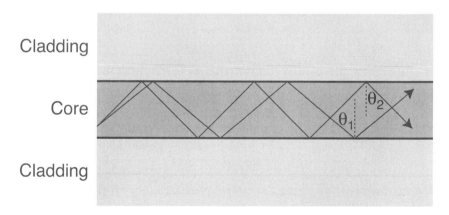

Figure 8.4 Cross section of a one-dimensional slab waveguide. Light is guided in the core region surrounded by lower index cladding material. Different modes propagate with differing reflection angles, θ.

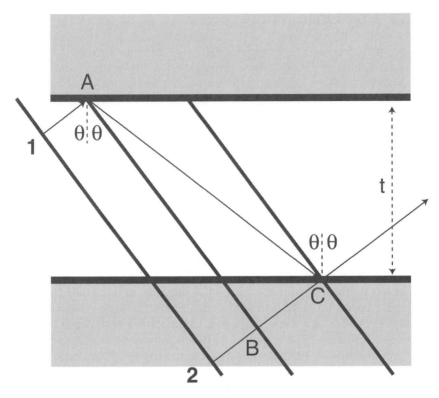

Figure 8.5 Geometric construction showing a wavefront propagating through a waveguide. The wavefronts starting from points 1 and 2 must be in phase after point C.

Consider the construction of Fig. 8.5, which shows a single-wavelength wave front moving through the waveguide core of thickness t. Two parts of the wave, labeled 1 and 2, are transmitted by two different paths. Portion 1 is reflected at A and C; portion 2 moves from B to C; both parts continue together after point C. To assure that the wave will continue to propagate, these two portions (1 and 2) must be in phase (i.e., have a relative phase shift of $2m\pi$ for integer m). For relative phase shifts that are not multiples of 2π, destructive interference effects would lead to an eventual loss of all optical intensity.

The path difference between 1 and 2, ΔL, is given by $\Delta L = AC - BC$, implying $\Delta L/l$ wavelengths. Elementary geometry leads to $\Delta L = 2t \cos(\theta)$ and thus a phase shift, due to the path length difference, of:

$$\frac{2\pi\Delta L}{\lambda} = \frac{4\pi t}{\lambda}\cos(\theta). \tag{6}$$

The ray from point 1 also undergoes two reflections, at A and B; these reflections result in a phase shift in ray 1 at the upper and lower interfaces, $\Delta\phi_{upper}$ and $\Delta\phi_{lower}$, respectively. Since the sum of these phase shifts must yield a total relative phase shift of an integer multiple of 2π, the in-phase condition for propagation in this waveguide is given by:

$$\frac{4\pi t}{\lambda}\cos(\theta) - \Delta\phi_{upper} - \Delta\phi_{lower} = 2m\pi. \tag{7}$$

We can now use the phase shifts on reflection presented in Section 8.2.1 for TE and TM polarizations, where n_g is the refractive index of the core region and n_s is the refractive index for both the upper and lower cladding regions. Plugging in these relations for $\Delta\phi$, the phase condition for TE propagation in the waveguide, after some algebraic manipulation, can be written as:

$$\tan\left(\frac{\pi t}{\lambda}\cos\theta - \frac{m\pi}{2}\right) = \frac{\sqrt{\sin^2\theta - \left(\frac{n_s}{n_g}\right)^2}}{\cos\theta}, \tag{8}$$

and for *TM* as:

$$\tan\left(\frac{\pi t}{\lambda}\cos\theta - \frac{m\pi}{2}\right) = \frac{\sqrt{\left(\frac{n_g}{n_s}\right)^2 \sin^2\theta - 1}}{\left(\frac{n_s}{n_g}\right)\cos\theta}. \tag{9}$$

These two expressions represent the characteristic equation for a symmetric waveguide, one for which the upper and lower claddings are identical.

The more general case of the asymmetric waveguide, where the upper cladding layer has index n_c (generally, c stands for cap), the lower cladding has index n_s (generally, s stands for substrate), and $n_c \neq n_s$, can be derived in the same way, except that now $\Delta\phi_{upper} \neq \Delta\phi_{lower}$. The resulting characteristic equations are slightly more complicated, and are given for TE by:

$$\tan\left(\frac{2\pi t}{\lambda}\cos\theta - m\pi\right) = \frac{\cos\theta\left[\sqrt{\sin^2\theta - \left(\frac{n_c}{n_g}\right)^2} + \sqrt{\sin^2\theta - \left(\frac{n_s}{n_g}\right)^2}\right]}{\cos^2\theta - \sqrt{\left(\sin^2\theta - \left(\frac{n_c}{n_g}\right)^2\right)\left(\sin^2\theta - \left(\frac{n_s}{n_g}\right)^2\right)}}, \quad (10)$$

and for TM by:

$$\tan\left(\frac{2\pi t}{\lambda}\cos\theta - m\pi\right) = \frac{\cos\theta\left[\frac{n_g}{n_c}\sqrt{\left(\frac{n_g}{n_c}\right)^2\sin^2\theta - 1} + \frac{n_g}{n_s}\sqrt{\left(\frac{n_g}{n_s}\right)^2\sin^2\theta - 1}\right]}{\cos^2\theta - \frac{n_g^2}{n_c n_s}\sqrt{\left(\left(\frac{n_g}{n_c}\right)^2\sin^2\theta - 1\right)\left(\left(\frac{n_g}{n_s}\right)^2\sin^2\theta - 1\right)}}.$$

(11)

Equations (10) and (11) represent the most general case. The modal behavior of a waveguide can thus be described using one of these characteristic equations; which one we use depends on the relevant polarization (TE or TM, or both) and whether the waveguide is symmetric or asymmetric.

Because the characteristic equations cannot be solved analytically, either a graphical or numerical solution is required. The equations are generally solved for θ as a function of waveguide thickness, wavelength, and refractive indices of the waveguide layers. The solutions for θ represent the allowed reflection angles that lead to the in-phase condition and thus propagation. Each value of θ represents a distinct propagating mode, and the number of allowed values of θ gives the number of allowed modes for the waveguide; modes are discussed in Section 8.2.3.

A typical result obtained in numerical solution of the characteristic equation is given in Fig. 8.6. This figure shows the allowed θ values for TE and TM polarizations in a symmetric waveguide as a function of the refractive index of the cladding layers. We see that the permitted values for θ, all greater than the critical angle θ_c, vary slowly with increasing cladding index and are slightly different for TE and TM. These solutions tell us that propagation down the waveguide can only occur for reflections at these given values of reflection angle; thus propagation is discrete, as was mentioned at the outset.

8.2.3 Modes and Propagation

For the example of Fig. 8.6, both the TE and TM polarizations have two solutions for any given refractive index combination, labeled TE0 and TE1 or TM0 and TM1. The existence of these two solutions implies that the waveguide supports two discrete modes in each polarization. These solutions were derived by solving the characteristic equations first with $m = 0$ and then with $m = 1$. No solutions exist for $m \geq 2$, implying that modes with $m \geq 2$ are cut off and thus do not propagate. Note that the reflection angle θ decreases as the mode number increases, so that θ approaches θ_c for higher order modes.

The maximum value for m, the mode index, is written m_{max}, and for a symmetric waveguide is given by:

$$m_{max} = \mathrm{int} \left[\frac{2t}{\lambda} \sqrt{1 - \frac{n_s^2}{n_g^2}} \right], \qquad (12)$$

where int represents the integer value of the expression in brackets. We see from this that the number of modes is primarily a function of the relationship between waveguide core thickness t and wavelength λ. Cutoff can be reached for a mode of a particular λ if the waveguide is too thin (i.e., t is too small); on the other hand, for a waveguide of a particular thickness t, cutoff can be reached if the wavelength λ is too long. Thus single-mode waveguides need to be as thin as possible or need to be operated at as long a wavelength as possible.

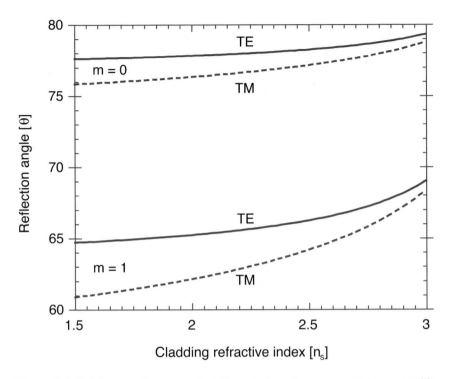

Figure 8.6 Solutions to the characteristic equations for a symmetric waveguide: The core has index $n_g = 3.35$, and the upper and lower cladding, n_s, vary from 1.5 to 3.0 along the abscissa. The waveguide has a thickness of 500 nm, and the calculations were done for $\lambda = 850$ nm. These parameter values are typical for a GaAs/AlGaAs-based waveguide.

From Eq. (12), the total number of allowed modes is then given by $m_{max} + 1$, since the lowest order (first) mode has the mode index $m = 0$. It should be noted that $m = 0$ is always allowed for a symmetric waveguide so that the lowest order mode for a symmetric waveguide is always guided.

Propagation of light in a waveguide is typically characterized by a propagation constant k, where:

$$k = \frac{2\pi}{\lambda} \qquad (13)$$

for the wavelength of light in the material λ, related to the vacuum wavelength λ_0 by $\lambda = \lambda_0/n$ for refractive index n. Thus propagation in vacuum is given by $k_0 = 2\pi/\lambda_0$.

The propagation in the direction of the waveguide is of interest in calculating the propagation properties of a mode; since the ray-optic model assumes that the wave bounces from sidewall to sidewall with an angle θ on its way down the guide, the relevant propagation parameter is the component that is parallel to the walls. As seen in Fig. 8.7, this component is in the z direction and proportional to $\sin\theta$. Since propagation takes place at discrete values of θ, we see that k also takes on discrete values.

The z component of k, k_z, is typically denoted β, so that:

$$\beta \equiv k_z = n_g \sin\theta k_0. \tag{14}$$

The factor $n_g \sin\theta$ represents a modification of the material refractive index n_g by a geometric factor corresponding to the mode that is propagating. Consequently, we often define an effective index for a waveguide, N, as:

$$N = n_g \sin\theta, \tag{15}$$

which includes both the material and mode characteristics. We can picture the mode, then, as propagating down a waveguide with index N, ignoring the details of reflection at the waveguide surfaces.

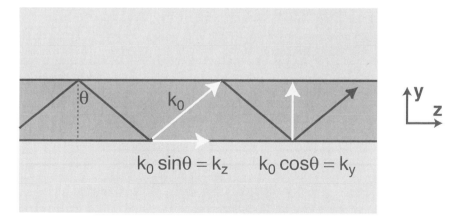

Figure 8.7 Components of the propagation vector k in a waveguide. Of primary interest is k_z, the propagation vector in the z direction, down the waveguide.

8.2.4 Electromagnetic Model

The ray-optic model described in Section 8.2.2 provided an intuitive picture of why propagation in a waveguide takes place with discrete modes. We can also describe this modal behavior using an electromagnetic model and essentially solving the wave equation in the waveguide. Applying the proper boundary conditions at the core/cladding interfaces, it can be shown that the same characteristic equations as derived above will result.[17] This more rigorous approach, then, confirms the validity of the more *ad hoc* ray-optic model.

The optical field, an electromagnetic wave, is characterized by an electric field E and magnetic field H. We solve the wave equation for electric field in a one-dimensional waveguide as shown in Fig. 8.8:

$$\frac{\partial^2 E}{\partial y^2} + k_{y,i}^2 E = 0, \qquad (16)$$

where the y direction is that normal to the plane of the waveguide, k_y is the vertical component of k, and i is an index corresponding to the layer in which we perform the solution. $i = c$ in the upper cladding or cap, $i = g$ in the waveguide core, and $i = s$ in the lower cladding or substrate. $k_{y,i}$ is given by:

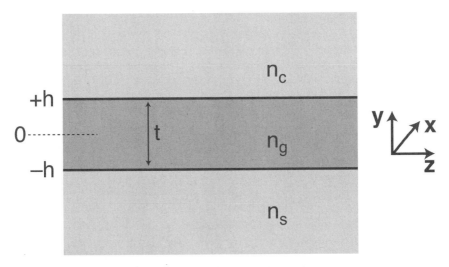

Figure 8.8 Cross section of a one-dimensional waveguide showing the three regions and coordinates for solving the wave equation.

$$k_{y,i}^2 = k_0^2 n_i^2 - \beta^2, \quad i = c, g, s. \tag{17}$$

Performing this solution in the three regions, we obtain, for an ideal guided mode, the following expressions for the variation of the x polarized electric field in the y direction:

Cap: $\quad E_x(y) = E_c \exp\left[-k_{yc}(y - h)\right].$ (18)

Core: $\quad E_x(y) = E_g \cos\left[k_{yg} y + \phi\right].$ (19)

Substrate: $E_x(y) = E_s \exp\left[k_{ys}(y + h)\right].$ (20)

The factors E_c, E_g, and E_s represent the peak magnitudes of the electric fields in the three regions, $2h = t$ denotes the waveguide thickness, and ϕ is an arbitrary phase shift, which may be zero.

For the above expressions, the y-directed propagation constants are given by

$$k_{yc} = \sqrt{\beta^2 - k_0^2 n_c^2}, \tag{21}$$

$$k_{yg} = \sqrt{k_0^2 n_g^2 - \beta^2}, \tag{22}$$

and

$$k_{ys} = \sqrt{\beta^2 - k_0^2 n_s^2}, \tag{23}$$

which are functions of the material refractive indices as well as, through β, the modal propagation characteristics.

The electromagnetic model thus tells us the form of the electric field distribution in the waveguide. In the core, we have an oscillatory solution, whose period is a function of the mode. The lowest order, or fundamental, mode has a single peak of the cosine function, the next higher order mode two peaks, and so forth. The number of peaks is primarily a function of β and thus is ultimately a function of the relative values of waveguide thickness t and wavelength λ.

We also see that the field outside of the core is exponentially decaying but is non-zero; this result contrasts with the assumptions of the ray-optic model, which assumed total internal reflection at the boundary and thus no penetration of the field into the cladding. The rate of decay of the exponential term is a function of k_{yc} or k_{ys}, and is thus again ultimately defined by β. If we define the $1/e$ penetration depth to be $1/k_{yc}$ or $1/k_{ys}$, we can show that the penetration depth increases for higher order modes, implying that these have increasing amounts of optical energy outside the waveguide core.

From knowledge of the electric fields in a waveguide, we can easily derive the energy of the optical field as well. The measurable time average of the optical intensity, $\langle S \rangle$ is given by:

$$\langle S \rangle = \frac{1}{2}\frac{1}{Z}E^2, \tag{24}$$

where Z is the impedance of the material,

$$Z = \frac{1}{n}\sqrt{\frac{\mu_0}{\varepsilon_0}} = \frac{1}{n}Z_0 = \frac{1}{n}377\Omega, \tag{25}$$

and Z_0 is the impedance of free space. Thus the energy, or measurable intensity, of the field is proportional to E^2. A typical result of the optical intensity distribution in a waveguide of rectangular cross section is shown in Fig. 8.9 for the fundamental mode and the first higher order mode.

8.2.5 Confinement Factor

We saw in Section 8.2.4 that the electric field (and thus the optical intensity, $\propto E^2$) is nonzero and decays exponentially outside the waveguide core. Thus a portion of the total energy of the optical mode is found outside the actual waveguide. The more energy exists outside of the core, the more poorly the mode is guided.

A measure for the fraction of optical power inside the waveguide core is the confinement factor Γ, defined as:

$$\Gamma = \frac{\int_{-h}^{h} E_x^2(y)dy}{\int_{-\infty}^{\infty} E_x^2(y)dy}, \tag{26}$$

for, in this case, an x polarized electric field and a waveguide bounded in the y direction, as shown in Fig. 8.8. The confinement factor is maximally unity, and a value for Γ close to unity implies little energy outside the core

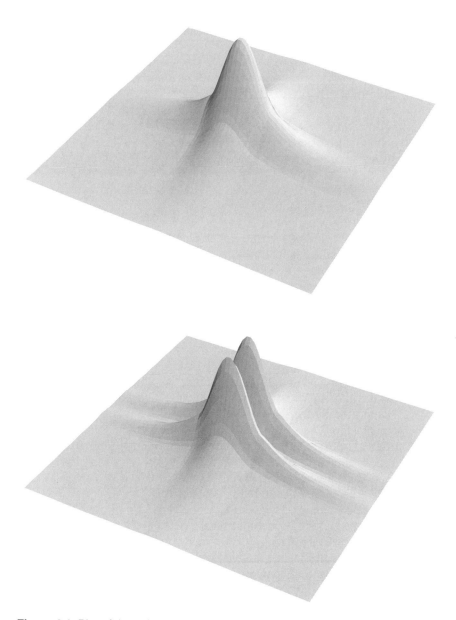

Figure 8.9 Plot of the calculated intensity distribution in a rectangular waveguide. The upper plot shows a single-mode characteristic; the lower has two transverse modes.

and thus good guiding. For some waveguide applications, for example, certain classes of integrated optical sensors, a small value for Γ is desired so that the optical field in the waveguide can interact with materials outside the waveguide core.

8.2.6 Solving a Waveguide

In the preceding sections, we have seen a selection of means to calculate the characteristics and behavior of modes in a waveguide. This section briefly summarizes how to approach a waveguide calculation.

1. Determine the necessary waveguide and operating parameters:
    - Waveguide thickness, t
    - Operating wavelength, λ
    - Refractive indices of the upper cladding (cap), core, and lower cladding (substrate)
    - Polarization of the optical field, TE or TM (or both)
2. Solve the appropriate characteristic equation for $m = 0$; the solution will give an allowed value of θ. Of Eqs. (8) through (11), choose:
    - The symmetric one, if the material of the cap and substrate are identical.
    - The asymmetric one, if they are different.
    - The TE or TM equation, depending on which polarization is desired; solve both if it is unclear which polarization is supported by the waveguide.
3. Repeat step 2 for m = 1,2,3... until no further solution is found. The total number of modes supported by the waveguide is then $m_{max} + 1$.
4. From the values of θ derived, calculate β and N for each mode from Eqs. (14) and (15), respectively.
5. From β, calculate $k_{y,c}$, $k_{y,g}$, and $k_{y,s}$ using Eqs. (21) through (23). Plug these into the expressions for the fields in the cladding and core, Eqs. (18) through (20). Each mode will have a separate set of field equations. One-dimensional plots of these field solutions will show the lateral distribution of the mode.
6. Using expressions for the optical power (Eq. 24) or the confinement factor (Eq. 26), these parameters can also be determined.

Although this approach will provide a simple and useful overview of the modal performance of a waveguide, quicker and more thorough calculations are typically performed using custom software packages, of which a wide variety is available commercially.[18]

8.3 Stripe Waveguides

The previous section presented an analysis of a slab waveguide structure, a simplified, one-dimensional structure that assumed infinite planes in two dimensions and reflections from boundaries in the third. Real waveguides, in contrast, are typically fabricated as stripe structures, where guiding takes place in two dimensions and propagation in the third; the two configurations are shown schematically in Fig. 8.10. Stripe waveguides, the optical "wire" of a photonic circuit, form the basis for most integrated optical functions.

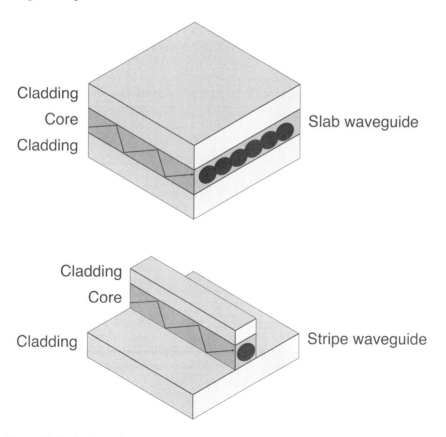

Figure 8.10 Schematic comparison of a slab waveguide with a stripe waveguide. The slab waveguide has many modes; the stripe waveguide may have only one.

8.3.1 Stripe Waveguide Structures

There are a number of approaches for confining light in two dimensions. We can, as with everything in this world, divide these concepts into two categories: (1) those that first fabricate a slab waveguide and, using one of a number of techniques, generate a stripe waveguide by removing or altering these layers; and (2) those that form a stripe waveguide structure directly. Which concept one uses is frequently a function of the materials used.

Waveguides in the first category start with a slab waveguide sequence, comprising at least a high-index waveguide core deposited on a substrate. Examples of the types of materials sequences used include the SiO_2/Si_3N_4 / air sequence or the $Al_{0.8}Ga_{0.2}As/Al_{0.3}Ga_{0.7}As/Al_{0.8}Ga_{0.2}As$ layers of Fig. 8.11. The former sequence represents a layer structure often used in Si-dielectric-based optical systems, and the latter sequence for III-V semiconductor integrated optics, including those for semiconductor lasers. In both cases, the waveguide layer is of course the one with the highest refractive index, namely, the Si_3N_4 layer or the $Al_{0.3}Ga_{0.7}As$ layer in the examples.

The waveguide stripe can be defined in a number of ways. The most straightforward approach is to remove the waveguide core in all regions outside the stripe, as shown in Fig. 8.12. This yields the so-called ridge waveguide structure. In the case where the upper cladding is air (such as the SiO_2/Si_3N_4/air example), the core layer is the topmost and can also be partially etched to form a waveguide; this configuration, termed a rib waveguide, is shown in Fig. 8.13. A strip-loaded waveguide, seen in Fig. 8.14, results if the upper cladding layer is partially or completely removed, while the core remains untouched. This approach has the advantage that etch processes used for removing the layers do not affect the core region, reducing the chances for damage-induced losses. A final scheme, shown in Fig. 8.15, uses implantation or diffusion to alter the core region outside the stripe; such an approach yields a planar surface, an advantage for some types of integrated optical structures.

The second stripe waveguide category, where the stripes are formed directly without the need to fabricate a slab waveguide first, represents the method predominantly used in glass-based integrated optics. With this approach, which yields the waveguide shown schematically in Fig. 8.16, the stripe is defined photolithographically and the waveguide is formed by a diffusion step. A typical example is the ion-exchange process used in the majority of glass-based waveguides. Using a stripe opened from a planar metal mask, a thermal diffusion process alters the refractive index of the

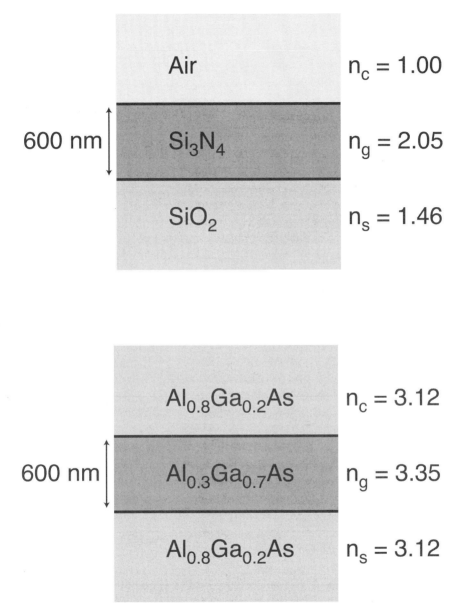

Figure 8.11 Typical layer sequences for slab waveguides before definition of the stripe structure; a silicon-based system and a III-V semiconductor example are shown.

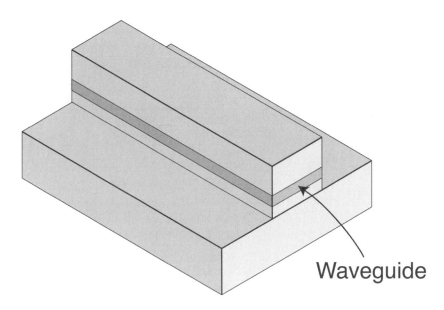

Figure 8.12 Schematic representation of a ridge waveguide, where the core layer is removed.

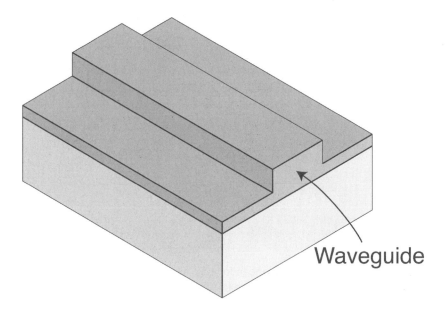

Figure 8.13 Schematic representation of a rib waveguide, where the core layer is partially removed.

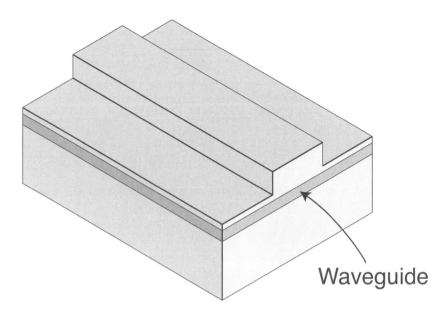

Figure 8.14 Schematic representation of a strip-loaded waveguide, where the top cladding layer is completely or partially removed.

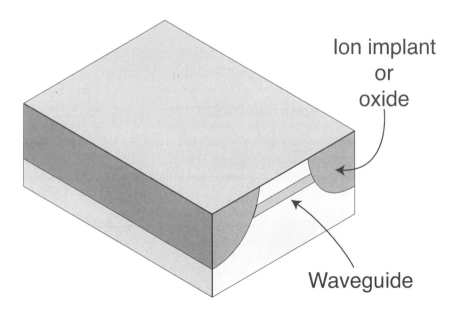

Figure 8.15 Schematic representation of an ion-implant defined waveguide, where the core layers outside the stripe are damaged or destroyed.

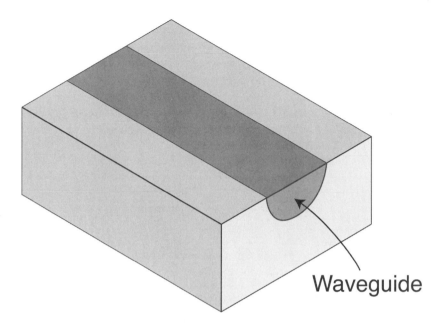

Figure 8.16 Schematic representation of a diffused waveguide, such as those made by ion-exchange, where a mask defines the waveguide diffusion.

exposed region such that the waveguide has a diffused index profile that changes more gradually than in the previous examples.

8.3.2 Stripe Waveguide Modeling

In contrast to the simple analytic models we used to describe the basic modal behavior of the one-dimensional slab waveguide, the two-dimensional stripe waveguide must in general be modeled numerically. A variety of approaches exist; some are available in commercial software packages.[18]

The mode structure of a stripe waveguide can be approximated for certain propagation conditions, namely for well-guided modes far from cutoff in waveguides with small Δn between core and cladding. The 2D effective index model,[19] for example, decomposes the waveguide into two orthogonal one-dimensional slab waveguides. Using the buried waveguide structure of Fig. 8.17, and assuming, for simplicity, only an x-directed electric field E_x, we first solve the waveguide one-dimensionally from bottom to top (in the y direction) using the appropriate charac-

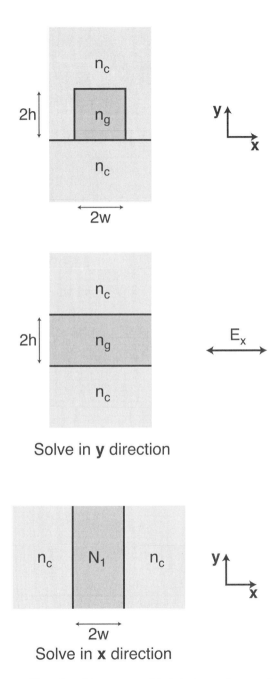

Figure 8.17 Decomposition of a stripe waveguide into two orthogonal slab waveguides for solution by the two-dimensional effective index model.

teristic equation from Section 8.2.2. In this case, we would use the asymmetric TE equation, Eq. (10), which can be rewritten as:

$$\tan\left(2k_{yg}h - m\pi\right) = \frac{k_{yc}}{k_{yg}} + \frac{k_{ys}}{k_{yg}}, \qquad (27)$$

where k_{yg}, k_{yc}, and k_{ys} are the propagation constants in the core, cap, and substrate, respectively. Solving this equation yields the one-dimensional effective index, $N = N_1$. We can subsequently solve the waveguide again, this time from left to right, in the x direction; instead of the refractive index n_g, we now use the calculated effective index N_1 in the core. Since we have rotated the guide 90 degrees, the symmetric TM equation is now applicable, written in the form:

$$\tan\left(k_{xeff}w - \frac{m\pi}{2}\right) = \left(\frac{N_1}{n_c}\right)^2 \left(\frac{k_{xL}}{k_{xeff}}\right), \qquad (28)$$

where k_{xeff} is the propagation constant in the core and k_{xL} is the propagation constant in the cladding, given by:

$$k_{xL} = \sqrt{\beta^2 - k_0^2 n_c^2}. \qquad (29)$$

This equation is solved for k_{xeff}, thereby yielding an expression for the two-dimensional effective index N_2:

$$N_2 = \sqrt{N_1^2 - \left(\frac{k_{xeff}}{k_0}\right)^2}. \qquad (30)$$

This derived value for the effective index is an approximation of the effective index of the two-dimensional waveguide. Figure 8.18 shows an example of the calculated two-dimensional effective index as a function of waveguide width for various core thicknesses. We see that $N_{eff(2D)}$ decreases for decreasing width, w as well as thickness, t. We stress again, however, that this approximation is only valid in a limited number of situations, for well-guided modes far from cutoff. The method is also limited, in practice, to well-defined refractive index sequences, and can only be applied to diffused structures with difficulty, if at all.

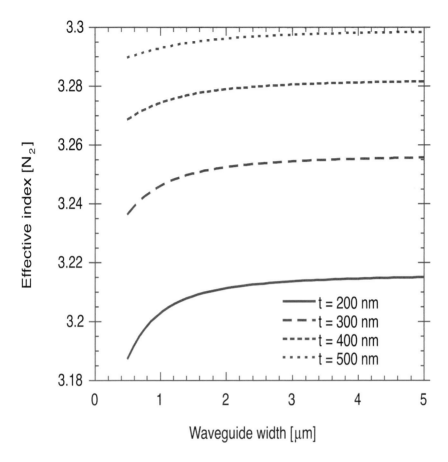

Figure 8.18 Two-dimensional effective index as calculated by the 2D effective-index model for a GaAs-based waveguide as a function of waveguide width w and core thickness t. The core has refractive index 3.35, the upper and lower claddings 3.12; the calculation was performed for $\lambda = 850$ nm.

8.4 Input/Output Coupling

With the exception of integrated optical circuits that include a monolithically integrated laser, most waveguide circuits used in a MEMS require light to be coupled from an external light source into the waveguide core. The waveguide begins and ends at an edge of the glass, semiconductor, or plastic substrate, and coupling techniques frequently use the resulting etched or cleaved facet as an access point for the core. The same techniques can be used for coupling light in as well as out.

The lateral dimensions of an optical waveguide, typically in the range 0.1 to 5 μm vertically and 2 to 10 μm laterally, result in considerable demands on positioning accuracy for coupling light into the core. Whereas numerous approaches are suitable for laboratory characterization, the incorporation of integrated optical chips into MEMS also requires coupling techniques that are robust as well as industrially viable.

8.4.1 End-Fire Coupling

Two coupling techniques of particular relevance are end-fire coupling and butt coupling. The former uses a lens structure to focus the optical field from a light source onto the facet of the waveguide chip, as shown in Fig. 8.19. The optical field is Gaussian,[20] with a radially symmetric intensity distribution of the form:

$$I = I_0 e^{-\frac{r^2}{\sigma^2}}, \qquad (31)$$

for $1/e$ width σ. The lens system should be configured to provide a Gaussian spot size on the waveguide facet of dimensions as closely matched to the core dimensions as possible; this mode-matching criterion assures optimal coupling into the waveguide.

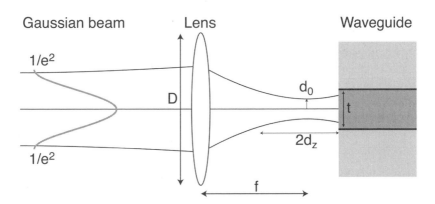

Figure 8.19 End-fire coupling an external optical field into a waveguide; the lens focuses the Gaussian beam onto the waveguide facet.

To avoid diffraction effects of the lens, its diameter, D, should be greater than 2σ, the $1/e$ width of the incoming beam. Given D and the focus, f, of the lens, the beam waist diameter d_0 and depth of focus d_z can be given as rules of thumb:[21]

$$d_0 = f \frac{4\lambda}{\pi}\left(\frac{f}{D}\right) \qquad (32)$$

and

$$d_z = \pm \frac{4\lambda}{\pi}\left(\frac{f}{D}\right)^2. \qquad (33)$$

The beam waist should be close to the waveguide dimensions, where the depth of focus provides an indication of the longitudinal positioning accuracy required to maintain a certain spot size focused onto the waveguide facet.

8.4.2 Butt Coupling

A simplified version of end-fire coupling eliminates the need for focusing optics by placing the optical source in very close proximity to the waveguide facet. This form of butt coupling is illustrated schematically in Fig. 8.20 and is frequently used to couple light from a fiber into a waveguide, from waveguide to waveguide, or from a semiconductor laser to a waveguide. The essential parameters leading to efficient butt coupling are smallest possible spacing, d, and best-possible mode-matching; coupling between a cleaved fiber and a Si dielectric waveguide, for example, will be limited in efficiency due to mode mismatch between the circular fiber core and the rectangular dielectric core.

Butt coupling is the technique of choice for permanent assembly of micro-optical and MEMS-based optical systems. Reduced system costs and reduced demands on assembly due to the absence of focusing optics simplify the total optical system. The various optical components are aligned for optimal butt-coupling efficiency and fixed using adhesives, welds, or mechanical alignment features. The losses due to Fresnel reflection at the facet can be reduced by antireflective surface coatings, thereby enhancing overall coupling efficiency.

A variety of further techniques for input and output coupling are available, although some require physical modification of the wave-

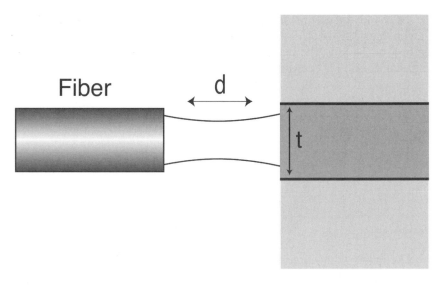

Figure 8.20 Butt coupling between an optical fiber and a dielectric waveguide. Coupling is enhanced for minimal d and optimized mode matching.

guide. By etching a periodic grating structure onto the surface of a waveguide, a grating coupler can be fabricated that allows access to the waveguide from the surface; no facet is thus required.[22] Light is coupled in at a particular angle from the surface normal (this time referring to the waveguide or substrate surface, not the facet); the angle depends on the grating period, light wavelength, and waveguide effective index. The wavelength dependence, while limiting for some applications, can be used to advantage in others. Related coupling techniques, such as prism coupling,[23] have a long history but are of limited utility in MEMS-type systems.

8.4.3 Numerical Aperture

Whereas most end-fire and butt-coupling approaches couple light under normal incidence, namely perpendicularly, into the waveguide, certain configurations may result in off-axis illumination of the waveguide facet. The maximum allowable coupling angle, θ_{max}, measured from the surface normal, is given by the numerical aperture (NA), where

$$NA = n_0 \sin(\theta_{max}) \qquad (34)$$

for refractive index of the outside material (usually air) n_0. For $n_0 = 1$, the maximum value of NA is 1, in which case $\theta_{max} = 90$ degrees, implying that all incidence angles are allowed.

In the case where a lightwave is coupled externally into a waveguide, as shown in Fig. 8.21, NA is defined as:

$$NA = \sqrt{n_g^2 - n_c^2},\qquad(35)$$

for waveguide core index n_g and cladding index n_c. For an asymmetric waveguide where $n_c \neq n_s$, the smaller of the two should be used. This last expression implies that the waveguide structure limits NA. As can be seen in Fig. 8.21, the coupling angle defines, through successive application of Snell's law, the reflection angle between waveguide core and cladding of the mode that begins to propagate in the waveguide.[24] Because this latter angle needs to be greater than the critical angle for that interface, θ_c, successful guiding of an optical field coupled into the facet will have a limiting maximum coupling angle. Values for NA vary strongly with material. A III-V semiconductor-based waveguide can have $NA = 1$, whereas an optical fiber typically has $NA \sim 0.25$, implying $\theta_{max} \sim 14$ degrees.

8.4.4 Tapers

Coupling efficiency is a critical function of mode matching between the optical source and the waveguide, as discussed above. The waveguide,

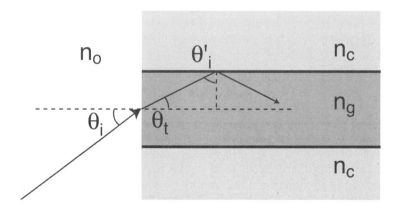

Figure 8.21 The numerical aperture limits the input angle, θ_i, for coupling into a waveguide.

due to its physical dimensions, often has a smaller mode size than the source. Matching of the mode profiles can be enhanced by the use of waveguide tapers, which allow an expansion of the waveguide mode size, either transversally (normal to the plane of the substrate) or laterally (parallel to the plane of the substrate), or both. The transverse mode size can be increased by gradually decreasing the waveguide thickness as it approaches the facet; as t decreases, confinement factor Γ decreases, and the transverse extent of the mode increases.[25] Such a taper is typically fabricated using advanced selective etch techniques.

In the lateral direction, the waveguide width can be gradually increased to allow the mode size to grow. A so-called adiabatic taper increases the width of a single-mode waveguide sufficiently slowly to ensure that only a single mode is maintained, despite the width of the guide, which would allow higher order modes to propagate.[26] The lateral taper can be fabricated more easily that the transverse one, as the former is defined photolithographically in the layout of the waveguide. Given the technology, however, a combination of the transverse taper (waveguide thickness decrease) and the lateral taper (waveguide width increase) can be used to generate a wide spectrum of mode sizes and shapes.

8.5 Waveguide Characterization

Once fabricated and combined with an appropriate excitation technique, a waveguide will typically require a modicum of characterization to ensure that it functions as desired in the design. Standard characterization procedures frequently include a determination of mode number and shape in addition to waveguide losses.

8.5.1 Modes

The modal behavior of a waveguide can be determined most easily by imaging the output at the facet. Using one of the approaches outlined in Section 8.4 for coupling light into the guide, a simple lens arrangement and imaging apparatus (typically a video or charge-coupled device (CCD) camera, depending on the wavelength regime) are used to image the optical intensity distribution at the output facet, yielding what is known as the near-field intensity distribution. A 20 to 40× microscope objective is usually sufficient for generating a reasonably sized image.

An example of the output typically derived from such a system is shown in Fig. 8.22. This series of images was taken at the facet of a series of GaAs/AlGaAs waveguides with identical layer structures but increas-

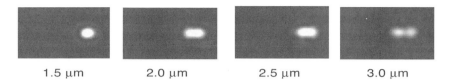

| 1.5 µm | 2.0 µm | 2.5 µm | 3.0 µm |

Figure 8.22 Near-field intensity distribution measured at the output of a series of GaAs/AlGaAs waveguides of varying widths. The transition from single-mode to multi-mode behavior is clearly shown.

ing widths. We can clearly see the transition from single-mode to multimode behavior: The 1.5-µm-wide waveguide is clearly single mode, yet the others show either the next higher order mode or a mixture of the two.

Related to the near-field distribution is the far-field distribution, which defines the optical intensity as a function of position at some distance from the facet. The distance from the facet at which the far-field becomes relevant is given roughly by:

$$L \geq \frac{\lambda^2}{d}, \tag{36}$$

for wavelength l and aperture (waveguide cross section, in this case) d. For typical waveguide structures and wavelengths in the visible or near-infrared range, L is on the order of several microns. For this reason, a movable photodetector placed several millimeters from the waveguide facet will most certainly measure the far-field distribution.

Many channel waveguide cross sections are asymmetric in shape. For example, a rectangular III-V-based waveguide can be 0.2×3 µm² and a dielectric waveguide 1×5 µm². The far-field distribution is essentially a diffraction effect from the aperture, so that a smaller aperture gives rise to a wider far-field pattern. As a result, the far-field pattern is likewise asymmetric, with a wider distribution in the direction of the smaller waveguide dimension, typically the transverse (normal to the substrate surface) direction. For the 0.2×3 µm² waveguide, as an example, the transverse far-field is considerably wider than the lateral far-field, as seen in Fig. 8.23; such a pattern is typical for semiconductor waveguides as well as lasers.

8.5.2 Losses

Optical loss is a further waveguide parameter that forms part of an essential characterization step. Every waveguide exhibits losses, which

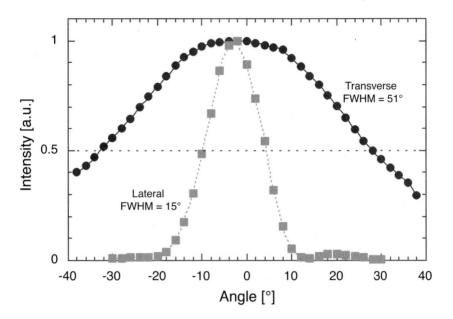

Figure 8.23 Typical far-field intensity distribution for a semiconductor waveguide; the transverse width is considerably greater than the lateral width.

can be due to a combination of factors including material absorption, scattering at sidewalls and interfaces, Rayleigh scattering and absorption due to doping, free carriers, or impurities. Depending on the material used for fabricating the waveguide, one or more of these effects can combine to give an overall optical loss for the waveguide, typically characterized by a loss parameter α with units [cm^{-1}]. α is also frequently quoted in units of [dB/cm], where the two are related as:

$$\alpha[dB/cm] = \frac{10}{\ln 10}\alpha[cm^{-1}]. \tag{37}$$

Given the loss parameter for a particular waveguide, the intensity I at a distance z from the input will be related to the input intensity, I_0, as:

$$I = I_0 e^{-\alpha z}. \tag{38}$$

This exponential relationship implies that the optical throughput of a waveguide will be a critical function of α, such that for an integrated optical system of any appreciable length, losses need to be well characterized and, if possible, reduced to a minimum.

The simplest, and intuitively most accessible, technique for measuring waveguide losses is the cutback method. This approach requires that light be coupled into a waveguide of length L_1 and the output intensity I_1 measured. The waveguide is then cut into two pieces of unequal length, L_2 and L_3; with the same light intensity coupled into the input (generally difficult to ensure), the output intensities, I_2 and I_3, of these shorter sections are measured. Using the length dependence of intensity as given in Eq. (38), we can then derive the loss parameter a from the ratio of the measured intensities, namely,

$$\alpha = \frac{1}{L_1 - L_2} \ln\left(\frac{I_2}{I_1}\right). \tag{39}$$

A larger number of different lengths and large length differences improve the accuracy of this approach.

A number of further approaches for measurement of waveguide loss have been developed. An approach that is conceptually similar to the cutback method uses a sliding prism coupler to couple light into the surface of a channel waveguide.[23] This approach functions for waveguides made from materials with a refractive index smaller than that of the prism (usually glass with $n \sim 1.5$). By moving the prism to various parts of the waveguide, with the output measured at a fixed position, various waveguide transmission lengths can be achieved, allowing calculation of the loss.

The light-scattering approach uses a fiber or small photodetector to measure the intensity of light scattered from the waveguide surface as a function of distance along the waveguide. This scattering takes place in waveguides with considerable surface or interface roughness and represents a significant loss factor. The scattered intensity, which is proportional to the light intensity in the waveguide, thus likewise decreases exponentially with distance. A plot of the scattered intensity as a function of position then yields an exponential curve with a characteristic a.

For semiconductor-based waveguides, which are defined by high-quality cleaved or etched facets that form a resonant cavity, the Fabry-Perot resonances can be used to determine loss.[27] In this more complex method, a tunable light source is coupled into the waveguide and the wavelength is swept across a small range. The waveguide cavity, formed by the two facets, is swept successively through resonance and anti-resonance, where the ratio of the two resulting intensities is given by C_R. Knowing the waveguide length L and the reflectance R of the facet, the waveguide loss in [cm^{-1}] can be determined from

$$\alpha = \frac{1}{L}\ln\left[\frac{1}{R}\frac{\sqrt{C_R}-1}{\sqrt{C_R}+1}\right]. \tag{40}$$

Although this approach becomes more accurate as waveguide loss decreases, it requires high-quality optical facets to achieve the necessary resonant cavity; it has been used successfully only with semiconductor-based waveguides.

8.6 Integrated Optical Devices

A wide variety of functional elements can be fabricated using waveguides as a basis. The passive devices include couplers and interferometers; these perform an optical operation with no external electrical input. In contrast, devices such as waveguide modulators, detectors, and lasers represent active elements, which function by using or generating electrical signals.

8.6.1 Couplers

In the design of a waveguide-based optical circuit, it is frequently necessary to split the optical signal into two or more parts, or to combine the signals from two or more waveguides and direct these into a single waveguide. A coupler performs this function.

Two families of waveguide couplers can be defined: the Y-coupler widens and divides the waveguide into two "arms," as shown in Fig. 8.24. In the transition region, two lateral modes are supported; these subsequently separate and divide into the two output arms. The Y-coupler can also be used in reverse, combining the inputs from the two arms into the single branch. The relative phases of the signals from the two arms define whether constructive or destructive interference takes place in the coupler. For two signals p out of phase, complete destructive interference implies that the output from the single output arm is zero.

The proximity coupler is based on the coupling between optical fields of two waveguides that are in close proximity but do not overlap. As shown in the schematic of Fig. 8.25, two waveguides sufficiently close together have an overlap in their evanescent fields. The penetration depth of the evanescent field, defined in Section 8.2.4, is a function of the propagation constants in the waveguide; the amount of overlap between the two waveguides fields is then a function of this and the waveguide spacing.

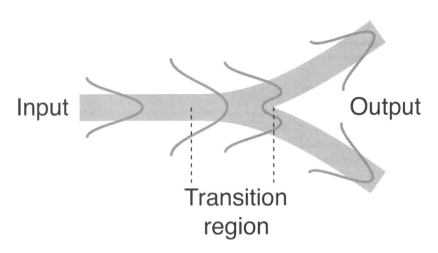

Figure 8.24 Schematic of a Y-coupler used to split the optical field from a single waveguide into two arms.

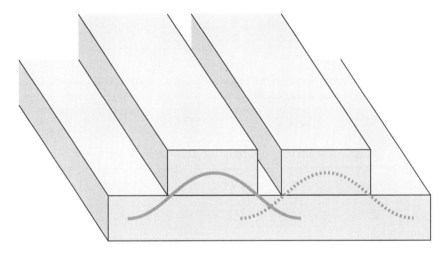

Figure 8.25 A proximity coupler using two closely spaced waveguides to couple light from one to the other.

As Fig. 8.26 shows, the coupling results from an exchange of energy between the waveguides as the modes propagate along the coupler. The spacing, w, is typically small (1 µm), and the length is typically 100 to 400 µm. The coupling length, L_c, is defined as the length of overlap for which all energy is coupled from one waveguide to the other;[28] this is the case shown in Fig. 8.26. If the coupler had a length $L_c/2$, only

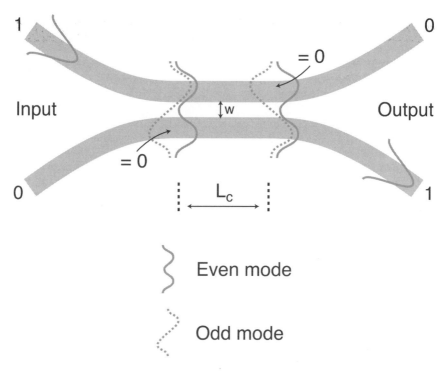

Figure 8.26 Coupling between the two waveguides of a proximity coupler. Coupling is a function of spacing, w, and coupling length, L_c. For the coupler as illustrated, $L = L_c$ so that all light is coupled from one waveguide to the other.

half the energy would be coupled; for this case, the power would be equally divided between the two output arms. For length $2L_c$, all the power would be coupled back into the original waveguide. Thus a well-designed (and fastidiously fabricated) proximity coupler can achieve an arbitrary power splitting ratio, defined by the length of the overlap region of the coupler.

A relatively new family of couplers is represented by the multimode interference coupler (MMI). The operation of this device, shown schematically in Fig. 8.27, is based on "self-imaging" of the interference pattern in a multimode waveguide due to the beating between the fundamental and higher order waveguide modes.[29] The intensity field is thus repeated at regular intervals. By choosing the correct multimode section length and properly positioning the output waveguides, the lateral intensity distribution can be chosen to achieve arbitrary coupling factors.

The MMI can be combined with modulator sections to variably couple or switch light between various output waveguides; the MMI can be

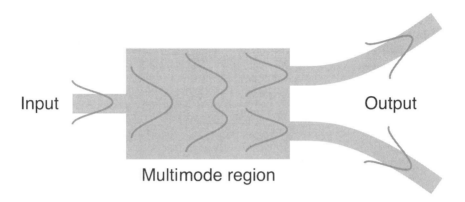

Figure 8.27 The multimode interference coupler uses a wide multimode region to couple light between single-mode waveguides. This example shows a 1 x 2 implementation.

implemented in $1 \times N$ or $2 \times N$ configurations, thus making it useful for a variety of optical switching applications. Since coupling can be achieved using shorter overall dimensions and with lower demands on fabrication tolerances than Y couplers or proximity couplers, the MMI is finding increasing use in complex optical microsystems.

8.6.2 Interferometers

One of the primary areas for which optics in MEMS is of interest is in the performance of optical measurement or sensing. In many of these cases, optical interferometers are a useful tool; these can easily be implemented in waveguide technology.

The Fabry-Perot interferometer can be realized using a straight waveguide delineated by two optical-quality facets. The resonant cavity can be used as a wavelength filter or for frequency stabilization of an external signal. In addition, exposing the waveguide core to the ambient surroundings can result in an environmentally dependent resonance and thus a high-resolution sensor.

The quality of the optical filter function achievable with the Fabry-Perot interferometer is a function of cavity finesse and is thus ultimately related to mirror quality and cavity length. The full-width half-maximum (FWHM) of the transmission peak can be written as

$$FWHM = \frac{c(1-R)}{2\pi Ln\sqrt{R}}, \qquad (41)$$

where R represents the mirror reflectance, L the cavity length, c the speed of light, and n the refractive index of the waveguide. Transmission recurs periodically in wavelength, and the free spectral range of the interferometer, Δv, defined as

$$\Delta v = \frac{c}{2Ln}, \qquad (42)$$

gives the peak spacing in frequency v.

For measurement of displacement, distance, and vibration, the Michelson interferometer is used both classically and in its integrated optical implementation;[30] the latter is shown in Fig. 8.28. Light coupled into the interferometer in the waveguide at the left is split in two using the proximity coupler; the length of the latter is chosen to be $L = L_c/2$, so that half the light is coupled into each output arm. The proximity coupler performs the function of the half-silvered mirror of the classical Michelson interferometer.

At one of the two output arms, a fixed mirror forms a reference reflection; at the other, the light is reflected from an external measurement object. The two resulting reflections, now moving to the left, recombine

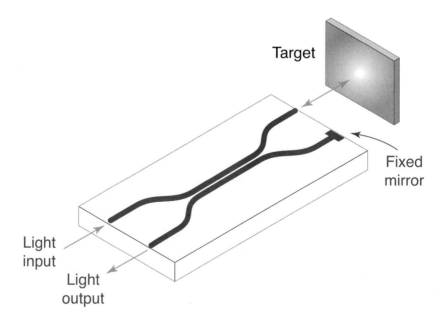

Figure 8.28 The waveguide implementation of a Michelson interferometer.

at the proximity coupler, and the resultant interference intensity is coupled to the output. The optical intensity at this output is then related to the movement of the measurement object, d, by

$$I = \frac{I_0}{2}\left[1 + V\cos\left(2\pi\frac{d}{\lambda}\right)\right], \qquad (43)$$

where I_0 is the initial intensity, V the contrast (≤ 1), and λ the operating wavelength. The resolution of the interferometer is thus on the order of a fraction of the wavelength used.

A third useful integrated optical interferometer is the Mach-Zehnder interferometer. Of particular utility in MEMS-based systems that use sensitive measurement of refractive index as a basis for an optical sensor, this interferometer configuration splits the optical field into two arms with separate propagation paths before recombining the optical signals, as shown in Fig. 8.29. The Y-couplers at the input and output replace the bulk beam splitters of the classical optical version.

This configuration is frequently used in chemical or physical sensing.[31] For the former, one interferometer arm can have the cladding removed from the waveguide, exposing the waveguide core to an external

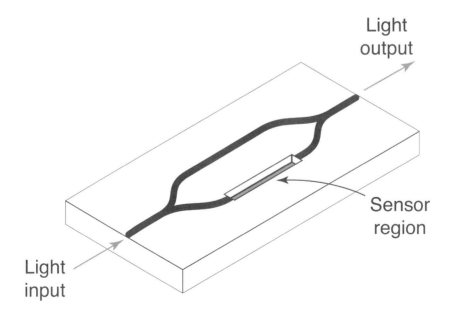

Figure 8.29 The waveguide implementation of a Mach-Zehnder interferometer.

medium. As the refractive index of this external medium changes, the effective waveguide index in the exposed arm also changes, giving rise to phase shift, $\Delta\phi$, with respect to the unchanged arm of the interferometer. This phase shift is again measurable through the resultant interference pattern following signal recombination in the second Y-coupler, where the intensity distribution is given by an expression similar to that for the Michelson interferometer, namely,

$$I = \frac{I_0}{2}\left[1 + V\cos(\Delta\phi)\right]. \tag{44}$$

If the phase shift is due to a change in the waveguide effective index, ΔN, over a section of length L_{ex} exposed to the external environment, the interference pattern can be expressed as

$$I = \frac{I_0}{2}\left[1 + V\cos\left(2\pi\frac{L_{ex}}{\lambda}\Delta N\right)\right]. \tag{45}$$

Alternatively, one waveguide arm can be directed over a physically deformable structure, such as a diaphragm or cantilever. The strain-induced change in refractive index will allow measurement of, for example, pressure or vibration in a MEMS.[32]

8.6.3 Active Optical Devices

The waveguide structures we have described also form the basis for active integrated optical devices; an active device is defined as one that has an electrical connection. This family of devices includes semiconductor lasers, LEDs, phase and intensity modulators, waveguide switches, and photodetectors. Because there is extensive literature available concerning these elements,[33] we provide only a very brief overview here.

In most cases, the active optical devices listed above have a waveguide structure similar to that of the passive devices. The waveguide layers are usually doped, forming a pn junction, and metallization for electrical contact is supplied on the top and bottom surfaces of the substrate. Depending on details of the structure and how it is biased and operated, the resultant device can emit, detect, modulate, or switch light. Since the physical dimensions and structure are usually similar to passive waveguides, mode sizes are likewise similar, facilitating coupling between device types. A MOEM system can thus be assembled from integrated optical components using various optical assembly and hybridization techniques.

A selection of active integrated optical devices includes the following:

- *Semiconductor lasers*: A waveguide-based resonant cavity, formed using cleaved mirrors or distributed Bragg mirrors, forms the basis of most semiconductor lasers. The materials used must be efficient emitters of light, such that compound semiconductors (III-Vs or II-VIs) are typically used.[34]
- *Modulators*: By applying a high electric field to the waveguide structure, the phase or intensity (or, typically, both) of the optical field in the waveguide can be modulated. Modulation is based on electro-optical effects in the material; the III-Vs or $LiNbO_3$ are frequently used. Thermal or carrier-concentration effects in Si-based waveguides can also be used.[35]
- *Switches*: Through combinations of waveguide couplers and modulators, the optical field can be switched between different waveguides. Most semiconductor materials would be suitable for a switch; Si seems to be the most popular.[36]
- *Photodetection*: Using materials that absorb at the wavelength of interest, a detector can be fabricated using a pn junction and electrical contacts. An entire length of waveguide can be used for absorption, such that highly efficient detectors can be designed. The materials used are again semiconductors, where Si is preferred for $\lambda \leq 1.1$ µm and the III-Vs are preferred for longer wavelengths.[37]

8.7 Materials

A variety of materials have been and continue to be used for fabrication of integrated optical devices and circuits. The choice of material is a function of required characteristics as well as external parameters such as MEMS microsystem compatibility and price. The required optical performance characteristics can include one or more of the following:

- Low optical absorption (low-loss waveguides)
- High optical absorption (detectors)
- Small or large mode size
- Light emission
- Electro-optic or thermo-optic effects
- Opto-mechanical effects.

As seen in Section 8.3.3, different materials are more or less able to fulfill these requirements. This section outlines some of the more important materials and provides a tabular listing of some of the relevant material parameters.

8.7.1 Silicon

Silicon, the classic MEMS material, is also of great utility for passive integrated optical and micro-optical components. Passive waveguides, couplers, splitters, and many types of free-space optical structures for MEMS are advantageously made using this material, not least due to the compatibility with established MEMS technology.[38] Si itself is most often used as a substrate, whereas most of the integrated optical components are made from SiO_2, Si_3N_4, or SiON. Data in Table 8.1 are partially from Zappe.[39]

8.7.2 GaAs

The compound semiconductor materials, of which the III-Vs are the most popular, and GaAs and InP are used as representative examples here, are used for all manner of active, integrated micro-optical structures. Efficient light emission, strong electro-optical effects, and the ability to make high-quality heterostructure waveguides have made this the material system of choice for laser diodes, modulators, and photonic integrated circuits. Whereas III-Vs have been micromachined,[40] most Si-based optical MEMS using III-V components are assembled using hybridization techniques. Data in Table 8.2 are partially from Zappe.[39]

Table 8.1 Selected Properties of Si, SiO_2, and Si_3N_4 Relevant to Integrated Optics

Parameter	Symbol	Units	Material Si	SiO_2	Si_3N_4
Refractive index	n		3.44	1.46	2.05
Temperature dependence of n	$\Delta n/\Delta T$	K^{-1}	2×10^{-4}	1×10^{-5}	
Energy gap	E_g	eV	1.12	~9	~5
Gap wavelength (= hc/qE_g)	λ_g	µm	1.11	~0.138	~0.248
Temperature dependence of E_g	$\Delta E_g/\Delta T$	eV/K	-2.8×10^{-4}		
Useful transmission range		µm	1.2 - 15	0.12 - 4.5	
Thermal conductivity	σ_{th}	W/cmK	1.45	0.014	0.19
Thermal expansion	$\Delta L/L\Delta T$	K^{-1}	2.6×10^{-6}	5×10^{-7}	
Electro-optics			weak	no	no
Light emission			no	no	no

Table 8.2 Selected Properties of GaAs and InP Relevant to Integrated Optics

Parameter	Symbol	Units	Material	
			GaAs	InP
Refractive index	n		3.36	3.45
Temperature dependence of n	$\Delta n/\Delta T$	K^{-1}	4.5×10^{-5}	2.3×10^{-4}
Energy gap	E_g	eV	1.42	1.35
Gap wavelength (= hc/qE_g)	λ_g	μm	0.870	0.919
Temperature dependence of E_g	$\Delta E_g/\Delta T$	eV/K	-4.2×10^{-4}	-2.9×10^{-4}
Useful transmission range		μm	1–15	1.1–15
Thermal conductivity	σ_{th}	W/cmK	0.46	0.68
Thermal expansion	$\Delta L/L\Delta T$	K^{-1}	5.7×10^{-6}	4.6×10^{-6}
Electro–optics			yes	yes
Light emission			yes	yes

8.7.3 Glass

Glass-based integrated optics are often limited to passive devices; well-established technology and low-cost fabrication techniques make glass integrated optical structures attractive. Low-loss waveguides, including couplers and interferometers, are frequently made using glass substrates. Although the sizes of optical systems tend to be larger than those made using semiconductor techniques, optical MEMS can be assembled hybridly using glass optical components. Data in Table 8.3 are from Smith.[41]

Table 8.3 Selected Properties of Three Glasses Used in Integrated Optics

Parameter	Symbol	Units	Material		
			BK 7	BaK 2	LaK 10
Refractive index (at $\lambda = 0.588$ μm)	n		1.5168	1.53996	1.72000
Thermal expansion	$\Delta L/L\Delta T$	K^{-1}	7.1×10^{-6}	8.0×10^{-6}	5.7×10^{-6}
Density	D	g/cm^3	2.51	2.86	3.81
Glass temperature	T_g	°C	559	562	620
Transmission through 25 mm (at $\lambda = 0.4$ μm)			0.991	0.974	0.91

8.7.4 Plastics

A wide variety of polymer and plastic materials have been developed for integrated optical applications. These materials are generally used for passive optical applications, although recent developments have led to polymers that exhibit light emission, detection, and electro-optical properties.[42] One of the advantages of plastics is the ability to structure them using embossing or injection molding techniques, such that they are potentially a source of low-cost components. Examples of the more popular materials include polycarbonate, polyimide, polymethylmethacrylate, and polystyrene. Data in Table 8.4 are from Gale.[43]

8.8 Applications

Integrated optical components and subsystems can be used for a variety of applications, either as standalone devices or as parts of a more comprehensive MEMS. Many of these application areas are developing rapidly. Optical communications applications currently dominate the market for integrated optical components, and research activities in sensor applications are very strong.

Table 8.5 summarizes a few of the more relevant fields of application. The companies listed under "Examples of Industry" represent a selection of firms that produce the integrated optical part listed.

8.8.1 Application Example: Monolithic Displacement Sensors

Contactless optical displacement is an industrially relevant application area for integrated-optics-based MEMS structures. As an example of what

Table 8.4 Selected Properties of Three Polymers Used in Integrated Optics

			Material		
Parameter	Symbol	Units	Polycarbonate	Polyvinyl-chloride	Polymethyl-methacrylate
Refractive index (at λ = 0.633 µm)	n		1.58	1.54	1.49
Glass temperature	T_g	°C	145	75–105	94–108
Useful transmission range		µm	0.38–1.6	0.4–2.2	0.4–1.1

Table 8.5 Some Relevant Fields of Application

Field	Application	Company	References
Communications	Waveguide switches for optical telecommunications networks	IMM (D) NTT (Japan) Nanovation (USA)	a b c
Communications	Fiber-pigtailed laser transceivers	NTT (Japan) Bookham (UK)	d e
Communications	Arrayed waveguide grating demultiplexers for wavelength division multiplexing	NTT (Japan) LETI (F) Bookham (UK)	f g e
Data storage	Monolithic CD reader head	–	h
Sensors	Michelson interferometers for displacement measurement	CSO (F)	i [30]
Sensors	Monolithic refractometers and bio/chemo-sensors	–	[31]
Sensors	Optical pressure sensors for automotive applications	Bookham (UK)	e
Sensors	Cantilever-based vibration sensors	LETI (F)	j

a www.imm-mainz.de/starlink/f_sl.html
b www.nel-world.com/products/photonics/thermo.htm
c www.nanovation.com
d www.nel-world.com/products/photonics/otm.htm
e www.bookham.com/productarena/productarena.html
f www.nel-world.com/products/photonics/awg.htm
g www.dta.cea.fr/leti/uk/pages/Optronique/m_opto42.htm
h S. Ura, T. Endoh, T. Suhara, and N. Nishihara, "Integrated optic head for sensing a two-dimensional displacement of a grating scale," *Appl. Opt.*, 35(31):6261-6266 (1996)
i www.cso.fr
j www.dta.cea.fr/leti/uk/pages/Optronique/m_opto45.htm

is possible with the technologies and devices discussed in previous sections, we will take a brief look at monolithic displacement measurement chips developed at two laboratories, which are described in detail by Zappe.[39]

The two integrated optical displacement sensors we consider are both interferometric, based on integrated optical Michelson interferometers. The first, developed at the University of Osaka, used a GaAs substrate, which allowed integration of a distributed feedback (DFB) laser as the light source. The entire chip had dimensions of 6.5 × 2.0 mm² and included the laser as well as coupling optics and photodetectors. The use of surface grating couplers allowed direct coupling of the measurement beam from the chip onto the measurement surface and coupling the reflected beam back into the interferometer. Operation of the sensor was at a wavelength of λ = 855 nm, and measurement at a distance of 20 mm led to a signal current amplitude of about 1 nA on a 10 nA offset, where the sinusoidal signal variation with translation had a period of $\lambda/2 \approx 427$ nm. Incorporation of the collimation function on-chip, in the final grating coupler, implied that this monolithic sensor did not require external optics for operation.

An alternative monolithic displacement sensor, based on a stripe-waveguide technology, was developed at the Centre Suisse d'Electronique et de Microtechnique in Zurich (CH). The chip consisted of a dual Michelson interferometer, with two reference arms and photodetectors, a distributed Bragg reflector (DBR) laser as mutual light source, and two phase shifters, one for each reference arm. The 2.6×0.3 mm^2 optical chip used selective quantum well intermixing for waveguide transparency, and phase modulation was through the quantum-confined Stark effect. A gradient-index (GRIN) lens was used at the measurement waveguide output for beam collimation. The optical measurement beam, at $\lambda = 820$ nm, was directed onto an external mirror, yielding a signal amplitude of about 1 nA on a 45 nA dc offset at a maximum measurement distance of 0.25 m. The use of two reference beams, one phase-shifted by $\pi/2$ with respect to the other, allowed unambiguous determination of movement amplitude and direction at a resolution of roughly 20 nm.

These two integrated optical sensor implementations serve as ideal examples of what can be accomplished using the technologies and devices described in this chapter. This type of highly integrated component will provide compact, low-power optical functionality for a variety of microsystem applications.

References

1. See, for example, *IEEE J. Selected Topics Quantum Electron.* (special issue on MOEMS), vol. 5(1), Jan./Feb. 1999
2. M. Tabib-Azar and G. Beheim, "Modern trends in microstructures and integrated optics for communication, sensing and actuation," *Opt. Eng.*, 36(5):1307–1318 (1997)
3. H.P. Herzig, ed., *Micro-optics*, Taylor & Francis, London (1997)
4. D.T. Neilson and R. Ryf, "Scalable micro-mechanical optical crossconnects," *Proceedings of the IEEE Laser and Electro-optical Society Meeting 2000*, Rio Grande, Puerto Rico (2000), pp. 48–49
5. L.J. Hornbeck, "Digital Light Processing™ and MEMS: Reflecting the digital display needs of the networked society," *Micro-optical technologies for measurement, sensors and microsystems*, SPIE Proc., 2783:2–13 (1996)
6. D. Leclrec, P. Brosson, F. Pommereau, R. Ngo, P. Doussière, F. Mallécot, P. Gavignet, I. Wamsler, G. Laube, W. Hunziker, W. Vogt, and H. Melchior, "High performance semiconductor optical amplifier array for self-aligned packaging using Si V-groove flip-chip technique," *IEEE Photonics Technol. Lett.*, 7(5):476–478 (1995)

7. O. Solgaard, M. Daneman, N.C. Tien, A. Friedberger, R.S. Muller, and K.Y. Lau, "Optoelectronic packaging using silicon surface-micromachined alignment mirrors," *IEEE Photonics Technol. Lett.*, 7(1):41–43 (1995)
8. M. Wu, "Micromachining for optical and optoelectronic systems," *Proc. IEEE*, 85(11):1833–1856 (1997)
9. H. Zappe, *Introduction to Semiconductor Integrated Optics*, Artech House, Boston (1995)
10. H. Nishihara, M. Haruna, and T. Suhara, *Optical Integrated Circuits*, McGraw Hill, New York (1989)
11. T. Hashimoto, Y. Nakasuga, Y. Yamada, H. Terui, M. Yanagisawa, Y. Akahori, Y. Tohmori, K. Kato, and Y. Suzuki, "Multichip optical hybrid integration technique with planar lightwave circuit platform," *IEEE J. Lightwave Technol.*, 16(7):1249–1258 (1998)
12. R.G. Hunsperger, *Integrated Optics: Theory and Technology*, 3rd ed., Springer, Berlin (1991)
13. J. Hecht, *Understanding Fiber Optics*, Prentice Hall, Upper Saddle River, NJ (1999)
14. J. Hecht, *The Laser Guidebook*, 2nd ed., McGraw-Hill, New York (1992)
15. B.A.E. Saleh and M.C. Teich, *Fundamentals of Photonics*, Wiley, New York (1991)
16. M. Born and E. Wolf, *Principles of Optics*, 6th ed., Pergamon, Oxford, Pergamon (1980), sec. 1.5
17. H. Zappe, *Introduction to Semiconductor Integrated Optics*, Artech House, Boston (1995), sec. 7.3
18. Software vendors include Apollo Photonics (USA), BBV Software (NL), Photon Design (UK), and RSoft Inc. (USA).
19. H. Kogelnik, "Theory of optical waveguides," in *Guided-Wave Optoelectronics* (T. Tamir, ed.), Springer, Berlin (1988), sec. 2.5.7
20. S.G. Lipson, H. Lipson, and D.S. Tannhauser, *Optical Physics*, Cambridge University Press, Cambridge (1995)
21. B.A.E. Saleh and M.C. Teich, *Fundamentals of Photonics*, Wiley, New York (1991), sec. 3.2
22. H. Nishihara, T. Suhara, and S. Ura, "Integrated-optic grating couplers," *Proceedings of the European Conference on Integrated Optics (ECIO)*, 1993, pp. 18–22
23. R.G. Hunsperger, *Integrated Optics: Theory and Technology*, 3rd ed., Springer, Berlin (1991), p. 97
24. E. Hecht, *Optics*, Addison-Wesley, Reading, MA (1987), sec. 5.6
25. T.L. Koch, U. Koren, G. Eisenstein, M.G. Young, M. Oron, C.R. Giles, and B.I. Miller, "Tapered waveguide InGaAs/InGaAsP multiple-quantum-well lasers," *IEEE Photonics Technol. Lett.*, 2(2):88–90 (1990)
26. R. Zengerle, H. Brückner, H. Olzhausen, and A. Kohl, "Low-loss fibre-chip coupling by buried laterally tapered InP/InGaAsP waveguide structure," *Electron. Lett.*, 28(7):631–632 (1992)
27. R.G. Walker, "Simple and accurate loss measurement technique for semiconductor waveguides," *Electron. Lett.*, 21:581–583 (1985); erratum, *Electron. Lett.*, 21:714 (1985)

28. A. Yariv, *Optical Electronics*, 4th ed., Saunders, Philadelpia (1991), chap. 13.8
29. L.B. Soldano and E.C.M. Pennings, "Optical multi-mode interference devices based on self-imaging: principles and applications," *IEEE J. Lightwave Technol.*, 13(4):615–627 (1995)
30. D. Hofstetter, H. Zappe, and R. Dändliker, "Optical displacement measurement with GaAs/AlGaAs-based monolithically integrated Michelson interferometers," *IEEE J. Lightwave Technol.*, 15(4:663–670 (1997)
31. B. Maisenhölder, H. Zappe, M. Moser, P. Riel, R.E. Kunz, and J. Edlinger, "Monolithically integrated optical interferometer for refractometry," *Electron. Lett.*, 33(11):986–988 (1997)
32. H. Porte, V. Gorel, S. Kiryenko, J.-P. Goedgebuhr, W. Daniau, and P. Blind, "Imbalanced Mach-Zehnder interferometer integrated in micromachined silicon substrate for pressure sensor," *IEEE J. Lightwave Technol.*, 17(2:229–233 (1999)
33. R.W. Waynant and M.N. Ediger, *Electro-optics Handbook*, 2nd ed., McGraw-Hill, New York (2000)
34. H.C. Casey and M.B. Panish, *Heterostructure Lasers*, parts A & B, Academic Press, New York (1978)
35. M. Fukuda, *Optical Semiconductor Devices*, Wiley, New York (1999)
36. J. Hecht, "All-optical networks need optical switches," *Laser Focus World*, May 2000, pp. 189–196
37. S. Donati, *Photodetectors: Devices, Circuits and Applications*, Prentice Hall, Upper Saddle River, NJ (1999)
38. A. Himeno, K. Kato, and T. Miya, "Silica-based planar lightwave circuits," *IEEE J. Special Top. Quantum Electron.*, 4(6):913–924 (1998)
39. H. Zappe, "Semiconductor optical sensors," *Sensors Update*, vol. 5 (H. Baltes, W. Göpel, and J. Hesse, eds., Wiley-VCH, Weinheim (1999), chap. 1
40. Y. Uenishi, H. Tanaka, and H. Ukita, "Characterization of AlGaAs microstructure fabricated by AlGaAs/GaAs micromachining," *IEEE Trans. Electron. Devices*, 41(10):1778–1783 (1994)
41. W.J. Smith, *Modern Optical Engineering*, McGraw-Hill, Boston (1990)
42. S. Bauer, "Poled polymers for sensors and photonic applications," *J. Appl. Phys.*, 80(10):5531–1558 (1996)
43. M.T. Gale, "Replication," in *Micro-optics* (H.P. Herzig, ed.), Taylor & Francis, London (1997), chap. 6

9 Microsensors for Magnetic Fields

Chavdar Roumenin

Bulgarian Academy of Sciences, Sofia, Bulgaria

9.1 Introduction

The devices and microsystems for measuring magnetic fields have the unique ability to reveal realities that cannot be perceived by the human senses. These transducers, which measure magnetic fields in a range not less than 15 to 16 orders of magnitude and are universal in their applications, are in continuous development. This chapter discusses recent progress in the most frequently used magnetic-field microsensors and MEMS, which are perfectly compatible with microelectronic technologies, most of which are silicon technologies. Up-to-date results are analyzed and abundant information is presented about the following topics: physical mechanisms of the origin of magnetosensitivity, device designs, sensor characteristics and the methods for their determination, biasing and interface circuits, the means for performance improvement and overcoming the basic transducer limitations, the most current and prospective applications, and development trends. Ample references to relevant literature are included for all modifications of Hall effect devices (orthogonal and parallel field); micromagnetodiodes; magnetoresistors (including feromagnetic versions such as giant magnetoresistance elements); microsensors based on $A^{III}B^{V}$ semiconductors; MOSFET, bipolar, CMOS, unijunction, and split-drain magnetotransistors and related devices; carrier-domain magnetometers; functional multisensors for the magnetic field, temperature, and light; 2-D and 3-D vector microsystems for the magnetic field; and magnetogradiometers and digital stochastic magnetotransducers.

9.2 Magnetic Fields for Different Applications

Contemporary microsystems for magnetic fields (hybrid or monolithic) must integrate at least two functions. One of them must be the sensing by an input transducer or sensor of the strength and the direction of this physical measurand; the other can be signal processing (when necessary, including a processor and the corresponding software) and/or an actuator. Elements that locally enhance, compensate, or change the direction of the external magnetic field such as ferrite flux concentrators or coils, can also be installed in the microsystem.[1-5] By using appropriate packages with small dimensions, these MEMS offer a variety of contactless sensing.

9.2.1 Methods for Sensing and Applications of Magnetic Fields

Two categories of applications of magnetic sensors and microsystems can be distinguished: direct and indirect.[1,4] Direct registrations are those in which only information about the magnetic field (strength and direction) itself is required. Indirect sensor systems use the field as an intermediary carrier to measure a nonmagnetic quantity (tandem transducers). Examples of direct applications include the following:

- Readout of information stored on disk, tape, and bubble memory
- Recognition of magnetic patterns on banknotes and credit cards
- Magnetometry: control of a magnetic apparatus such as classical and superconducting electromagnets; instrumentation for particle accelerators as well as determination of the full magnetic field vector, its direction, and its gradient by detecting two or all three vector components
- Magnetic levitation control
- Earth magnetic-field measurement and the electronic compass
- Geomagnetic remote sensing for geological and volcanic surveying
- Attitude control for satellites
- Positioning of aircraft, ships, missiles, projectiles, or submarines by the perturbations they cause in the geomagnetic field, and for the development of global navigation systems
- Biomagnetometry: obtaining diagnostic data by cardiomagnetism, myomagnetism, and neuromagnetism to map the functions of the heart, muscles, nerves, and brains of humans and animals

Indirect applications are much more common. Examples include the following:[1-14]

- Distance (linear and angular), velocity, speed, and vibration measurements

- Position detection
- Rotation and direction of rotation, e.g., for tachometry
- Collectorless DC motor control
- Keyboards and proximity switches
- Microphones
- Angular displacement detection and angle decoders and synchro resolvers
- Linear and rotary potentiometers, and crankshaft position transducers in automobile ignition control
- Automotive anti-skid breaking systems
- Nondestructive magnetic methods, including characterization of materials and metal detection
- Electrical current and power measurements (watt-hour meters) that do not interrupt the current-carrying conductor
- Analog multiplication
- Galvanic isolation
- Traffic detection when a ferromagnetic body is passing
- Measurement of mechanical and chemical parameters, pressure, mass filters, and so on, using suitable magnetomodulating systems that contain permanent magnets

These incomplete lists of magnetic sensor and microsystem applications should suffice to show how universal they are. From the viewpoint of the achieved accuracy, these are some of the most precise instruments available in control-measurement technology.

The range of operation of various magnetic sensors and microsystems is systematized in Fig. 9.1. The applications selected above clearly show the wide variety of fields met by magnetic (micro) sensors—from the

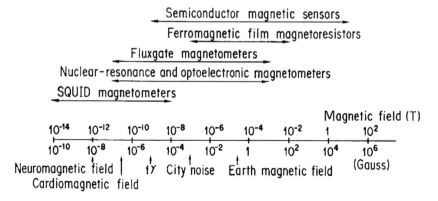

Figure 9.1 Field range of different types of magnetic transducers and microsystems.

biosignals, which are lower than 10 fT; passing through the variations of the geomagnetic field, which is about 0.05 mT; through the most important large-scale applications, using permanent magnets with induction fields at about 50 mT and with recording of magnetic fields of the range from 0.01 to 10 mT; and reaching to flux densities of several teslas in the apparatus for nuclear physics and the colossal inductions of millions of teslas in the new stars and black holes of the Universe. Therefore there is a dynamic range of fields not less than 15 to 16 orders of magnitude! As seen from Fig. 9.1, at least five principally different types of devices are capable of measuring this unique range of magnetic fields: superconductor magnetic sensors such as superconducting quantum interference devices (SQUIDs), nuclear-resonance and optoelectronic magnetic transducers, flux gate magnetometers, ferromagnetic thin-film magnetoresistors, and semiconductor magnetic devices. That is why this broad range cannot be covered by only one type of magnetic sensor microsystem.[1-14]

9.2.2 (Micro) Sensors for a Magnetic Field

This chapter discusses the most frequently used magnetic-field microtransducers, which according to Fig. 9.1 cover the widest interval of fields of 0.01 mT to 10^2 T. The concrete selection is realized according to my viewpoint, which conforms with my personal forecast concerning the future of this class of sensors and MEMS as well as their perfect compatibility with microelectronics technologies.[1-4]

Microsensors for a magnetic field are basically modulating transducers (all semiconductor sensors and all silicon sensors). They convert the magnetic field, whether it is constant or variable, or even of biological origin, if possible with a maximal degree of accuracy and reliability, into an output electrical signal (current, voltage, or frequency) with high fidelity.[1-4]

The output energy of the modulating sensors is fed by an external power source through an additional input.[1,4] More specifically, these microdevices are fabricated using standard semiconductor IC technologies (most frequently, silicon processing technologies). In this chapter, besides the different modifications of magnetosensitive transducers, we have included abundant information about their device designs, the sensor characteristics, the concrete circuitries, the means for improving and overcoming their basic limitations, the most frequently used magnetomodulation systems for the indirect applications, ample references to relevant literature, and other useful data for both industry-based engineers and researchers.

9.3 Main Figures of Merit of Magnetic Microsensors

9.3.1 Classification of Magnetic Sensors: Figures of Merit

The great variety of applications, the peculiarities of transducer action and reliability, the growing research activity and the necessity to compare the results, the demanding requirements concerning the quality of used materials, and the tendency for unification necessitate the development of clear criteria for evaluation of the different sensors, including the magnetic sensor microsystems. We believe[4] that the consideration of all figures of merit can be made more comprehensible if they are classified into three groups, as follows:

1. The change of the output (OUT) in a magnetic field B, keeping (or leaving) constant other possible external influences, e.g., temperature T, pressure P, and radiation Φ, summarized here with the symbol C. The denotation of these characteristics is $\text{OUT}(B)_C$.
2. The behavior of the output signal OUT as a function of all possible external influences C at a constant (or absent) magnetic field B, denoted $\text{OUT}(C)_B$
3. The parameters describing the magnetosensor as a circuit device, denoted SD (system descriptors)

In Table 9.1, the principal figures of merit of magnetic-field microsensors are presented according to this classification.

9.3.2 Characteristics Related to OUT(B)$_C$

Magnetosensitivity. By definition, the sensitivity S or the transduction efficiency is the ratio of the change of the output signal (current, voltage, or frequency) to the variation of the external magnetic field B at a constant T, P, Φ, etc. Both an absolute sensitivity and a relative sensitivity of the modulating magnetosensors can be defined.[1,2,4,6,15] Equations (1a), (1b), and (1c) define the absolute sensitivity when the output is a current I, voltage V, or frequency f, respectively:

$$S_A^{(V)} \equiv \left|\frac{\partial V}{\partial B}\right|_C, \; [\text{VT}^{-1}]; \tag{1a}$$

Table 9.1 Classification of the Principal Figures of Merit of Magnetic Sensors

OUT(B)$_{C=const}$	OUT(C)$_{B=const}$	System Descriptors
Sensitivity	Noise	Excitation
Nonlinearity	Offset	Input impedance
Calibration	Cross-sensitivity	Output impedance
Range	Drift	Size
Frequency response	Creep	Weight
Directivity	Temperature error	Packaging
Resolution	Operating life	Sensor material
Accuracy	Reliability	Environmental conditions
Hysteresis	Long-term stability	Device design
Error (reversibility error)	Response time	
Output form		
Repeatability		

$$S_A^{(I)} \equiv \left|\frac{\partial I}{\partial B}\right|_C, \; [\text{AT}^{-1}]; \tag{1b}$$

$$S_A^{(f)} \equiv \left|\frac{\partial f}{\partial B}\right|_C, \; [\text{HzT}^{-1}]. \tag{1c}$$

The relative magnetosensitivity is determined by the ratio of the absolute sensitivity to the supply current or voltage applied to the additional transducer input. The figure of merit is the current-related sensitivity S_{RI} [Eq. (2)] when the additional input is fed by the supply current I_S, and the voltage-related sensitivity S_{RV} [Eq. (3)] when the additional input is fed by a supply voltage V_S:

$$S_{RI}^{(V)} = \frac{S_A^{(V)}}{I_S} \equiv \left|\frac{1}{I_S}\frac{\partial V}{\partial B}\right|_C, \; [\text{VA}^{-1}\text{T}^{-1}];$$

$$S_{RI}^{(I)} = \frac{S_A^{(I)}}{I_S} \equiv \left|\frac{1}{I_S}\frac{\partial I}{\partial B}\right|_C, \; [\text{T}^{-1}];$$

$$S_{RI}^{(f)} = \frac{S_A^{(f)}}{I_S} \equiv \left|\frac{1}{I_S}\frac{\partial f}{\partial B}\right|_C, \; [\text{HzA}^{-1}\text{T}^{-1}]; \tag{2}$$

$$S_{RV}^{(V)} = \frac{S_A^{(V)}}{V_S} \equiv \left|\frac{1}{V_S}\frac{\partial V}{\partial B}\right|_C, \ [\text{T}^{-1}];$$

$$S_{RV}^{(I)} = \frac{S_A^{(I)}}{V_S} \equiv \left|\frac{1}{V_S}\frac{\partial I}{\partial B}\right|_C, \ [\text{AV}^{-1}\text{T}^{-1}];$$

$$S_{RV}^{(f)} = \frac{S_A^{(f)}}{V_S} \equiv \left|\frac{1}{V_S}\frac{\partial f}{\partial B}\right|_C, \ [\text{HzV}^{-1}\text{T}^{-1}]. \tag{3}$$

The relative sensitivity should be suitable to the comparative analysis of magnetic microdevices, while in the purely practical applications, the absolute sensitivity is preferred.

Nonlinearity. When the ideal output characteristic of the magnetosensor is a straight line, the deviation of the real output characteristic from it is a relative error that is termed a nonlinearity (NL). This parameter is determined by the expression

$$\text{NL} \equiv \frac{|\text{OUT}_1 - \text{OUT}_2|}{(\text{OUT})_2} 100\%, \tag{4}$$

where OUT_1 is the measured value of the output signal at a given fixed magnetic induction B_o, and OUT_2 is the corresponding value from the straight line at a field B_o. This line is the best fit to the measured output values. Another convenient expression for NL is

$$\text{NL} = \left\{1 - \frac{[\text{OUT}(B)/B]}{[\text{OUT}(0.1)/0.1]}\right\} 100\% \tag{5}$$

The NL is expressed as within ± ... percent of full-scale output.[1-4,6,15]

Calibration. This figure of merit is necessary to determine the real output signal $\text{OUT}(B)$ of any magnetosensor. The calibration is a test during which known values of the magnetic induction (taking into account the sign of B) are applied to the microtransducer and the corresponding output readings are recorded.[4,15]

Range of Magnetic Fields. Each magnetic microdevice is designed to measure a certain optimum range of the magnetic-field magnitude: B_{lower} and B_{upper}. The most frequently used magnetic terms and units are systematized in Table 9.2.[4,15,16]

Table 9.2 Magnetic Terms and Units

Term, Quantity, Symbol	MKSA Unit	Subunits	CGS Unit	Conversion Factors
Magnetic field strength, H	$A\,m^{-1}$	$1\,A\,cm^{-1} = 10^2\,A\,m^{-1}$ $1\,mA\,cm^{-1} = 0.1\,A\,m^{-1}$ $1\,kA\,m^{-1} = 10^3\,A\,m^{-1}$	Oe (Oersted)	$1\,Oe = 79.58\,A\,m^{-1}$ $= 0.796\,A\,cm^{-1}$
Magnetization, M	$A\,m^{-1}$			
Magnetic induction, B (flux density)	$T\,(Tesla) = V\,s\,m^{-2}$	$1\,mT = 10^{-3}\,T$ $1\,\mu T = 10^{-6}\,T$ $1\,nT = 10^{-9}\,T$ $1\,pT = 10^{-12}\,T$ $1\,fT = 10^{-15}\,T$ $1\,\gamma = 1\,nT$	G (Gauss)	$1\,G = 10^{-4}\,T$ $1\,kG = 0.1\,T$ $1\,mG = 10^{-7}\,T$ $1\,\gamma = 10^{-5}\,G$
Magnetic flux, Φ	$Wb\,(Weber) = V\,s$		Mx (Maxwell)	$1\,Mx = 10^{-8}\,Wb$
Magnetic polarization, J	$T\,m\,A^{-1}$		$G\,Oe^{-1}$	
Permeability absolute, μ				
Permeability of vacuum, μ_0	$4\pi \times 10^{-7}\,T\,m\,A^{-1}$	$0.4\pi \times 10^{-4}\,T\,cm\,A^{-1}$	1	

Sensor Excitation. An important feature of the magnetosensitive device is the manner of their excitation by the field B, i.e., whether the vector B is perpendicular or parallel to the active surface of the structure.[4,15] Depending on the orientation of B with respect to the plane of the device, we have divided the magnetic microsensors known so far into orthogonal and parallel-field transducers.

Frequency Response. The dynamic behavior of the magnetic-field sensors is of primary importance in the measurement of fast variations of the field B. The frequency response is the dependence of the amplitude ratio of the output signal to the ac field B on the frequency of the external sine wave of the ac field B within a specified frequency interval. This figure of merit is usually specified as within ± ...% (or ± ...db) from ... to ... Hz. It is generally accepted that the dynamic degradation of a microtransducer in an ac field B starts with a reduction of the output by a factor $1/\sqrt{2}$, i.e., by 3 db.[4,15]

Resolution. This characteristic determines the smallest possible step change of the magnetic induction, which can be detected on the sensor output.[4,15]

Error. This figure of merit is the algebraic difference between the magnetic field recorded at the sensor output and its exact value. The error is expressed as a percentage of the full-scale output. The error curve is a graphical representation of the errors obtained from a certain number of calibration cycles. There is a slight error given in percent termed reversibility error. It expresses the difference in the output readings of the magnetotransducer at fixed value B_o with changes in the magnetic-field B direction (from positive to negative and vice versa). This error strongly depends on the difference in the contact area and nonuniformity of the sensor material used.[1,4,15]

Accuracy. This characteristic is defined as a ratio of the error to the full output scale expressed as a percentage. The accuracy can be presented in the same units by which the magnetic field is measured as within ± ...% of the full scale output.[1,4,15]

Hysteresis. This parameter determines the maximal change in the output signals for any fixed value B_o of the magnetic induction within the specified range. The value B_o is reached first by increasing and then by decreasing the external field B. The hysteresis is expressed as a percentage of the full-scale output during any calibration cycle.[1,4,15]

Repeatability. The ability of a magnetic microsensor to give the same output when the same measurand value B_o is applied. It is expressed as a percentage of the full-scale output. Many calibration cycles are needed to determine the repeatability.[1,4,15]

9.3.3 Characteristics Related to OUT(C)$_B$

Noise. The output electrical noise is a fundamental property that determines the lowest detected value of the magnetic field B_{min}.[1-4,6,15] In this output signal (voltage or current), with a random amplitude and random frequency, which has nothing in common with the measurand **B**, the most disturbing component is $1/f$ noise. By selection of a more qualitative semiconductor material and more sophisticated technological steps, the $1/f$ noise can be significantly reduced. Often this figure of merit is presented as an equivalent magnetic-field noise at a signal-to-noise ratio equal to one.

Offset. The offset is a parasitic output signal in magnetic microsensors of a modulating type (in most cases with a differential output) in the absence of a magnetic field **B**. Without any supplementary information about the value of induction B, the offset cannot be distinguished from the useful output signal (voltage, current, or frequency):

$$\text{OUT}(B) = SB + \text{OUT}'(B = 0), \tag{6}$$

where $\text{OUT}'(B = 0)$ is the offset that is most frequently the result of a structural or electrical asymmetry of the sensor.

This error is often expressed as within + ... % of the full-scale output.[1,4,6,15] A more relevant definition of the offset is its expression as a signal of an ideal sensor (without offset) generated by an equivalent magnetic induction $B_{\text{off, eq}} \equiv B_{\text{off}}$:

$$B_{\text{off},eq} \equiv \frac{V_{\text{off}}}{S_A^{(V)}}; \quad B_{\text{off},eq} \equiv \frac{I_{\text{off}}}{S_A^{(I)}}; \quad B_{\text{off},eq} \equiv \frac{f_{\text{off}}}{S_A^{(f)}}. \tag{7}$$

Cross-Sensitivity and Temperature Error. The influence of one or more measurands as pressure P, light Φ, temperature T, and so forth, on the magnetosensitivity is an undesired parasitic output signal termed cross-sensitivity (CS).[1-4,6,15] The figure-of-merit CS is expressed as $\text{CS} \equiv (1/S\ \partial S/\partial C)$, where S is the sensor magnetosensitivity [Eqs. (1) through (3)] and C is the source of perturbation. Most frequently this is the temperature T; consequently,

$$\text{TC} = (1/S)\ (\partial S/\partial T)100\%. \tag{8}$$

In this case, TC is the temperature coefficient of magnetosensitivity, measured as % K^{-1} or % °C^{-1}. The temperature interval (T_{min}, T_{max}) within which the value of the TC is constant should be specified.

Drift and Creep. The drift is an undesirable slow change in time of the output signal at a constant magnetic field *B*. The drift is in no way connected with the measurand *B*. Drift can be caused by temperature, pressure, light, and so on. The creep, which is also a parasitic magnetosensor fluctuation, is a weak and continuous change at the output at constant magnetic field *B* and all other environmental parameters such as T, P, and Φ.[4,15]

Response Time. This figure of merit determines the time it takes for the sensor output signal to reach its final value as a result of a step change of the magnetic field. This parameter is indicated in the handbooks, for example as "95% response time ... µs".[1,4,15]

9.3.4 Characteristics Related to the SD

Electrical Excitation. This is the external electrical voltage and/or current applied to a magnetic modulating microsensor for its proper operation.

Input and Output Impedance. The input impedance presented to the power supply is measured between the excitation terminals. The output impedance is measured between the output leads of the magnetosensor under conditions with open-circuited additional input terminals.

Environmental Conditions. Magnetic-field microsensors most frequently function under the following environmental conditions: temperature $(25 \pm 10)°C$ or $(77 \pm 18)°F$, relative humidity 90% or less, and barometric pressure 26 to 32 inches Hg.

The remaining figures of merit of the magnetodevices from Table 9.1, as well as other characteristics, methods, and circuits for their determination, can be found in the literature[1-4,13,15] and the references therein.

9.4 Hall Microsensors

9.4.1 The Lorentz Force

With the movement of a charge carrier with an average drift velocity *v* (under the action of an applied electric field $E = qv$) and charge *q* in a direction perpendicular to the vector *v* and the external magnetic field *B*, a transverse electromotive force F_L appears, termed the Lorentz force:

$$F_L = q(v \times B). \tag{9}$$

What is unique about this force F_L is that it is perpendicular both to the velocity v and to the field B, deflecting the trajectories of the moving carriers.[1-4,6,7,9,13,16] In a solid, the force F_L can cause the following fundamental phenomena:

a. In the collisions of the carriers with the atoms of the lattice, the carriers transmit pulses to the conductor (semiconductor). The summation of these pulses results in the appearance of a movement of the sample itself with respect to the magnetic flux lines. Such a lateral shift will also occur if a previously fixed conductor is released.
b. Depending on the geometry of the concrete structure (long or short), the electromotive force F_L can generate a transverse voltage across the sample or a current deflection in it (when the output is short-circuited).

9.4.2 Hall Effect

The genesis of these phenomena was discovered in 1879 by Edwin Herbert Hall, a 24-year-old Ph. D. student at Johns Hopkins University.[17] Although Hall first, by a purposeful experiment, proves the hypothesis about a link between the described important physical phenomena discussed in Section 9.3.1(a) and (b), it is accepted for the Hall effect to refer only to the manifestations of the force F_L from Section 9.3.1(b). However, we believe that the Hall effect is in the unity of the key properties described in Section 9.3.1(a) and (b). The devices in which the action of the electromotive force F_L is optimized so that the magnetic field B maximally changes their characteristics, generating transverse voltage (or current), form the most widely used class of galvanomagnetic transducers: the Hall (micro) sensors. Other generally used terms for Hall devices are Hall plate, Hall generator, Hall transducer, Hall cell, Hall slab, and Hall element.[4]

9.4.3 Hall Effect as Sensor Action

The shape of the classical Hall sensor is similar to the shape of the device used by Hall when he discovered his effect.[17] Such a structure, presented in Fig. 9.2, is a long, rectangular n-type resistive layer with length l, width w, and thickness t, $l > w$. The material is a semiconductor with high carrier mobility. Four ohmic leads are attached along the periphery. Two of them are bias input contacts C_1 and C_2; the other two are dif-

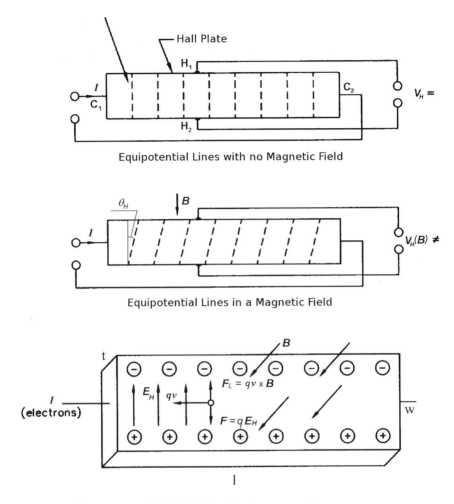

Figure 9.2 Occurrence of the Hall effect in a long sample.

ferential point output contacts H_1 and H_2. The normal component of a homogeneous magnetic field B is perpendicular to the plate plane. According to the excitation, this Hall sensor is orthogonal. If a current I is sent through the layer (slab), the deflection of the trajectories of the carriers under the influence of the force F_L [Eq. (9)] results in a redistribution of the current I along the device section at an angle Θ_H between the current density j and the electric field E vectors.

The Hall angle Θ_H is given by $\Theta_H = \text{arctg}(\mu_H B)$, where μ_H is the carrier mobility determined in a magnetic field (for electrons n or holes p, respectively, depending on the semiconductor type). An uneven charge distribution results, with the electrons accumulated in the upper edge of the plate where the potential is negative (Fig. 9.2). The opposite edge will become depleted of electrons and will therefore be positively charged. This process will continue until an electric field (Hall field) E_H across the structure occurs, exerting a force $F = qE_H$, opposite in direction to the Lorentz force F_L. At equilibrium, the resultant forces balance. This electric field E_H or its corresponding voltage $V_H = wE_H$ (Hall voltage) is the most important widespread criterion for the presence of the Hall effect. It can be said the Hall voltage stems from excessive charges on the two opposite edges, separated by the Lorentz force F_L. The obligatory condition for appearance of the Hall effect is that the biasing and sense contacts must have an altering arrangement.

9.4.4 Hall Voltage Mode of Operation

According to the theory of the Hall effect, the voltage V_H is proportional to the product of the input current I and the normal component of the induction B, and is inversely proportional to the layer thickness t and the carrier concentration n (or p):

$$V_H = GR_H IB\sin\varphi/t = G(r_n/qn)(1/t)\, IB\sin\varphi, \qquad (10)$$

where $G \leq 1$ is the geometrical correction factor, a dimensionless quantity that determines the reduction of the Hall voltage due to the imperfect bias current confinement and depends mainly on the ratio l/w; $R_H = r/qn$ is the Hall coefficient that determines the Hall effect intensity in a specific material; r_n is the electron scattering factor; and φ is an angle in degrees of the incident magnetic field from a line drawn parallel to the Hall plate.[1-4,6,18]

The voltage V_H is proportional to the product of the carrier velocity v and the width w of the structure.

The Hall coefficient R_H for bipolar semiconductor materials (with both types of carriers) is given by the relation

$$R_H = \frac{1}{q}\frac{r_p p - b^2 r_n n}{(p+bn)^2}$$

and

$$b \equiv \mu_n/\mu_p, \qquad (11)$$

where r_n and r_p denote the scattering factors, μ_n and μ_p are the mobilities, and n and p are the concentrations of electrons and holes, respectively.

The scattering factor r varies between 1 and 2; it depends on the kind of semiconductor and the temperature T, and plays an important role in the temperature behavior of the Hall sensors. In most practical cases, the value of r is close to 1, $r \approx 1$. When extrinsic conductivity dominates, R_H is given by the relations $R_H = -r_n/nq$ for n-type and $R_H = +r_p/pq$ for p-type.

The important quantities and constants used in galvanomagnetic sensors are summarized in Table 9.3.[16]

One particularly useful aspect of the Hall voltage is that it is directly proportional to the number of flux lines passing through the active area of the sensor. This means that with vw = const at l/w = const, the voltage V_H is proportional to the active sensor surface. The Hall effect is odd: The polarity of the voltage V_H reverses when the direction of the flux lines or the bias current reverses. Through the voltage V_H, both static (dc) and alternating (ac) magnetic fields can be detected.

The Hall sensor output is frequently operated with a constant voltage $V_S = lE_x$. In this case, the expression for the voltage V_H is

$$V_H = \mu_H (w/l) G V_S B. \tag{12}$$

Equation (12) categorically indicates the importance of high carrier mobility μ_H materials for the Hall devices. Table 9.4 contains the most important semiconductor parameters for magnetosensitive devices.[1-4,6,16,19,20]

According to Table 9.4, $A^{III}B^V$ compounds such as n-GaAs, n-InAs, and n-InSb are preferred for the development of Hall elements.

The maximum value of V_H with a fixed product $V_S B$ is given by the expression $V_H \approx 0.742 \, \mu_H V_S B$. For example, the maximum Hall voltage V_H at T = 300 K, which can be generated by an n-Si sensor, is about 12% of the supply voltage V_S at B = 1 T. For an n-GaAs Hall element, the corresponding value under the same conditions is about 67% of the voltage V_S. The reason for this is a 5.5 times higher mobility in n-GaAs.

The relationship between the Hall voltage V_H and the dissipated input power[4,6] $W_{in} = I_S V_S$ is

$$V_H = G \left(\frac{w}{l}\right)^{\frac{1}{2}} r_H \left(\frac{\mu_H}{qnt}\right)^{\frac{1}{2}} W_{in}^{\frac{1}{2}} B \tag{13}$$

In fact, n-type semiconductors are preferred because the mobility of electrons is larger than that of the holes. This reduces the supply power W_{in} at the same sensitivity S.

The shunting of the Hall voltage V_H by the ohmic contacts C_1 and C_2 results in a flow of Hall current, which reduces V_H near C_1 and C_2. In these

Table 9.3 Important Quantities and Constants Used in Galvanomagnetism

No.	Physical Quantity and Constants	Symbol	SI Unit	CGS Unit and Conversion
1	Conductivity	σ	S m^{-1}	$10^{-2} \times \Omega^{-1}$ cm^{-1}
2	Resistivity	ρ	Ω m	$10^{2} \times \Omega$ cm
3	Hall coefficient or constant	R_H	m^3 C^{-1}	$10^{6} \times$ cm^3 C^{-1}
4	Carrier concentration	n, p	m^{-3}	$10^{-6} \times$ cm^{-3}
5	Carrier mobility and Hall mobility	$\mu_n, \mu_p; \mu_{Hn}, \mu_{Hp}$	m^2 V^{-1} s^{-1}	$10^{4} \times$ cm^2 V^{-1} s^{-1}
6	Elementary charge	q	1.602×10^{-19} C	4.8×10^{-10} statcoulombs
7	Electron rest mass	m_0	9.109×10^{-31} kg	9.109×10^{-28} g
8	Boltzmann constant	k_B	1.380×10^{-23} J K^{-1}	1.380×10^{-16} erg °C^{-1}
9	Thermal energy ($T = 300$ K)	$k_B T$	4.14×10^{-21} J	4.14×10^{-14} erg
10	Thermal voltage at $T = 300$ K	$k_B T/q$	0.0259 V	
11	Planck's constant	h	6.626×10^{-34} J s	6.626×10^{-27} erg s
	Reduced constant	$h/2\pi$	1.054×10^{-34} J s	1.054×10^{-27} erg s
12	Dielectric constant (permittivity of vacuum)	ε_0	8.854×10^{-12} F m^{-1}	1

Table 9.4 Important Material Parameters of Selected Semiconductors for Galvanomagnetic Sensors at $T = 300$ K; adapted from Roumenin[4]

No.	Material	E_g (eV)	μ_n (m² V⁻¹ s⁻¹)	μ_p (m² V⁻¹ s⁻¹)	ρ (Ω cm)	$-R_H$ (cm³ C⁻¹)	n (cm⁻³)	$\alpha(RT)$ $\mu_H \rho^{1/2}$	(K⁻¹)	Remarks
1	Ge	0.67	0.39	0.19	45	9×10^4	2.4×10^{13}			Intrinsic
2	Si	1.12	0.14	0.05	2.3×10^5	4×10^8	1.5×10^{10}		$+5 \times 10^{-3}$	Intrinsic
3	Si	1.12	0.13		2	2.5×10^3	2.5×10^{15}	1840		n-type
4	GaAs	1.42	0.85	0.045	7.8×10^7	6×10^{11}	9.2×10^6		$+8 \times 10^{-4}$	Intrinsic
5	GaAs	1.42	0.64		0.32	2.1×10^3	3×10^{15}	3680		n-type
6	InAs	0.36	2.5	0.046	0.08	2.1×10^3	3×10^{15}		$+1 \times 10^{-3}$	Nearly intrinsic
7	InAs		2.2		0.006	125	5×10^{16}	1700		n-type
8	InSb	0.17	7.5	0.075	0.005	380	2×10^{16}		-2×10^{-2}	Nearly intrinsic
9	InSb		5.5		0.0013	70	9×10^{16}	1990		n-type

areas, the force F_L is not balanced by the field E_H, and the dimensions of these zones on the x axis are 1.5 w. Consequently, the Hall contacts H_1 and H_2 must not be placed in these regions; i.e., the Hall samples must be designed with dimensions satisfying $l \geq 3w$ (Fig. 9.2). By reducing the dimensions of the Hall device, the magnetosensitivity S is also decreased, mainly due to short-circuiting effects.

9.4.5 Hall Current Mode of Operation

The Hall current I_H in short samples ($l < w$) is an equivalent alternative to the Hall voltage V_H and can also be used as a sensor signal to detect the field $\boldsymbol{B}$.[4,6] The device structure of a Hall sensor with a current output is presented in Fig. 9.3. The Hall current mode of operation arises when the boundary surfaces are electrically shunted and the Hall field $E_H \approx 0$. When crossed electric and magnetic fields are applied to the short sample, the carriers are deflected by the force $\boldsymbol{F}_L$. The Hall current is given by the relationship

$$I_H = (l/w)\mu_H B I_S. \qquad (14)$$

The version in Fig. 9.3(b) is more sensitive than the one in Fig. 9.3(a) because of increased relative variation of the useful output signal $\Delta I(B)$ in the expression $S_{RI} = \Delta I(B)/(I_1(B) + I_2(B))$.

Long samples ($l > w$) can operate in the Hall current mode too. For this purpose, the voltage V_H must be shunted by a measuring instrument with a very low resistance. In this case, the generated output current is $I_H = 0.742 \mu_H B I_S$.

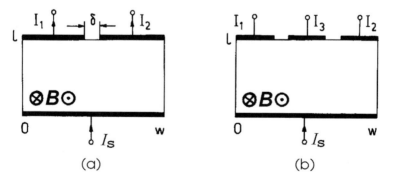

Figure 9.3 Hall current sensors: (a) with dual leads; (b) with triple terminals. The output is the differential current $\Delta I(\boldsymbol{B}) = I_1(\boldsymbol{B}) - I_2(\boldsymbol{B})$.

9.4.6 Diode Hall Effect

The existence of the Hall effect in diode and transistor structures is a recently discovered property.[21,22] This phenomenon consists of the appearance of a Hall effect (Hall voltage or Hall current) generated by the majority carriers in the base region of diodes and transistors in a magnetic field. The original cause of the diode Hall effect is also the Lorentz deflection. The paradox is that according to [Eq. (11)], under the conditions of nonequilibrium bipolar conductivity, when the concentrations of the electrons and the holes are equal, i.e., $n \approx p$, the Hall voltage V_H in diodes and transistors in a first approximation is absent. Theoretically and experimentally, it is established that if the length of the base region l, in which minority carriers are injected, exceeds their diffusion length L_D, $l > L_D$, the drift component of the velocity of the majority carriers generates a Hall effect irrespective of the mixed conductivity.[21,22] The behavior of the diode Hall effect is the same as the classical one. In the literature, the diode Hall effect is also associated with the author's name.[23]

9.4.7 Hall Effect Devices

Orthogonal Hall Sensors. In Fig. 9.4, many possible device designs of orthogonal Hall sensors are collected. The peculiarities of their actions and process fabrication are described in detail in the literature.[1-4,6,13] The functional relations of the different geometrical shapes of Hall transducers at factor $G = 1$ show that it is impossible to improve the magnetosensitivity of any of the Hall devices by changing their geometry. In this aspect, the Hall elements from Fig. 9.4 are equivalent.[4,24,25]

The orthogonal bulk versions of Hall sensors from Fig. 9.4 are based on high-mobility compound semiconductors of the type $A^{III}B^V$. Parameters of selected Hall sensors are presented in Table 9.5.[4,26] One of the goals of sensor development is to reduce the size of the magnetotransducers; this is realized by IC technology. Silicon, which is used most of all, unfortunately does not possess a high mobility of the n-carriers. The availability of sophisticated IC processing steps, and the material, technological, and electrical compatibility between silicon integrated circuits and Hall plates are nevertheless excellent reasons to use Si for the fabrication of Hall microsensors.

The first integrated Hall microtransducer was realized in a MOS process.[27] Its greatest advantage is that it is extremely thin in the *B* field direction. The n-type inversion layer of the MOSFET transistor is used as an active zone, to which Hall leads are attached (Fig. 9.5). Besides this classical design, two- and three-drain structures are used, as well as split-

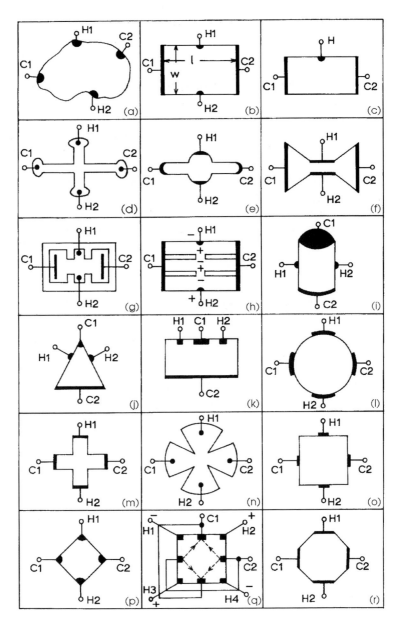

Figure 9.4 Various shapes of orthogonal Hall devices. Supply contacts are denoted by C_i, Hall leads by H_i.

Table 9.5 Parameters of Selected Orthogonal Bulk Hall Sensors (device structures shown in Fig. 9.4); adapted from Roumenin[4]

No.	Structural Material	l/w	n⁺ Hall Region	Epi-Doping, N_D (cm^{-3})	Epi-Thickness, t (μm)	Sensitivity, S	Equivalent Magnetic-Field Noise (mT)	Equivalent (mT)	TC of Offset (mT K^{-1})	TC of Sensitivity (% K^{-1})	Nonlinearity (%)	Frequency Response (kHz)
1	(b), Si	2	epi	10^{15}	10	0.6 V T^{-1}	7×10^{-4}	17				15
2	(b), Si	1	epi	10^{15}	8	0.5 V T^{-1}		80				
3	(b), Si	1	epi	10^{15}	10	0.4 V T^{-1}		25				10
4	(b), Si	1.2	epi	10^{15}	10	B_{switch} = 50 mT				-0.6		
5	(q), Si	1	epi	2.5×10^{15}	12	7.2% T^{-1}		10	0.025	-0.45		
6	(b), Si	1.2	epi + implementation	10^{16}	15	7.6% T^{-1}		10				
7	(b), Si Hall IC TL 173	1	epi			15 V T^{-1}		7	0.2	-0.18		10
8	(b), Si Hall IC SAS 231		epi			100 V T^{-1}		35	0.4		±2	
9	(p), Si	1	epi			300 V A^{-1} T^{-1}						
10	(m), Si	1	epi + impl.	4×10^{15}	14	290 V A^{-1} T^{-1}					comp.	
11	(m), Si	1	epi + impl.	1.5×10^{15}	14	3300 V A^{-1} T^{-1}					comp.	
12	(b), Si		epi	5×10^{14}		1 V T^{-1}; 7% T^{-1}				-0.05		
13	(b), GaAs	2	epi	5×10^{14}	4	22 V T^{-1}				-0.04		
14	(b), GaAs	1	epi + impl.	5×10^{17}	0.13	0.17 V T^{-1}		8	0.5	-0.05	±0.1	
15	(m), GaAs	1	epi + impl.	10^{17}		500 V A^{-1} T^{-1}						
16	(m), GaAs	1	epi			15 V T^{-1}				-0.03	±0.2	
17	(o), GaAs	1	epi + impl.	$(1.1-3.8)\times10^{17}$	0.25	65 V A^{-1} T^{-1} (350-1100) V A^{-1} T^{-1}	4×10^{-3}	comp.				
18	(m), GaAs with concentr.	1					7×10^{-5}			-0.08		10^4
19	(b), GaAs	4	epi + MOCVD + implementation	10^{17}	0.35	220 V A^{-1} T^{-1}		3.4			±2	
20	(b), InSb		epi, MBE	2×10^{18}		1 V T^{-1}				-0.01		

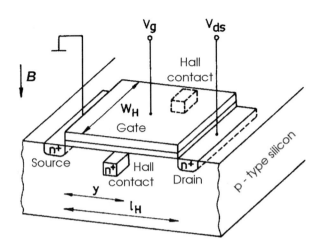

Figure 9.5 N-channel MOS Hall microsensor.

source configurations, to improve the magnetosensitivity.[4,6,28-30] The Hall microdevices are also realized by using bipolar IC technology.

The n-type sensor region is isolated from the p-type semiconductor material by a depletion layer. In a CMOS process, the n-well satisfies both important requirements for reaching high sensitivity in the Hall sensors: a low value of the thickness t and a low concentration of the n-region (Fig. 9.6).[31]

The available parameters of typical MAGFET Hall sensors are generalized in Table 9.6.[4] The overcoming of some limitations in the sensitiv-

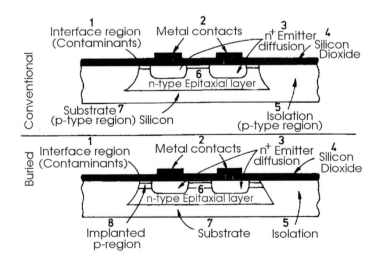

Figure 9.6 Cross section of a buried Hall microdevice realized using IC technology.

Table 9.6 Parameters of Selected Orthogonal MAGFET Hall Sensors; adapted from Roumenin[4]

Type	l/w	Sensitivity, S	Equivalent Noise ($\mu T\,Hz^{-1/2}$)	Equivalent Offset (mT)	TC of S (% K^{-1})
n-channel, Si	0.4	3.1% T^{-1}			
n-channel, Si	1	3.5% T^{-1}		20	
n-channel, Si	1.2	6.4% T^{-1}		14	−0.4
n-channel, Si	0.1	14% T^{-1}			
CMOS differential amplifier, split-drain, Si	0.3	1.2 V T^{-1} 1.2×10^4 V A^{-1} T^{-1}			
n-channel, Si, split-drain	1	9.6% T^{-1}	7		
n-channel, Si, open split-drain	1	172 V A^{-1} T^{-1} 109 mV T^{-1}	0.15		
n-channel, Si, split-drain	1	10% T^{-1}	10		
n-channel, GaAs, split-drain	4	70% T^{-1}	25		
n-channel, GaAs, open split-	4	180 mV T^{-1}	8		
n-channel, Si, split-drain	3	7 mV T^{-1}	10		
n-channel, Si, split-drain	1	2.5 V T^{-1}			−1
p-channel, Si, split-drain	1	2.2 V T^{-1}			
n-channel, Si, 3-drain	1	6.2% T^{-1}	0.76		
n-channel, Si, 3-drain	3	20 nA T^{-1}	22		
n-channel, GaAs, 3-drain	4	24 μA T^{-1}	10		
n-channel, GaAs, 3-drain	4	27 μA T^{-1}	2.3		

ity, offset, long-term stability, noise, stress and aging, temperature cross-sensitivity, enlargement of the temperature operation range, and so on, can be found in the literature.[4,6,13] The necessary conditions for improving the magnetosensitivity in Hall microsensors (principally high mobility) are best accomplished by heterojunction structures with a two-dimensional electron gas (2-DEG).[4,32] The molecular-beam epitaxial growth technique is used for the fabrication of the modulation-doped (AlAs/GaAs)$_{SL}$ GaAs lattice Hall devices.

The 2-DEG occurs in a two-dimensional quantum well with an active thickness of about 10 nm. The energy-band diagram of a superlattice structure and the cross section of the superlattice Hall microsensor are shown in Fig. 9.7. At $T = 300$ K, a very high sensitivity of about 1200 V/AT is reached. Quantum-well Hall microsensors are also developed, using the InAs/AlGaSb system.[33,34]

Parallel-Field Hall Devices. Parallel-field devices, designed for the first time in 1983 at the Bulgarian Academy of Sciences,[35] are a new and promising trend in the development of Hall microsensors. It is regrettable that for more than 100 years, the Hall effect was investigated, analyzed, and used in sensor electronics, only in elements with an orthogonal applied magnetic field. The invariance of a conventional Hall plate is proved with respect to its shape by conformal mapping theory. The upper complex half-plane appears as the basic connecting link in the conformal transformation (Fig. 9.8; see Roumenin[36] and references therein). The new class of Hall microsensors with parallel-field sensitivity can be considered as a materialization of device shapes from the upper complex half-plane Im(w) > 0, which should be activated with a magnetic field parallel

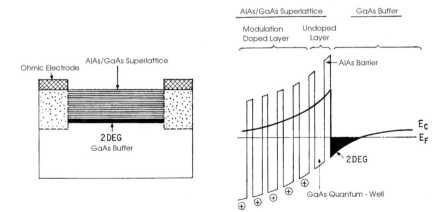

Figure 9.7 Cross section of the superlattice two-dimensional electron gas (2-DEG) GaAs Hall device and energy-band diagram of a superlattice structure.

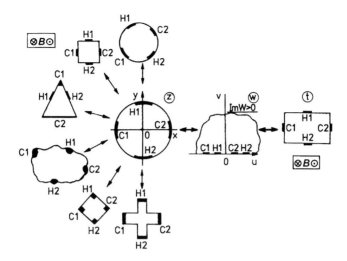

Figure 9.8 The connection of selected orthogonal Hall plates with the classical rectangular Hall device via conformal mapping.

to them.[4,36] The main advantages in comparison with the orthogonal versions are as follows:

- They are fully compatibility with the IC processes because all or almost all contacts are planar, i.e., located on the same side of the silicon plate.
- There is the possibility of creating two-dimensional and three-dimensional integrated vector magnetometers.
- It is possible to compensate electrically for the current and potential asymmetry inside the structures, which eliminates the offset
- The necessity of growing an n-type epitaxial layer of high quality (with high mobility) drops off because the devices are bulk devices; i.e., the active sensor zone is a part of the semiconductor substrate itself.

The trajectory of the majority carriers in the volume of such structures is curvilinear. The trajectories start and end on the metal or heavy doped n^+ contacts on the upper surface of the slabs.[4,36] Figure 9.9 demonstrates all currently known device designs of parallel-field Hall microsensors.[4,6,9,13,35-43] A specific feature of these microsensors is that at a field B, the Lorentz force acts simultaneously on the lateral and vertical components of the drift velocity $v_{dr} = v_x + v_y$. As a result, a Hall field will be gen-

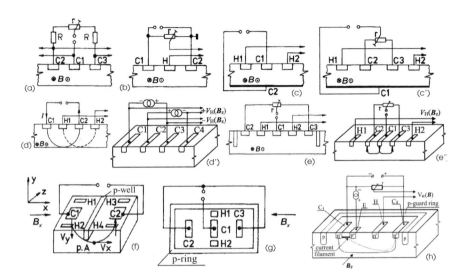

Figure 9.9 Device structures of parallel-field Hall microsensors known so far; r is the trimmer for offset regulation.

erated on the upper surface containing the contacts (Fig. 9.9). The voltage V_H will be registered by terminals H_1 and H_2. Figure 9.9(h)[43] shows how the geometrical factor G can be increased simultaneously with a decrease of the parameter t in Eq. (10), leading to an overall increase of V_H. By a supplementary emitter E and a current filament of high conductivity, the effective trajectory of the carriers maximally approaches the upper surface of the chip. Thus the conditions for a higher Hall voltage are optimized [Eq. (10)] The Hall signals $V_{C1,4}(B)$ and $V_{C2,3}(B)$ of the device in Fig. 9.9(d′) are of opposite signs, the magnetosensitivities of the two outputs are equal at currents $I_1 = I_2$, and the two offsets are almost equal too. Parameters of Si parallel-field Hall devices are generalized in Table 9.7.[4,39,43] The results obtained are not worse than for the orthogonal counterparts.

9.5 Magnetoresistors

Magnetoresistance is the change in the resistivity ρ of a material when it is subjected to a magnetic field. Usually the resistivity increases quadratically with the induction B, and ρ is not sensitive to the sign of the field $\boldsymbol{B}$.[1-4,18,44-46]

Table 9.7 Parameters of Silicon Parallel-Field Hall Microsensors (device structures shown in Fig. 9.9); adapted from Roumenin[4]

No.	Device Structure	Output	Sensitivity, S	Equivalent Magnetic-Field Noise Density ($T^2\ Hz^{-1}$)	Equivalent Offset (mT)	Linear Range (T)	NL (%)	Temperature Coefficient of S (% K^{-1})
1	(a), n-substrate as active reg.	Differential	70 mV T^{-1}, I = 10 mA, $l_{C1,2}$ = 300 µm		5	±1	1	0.1
2	(b), n-substrate as active reg.	Single, after magnetoresistance compensation	75 mV T^{-1}, I = 10 mA, $l_{C1,2}$ = 200 µm			±1	1	0.1
3	(b), n-substrate as. active reg.	Single	DFM ~ 4, $\xi B \xi$ = 1 T					
4	(c), n-substrate as active reg.	Differential	100 mV T^{-1}, I = 10 mA, $l_{C1,2}$ = 200 µm		6	±1	1	0.1
5	(d), n-substrate as active reg.	Differential	120 mV T^{-1}, I = 9 mA, $l_{C1,2}$ = 300 µm		10	±1	1	0.1
6	(d), n-substrate, confin. by p-ring	Differential	300 V A^{-1} T^{-1}					
7	(e), n-substrate, confin. by p-ring	Differential	450 V A^{-1} T^{-1}	3×10^{-13} (f = 100 Hz), 10^{-15} (f = 100 kHz)			1	

(continued)

Table 9.7 Cont.

No.	Device Structure	Output	Sensitivity, S	Equivalent Magnetic-Field Noise Density ($T^2\ Hz^{-1}$)	Equivalent Offset (mT)	Linear Range (T)	NL (%)	Temperature Coefficient of S (% K^{-1})
18	(e), n-substrate as active reg.	Differential	190 mV T^{-1}, I = 10 mA		7	±1	1	0.1
9	(e), n-substrate as active reg.	Differential	24.7 mV T^{-1}, I = 1.0 mA, $l_{C1,2}$ = 80 µm			±1	2	
10	(e), n-substrate as active reg.	Differential	66.6 mV T^{-1}, I = 1.0 mA, $l_{C1,2}$ = 80 µm			±1	2	
11	(e), n-epilayer	Differential	101 mV T^{-1}			±0.4		
12	(g), n-epilayer	Differential	75 V $A^{-1}\ T^{-1}$					0.1
13	(g), n-epilayer	Differential	1243 V $A^{-1}\ T^{-1}$					
14	(g), n-epilayer	Differential	41 V $A^{-1}\ T^{-1}$	10^{-10} (f = 40 Hz)		±1	1	0.1
15	(h), n-bulk	Single	190 mV T^{-1}					
16	(e), CMOS n-dif. layer	Differential	127 V $A^{-1}\ T^{-1}$ 400 V $A^{-1}\ T^{-1}$	50 nT $Hz^{-1/2}$	30	±2	0.04	0.04

9.5.1 Physical Magnetoresistance Effect

This phenomenon appears in homogeneous semiconductor systems and structures, maximally optimized for the generation of Hall voltage, $l > w$.[1,2,4,18,44] In these structures, the velocity v of electrons is not the same for all carriers; that is, it has a Boltzmann distribution about a mean value v_o. However, the electrostatic force $F = qE_H$, caused by the Hall electric field E_H and balancing the Lorentz force F_L [Eq. (9)], is the same for all electrons. For the dominating part of the carriers (Fig. 9.10), moving with the same velocity v_o, these two forces are mutually balanced. The electrons that have bigger and smaller velocities than the mean value v_o experience deflections from the forces F_L and F (the field E_H), respectively, which elongates their path in a magnetic field. This is the origin of the physical magnetoresistance effect. The resistivity ρ of a semiconductor material only increases in a magnetic field, and for unipolar conductivity is given by

$$\Delta\rho/\rho = \mu^2 r_M^2 B^2, \tag{15}$$

where μ is the carrier mobility and r_M is a coefficient that depends on the scattering mechanism.[4] This effect is too small to be used practically at room temperature for sensor applications.

9.5.2 Geometrical Magnetoresistance Effect

If the width, w, of the semiconductor sample is greater than the length, l, $w > l$ (short slab, Fig. 9.10), the Hall voltage is practically absent and

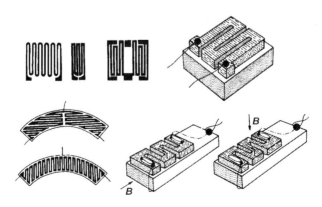

Figure 9.10 Layout and examples of meander-type semiconductor magnetoresistors.

the Lorentz deflection for all carriers dominates.[1-4,6,18,44-46] Consequently, in a magnetic field **B**, the effective length of the current path increases proportionally to the induction B and the mobility μ. This geometrical magnetoresistance effect is most strongly expressed in semiconductors such as n-InSb, n-GaAs, and n-InAs, and in elements with a circular shape (Fig. 9.11) (Corbino disk).[45] In general, the resistance dependence on induction B is quadratic and does not depend on the sign of the field **B**. It is established that in the case of samples with $l = w$, the resistance $R(B)$ dependence on the induction B is linear, i.e., $R(B) = R_H B/t$, where R_H is the Hall coefficient and t is the thickness of the structure.[46] By the geometrical magnetoresistance effect, an increase in the resistance of several hundred percent is possible, which makes this mechanism very suitable for sensor applications.

9.5.3 Semiconductor Magnetoresistors

This class of sensors includes thin plates or films of high-mobility material fitted with two ohmic (supply) contacts, subject to an orthogonal magnetic field **B**. To obtain high accuracy, an increase of the otherwise low onset resistance of the individual magnetoresistor is necessary. This is achieved using suitably constructed resistor cells connected in series and produced by common technological processes. Some typical structures of magnetoresistors are shown in Fig. 9.10.[4,45,47]

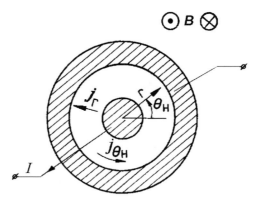

Figure 9.11 Corbino disk; the current density j is tilted relative to the electric field by the Hall angle Θ_H.

It should be noted that the linear magnetic sensitivity in square ($w = l$) structures from InAs and InSb is equal to that of Hall elements of the same material and of the same thickness, t.[46] The high nonlinearity and temperature dependence of magnetoresistor sensitivity can be overcome in software by smart sensor systems. Hybrid MEMS that contain magnetoresistors recently found an application as precise field gradient instruments. Equipped with permanent magnets, these sensors are widely used in different modifications of proximity detectors and many other contactless devices.

9.5.4 Spin-Dependent Magnetoresistance

In 1988, a new class of magnetic structures, quite suitable for sensor applications, was grown. These metallic multilayers are magnetic layers separated by nonmagnetic layers, e.g., Fe/Cr, Co/Ag, or Co/Cu. For field $B = 0$, if the adjacent magnetic layers are antiparallel or randomly oriented, the application of field B, which orients their magnetic moments in parallel, leads to a very large decrease in the resistance.[48] This phenomenon is called giant magnetoresistance (GMR) or, more exactly, giant negative magnetoresistance. The GMR effect is generally explained by spin-dependent electron scattering and is related to different shifts of energy levels between spinup and spin-down conductivity-band electrons. As a result of the exchange interaction in the case that the ferromagnetic layers have parallel magnetization directions, only one type of carrier is scattered strongly, resulting in a relatively low resistivity. For an antiparallel configuration of the magnetization directions, both types of electrons are scattered appreciably; thus the resistivity is high. The conditions for GMR occurring are: (1) There is a possibility for change of the magnetization direction in the neighbor magnetic layers from antiparallel to parallel. (2) The magnetic and nonmagnetic layers must be much smaller than the mean free path of the conductivity electrons, i.e., within the range of 10 to 30 Å (see Lenssen[49] and references therein). The resistivity is dependent on the relative angle φ between the supply current and the magnetic moments:

$$\rho(\varphi) \approx \rho(0°) + \Delta\rho[1 - \cos\varphi]/2. \qquad (16)$$

The quantity $\rho(\varphi)/\rho(0°)$ is called the GMR ratio and is about 10%. (There is evidence that it can reach values several times larger at room temperature.) According to Eq. (16), the GMR effect has a period of 360 degrees. The GMR does not depend on the current direction.[49]

9.5.5 GMR Sensors

Up until now, the GMR effect has been successfully used in read heads for hard disk drives.[2] The larger output signal, the miniaturization opportunities, and the possibility to register the angle φ within the interval $0° \leq \varphi \leq 360°$, independent of magnetic induction over a broad field range, are advantageous properties of GMR for application in magnetic sensor microsystems. Wheatstone bridge versions, simple resistors, and half-bridge configurations have already been created. The first commercial GMR bridge transducer has been realized by shielding two elements (serving as reference resistors) among four elements by a soft magnetic shield, which essentially performs simultaneously the function of a flux concentrator for the other two (active) magnetosensitive devices. In this structure, an output for a 10% change corresponds to about a 5% change of the supply voltage to the bridge.[49-52] The GMR bridge sensitivity can be controlled by the lengths of the flux concentrators. In this way the GMR material, which saturates at about 300 Oe, can be used to build transducers that are saturated at 20, 50, and 80 Oe. The integrated GMR sensors are compatible with IC processes and have potential as MEMS. In this way a magnetic and digital switch microsystem has been successfully realized. Also, on the basis of the bridge version, a GMR contactless potentiometer has been constructed. A very high-potential application is the galvanic separation in signal processing by a GMR magnetocoupler.[49-52] However, the problems of overcoming both the strong temperature dependence of the characteristics and the relatively high level of internal noise are still to be solved. In comparison with the other magnetic sensors, GMR transducers have higher sensitivity at low fields. This is because of their role as flux concentrators. There is also a possibility for linearization of the output.

9.6 Magnetodiodes

9.6.1 Magnetoconcentration and Magnetodiode Effects

If a parallelepiped semiconductor slab with quasi-intrinsic conductivity $n \approx p \approx n_i$ is activated by an external electric field $\boldsymbol{E}_x$ directed parallel to the long side (the x axis) and a magnetic field $\boldsymbol{B}_z$ directed parallel to the short side (the z axis), as a result of a Lorentz force $\boldsymbol{F}_L$ action, the quasi-free electrons and holes deflect toward the same surface in the direction of the y axis. As the Hall voltage is prevented due to electrical neutrality, the concentrations n and p grow at one side of the device. An important condition for the occurrence of the magnetoconcentration effect or Suhl-Shockley effect is that the surface recombination rates s_1 and s_2 on the two surfaces Σ_1 and Σ_2 toward which Lorentz force F_L can deflect the carriers are to be strongly different, for example, $s_1 \gg s_2$. In this case, if the force F_L concentrates the holes and electrons toward Σ_1, their recombination significantly grows and the sample conductivity significantly decreases. (The resistance increases.) In the opposite direction of the force F_L, as a consequence of the low s_2 value, the concentrations of the electrons and the holes will be increased in comparison with their equilibrium bulk value, which will result in a considerable increase in the conductivity of the slab.[4] Under such conditions, the current-voltage characteristic of the sample, which is linear when $B = 0$, becomes nonlinear and resembles a "diode" curve. This behavior is suitable for the development of magnetic sensor devices, but the instability of the parameters in a wide temperature interval, because of the intrinsic conductivity, is a limiting condition. In its physical essence, the magnetodiode (MD) effect is a superposition of magnetoconcentration and injection.[4,53-57] In the structures where this phenomenon is observed, there is a highly nonequilibrium electron-hole plasma obtained by injection at one or two junctions (p^+-n and n^+-n) in semiconductor structures with a low doping impurity concentration of the base region. The conductivity modulation by a magnetic field B allows control of the nonequilibrium electrons and holes (as a consequence of the strongly different rate of the surface recombination s_1 and s_2) representing the MD effect; it is at the bottom of the highly magnetosesitive microsensors class called magnetodiodes. By means of this transduction mechanism, the sign of the magnetic field B can be detected as well. Most strongly expressed, the MD effect occurs in diode structures with two injecting contacts and an operating condition that generates a high injection level.[1,2,4]

9.6.2 Magnetodiode Microsensors

The MD devices in their essence are two-terminal resistors and are activated with an orthogonal and a parallel magnetic field. Various shapes of MD sensors are presented in Fig. 9.12.[4] These devices can be realized on the basis of Ge, GaAs, InSb, InAs, Si, and so on. The MD transducers must meet two important requirements:

1. The diffusion length of the injected minority carriers must be smaller than the diode base length l, $l > L_D$ (long diode); the optimal ratio is $l/L_D \geq 3$.
2. The sample thickness t must be similar to the length L_D, $t \approx L_D$.[4]

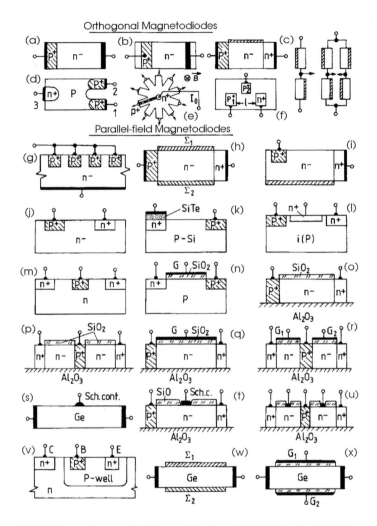

Figure 9.12 Various shapes of orthogonal and parallel-field magnetodiode sensors.

The silicon-on-sapphire (SOS) structure is of special interest. The crystallographic and thermal properties of Si and Al_2O_3 produce a high density of defects and recombination centers at the Si-Al_2O_3 interface [Fig. 9.12(o) through (r)] Because of this asymmetry, SOS is the best system for the miniaturization of MDs and their use in MEMS. Another integrated MD microsensor has been manufactured by a standard bulk CMOS process. Table 9.8 presents the sensor characteristics of selected MD devices.[4] These highly sensitive transducers have severe disadvantages.

Table 9.8 Parameters of Magnetodiode Sensors (device structures shown in Fig. 9.12); adapted from Roumenin[4]

Device Structure Material	Doping n_0 (cm^{-3}), Resistivity, ρ (Ω cm)	Dimensions, $l \times w \times t$	Equivalent Sensitivity, S	Magnetic-Field Noise	TC of S (% K^{-1})	Frequency Response (kHz)
(a,b) Ge	50 Ω cm	4×1×0.2 mm	0.5×10^3 V A^{-1} T^{-1}			10
(c) Ge	50 Ω cm	3×0.6×0.4 mm	10 V T^{-1}		−0.11	10
(c) Si	20 kΩ cm	1×0.5×0.5 mm	30 V T^{-1}			1
(a) Si S-type	2 kΩ cm	0.15×0.55×− mm	50 V T^{-1}			
(a) GaAs S-type			10^8 V A^{-1} T^{-1}			
(d) InSb S-type	10^2 Ω cm	1×0.05×− mm	12 mA mT^{-1}			
(f) Si	55 Ω cm	0.1×−×− mm	2 V T^{-1}			
(g) Ge	50 Ω cm		10 V T^{-1}			15
(h) Ge	50 Ω cm	7×1×0.25 mm	60 V T^{-1}	10^{-13} V^2 Hz^{-1}, $f \leq 1$ kHz		100
(i) Ge	50 Ω cm	5×2×1 mm	300 V T^{-1}			100
(h) Si	20 kΩ cm	400×−×− μm	40 V T^{-1}	10^{-13} V^2 Hz^{-1}		10
(j) Si S-type	20 kΩ cm	100×−×30 μm	10 V T^{-1}			10^3
(j) Si	20 kΩ cm	3×1×0.6 mm	80 V T^{-1}			
(m) Si	10^{15} cm^{-3}	0.4×0.15×0.3 mm	10^2 V A^{-1} T^{-1}		−0.2	100
(o) Si SOS	10^{15} cm^{-3}	10×50×7 μm	10 V T^{-1}, 150 mA T^{-1}	$B_{min} = 10^{-7}$ T	−0.1	100 MHz
(o) Si SOS	7×10^{15} cm^{-3}	31×20×4 μm	4.9 V T^{-1}			
(o) Si SOS	5×10^{15} cm^{-3}	50×50×0.65 μm	14 V T^{-1}, 2.2 mA T^{-1}	1 μV^2 Hz^{-1}	−0.1	100 MHz
(q) Si SOS	10^{15} cm^{-3}	15×100×0.65 μm	10% T^{-1}	$B_{min} = 5$ μT	−0.1	100 MHz
(s) G.MD	60 Ω cm	2×3×1 mm	250 V T^{-1}	$B_{min} = 10^{-7}$ T		100
(t) Si SOS	5×10^{15} cm^{-3}	20×50×0.65 μm	20 V T^{-1}	$B_{min} = 10^{-9}$ T	−0.1	100 MHz
(v) Si CMOS	4×10^{15} cm^{-3}	126×48×10 μm	25 V T^{-1}			
(w) Ge	40 Ω cm	8×0.6×0.2 mm	50 V T^{-1}		−0.5	1
(x) Ge	40 Ω cm	8×0.6×0.2 mm	15 V T^{-1}		−0.5	1

Their output characteristics are nonlinear, especially at high values of the induction B, and the transduction efficiency is also dependent on the field B direction. The main drawback is connected with the strong temperature dependence of the output signal. The device designs presented in Fig. 9.12(m), (p), (r), (t), and (u) are differential.[58] In this way, the temperature drift of the output is drastically reduced, the linear range is extended, and the sensitivity is increased by a factor of almost two.

Detailed information about the operation of MD sensors and the peculiarities of their technological fabrication is provided by Roumenin.[4]

9.7 Magnetotransistors and Related Microsensors

9.7.1 General Approach to Bipolar Magnetotransistor Design

Concerning their electrical principle of operation, the bipolar magnetotransistors (BMTs) are active devices with a current output in contrast to the classical Hall plates, where the output signal is a potential difference.[4,59,60] Irrespective of their variety, each BMT has as its essential component a current source in the form of a forward-biased *p-n* junction, injecting minority carriers into the base region, and one or more reverse-biased *p-n* junctions as collectors that pick up the useful signal. As a BMT is realized on the basis of a nonmagnetic substrate (such as Si, Ge, or GaAs), the magnetic field B influences only the kinetic processes in different regions of the transistor (emitter, base, or collectors). As a result, the changes in the output collector current $I_C(B)$ or a current difference $\Delta I_{C1,2}(B)$ (at dual collector devices) are a measure of the direction and strength of the field B. Because of the effect of transistor amplification, the change in the output voltage may substantially exceed that of Hall elements and MDs. Depending on the type of carriers injected into the base, the BMT is of the *p-n-p* or *n-p-n* type. According to the geometry of excitation by an external magnetic field with respect to the chip surface, there are two methods of activation: parallel-field and orthogonal. These two main types of BMTs are presented in Figs. 9.13 and 9.14.[4,59] If the current moves parallel to the chip surface, the BMT is lateral [for example, Fig. 9.14(h) and (j)] If the direction of the carriers is perpendicular to the surface, the sensor is vertical [Fig. 9.14(c), (d), and (f)] When the substrate serves as an active sensor zone, the BMT is a bulk device; when the base region is a part of the epitaxial layer, it is an epitaxial BMT. This class of transducers includes single collector elements [Fig. 9.14(m)], differential or dual collector sensors (Fig. 9.13), and multicollector devices

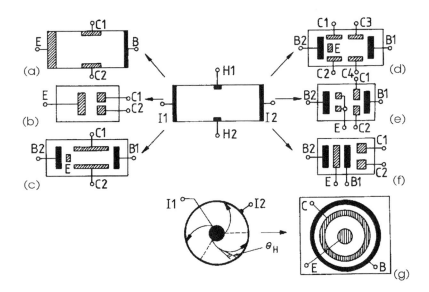

Figure 9.13 Classification of orthogonal BMTs: (a) through (f) show the respective classical Hall element; (g) shows the evolution from the Corbino disk into a circular BMT.

for measuring the magnetic vector components; each different x-, y- or z-channel represents a differential version. In general, the trajectory of movement of the carriers in BMTs is not a straight line, and it has both a vertical and a lateral component, subject to Lorentz force deflection.

BMTs are most frequently realized by conventional IC technology (CMOS or bipolar), with some specific steps such as anisotropic etching techniques or micromachining so that part of the substrate is removed to increase the magnetosensitivity, improve the remaining sensor characteristics, and so on.[61-64] The mobility of carriers is of great importance for the transduction efficiency because of the Lorentz force that determines the action of the carriers.[4,59-66] To increase the drift velocity, a supplementary electric field in the base can be used to force a current through an additional base contact, the so-called drift-aided BMTs [Fig. 9.13(c) through (e) and Fig. 9.14(j) through (m)] At present, BMTs are mainly fabricated by silicon IC technology; the use of GaAs is being considered as a future alternative.[63,64] In addition to the BMT versions, MOSFET magnetotransistors (MAGFETs) are often added to this class of sensors. However, we believe that these transducers are pure Hall-effect devices.

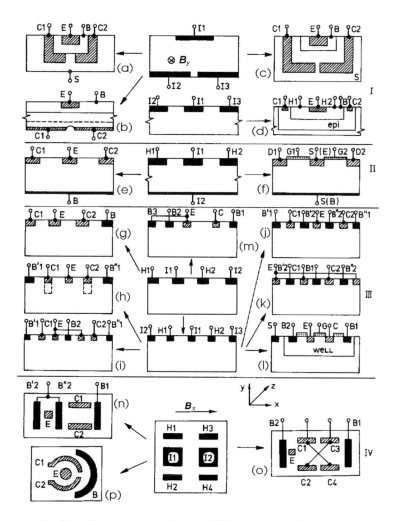

Figure 9.14 Classification of parallel-field BMT sensors, showing the respective parallel-field Hall devices.

9.7.2 Principles of BMT Operation

As a result of the variety of device structures, specific boundary conditions, and modes of operation, the sensor mechanisms of magnetic control of the output current $I_C(B)$ or $\Delta I_{C1,2}(B)$ can be attributed to one of the following:

- Emitter-region magnetosensitivity connected with the inhomogeneous in-depth distribution of doping impurities in the

emitter junction and the occurrence of a circulation Hall current in a field B. This current causes asymmetric injection, which leads to an increase of one collector current and a simultaneous decrease of the other collector current in the differential BMT structures.[4,9,59,60,65]
- Base-region magnetosensitivity related to the occurrence of the diode Hall effect (Hall voltage and Hall current); Lorentz deflection of minority carriers and the magnetoconcentration and MD effects, associated with the possibility of magnetic modulation of the base resistance by surface conditions at high emitter injection levels. Besides these mechanisms, newly described effects also operate in the base region. These include electrical control of the sign of magnetosensitivity, filament magnetosensitivity, and emitter-injection modulation.[4,23,59,60] The combined effect of all these phenomena can lead to record values of the sensitivity, reaching, for example, up to 10^6 %/T in the BMT of Fig. 9.14(m).[4,60]
- Magnetosensitivity associated with the collector-base depletion region. The electric field E_C in the collector-base depletion layer could contribute to the sensitivity in some types of BMTs [Fig. 9.14(a) through (d)] Essentially, the mechanism is identical in origin to the Hall current in short samples: Its fundamental cause is the majority-carrier deflection in the depletion region. All mechanisms act synergistically, and their identification experimentally is not a trivial problem.[4]

9.7.3 BMT Microsensors

The figures of merit of BMTs, particularly the sensitivity, the noise, and the temperature coefficient of magnetosensitivity, are quite purposeful in building MEMS. There are no problems connected with the sensor characteristics of BMTs that cannot be overcome in the same way and applying the same logic as was done for pure Hall effect devices. However, the bipolar conductivity is not at all a shortcoming; we believe that it gives more flexibility to the optimization of sensor action, particularly because BMTs can be considered as functionally integrated transistors and Hall elements. This connection is clearly proven in Figs. 9.13 and 9.14 for orthogonal and parallel-field versions. There are modes of operation in which the output characteristics of the BMTs are linear and odd functions of the field B, just like Hall sensors. In Table 9.9, the parameters of selected orthogonal and parallel-field BMTs are generalized.[4]

Table 9.9 Parameters of Selected Silicon BMTs (device structures shown in Figs. 9.13 and 9.14); adapted from Roumenin[4]

No.	Type	Device Structure	Output	Sensitivity S (% T^{-1})	Equivalent Noise (T), Δf	Equivalent Offset (mT)	Linear Range (T)	Nonlinearity (%)	TC of S (% K^{-1})	Frequency Response (MHz)	Sensor Mechanism
Orthogonal BMT Sensors (Fig. 9.13)											
1	n–p–n	(b)	Differ.	56	5×10^{-5}, 1 MHz	7	±0.2			5	Deflection
2	p–n–p	(c)	Differ.	7	2×10^{-5}, 5 Hz	36	±0.3	5	−0.7	10	Deflection
3	p–n–p	(c)	Differ.	2.6							Deflection
4	p–n–p	(c)	Differ.	1.77		45			−0.5		Deflection
5	p–n–p	(c)	Differ.	0.6			±1	5			Deflection
6	p–n–p	(d)	Differ.	500							Filament effective sensitivity
7	n–p–n	(g)	Single	60 (77 K)							Deflection
8	n–p–n	–	Differ.	400							Emitter-injection modulation
Parallel-Field BMT Sensors (Fig. 9.14)											
1	p–n–p	6.8 (c)	Differ.	3							Deflection
2	n–p–n	(b)	Differ.	8							Deflection
3	n–p–n	(c)	Differ.	5	1×10^{-5}, 2 Hz	50	±1	0.5	−0.62	1	Deflection
4	n–p–n	(d)	Differ.	6	2×10^{-5}, 5 MHz						Deflection
5	n–p–n	(d)	Differ.	1.8			±0.2	1			Deflection
6	n–p–n	(d)	Differ.	0.8							Deflection
7	n–p–n	(h)	Differ.	7.6	1×10^{-5}, 1 MHz					50	Emitter-injection modulation

#	Type	Ref	Config	Val1	Val2	Val3	Val4	Val5	Val6	Val7	Description
8	p–n–p	(h)	Differ.	7.3					−0.34	0.2	Deflection
9	p–n–p	(h)	Differ.	1.8	3×10^{-7}, 1 kHz	180	±1	1	−0.5	0.21	Magnetoconcentration deflection
10	p–n–p	(i)	Differ.	400				5			Emitter-injection modulation
11	p–n–p	(i,l)	Single	3000							Emitter-injection modulation
12	n–p–n	(j)	Differ.	3050			±0.03				Double deflection
13	p–n–p	(j,k)	Differ.	40					−0.4		Deflection
14	p–n–p	(k)	Single	30000					−0.4		Filam. eff. sensit.
15	n–p–n	(l)	Single	150	4×10^{-5}, 10 Hz		±0.1		−0.5		Deflection
16	p–n–p	(m)	Single	10^6							Filament effective sensitivity emitter-injection modulation
17	p–n–p	(n)	Differ.	3.6			±1		−0.4		Deflection
18	p–n–p	(o)	Differ.	12		100	±1	1.5	−0.5		Deflection
19	n–p–n	(p)	Differ.	8	3.3×10^{-5}, 1 MHz		±0.058			1	Deflection

9.7.4 Sensors Related to the BMTs

The circular carrier-domain magnetometers (CDMs) are equipped with a digital output proportional to the magnitude of the field B.[67,68] The position of a strongly localized current filament or electron-hole plasma domain can be observed as a function of time by means of symmetrically placed collectors. The formation of the filament is associated with operation in the negative resistance region of the S-type characteristic of a central emitter.[4] In the absence of a field $B = 0$, the domain is most probably in a neutral equilibrium. When $B \neq 0$, the Lorentz force exerts a tangential influence on the horizontal velocity component of the filament and it acquires a circular motion. The important condition for this sensor mechanism is the absence of structural boundaries in the direction of the force F_L action. These attractive transducers can be three-layer p-n-p or n-p-n devices and four-layer magnetometers such as n-p-n-p.[69] The current pulses are generated at the collectors with a frequency proportional to the induction B. A relatively high sensitivity of 250 kHz/T is reached. One of the CDMs is shown in Fig. 9.15.[70] A problem in the CDMs that should be overcome is the strong temperature dependence of the sensor parameters.

Thyristor-like magnetosensitive transducers (magnetothyristors) are related to the BMT devices. Their action is also based on magnetic control of a current filament as a result of negative resistance. In this way, the Lorentz force modulates the turn-on voltage, with a linear dependence on the induction B. Data are also available about unijunction MTs in which the current-voltage curves show negative resistance of S- and N-types and a current filament occurs as well. By magnetic control of the peak voltage, proximity switch devices can be realized.[4]

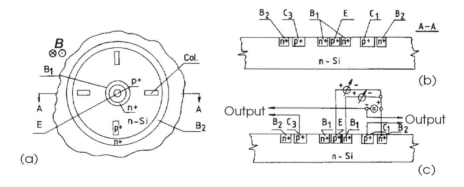

Figure 9.15 Device structure of a CDM with frequency output: (a) horizontal plane; (b) cross section; (c) biasing circuitry.

9.8 Magnetic Field-based Functional Multisensors

9.8.1 Functional Approach to Multisensors

If a multisensor IC for the measurement of more than one nonelectrical quantity is to be developed, individual sensor regions for the detection of parameters such as magnetic field, pressure, temperature, and light must be formed in the substrate.[71] The integration of several hundred thousand discrete elements on a single chip is a very serious technological task; the layout, physical limitations, electromagnetic compatibility, and so on are factors of paramount importance. The functional approach to the development of multisensors lies in the planned exploitation of physical phenomena and effects in solid-state materials that coexist in a structure (in a single active zone) and generate definite sensor properties. By this principle it is possible, simultaneously and independently, to detect more than one nonelectric measurand, such as the strength and direction of the magnetic-field components B_x, B_y, and B_z; the magnetic-field gradient ∇B; the temperature of the environment T; the light intensity Φ_l; and so on. Some unexpected new results in sensor electronics led to the question of what, in addition to the magnetic signal conversion into an electric one, is "hidden" in the active region of solid-state transducers, what parallel mechanisms take place that might yield new information by signal processing or by the formation of additional output leads. This approach stimulated the creation of functional multisensors for the magnetic field, temperature and light, 2-D and 3-D vector magnetometers, high-resolution magnetic-field gradiometers, and so on.[4,7,9,60]

9.8.2 Linear Multisensors for Magnetic Field and Temperature

Figure 9.16 schematically illustrates some of the most frequently quoted versions of differential BMTs. The onset circuit of the device is shown in Fig. 9.16(a). The analysis of these microsensors has shown that they contain a diode (p-n) thermometer.

The linear dependence of the emitter junction forward voltage drop $V_{EB}(T)$ of a diode is well known as a function of the parameter T at a constant supply current I_{EB} = const, due to the temperature dependence of the semiconductor bandgap. In this case, the voltage of the space charge

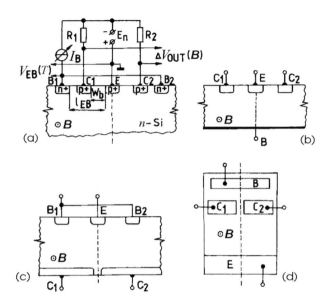

Figure 9.16 Dual-collector BMTs for multisensing.

region of the emitter p^+-n junction decreases linearly with an increase in the temperature T:

$$V_{p-n} = \frac{\Delta E_g}{q} - \left[\left(\frac{k_B T}{q}\right)\ln\left(\frac{I_{S\infty}}{I_{EB}}\right)\right] = \varphi_0 - \left[\frac{T}{11600}\ln\left(\frac{I_{S\infty}}{I_{EB}}\right)\right], \quad (17)$$

where $\varphi_0 = \Delta E_g / q$ is the bandgap in Si, $I_{S\infty}$ is the saturation current at $T \to \infty$, $I_{EB} \ll I_{S\infty}$, and k_B is the Boltzmann constant. The other notations are as usual.

According to Eq. (17), the coefficient of thermosensitivity of the p-n junction $K_{p-n} = \partial V_{p-n} / \partial T$ is negative, i.e., $K_{p-n} = -(k_B/q)\ln(I_{S\infty}/I_{EB})$. The voltage drop over the resistance R_B of the base region can be determined by the dependence $V_B = I_{EB} R_B$, as $R_B \sim \rho_B \sim 1/qn_0\mu_n$. In fact, the resistance R_B increases linearly with temperature T because of the decreasing carrier mobility μ_n. The temperature coefficient of the base K_B is positive. Therefore, the resulting coefficient $K_{EB} = \partial V_{EB} / \partial T$ of the differential diode B_1-E-B_2 in Fig. 9.16(a), which depends on the distance $l_{EB1} = l_{EB2} = l_{EB}$ and the supply current I_{EB}, is smaller than the standard value of $K_{p-n} = -2$ mV/°C for a silicon p-n junction. Hence the temperature output is a linear function of the parameter T.

The conversion of the homogeneous magnetic field B into an electrical signal by the BMTs of Fig. 9.16 was been described in Section 9.7.2. A determining condition is the sensor mechanism for generating a linear output signal $\Delta I_{C1,2}(B)$. The metrological qualities of the suggested functional multisensor are intrinsically connected by cross-sensitivity effects, mainly due to the influence of the field B on the thermodependent signal $V_{EB}(T)$. The magnetoresistance, the magnetoconcentration, the magnetodiode effect, and the emitter magnetoinjection modulation, which in principle are present in the microsensors of Fig. 9.16, can have a disturbing influence on the voltage $V_{EB}(T)$. However, the nonequilibrium bipolar conductivity and the diffusion processes with low mobility at $l_{EB1} = l_{EB2} \sim L_P$ and $B \leq 1T$ cause a small change of $V_{EB}(T)$ at $T = 300$ K of no more than 0.4%. (L_P is the diffusion length.) The influence of the temperature T on the magnetosensitivity is stable with a constant coefficient (TC) in a wide temperature range ΔT. The multisensor characteristics $V_{EB}(T)$ and $\Delta I_{C1,2}(B)$ of the silicon multisensor from Fig. 9.16(a) are presented in Fig. 9.17.[4,7] Multisensors for a magnetic field and temperature are also successfully realized by differential diode structures similar to the device shown in Fig. 9.16(a). In these devices, the conventional reverse bias of the collector p-n junction is removed because the magnetosensitive output is amperometric.[72] Ohmic n^+-n contacts have been used, besides floating p-n junctions, as differential output. These n^+-n contacts detect the magnetic control of the base currents I_{B1} and I_{B2}.[22,73,74] The highly reduced internal noise, the low-resistance output, the substantial low-power dissipation, and so on, are advantages of these amperometric multisensors.

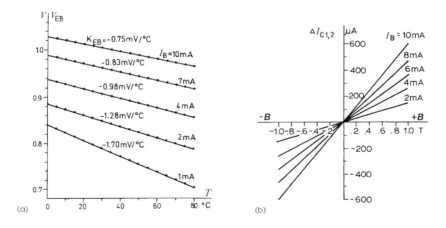

Figure 9.17 Multisensor characteristics: (a) temperature dependence of the bias $V_{EB}(T)$; (b) output magnetosensitive current.

9.8.3 Linear Multisensor for Magnetic Field, Temperature, and Light

The next step in the development of the multisensing functional principle is the increase in the number of measurands. It has been established that the transducing zone of the device in Fig. 9.16(a) can be used to measure a radiant flux Φ_l in addition to magnetic fields and temperature. The useful information about the flux Φ_l can be derived by means of simple signal processing. The circuit, including a device similar to that in Fig. 9.16(a), is presented in Fig. 9.18. The measurements of the homogeneous magnetic field B and the temperature T are identical with the ones described in Section 9.8.2. The principle of light detection is analogous to the mechanism of the well-known photodiode sensors operating in the reverse collector (C_1 and C_2) bias mode.[1] The radiant flux Φ_l arriving at the top surface will generate a nonequilibrium electron-hole plasma. As a result, both collector currents $I_{C1}(\Phi_l)$ and $I_{C2}(\Phi_l)$ are proportional to the concentration of excess minority carriers, and the two collector currents are a linear function of the flux Φ_l. The bandgap of Si is $Eg \sim 1.12$ eV; this value corresponds to the absorption of visible light whose wavelength is within the interval 0.4 µm $\leq \lambda \leq 0.76$ µm.[1] The suggested circuitry of

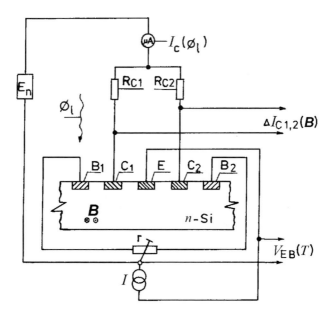

Figure 9.18 Device structure and circuitry of the multisensor for magnetic field, temperature, and radiant flux.

Fig. 9.17 adds the two collector currents. Thus, information concerning the radiant flux is represented by the difference between the total collector current $I_C(\Phi_l) = I_{C1}(\Phi_l) + I_{C2}(\Phi_l)$ and its dark value $I_C(0) = I_{C1}(0) + I_{C2}(0)$ at $\Phi_l = 0$.[4] The cross-sensitivity of these devices is discussed by Roumenin.[4,7]

9.8.4 Functional Gradiometer Microsensors

BMT Gradiometers. BMT gradiometers use the recently discovered magnetogradient effect (MGE) arising in linear output differential BMTs.[75] A possible device design and the circuitry used to observe the MGE are identical to the system shown in Fig. 9.18 except for the detection of the radiant flux. At field $B = 0$, the offset is adjusted to zero by the trimmer r to obtain sensitivities $|S_L| = |\text{-}S_R|$ on the left and right of the emitter E. The MGE linearly controls the sum of the two collector currents $I_C = I_{C1} + I_{C2}$ by the gradient ∇B in a nonuniform magnetic field when the gradient $\partial B_z/\partial x$ is oriented along the length $l_{C1,2}$ of the BMT. If the sign of the gradient changes, the polarity of $\Delta I_C(B)$ also changes. Assuming that the magnetosensitivities S_L and S_R are equal (which is achieved by the adjustment of the offset to zero) and that the induction B along the x axis has a linear distribution, the parameter $\partial B_z/\partial x$ is given by the expression

$$\Delta B_z/\Delta x = \Delta I_C(B) / S_{L,R} l_{C1,2}, \tag{18}$$

where $S_{L,R} = S/2$, $|S_L| = |\text{-}S_R|$, and S is the BMT magnetosensitivity determined in a homogeneous magnetic field.

The change of the current ΔI_C is a linear function of ∇B. The high resolution of this BMT gradiometer is determined by the distance $l_{C1,2}$ between the two collectors, which can be 50 to 80 μm in IC versions. There are also other modifications of BMT gradiometers.[7] In addition to the magnetic-field gradient measurement, the sensor described above can be used to determine the homogeneous field B using the differential signal $\Delta I_{C1,2}(B)$ and the environmental temperature T by the voltage $V_{EB}(T)$.

Hall Effect Gradiometer Microdevice. A circuit for the measurement of the magnetic-field gradient by a classical Hall microsensor ($l > w$) is presented in Fig. 9.19.[7] The distribution of the magnetic induction along the x axis is a linear function. The useful signal proportional to the gradient $\partial B_z/\partial x$ is the difference $(V_{H1} + V_{H2}) - V_{1,2}$, where $V_{1,2}$ is a voltage at the supply contacts C_1 and C_2. The circuit from Fig. 9.19 realizes the summation of the voltages V_{H1} and V_{H2} and the voltage $V_{1,2}$. Simultaneous with the measurement of the gradient ∇B, information about the strength and

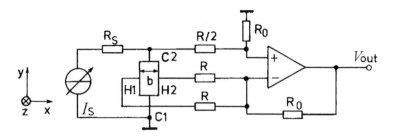

Figure 9.19 Circuit for magnetic-field gradient measurement by means of a classical Hall device.

direction of the homogeneous magnetic field by the classical Hall voltage $V_H(\mathbf{B})$ is obtained as well.

9.8.5 2-D and 3-D Vector Microsystems for Magnetic Fields

Methods for Measuring the Magnetic-Field Components. Functional integration allows the measurement with the same microstructure fixed in space of the different components of the vector of the magnetic field $\mathbf{B}$: B_x, B_y, and B_z. This approach has several advantages: extremely high resolution (the possibility of detecting the vector components at a "spot"); improved orthogonality when measuring the vector component because of the precision of planar technology steps; the position of the 3-D sensor with respect to the magnetic source is not as critical as it is in the case of 1-D transducers; optimum electrical, thermal, technological, electromagnetical, and signal processing compatibility of B_x, B_y, and B_z channels; etc. Two mutually complementary techniques (simultaneous and successive) can be used for detection of the magnetic components of the vector $\mathbf{B}$. The first technique suggests the availability of two or three sensor channels in the IC vector magnetometer, the information about B_x, B_y, or B_z coming on-line. The second technique is based on the time-extension of the metrological task. Constancy of the field $\mathbf{B}$ is necessary during the measurements. In such devices, a single structure with a fixed spatial position is used for successive detection of the orthogonal components B_x, B_y, and B_z, changing only the mode of operation.[1,4,76]

2-D and 3-D Vector Magnetometers. The synthesis of 2-D and 3-D magnetic-field microsystems for simultaneous sensing of the components of $\mathbf{B}$ is based on the functional integration of orthogonal and parallel-field Hall effect devices and BMTs. Because of the planar positions of the contacts, the trajectories of the carriers are curves; therefore the velocity has both a lateral and a vertical component. These two velocities, and the effect of the respective Lorentz deflections in magnetic fields B_x, B_y, and B_z, are used in the concrete solutions for detecting the different vector components. As an illustration, the device designs of 2-D and 3-D Si vector devices using the Hall effect are shown in Fig. 9.20.[4,36,42,77] The device designs generally contain two parallel-field devices, which are oriented to each other at 90 degrees and have a common central contact. The region in which the conversion of the different magnetic vectors in an electrical signal occurs is a single zone. The device designs of the multidimensional magnetometers based on BMTs are analogous. A very promising 2-D sensor, free from cross-sensitivity, has been made by differential SOS MDs.[58] Additional processing steps, including micromachining, are used

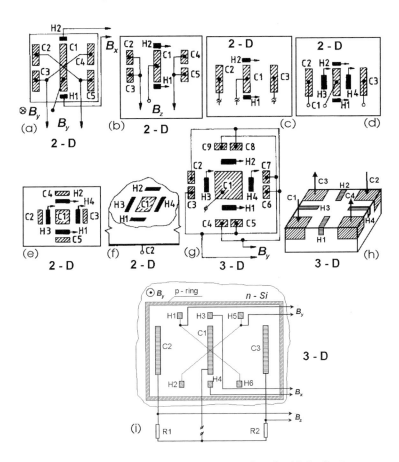

Figure 9.20 2-D and 3-D vector microsensors using the Hall effect.

to minimize the channel cross-sensitivities and increase the transducer efficiency of the different outputs. By amperometric output mode of operation, keeping equal voltage conditions on the top of the chip with and without magnetic field, the cross-sensitivity is also strongly reduced. Detailed information on the 2-D and 3-D multisensors can be found in the literature.[4,78-85] The development of vector magnetometers is one of the most promising trends in the area of magnetic sensor microsystems.

9.9 Interfaces and Improvement of Characteristics of Magnetic Microsensors

9.9.1 Biasing Circuits and Signal Processing Electronics

Generally, low resistance of voltage outputs and high resistance of current outputs are needed when output efficiency and interfacing convenience are desired. Two modes of biasing magnetic microsensors are possible, depending on the particular transducer: constant voltage and constant current.

Hall Microsensors. Hall sensors can operate in both supply modes. Certain peculiarities occur: The temperature coefficient of magnetosensitivity TC is negative at V = const, while at I = const, the TC is positive. The typical values of the output resistance of the Hall devices are of the order of 500 Ω to 5 kΩ. When using a differential amplifier, which is the most convenient solution, the output voltage of the Hall transducer is efficiently used because the input resistance of the amplifier is sufficiently large. For practical purposes, the interfaces presented in Fig. 9.21 are the most suitable.[1,4,29,45,47,86] Standard differential amplifiers are used here.

One shortcoming of the circuitry from Fig. 9.21(a) is the occurrence of a nonlinearity at the output at induction B ≥ 0.3 T, which is due to an amplifier nonlinearity.

The interface of Fig. 9.21(b) has an analog output voltage and very frequently is used in IC realization. The emitter resistors R_{E1} and R_{E2} reduce the nonlinearity, and the operation point of the circuit is improved by the symmetry of the scheme. The possibility of compensation of the offset, the stabilization of the bias circuit, and the drift minimization are some of the advantages of the interface from Fig. 9.21(c). The current output is analog. For the purposes of switching IC devices, the interfaces from Fig. 9.21(d) and (e) are particularly suitable. They have hysteresis circuitry and an amplifier. On the Fig. 9.21(d) diagram, the hysteresis is obtained by positive feedback through the additional contacts f_1 and f_2. When the differential current from split-drain MAGFETs or other analo-

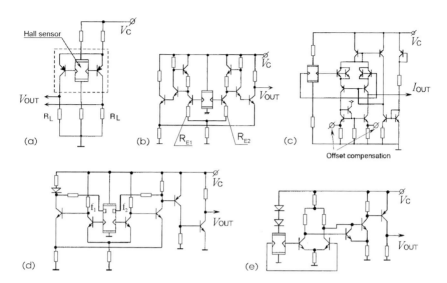

Figure 9.21 Circuit electronics for Hall sensors.

gous devices that operate in a Hall-current mode need to be measured, the most suitable scheme is that of Fig. 9.22. The well-known nulling potentiometric circuit in Fig. 9.23 is an alternative solution for the measurement of the Hall current. A digitally offset-trimmable MAGFET array has been described by Ning and Bruun.[87]

Magnetoresistors and MD Microsensors. The best metrological characteristics of the signal processing for magnetoresistors and magnetodiodes are obtained by Wheatstone-bridge circuits (Fig. 9.24). If two opposite resistors are magnetically activated, the output is doubled and becomes $V_x/2$ [Fig. 9.24(b)] If the bridge is built of four magnetoresistivity-type sensors (magnetoresistors or magnetodiodes) in a magnetic field, two of them grow and two of them decrease in value, so that the output signal will be maximal [Fig. 9.24(c)] A magnetosensor bridge amplifier with a differential stage is shown in Fig. 9.24(d).[1,4]

BMT Sensors. Figure 9.25 presents one of the most widespread BMT circuits. At $R_{C1} = R_{C2}$, the offset is adjusted to zero by the trimmer r. Figure 9.26 presents signal processing circuitry for a dual-collector BMT. This scheme provides a stable and easily adjustable bias; it measures, compensates the offset signal, and transforms the collector-current difference into an output voltage independent of the bias condition. Drift-aided BMTs and the so-called suppressed-side-wall injection BMTs can use the interface of Fig. 9.27.[4,86] The biasing circuitry of the BMT with a negative S-type resistance is shown in Fig. 9.28. The unijunction magnetotransistor B_1-E-B_2 acts as a relaxation generator. The oscillation period is

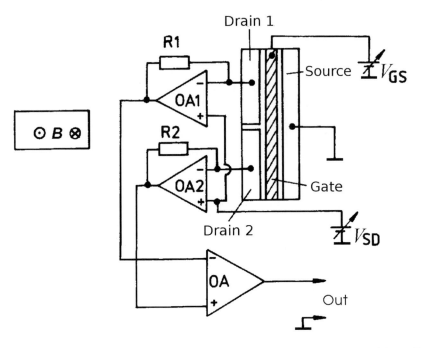

Figure 9.22 Interface for Hall current measurement using low-noise and low-offset operational amplifiers.

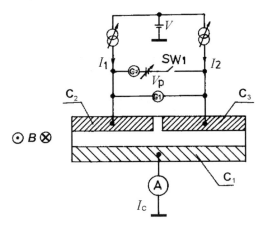

Figure 9.23 Nulling potentiometric interface for Hall current measurement. A three-terminal Hall device has been taken as an example.

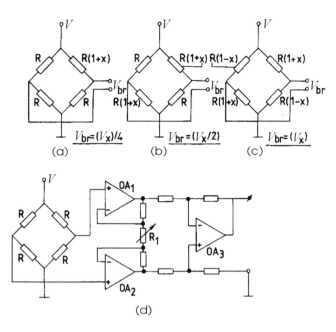

Figure 9.24 Bridge output voltage: (a) one, (b) two, and (c) four varying magnetoresistor-type sensors; (d) magnetosensor bridge amplifier with a differential stage.

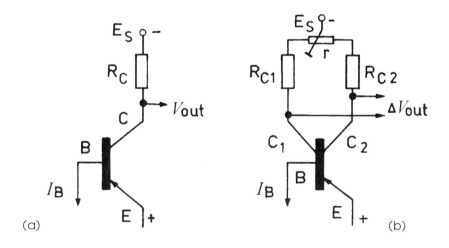

Figure 9.25 BMT circuits: (a) for a single collector; (b) for a dual-collector version.

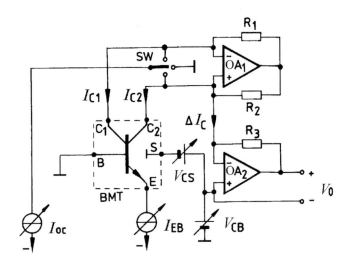

Figure 9.26 Bias and measurement interface for differential BMTs using op-amps.

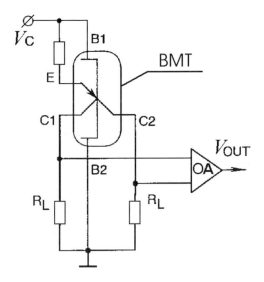

Figure 9.27 Interface for suppressed sidewall-injection magnetotransistors (SSIMT) and drift-aided BMT with op-amp.

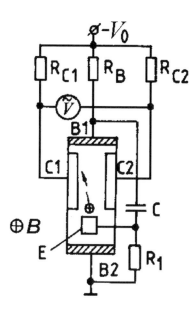

Figure 9.28 Circuit of S-type *p-n-p* BMT with frequency output.

determined by the capacitor C. At a magnetic field +*B*, the output pulsations are positive; at −*B*, the amplitude of the output signal is negative.[4] In the CDMs, it is necessary to keep the emitter at a constant current. In this way a stable working point is adjusted.[4,67-70,88]

9.9.2 Improvement of Magnetosensor Characteristics

Offset Reduction. The circuits most widely used for offset compensation of Hall sensors are demonstrated in Fig. 9.29. In general, at constant supply current I_S = const,, the offset is proportional to the input voltage, i.e., $V_{off} \approx kV_S$, with a proportionality constant k. The schemes of Fig. 9.29(a), (b), and (d) are traditional and are intended for four-lead Hall devices. The other solutions are meant for five terminal microsensors. The offset in parallel-field Hall transducers can be successfully eliminated by schemes shown in Fig. 9.29(a), (b), (d), and (e).[4] Systematic offsets can be minimized by means of symmetric device design. An effective technique for random offset compensation in a Hall element is the cross-coupled con-

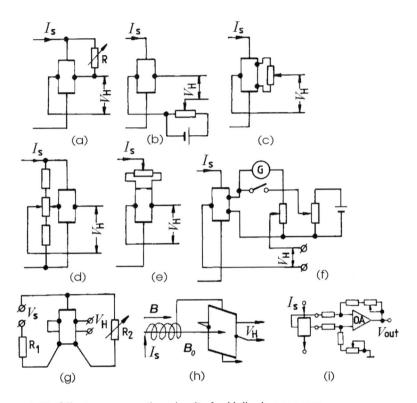

Figure 9.29 Offset-compensation circuits for Hall microsensors.

figuration. This approach is suitable in IC versions (Fig. 9.30). A very promising spinning-current offset reduction method is proposed: The supply current is applied in all directions (for example, four), and the resulting Hall voltages are averaged.[89-94] This is the best entirely IC-compatible solution. In this way, the $1/f$ noise of the Hall sensor is removed. Compensation of the temperature-dependent offset drift of a Hall device is discussed by Blanchard et al.[95] The industry's main problem is the stress-

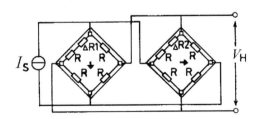

Figure 9.30 Orthogonally switched on-chip Hall plates to minimize offset errors.

related offset reduction. Sources of mechanical stress are introduced during fabrication due to intrinsic stress caused by overlying dielectric and conducting thin films[96,97 and references therein] and by the packaging process. A powerful method for dynamic stress-offset compensation of Hall devices has been developed.[98]

The offset reduction in the differential BMTs is analogous to Hall microsensors. One of the most widespread approaches for this purpose is to trim using an additional resistor connected to the collector resistor.[4] Two versions of offset compensation in drift-aided BMTs are shown in Fig. 9.31. For BMTs, a possibility is to use a fully IC-based method called sensitivity-variation offset reduction, which is based on the change of the device magnetosensitivity at constant offset.[99] The threshold offset-frequency problem in CDMs can be overcome by the use of a small biasing magnet close to the sensor. The MDs and magnetoresistors usually have a large quiescent signal. If a bridge-like circuit is used for these transducers, the quiescent offset is reduced. The MDs with differential design are also preferred because a stable behavior of a vanishing offset at the output is reached in these devices.[4]

Cross-Sensitivity Reduction. A few common compensation techniques for magnetic sensors are described here. They concern the drift of the offset and the temperature dependence of the magnetosensitivity.

a. *Compensation of the temperature drift of the offset.* Figure 9.32 shows a passive offset temperature compensation network that contains a Hall sensor, a thermistor $R_T(T)$, and two resistors R_1 and R_2. An active circuit for temperature compensation of the Hall sensor offset includes operational amplifiers

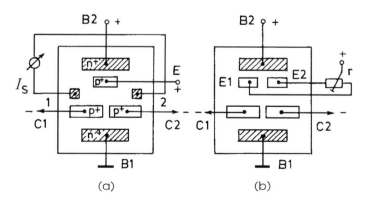

Figure 9.31 Circuitries for offset compensation in drift-aided BMTs: (a) with additional ohmic contacts 1 and 2; (b) with two emitters E_1 and E_2.

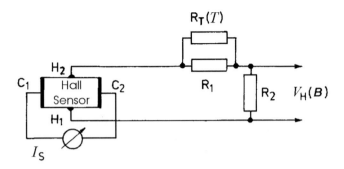

Figure 9.32 Temperature compensation of the offset using a thermistor $R_T(T)$.

(op-amps) and a temperature sensor (TS) (Fig. 9.33). These schemes can also be used successfully for BMTs.

b. *Compensation of the temperature dependence of magnetosensitivity.* Figure 9.34 shows a passive circuit for compensation of the TC of the magnetosensitivity. This scheme is suitable for Hall devices, BMTs, MDs, and other related magnetosensors. The circuitry is analogous to that shown in Fig. 9.32. An active scheme for TC compensation, shown in Fig. 9.35, contains an op-amp and is appropriate for differential magnetoresistors. An effective autocalibration method of eliminating all sensor defects in Hall devices (including the TC of magnetosensitivity) has been proposed.[100]

The TC of sensitivity of the BMTs can most effectively be compensated on the basis of the multisensor principle described in Section 9.8.2 (Fig. 9.36). The thermodependent output signal of the magnetotransistor,

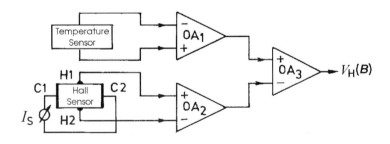

Figure 9.33 Interface for active offset thermocompensation in a Hall sensor.

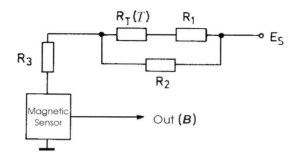

Figure 9.34 Compensation of TC of magnetosensitivity using a thermistor $R_T(T)$.

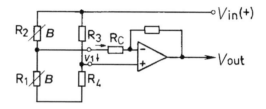

Figure 9.35 Circuit for temperature compensation of a differential magnetosensor.

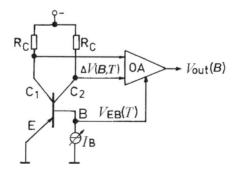

Figure 9.36 Interface for reducing the TC of sensitivity for BMTs using the multi-sensor principle.

$\Delta I_{C1,2}(B,T) \sim \Delta V(B,T)$, is supplied to the input of an op-amp OA with a controllable gain coefficient. A linear thermodependent bias $V_{EB}(T)$ is fed to its controlling input. The influence of temperature on the sensor output voltage $\Delta V(B,T)$ is compensated by the respective variation of the gain coefficient control of the op-amp, proportional to the temperature T. As a result, the output signal of the op-amp $Vout(B)$ is a function of the external magnetic field only. This method provides for the complete and precise reduction of the temperature dependence of the magnetosensitivity within a wide temperature range. A significant advantage of this method is the fact that the thermometric control signal $V_{EB}(T)$ to the op-amp OA is taken directly from the same crystal (chip) and reflects its actual temperature T very well. Thus, no additional forming of a thermodependent element in the structure is needed. This is a completely integrated solution.[4,22]

In the literature, schemes and technological steps are described by which the nonlinearity of magnetosensors can be compensated.[6,13] It must be remembered that digital signal processing using a microprocessor is a sophisticated approach to the compensation of all types of cross-sensitivities; this digital approach is of paramount importance in building smart magnetic microsystems.

9.9.3 Magnetic Systems for Contactless Measurements

The magnetic sensor is very rarely used alone, except in cases when the final aim is the measurement of the magnetic field itself. It is necessary, especially in weak magnetic fields, to use flux concentrators, coils, feedbacks, and other findings to increase the density of the magnetic flux in the active transducer zone of the sensor. Essentially, contactless galvanomagnetic instruments imply the development of magnetic systems or magnetic circuits containing constant magnets in various configurations. The goal is the generation of a suitable gradient for the measurement of different mechanical, electrical, chemical, and other quantities. Various configurations, including a magnetic-field sensor and magnetic systems, are classified in Fig. 9.37.[4]

As is clearly seen, the form of the output signal is determined by the corresponding configuration. [The relative variation of the distance between the sensor and the magnetic system is denoted by (d).] The arrangements in Fig. 9.37 can be used quite successfully for registering the magnetically conductive cores (a2, b2, c1-c3, d1, g3, g4, h1, h2); digital position instruments (a1, b1, b2, c1-c3, e2); brushless DC motors,

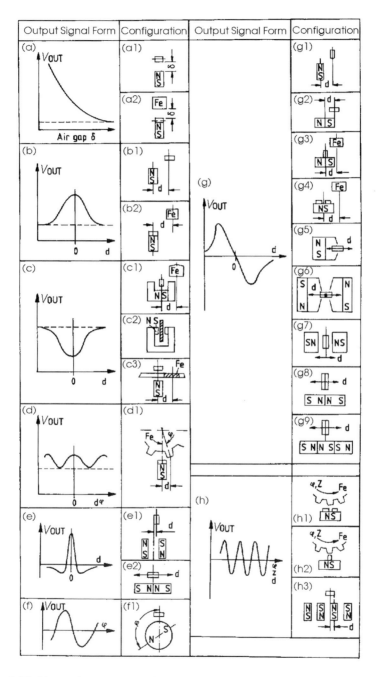

Figure 9.37 Magnetic systems with magnetic-field microsensors: configurations and forms of output signals.

angle detectors and decoders, synchros, resolvers, tachometers, and crankshaft position devices in car ignition control (d1, f1, h1, h2); linear instruments for displacement sensing within the 3 to 4 mm range (g1-g9), and so forth. The magnetic circuits shown in Fig. 9.37 (g6-g9) exhibit the highest transducer efficiency.[101] If the measuring system operates at high temperature, GaAs microtransducers are indispensable. For improvement of low-field measurement (avoiding the noise disturbances), a digital stochastic magnetic-field sensing method has been developed.[102] The fluxgate devices are a new and prospective trend in magnetic-field measurement technology.[103] Their high sensitivity and resolution, reliability, low power dissipation, long-term stability, low noise level, and so forth, allow a wide range of sophisticated applications, including antiterrorism activities, to be addressed.

9.10 Conclusions and Outlook

The short analysis of contemporary magnetic-field microsensors contained in this chapter shows clearly that they have great future potential. The magnetodevices have the unique ability to reveal realities that cannot be perceived by the human senses. With IC technology, these universally applicable transducers are completely compatible with MEMS.

The following observations and conclusions, which summarize some practical experience gathered over many years in the laboratory, may be useful to future engineers.

1. There are very few cases in which magnetic sensor microsystems function in conditions where the gradient of the measured magnetic field is effective over distances comparable to or smaller than the transducer region. In most applications, the field B is homogeneous in a range of several millimeters. Therefore, in most of the devices, the microdimensions that are typical for the sensors are hardly beneficial. This is why, for reaching high values of the absolute magnetosensitivity S, the most popular magnetic transducers, i.e., the Hall elements, must have a relatively large surface. Larger dimensions lead to more flux lines passing through the active sensor zone; a larger supply power, without heating the sensor above the permissible temperature; more effective thermal energy dissipation; reduction of the $1/f$ noise and offset; a decrease in the negative influence of short-circuited effects of the contacts on

the device's output, and so forth. Using this surface-scaling factor, the actual Hall signal can be increased many times. This approach is very useful in low-field magnetometry of the geomagnetic field, for compasses, and for hybrid MEMS. Therefore, the Hall sensor dimension is a tradeoff, depending on the particular task.

2. Enhancement of the effective magnetic field in the active zone of a magnetosensor's (such as Hall devices, BMTs, magnetoresistors, and MDs) can be attained with a magnetic flux concentrator. Even a ferromagnetic concentrator (hybrid or integrated) is able to boost the output signal at least 5 to 7 times when used for this purpose. This is why the most sensitive Hall sensors must have a large effective surface and must use flux concentrators.

3. If plastic-encapsulated packages are *not* used, the inevitable instability in the characteristics of all microsensors can be strongly reduced and the well-known piezoresistance effect and packaging stress can be eliminated. The best solutions are to cover the sensor only with nontransparent silicone glue or to use a ceramic carrier. In this way, the characteristics of magnetic devices (including their temperature behavior) can be reproduced over time and the hysteresis is drastically decreased. A good result for long-term stability can be obtained if the silicon sensors are artificially aged with a thermal treatment: 2 hours at 100°C, and 1 hour at -50°C. The temperature rate must be a degree per 2 to 3 minutes. Some stabilization of the characteristics can be reached if the supply current exceeds its nominal value by 100% for 2 to 3 hours when the sensor is switched on for the first time.

4. The multisensor approach is preferable in most of the applications. In this case, information about both the magnetic field and the temperature can be obtained by the same sensor zone in the chip. This type of transducer seems to give the most accurate and precise solution of the temperature dependence of the magnetosensitivity and the offset drift. Independent of the output type, amperometric or potentiometric, the multisensors guarantee the positive result needed.

5. If plastic-encapsulated packages are not used, all silicon microsensors for magnetic fields are efficient in a wide temperature range: 50 to 400 K. It is very important that the

cooling to cryogenic temperatures ($T = 77$ K) be quite slow: 20 minutes or longer. The samples may change their characteristics once, in the first heating/cooling cycle, but their parameters are well reproducible thereafter. The cryogenic temperatures increase the magnetosensitivity of silicon Hall sensors several times and with more than one order of magnitude for BMTs and MDs.

6. $1/f$ noise reduction in Hall sensors is still not universally recognized, despite the existence of some data. However, in BMTs and similar devices that operate in the amperometric mode, there is always a significant $1/f$ noise reduction on the output terminals in the absence of an initial current at field $B = 0$. In this situation the noise reaches levels lower even than the noise level in Hall sensors. The resolution is also highly improved.
7. A very good tip is to test the MEMS interface before correcting the sensor's defects. When the test gives the parameters needed, the magnetosensor characteristics may be improved.

The development of magnetic-field sensors is therefore likely to continue to progress on many fronts.

References

1. S. Middelhoek and S.A. Audet, *Silicon Sensors*, Academic Press, London, UK (1989)
2. R. Boll and K.J. Overshott (eds.), Magnetic Sensors, in *Sensors, A Comprehensive Survey*, vol. 5, VCH, Weinheim, Germany (1989)
3. S.H. Sze, *Semiconductor Sensors*, Wiley, New York (1994)
4. C.S. Roumenin, *Solid State Magnetic Sensors*, Elsevier, Amsterdam, The Netherlands (1994)
5. J.E. Brignell and N. White, *Intelligent Sensors Systems*, IOP Publishing, Bristol, UK (1994)
6. R.S. Popovic, *Hall Effect Devices*, Adam Hilger, Bristol, UK (1991)
7. C. Roumenin, "Functional magnetic-field microsensors: An overview," *Sensors and Materials*, 5(5):285 (1994)
8. S. Middelhoek, A.A. Bellekom, U. Dauderstadt, P.J. French, S.R. in t'Hout, W. Kindt, F. Riedijk, and M.J. Vellekoop, "Silicon sensors," *Meas. Sci. Technol.*, 6:1641 (1995)
9. C.S. Roumenin, "Magnetic sensors continue to advance toward perfection," *Sensors and Actuators*, A46-47:273 (1995)
10. W.H. Ko, "The future of sensor and actuator system," *Sensors and Actuators*, A56:193 (1996)

11. H. Baltes, "Future of IC microtransducers," *Sensors and Actuators*, A56:179 (1996)
12. H. Fujita, "Future of actuators and microsystems," *Sensors and Actuators*, A56:105 (1996)
13. R.S. Popovic, J.A. Flanagan, and P.A. Besse, "The future of magnetic sensors," *Sensors and Actuators*, A56:39 (1996)
14. K. Mohri, T. Uchiyama, and L.V. Panina, "Recent advances of micro magnetic sensors and sensing application," *Sensors and Actuators*, A59:1 (1997)
15. H.N. Norton, *Sensor and Transducer - Selection Guide*, Elsevier, Oxford, UK (1992)
16. H.J. Gray and A. Isaacs (eds.), *Dictionary of Physics*, 3rd ed., Longman Ltd., London, UK (1990)
17. E.H. Hall, "On a new action of the magnet on electric current," *Am. J. Math.*, 2:287 (1879)
18. E.H. Putley, *The Hall Effect and Semiconductor Physics*, Dover, New York (1960)
19. A.S. Grove, *Physics and Technology of Semiconductor Devices*, Wiley, New York (1967)
20. R. Mueller, *Grundlagen der Halbleiter-Elektronik*, Springer Verlag, Berlin (1971)
21. C.S. Roumenin, "Hall effect in diode structures," *Compt. rendus ABS*, 38(11):1501 (1985); "Optimized emitter-injection modulation magnetotransistor," *Sensors and Actuators*, 6:19 (1984)
22. C.S. Roumenin, A. Ivanov, and P. Nikolova, "A new class of multisensors for magnetic field and temperature based on the diode Hall effect," *Sensors and Actuators*, A85:163 (2000); "The diode Hall effect and its sensor applications—an overview," *Sensors and Materials*, 13(1):1 (2001)
23. A. Nathan, K. Maenaka, W. Allegretto, H. Baltes, and T. Nakamura, "The Hall effect in integrated magnetotransistors," *IEEE Trans. Electron. Devices*, ED-36(1):108 (1989)
24. R.F. Wick, "Solution of the field problem of the germanium gyrator," *J. Appl. Phys.*, 25(6): 741 (1954)
25. W. Versnel, "Analysis of a circular Hall plate with equal finite contacts," *Solid-State Electron.*, 24:63 (1981)
26. M. Oszwaldowski, "Hall sensors based on heavily doped n-InSb thin films," *Sensors and Actuators*, A68:234 (1998)
27. R.C. Gallagher and W.S. Corak, "A metal-oxide-semiconductor (MOS) Hall element," *Solid-State Electron.*, 9:571 (1966)
28. J. Lau, C. Nguyen, P.K. Ko, and P.C.H. Chan, "Minimum detectable signals of integrated magnetic sensors in bulk CMOS and SOI technologies for magnetic read heads," in *Proc. 8th Intern. Conf. on Solid-State Sensors and Actuators, and Eurosensors IX*, Stockholm, Sweden (1995), p. 257
29. S. Kawahito, K. Hayakava, Y. Matsumoto, M. Ishida, and Y. Tadokoro, "An integrated MOS magnetic sensor with chopper-stabilized amplifier," *Sensors and Materials*, 8(1):1 (1996)
30. D. Killat, J. v. Kluge, F. Umbach, W. Langheinrich, and R. Schmitz, "Measurement and modelling of sensitivity and noise of MOS magnetic field-effect transistors," *Sensors and Actuators*, A61:346 (1997)

31. "Improved Hall devices find new uses: orthogonal coupling yields sensitive products with reduced voltage offset and low drift," *Electron. Week*, 29:59 (April 1985)
32. Y. Sugiyama, H. Soga, and M. Tacano, "Highly-sensitive Hall element with quantum-well superlattice structures," *J. Crystal Growth*, 95:394 (1989)
33. M. Behet, J. de Boeck, G. Borghs, and P. Mijlemans, "Comparative study on the performance of InAs/Al$_{0.2}$Ga$_{0.8}$Sb quantum well Hall sensors on germanium and GaAs substrates," *Sensors and Actuators*, A79:175 (2000)
34. M. Behet, J. Bekaert, J. de Boeck, and G. Borghs, "InAs/Al$_{0.2}$Ga$_{0.8}$Sb quantum well Hall effect sensors," *Sensors and Actuators*, A81:13 (2000)
35. C.S. Roumenin and P.T. Kostov, "Planar Hall-effect device," Bulgarian Patent no. 37208, December 26, 1983
36. C.S. Roumenin, "Parallel-field Hall microsensors: an overview," *Sensors and Actuators*, A30:77 (1992)
37. R.S. Popovic, "The vertical Hall-effect device," *IEEE Electron. Device Lett.*, EDL-5:357 (1984)
38. K. Maenaka, T. Ohgusu, M. Ishida, and T. Nakamura, "Novel vertical Hall cells in standard bipolar technology," *Electron. Lett.*, 23:1104 (1987)
39. E. Schurig, M. Demierre, C. Schott, and R.S. Popovic, "A vertical Hall device in CMOS high-voltage technology," *Sensors and Actuators*, A97-98:47 (2002)
40. C.S. Roumenin and P.T. Kostov, "Optimized parallel-field Hall microsensor," *Compt. rendus ABS*, 40:51 (1987)
41. S.R. in't Hout and S. Middelhoek, "A 400°C silicon Hall sensor," *Sensors and Actuators*, A60:14 (1997)
42. R.S. Popovic, "Not-plate-like Hall magnetic sensors and their applications," *Sensors and Actuators*, A85:9 (2000)
43. C.S. Roumenin, D. Nikolov, and A. Ivanov, "Enhancing the sensitivity of Hall microsensor by minority carrier injection," *Electron. Lett.*, 36:1375 (2000)
44. R.A. Smith, *Semiconductors*, 2nd edition, Cambridge Univ. Press, Cambridge, UK (1977)
45. H. Weiss, *Structures and Application of Galvanomagnetic Devices*, Pergamon, Oxford, UK (1969)
46. S. Kataoka, "Recent developments of magnetoresistive devices and applications," *Circulars Electrotech. Lab.*, no. 182, Electrotech. Lab., Tokyo, Japan (1974)
47. *Sensoren*, Magnetfeldhalbleiter, Teil 1, Siemens AG Aktiengesellschaft, Munchen, Germany (1997)
48. M.N. Balbich, J.M. Broto, A. Fert, F. Nguyen Van Dau, F. Petroff, P. Etienne, G. Creuzet, A. Friederich, and J. Chazelas, "Giant magnetoresistance of (001)Fe/(001)Cr magnetic superlattices," *Phys. Rev. Lett.*, 61:2472 (1988)
49. K.-M.H. Lenssen, D.J. Adelerhof, H.J. Gassen, A.E.T. Kuiper, G.H.J. Somers, and J.B.A.D. van Zon, "Robust giant magnetoresistance sensors," *Sensors and Actuators*, A85:85 (2000)
50. http://www.semiconductors.philips.com/discretes
51. http://www.nve.com
52. A. Barthelemy, A. Fert, and F. Petroff, *Handbook of Magnetic Materials*, vol. 12 (K.H.J. Buschow, ed.), Elsevier, Amsterdam, The Netherlands (1999), Chap. 1

53. H. Pfleiderer, "Magnetodiode model," *Solid-State Electron.*, 15:335 (1972)
54. E.I. Karakushan and V.I. Stafeev, "Magnetodiodes," *Sov. Phys.-Solide State*, 3:493 (1961)
55. S. Cristoloveanu, "L'effet magnetodiode et son application aux capteurs magnetiques de haute sensibilite," *L'Onde Electrique*, 59:68 (1979)
56. A. Chovet, S. Cristoloveanu, A. Mohaghegh, and A. Dandache, "Noise limitation of magnetodiodes," *Sensors and Actuators*, 4:174 (1983)
57. O.S. Lutes, P.S. Nussbaum, and O.S. Aadland, "Sensitivity limits in SOS magnetodiodes," *IEEE Trans. Electron. Devices*, ED-27:2156 (1980)
58. C.S. Roumenin, "2-D magnetodiode sensors based on SOS technology," *Sensors and Actuators*, A54:584 (1996)
59. A.W. Vinal and N.A. Masnari, "Operating principles of bipolar transistor magnetic sensors," *IEEE Trans. Electron. Devices*, ED-31:1486 (1984)
60. C.S. Roumenin, "Bipolar magnetotransistor sensors: an invited review," *Sensors and Actuators*, A24:83 (1990)
61. R. Castagnetti, M. Schneider, and H. Baltes, "Micromachined magnetotransistors," *Sensors and Actuators*, A46-47:280 (1995)
62. Y.A. Chaplygin, A.I. Galushkov, I.M. Romanov, and S.I. Volkov, "Experimental research on the sensitivity and noise level of bipolar and CMOS integrated magnetotransistors and judgment of their applicability in weak-field magnetometers," *Sensors and Actuators*, A49:163 (1995)
63. R. Castagnetti, *Integrated Magnetotransistors in Bipolar and CMOS Technology*, Phys. Electr. Lab., Zurich, Switzerland (1994), ETH-Hoenggerberg, HPT, ISBN 3-907574-05-2
64. C. Riccobene, *Multidimensional Analysis of Galvanomagnetic Effects in Magnetotransistors*, Phys. Electr. Lab., Zurich, Switzerland (1995), ETH-Hoenggerberg, HPT, ISBN 3-907574-08-7
65. Uk-Song Kang, Seung-Ki Lee, and Min-Koo Han, "Highly sensitive magnetotransistor with combined phenomena of Hall effect and emitter injection modulation operated in the saruration mode," *Sensors and Actuators*, A54:641 (1996)
66. A. Nagy and H. Trujillo, "Highly sensitive magnetotransistor with new topology," *Sensors and Actuators*, A65:97 (1998)
67. B. Gilbert, "Novel magnetic field sensors using carrier domain rotation: proposed device design," *Electron. Lett.*, 12:698 (1976)
68. J.I. Goicolea, R.S. Muller, and J.E. Smith, "Highly sensitive silicon carrier domain magnetometer," *Sensors and Actuators*, 5:174 (1984)
69. G. Persky and D.J. Bartelink, "Controlled current fillaments in PNIPN structures with application to magnetic field detection," *Bell Syst. Tech. J.*, 53:467 (1974)
70. C.S. Roumenin and V. Dobovska, "Circular magnetometer," *Sensors and Actuators*, A32:661 (1992)
71. D.L. Polla, R.S. Muller, and R.M. White, "Integrated multisensor chip," *IEEE Electron. Device Lett.*, EDL-7:254 (1986)
72. C. Roumenin, P. Nikolova, and A. Ivanov, "Magnetotransistor sensors with amperometric output," *Sensors and Actuators*, A69:16 (1998)
73. C.S. Roumenin, A. Ivanov, and P. Nikolova, "A linear multisensor for temperature and magnetic field based on diode structure," *Sensors and Actuators*, A72:27 (1999)

74. C.S. Roumenin, A. Ivanov, and P. Nikolova, "A new anisotropy effect in the amperometric magnetic-field microsensors," *Sensors and Actuators*, A77:195 (1999)
75. C.S. Roumenin, "Magnetogradient effect in differential bipolar magnetotransistors," *Compt. rendus ABS*, 42(12):63 (1989)
76. C.S. Roumenin, "Three-dimensional magnetic field vector sensor," *Compt. rendus ABS*, 41(1):59 (1988)
77. C.S. Roumenin, K. Dimitrov, and A. Ivanov, "Novel integrated 3-D silicon Hall magnetometer," in *Proc. 14th Europ. Confer. on Solid-State Transducers - Eurosensors XIX*, Copenhagen, Denmark (2000), p. 759
78. M. Paranjape, L.M. Landsberger, and M. Kahrizi, "A CMOS-compatible 2-D vertical Hall magnetic-field sensor using active carrier confinement and post-process micromachining," *Sensors and Actuators*, A53:278 (1996)
79. M. Paranjape and I. Filanovsky, "A 3-D vertical Hall magnetic field sensor in CMOS technology," *Sensors and Actuators*, A34:9 (1992)
80. L. Zongsheng, W. Tianping, and W. Guangli, "A new integrated JFET 3-D magnetic-field sensor in VIP technology," *Sensors and Actuators*, A35:213 (1993)
81. Lj. Ristic and M. Paranjape, "Hall devices for multidimensional sensing of magnetic field," *Sensors and Materials*, 5:301 (1994)
82. K. Maenaka, Y. Shimizu, M. Baba, and M. Maeda, "Application of multidimensional magnetic sensors - position and movement detection," *Sensors and Materials*, 8:33 (1996)
83. H. Lin, T. Lei, Jz-Jan Jeng, Ci-Ling Pan, and C. Chang, "A novel structure for three-dimensional silicon magnetic transducers to improve the sensitivity symmetry," *Sensors and Actuators*, A56:233 (1996)
84. A. Nagy and H. Trujillo, "3D magnetic-field sensor using only a pair of terminals," *Sensors and Actuators*, A58:137 (1997)
85. Ch. Schott, D. Manic, and R.S. Popovic, "Microsystem for high-accuracy 3-D magnetic-field measurements," *Sensors and Actuators*, A67:133 (1998)
86. K. Maenaka and M. Maeda, "Interface electronics for integrated magnetic sensors," *Sensors and Materials*, 5:265 (1994)
87. F. Ning and E. Bruun, "An offset-trimmable array of magnetic-field-sensitive MOS transistors (MAGFETs)," *Sensors and Actuators*, A58:109 (1997)
88. S. Kirby, "Temperature compensation technique for the carrier domain magnetometer," *Sensors and Actuators*, 3:373 (1983)
89. P. Munter, *Spinning-Current Method for Offset Reduction in Silicon Hall Plates*, Delft Univ. Press, Delft, The Netherlands (1992), ISBN 90-6275-780-4/CIP
90. S. Bellekom, *Origin of Offset in Conventional and Spinning-Current Hall Plates*, Delft Univ. Press, Delft, The Netherlands (1998), ISBN 90-407-1722-2/CIP
91. R. Steiner, C. Maier, A. Haberli, F.-P. Steiner, and H. Baltes, "Offset reduction in Hall devices by continuous spinning current method," *Sensors and Actuators*, A66:167 (1998)

92. S. Bellekom and P.M. Sarro, "Offset reduction of Hall plates in three different crystal planes," *Sensors and Actuators*, A66:23 (1998)
93. S. Bellekom, "CMOS versus bipolar Hall plates regarding offset correction," *Sensors and Actuators*, A76:178 (1999)
94. C. Muller-Schwanneke, F. Jost, K. Marx, S. Lindenkreuz, and K. von Klitzing, "Offset reduction in silicon Hall sensors," *Sensors and Actuators*, A81:18 (2000)
95. H. Blanchard, C. de Raad Iseli, and R.S. Popovic, "Compensation of the temperature-dependent offset of a Hall sensor," *Sensors and Actuators*, A60:10 (1997)
96. Y. Kanda and Migitaka, "Effect of mechanical stress on the offset voltage of Hall devices in Si IC," *Phys. Status Solidi* (A), 35:115 (1976)
97. R. Steiner, C. Maier, M. Mayer, S. Bellekom, and H. Baltes, "Influence of mechanical stress on the offset voltage of Hall devices operated with spinning current method," *J. Microelectromech. Syst.*, 8(4):466 (1999)
98. R.S. Vanha, *Rotary Switch and Current Monitor by Hall-Based Microsystems*, Phys. Electr. Lab., ETH Zurich, Switzerland (1999), ISBN 3-89649-446-5
99. S. Kordic, V. Zieren, and S. Middelhoek, "A novel method for reducing the offset of magnetic-field sensors," *Sensors and Actuators*, 4:55 (1983)
100. P.L. Simon, P.H.S. Vries, and S. Middelhoek, "Autocalibration of silicon Hall devices," in *Proc. 8th Intern. Conf. on Solid-State Sensors and Actuators, and Eurosensors IX*, Stockholm, Sweden (1995), p. 237
101. C.S. Roumenin, "New magnetomodulating systems for contactless measurement of linear displacements," *Compt. rendus ABS*, 41(4):25 (1988)
102. S. Hentschke, "Digital stochastic magnetic-field detection," *Sensors and Actuators*, A57:1 (1996)
103. P. Ripka, *Magnetic Sensors and Magnetometers*, Artech House Books, London, UK (2001)

10 Mechanical Microsensors

Franz Laermer

Robert Bosch GmbH, Stuttgart, Germany

10.1 Introduction

Micromachining technologies have enabled a reduction in the size of mechanical sensors and an increase in their functionality to unprecedented levels of miniaturization. In many applications, precision-machined devices already existed when micromachined solutions entered the market. To replace established solutions, mechanical microsensors had to prove their competitiveness with respect to cost, size, and performance. Success stories were due to enhanced functionality, increased accuracy and performance, and higher reliability, at lower device, packaging, and mounting costs. In many applications, such as the automotive area, which was and still remains the strongest driver for MEMS-based sensor systems, sensor cost is one of the most important factors deciding the success and the degree of market penetration of a new system. Starting with the replacement of older sensor generations established in already-existing systems like the airbag, mainly for cost reasons, mechanical microsensors are currently enabling completely new systems that critically rely on them. One well-known example is the Electronic Stability Program (ESP) or Vehicle Dynamics Control (VDC) system, which would not have been affordable and would not have reached today's performance if it had to rely on classical mechanical sensor approaches. It becomes more and more obvious that the exclusivity of systems like ESP/VDC depends on the availability of high-performance sensors, in this case mainly inertial sensors, around which hardware and software algorithms are developed.

As for all MEMS devices, a sufficiently high production volume is a must to achieve a low cost. Factory costs easily reach $50 to 100 million because the production of MEMS devices requires equipment, tools, processes, and a processing environment close to semiconductor standards, such as a clean-room infrastructure, photolithography, deposition and etch tools, and wafer handling and cleaning systems. These high infrastructure costs can only be recovered by a production volume that is sufficiently high, i.e., several million (or, better, tens of millions of) sensors per year. Even then, the enormous initial investments are only bearable by large enterprises. Quite often, semiconductor manufacturers who have the chance to share part of their semiconductor equipment for the fabrication of MEMS have become key players in the new field of mechanical microsensors. Because of the high infrastructure costs, only a few developments have led to successful large-volume products. Important markets that offer the high-volume opportunities needed for the success of mechanical microsensor products are automotive, computers and peripherals, consumer products, and medical and biological applications.

Other application fields like space, avionics, naval, the military sector (intelligent bombs, missiles, smart ammunition), and the gas and petroleum industry (hydrophones, geophones) are comparatively small in volume and require highly specific sensors with dedicated performance features, enhanced robustness, and shock and vibration tolerance, often combined with the capability to function under harsh environmental conditions. These constitute attractive niches for specialized MEMS players rather than a true mass market. The highly specialized sensors are manufactured by a number of highly specialized companies at an elevated cost, mostly through foundry access granted by larger production facilities. The task of these companies is not running their own technological infrastructure, but organizing and coordinating the foundry services network needed to fabricate their devices, which often extends to more than one location.

10.1.1 Automotive

The automotive sector is one of the most important drivers in the field of mechanical microsensors. The penetration of the car by electronic systems is rapidly increasing, to improve the comfort, safety, and economy of driving, as well as for environmental reasons. In the year 2000, about 17% of the total value of a new car was electronic systems. In 2010, this fraction is expected to reach about 24%. Most of these electronic systems require the input of physical parameters. This input is provided by a rising number of sensors. Mechanical microsensors in modern cars are mainly pressure sensors, inertial sensors such as accelerometers and gyroscopes, and sensors for steering angle and torque. The automotive sector is characterized by its requirement for high precision and high reliability, with a guaranteed lifetime of more than 10 years, which is close to the high end of MEMS sensor technology. Strong pressure is also exerted on sensor costs. Under these requirements, MEMS-based sensors have been able to achieve a strong position in this field over the past few years. Automotive siliconbased sensors reached a total annual market volume of about $1.3 billion in 2003. This number is expected to grow to $2.7 billion in 2010.

10.1.2 Computers and Peripherals

Computers and peripherals are another important field for mechanical microsensors. One of the first successful MEMS devices that conquered a mass market was the ink-jet printing head (see Chapters 12 and 14). An astonishingly wide spectrum of technologies is covered by the diverse printing head generations on the market, ranging from devices fabricated in nickel using the LIGA (from German: Lithographie, Galvanik, und Abformung) process, to purely silicon-based devices fabricated by deep reactive ion etching (DRIE), together with bonding techniques. The ink-jet printing head success story created a strong awareness of the new MEMS-related capabilities within the computer society. Presently, inertial sensors are found in hard disk drives for position control of the read-write head, in wireless computer mice to be used as a pointer to control curser position and program actions, and in virtual reality head gear where both accelerometers and gyroscopes pick up the position and motion of the player's head as inputs to scientific software (e.g., micromolecular structure visualization, 3D animation) and to advanced computer games guiding the player through virtual worlds. These fascinating perspectives are being realized by the computer and software industries, and a huge market potential is expected for these applications. Sensor requirements are in general at the very low end of the technical spectrum; however, extreme cost pressure is pushing down sensor prices to between $2 and $3.

10.1.3 Consumer Products

The consumer products sector is an area of strong growth for MEMS-based devices, including mechanical sensors. One of the more spectacular achievements in the past few years in the field of MEMS consumer applications is the digital mirror device (DMD), from Texas Instruments, for high-resolution projection displays. Once again, this is not a sensor, but another success story that creates awareness and opens the field for MEMS in general. More than 1.5 million tiny mirrors made from aluminium on top of an integrated circuit allow projecting the television screen onto m^2 of white wall with unprecedented brilliance. If prices come down further by a factor of 5 to 10 from today's levels, which is expected with rising production volumes and increasing wafer sizes, this will be an attractive solution for large-screen television, enabled by a very smart device. Inertial sensors, mainly gyroscopes, found their way into camcorders and electronic cameras, primarily for blur compensation. Acoustic devices such as silicon microphones are making their way into high-fidelity equipment, cellular phones, and electronic cameras. In the latter application, the purpose is ultrasonic distance measurement.

Household goods including dishwashers, washing machines, and vacuum cleaners are upgraded by microcontrollers combined with sensors, e.g., pressure sensors that control the amount of water used or the efficiency of cleaning.

Packages for fragile or precious goods are equipped with accelerometers that monitor shocks during transport and document peak values in a nonvolatile memory. These data serve as a proof for the carrier with regard to later actions for damages.

Toys equipped with inertial sensors also enjoy increasing popularity. One example is model helicopters that use gyroscopes to stabilize their flight operation.

The first handheld navigation systems have appeared on the market, combined with cellular phone functions. They both rely on periodic satellite contact to update their position accurately and on the output of gyroscopes to cover the time between satellite contacts. These systems are similar to car navigation systems, but smarter. The commercial sector spans a wide range of sensor specifications, most of them low-end and low-price, but in some applications, high performance is needed in a mid-price range.

10.1.4 Medical and Biological Applications

Much attention has been drawn to biomedical applications, where the combination of high-performance sensing and a high degree of miniaturization through MEMS techniques offers unique opportunities to improve the quality of life for the ill and the disabled. Microdosing systems combined with microsensors to control the supply of drugs such as insulin to the bloodstream are close to maturity. Pressure sensors are widely used in electronic blood pressure controllers. Implants for a permanent monitoring of blood or ocular pressure help to optimize medical treatment schemes. Inertial sensors attached to the legs of people suffering from certain types of brain damage help improve their walking capabilities. Accelerometers that monitor motion activity of the body yield input to cardial stimulators, to provide the right amount of electrical stimulation needed at the current level of activity. The biomedical field is certainly one of the most attractive for MEMS-based sensors, for its high volumes and very strong growth. Sensor specifications are comparatively high-end in a mid- to high-price regime.

This short overview of existing and potential markets for mechanical microsensors does not claim to cover the full spectrum of applications. The intention was just to show some important fields in which large markets or market potentials exist today and which were identified as the driving forces for present mechanical microsensors. In the future, it may well be that completely new markets for MEMS sensors will emerge, given their enormous capabilities with respect to performance, reliability, miniaturization, low power consumption, and low price. (See also Chapters 12 and 13.)

10.2 Inertial Sensors

Inertial sensors measure physical quantities of motion of a solid object. The accessible physical quantities are acceleration and angular velocity. Acceleration of one point of a body and its rate of rotation around three orthogonal axes provide a full description of the body's motion. Inertial data are most relevant to control moving objects, whether it be an airplane, a marine vessel, or an automobile. The automotive field exerts the strongest push on micromechanical inertial sensor technology, for passenger safety systems like front- and side-airbag, rollover protection, vehicle dynamics control, and navigation systems. Strong innovation pressure comes from vehicle dynamics control,[1] a system that prevents skidding of the car in difficult situations by selective braking actions to certain wheels, with ever-improving system capabilities with each new generation of gyroscopes. Here the correlation between system performance and enabling sensor performance is most obvious.

Inertial sensors detect forces resulting from motion. These forces act on inertial masses within the sensor core and form part of the sensor, without needing any contact with an outside medium. Inertial sensors are generally sealed and isolated from the environment and the surrounding media by a hermetic package, e.g., a welded steel housing, a ceramic package, or a silicon cap soldered onto the sensor chip. This perfect isolation has a significant impact on the system design, since contact with a medium does not have to be considered. For example, the question of whether monolithic or hybrid integration is the best approach is strongly affected by the exclusion of the ambient. This question must be considered for each type of sensor and answers will differ, depending on the application.

There are two important inertial sensors: accelerometers and gyroscopes. These MEMS-based sensors are discussed below.

10.2.1 Accelerometers

10.2.1.1 Fundamentals

Mechanical accelerometers consist of a spring-mass system, with a seismic mass carried by elastic tether beams. Acceleration along the sensitive axis leads to a deflection of the seismic mass, with elastic forces from the tether beams balancing the external forces. With m and k the seismic mass and the spring constant of the spring-mass system, the resonance frequency of the spring mass system f_0 is:

$$f_0 = \frac{1}{2\pi}\sqrt{\frac{k}{m}},$$

and the quasistatic sensitivity s, i.e., the deflection per acceleration unit at low frequencies, is:

$$s = \frac{1}{\omega_0^2} = \frac{1}{4\pi^2 f_0^2}.$$

There is a fundamental conflict between sensitivity of the mechanical sensor element and the dynamic range, which is limited by f_0. The higher the sensitivity of the accerometer, the slower its response, and vice versa.

The deflection of the mass itself is detected by means of capacitive elements, the capacitances of which change with deflection, or by piezoresistive elements that detect strain induced by the motion of the seismic mass through a change in resistor values. Frequency analog sensors convert induced stress σ into a change of resonance frequency of vibrating elements:

$$\Delta f \approx f_0 \cdot \frac{\sigma}{2\sigma_{crit}},$$

where σ_{crit} denotes the critical stress at which the vibrating beam buckles.

Piezoelectric sensors convert strain into an electrical output voltage through the piezoelectric effect. Since silicon is not piezoelectric, the latter variant needs additional piezoelectric elements made, for example, from zinc oxide (ZnO),[2] lead zirconate titanate (PZT), or aluminum nitride (AlN),[3] which are deposited as thin films onto the silicon structures. As an alternative, the sensors can be fabricated completely from PZT[4] or other piezoceramic materials by molding and subsequent firing of the raw ceramics, or from quartz material.[5]

Apart from these detection methods, optical detection[6] and tunneling current sensing have also been used in more exotic applications, as have dynamic effects in resonant sensors.[7] There is one big supplier of a resonant type of accelerometer on the market with frequency-analog output, namely, SensoNor in Horten, Norway, with its SA30 accelerometers, among other products. A frequency analog sensor has the major advantage that its oscillating parts represent a way of permanent self-test, with the resonance frequency being an indicator for mechanical integrity of the whole sensor, which would not be achievable to this extent with other sensing principles. The disadvantage is an increased circuit effort for undamping the resonator and frequency-to-voltage conversion (phase-locked-loop or timer), and the lack of anti-aliasing capabilities in the primary mechanical conversion of acceleration into resonance frequency change. In addition, permanently vibrating elements need a drive structure, which is achieved, in the case of the SA30, by a thermoelectric drive through heated resistors powered by an alternating current of corresponding frequency. Alternatively, a piezoelectric drive can be implemented by a piezoelectric thin film deposited on top of the vibrator, for electromechanical conversion.

Today, however, the most widespread MEMS-based accelerometers rely either on piezoresistive detection or on capacitive detection principles.

10.2.1.2 Piezoresistive Accelerometers

The Piezoresistive Effect

Piezoresistive accelerometers consist of a seismic mass carried by one or several tether beams, with at least one of the tether beams containing piezoresistors, in most cases four piezoresistors in a full Wheatstone-bridge arrangement. If the mass is deflected by external forces, the piezoresistive elements are strained and change their resistance values, leading to imbalance of the bridge and appearance of a bridge voltage if fed by an electrical current. The piezoresistive effect is commonly described by:

$$\frac{\Delta R}{R} = \Pi_l \cdot \sigma_l + \Pi_t \cdot \sigma_t$$

where Π_l and Π_t denote the longitudinal and transversal piezoresistive coefficients, respectively; σ_l and σ_t are the corresponding stress values caused by mechanical strain of the structure; and $\Delta R/R$ denotes the relative resistance change.[8]

Fabrication Aspects

Piezoresistors are either implanted and diffused into silicon by dopant atoms, or deposited on top in the form of metal, polysilicon, or single crystalline conducting lines (see Chapter 15). These processing steps are common in semiconductor technology and represent no deviation from standards. A more detailed discussion on piezoresistive sensing, its advantages and drawbacks, is given in Section 10.3. All questions related to the detection principle are equivalent for both piezoresistive pressure and acceleration sensors. As with pressure sensors, piezoresistive detection appears as the detection principle of choice with bulk micromachined sensors. The main reason for this is that practically all of the bulk micromachining steps can be restricted to the wafer backside. In general, backside etching is done in alkaline solutions, with the finished wafer frontside protected by an etchbox, which also provides electrical contacts to the wafer in case an electrochemical etch stop (pn junction) is used in the process. At all times, closed silicon membranes separate the wafer frontside from the backside of the wafer, which is KOH-etched. Instead of a simple cavity defining a membrane, more complicated geometries need to be etched to preserve seismic masses of full wafer thickness within the cavity, for increased inertial mass and higher sensitivity toward external acceleration compared with a mass of only membrane thickness. The mass is tethered to the substrate through appropriate design and processing. Since the mass represents a convex structure, the corners of which are subject to a substantial attack by the etchant solution, corner compensation structures[9] discussed in Chapter 15 must be applied. Otherwise, a significant fraction of the seismic mass as defined by the etch-mask would be removed by the poorly controlled under-etching of the mask corners during the wet etching. This is highly undesirable.

In contrast to "dirty" backside processing, the preceding frontside processes (oxidation, deposition steps, implantation of dopants, metallization) are "clean" and remain close to or within semiconductor standards. Only at the very end of the fabrication process, after finishing both frontside and subsequent backside processing with an intact membrane still in between both wafersides, is plasma etching (RIE, DRIE) through the membrane performed from the wafer frontside, to separate the tether beams and the movable mass out of the membrane. This last step departs from pressure sensor bulk-micromachining: for an accelerometer process, the membrane must finally be structured to yield free-standing beams and a mass free to move. After the final release step, in general, the sensor wafer is bonded to a carrier wafer (pyrex glass or a second silicon wafer), sometimes also to a capping wafer, before the wafer sandwich is diced into device chips. This provides for both mechanical damping and overload protection, at least in one direction. For bulk-micromachined accelerometers, due to the comparatively large mass and spring constants, gas damping by the enclosed atmosphere[10] is normally insufficient to suppress gain enhancement around the fundamental resonance of the spring-mass system to values below 10, which must be taken into account, for example when designing the electronics.

Monolithic Integration

Monolithic integration is comparatively easy to achieve with this type of sensing principle. It is not a must, however, thanks to the low-impedance raw signals from the resistive bridge, which permits even long wires to external circuitry with no significant drawbacks. No process compatibility problems appear for fabricating the piezoresistors and the electronics on the same wafer side. This statement also holds for the final DRIE membrane separation step, which is compatible with both CMOS and bipolar circuit technology.

Looking at the advantages of a monolithically integrated accelerometer in the piezoresistive case, there is potentially an increase in reliability (no bond-wires, lower EMI sensitivity), better exploitation of surface area already consumed by the anisotropic backside etching (which is lost otherwise), and easier temperature compensation and calibration by the electronics integrated with the sensing element on the same chip (negligible temperature difference between the sensor and the electronics). However, on the market, monolithically integrated piezoresistive accelerometers have no significant impact, mainly because of the higher process complexity of an integrated process, lower yield, and increased sensor cost by the relatively large chip sizes compared to the electronic circuit sizes: The sensing elements are generally so large (10 mm^2 or even larger) that the fabrication of sensor and electronics chips in separate processes on separate wafers is more cost-efficient than an integrated process.

Conclusions

Piezoresistive micromechanical accelerometers with hybrid electronics, fabricated as described here, have been on the market since the early 1980s from suppliers such as SensoNor, Denso, and Hitachi. They have proven very successful in the automotive and the avionics field. The main disadvantages are the need for extensive calibration and temperature compensation of offset and sensitivity, since the piezoresistors change their resistance values (and leakage currents, in the case of diffused piezoresistors) by more than an order of magnitude between −20 and +140°C, and the lack of an integrated self-test feature: In any safety-critical application, periodic self-tests are requested to check regularly for the integrity of the sensor elements and the functionality of the electronic circuit, which means inducing mechanical deflection of the sensing element and subsequent evaluation of the resulting electrical signals. In a piezoresistive-type accelerometer, substantial actuation is not intrinsic to the detection principle and requires additional, appropriate actuators, for example an electrostatic, electrothermal,[11] or piezoelectric actuator affecting the seismic mass. This entails additional effort and costs and represents a serious drawback of piezoresistive accelerometers in safety-critical applications.

10.2.1.3 Capacitive Accelerometers

Working Principles

Capacitive micromechanical accelerometers appeared in the late 1980s.[12] They rapidly established themselves as preferred devices, mainly because of their increased accuracy, linearity, and temperature stability, and the reduced calibration effort. In general, no temperature compensation is needed in capacitive sensing, and a full-range adjustment is the only calibration factor needed. In addition, capacitors serve both as sensing and electrostatic actuation elements: If a voltage U is applied to the capacitive structure, electrostatic forces proportional to U^2 act on the movable structure. This dual nature of capacitive detection and electrostatic driving is one of the key advantages of this type of accelerometer, which makes, for example, a self-test option easy to implement into the evaluation algorithm. In addition, force-feedback in a closed-loop evaluation circuit is possible, which keeps the seismic mass at a nearly fixed position by compensating external forces with corresponding forces from the electrostatic actuator. This reduces the effective deflection of the seismic mass by the loop gain factor, thus improving linearity, noise behavior, and dynamic response of the system in comparison to open-loop operation.

In capacitive sensors, deflection of the seismic mass is usually transformed into corresponding capacitance changes in a differential capacitor arrangement. As the seismic mass approaches one detecting surface, thereby increasing the corresponding capacitance value, it simultaneously moves away from the other detecting surface, thereby decreasing the corresponding capacitance value. There are a number of electronic circuit principles enabling these capacitance changes to be transformed into an electrical voltage, in so-called C/V converters: switched capacitor circuits, ΣΔ-converters, pulsewidth converters, carrier-frequency-based demodulators, and so on, each with or without force-feedback (closed loop or open loop, respectively). In most cases, the electrical output voltage of the C/V converter is proportional to:

$$\frac{C_1 - C_2}{C_1 + C_2} = \frac{\Delta x}{x_0},$$

with Δx the deflection of the seismic mass, x_0 the distance between capacitor plates in the undeflected state, and C_1 and C_2 the individual capacitance values in the differential capacitor arrangement, respectively. In the undeflected state, $C_1 = C_2 = C$ holds in most cases. For the mechanical sensitivity $s = \Delta x/a$, the same relation as before links deflection per acceleration unit with resonance frequency of the spring-mass-system:

$$s = \frac{1}{\omega_0^2} = \frac{1}{4\pi^2 f_0^2},$$

i.e., sensitivity is gained at the cost of dynamic response and vice versa. In contrast to the piezoresistive case however, closed-loop operation of a capacitive micromechanical accelerometer opens the way to increasing sensing speed without sacrificing sensitivity: Additional stiffness is added by the "electrostatic spring" to the mechanical spring, resulting in improved dynamic response. This corresponds to an increase of resonance frequency significantly above f_0 by a factor equal to the square root of the closed-loop gain factor. In addition, the desired damping characteristics of the sensing element can be adjusted by adding a differential behavior to the loop characteristics: A proportional loop regulator (p type) is thus transformed into a proportional-differential loop regulator (pd type).

Fabrication Aspects

Capacitive sensing appears as a technology of choice for surface-micromachined accelerometers, although examples based on bulk micromachining do exist. Because of the very small sizes for spring constants and seismic masses, inertial forces that appear during operation of a surface-micromachined accelerometer stay well within the range covered by electrostatic forces from the capacitive pickups. The intrinsic dualism of sensing and actuation allows self-test and force-feedback in closed-loop operation mode to be achieved for reasonably small voltage levels, e.g., below 4 V. In general, low-g (0.1 to 10 g) accelerometers are preferably used in closed-loop operation, while high-g (>10 g) accelerometers are preferably operated in open-loop mode. These preferences account for the limited electrostatic feedback forces available with low voltages.

In surface micromachining (see Chapter 15), all processing steps are performed on the wafer frontside. A sequence of layers, namely functional polysilicon layers and sacrificial silicon oxide layers, is deposited and structured into functional elements. The moving structures are finally released from the substrate by a sacrificial etching process, removing silicon oxide underneath or in between the polysilicon layer(s) in an HF-based (either wet or vapor phase) etchant. Great care must be taken to prevent mechanical sensor elements from sticking during or immediately after the release etch, and during sensor operation: Because mechanical forces (i.e., pullback forces from the mechanical springs), electrostatic forces, Van der Waals adhesion forces, and chemical bonding forces appear on the same scale of sizes, extended parts that enter into contact with each other are hard to release again. Drying liquids, through their surface tension, tend to pull structures together and bring them into unwanted contact. To prevent stiction during or immediately after the release etch, dedicated techniques had to be developed.

Vapor-phase etching in hydrofluoric acid vapor as an alternative to liquid-phase etching prevents the formation of droplets underneath the released structures and completely avoids any need for subsequent drying.[13] Wet etching in hydrofluoric acid liquid requires either mechanical fixations of the sensor structures in the form of photoresist spacers, for example, which can be plasma-ashed in an oxygen plasma stripper after the etch-and-dry procedure,[14] the use of supercritical drying like CO_2,[15] or freeze-and-dry procedures based on the sublimation of organic solvents,[16] thereby avoiding going through a liquid phase during the final drying procedure.

During operation, sticking can be prevented by mechanical stoppers restricting the surface areas that could possibly get in contact with each other, for example in an overload situation. An alternative is provided by appropriate coatings such as plasma-deposited teflon-like layers or self-assembled monolayers (SAMs) exhibiting very low surface energy, thus reducing the adhesion forces by several orders of magnitude.[17] If structures get into contact, the remaining adhesion forces are easily overcome by the pullback forces from the mechanical springs.

The sensing direction can be in-plane, with lateral interdigitated comb structures forming a differential capacitor arrangement. Today a preferred technology[18] uses about 10- to 20-µm-thick poly- or monocrystalline silicon for the active structures. The mass and comb-structure arrangement is fabricated by DRIE, which has reached a high level of accuracy and maturity over the past few years, driven mainly by applications such as the accelerometer and gyroscopes. Figure 10.1 shows a secondary electron microscope (SEM) picture of a front-airbag accelerometer from Robert Bosch GmbH fabricated using 10-µm-thick polysilicon technology.[19] The gaps between the interdigitated fingers are only 2.5 µm wide, resulting in a base capacitance of the differential capacitor arrangement on the order of 0.5 to 1 pF. Capacitance change from full-range acceleration is typically an order of magnitude smaller. Figure 10.2 illustrates the working principle of this type of accelerometer: Acceleration moves the seismic mass out of position and changes the gaps of the interdigitated differential comb capacitor arrangement. The resulting capacitance changes are converted into an electrical signal.

Alternatively, for out-of-plane sensing, a seismic polysilicon mass is placed and free to move vertically between two capacitor plates, which are also made from polysilicon.[20] Three polysilicon layers are sequentially deposited and structured during the process, with silicon oxide layers in between; finally, the mass is released by a sacrificial etching step to remove the oxides. Figure 10.3 shows a schematic of an out-of-plane sensor with a vertical sensitivity axis, made from three polysilicon layers.

For sensors of the lateral type, manufacturers like Analog Devices and Robert Bosch GmbH are important suppliers. Vertical sensors are manufactured by Motorola SPS, for example.

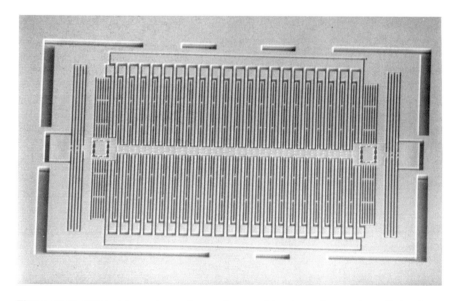

Figure 10.1 SEM picture of a surface-micromachined accelerometer with in-plane sensitivity and capacitive evaluation through comb capacitors, mainly for airbag applications (courtesy of Robert Bosch GmbH).

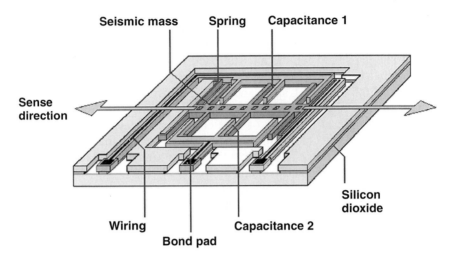

Figure 10.2 Working principle of capacitive accelerometer structure with comb capacitors and in-plane sensitivity axis.

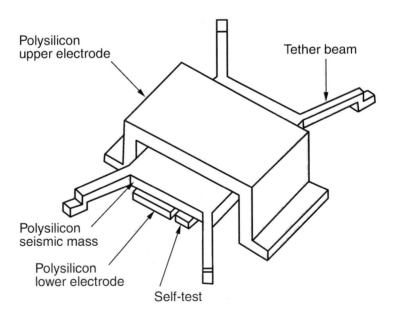

Figure 10.3 Sketch of surface-micromachined capacitive accelerometer made from three polysilicon layers and vertical sensitivity axis (e.g., Motorola SPS).

Packaging Aspects

Both sensor types need long-term stable hermetic encapsulation, since dust and humidity would otherwise prevent reliable operation. One packaging solution is to mount the sensor chips into hermetic housings, such as Cerdip packages or metal TO housings, and weld them while still in the cleanroom. This corresponds to an outer hermetic sensor package. An alternative solution is by wafer-level encapsulation, either by bonding a capping wafer onto the sensor wafer even before dicing the wafer stack into single chips, or by thin-film coverage of the sensor cores, for example by a polysilicon capping layer. The latter approaches correspond to an inner hermetic sensor package, with a hermetic encapsulation provided for each device at the chip level. The final package may then be achieved by plastic molding (for example, PLCC). Plastic molding in combination with an inner hermetic package is more cost-effective than expensive hermetic housings, if stress levels involved do not exclude this type of package in a particular sensing application.

Electronic Readout

In capacitive sensing, especially with surface-micromachined sensors, working capacitances are quite small (typically in the range of 0.1 to 1 pF, depending on the sensor technology). Electronic evaluation of capacitance changes on the order of femto- or even atto-farads is not easy at all. If a two-chip solution with separate electronics and sensor chips is chosen, large parasitic capacitances from large bondpads, bonding wires, and the EMI protection of the electronic circuit input totalling an order of 10 pF are adding onto the already small sensor capacitances, making things even more difficult. For this reason, monolithic integration was early considered a promising option with surface-micromachined accelerometers. In fact, the first product introduced to the market in 1994 by Analog Devices was a fully monolithically integrated acceleration sensor in BiCMOS technology. Other manufacturers, including Bosch and Motorola, decided to stay with hybrid integration in the form of two chips mounted together in the same housing. They could rely on a base capacitance of their sensors a factor of 5 higher than that of Analog Devices, thanks to the larger polysilicon thickness in the laterally sensing devices (10 µm thick instead of 1.5 µm) or thanks to the larger capacitor plates separated by only very small gaps in the vertically sensing devices. The question of whether monolithic integration makes sense or not in a particular application depends on various factors, and no general answer can be given. The sensor technology is obviously an important criterion. It determines the sensor size, the value of the available base capacitance, and the compability with IC technology. Compatibility is not so much an issue with surface micromachining, since this technology remains close to IC standards except for some dedicated steps, like the release etch, which may raise potential compatibility problems.

Integration Concepts

With a base capacitance as low as 0.1 pF, monolithic integration seems to be a good choice for evaluation, with high sensitivity and resolution, to reach the required sensor performance. This decision would be even more strongly supported by a sensor core as small as possible, e.g., 50×50 μm^2 or preferably even less, to restrict additional surface area consumption in the IC process to a minimum value. Note that in an IC process, surface area is getting expensive not for the cost of silicon wafers, which is negligible in the end, but due to the accumulation of process costs through as many as 25 masking levels. Chip yield, i.e., the total number of chips and fraction of functional chips per wafer, directly determines the final chip price. Integrating a sensor process into an IC process necessarily adds to the chip size, reduces overall yield (the larger the size, the higher the chip failure risks), and increases total process complexity and costs. The resulting higher costs must compete with the extra costs for the mounting and wire bonding efforts in a two-chip solution, provided the two-chip solution can fulfill the same performance requirements as the integrated solution.

There are some advantages to hybrid integration concepts. Hybrid integration is a modular approach; both sensor and electronics can be developed and processed separately, which reduces development time and costs, risk, and time-to-market. Having both processes decoupled provides the flexibility of easily switching over to a new generation of sensor elements or to a new IC process generation. In addition, flexibility with respect to sensor packaging is high in a two-chip hybrid solution, offering a number of feasible approaches for both inner and outer sensor packaging concepts. The choice is much more restricted if compatibility with a surrounding delicate circuit area must be maintained. Here, outer sensor packaging by expensive hermetic housings or thin-film hermetic inner sensor packaging appears feasible. Thin-film inner packages are being developed in several places worldwide in the frame of MEMS-first integration concepts; however, their process complexity is very high.

In summary, monolithic integration appears to be a good choice compared to hybrid integration if market volumes are extremely high (tens of millions of pieces per year) and if integration yields enhanced performance levels that could hardly be reached otherwise in a hybrid solution. Most important is whether these performance enhancements really pay off, i.e., are appreciated by the market in the form of acceptance of higher sensor prices. Gyroscopes may potentially emerge as one case where monolithic integration is beneficial, at least for high-end applications.

10.2.2 Yaw-Rate Sensors

10.2.2.1 Fundamentals

While accelerometers are passive devices that react only to external forces, gyroscopes for yaw-rate detection create their own internal forces if they are subjected to an external rotation. Soederkvist[21] and Funk[22] discuss gyroscopes in detail; Funk focuses on surface-micromachined yaw-rate sensors. Mechanical gyroscopes contain at least one moving element that responds to Coriolis forces if its motion is disturbed by a forced rotation around a sensitive axis. The size of the Coriolis force generated represents a direct measure for the applied rate of turn. In a classical precision-machined rotating-type gyroscope, Coriolis forces are, for example, measured and compensated in the bearings of a spinning rotator, by magnetic force-feedback, and so forth. Such classically machined gyroscopes are high-precision instruments dedicated to aircraft or naval navigation purposes and cost several thousand dollars per piece. Only with micromachining techniques has it been possible to bring sensor costs down into a range where their wide use became economically meaningful in consumer products such as video cameras, virtual reality heads, or airplane models, or in the automotive field for car navigation, vehicle dynamics control systems, and rollover protection systems. Originally, micromachined gyroscopes were manufactured from quartz oder piezoceramics, e.g., PZT, taking advantage of the piezoelectric effect both for driving and detection of motion. Recently, micromachined silicon devices have entered the market, most of them based on surface-micromachining or surface-near-bulk-micromachining technology. Because silicon is a nonpiezoelectric material, electromechanical conversion has to be performed either by magnetic drive and magnetic detection, electrostatic drive and capacitive detection, or mixed magnetic drive and capacitive detection of motion. Magnetic principles need a supplementary external magnetic field provided, for example, by permanent magnets close to the sensor structure.

The fundamentals of vibrating micromechanical gyroscopes in their different technologies are very similar and rely on the generation of

Coriolis forces in response to an external rate of rotation, irrespective of the chosen sensor materials and electromechanical conversion principle, whether it is piezoelectric, magnetic, or electrostatic/capacitive. Although other technologies based on quartz[23] or PZT[4,24] exist, silicon micromachining is promising the largest cost reduction potential and will certainly dominate in a growing number of applications.

Because no solution exists to the fabrication of high-quality bearings in micromechanics, in contrast to precision mechanics, MEMS gyroscopes are mostly based on a periodic vibratory motion, not a unidirectional rotation of mass(es), like a spinning-wheel gyroscope. Bearings are replaced by a spring-mass system. Examples are a periodic linear vibration, a ring vibration, or a periodic rotational vibration (rotational-type oscillator). In all cases, their action can be understood in terms of a forced change of an angular momentum $\vec{L}$, which induces a corresponding reaction of the mechanical element. The resulting mechanical torque $\vec{M}_C$ is:

$$\vec{M}_C = \frac{d\vec{L}}{dt} = \vec{\Omega} \times \vec{L}$$

and the resulting Coriolis force $\vec{F}_C$ is:

$$\vec{F}_C = \vec{\Omega} \times \vec{v},$$

where $\vec{\Omega}$ and $\vec{v}$ denote the angular rotation speed and the speed of linear motion, respectively.

Given a vibrating spring-mass arrangement, measuring the Coriolis forces is then similar to acceleration sensing, which was discussed in Section 10.2.1.

Two orthogonal degrees of motion are relevant for the operation of a micromechanical gyroscope: the base vibrating mode, which is kept under permanent oscillation at constant amplitude, and the detection mode, where motion appears in the ideal case only if the rate of rotation is applied to the gyroscope around its sensitive axis. "Orthogonal" means in this case that the motion axes in the drive and detect modes include a 90-degree angle. The Coriolis force $\vec{F}_C$ or Coriolis-induced torque $\vec{M}_C$ appears in a direction perpendicular to both the axis of external rotation $\vec{\Omega}$ and the axis of the inner motion $\vec{v}$ or angular momentum $\vec{L}$, respectively. Each of the two degrees of motion has its own resonance frequency, quality factor, and anharmonicity. In practice, the lowest eigenfrequencies in both modes are the most relevant ones.

Since the generated Coriolis force amplitude is proportional to the amplitude v_0 of velocity in the base vibration mode, it is highly desirable to achieve a motion amplitude x_0 as large as possible. Note that for a harmonic motion at frequency ω, $v_0 = \omega \cdot x_0$. For a micromechanical sensor, given the limited driving capabilities of available microactuators, this strongly suggests resonant excitation in an eigenmode of the base vibration, to profit from the quality factor enhancement at resonance. Practically all existing micromechanical gyroscopes use resonant excitation, which makes the driving frequency equal to an eigenfrequency ω_0 (generally the first or second eigenfrequency) of the base vibration. For slowly varying angular rate Ω, this leads to a Coriolis force $F_C(t)$ of the same frequency ω_0, and in phase with the velocity $v(t)$. In other terms, the Coriolis force precedes the motion $x(t)$ by a phase angle of $\pi/2$:

$$x(t) = x_0 \cdot \sin(\omega_0 t) \Rightarrow F_C(t) = \Omega \cdot v(t) = \Omega \cdot x_0 \omega_0 \sin(\omega_0 t + \pi/2)$$

The task for the accelerometer part of the gyroscope is now to measure as accurately as possible this Coriolis force, which appears at a frequency ω_0. This introduces the resonance frequency of the detection mode as a second important parameter. The usable bandwidth of an acceleration sensor is generally limited by its first eigenfrequency ω_1: below ω_1, the induced mechanical motion follows the acceleration signal, and the resulting electrical signal from the detector is a true representation of the acceleration signal with respect to amplitude and phase. At resonance ω_1, resonant enhancement amplifies the measured signal by the quality factor, which can be very large (100 to 10,000) at resonance, and the measured phase is lagging behind the phase of the acceleration by an angle of $\pi/2$. For acceleration signals of frequency ω above the detection mode resonance ω_1, the measured signal falls off by a factor of $(\omega-\omega_1)^{-2}$ with increasing frequency ω, and the phase lag is approaching π. As already mentioned, Coriolis force-related signals in a gyroscope appear at fixed frequency $\omega = \omega_0$, the base vibration frequency. As an example, the layout of a surface-micromachined gyroscope based on the Coriolis force, with in-plane base vibration and orthogonal in-plane detection mode, is shown in Fig. 10.4.

10.2.2.2 Single- or Double-Resonant Operation

Single Resonance

To gain a true representation of the Coriolis forces from the accelerometer output, only two situations are of practical relevance with respect to the gyroscope's eigenfrequencies: either ω_1 is higher than ω_0 (in practice, significantly higher), or ω_1 is equal to ω_0. The first option is very

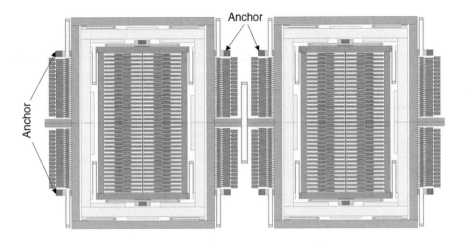

Figure 10.4 Structure of surface-micromachined gyroscope with in-plane base vibration and orthogonal in-plane detection mode (courtesy of Robert Bosch GmbH).

common to all kinds of linear or rotational type vibrating micromechanical gyroscopes. The base vibration frequency is kept well below the first detection eigenfrequency, and the measured acceleration signals are true with respect to their amplitude and phase. No special attention has to be drawn toward the exact position of resonance frequencies, and mechanical frequency tuning is not required even for manufacturing tolerances. For a fixed base vibration amplitude, sensitivity scales with ω_0/ω_1^2.

Double Resonance

The double-resonant option is more difficult to achieve and to maintain, since two different mechanical motion modes must be tuned and kept at the same resonance frequencies. To better understand the difficulties, it is useful to think of a spring-mass system with an in-plane base vibration mode and an out-of-plane detection mode, or a spring-mass system with an in-plane base vibration mode and a detection mode in a second orthogonal in-plane direction. The geometrical dimensions and their manufacturing tolerances deciding on resonance frequencies and their resulting statistical distributions are quite different for these modes. In practice, eigenfrequencies either have to be adjusted carefully by laser trimming, for example, or by a permanent electrical calibration through nonlinear electrostatic forces applied via separate control structures.

Example: Ring Resonators

The ring resonator is probably the best-known example of a double- or multi-resonant gyroscope, the operation of which depends on a complete degeneracy of its ring vibration frequencies. In a perfectly isotropic ring, for any orientation of the vibration axis in the plane of the ring, vibration frequency is exactly the same and the vibration mode shape is free to move along the ring into any position. For a given excitation geometry, whatever the actuator may be, a well-defined ring vibration undamped by the pickup-and-drive method, with four stationary vibration nodes along the ring, has to be guaranteed. If the ring is rotated with an angular rate Ω, the vibration mode shape tends to remain stationary in space despite the forced rotation of the ring. This leads to a motion of the four nodes along the ring with angular velocity Ω. The motion of the nodes is detected and compensated by force-feedback, with the size of the feedback signal needed for full compensation used as a measure for the applied angular rotation rate Ω. This is the general working principle of ring gyroscopes, irrespective of the applied electromechanical conversion principle. One ring gyroscope that is manufactured by Sumitomo Precision Products and British Aerospace (SPP/BASE) uses electromagnetic drive and detection in an external magnetic field from a small permanent magnet. In a publication by the University of Michigan, a metal ring gyroscope was described that uses electrostatic forces combined with capacitive detection to drive the ring vibration and to detect and annihilate motion in the vibration nodes.[25]

The necessary rotational symmetry of the ring is achieved by a careful and extensive frequency-tuning procedure like laser-trimming, or by introducing nonlinear electrostatic forces through appropriate balancing electrodes. The key advantage of a double-resonant gyroscope is the related gain in sensitivity and signal-to-noise ratio. First, with respect to the operating frequency ω of the gyroscope, sensitivity scales with $1/\omega$, which is better than for two different frequencies $\omega_0 < \omega_1$. Second, and most important, quality factor enhancement amplifies the detection amplitudes in the accelerometer, which is excited by Coriolis forces exactly in its resonance. Quality factors in typical accelerometer structures can reach values of 100 to 10,000 under vacuum operating conditions. If quality factor amplification were used directly for signal increase, bandwidth would shrink by the same factor, which is undesired in most cases. Therefore, electronic closed-loop operation similar to that already described for suppressing motion in the vibration nodes of a ring gyroscope is used also in most other double-resonant gyroscopes. Closed-loop operation works by compensating the quality factor enhancement of the vibration amplitude by an electrical damping force, thereby reducing the noise level in the loop by a corresponding factor. The latter evaluation method allows a sufficiently large bandwidth to be maintained and increases signal-to-noise ratio by lowering the noise, rather than by increasing the measured signals. This is the way electronics work for most double-resonant gyroscopes.

The question of which approach is preferable depends very much on the performance required. Double-resonant operation yields a higher performance but requires a higher effort for adjustment and trimming. In contrast, two different resonance frequencies for excitation and detection represent a lesser effort in device operation at the cost, however, of lower performance.

10.2.2.3 Electronic Readout

Quadrature Discrimination

The need for a true representation of Coriolis forces in the signal output with respect to amplitude and phase becomes clearer when disturbances that are always present from manufacturing tolerances and structural imperfections are taken into account. Part of the base vibration amplitude is coupled into the "orthogonal" detection mode through these imperfections, giving rise to a motion signal proportional and in phase with the base vibration. This so-called quadrature component, which follows $x(t)$, is phase-shifted by $-\pi/2$ with respect to the "true" Coriolis signal $F_C(t)$, which follows $v(t)$. Only if the detector electronics show well-defined phase characteristics is discrimination of the Coriolis signal from the quadrature component possible by phase and frequency-selective demodulation, using the base vibration motion as a reference input to the demodulator. The importance of an effective discrimination of the quadrature component is even more obvious given the fact that the quadrature signal amplitude is typically several times the full-range signal amplitude for a typical measuring range of ±100 degrees/sec. With a requested resolution limit in typical applications between 0.02 to 0.5 degree/sec, the only chance for true signal extraction from large disturbances is to maintain proper phase relationships in the detector output to allow for phase-sensitive discrimination. Phase-selective demodulation, with the relative phases carefully adjusted, is capable of suppressing the quadrature signal to a large extent. However, with increasing quadrature signal amplitude, phase noise or phase jitter occuring anywhere in the demodulation path is transformed into amplitude noise of corresponding strength. Amplitude noise, irrespective of its source, is a limiting factor for sensor resolution.

Signal Evaluation

Electronic evaluation of a high-performance gyroscope is quite a challenging task. A detailed description is given by Funk.[22] Figure 10.5 shows a typical gyroscope readout circuit. One circuit block is responsible for undamping the vibrating structure, keeping it permanently vibrating at a fixed amplitude. The high-impedance raw signals from the accelerometer stage are converted by a low-noise primary amplifier to a reasonably low impedance output voltage. This contains the true Coriolis signals superimposed by quadrature contributions. The output of the low-noise impedance converter is fed into a multiplier (mixing stage) and demodulated with the speed signal derived from the base vibration of the sensor element as a reference. A phase shifter is needed for proper phase adjustment, to achieve the most effective quadrature suppression. The low-pass filter following the frequency- and phase-selective demodulator eliminates higher frequency mixing products and determines the bandwidth of the sensor output. For typical applications, the filter transition frequency is in a range between 10 to 100 Hz.

Examples

The first gyroscope[26] consists of a vibrating structure manufactured from a bulk-silicon membrane by Bosch DRIE. Its operating principle and a photograph of a production wafer combined with a SEM picture of a device are shown in Figs. 10.6 and 10.7, respectively. Two masses are car-

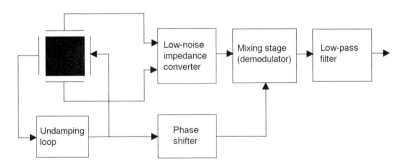

Figure 10.5 Block diagram of gyroscope evaluation circuit. The raw sensor signal is amplified by an impedance converter and mixed with a base vibration speed signal from the undamping loop. Low-pass filtering after signal mixing provides the desired angular rate signal.

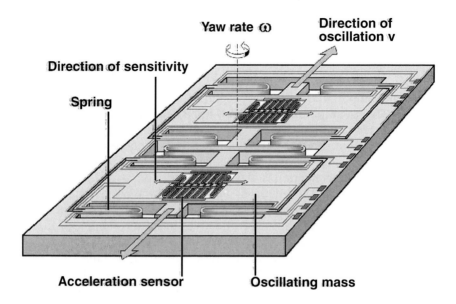

Figure 10.6 Working principle of surface-near bulk-micromachined gyroscope. The bulk-silicon structure is kept at permanent in-plane oscillation by electromagnetic drive. Coriolis forces are detected by accelerometers on top of the bulk vibrators with lateral sensitivity perpendicular to the base vibration (courtesy of Robert Bosch GmbH).

ried and coupled by a spring mass arrangement to split the degenerate base vibration modes of the two masses into two eigenmodes of different frequencies. The lower frequency is for the two masses moving in phase; the higher frequency is for the two masses moving at 180-degree phase difference. Each of the two masses is carrying a surface-micromachined accelerometer with capacitive comb structures to measure Coriolis forces induced by a rotation around a perpendicular axis, i.e., out of plane. The two masses are driven into their counterphase vibration mode by electromagnetic forces using the Lorentz force on current-carrying conducting lines in an external magnetic field. Counterphase motion yields Coriolis forces to the accelerometers, which are also in counterphase. This provides a first means to discriminate true internal force signals from external acceleration signals by subtracting the two accelerometer outputs. Frequency and phase-selective detection yields a signal proportional to the angular rate Ω.

The second example is a rotational type of vibrator that is manufactured in surface-micromachining technology.[27] Figure 10.8 illustrates the operating principle of the gyroscope, while Fig. 10.9 shows a SEM micro-

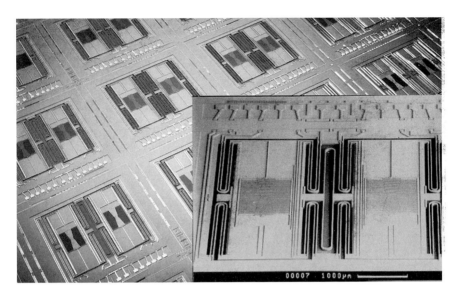

Figure 10.7 Photograph of a wafer and SEM picture of gyroscopes made from surface-near bulk-micromachining. The base vibrator is fabricated by Bosch DRIE from a 70-µm-thick silicon membrane (KOH etched from the wafer backside). Accelerometers on top are made by surface-micromachining of 10-µm-thick epitaxially deposited polysilicon (EpiPoly).

graph of the device with a detailed view of the comb drives. The in-plane vibrating mass is carried by a central tether beam structure. A yaw rate applied to any in-plane axis leads to a corresponding out-of-plane motion of the vibrating mass, adding an out-of-plane component to the otherwise purely in-plane oscillation. The out-of-plane vibration is detected by capacitor plates underneath the vibrating structure, which form a differential capacitor arrangement together with the mass. The drive is achieved by electrostatic forces through comb structures. All detection is done capacitively, either by comb structures or capacitor plates. With two orthogonal pairs of capacitor plates placed underneath the mass, this type of device is intrinsically a two-axis gyroscope.

The third example is a linear-linear type of vibrator with two orthogonal in-plane modes for drive and detect, also made in surface-micromaching technology.[28] A SEM micrograph of the sensor is shown in Fig. 10.10. A seismic mass is vibrating in one direction in-plane, driven by electrostatic forces from comb structures. Angular rotation around an out-of-plane axis leads to Coriolis forces that excite the structure into motion in the perpendicular direction in-plane. This second motion orthogonal to both the angular rotation axis and base vibration direction is again

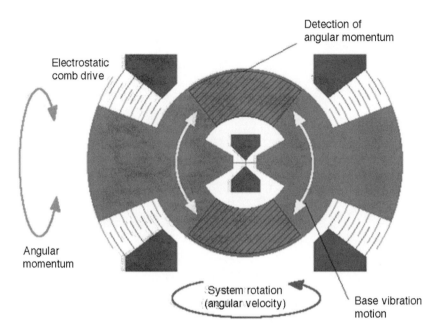

Figure 10.8 Working principle of a rotational type of surface-micromachined gyroscope. The base mode is an in-plane rotation that is undamped electrostatically using comb drives; the detection mode is an out-of-plane torsion. Capacitor plates under the rotationally vibrating structure pick up capacitance changes in response to out-of-plane torsion.

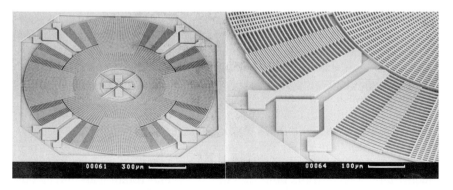

Figure 10.9 SEM picture of a rotational type of surface-micromachined gyroscope showing an overview of the sensor structure and details of the comb drives.

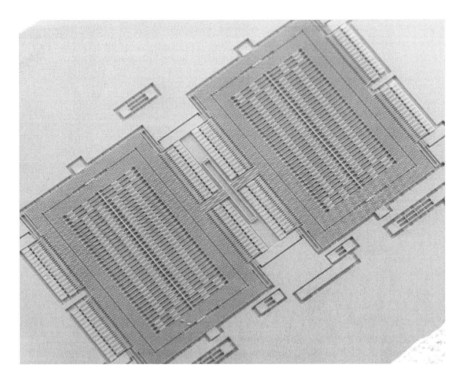

Figure 10.10 SEM picture of a linear type of surface-micromachined gyroscope. The base mode is a linear in-plane vibration, excited by electrostatic drive through comb structures. The detection mode is an orthogonal, linear in-plane motion (inverse tuning fork principle).

detected by comb structures that form a differential capacitor arrangement. Care must be taken to prevent coupling of moments from the electrostatic driving combs into the detection mode, as the electrostatic instability of the drive structure acts in the same direction as the detection degree of freedom. From the electrostatic drive, the detection mode is decoupled by a tether-beam construction. Electronic readout of this type of gyroscope follows the same basic principles as described above.

Monolithic Integration

All three examples of micromechanical gyroscopes use capacitive detection. They are currently built as hybrid solutions, with sensor and electronics integrated as two separate chips. Because Coriolis forces are very small, the electronic evaluation circuit has to resolve capacitance changes on the order of a few tens of atto-farads (aF). These extremely small capacitance changes are covered by parasitic capacitances on the order of 10 pF, mainly from conducting lines, large bondlands, bond wires, and the EMI protection diodes for the amplifier inputs. Obviously, these large parasitics represent a significant limitation to the achievable performance of the gyroscope. Monolithic integration offers the opportunity to reduce parasitics by more than an order of magnitude, to below 1 pF, with a corresponding increase in sensor performance (omission of bonding pads, wires, and input protection network), mainly with respect to signal-to-noise ratio. For purely surface-micromachined gyroscopes like the latter two examples, given their small sizes, monolithic integration is a potential option to be considered, mainly for higher achievable performance. This is a reasonable approach for applications where sensor performance is a priority issue.

10.3 Pressure Sensors

10.3.1 Fundamentals

In contrast to inertial sensors, which receive their information on the relevant physical quantities of motion in a perfectly shielded, isolated world, with their package separating them completely from the environment, pressure sensors must be in contact with the medium, the pressure of which is to be measured. This may sound trivial; however, it involves new challenges with regard to the passivation of the sensor and the approaches to protecting the device from environmental influences to which it necessarily has to be exposed during operation. This is why a larger variety of technologies exists for the different pressure-sensing fields, for example, low-pressure or high-pressure range, low or high operating temperature, gaseous or liquid media, aggressive or nonaggressive ambient. Nevertheless, pressure sensors were the first mechanical microsensors, reported as early as the 1970s.

The key elements of many pressure sensors are a membrane that is elastically deformed by the applied pressure and components that are able to detect this deflection either by the deflection itself or through the strain occuring as a result of the deflection.[29] Another way to measure pressure is by thermal sensing (so-called thermovacs; see Chapter 5). Thermal-type vacuum pressure sensors (ranging from 0.1 to 10,000 Pa) rely on heat transfer to the surrounding gas atmosphere, with the amount of electrical heating power required to maintain a constant temperature on the sensor element as a possible measure for ambient pressure if the heat conductivity of the ambient gas is known. A different measuring principle is by evaluation of viscous damping of vibrating structures or surface acoustic waves by the ambient gas atmosphere, with the resultant quality factor of vibration yielding the ambient pressure for known damping characteristics of the ambient gas (ranging typically from 1 to 1,000 Pa). Ion current pressure sensors for high vacuum (so-called ion-gauges, or "ionivacs") take the electrical current for a fixed high voltage between a heated filament cathode and a grid anode as a measure for ambient pressure (ranging below 0.1 Pa). This chapter considers only membrane-type pressure sensors.

10.3.2 Bulk-Micromachined Pressure Sensors

The deflectable membrane of a silicon micromechanical pressure sensor is fabricated either by bulk or surface micromachining. In bulk micromachining technology, predominantly wafers with (100) crystal orientation are etched from the backside to a predetermined depth, which leaves a remaining membrane of the desired thickness at the frontside of the wafer. Wet etching in KOH solutions is one technology of choice for etching nearly through the full wafer thickness. For membrane thickness control, either a time-determined etch or etchstop techniques (p^+-etchstop, electrochemical etchstop on reversely polarized pn junction, or dielectric etchstop for SOI-wafers) can be applied. Wet etching leads to tilted sidewalls of the etched cavity, which represent (111)-crystal planes. As a consequence of this anisotropic etching behavior, the opening of the backside cavity is much larger than the membrane size itself, thus consuming additional wafer area. Sharp corners appear at the transitions from the 111-oriented sidewalls to the 100-oriented bottom of the membrane. Sharp transitions may cause early fracture of the sensor membrane under an overpressure load. This is one of the reasons why, as an alternative to wet etching, DRIE of silicon is becoming more and more popular and expanding over traditional KOH-dominated application fields. In DRIE-fabricated devices, the transition from the sidewalls to the bottom of the membrane can be made smooth, yielding enhanced tolerance for high-pressure loading. One more advantage of DRIE is that the geometry of the cavities and thereby of the membranes can be arbitrarily chosen, irrespective of crystal orientation (for example, round cavities and membranes instead of squares or rectangles). Sidewalls are vertical, resulting in minimum surface-area consumption. As a disadvantage, no electrochemical nor p^+-etchstop exists for DRIE. The membrane thickness is either defined through a time-controlled etch, or an in-situ etch-depth measuring system (laser interferometry) is needed to monitor the progress of the etch front. A third alternative is to use SOI wafers, which provide a dielectric etch-stop at the buried oxide of the SOI-wafer stack.

A general feature of bulk-micromachined pressure sensors is a clear separation between frontside and backside processing. As a rule, semiconductor-type process steps are performed on the wafer frontside, for example, diffusions, oxidation, layer deposition, and etching, up to a full IC process in the case of an integrated sensor. The "dirty" micromachining steps are performed afterward on the backside of the wafer, following the IC fabrication process, with the finished frontside being protected by appropriate measures.

This approach represents both advantages and disadvantages. The main advantages are that compatibility requirements between frontside and backside processing are weak, since the membrane is always in between. This makes it possible to integrate a full electronic circuit at the wafer frontside and to perform the KOH or DRIE micromachining at the wafer backside with a low risk of interfering with the frontside. During wet etching, the wafer frontside is normally protected by an etch box; during DRIE, the frontside is protected by its passivation layers and the substrate electrode of the DRIE tool. Most important, during operation of the sensors, medium contact can be easily restricted to the wafer backside, with no access of the media to the delicate frontside features, once again having the membrane as a separation in between. Figure 10.11 shows a cross section and packaging concept of a pressure sensor and illustrates the separation of the inner sensor part from the outer world. Figure 10.12

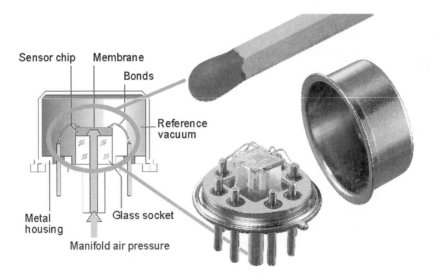

Figure 10.11 Cross section and encapsulation of pressure sensor. Contact with the outside world is only to the back side of the sensor membrane, hermetically isolating the front side containing the electronics from the environment.

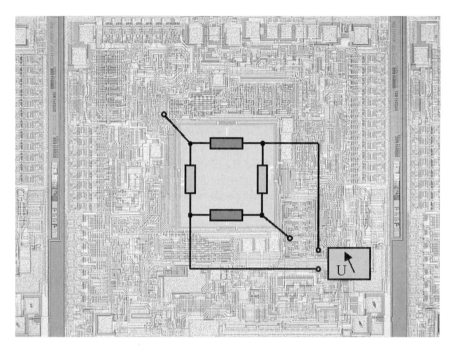

Figure 10.12 Top-view photograph of an integrated pressure sensor chip with piezoresistor positions indicated schematically.

shows a top-view photograph of the integrated pressure sensor chip, with the position of the piezoresistors indicated schematically.

The main disadvantages of this concept are that backside processing is nonstandard in semiconductor technology, requiring dedicated equipment and specialized wafer handling, and that backside quality and surface finish, together with thickness and thickness variations of the wafers, must be carefully monitored. A second drawback is that the wafers tend to be fragile after cavity etching, with negative impact on overall yield.

10.3.3 Surface-Micromachined Pressure Sensors

In contrast to bulk-micromachining, which is a subtractive microstructuring technology, surface micromachining is an additive microstructuring technology that is based on the deposition or bonding of additional layers onto the wafer frontside. Processing steps take place exclusively on the wafer frontside, which moves this technology very close to semiconductor standards. Surface-micromachined pressure sensors are fabricated by depositing the membrane material (in most cases, either poly- or single-crystalline silicon) over a so-called sacrificial layer, for example, silicon oxide or porous silicon. This sacrificial layer is later removed either by selective etching (for example, by hydrofluoric acid) through membrane perforations, which have to be closed by a subsequent deposition process, or by thermally collapsing the sacrificial layer in the case of porous silicon. In SOI-type approaches, a first wafer containing the cavities is bonded by silicon fusion bonding to a second wafer, which is then etched or ground back to the desired membrane thickness. In all cases, the result is a membrane overspanning a cavity and enclosing a reference pressure, most often a reference vacuum. The main advantages of surface-micromachined pressure sensors are the minimum surface area consumption; independence from wafer thickness and thickness variations; no requirements with respect to the backside quality or surface finish, since the backside is not processed at all; ease of mounting the devices; robustness of the wafers throughout the whole manufacturing process; and standard processing and handling. A potential disadvantage of surface micromachined pressure sensors is the process restrictions resulting from compatibility requirements between MEMS and non-MEMS processes performed on the same wafer side. Most important is the fact that during sensor operation, pressure has to be applied to the frontside membrane, exposing the whole wafer frontside to the medium whose pressure is to be measured. If the sensor application involves only nonaggressive media, silicone gel coverage is in many applications sufficient to protect the delicate wafer frontside from humidity and dust. In applications involving aggressive media, however, special care has to be taken for a proper passivation of the wafer frontside, including means to prevent a direct contact with the medium through appropriate packaging. Possible solutions include the isolation by silicone oil and a diaphragm,[31] or an additional macromechanical membrane, often made from steel, to separate the micromachined sensor element from the outer world as part of the package.[32,33] These measures may represent a considerable extra effort and a severe complication of the sensor device. They all impose significant additional cost.

10.3.4 Signal Generation

The question of how to detect the deflection of the membrane, i.e., the applied pressure, is not fully independent from the fabrication technology of the sensor element. Most bulk-micromachined pressure sensors use a piezoresistive type of detection, which is most appropriate and easy to realize for this particular technology. Although solutions also exist for capacitive evaluation of bulk-micromachined pressure sensors,[34] in most cases these involve complex processes like direct bonding of several structured wafers. One wafer contains the membranes, and a second wafer provides the required counter-electrodes opposite to the membranes. Electrical feedthroughs required to contact both wafers from the top side of the bonded wafer stack by bond wires represent another significant complication in all of these approaches. Capacitive detection is much simpler with surface-micromachined pressure sensors.[35] Here, the additive and subsequent deposition of several layers on the wafer front side offer the opportunity to create capacitor plates underneath the freestanding membranes. For example, after deposition of a first insulating oxide, a first polysilicon layer is deposited and structured for the buried capacitor plates. On top, the sacrificial oxide is deposited, which is to be removed later in the process. On top of this, a second polysilicon layer is deposited to form the membrane layer. This makes capacitive detection a preferred choice with surface-micromachined pressure sensors. However, piezoresistive evaluation is also possible, in principle, with surface-micromachined devices, especially if single-crystalline membranes can be provided by the sensor fabrication technology.

10.3.4.1 Piezoresistive Evaluation in Bulk-Micromachined Pressure Sensors

In pressure sensors fabricated using bulk-micromachining,[36] the preferred method for response evaluation is based on the piezoresistive effect, i.e., the strain and/or stresses induced by the deflection of the membrane are evaluated as a measuring signal. Piezoresistive elements, normally four piezoresistors in a full Wheatstone bridge arrangement, are placed in or close to locations on the membrane where the maximum stress levels appear under deflection. Preferably, two pairs of resistors are placed in areas of opposite stress, i.e., one pair of resistors is exposed to tensile stress under a given deflection of the membrane, and the other pair is exposed to compressive stress, or vice versa. This leads to an asymmetric change of their resistivity, i.e., resistance increases for one pair of resistors and decreases for the other pair of resistors. The asymmetry thus introduced into the Wheatstone bridge yields a corresponding bridge voltage that depends to first order only on the resistance changes, and is proportional to the applied external pressure or the resulting deformation of the membrane.

The piezoresistive effect leads to a resistivity change as a function of a geometrical change of the individual resistor, which can be written in the form:

$$\frac{\Delta R}{R} = K \cdot \varepsilon,$$

with K the piezoresistive factor and ε the strain. For metal resistors, K is close to a value of 2. In a semiconductor material, electronic band deformation effects are dominating, which leads to K-factor values of typically around 30 for monocrystalline silicon. The change in resistivity in response to transversal and longitudinal stresses σ_l and σ_t was described in Section 10.2.1.2.[8]

Piezoresistors on a membrane can be fabricated in different ways. They can be deposited as a metal or a polysilicon layer on top of the wafer surface, preferably with an insulating oxide between the membrane silicon and the resistor material, and structured into the desired resistor geometries. In SOI technology, a single-crystalline silicon layer is deposited on top of a silicon oxide layer by silicon fusion bonding, then structured into the resistor geometries. This provides single-crystalline piezoresistors with dielectric insulation, which yields high-temperature capabilities. Single-crystalline resistors show exceptional stability with respect to aging and drifts, even under extended high-temperature loading. The easiest approach is certainly diffusing the resistors into the silicon, for example by implantation of n-type dopants into p-type bulk silicon or vice versa. These diffused resistors share the exceptional long-term stability and very small drift of the SOI-approach. However, their temperature range is restricted to a maximum temperature of 120 to 140°C because of leakage currents through the pn junction at higher temperatures.

Due to the dependence of the piezoresistive coefficients on temperature, which is a very strong effect with all except strongly doped semiconductor-type piezoresistors, temperature compensation is needed both for bridge offset and sensitivity. Low-doped silicon resistors change in resistance by an order of magnitude over a temperature range of 100°C. This has a corresponding effect on the sensitivity of the same order of magnitude, which makes correction necessary. Offset changes are mainly caused by temperature-dependent leakage currents occuring through pn junctions or dielectrics. One method of temperature compensation is to feed a constant current into the full Wheatstone bridge and measure both the bridge voltage and the difference voltage proportional to bridge asymmetry (the pressure signal). The full bridge voltage is used as a temperature gauge and an input to the temperature calibration algorithm, to correct the bridge asymmetry voltage, for example through a lookup table.

Electrical currents needed for evaluation of the bridge resistors lie in the range of 0.1 to 1 mA, which is a comparingly large value. In contrast to capacitive sensing, which is very low in power consumption, piezoresistive sensing consumes significant power, thereby excluding applications where the power supply is restricted (telemetric sensing, autonomous power supply).

The methods discussed above for fabricating piezoresistors in or on top of silicon fit very well into IC processing, requiring only processing steps and dopant levels common in semiconductor manufacturing. One exception is direct-fusion-bonded SOI approaches. These do not really fit in, but they represent special solutions dedicated to high-temperature applications. IC compatibility makes the fabrication of piezoresistive pressure sensors with standard semiconductor tools in a more or less standard semiconductor process possible, with the MEMS etching steps added at the back end. This is why, in many cases, monolithic integration is found with piezoresistive pressure sensors,[30,36] although the low source impedance of the Wheatstone bridge does not necessarily imply integrated solutions. Integration in the case of piezoresistive sensors has some striking benefits. (1) Mainly, the processes fit perfectly together, and complexity added to the IC process by the MEMS is low. (2) The large surface area consumed by wet etching from the wafer backside is no longer lost; it is used by circuitry surrounding the membrane on the wafer frontside. (3) Temperature compensation can be achieved easily by the electronics located on the same chip as the sensor membrane and the piezoresistors. (4) The individual sensor calibration data can be stored on-chip, within its electronic circuit part by Zener or thyristor zapping or E^2PROM for the storage of calibration data.

10.3.4.2 Capacitive Evaluation in Surface-Micromachined Pressure Sensors

In capacitive pressure sensors,[35] the membrane forms part of a capacitor arrangement with a second counter-electrode opposite the membrane. If the membrane is deflected toward the counter-electrode or away from it, the capacitance of the arrangement formed by both membrane and counter-electrode changes correspondingly. An electronic C/V converter provides an output voltage proportional to the applied pressure. Capacitive pressure sensors have the advantage of high accuracy and show only a small temperature dependence of offset and sensitivity, which is inherent to this measuring principle. In many cases, this makes it possible to disregard temperature compensation. Another advantage is the very low power consumption, especially in combination with CMOS ciruitry, which makes this sensor an ideal choice for telemetric applications with an autonomous power supply, e.g., remote pressure control and tire pressure monitoring. A potential drawback of capacitive pressure sensing is its higher nonlinearity and that integration of micromechanical sensor functions and evaluation electronics is more or less a must. This leads to a high process complexity and raises the full range of compatibility issues mentioned in Section 10.2.1.3. The reason that monolithic integration is hard to avoid in this case is the small capacitance values, on the order of only 1 pF, and their even smaller changes, on the order 0.1 pF maximum for a linear response, combined with the application of pressure to the frontside of the sensor chip leading to medium contact. The medium would impact more or less directly onto the bond wires and affect their parasitic capacitances in an unpredictable way. Therefore, the first electronic amplifier stage to transform from the high source impedance of the membrane capacitor arrangement to a low-impedance output signal should be as close as possible to the sensing area (i.e., monolithically integrated with the sensor).

10.3.4.3 Resonant Sensors

Other, more exotic sensing principles make use of resonant structures,[37,38] or exploit the propagation of acoustic waves on the membrane surface or through auxiliary waveguides connected to the membrane, such as surface acoustic wave (SAW) devices.[39] The basic principle of resonant sensing is that mechanical deflection and ensuing stress shift the mechanical resonance frequencies of the eigenmodes of a structure that is excited into permanent vibration.

Bulk Acoustic Wave Sensors

The simplest mechanical resonator for a pressure sensor is the membrane itself, with its eigenfrequencies changing with the square of the applied pressure. Beam-on-diaphragm (BOD) structures have clamped-clamped silicon or dielectric beams connected to the membrane in the locations of maximum stress on the deflected membrane. When the membrane is deflected, the beams are strained and their eigenfrequency is detuned correspondingly. BOD structures are similar to piezoresistive sensing in that they transform membrane deflection into an electrical signal, in this case the resonance frequency proportional output of an f/V converter, which is usually a phase-locked-loop demodulator. Similarly, bridging of several vibrating beams strained in opposite directions by the applied pressure is done in a Wheatstone-bridge-like arrangement, to isolate the frequency change Δf from the base vibration frequency f_0 through frequency subtraction. (See the discussion of piezoresistive pressure sensing in Section 10.3.4.1.)

Surface Acoustic Wave Sensors

Surface acoustic waves (SAWs) are periodic strain fields propagating along material surfaces. The phase velocity of SAWs depends on the surface strain, which in turn varies with the external load. In SAW sensors, this relationship is exploited to measure externally applied forces. In most cases, this is achieved by transforming phase velocity into a corresponding resonance frequency of a closed-loop oscillator. The resonance frequency of the loop is dependent on phase velocity of the wave along the probe path and thus varies with the applied pressure through the corresponding strain induced into the SAW probe path.

Resonant Sensing

All resonant structures have in common that they need mechanical actuation, i.e., excitation of a sensitive vibration eigenmode. This excitation can be achieved, for example, by electrostatic driving forces [realized in a straightforward way in a BOD structure by applying an external voltage between the beam(s) and the membrane], by electrothermal excitation using heating resistors, or by piezoelectric actuation. Electrothermal actuation by periodic heating using resistors on or in the membrane, or in the beams of a BOD, is limited to a driving frequency of up to about 1 MHz, but can be easily integrated into a sensor process close to IC processing standards. Piezoelectric actuation needs deposition of additional piezoelectric thin-film layers, such as zinc oxide, aluminium nitride, or PZT. This adds to the complexity of the manufacturing process and gives rise to problems due to different thermal expansion coefficients between silicon and the piezoelectric material, leading to strong temperature-induced drifts and the need for temperature compensation. The advantage of piezoelectrics is that driving frequencies can be extremely high, up to several hundreds of MHz or even in the lower GHz range, which is needed in SAW-type sensing applications. In general, resonant sensing principles are not very favored with silicon pressure sensors and are only found in some dedicated niche applications where very high accuracy is required at the cost of a slow dynamic response. If, however, a material other than silicon (such as quartz, which is piezoelectric) is used as the basic sensor material, resonant sensing principles offer some potential. However, quartz micromachining is lagging considerably behind silicon micromachining and is not likely to reach a similar level of maturity, mainly due to the lack of plasma processes with capabilities similar to those for silicon microstructuring, and to the restricted availability of large quartz substrates. This leads to an on-going substitution of silicon for quartz in micromechanical sensors.

10.3.4.4 Nonsilicon Pressure Sensors

In special applications, including hydraulics and diesel injection systems, which imply very high pressures up to more than 2000 bars, silicon is no longer a good choice as a base material for the sensor element. Sensor elements fabricated from steel[32,33] have the advantage that classical machining methods, like welding or applying threads to the steel body, provide robust connectors to the outer world that are able to withstand these pressures. For silicon, no connecting nor mounting techniques are available for pressure loads in this range. However, even with sensor bodies made from steel, the sensing technology itself is in most cases still MEMS-based. Piezoresistive elements are deposited on top of the steel membrane, with an insulating silicon oxide layer in between, either as metal thin film (for example, sputtered Ni or NiCr) resistors, polysilicon gauges, or single-crystalline silicon gauges transferred onto the steel membrane and structured by processes very similar to SOI technology. Regardless of the different base material, the principles of piezoresistive sensing still apply.

10.4 Force and Torque Sensors

Force and torque sensors can be used in a large number of applications. Whenever mechanical loads are being moved or lifted by machinery, information on the resulting forces and torques is of high relevance for safe and proper operation, for the correct adaptation of the driving mechanical moments to the load, and, as a consequence, for efficient use and the long lifetime of the mechanical construction. For example, an automatic gearbox controller needs feedback on torque moments that occur in the power train to choose the right gear. Another example is the steering column in a car, where both steering angle and steering torque are measured as inputs to the vehicle dynamics control system, and for future applications like steer-by-wire. The latter system uses magnetic or optical absolute- and relative-position sensors attached close to the steering column to measure both angle and torsion of the column, in particular of torsion bars integrated within the column. Optical and magnetic sensors are discussed in Chapters 8 and 9, respectively.

10.4.1 Linking the Macro World to the Micro World

Microsensors currently used to measure deflections or stresses in mechanical parts such as torque shafts, crankshafts, gears, bearings, and balances, acting from the outside world, like from a driving engine, human interaction, or just gravity, are basically not different from those discussed in Sections 10.2 and 10.3. The problem of detecting either stress or deflection or both is in principle the same as in the cases discussed in this section. Piezoresistive or resonant-type strain gauges provide appropriate solutions. The major difference between the cases discussed in Sections 10.2 and 10.3 and those discussed here is that here the mechanical microsensor is no longer restricted to its own microworld; it is and must be closely linked to macromechanical parts. The combination of both the mechanical microsensing element and the macromechanical part constitutes the sensor.

Interaction with the outside world is the strongest compared to the sensor examples discussed earlier in this chapter. An inertial sensor has no connection to the outer world at all; it is hermetically isolated by its package. It can be used as a plug-in device for many different applications, and mounting, calibration, and testing is usually done before it is mounted in its application site and often without knowledge about the particular intended application. In contrast, pressure sensors need some restricted contact with the external world to determine the pressure of a medium; apart from this, they constitute self-contained devices that can be individually mounted, calibrated, and tested. In contrast, a micromechanical force sensor can only fulfill its task in most cases together with a macromechanical construction, as a macroscopic unit to determine the relevant physical quantities. This has a strong impact on basic issues such as mounting and reliability, long-term stability and drifts, hermeticity of microencapsulation and macroencapsulation, calibration, temperature compensation and testing, and overload protection.

10.4.2 Fabrication, Protection, Test, and Calibration

10.4.2.1 Mounting of the Sensor Element

As an example, consider a micromechanical force sensor in a balance: The mechanical microsensing device might be a piezoresistive strain gauge mounted onto a lever construction, or a resonant sensor to convert mechanical strain into a change in resonance frequency of a vibrating beam.[40] After fixing the microsensor to the macrodevice, some passivation must be applied to protect the sensor and its electrical connections (bonding areas) from corrosion or other detrimental environmental effects, such as humidity.

10.4.2.2 Calibration

The glue used for mounting and the passivation materials have a decisive influence on the reliability of the microsensing device and its accurate operation. Calibration of the microsensor makes no sense at the microelement level, since the mounting and passivation procedure introduces significant extra stresses onto the microdevice and affects the electrical output from the sensing element. With the use of glues such as epoxy glue, which shows pronounced volume shrinkage during polymerization, extensive burn-in and aging procedures are needed before a stable situation is reached. Temperature compensation data are only of value if they are obtained for the whole device, taking into account all the different materials involved (such as silicon, metal, and epoxy glue) with their different thermal expansion coefficients. For temperature compensation, the complete macrodevice is placed in an oven and exposed to temperature cycling to determine the required data maps for correction of temperature-induced signal alterations.

10.4.2.3 Testing

The same holds true for testing: The mounting procedure adds numerous failure and reliability risks, so that testing must be carried out at the macrodevice level, for the completed lever, including the microsensing cell. This limits the scope of automatic wafer-level testing and implies that costly manual testing is necessary, which requires the handling of large parts, with routines going beyond purely electrical testing.

10.4.2.4 Protection and Reliability

Finally, to ensure safe and reliable operation, overload and overrange protection must be implemented in the macro-construction so that the microdevice will ultimately be protected. This cannot be achieved on the microscopic level, since forces in overload situations easily exceed the mechanical robustness of micromechanical stoppers, nor can macroscopic overrange protection exclude the appearance of overload situations at the microscale under all circumstances. Again, using the example of the balance, macroscopic stoppers can well limit the way of the lever. However, to prevent destruction of a microsensor fixed to the lever, this may not be enough. Despite properly limiting macroscopic deflections, vibrations induced from shocks to the lever are easily capable of destroying a microsensing element connected to it.

10.4.3 Conclusions

These sensor applications that so closely link the microworld and the macroworld involve the largest complexity of all the applications discussed, which is in many cases a true obstacle preventing the successful replacement of classical sensing elements by microsensing devices. Another difficulty is that the microsensing cell is often fixed to a rotating part (such as a torque shaft), with additional issues raised by the supply of energy from and the transfer of information to the environment. Telemetric systems based on electromagnetic fields for both wireless energy transfer and information exchange are promising solutions to sensing on moving objects and are attracting growing attention from MEMS developers. However, until today, the complexity of combined microsystems and macrosystems has put effective barriers against a stronger penetration of MEMS into these application fields. As technical issues resulting from this complexity are being increasingly analyzed and mastered, much progress in these fields can be expected to occur in the near future.

References

1. A. Von Zanten, R. Erhardt, and G. Pfaff, "FDR—Die Fahrdynamikregelung von Bosch," *ATZ—Automobiltechnische Zeitschrift*, 96:674 (1994)
2. C. J. Van Mullem, F. R. Blom, J. H. J. Fluitman, and M. Elwenspoek, "Piezoelectrically driven silicon beam force sensor," *Sens. Actuators*, A25–27:379 (1991)
3. U. Nothelfer, "Entwicklung und Anwendung neuer Techniken zur Steigerung der Leistungsfaehigkeit piezoresistiver Siliziumdrucksensoren," Ph.D. thesis, Technical University of Braunschweig (1995)
4. S. Fujishima, T. Nakamura, and K. Fujimoto, "Piezoelectric vibratory gyroscope using flexural vibration of a triangular bar," *Proc. 45th Symposium on Frequency Control* (1991), p. 261
5. T. Ueda, F. Kohsaka, D. Yamazaki, and T. Lino, "Quartz crystal micromechanical devices," *Proc. 3rd Int. Conf. Solid-State Sensors and Actuators*, Philadelphia, 1985, p. 113

6. H. Guckel, M. Nesnidal, J. D. Zook, and D. W. Burns, "Optical drive/sense for high Q resonant microbeams," *Proc. 7th Int. Conf. Solid-State Sensors and Actuators* (Transducers'93), Yokohama, Japan (1993), p. 686
7. R. A. Buser, "Theoretical and experimental investigations on silicon single crystal resonant structures," Ph.D. thesis, Université de Neuchâtel, 1989; S. Buettgenbach, "Frequenzanaloge Quarzsensoren," *Hard and Soft* (Oct. 1988), Fachbeilage Mikroperipherik; C. Burrer, "Design, fabrication and characterization of resonant silicon accelerometers," Ph.D. thesis, Universitat Autònoma de Barcelona, 1995
8. Y. Kanda, "A graphical representation of the piezoresistance coefficients in silicon," *IEEE Trans. Electron. Devices*, ED-29 (1982), p. 64
9. M. M. Abu-Zeid, "Corner undercutting in anisotropically etched isolation contours," *J. Electrochem. Soc.*, 131:2138 (1984); H. Seidel, H. Csepregi, A. Heuberger, and H. Baumgartner, "Anisotropic etching of crystalline silicon in alkaline solutions," *J. Electrochem. Soc.*, 137:3612–3632 (1990)
10. P. Barth, F. Pourahmadi, R. Mayer, J. Poydock, and K. Petersen, "A monolithic silicon accelerometer with integral air damping and overrange protection," *Tech. Digest IEEE Solid-State Sensor and Actuator Workshop*, Hilton Head Island, USA (1988), p. 35
11. F. Pourahmadi, L. Christel, and K. Petersen, "Silicon accelerometer with new thermal self-test mechanism," *Tech. Digest IEEE Solid-State Sensor and Actuator Workshop*, Hilton Head Island, USA (1992), p. 122
12. F. Rudolf, "A micromechanical capacitive accelerometer with a two-point inertial-mass suspension," *Sens. Actuators*, 4:191 (1983)
13. M. Offenberg, B. Elsner, and F. Laermer, "Vapor HF etching for sacrificial oxide removal in surface-micromachining," *Electrochem. Soc. Fall Meeting*, Miami Beach, FL, USA (1994), abstract no. 671, p. 1056
14. T. Core and R. Howe, *Analog Devices*, WO 93/21536
15. G. T. Mulhern, D. S. Soane, and R. T. Howe, *Proc. 7th Int. Conf. on Solid-State Sensors and Acutators* (Transducers'93), IEEE Electron Devices Society, Yokohama, Japan (1993), p. 296
16. H. Guckel, J. J. Sniegowski, T. R. Christenson, and F. Raissi, *Sens. Actuators*, A21:346 (1990)
17. R. W. Ashurst, C. Carraro, R. Maboudian, and W. Frey, "Wafer level antistiction coatings for MEMS," *Sens. Acutators*, A104:213 (2003)
18. M. Offenberg, F. Laermer, B. Elsner, H. Muenzel, and W. Riethmueller, "Novel process for a monolithic integrated accelerometer," *Transducers'95*, Stockholm (1995), 148-C4, p. 589
19. M. Offenberg, H. Muenzel, D. Schubert, O. Schatz, F. Laermer, E. Mueller, B. Maihoefer, and J. Marek, "Acceleration sensor in surface-micromachining for airbag applications with high signal/noise ratio," *Robert Bosch GmbH*, SAE 960758 (1996)
20. Lj. Ristic, D. Koury, E. Joseph, F. Schemansky, M. Kniffin, and L. Cergel, "A two chip accelerometer system for automotive applications," *Micro System Technologies Conference'94*, Berlin, Germany (1994), p. 77
21. J. Soederkvist, "Micromachined gyroscopes," *Sens. Actuators*, A43:65 (1994)

22. K. Funk, "Entwurf, Herstellung und Charakterisierung eines mikromechanischen Sensors zur Messung von Drehgeschwindigkeiten," Ph.D. thesis, Technical University of Munich, Germany, 1998
23. J. Soederkvist, "Piezoelectric beams and vibrating angular rate sensors," *IEEE Trans. Ultrasonics, Ferroelectrics and Frequency Control*, 38(3):271 (1991)
24. H. Shimizu, T. Yoshida, and C. Mashiko, "Gyroscope using circular rod type piezoelectric vibrator," European patent appl. no. 488 370 (1991)
25. M. W. Putty and K. Najafi, "A micromachined vibrating ring gyroscope," Solid-State Sensors and Actuators Workshop, Hilton Head, USA (1994), p. 213; Putty, M.W., Ph.D. thesis, University of Michigan, USA, 1995
26. M. Lutz, W. Golderer, J. Gerstenmeier, J. Marek, B. Maihoefer, S. Mahler, H. Muenzel, and U. Bischof, "A precision yaw rate sensor in silicon surface micromaching," *Proc. Transducers'97*, Chicago, USA (1997)
27. K. Funk, A. Schilp, M. Offenberg, B. Elsner, and F. Laermer, "Surface-micromachining of resonant silicon structures," *Transducers'95*, Stockholm, (1995), Late News, p. 50; R. Schellin, A. Thomae, M. Lang, W. Bauer, J. Mohaupt, G. Bischopink, L. Tanten, H. Baumann, H. Emmerich, S. Pinter, J. Marek, K. Funk, G. Lorenz, and R. Neul, "A low cost angular rate sensor for automotive applications in surface micromachining technology," *Advanced Microsystems for Automotive Applications 99*, Berlin, Germany (1999), p. 239
28. R. Willig and M. Moerbe, "New generation of inertial sensor cluster for ESP and future vehicle stabilizing systems in automotive applications," SAE paper 2003-01-0199
29. E. Obermeier, "Polysilicon layers lead to a new generation of pressure sensors," *Transducers'95*, Philadelphia (1985), p. 430
30. S. Armbruster, F. Schaefer, G. Lammel, H. Artman, C. Schelling, H. Benzel, S. Finkbeiner, F. Laermer, P. Ruther, and O. Paul, "A novel micromachining process for the fabrication of monocrystalline Si-membranes using porous silicon, *Transducers '03*, boston, MA (2003), p. 246
31. S. Otake, M. Onoda, and K. Nagase, "Fuel pressure sensor," Spec. Publ. SAE 1998, SP-1312 (1998), p. 61
32. Y. Suzuki, H. Tanaka, M. Imai, M. Harrison, and N. Oba, "Common rail pressure sensor," Spec. Publ. SAE 2002, SP-1076 (2002)
33. Data sheet on "Piezoresistive absolute pressure sensors 4043A...," Kistler Instrumente AG, CH-8408, Winterthur, Switzerland
34. C. Sander, J. Knutti, and J. Meindl, "A monolithic capacitive pressure sensor with pulse-period output," *IEEE Trans. Electron Devices*, 27(5):927 (1980)
35. Y. Manoli, W. Mokwa, and E. Spiegel, "Surface micromachined pressure sensors with integrated CMOS read-out electronics," *Sens. Actuators*, Phys. (Switzerland), A43(1–3):157 (1994); T. Scheiter, K.-G. Oppermann, M. Steger, C. Hierold, W.M. Werner, and H.-J. Timme, "Full integration of a pressure sensor system into a standard BiCMOS-process," *Proc. Eurosensors XI*, Warsaw, Poland (1997), p. 1595
36. H.-J. Kress, J. Marek, M. Mast, O. Schatz, and J. Muchow, "Integrated silicon pressure sensor for automotive application with electronic trimming," SAE Tech. Paper Series 950533 (1995)

37. J. Smits, H. Tilmans, and T. Lammerink, "Pressure dependence of resonant diaphragm pressure sensor," *Transducers'85*, Philadelphia, USA (1985), p. 93
38. T. Lammerink and W. Wlodarski, "Integrated thermally excited resonant diaphragm pressure sensor," *Tranducers'85*, Philadelphia, USA (1985), p. 97
39. H. Wohltjen, "Surface acoustic wave microsensors," *Proc. 4th Int. Conf. Solid-State Sensors and Actuators*, Tokyo, Japan (1987), p. 471
40. F. R. Blom, S. Bouwstra, J. H. J. Fluitman, and M. Elwenspoek, "Resonating silicon beam force sensor," *Sens. Actuators*, 17:513 (1989)

11 Semiconductor-Based Chemical Microsensors

Andreas Hierlemann and Henry Baltes

ETH, Zürich, Switzerland

11.1 Introduction

The detection of molecules or chemical compounds is a general analytical task in the efforts of chemists to obtain qualitative and/or quantitative time- and spatially resolved information on specific chemical components.[1] Examples of qualitative information include the presence or absence of certain odorant, toxic, carcinogenic, or hazardous compounds. Examples of quantitative information include concentrations, activities, or partial pressures of such specific compounds exceeding, e.g., a certain threshold-limited value or the lower explosive limits of combustible gases.

All this information can, in principle, be obtained from either a chemical analysis system or alternatively by using chemical sensors. In both cases, sampling, sample pretreatment, separation of the components, and data treatment are the tasks to be fulfilled. The main components of a state-of-the-art chemical analysis or sensor system are depicted in Fig. 11.1.

It is not easy to distinguish clearly between a chemical sensor and a complex analytical system. Integrated or miniaturized chromatographs or spectrometers may also be called chemical sensors. However, a typical chemical sensor in most cases is a cheaper, smaller, and less complex device compared to miniaturized analytical systems. The International Union of Pure and Applied Chemistry (IUPAC) provides the following draft definition of a chemical sensor:[2] "A chemical sensor is a device that transforms chemical information, ranging from the concentration of a specific sample component to total composition analysis, into an analytically useful signal." This rather wide definition does not require the sensor to be continuously operating and the sensing process to be reversible. But intermittently operating devices exhibiting irreversible characteristics are usually referred to as dosimeters.[3] In this context, it is useful to introduce some important key words used extensively throughout the chemical sensor literature.[1,4-11]

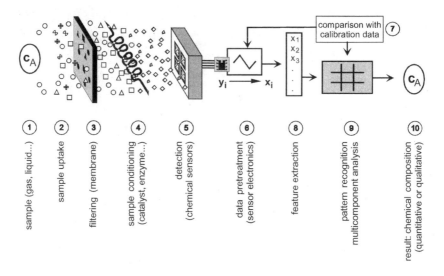

Figure 11.1 Components of a chemical analysis or sensor system. Adapted from Hierlemann *et al.*[15] with permission.

Reversibility

Thermodynamic reversibility, strictly speaking, requires that the sensor measurand be related to a thermodynamic state function. This implies that, e.g., a certain sensor response unequivocally corresponds to a certain analyte concentration. (Analyte here denotes the chemical compound to be monitored.) The sensor signal may not depend on the history of previous exposures or on how a certain analyte concentration is reached (no memory effects or hysteresis). More details are given in Section 11.2.

Sensitivity and Cross-Sensitivity

Sensitivity is usually defined as the slope of the analytical calibration curve, i.e., to what extent the change in the sensor signal depends on a certain change in the analyte concentration. Cross-sensitivity therefore refers to the contributions of compounds, other than the desired compound, to the overall sensor response.

Selectivity/Specificity

Selectivity can be defined according to Janata[4] as the ability of a sensor to respond primarily to only one species in the presence of other species (usually called interferants).

Limit of Detection and Limit of Determination

The limit of detection (LOD) corresponds to a signal equal to k times the standard deviation of the background noise (i.e., k represents the signal-to-noise ratio), with a typical value of $k = 3$. Values above the LOD indicate the presence of an analyte; values below the LOD indicate that no analyte is detectable.

The limit of determination implies qualitative information, i.e., that the signal can be attributed to a specific analyte. This in turn requires more information; therefore, the limit of determination is always higher than the limit of detection.

Transducer

The word *transducer* is derived from the Latin *"transducere,"* which means "to transfer or translate." Therefore, a device that translates energy from one kind of system (e.g., chemical) to another (e.g., physical) is called a transducer.

Biosensor

Biosensors are usually considered a subset of chemical sensors that make use of biological or living material for their sensing function.[11] Since the respective transducers and principles are similar, this chapter does not further diversify into chemosensors and biosensors. (See Chapter 13 for more information on biosensors.)

Using these definitions, chemical sensors usually consist of a sensitive layer or coating and a transducer. Upon interaction with a chemical species (such as absorption, chemical reaction, or charge transfer), the physicochemical properties of the coating (such as its mass, volume, optical properties, or resistance) reversibly change (Fig. 11.2). These changes in the sensitive layer are detected by the respective transducer and are translated into an electrical signal such as frequency, current, or voltage, which is then read out and subjected to further data treatment and pro-

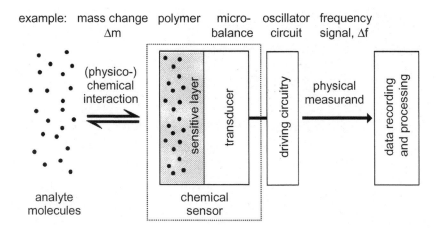

Figure 11.2 Definition of chemical sensors and components exemplified for the mass-sensitive principle (top).

cessing. In Fig. 11.2, this is exemplified for the mass-sensitive principle. Analyte molecules are absorbed into a coating material (polymer) to an extent governed by intermolecular forces. The change in mass of the polymeric coating in turn causes a shift in the resonance frequency of the transducer, e.g., a quartz microbalance. This frequency shift constitutes the electrical signal used in subsequent data processing.

To supply the different needs in chemical sensing, a variety of transducers based on different physical principles have been devised. Following the proposal of Janata et al.[4,5], chemical sensors can be classified into four principal categories according to their transduction principles:

1. Chemomechanical sensors (e.g., mass changes due to bulk absorption)
2. Thermal sensors (e.g., temperature changes through chemical interaction)
3. Optical sensors (e.g., changes of light intensity by absorption)
4. Electrochemical sensors (e.g., changes of potential or resistance through charge transfer)

These four categories of chemical sensors are treated in detail in Sections 11.3 to 11.6, respectively. A general overview of the recent literature on chemical sensors with regard to different transduction principles is given by Janata et al.[5]

Various inorganic and organic materials serve as chemically sensitive layers that can be coated onto the different transducers. Typical inorganic materials include electron-conducting oxides such as tin dioxide (SnO_2) for monitoring reducing gases such as hydrogen or carbon monoxide. Ion-conducting oxides such as zirconium dioxide (ZrO_2) are applied to detect oxygen; they are also used for nitrogen oxide and ammonia. Organic layers consisting mostly of polymers such as polysiloxane and polyurethane derivatives are used to monitor hydrocarbons, halogenated compounds, and different kinds of toxic, volatile organics. Table 11.1 surveys typical chemically sensitive materials and their applications. Further information on the coating materials is provided in the context of the different transducers in Sections 11.3 through 11.6.

Current research and development efforts in chemical sensors and sensitive materials follow two different strategies:

1. Searching for highly selective (bio)chemical layer materials (molecular recognition, key-lock-type interactions)
2. Using arrays of sensors exhibiting different partial selectivity (polymers, metal oxides) and applying pattern recognition (odors, aromas) and multicomponent analysis methods (mixtures of gases and liquids)

The latter strategy has grown very popular,[12-17] especially since compact sensor arrays can presently be fabricated at low cost, and interferants, which are present in almost any practical application, can be handled.

Meanwhile, chemical sensors have also reached the stage of exploratory use in a variety of industrial and environmental applications. Some examples are quality control or on-line process monitoring in the food industry as well as preliminary tests in the areas of medical practice and personal (workplace) safety.[18] In environmental monitoring, in particular, there is an urgent need for low-cost sensor systems that detect various pollutants at the trace level.

Key requirements for a successful chemical sensor include:

- High sensitivity and low LOD
- High selectivity to the target analyte and low cross-sensitivity to interferants
- Short recovery and response times
- Large dynamic range
- Reversibility
- Precision and reproducibility of the signal
- Long-term stability and reliability (self-calibration)
- Low drift
- Low temperature dependence or temperature compensation mechanisms
- Ruggedness

Table 11.1 Typical Sensor Materials and Applications

Materials	Examples	Applications
Metals	Pt, Pd, Ni, Ag, Sb, Rh, ...	Inorganic gases (CH_4, H_2,...)
Ionic compounds	Electronic conductors (SnO_2, TiO_2, Ta_2O_5, In_2O_3, $AlVO_4$,...) Mixed conductors ($SrTiO_3$, Ga_2O_3, perowskites,...) Ionic conductors (ZrO_2, LaF_3, CeO_2, nasicon,...)	Inorganic gases (CO, NO_x, CH_4,...), exhaust gases, oxygen, ions in water,...
Molecular crystals	Phthalocyanines (PCs), (PbPc, $LuPc_2$,...)	Nitrogen dioxide, volatile organics
Langmuir-Blodgett films	Lipid bilayers, polydiacetylene,...	Organic molecules in medical applications, biosensing,...
Cage compounds	Zeolites, calixarenes, cyclodextrins, crown ethers, cyclophanes,...	Water analysis (ions), volatile organics,...
Polymers	Nonconducting: polyurethanes, polysiloxanes,... Conducting: Polypyrroles, polythiophenes, nafion,...	Detection of volatile organics, food industry (odor and aroma), environmental monitoring in gas and liquid phases,...
Components of biological entities	Synthetic: phospholipids, lipids, HIV-epitopes,... Natural: enzymes, receptors, proteins, cells, membranes,...	Medical applications, biosensing, water and blood analysis, pharmascreening,...

- Low costs (batch fabrication) and low maintenance
- Ease of use

Semiconductor technology provides excellent means to meet some of these criteria (low cost, batch fabrication) and offers additional features such as small size and on-chip signal processing. The rapid development of integrated-circuit (IC) technology has initiated many initiatives to fabricate chemical sensors consisting of a chemically sensitive layer on a signal-transducing silicon chip.[19,20] The earliest types of chemical sensors realized in silicon technology were based on field-effect transistors (FETs).[21,22] Reviews of silicon-based sensors (not only chemical sensors) are given by Middelhoek,[23] Sze,[24] and Gad-el-Hak.[25]

The largely two-dimensional IC and chemical sensor structures processed by combining lithographic, thin-film, etching, diffusive, and oxidative steps have been recently extended into the third dimension using micromachining technologies – a combination of special etchants, etch stops, and sacrificial layers. (See Chapters 1, 2, and 15.) A variety of micromechanical structures including cantilever beams, suspended membranes, freestanding bridges, gears, rotors, and valves have been produced using micromachining technology (microelectromechanical systems, or MEMS). (See Chapters 1 and 15.)[26-29] MEMS technology thus provides a number of key features that can serve to enhance the functionality of chemical sensor systems.[9,11,26,29-34] In a next step, microelectronics and micromechanics (MEMS structures) have been realized on a single chip, allowing for on-chip control and monitoring of the mechanical functions as well as for data preprocessing such as signal amplification, signal conditioning, and data reduction.[29-34] In this context, two more keywords are introduced: integrated sensor and smart or intelligent sensor.

Integrated Sensor

A sensor is called an integrated sensor if the chemical sensing operation is based on a direct influence on an electric component (resistor, transistor, capacitor) integrated in silicon or another semiconductor material.[11]

Smart or Intelligent Sensor

The combination of interface electronics and an integrated sensor on a single chip results in a so-called smart sensor. At least some basic signal conditioning is usually carried out on the chip. The major advantage of smart sensors is the improved signal-to-noise and electromagnetic interference characteristics.[11]

The treatment of the different transducers and chemical sensors in Sections 11.3 through 11.6 focuses exclusively on semiconductor-based and MEMS-based devices.

11.2 Thermodynamics of Chemical Sensing

The interaction of a chemical species with a chemical sensor can either be confined to the surface of the sensing layer or can take place in the whole volume of the sensitive coating. Surface interaction implies that the species of interest is *adsorbed* at the surface or interface (gas/solid or liquid/solid) only, whereas volume interaction requires the *absorption* of the species and a partitioning between the sample phase and the bulk of the sensitive material. The different types of chemical interactions involved in a sensing process range from very weak physisorption through rather strong chemisorption to charge transfer and chemical reactions.

Physisorption in this context implies that the compound is only physically absorbed/adsorbed (London or Van-der-Waals dispersion forces) with an interaction energy of 0 to 30 kJ/mol. In the case of the much stronger chemisorption (interaction energy >120 kJ/mol), the particles stick to the surface by forming a chemical (usually covalent) bond. In most cases, charge transfer and chemical reactions involve interaction energies comparable to those of chemisorption and higher. Some of the most common interaction mechanisms and their energies are listed in Table 11.2; Atkins[35] provides further details.[35]

High chemical selectivity and rapid reversibility place contradictory constraints on desired interactions between chemical sensor coating materials and analytes. Low-energy, perfectly reversible (physisorptive) interactions generally lack high selectivity, while chemisorptive processes, the strongest of which result in the formation of new chemical bonds, offer selectivity but are inherently less reversible. A practicable compromise has to be achieved with due regard to the specific application. In this context, the commonly accepted limit of reversibility up to 20 kJ/mol refers to room temperature and will not apply to the case of, e.g., tin-dioxide-coated semiconductor sensors operated at 300 to 400°C. On the other hand, spontaneous chemical reactions occurring at room temperature often require a tedious regeneration of, e.g., biological recognition units (enzymes).

Any interaction between a coating material and an analyte is governed by chemical thermodynamics and kinetics. Thus a fundamental thermodynamic function, the Gibbs free energy, G [J], is the most important

Table 11.2 Typical Interactions and Energies

Interaction Type	Typical Energy (kJ/mol)	Comment
Covalent bond	120 to 800	Chemical reaction
Ion-ion	250	Only between ions
Coordination, complexation, charge-transfer	8 to 200	Weak "chemical" interaction
Ion-dipole	15	
Hydrogen bond	20	Hydrogen bond: A–H···B
Dipole-dipole	0.3 to 2	Between polar molecules
London (dispersion) (induced dipole-induced dipole)	0.1 to 2	Physical interaction between any type of molecules

descriptor in all chemical sensing processes: The direction of spontaneous reactions is always toward lower values of G (minimization of the Gibbs energy). The Gibbs free energy is a state function in the thermodynamic sense, i.e., its value depends only on the current state of the system and is independent of how that state has been prepared. This implies that any chemical (sensing) process described by a Gibbs energy function moves toward a dynamic equilibrium ($\Delta G = 0$, G minimal), in which both reactants and products are present but have no tendency to undergo net change. This equilibrium is reversible, i.e., an infinitesimal change in the conditions in opposite directions results in opposite changes in its state. The interaction equilibrium of an analyte, A, with a sensor coating, S, can thus be represented by

$$A + S \underset{\overleftarrow{k}}{\overset{\vec{k}}{\rightleftharpoons}} A \cdots S. \tag{1}$$

Here, $\vec{k}$ and $\overleftarrow{k}$ denote the rate constants of the forward reaction and the reverse reaction, which are detailed below. Such equilibrium can be described by an equilibrium constant, K, which relates the activity, a, of reaction products (A···S) to those of the reactants (A and S). This constant is thus a characteristic value for the progression of the reaction ($K \leq 1$: no reaction takes place); its numerical value depends on the system temperature.

$$K = \frac{a_{A \cdots S}}{a_A \cdot a_S} \quad \text{and, in general,} \quad K = \prod_i a_i^{n_i} \quad (2)$$

The index i denotes the chemical substance, and n_i are the corresponding stoichiometric numbers in the chemical equation. This expression signifies that each activity (or fugacity) is raised to the power equal to its stoichiometric number, then all such terms are multiplied together. Stoichiometric numbers of the products are positive and those of the reactants are negative, i.e., reactants appear as the denominator and reaction products as the numerator. The activity, a (fugacity, f, for gases), which denotes the effective quantity of a compound i participating in, e.g., a chemical reaction, is related to the mole fraction, x (partial pressure, p, for gases), of a species via $a_A = \gamma \cdot x_A$ with $\gamma \leq 1$. The constant γ takes into account all kinds of nonideal behavior. The equilibrium constant, K, is also related to kinetics. For the simple reaction in Eq. (1), we can define two kinetic constants: $\vec{k}$ for the reaction leading to the product A$\cdots$S, and $\overleftarrow{k}$ for the reaction in the opposite direction.

$$\frac{da_A}{dt} = -\vec{k}\, a_A a_S + \overleftarrow{k}\, a_{A \cdots S} \quad (3)$$

K then represents the ratio of those two kinetic constants in the equilibrium state.

$$K = \frac{\vec{k}}{\overleftarrow{k}} \quad (4)$$

Both thermodynamics and kinetics therefore affect the progress of any chemical process or reaction. Thermodynamics, namely the Gibbs free energy (minimum) or the equilibrium constant, can tell us the direction of spontaneous change and the composition at the equilibrium state, whereas kinetics tell us whether a kinetically viable pathway exists for that change to occur and how fast an equilibrium state will be achieved. Kinetics is important in the context of chemical sensors because chemical processes exist, the activation barrier of which is too high to get a reaction going, although the Gibbs free energy of the products would be below that of the reactants. Such effects can be used to advantage in tuning the selectivity of, e.g., catalytic chemical sensors.

A chemical potential has been introduced in thermodynamics. The chemical potential shows how the Gibbs energy of a system changes when a portion of a specific chemical compound is added to it or removed from it. The chemical potential of the ith component, μ_i, is defined as

$$\mu_i = \left(\frac{\partial G}{\partial n_i}\right)_{p,T}$$

and

$$\Delta G = \sum_i \mu_i \, dn_i. \tag{5}$$

Here n_i denotes the stoichiometric number or the amount of substance in moles. Again, the stoichimetric numbers of the products are positive and those of the reactants are negative. The pressure, p, and the temperature, T, are kept constant. The chemical potential can be expressed in terms of mole fractions, x, or activities, a, in liquids, and partial pressures, p, or fugacities, f, in the gas phase:[35]

$$\mu_i = \mu_i^0(p,T) + RT \ln a_i$$

or

$$\mu_i = \mu_i^0(p,T) + RT \ln f_i. \tag{6}$$

Here, $\mu_i^0(p,T)$ denotes the chemical potential of a (hypothetical) reference state, e.g., infinite dilution or pure compound; R is the molar gas constant (8.314 J/Kmol); and T denotes the temperature (in K). So Eq. (6) has two terms: a reference term and an activity-dependent term. Plugging the terms of Eq. (6) into Eq. (5), the reference terms (μ_i^0) can be subsumed into ΔG^0 as shown in the following equation:

$$\Delta G = \sum_i n_i \mu_i^0(p,T) + RT \sum_i n_i \ln a_i = \Delta G^0(p,T) + RT \ln \prod_i a_i^{n_i}. \tag{7}$$

Both ΔG and K are characteristic descriptors for the direction of a chemical reaction. Comparing Eq. (2) with Eq. (7) shows that ΔG^0 and K in a thermodynamic equilibrium state ($\Delta G = 0$) are interrelated via the following equation: (For details, see Atkins.[35])

$$\ln K = -\frac{\Delta G^0}{RT}. \tag{8}$$

The more negative ΔG^0 is, the larger K is. In other words, the higher the chemical potential of the reactants is with regard to the products, the larger the reaction extent is and the more spontaneously the reaction will occur provided that kinetic factors already discussed do not upset such predictions. According to the Gibbs fundamental equation, ΔG^0 is composed of an enthalpy term, ΔH^0, representing the reaction heat at constant pressure, and an entropy term, ΔS^0, representing the degree of "disorder" or, to put it more precisely in thermodynamic terms, the number of different ways in which the energy of a system can be achieved by rearranging the atoms or molecules among the states available to them: (For details, see Atkins.[35])

$$\Delta G^0 = \Delta H^0 - T\Delta S^0. \tag{9}$$

For spontaneous reactions (ΔG^0 negative), the entropy increases and/or the enthalpy term is negative, i.e., heat is released during the chemical reaction.

The thermodynamics of three prototype reactions of chemical sensors are discussed below.

Simple Adsorption/Absorption

At thermodynamic equilibrium state, the free species and the adsorbed/absorbed species are in dynamic equilibrium, i.e., the chemical potentials of a certain compound A in the gaseous phase and the polymeric phase are identical: $\mu_A^{gas}(p,T) = \mu_A^{polymer}(p,T)$ [Eq. (6)]. Absorbing all the constant terms (μ_i^0, R, T) into a sorption constant, $K_{sorption}$, or so-called partition coefficient, the equilibrium state can be described by

$$K_{sorption} = \frac{a_A^{sorbed}}{a_A^{free}}. \tag{10}$$

The partition coefficient is a dimensionless "enrichment factor" relating, e.g., the activity of a compound in the sensing layer a^{sorbed} to that in the probed gas or liquid phase a^{free}; it also represents a thermodynamic equilibrium constant, which is related to ΔG^0 via Eq. (8).

For surface adsorption, it is more common to relate the fractional coverage of the surface, θ, to the concentration of the analyte in the probed phase and to use different types of adsorption isotherms such as Langmuir, Freundlich, or Brunauer-Emett-Teller (BET) isotherms.[35]

Chemical Reaction

In this case, Eq. (2) can be applied in principle. It has to be modified with regard to the respective reaction mechanism occurring. For a simple reaction like $n_A A + n_B B \leftrightarrow n_C C + n_D D$, the equilibrium constant is given in analogy to Eq. (2):

$$K = \prod_i a_i^{n_i} \quad \text{and in particular:} \quad K = \frac{a_C^{n_C} \cdot a_D^{n_D}}{a_A^{n_A} \cdot a_B^{n_B}}. \qquad (11)$$

The chemical potentials as defined in Eq. (6) can be used, and Eq. (8) holds. As already mentioned, the interaction leading to a true chemical reaction may be too strong to be reversible.

Charge Transfer and Electrochemical Reaction

For a reaction of type $A^+ + e^- \leftrightarrow A$, an electrochemical potential has to be introduced. The contribution of an electrical potential to the chemical potential is calculated by noting that the electrical work, W_e, of adding a charge, $z \cdot e$ (z denotes the number of elementary charges, e), to a region where the potential is ϕ (ϕ denotes the *Galvani* potential, which represents the bulk-to-bulk inner contact potential of two materials and is defined as the difference of the Fermi levels of these two materials), is

$$W_e = z \cdot e \cdot \phi;$$

hence, the work per mole is

$$W_e = z \cdot F \cdot \phi. \qquad (12)$$

F here denotes the Faraday constant, 96485 C/mol, which is equivalent to one mole of elementary charges. Hence, the electrochemical potential is [compare Eq. (6)]

$$\mu_i = \mu_i^0(p,T) + RT \ln a_i + zF\phi. \qquad (13)$$

When z = 0 (a neutral species), the electrochemical potential is equal to the chemical potential [Eq. (6)]. Rewriting Eq. (7) for the electrochemical potentials leads to

$$\Delta G = \Delta G^0(p,T) + RT \ln \prod_i a_i^{n_i} + zF \cdot \Delta\phi. \qquad (14)$$

In the equilibrium state ($\Delta G = 0$), Eq. (13) can be expressed in terms of K [Eq. (2)]. By replacing E, the "electromotive force," for $\Delta\phi$, and by replacing E^0, the standard cell potential, for $-\Delta G^0/zF$ (a positive voltage per convention always corresponds to a negative ΔG: spontaneous reaction), the so-called Nernst equation results:

$$E = E_0 - \frac{RT}{zF} \ln K. \qquad (15)$$

The Nernst equation now can be used to derive an expression for the potential of any electrochemical cell or, in our case, electrochemical sensor. Electrochemical reactions can be triggered by applying currents or voltages via electrodes to a sensing layer.

The different sensor principles and characteristic applications are detailed in the remainder of this chapter.

11.3 Chemomechanical Sensors

The change in mechanical properties (e.g., mass) of a sensitive layer upon interaction with an analyte can be conveniently recorded with micromechanical structures. Any species that can be immobilized on the sensor can, in principle, be sensed. As with most of the chemical sensors (excluding thermal sensors), the measurements are performed at a thermodynamic equilibrium state [Eqs. (1) through (4) and (10)] defined by the Gibbs energy of the system. In the simplest case, such chemomechanical sensors are gravimetric sensors that respond to the mass of species accumulated in a sensing layer.[36-38] Some of the sensor devices also respond to changes in a variety of other mechanical properties of solid or fluid media in contact with their surface such as polymer moduli, liquid density, and viscosity,[36-38] which are not discussed here.

The high sensitivity of gravimetric sensors provides good chemical sensitivity: Mass changes in the picogram range or lower can be detected and ppm (parts per million) to ppb (parts per billion) detection levels have been reported for, e.g., gas and vapor sensors.[36-38] The large number of chemical species that can be present in the environment, and the difficulty in selectively and, at the same time, in reversibly sorbing these species on the sensor, however, make specific detection difficult.

Most of the gravimetric sensors rely on piezoelectric materials such as quartz, lithium tantalate or niobate, aluminum nitride, and zinc oxide. Piezoelectricity results in general from coupling of electrical and mechanical effects. The prerequisite is an anisotropic, noncentrosymmetric crystal lattice. Upon mechanical stress, charged particles are displaced and thus generate a measurable electric charge in the crystal. In turn, mechanical deformations can be achieved by applying a voltage to such a crystal. (For details, see Janata et al.[4]) Using an alternating current (AC), the crystals can be electrically excited into a fundamental mechanical resonance mode. The resonance frequency, which is the recorded sensor output in most cases, changes in proportion to the mass loading on the crystal or device. The more mass (analyte molecules) is absorbed, e.g., in a polymer coated onto a piezoelectric substrate or transducer, the lower is the resonance frequency of the device:

$$\Delta f = -C f_0^2 \Delta m / A. \qquad (16)$$

This equation was published by Sauerbrey in 1959.[39] Here, Δf denotes the frequency shift due to the added mass (in Hz), C is a constant, f_0 is the fundamental frequency of the quartz crystal (in Hz), and $\Delta m/A$ is the surface mass loading (in g cm^{-2}). The following equation describes the relationship between analyte gas-phase concentration change, Δc_A, and the responses of mass-sensitive sensors:

$$\Delta f_A = \Gamma \cdot M_A \cdot K \cdot \Delta c_A. \qquad (17)$$

Here, Δf_A (Hz) denotes the frequency shift (sensor response) measured upon exposure to analyte at a concentration c_A (mol/L). M_A (kg/mol) is the molar mass of the analyte vapor, K is the partition coefficient [Eq. (10)], and Γ is a gravimetric constant (L/kg·s) that includes, for example, the frequency shift measured upon initial deposition of the sensitive layer, the coating density, and the transducer dimensions. Figure 11.14 displays a typical signal of a gravimetric sensor, showing the frequency shifts of a resonant cantilever coated with poly(etherurethane) (PEUT) upon exposure to various concentrations of n-octane. At low analyte concentrations (trace level), a linear correlation between the frequency shift due to analyte absorption and the corresponding analyte concentration in the gas phase is usually observed (Fig. 11.3), provided that the sensing film on the transducer moves synchronously with the oscillating crystal surface. Significant deformations across the film thickness result in a more complex relationship between mass changes and resonant frequency due to, e.g., viscoelastic effects (see the concept of "acoustically thin and thick" films as detailed by Martin et al.[40]).

The most common devices are the thickness shear mode resonator (TSMR) or quartz microbalance (QMB), which is a bulk resonator, and

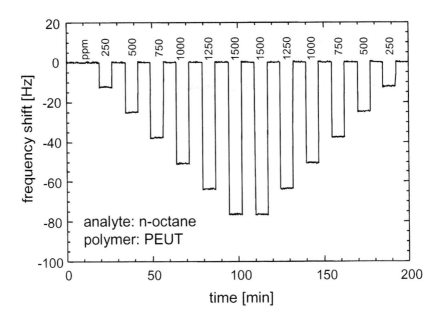

Figure 11.3 Typical response of mass-sensitive sensors. Frequency shifts of a polymer-coated [poly(etherurethane), PEUT] cantilever upon exposure to different concentrations of an organic volatile: n-octane.

the Rayleigh surface acoustic wave (SAW) device, both based on quartz substrates. The QMB was demonstrated to function as an organic vapor sensor by King in 1964;[41] the SAW device became popular after interdigital transducers were used to fabricate acoustic sensors in 1970.[42] Since the TSMR is not semiconductor-based and is not compatible with IC technology, it is not treated here. Shear horizontal acoustic plate mode (SH-APM) devices, shear transverse wave (STW) devices, and Love wave devices require quartz, lithium niobate, or lithium tantalate substrates [36-38] and therefore are also not dealt with here. A wealth of literature provides details and further information.[36-38,43-45]

Silicon is not a piezoelectric material. The realization of silicon-based piezoelectric transducers therefore requires an additional piezoelectric layer to be patterned on the silicon. Different materials have been used such as cadmium sulfide,[46] aluminum nitride,[47,48] and in particular zinc oxide (ZnO)[25,49-51]; these are discussed in Sections 11.3.1 and 11.3.2.

Three semiconductor-technology-compatible types of mass-sensitive devices are discussed in Sections 11.3.1 through 11.3.3, respectively: (1) Raleigh SAW devices on Si-substrates with piezoelectric overlay, (2) flexural-plate-wave devices (FPWs), and (3) micromachined resonating cantilevers. Operability in gas or liquid media, typical coating materials, target analytes, and applications are discussed in the context of each transducer. Overviews of micromachined resonant sensors are given by Martin et al.[45] and Brand et al.[52]

11.3.1 Rayleigh SAW Devices

Transduction Principle and Sensing Characteristics

Interdigital transducers can be used to launch and detect a surface acoustic wave on a piezoelectric substrate,[42] as shown in Fig. 11.4. By applying an AC voltage to a set of interdigital transducers patterned on a piezoelectric substrate with appropriate orientation of the crystal axes, one set of the fingers moves downwards, the other upwards, thereby creating an oscillating mechanical surface deformation. This surface deformation generates an acoustic wave that propagates along the surface and is converted back into an electrical signal by deforming the surface in the region of the receiving transducer. The electrical signal of the receiving transducer is recorded and represents the sensor signal.

For a given piezoelectric substrate, the acoustic wavelength and thus the operating frequency of the SAW are determined by the transducer periodicity, which is equal to the acoustic wavelength at the transducer center frequency. Typical frequencies range between 100 and 500 MHz.[36-38] Such frequencies require a sophisticated, high-frequency circuit design.

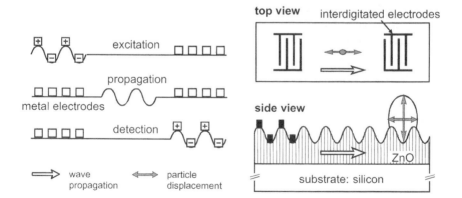

Figure 11.4 Launching, propagation, and detection of a Rayleigh-type surface acoustic wave by interdigitated transducers. The top view shows the electrode configuration and the wave propagation. The side view shows the elliptical particle displacement.

Therefore, a bare reference oscillator is operated together with the sensor in many cases, and the outputs are mixed to produce a difference frequency, with values in the kHz range, that is recorded.[49,51]

The acoustic wave is confined to a surface region of approximately one acoustic wavelength thickness. The velocity and damping characteristics of the acoustic wave are therefore extremely sensitive to changes at the transducer surface. When used in an oscillator circuit, relative changes in the wave velocity are reflected as equivalent changes in fractional oscillation frequency. A change in mass due to, e.g., absorption of a gaseous analyte in a polymeric sensing layer thus changes the device frequency according to Eq. (16).

The acoustic (Rayleigh) wave causes an elliptical particle movement at the transducer surface (Fig. 11.4), i.e., the sensitive films deposited on top of the transducers and the piezoelectric substrate are severely deformed. Thus, additional effects such as changes in viscoelastic properties of the sensing layer can affect the sensor response.[40]

Fabrication

Since a variety of custom-designed semiconductor and silicon processes exist in the literature, only the additional fabrication steps after semiconductor processing are mentioned in Sections 11.3 through 11.6. However, industrial standard IC processes such as CMOS fabrication are explicitly identified. Chapters 15 and 16 provide details of semiconductor and MEMS process steps.

- Optional back-etching using potassium hydroxide (KOH) or ethylenediamine-pyrocatechol (EDP) to achieve a membrane structure [49,53])
- Zinc-oxide processing: Deposition mainly by sputtering techniques at 423 to 723 K; highly oriented layers of 5 to 50 µm with a high degree of surface flatness.[25,49]
- Electrode processing: Vacuum evaporation of aluminum or gold, with a layer thickness >200 nm
- Sensitive layer: Spin or spray coating of polymers, organic layers, or biological entities

Applications

Since surface-normal particle displacements occur (Fig. 11.3), and the acoustic wave velocity is larger than the compressional velocity of sound in water, the device radiates compressional waves in the liquid phase, which causes severe attenuation. Rayleigh SAW devices therefore cannot be used in liquids.[36,38]

Typical application areas are environmental monitoring and personal safety devices. These applications include the detection of different kinds of organic volatiles (e.g., hydrocarbons, chlorinated hydrocarbons, alcohols) by using polymeric layers[49] or porphyrins,[54] and the detection of nitrogen dioxide using phthalocyanines.[51] In most cases, the interaction mechanisms involve fully reversible physisorption and bulk/gas phase partitioning [see Eqs. (10) and (17)].

Integrated Gallium Arsenide SAW Sensor

Gallium arsenide (GaAs) is a well-developed semiconductor device material for fabricating high-frequency integrated circuits; GaAs is piezoelectric. The piezoelectric properties of GaAs and, hence, the device characteristics are similar to those of quartz except for the strong temperature dependence. An integrated GaAs SAW sensor is shown in Fig. 11.5.[55] It consists of a 470-MHz SAW device along with a multistage amplifier (four gain

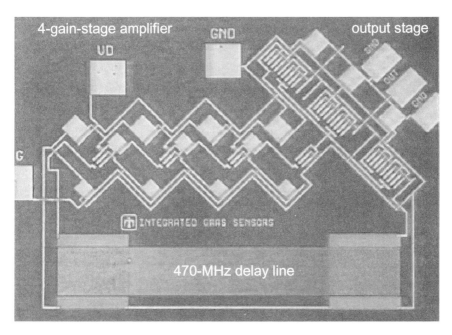

Figure 11.5 Micrograph of a monolithically integrated GaAs surface acoustic wave device showing the delay line, the amplifier, and the output stage. Reprinted from Casalnuovo *et al.*[55] with permission.

stages and an impedance-matching output stage), forming a monolithic oscillator circuit, thus eliminating the need for high-frequency interconnections.[55]

11.3.2 Flexural-Plate-Wave or Lamb-Wave Devices

Transduction Principle and Sensing Characteristics

Flexural-plate-wave devices were introduced in 1988.[56] Their chief advantage is their high sensitivity to added mass at a low operating frequency (typically 3 to 10 MHz).[57] Flexural-plate-wave devices feature plates that are only a few percent of an acoustic wavelength thick (typically 2 to 3 µm). The plates are composite structures (Fig. 11.6), consisting of a silicon nitride layer, an aluminum ground plane, and a sputtered zinc-oxide piezoelectric layer, all supported by a silicon substrate.[36,38,56,58]

The interdigital transducers (IDTs) on these devices generate flexural waves (Lamb waves; Fig. 11.6) with retrograde elliptical particle motions, as in the SAW devices. However, the velocity in the membrane is much less than in a solid substrate, and the operating frequency for a given transducer periodicity is hence considerably lower.[36-38,45,56] The Lamb waves give rise to a series of plate modes, one of which has a frequency that is much lower than those of the other possible modes. The velocity of this unique wave decreases with decreasing plate thickness. The entire thickness of the plate is set in motion like the ripples in a flag.[36-38,45,56,58] The confinement of acoustic energy in the thin membrane results in a very high mass sensitivity. The sensor response (frequency shift) is proportional to the mass loading [Eq. (16)].

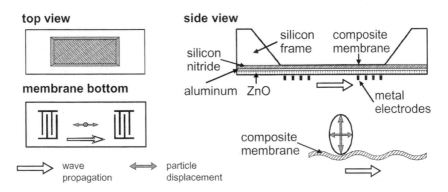

Figure 11.6 Schematic of a flexural-plate-wave device. The side view shows the different layers and the membrane movement. Interdigitated electrodes are used for actuation.

Since the Lamb wave causes an elliptical particle movement at the transducer surface (Fig. 11.6), the sensitive films are deformed, just as they are in the case of the SAW. Because the frequency is much lower, however, changes in the viscoelastic properties of the sensing layer do not severely affect the sensor response.

The sensitive layer can be deposited on either side of the membrane. Deposition on the backside (nonprocessed side of the wafer) has the advantage that on-chip circuitry will not be exposed to chemicals.[56-59]

Transducer Modification: Magnetically Excited FPW

Although magnetic excitation requires an externally applied magnetic field, it eliminates the need for a piezoelectric layer, which frequently contains elements (e.g., Zn) that pose contamination problems in IC fabrication. The FPW device consists of a silicon nitride membrane suspended in a silicon frame. A metal meander-line transducer is patterned on the membrane surface (Fig. 11.7). Alternating current flowing in the transducer interacts with a static in-plane magnetic field to generate time-varying Lorentz forces (Fig. 11.7). These deform the membrane, exciting it into a

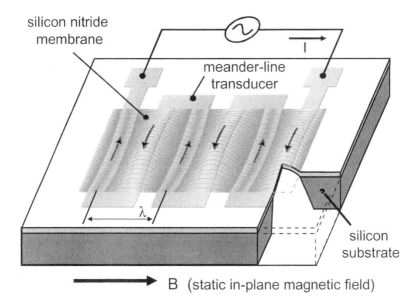

Figure 11.7 Schematic representation of the magnetically excited flexural-plate-wave device. Lorentz forces are generated between an impressed alternating current in a serpentine conductor and a static in-plane magnetic field. Reprinted from Martin *et al.*[60] with permission.

resonant mode.[45,60] To excite the mode efficiently, the current lines of the transducer must be positioned along lines of maximum mode displacement (Fig. 11.7). This requires a critical alignment between the top metallization pattern and the backside etch mask.[45,60]

Fabrication

Additional fabrication steps after semiconductor processing include the following:

- Evaporation of Al and Si-nitride (low-pressure chemical-vapor deposition, LPCVD)[56]
- Back-etching (KOH or EDP) to achieve a membrane structure[56,58]
- Zinc-oxide or lead zirconate titanate (PZT)[61] processing, if necessary; see Section 11.3.1
- IDT processing: vacuum evaporation of aluminum or gold; see Section 11.3.1
- Sensitive layer: spin or spray coating of polymers, deposition of biological entities

IC process-compatible fabrication sequences for monolithic integration of the Lamb device with electronics are detailed in Vellekoop *et al.*[58,62]

Applications

Surface-normal particle displacements occur (Fig. 11.6), but the acoustic-wave velocity is much less than the compressional velocity of sound in water. FPW devices thus can be used in the liquid phase.[36,38,58,63]

Typical application areas are environmental monitoring (gas and liquid phases) or biosensing in liquids. These applications include the detection of different organic volatiles in the gas phase (hydrocarbons, chlorinated hydrocarbons, alcohols, etc.) by using polymeric layers,[56,59,64-66] the detection of the weight percentage of alcohol/water[58] or glycol/water[63] mixtures, and the use of an FPW-based immunoassay for the detection of breast cancer antigens.[67]

The interaction mechanisms involve reversible physisorption and bulk/gas phase partitioning [see Eqs. (10) and (17)] as well as antigen/antibody binding.[67]

11.3.3 RESONATING CANTILEVERS

Transduction Principle and Sensing Characteristics

Micromachined cantilevers commonly used in atomic force microscopy (AFM) constitute a promising type of mass-sensitive transducer for chemical sensors.[68-82] The sensing principle is simple. The cantilever is a layered structure (Fig. 11.8) composed of, e.g., the dielectric layers of a standard CMOS process, silicon, metallizations, and eventually, zinc oxide. The cantilever base is firmly attached to the silicon support. The freestanding cantilever end is coated with a sensitive layer.

There are two fundamentally different methods of operation:

1. Static mode: measurement of the cantilever deflection upon stress changes in the sensitive layer by means of, e.g., laser light reflection [75,78-80]
2. Dynamic mode: excitation of the cantilever in its fundamental mode and measurement of the change in resonance frequency upon mass loading[68-70,76,77] is analogous to other mass-sensitive devices [Eqs. (16) and (17)]

The two methods impose completely different constraints on the cantilever design for maximum sensitivity. Method 1 requires long and deformable cantilevers to achieve large deflections, whereas method 2

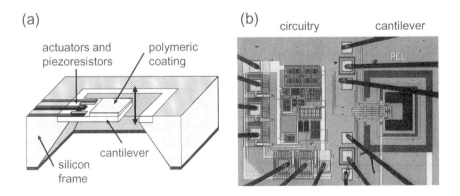

Figure 11.8 (a) Schematic representation and (b) micrograph of an integrated CMOS cantilever with on-chip circuitry. The cantilever is thermally actuated; the vibration is detected by piezoresistors. Reprinted from Hierlemann *et al.*[70] with permission.

requires short and stiff cantilevers to achieve high operational frequencies. Method 2 is preferable with regard to integration of electronics and simplicity of the setup (feedback loop).[68-70,76,77,81,82] Method 1 can be applied in liquids as well,[76,79] which is more difficult using the dynamic mode. The excitation of a cantilever in the resonant mode is usually performed by applying piezoelectric materials (ZnO)[81] or by making use of the bimorph effect, i.e., the different temperature coefficients or mechanical stress coefficients of the various layer materials forming the cantilever.[68-80] This difference in material properties gives rise to a cantilever deflection upon heating. Periodic heating pulses in the cantilever base thus can be used to thermally excite the cantilever in its resonance mode up to 500 kHz.[68-70,81] The frequency changes can be detected by embedding piezoresistors in the cantilever base,[68-70,76,77] by measuring motional capacitance changes,[82] or by using optical detection by means of laser light reflection on the cantilever.[72-75,78-80] The mass resolution of the cantilevers is in the range of a few picograms.[68-70,71,78-80] This high mass-sensitivity does not necessarily imply an exceptionally high sensitivity to analytes since the area coated with the sensitive layer is usually very small (on the order of 100×150 μm^2).[68]

The sensing layer is deformed upon motion of the cantilever; therefore, modulus effects are expected to contribute to the overall signal, especially since the coating thickness may exceed the thickness of the cantilever.

Fabrication

Additional fabrication steps after semiconductor processing include the following:

- Eventually, additional Al and Si-nitride (LPCVD)[68-80]
- Back-etching (KOH or EDP) to achieve a membrane structure[68-82]
- Zinc-oxide processing, if necessary[81]
- Release of the cantilevers by frontside reactive-ion etching[68-78]
- Sensitive layer: spray or drop coating of polymers, deposition of biological entities

IC process-compatible and CMOS-compatible fabrication sequences for monolithic integration of the cantilevers with electronics are detailed in the literature.[68-70,76,77,82]

Applications

The application of the dynamic mode is restricted to the gas phase, whereas the static mode can be used to detect analytes in the liquid phase as well. Due to the bimorph effect (cantilever deformation upon heat generation), cantilevers have also been used in microcalorimetric applications.[78,83]

Typical applications are environmental monitoring (gas and liquid phases) or biosensing in liquids. These applications include the detection of different kinds of organic volatiles (hydrocarbons, chlorinated hydrocarbons, alcohols, etc.) or humidity in the gas phase by using polymeric layers,[68-75,77,80] the detection of alcohol in water,[76] and the hybridization and detection of complementary strands of oligonucleotides.[79]

The interaction mechanisms involve reversible physisorption and bulk/gas phase partitioning [see Eqs. (10) and (17)], as well as receptor-ligand binding.[79]

11.4 Thermal Sensors

Calorimetric or thermal sensors rely on determining the presence or concentration of a chemical by measurement of an enthalpy change produced by the chemical species to be detected.[1,4,34,84] Any chemical reaction [Eqs. (1) and (11)] or physisorption process [Eqs. (1) and (10)] releases or absorbs from its surroundings a certain quantity of heat [enthalpy term, ΔH^0, in Eq. (9)]. Reactions that liberate heat are called exothermic; reactions that abstract heat are called endothermic. This thermal effect shows a transient behavior: Continuous heat liberation/abstraction occurs only as long as the reaction proceeds. This implies that only a steady-state situation can be achieved: A chemical reaction is proceeding at a constant rate and is thus releasing/abstracting permanently a constant amount of heat. However, there will be no heat production, and hence no measurable signal, at thermodynamic equilibrium ($\Delta G = 0$), in contrast to mass-sensitive, optical, or electrochemical sensors.

Conflicting constraints are imposed on the design of a thermal sensor: The sensor has to interact with the chemical species (exchange of matter), and thus constitutes a thermodynamically open system; at the same time, the sensing area should be as thermally isolated as possible.

The liberation or abstraction of heat is conveniently measured as a change in temperature, which can be easily transduced into an electrical signal. All sensors aimed at thermal infrared radiation detection (Chapter 5) can in principle be used as chemical sensors as well. The various types of calorimetric sensors differ in the way that the evolved heat is transduced. The catalytic sensor (often called a pellistor) uses platinum resistance thermometry,[85-97] the thermistor uses composite oxide resistance thermometry,[98-104] and the pyroelectric[105,106] and Seebeck-effect[107-119] sensors use the respective effects to measure the temperature change. There are also thermal (flow) sensors based on the different thermal conductivity of gaseous analytes.[120,121] A micromachined cantilever that enables microthermal analysis due to the bimorph effect[78,83] was mentioned in Section 11.3.3.

Catalytic sensors, Seebeck-effect or thermoelectric sensors, pyroelectric sensors, and thermal conductivity sensors are semiconductor-technology-compatible. The latter is not discussed further here because it is essentially a flow sensor (see Chapter 5) responding to the thermophysical properties of a gas (thermal conductivity, heat capacity).[120,121]

Sensors based on the pyroelectric effect (anisotropic, noncentrosymmetric crystal lattice, permanent polarization, creation of macroscopic charges due to thermal stress in the crystal) require the deposition of pyroelectric material (lithium tantalate, zinc oxide,[105] polycyclic organic compounds[106]) on the silicon chip as for piezoelectric sensors. Because there are very few chemical-sensing applications reported on in the literature,[34,105,106] this class of sensor is not discussed here.

Thermistors are temperature-sensitive bead resistors composed of either oxide semiconductors with a negative temperature coefficient (NTC) or a positive temperature coefficient (PTC); that is, their resistance decreases or increases nonlinearly with temperature. PTC resistors are made, e.g., from barium or lead titanate, while NTC resistors are made from sintered transition metal oxides (titanium oxide) doped with aliovalent ions. The beads are contacted via two metallic (platinum) leads and coated with glass for chemical inertia.[4,98] The thermistors themselves have not been fabricated in planar semiconductor technology yet, but they have been integrated into, e.g., silicon-based biosensors due to their small size.[99,100] They have been used as thermal biosensors in strongly exothermic enzymatic reactions to detect urea,[100] glucose,[101-103] uric acid,[104] and other compounds of relevance in blood analysis.

The remainder of this section details catalytic thermal and thermoelectric sensors.

11.4.1 Catalytic Thermal Sensors (Pellistors)

Transduction Principle and Sensing Characteristics

The development of the catalytic sensor is derived from the need for a handheld detector for methane to replace the flame safety lamp in coal mines. The catalytic device measures the heat evolved during the controlled combustion of flammable gaseous compounds in ambient air on the surface of a hot catalyst by means of a resistance thermometer in proximity with the catalyst. This method is therefore calorimetric. A catalyst is a chemical compound [often a noble metal like platinum (Pt)] that enables or accelerates a chemical reaction by providing alternative reaction paths involving intermediates with lower activation energies than the uncatalyzed mechanism. The catalyst itself is not permanently altered by the reaction. The heated catalyst here permits oxidation of the gas at reduced temperatures and at concentrations below the lower explosive limit. Three elements are necessary for this method: a catalyst, a method to heat it, and a means to measure the heat of the catalytic oxidation. The term *pellistor* originally refers to a device consisting of a small platinum coil embedded in a ceramic bead impregnated with a noble metal catalyst.[85] A ceramic bead is used since the rate of reaction (and thus the sensor signal) is directly proportional to the active surface area.

Figure 11.9 shows two different micromachined designs to realize a catalytic calorimetric sensor: a meander structure on a micromachined membrane[86] and a freestanding, Pt-coated polysilicon microfilament (10 μm wide, 2 μm thick) separated from the substrate by a 2 μm air gap.[87,88] Heat losses to the silicon frame are minimized in these designs. By passing an electric current through the meander, the membrane/microbridge is heated to a temperature sufficient for the Pt surface to oxidize the com-

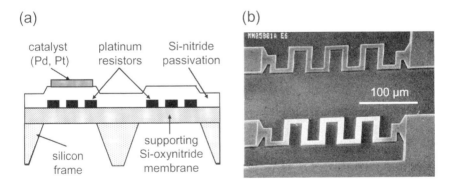

Figure 11.9 (a) Cross section of side-by-side micro hotplates composed of a dielectric membrane on an etched silicon wafer and platinum resistors/heaters. The device on the left has a deposited catalyst, making it the active element. Adapted from Zanini et al.[86] with permission. (b) Micrograph (SEM) of two meandered polysilicon microbridges. The lower meandered bridge is coated with a thin (approx. 0.1 μm) layer of platinum (CVD). In a differential gas-sensing mode, the upper, uncoated filament acts to compensate changes in the ambient temperature, thermal conductivity, and flow rate, while the lower filament is used to calorimetrically detect combustible gases. Reprinted from Manginell et al.[88] with permission.

bustible mixture catalytically; the heat of oxidation is then measured as a resistance variation in the Pt. The combustion of methane, for example, generates 800 kJ/mol heat, which translates into a corresponding temperature change. The structures described here are very similar to hotplate structures discussed in Section 11.6.3.1.2, which deals with electrochemical sensors.

The temperature change of the sensor element is proportional to the combustible concentration when the device is operated in excess oxygen and in the mass-transfer-limited regime.[89,90] The combustion of hydrogen in dry air is exemplified in Fig. 11.10.[87,88] The circuit maintains a constant sensor temperature by adjusting the supplied current to keep the filament resistance at a constant value. The sensor response is measured at steady state, i.e., continuous combustion. In most realizations, the measuring resistor forms part of a Wheatstone bridge configuration.[86,89] Temperature-modulated operation has been reported by Aigner et al.[91,92]

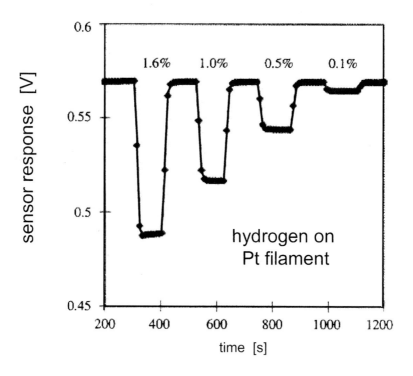

Figure 11.10 Sensor response of a Pt-coated filament exposed to various concentrations of hydrogen in synthetic air. Reprinted from Manginell et al.[87] with permission.

Fabrication

Additional fabrication steps after semiconductor processing include the following:

- Back-etching (KOH or EDP) for membranes[86,91,93]
- Surface micromachining: sacrificial layer etching (hydrogen fluoride, HF) for the bridges[87,88]
- Pt or catalyst processing: sputtering,[91,92] evaporation,[93-95] LPCVD[87,88]

Processing sequences for microbridges are given in Mastrangelo[96] and in Accorsi *et al.*[97]

Applications

Main applications include monitoring and detection of flammable gas hazards in industrial, commercial, and domestic environments. The lower explosive limit is the concentration of gas in air below which it cannot be ignited. Target gases include methane,[89,93,97] hydrogen,[86-90,93] propane,[86] carbon monoxide,[86,94] and organic volatiles.[91,92,95] The detectable gas concentrations usually range between a few percent (1 to 5%) and the region of some hundreds of ppm. The interaction process is an irreversible chemical combustion reaction at high temperature that liberates the respective reaction enthalpy [Eqs. (9) and (11)].

11.4.2 Thermoelectric or Seebeck-effect Sensors

Transduction Principle and Sensing Characteristics

This type of sensor relies on the thermoelectric or Seebeck effect. If two different semiconductors or metals are connected at a hot junction and a temperature difference is maintained between this hot junction and a colder point, an open-circuit voltage is developed between the different leads at the cold point. This thermovoltage is proportional to the difference of the Galvani potentials (inner contact potential, the difference of the Fermi levels of the two materials) at the two temperatures and is thus proportional to the temperature difference itself.[107] This effect can be used to develop a thermal sensor by placing the hot junction on a thermally isolated structure, such as a membrane or a bridge, and the cold part on the bulk chip with the thermally well-conducting silicon underneath.[70,108,109,110] To achieve a higher thermoelectric voltage, several thermocouples are connected in series to form a thermopile. The membrane structure featuring the hot junctions is covered with a sensitive or chemically active layer that liberates or abstracts heat upon interaction with an analyte. The resulting temperature gradient between hot and cold junctions then generates a thermovoltage, which can be measured.

Figure 11.11 displays the schematic of a CMOS thermopile. The overall sensor system includes two 700 by 1500 μm dielectric membranes with 300 polysilicon/aluminum thermocouples each (Seebeck coefficient: 111 μV/K) and an on-chip amplifier. One of the membranes is coated with a gas-sensitive polymer; the uncoated one serves as a reference.[70,110,111]

The detection process includes four principal steps: (1) absorption and partitioning or chemical reaction, (2) generation of heat, which causes (3) temperature changes to be transformed into (4) thermovoltage changes. (See, for example, Hierlemann[70] and Van Herwaarden et al.[109]) Each of the four steps contributes to the overall sensor signal. The thermovoltage change ΔU [V] is proportional to the derivative of the analyte concentration as a function of time dc_A/dt (mol/m³s):

$$\Delta U = A \cdot B \cdot V_{poly} \cdot \Delta H \cdot K \cdot \frac{dc_A}{dt} \tag{18}$$

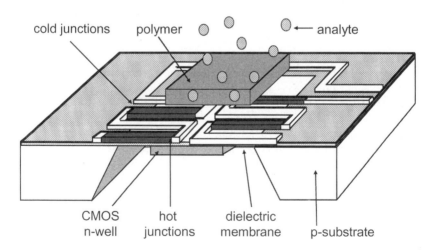

Figure 11.11 Schematic of a thermoelectric sensor. Polysilicon/aluminum thermopiles are used (hot junctions on the membrane, cold junctions on the bulk chip) to record temperature variations caused by analyte sorption in the polymer.

Here A [K·s/J] and B [V/K] are device- and coating-specific constants that describe the translation of a generated molar absorption/reaction enthalpy ΔH [J/mol] via a temperature change into a thermovoltage change. V_{poly} denotes the sensitive-polymer volume, and K is the partition coefficient [Eq. (10)] or reaction equilibrium constant [Eqs. (2) and (11)].

The calorimetric sensor only detects changes in the heat budget at non-equilibrium state (transients) due to changes in the analyte concentration (Fig. 11.12). Thus, the sensor provides a signal upon absorption (condensation heat) and desorption (vaporization heat) of gaseous analytes into the polymer[109-113] or during chemical reactions of an analyte with the sensing material.[114-116]

Figure 11.12 (a) and (b) shows the output voltage of the microsystem while switching from synthetic air (a nitrogen/oxygen mixture without humidity) to toluene (4000 ppm) and back to air at a temperature of 301 K.[70,111] Enthalpy changes can be approximated by integration over the peak area of the sensor signals.[110]

Fabrication

Additional fabrication steps after semiconductor processing include the following:

- Back-etching (KOH or EDP) for membranes[70,108,110,117]
- Processing of the sensitive layer: airbrush,[70,110] dispensing, spin coating, or enzyme immobilization methods[114-116]

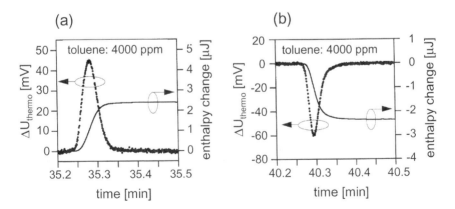

Figure 11.12 Output voltage of the integrated thermoelectric sensor while switching (a) from synthetic air to toluene (4000 ppm) and (b) back to air, at a temperature of 301 K and a total flow of 200 ml/min. Reprinted from Hierlemann et al.[70] with permission.

Processing sequences for the integration of thermoelectric sensors with circuitry in a CMOS standard process are detailed in Hierlemann et al.,[70] Kerness et al.,[110] and Sarro et al.[117] Sensors are commercially available from Xensors Integration.[118]

Applications

Typical application areas are environmental monitoring (gas and liquid phase) or biosensing in liquids. These applications include the detection of different kinds of organic volatiles in the gas phase (such as hydrocarbons, chlorinated hydrocarbons, and alcohols) by using polymeric layers,[70,110-113] or metal oxides,[119] the monitoring of acid/base neutralization,[114,116] and the biosensing of glucose, urea, and penicillin in the liquid phase by using suitable enzymes.[109,114-116]

The interaction mechanisms involve reversible physisorption and bulk/gas phase partitioning [see Eq. (10)] as well as enzymatic chemical reactions [Eq. (11)].[109,114-116]

11.5 Optical Sensors

Light can be considered to consist of either particles (photons) or electromagnetic waves, according to the principle of duality. The characteristic properties of the electromagnetic waves such as amplitude, frequency, phase, and/or state of polarization can be used to devise optical sensors.[1,4,6,10,122-124] The energy, E, of an electromagnetic wave is quantized, a quantum being called a photon (h is Planck's constant, $6.626 \cdot 10^{34}$ Js; v denotes the frequency):

$$E = h \cdot v \tag{19}$$

When light interacts with matter, several processes can take place, sometimes simultaneously.

Absorption

If a sample is irradiated with visible light or electromagnetic waves, the radiation can be absorbed, which results in a decrease of the intensity in the detected radiation as compared to the primary beam (Fig.11.13). Alternatively, the radiation can be transmitted without attenuation. A pre-

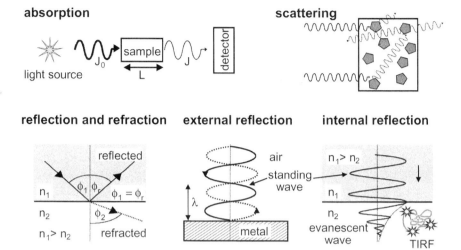

Figure 11.13 Schematic representation of the different processes taking place upon interaction of light with matter: absorption, scattering, reflection/refraction, external and internal reflection. n denotes the refraction index, J the light intensity, λ the wavelength, and ϕ the angle. Details are given in the text and by Wolfbeis et al.,[122] Gauglitz,[123] and Brecht et al.[124]

requisite for absorption is that the absorbing matter (atom, molecule, etc.) exhibit unoccupied energy states with an energetic difference exactly equal to or less than the energy of the incoming radiation quanta. The matter then absorbs the radiation energy by transition into a so-called excited state with higher internal energy. The absorption of radiation forms the base for most traditional spectroscopic methods, which are usually distinguished by the different radiation wavelengths or frequency ranges as given in Table 11.3.[35]

The absorption of monochromatic radiation (only one selected wavelength) can be quantitatively determined using the well-known Lambert-Beer relation:

$$I = I_0 \cdot e^{-\varepsilon_\lambda c_A l} \tag{20}$$

Here I denotes the transmitted radiation intensity at the detector, I_0 the intensity of the incident radiation, ε_λ the molar absorptivity at the measured wavelength, c_A the analyte concentration, and l the optical path length in the probed volume.

Table 11.3 Different Spectroscopy Methods and Radiation Energies

Radiation	Energy (J/mol)	Wavelength	Transition
γ-radiation	10^9 to 10^{11}	10^{-13} to 10^{-11} m	Nucleus excitation
X-rays	10^7 to 10^9	10^{-11} to 10^{-9} m	Core electron excitation
Ultraviolet (UV)	10^6 to 10^7	10^{-9} to 10^{-7} m	Shell electron excitation
Visible (VIS)	10^5 to 10^6	400 to 800 nm	Shell electron excitation
Infrared (IR)	10^2 to 10^5	10^{-6} to 10^{-4} m	Vibrational states
Microwaves	10^{-2} to 10^2	10^{-4} to 1 m	Rotation states
Radio waves	$<10^{-2}$	>1 m	Electron spin, nuclear spin

Scattering

Changes of the direction and/or frequency of light are commonly called scattering (Fig. 11.13). Scattering of light does not necessarily involve a transition between quantized energy levels in atoms or molecules. A randomization in the direction of light radiation occurs. Particles with sizes that are small compared to the wavelength of radiation give rise to Rayleigh scattering; particles that are large compared to the wavelength give rise to Mie scattering.[122,123] In both processes, the particle polarization is unaltered. However, the incident radiation can promote vibrational changes (energy quantum absorption) that can alter the polarization of the irradiated particle/molecule. The frequency of the light scattered by these molecules will be different from that of the incident light and much less intense. Such a phenomenon is known as Raman scattering.[6,35,125]

Fluorescence and Phosphorescence

The mechanism of fluorescence and phosphorescence is an absorption-emission process. The wavelength or energy of the incident radiation is absorbed and promotes changes in the molecular energy states. The resulting excited state is unstable, and the molecule dissipates some of its energy to rotational and/or vibrational energy states. The molecule then can return to the ground state by emitting light at a lower frequency than the incident radiation. This process is called fluorescence. If a more complex and slower intersystem crossing process into a triplet state occurs, followed by a radiative transition from there to the ground state, the process is called phosphorescence. For details and the respective Jablonski diagrams, which represent simplified portrayals of the relative positions of the electronic energy levels of a molecule, are given in the literature.[4,35,122] Both processes, fluorescence and phosphorescence, are sometimes subsumed under the term *luminescence* processes, but here *luminescence* is only used for chemiluminescence, detailed below.

Fluorescence processes are extensively used in gene analysis techniques, where a defined array of single-stranded deoxyribonucleic acid (DNA) fragments is hybridized with the respective complementary strands labeled with a fluorescent marker. By illuminating the array with a laser, the sites, where the labeled DNA fragments are bound by interaction between the two complementary strands, can be detected by their positive fluorescence response.[124] This technique has been commercialized by several companies.[126]

Chemoluminescence

The excited state of a molecule (C*) is created by a chemical reaction;[4,10,127] the molecule emits light during transition to the ground state according to

$$A + B \Rightarrow C^* \Rightarrow C + h \cdot v. \tag{21}$$

Chemical energy is thus directly converted into light energy, in most cases without additional heat generation (cold luminescence). In the biological domain, this process is called bioluminescence. It takes place, for example, in glowworms.

Reflection and Refraction

Reflection and refraction take place when light infringes on a boundary surface between two media of distinct optical properties (refraction index). The light can either be reflected back into the original medium or refracted (transmitted) into the adjacent medium (Fig. 11.13).

Several distinct types of reflection are possible. The first is a mirror type or specular external reflection (Fig. 11.13) occuring at, e.g., a metal surface or generally at interfaces of media with no transmission through them. (Evanescent waves are treated in the context of refraction, below.) Another type is diffuse reflection, where the light penetrates the medium and subsequently reappears at the surface after partial absorption and multiple scattering within the medium. The optical characteristics of diffusely reflected radiation provide information on the composition of the reflecting medium.[6]

Thin films (10 μm and less) on a surface can strongly affect the propagation of incident light due to reflection at each of the thin-film interfaces, causing a multitude of reflected, coherent beams with small phase shifts. Sensor techniques to interrogate such thin-film structures include ellipsometry[128,129] and thin-film reflectometric interference spectroscopy (RIFS), which is based on spectral modulation of the reflectance of a thin film under white-light illumination without using the polarization information. The spectral characteristics are a function of the film thickness; therefore, any adsorption/absorption of organic matter leads to changes in the interferograms of the reflected beams.[130,131]

A variety of transducers respond to changes in the refractive index in the immediate vicinity of the device surface. The propagation behavior of a wave guided by nonmetallic total internal reflection (Fig. 11.13) in a medium of high refractivity depends on the dielectric characteristics of the surrounding medium. This effect is mediated by the evanescent field, which penetrates from the optically denser guiding medium a few hundred nanometers into the optically rarer environment.[124,132,133] (More details are given in Section 11.5.1, in Fig. 11.14, and in Chapter 8.)

If the environment absorbs, energy will be transferred from the evanescent wave to the environment, and attenuation of the traveling wave will occur; this is called attenuated total reflection (ATR).[122,123] The

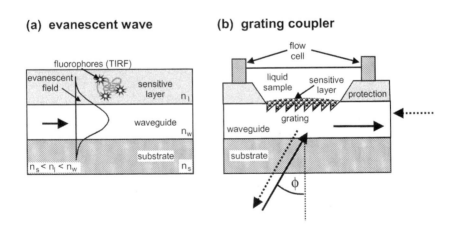

Figure 11.14 (a) Schematic of an evanescent wave in an optical waveguide. If fluorophores are in the range of the evanescent wave, they can be excited and the fluorescence can be detected (TIRF). (b) Schematic of a grating coupler. A periodic grating on the surface of the waveguide is used for in- or out-coupling of radiation to/from the waveguide. The deflection angle depends on the light wavelength and the grating period and is altered by binding of an analyte on the grating. Adapted from Lukosz[133] with permission.

energy of the evanescent wave can also be used to initiate fluorescence; this is called total internal reflectance fluorescence (TIRF).[132] Light propagating in waveguide structures without absorbing cover layers is not attenuated by environmental influences; this is called frustrated total reflection (FTR).[122,123] However, its propagation velocity changes depending on the refractive index in the vicinity of the waveguide. Several setups have been proposed and are discussed in Section 11.5.1.[122,124,133]

Surface plasmon resonance (SPR) is based on collective fluctuations in the electron density at the surface of thin films, typically gold or silver, on a waveguide. Surface plasmon waves show the maximum of the electrical field distribution located at the waveguide/metal interface, which is exponentially decaying into the metal and the adjacent medium. SPR is detected as a strong attenuation of the reflected light beam sensitive to the medium adjacent to the metal film.[134-136] SPR is described in more detail in Section 11.5.4.

In comparison to other chemical sensing methods, optical techniques offer a great deal of selectivity already inherent in the various transduction mechanisms. Characteristic properties of the electromagnetic waves, such as amplitude, frequency, phase, and/or state of polarization, can be used to advantage. The wavelength of the radiation, e.g., can be tuned to specifically match the energy of a desired resonance or absorption process. Geometric effects (scattering) can provide additional information. Moreover, optical sensors, like any other chemical sensor, can capitalize on all the selectivity effects originating from the use of a sensitive layer.

Optical sensors and classical spectroscopy methods are often very similar in methodology but differ in the arrangement of the experiment and equipment. In particular, the introduction of fiber-optic techniques has promoted the development of comparatively inexpensive optical sensor setups. Sections 11.5.1 through 11.5.4 focus exclusively on semiconductor-based and MOEMS-based (micro-opto-electro-mechanical[137]) transducers and their respective mechanisms. The wide field of glass-based and fiber-optical techniques such as optodes[138] or microoptodes[139,140] is not covered. Reviews and articles are provided by Wolfbeis and others.[122,123,141-145] The light-addressable potentiometric sensor (LAPS)[146] is discussed in Section 11.6.2.2.5.

11.5.1 Integrated Optics

Transduction Principle and Sensing Characteristics

The generation of light in silicon devices is very difficult since there is no first-order transition from the valence band to the conduction band without the involvement of a phonon (lattice vibrations).[147] Only direct bandgap semiconductors such as gallium arsenide (GaAs) or indium phosphide (InP) show first-order radiative electron-hole recombinations with high quantum efficiency. (See the discussion of GaAs-based devices, below.) The detection of light is possible with either silicon-based devices (photodiodes) or other semiconducting materials.

Integrated optical (IO) sensors use guided waves or modes in planar optical waveguides. (See Chapter 8.) The waveguide materials usually include high-refractivity silicon dioxide, titanium dioxide, or silicon nitride films on oxidized silicon wafer substrates. The guided waves or modes in planar optical waveguides include the TE (transverse electric or s-polarized, surface-normal) mode and the TM (transverse magnetic or p-polarized, surface-parallel) mode. Changes in the effective refractive index of a guided mode are induced by changes of the refractive index distribution in the immediate vicinity of the waveguide surface, i.e., within the penetration depth (some hundred nanometers) of the evanescent field into the sample [Fig. 11.14(a)].[124,133] The evanescent field decays exponentially with increasing distance from the waveguide surface. Changes in the effective refractive index can be induced by absorption of an adlayer onto the surface of the waveguide from the gas or liquid phase,[133,148,149] by interaction of an analyte molecule with a recognition structure immobilized on the waveguide surface,[133,150-154] or by changes of the refractive index of the medium adjacent to the waveguide in a flow-through configuration.[133,150,151] In the case of microporous waveguides, analyte molecule absorption or desorption directly into the pores of the waveguiding film itself can change the waveguide refractive index.[133]

A number of different integrated optical sensors have been developed to transform the changes of the effective refractive index into readily measurable physical quantities.

Grating Couplers

Among the first integrated optical devices were grating coupler structures embossed with a monomode film waveguide as proposed by Lukosz.[133,149-151,155] A periodic grating on the surface of the waveguide can be used for in- or out-coupling of radiation (TE and TM modes) from the waveguide. In-coupling and out-coupling are governed by the same physical laws, since the reciprocity theorem permits the reversal of the propagation direction of all light waves. The deflection angles (coupling angles) of the TE and TM modes depend on the light wavelength and the grating period [Fig. 11.14(b)]. Binding of an analyte on the grating alters the coupling angle, which can be detected using position-sensitive detectors. The sensitivity of the grating coupler is related to the lateral dimensions of the grating region interacting with the sample.

Prism Couplers

Prism couplers[155] are not treated here.

Difference Interferometer

In a planar waveguide, the TE and TM modes are coherently excited by a laser. Both propagate along a common path down the same waveguide and interact with the sample within a certain length of the waveguide. The polarization-dependent interaction induces a phase difference between the two modes, which can be measured using a dedicated interferometer setup.[133,148,149,156,157]

A variant of this method involves a Zeeman laser to generate two orthogonally polarized modes in a silicon nitride waveguide.[158, 159]

Two-Beam Interferometers

Mach-Zehnder integrated optical devices [Fig. 11.15(a)] are monomode channel waveguides (TE or TM mode) that allow for a straightforward implementation of an interferometer structure.[133,152-154,160]

(a) Mach-Zehnder interferometer   (b) integrated GaAs interferometer

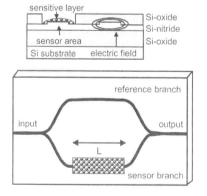
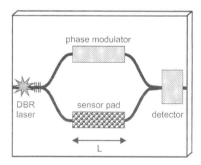

Figure 11.15 Schematic of (a) a conventional Mach-Zehnder interferometer and (b) an integrated Mach-Zehnder interferometer (light source and detector on chip) in GaAs-technology. The cross section shows the separate sensor (left side, open) and reference (right side, covered) branches. Adapted from Maisenhölder et al.[166] with permission.

A waveguide is split into an open measurement path and a protected reference path, which are recombined after some distance. The phase difference, introduced by analyte interaction (refractive index change) in the sensing path, is detected by interference effects. Detection limits are in the range of some picograms/mm^2.[161]

Integrated Waveguide Absorbance Optodes

A membrane inserted between two micromachined waveguides acts simultaneously as the light-guiding medium and the sensing element, and therefore changes its spectral properties while interacting with an analyte. The first results with potassium-selective optode membranes have been reported on by Puyol et al.[162]

Fabrication

Additional fabrication steps after semiconductor processing include the following:

- Patterning of silicon nitride as a waveguide (LPCVD, RIE, lithography)[133,152-154]
- Deposition of a silicon-oxide cladding layer (plasma-enhanced chemical-vapor deposition, PECVD)[133,152-154]
- Deposition of the chemically sensitive layer, immobilization of biological entities[150-154]

Electro-optical modulation techniques of the sensor signal, when using zinc oxide as the optical waveguide on a chip fabricated in silicon oxinitride technology, have been reported by Heideman et al.[163,164]

Applications

Typical applications are humidity sensors,[133,148,149] gas sensors[159,160] (adsorption on the device or absorption in a microporous waveguide), and environmental monitoring or biosensing. Examples include the detection of different organic solvents in the liquid phase (hydrocarbons, alcohols, etc.), the monitoring of sucrose and buffer solutions,[133,148,150,151] and biotin/streptavidin-mediated immunosensing (Fig. 11.16)[153] involving antibody-antigen binding experiments.[133,153,154]

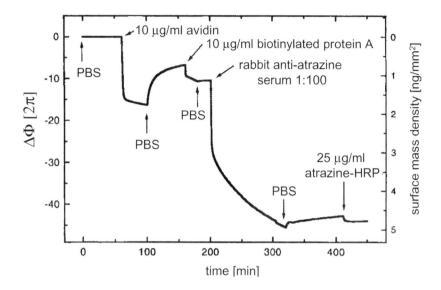

Figure 11.16 Sensor response (phase shift, ΔΦ) of a difference interferometer used for biosensing. (1) Adsorption of avidin at the surface; (2) affinity binding of biotinylated protein A to the avidin layer; (3) binding of the rabbit anti-atrazine serum to the protein A layer; (4) immunoreaction of the immobilized anti-atrazine antibodies with atrazine (atrazine-HRP). PBS denotes phosphate buffer solution washing steps. Reprinted from Lukosz et al.[148] with permission.

The interaction mechanisms include reversible physisorption [Eq. (10)] as well as biochemical affinity reactions [Eq. (11)].[133,150-154]

Gallium-Arsenide-Based Devices

Due to their direct bandgap, III-V-semiconductors offer the opportunity for fabricating and integrating lasers, waveguides, phase modulators, and waveguide detectors on the same chip. (See Chapter 8 for details.) GaAs/AlGaAs-based Mach-Zehnder devices with integrated light sources and detectors have been developed, as shown in Fig. 11.15(b).[165-167] The light source is a distributed Bragg reflector (DBR) laser, fabricated with a simplified grating recess technology,[168,169] which is operating on a single mode. A dielectric waveguide pad (silicon oxide, tantalum oxide) is integrated in the measurement arm of the interferometer.[165-167] The detector is a long-absorbing-length photodiode with high quantum efficiency.[168]

A more recent development for optical gas sensing is the vertical-cavity, surface-emitting laser (VCSEL). The cavity is formed vertically on the wafer surface. Epitaxially grown Bragg mirrors serve as distributed reflectors above and below the laser's very short active region. GaAs/AlGaAs-based VCSELs emit in the near-infrared region. Oxygen sensing at 762 nm (absorption due to magnetic dipole transitions in the gas molecule without interference from other gases) was demonstrated in first experiments.[170,171]

11.5.2 Microspectrometers

11.5.2.1 Fabry-Perot-Type Structures

Transduction Principle and Sensing Characteristics

A Fabry-Perot interferometer (FPI) is an optical element consisting of two partially reflecting, low-loss, parallel mirrors separated by a gap. The optical transmission characteristics through the mirrors consist of a series of sharp resonant transmission peaks occurring when the gap width is equal to multiples of a half wavelength of the incident light. These transmission peaks are caused by multiple reflections of the light in the cavity. By using highly reflective mirrors, small changes in the gap (width, absorptivity) can produce large changes in the transmission response. Even though two reflective mirrors are used, transmission through the element at the peak wavelengths approaches unity. The transmission is a function of both the gap spacing and the radiation wavelength. The devices can be used as a wavelength selector or monochromator by adjusting the gap width to achieve the desired wavelength. Tunable devices with a gap width variable by electrostatic actuation using electrodes on movable micromachined parts have been reported[172-175] [Fig. 11.17(a)]. Such devices operate preferably in the near-infrared region at wavelengths larger than 1 μm, where silicon substrates become transparent.[172]

A single-chip CMOS optical microspectrometer based on FPI and operating in the UV/visible region has been reported by Correia et al.[176] and Aratani et al.[177] It contains an array of 16 addressable Fabry-Perot etalons (500×500 μm^2), each with a different resonant cavity length realized as a PECVD silicon-oxide layer sandwiched between an aluminum and a silver layer and placed on top of an array of vertical pnp phototransistors[178] [Fig. 11.17(b)]. It also includes circuits for readout, multiplexing, and driving a serial bus interface.

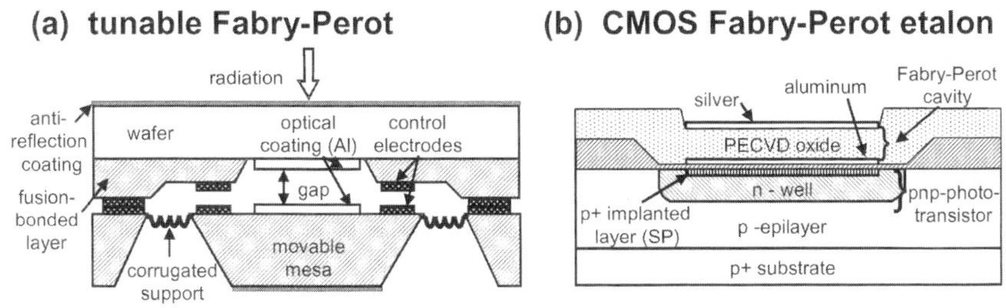

Figure 11.17 (a) Schematic of a tunable Fabry-Perot interferometer. The wavelength-defining gap width can be changed by applying DC to the control electrodes. Aluminum is used as optical coating material. Adapted from Jerman and Clift[172] with permission. (b) Cross section of an integrated Fabry-Perot etalon. The gap width is determined by the thickness of the PECVD oxide; silver and aluminum are used as the optical coating. A pnp-phototransistor (p+ implanted layer/n-well/p-epilayer) is located directly underneath the etalon. Adapted from Correia et al.[176] with permission.

Fabrication

Additional fabrication steps after semiconductor processing include the following:

- Surface micromachining techniques to achieve an air gap (HF)[175,177]
- Deposition of the lower mirror (evaporation and liftoff of aluminum)[176]
- Deposition of a silicon-oxide layer of defined thickness (PECVD)[176]
- Deposition of the upper mirror layer (silver)[176]

Some devices are derived from air gap pressure sensors,[173-175,179,180] so fabrication sequences already described in this context (Chapter 10) have been applied. The same holds for wafer-stacking techniques to achieve the FPI cavities.[173,174] The complete fabrication sequence for the monolithic CMOS VIS spectrometer is given by Correia et al.[176]

Applications

Typical applications include gas sensors.[173-175,178-182] The characteristic absorption wavelengths of carbon monoxide, carbon dioxide, and methane or hydrocarbons are 4.7 µm, 4.2 µm, and 3.3 µm (IR region, molecular vibrations), respectively. Polymeric coatings in the FPI cavity have been used to detect iodine.[179,180] Carbon dioxide sensors based on tunable FPIs (dual-wavelength measurements) are commercially available from Vaisala.[182] The radiation source in most cases is a light bulb or a light-emitting diode (LED).

11.5.2.2 Grating-Type Structures

Transduction Principle and Sensing Characteristics

Micromachined diffraction gratings have been used in combination with imaging devices to set up microspectrometer arrangements; Fig. 11.18(a) provides a schematic. Two basic types of gratings can be micromachined easily: (1) amplitude gratings, by blocking out light with an array of opaque and transparent sections; (2) phase gratings, where the light phase is modulated by variations in the grating shape.[183,184] In the example displayed in Fig. 11.18(a), a phase grating with a fine grating pitch creating high dispersion angles was etched into a quartz wafer. The transmission grating was then mounted directly over a charge-coupled device (CCD) imager[183] or a CMOS imaging chip consisting of an array of custom-designed photodiodes.[185] A glass spacer was placed in the optical path between the grating and the detector. In another approach, two

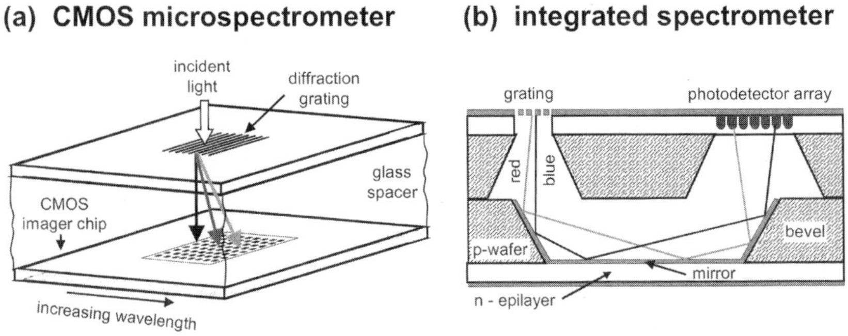

Figure 11.18 (a) CMOS microspectrometer consisting of a grating, a glass spacer, and a CMOS imager chip. Different wavelengths are diffracted at different angles from the surface normal. Adapted from Yee et al.[185] with permission. (b) Schematic of an integrated spectrometer using two bonded chips; one of them carrying a grating and a photodetector array. Adapted from Kwa and Wolffenbuttel[186] with permission.

wafers are micromachined such that an optical path of about 4 mm in length is obtained. Light dispersed by a 32-slit diffraction grating travels along the optical path and is directed to an array of photodiodes [Fig. 11.18(b)].[186] The interior of one of the wafers is coated with a reflective film. The grating and the diode array are integrated in the second wafer, which remains uncoated. The wafers are bonded together by silicon/silicon fusion bonding. The performance of such micromachined spectrometers is comparable to that of low-end, bench-top spectrometers.

Diffractive optical elements, which produce synthetic analyte infrared spectra of compounds such as hydrogen fluoride (HF) for chemical sensor systems based on correlation spectroscopy, are reported on in Sinclair et al.[187]

Fabrication

Additional fabrication steps after semiconductor processing include the following:

- Patterning of gratings, quartz micromachining[183,186]
- Back-etching using electrochemical etch stop[186]
- Deposition of reflective layers[186]
- Wafer fusion bonding[186]

Applications

Typical applications include biochemical and chemical analysis. Emission gas spectra have been recorded for carbon dioxide and helium. Fluorescence of a dye (fluorescein) was induced by a laser and detected with the microspectrometer.[183]

11.5.3 Bioluminescent Bioreporter Integrated Circuits

Transduction Principle and Sensing Characteristics

This technique uses bioluminescent bacteria placed on an application-specific optical integrated circuit (standard CMOS).[188-190] The bacteria have been engineered to luminesce [Eq. (21)] when a target compound such as toluene is metabolized. The integrated circuit detects, processes, and reports the magnitude of the optical signal. The microluminometer uses the p-diffusion [source and drain diffusions of p-channel MOS field-effect transistors (MOSFETs)] as the photodiode. The shallow p-diffusion has a strong response to the 490-nm bioluminescent signal. The entire sensor, including all signal processing and communication functions, can be realized on a single chip. The integrated circuit contains the subunits needed to detect the optical signal, to distinguish the signal from noise, to perform analog or digital signal processing, to communicate the results, and to perform auxiliary functions (temperature, position measurement) (Fig. 11.19).

Many types of bioluminescent transcriptional gene fusions have been used to develop light-emitting bioreporter bacterial strains to sense the presence, bioavailability, and biodegradation of different kinds of pollutants. The cells here were entrapped on the chip by encapsulation in natural or synthetic polymers that provided a nutrient-rich, hydrated environment.[188-190]

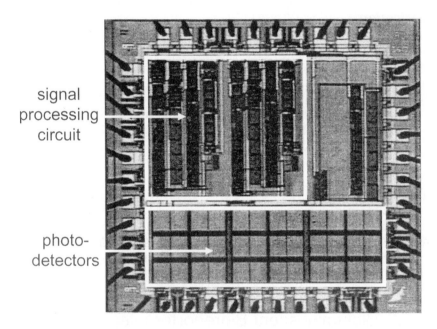

Figure 11.19 Micrograph of a bioluminescent bioreporter integrated circuit (BBIC). Details are given in the text. Reprinted from Simpson *et al.*[189] with permission.

Fabrication

Additional fabrication steps after semiconductor processing include the following:

- Silicon nitride protective coating using a jet vapor-deposition technique[191]

- Deposition of the cell-containing polymer (drop coating, spraying, beads)[188-190]

Applications

Typical applications include chemical analysis in the gas or liquid phase. Depending on the integration time of the device, trace amounts of toluene and naphthalene were detected in the gas phase using the appropriate cell colonies (*Pseudomonas putida*).[188-190] The interaction mechanism is a chemical reaction [Eqs. (11) and (21)].

11.5.4 Surface Plasmon Resonance Devices

Transduction Principle and Sensing Characteristics

The quantum-optical/electronic basis of surface plasmon resonance is due to the fact that the energy carried by photons of light can be "coupled" or transferred to electrons in a metal.[134] The wavelength of light at which coupling (i.e., energy transfer) occurs is characteristic of the particular metal and the environment in which the metal surface is illuminated; gold is the preferred metal. The coupling can be observed by measuring the amount of light reflected by the metal surface. All the light is reflected except at the resonant wavelength, where almost all the light is absorbed (Fig. 11.20). The coupling of light into a metal surface results in the cre-

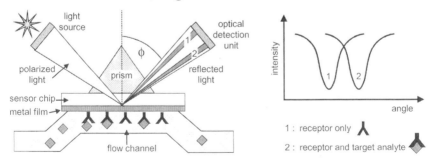

Figure 11.20 Surface plasmon resonance (SPR) principle and a typical sensor response diagram: intensity versus angle. Monochromatic light is used to excite the surface plasmon, which leads to a drastic intensity decrease at a defined reflection angle. Adsorption of an analyte changes this angle.

ation of a plasmon, a group of excited electrons, that behave like a single electrical entity. The plasmon, in turn, generates an electrical field, which extends about 100 nm above and below the metal surface.[134] The characteristic of this phenomenon, which makes SPR an analytical tool, is that any change in the chemical composition of the environment within the range of the plasmon field causes a change in the wavelength of light that resonates with the plasmon. That is, a chemical change results in a shift in the wavelength of light, which is absorbed rather than reflected, and the magnitude of the shift is quantitatively related to the magnitude of the chemical change.[136,192] The phenomenon of SPR is nonspecific. Different chemical changes cannot be distinguished.

A hybrid SPR system consisting of two micromachined silicon layers, one micromachined glass layer, and an alumina substrate is shown in Fig. 11.21.[193,194] Bonding was achieved using a low-temperature curing polyimide and a solder sealing. The micromachined silicon layers contain a torsional, all-silicon micromirror, V-grooves for optical fiber and lens, and a position-sensing photodiode (PSD). The silicon micromirror was electrostatically deflected through 9 to 10 degrees to direct the light beam emitted from the end of a fiber through a range of angles incident onto the metal film, setting up a surface plasmon. The position and intensity of the reflected beam were recorded with the position-sensing photodiode. The microsystem measures approximately $1 \times 2 \times 0.2$ cm^3.[193,194]

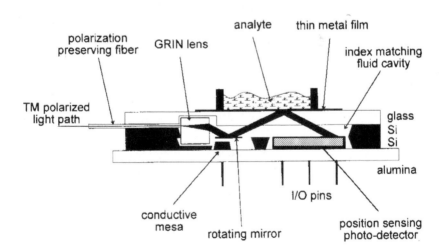

Figure 11.21 Cross-sectional diagram of the SPR microsystem, labeling individual components and showing the light path. Reprinted from Garabedian et al.[193] with permission. (Color image is available at *http://www.ece.ucdavis.edu/misl/web/pages/projects/plasmon.html*.)

Fabrication

Additional fabrication steps after semiconductor processing include the following:

- Alumina substrate: laser drilling, screen printing of thick-film conductors[193]
- Anisotropic etching of the silicon micromirror and the fiber-to-lens alignment groove[193]
- Filling the interior of the assembly with index-matching fluid[193,194]
- Glass slide: deposition of the sensitive layer (drop coating, spraying, beads)[135,136]

Small SPR instruments are commercially available from Texas Instruments.[195]

Applications

Any pair of molecules that exhibit specific binding can be adapted to SPR measurement. These may be an antigen and an antibody, a DNA probe and a complementary DNA strand, an enzyme and its substrate, or a chelating agent and a metal ion. Typical application areas are environmental monitoring (gas sensing[136,193]) or biosensing and immunosensing in liquids.[135,136] The surface adsorption of bovine serum albumin (BSA) was tested using the micromachined instrument.[193,194] The interaction mechanisms involve, in most cases, specific biochemical reactions [Eq. (11)].[135,136]

11.6 Electrochemical Sensors

Electrochemical sensors constitute the largest and oldest group of chemical sensors.[4,6,8-11,23,26,34] They make use of electrochemical or charge-transfer reactions: $A^+ + e^- \leftrightarrow A$. [See Eqs. (12) through (15) and related text in Section 11.2.] Electrochemistry includes charge transfer from an electrode to a solid or liquid sample phase and vice versa. Chemical changes take place at the electrodes or in the probed sample volume, and the resulting charge or current is measured. Electrode reactions and charge transport in the sample are both subject to changes by chemical processes and are, hence, at the base of electrochemical sensing mechanisms.[4]

A key requirement for electrochemical sensors is a closed electrical circuit, although there may be no current flow. (See the following discussion of potentiometry.) An electrochemical cell is always composed of at least two electrodes with two electrical connections: one through the probed sample, the other via the transducer and measuring equipment. The charge transport in the sample can be ionic, electronic, or mixed; that in the transducer branch is always electronic.

Electrochemical sensors are usually classified according to their electro-analytical principles.[4,8,11]

Voltammetry

Voltammetric sensors are based on the measurement of the current-voltage relationship in an electrochemical cell comprising electrodes in a sample phase. A potential is applied to the sensor, and a current proportional to the concentration of the electro-active species of interest is measured. Amperometry is a special case of voltammetry, where the potential is kept constant as a function of time.

Potentiometry

Potentiometric sensors are based on the measurement of the potential at an electrode, which is, in most cases, immersed in a solution. The potential is measured at an equilibrium state, i.e., no current is allowed to flow during the measurement. According to the Nernst equation [Eq. (15)], the potential is proportional to the logarithm of the concentration of the electro-active species. (Work function sensors are discussed in the context of field-effect devices in Section 11.6.2.2.4.)

Conductometry

Conductometric sensors are based on the measurement of a conductance between two electrodes in a sample phase. The conductance is usually measured by applying an AC potential with a small amplitude to the electrodes to prevent polarization. The presence of charge carriers determines the sample conductance. In contrast to conductometry, AC-impedance measurements are not really used for analytical applications. Impedance measurements are of special interest for the study of membrane processes and electrode and electrolyte characterization. The goal of impedance measurements is to find an equivalent electronic circuit model and to correlate that model with electrochemical phenomena.

Another categorization method relies on discerning the electronic components.[9,34] There are chemoresistors, chemodiodes, chemocapacitors, and chemotransistors. This chapter uses the electroanalytical principles as the superordinated scheme and uses the component notation within this scheme.

The discussion is restricted to semiconductor-based systems, omitting a wealth of literature on other designs of ion-sensitive electrodes.[196] Silicon-carbide-based devices operating at extremely high temperatures[197-200] are also omitted. Further details are provided by review articles available in the literature.[4,6,8-11,23,26,34,201-206]

11.6.1 Voltammetric Sensors

Transduction Principle and Sensing Characteristics

Voltammetry in general is the measurement of the current that flows at an electrode as a function of the potential applied to the electrode. The result of a voltammetric experiment is a current/potential curve. Amperometry, which is more frequently applied in chemical sensors, usually provides a linear current/analyte concentration relationship at a constant potential, predefined with regard to the target analyte.

Two different electrode configurations are typically used. The two-electrode configuration[4,8,11] consists of a reference electrode (RE) and a working electrode (WE) [Fig. 11.22(a)].[11] The disadvantage of this method is that the RE carries current and may become polarized if it is less than 100 times the size of the WE. Material consumption due to the current in the RE is another problem. A better approach is, therefore, the use of a three-electrode system[4,8,11] in a potentiostatic configuration. An additional auxiliary electrode [AE, sometimes called a counter-electrode (CE)] is introduced for current injection in the analyte [Fig. 11.22(b)].[11] The reference electrode is now a true RE with a well-defined potential, since no current is flowing through the RE. The potentiostat controls the current at the auxiliary electrode as a function of the applied potential. This is realized in practice with an operational amplifier (opamp).[11] The potential is applied to the positive input of the opamp. The RE is connected to the negative input and measures the potential in the solution. The AE is connected to the output. The opamp injects a current into the solution through the AE. Due to the feedback mechanism, the current is controlled in such a way that the potential at the negative input equals the potential at the positive input. The potential difference between WE and RE therefore equals the applied potential. No current is flowing through

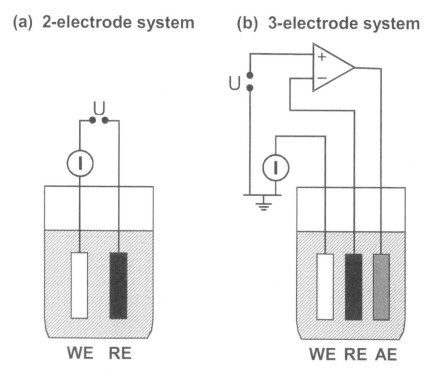

Figure 11.22 Schematic of (a) two-electrode and (b) three-electrode configurations used for voltammetric measurements. Details are given in the text. Adapted from Lambrechts and Sansen[11] with permission.

the RE since the opamp has a very high input impedance. The sensor signal current is measured at the working electrode.

The measured current at any given potential difference depends on the material properties, the composition and geometry of the electrodes, the concentration of the electro-active species (presumably the target analyte), and the mass-transport mechanisms in the analyte phase.[4,8,11,34] Among these are *migration*, the movement of charged particles in an electric field; *convection*, the movement of material by forced means such as stirring or as a consequence of density or temperature gradients; and *diffusion*, the movement of material from high-concentration regions to low-concentration regions. The electrochemical reactions at the electrodes are normally fast in comparison to the transport and supply mechanisms. Diffusion is normally regarded to be the dominant mechanism since convection in the electrode vicinity is avoided, and migration is suppressed by, e.g., a large excess of electro-inactive salts (electro-inactive at the respective applied potential). There are two components to the measured current: a capacitive component, resulting from redistribution of charged and polar particles in the electrode vicinity, and a component called faradaic current that results from the electron exchange between the electrode and the redox species (analyte).[4,8,11,34] The faradaic component is the important measurand and is, for the case of diffusion-limited conditions, directly and linearly proportional to the target analyte concentration. The limiting current (all analyte ions are immediately charged or discharged upon arrival at the electrode) is then given by the Cottrell equation:[4,8,11,26,34]

$$I_\infty = n_e \cdot F \cdot A \cdot D_{diff} \frac{c_A}{L_{diff}}. \qquad (22)$$

Here n_e denotes the number of electrons, F is the Faraday constant, A is the effective electrode area, D_{diff} is the diffusion coefficient, c_A is the target analyte concentration, and L_{diff} is the diffusion length. (The literature provides more mechanistic details.[4,8,11,26,34]) Correction terms to this equation for small electrodes have to be introduced to take into account the electrode geometry.[4,8,26]

Cyclic Voltammetry

A stationary WE is used, and a cyclic ramp potential versus some RE is applied. The potential versus time is triangular: It increases at a rate linear with time, then reverses and decreases at the same rate. The current flows as a consequence of the applied potential between the WE and an AE. This technique is used to study the electrode/sample interface.[4,8,11,207-210]

Stripping Voltammetry

This method is used to detect heavy-metal ions at trace level by means of a mercury electrode.[211-214] The method involves an initial preconcentration phase, in which the array is held at a cathodic potential such that the metal ions from solution are reduced and amalgamated into the mercury. Then the electrode potential is reversed to anodic, and the metals in the mercury are reoxidized and stripped from the mercury into the solution. The charge required to strip a given metal completely from the mercury is proportional to its initial concentration in the test solution.[211-214]

Fabrication

Additional fabrication steps after semiconductor processing include the following:

- Additional silicon nitride as a protective coating
- Optional back-etching, membrane formation, and perforation for liquid-electrolyte access to membrane-covered sensitive electrodes or gas permeation[215,216]
- Deposition/patterning of metal electrodes (liftoff, thermal evaporation, sputtering)[8,11,215-231]
- Deposition of electro-active polymers, membrane materials, hydrogels (spin-casting, spraying, screen printing)[215-232]

Sensor processing sequences are given in the literature.[11,215,216,226,228] The fabrication of electrochemical sensors integrated with CMOS circuitry components is described by Lambrechts et al.[11] and Sansen et al.[226] Microsensor arrays with up to 1024 individually addressable elements have been reported by Hinkers et al.[229] and Hermes et al.[230] A picture and a schematic of a CMOS-based, three-electrode amperometric sensor is shown in Fig. 11.23.[226]

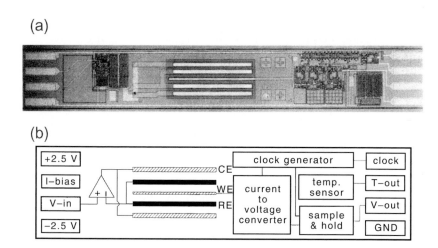

Figure 11.23 (a) Micrograph and (b) layout of a CMOS-based three-electrode amperometric sensor. Reprinted from Sansen et al.[226] with permission.

Applications

Typical applications include chemical analysis in the gas or liquid phase. If the target analyte is not an electro-active species such as glucose, oxygen, or carbon dioxide, polymer electrolytes or enzymes (glucose oxidase) that produce analyte-related ionic species are used as components of the sensitive electrode coatings. Typical target analytes in the gas phase are nitrogen oxides, hydrogen sulfide using Nafion polymer electrolyte,[216-218] oxygen,[223,228,230] and carbon dioxide using liquid electrolytes.[215] Target analytes in the liquid phase comprise dissolved oxygen,[210,226] glucose,[220,221,226] hydrogen peroxide,[215,222] and chlorine in drinking water[219,224] (Fig. 11.24). One of the best known voltammetric cells is the Clark cell, which is based on a two-step reduction of oxygen via hydrogen peroxide to hydroxyl ions in aqueous solution. The Clark cell is used to measure dissolved oxygen in blood and tissue.[233] The reference electrodes in the liquid phase are, in most cases, silver/silver-chloride elements.

The interaction mechanism in all cases is an electrochemical redox reaction [Eqs. (12) through (15)].

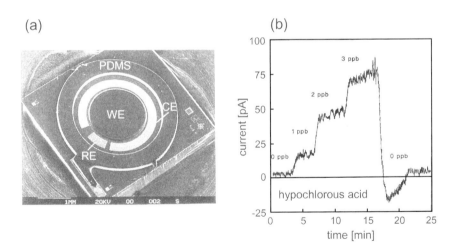

Figure 11.24 (a) Micrograph of an amperometric three-electrode free-chlorine sensor with central membrane-covered working electrode (WE) surrounded by the counter-electrode (CE, ring) and reference electrode (RE, ring segment). A polysiloxane (PDMS) encapsulation ring guides the liquid sample phase. (b) Sensor response upon exposure to chlorine from hypochlorous acid near the limiting threshold (ppb range). Reprinted from Van den Berg et al.[224] with permission.

11.6.2 Potentiometric Sensors

Potentiometry is the direct application of the Nernst equation [Eq. (15)] through measurement of the potential between nonpolarized electrodes (WE and RE) under conditions of zero current. The measurement is carried out at thermodynamic equilibrium.

The following discussion distinguishes between two different types of potentiometric devices: electrochemical cells with metal electrodes and field-effect semiconductor devices.

11.6.2.1 Electrochemical Cell

Transduction Principle and Sensing Characteristics

The electrochemical cell used for potentiometric microsensors consists of two metal electrodes; a WE covered with an ion-selective membrane or gel, which preferably hosts a specific target ion; and an RE, which, in most cases, is a silver electrode covered by a thin silver-chloride film.[234, 235] The WE is called an ion-selective electrode (ISE). Both electrodes are on the same chip and are simultaneously exposed to the analyte phase [Fig. 11.25(a)]. The ISE is, in principle, the oldest solid-state chemical sensor.[236]

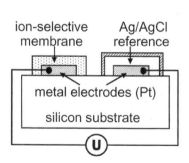
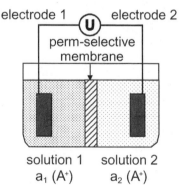

Figure 11.25 (a) Schematic of a potentiometric nonsymmetrical electrochemical cell with metal electrodes. The ion-selective electrode (ISE) and RE are on the same chip exposed to the analyte. The RE is protected by a membrane. (b) Schematic of a classical symmetrical potentiometric concentration cell. The potential is measured between two half-cells containing different activities/concentrations of the same analyte (A^+).

The design of modern micromachined nonsymmetrical ISEs, however, is completely different from that of conventional symmetrical ISEs such as a pH glass electrode or a lanthanum fluoride membrane electrode. The traditional symmetric arrangements exhibit a liquid-phase reservoir separated by a permeation-selective membrane from the analyte phase and thus essentially constitute electrochemical concentration cells [Fig. 11.25(b)].

The charge-transfer processes at all interfaces (solution/solid electrolyte, solid electrolyte/metal, ionic conductor/electronic conductor) of a modern, nonsymmetrical ISE must be carefully designed. If the exchange current density of the charged species of interest is sufficiently high (>10^{-3} A/cm^2), such interfaces are well defined and the devices are stable.

The exchange current is due to significant and continuous movement of charge carriers in both directions through the interphase region at an electrode at equilibrium (dynamic equilibrium). The magnitude of these mutually compensating currents (no net current) flowing at any zero-current potential is called exchange current. The Nernst equation [Eq. (15)] can be applied to calculate the electrochemical potential caused by a certain analyte concentration.

The design of metal-electrode potentiometric sensors is very similar to that of voltammetric sensors (two-electrode configuration) described in Section 11.6.1. In comparison to voltammetric techniques, the concentration dependence of the measured potential is logarithmic [Eq. (15)]. Using amperometric techniques, the target ions can be selected by careful tuning of the appropriate redox potential. In potentiometry, all ions in the sample that exhibit comparable exchange current density contribute to the measured overall potential. To achieve selectivity to one specific ion, this target ion must provide a significantly higher exchange current density than the other interfering ions. This condition can be achieved by incorporating selective binding sites or ionophores in a membrane or gel material. The sensitive coating therefore has to provide selectivity.

Fabrication

Because the fabrication is very similar to that of voltammetric devices, it is not further specified here. Sensor processing sequences are given in the literature.[234,237,238] The fabrication of potentiometric sensors integrated with CMOS circuitry components is described by Goldberg et al.[238] and Lauks et al.[239]

Applications

Prototype applications include chemical analysis in the gas or liquid phase. Typical target analytes in the gas phase are nitrogen oxide, sulphur dioxide, and carbon dioxide, using sintered ceramic electrolytes (sodium, barium, and silver sulfate or Nasicon).[237,240] Target analytes in the liquid phase comprise all kinds of ionic species such as hydrogen (pH, *Potentia hydrogenii*, the negative decadic logarithm of the hydrogen ion concentration), potassium, ammonium, calcium, chloride, cyanide, or nitrate, using ionophores in polymeric membranes[234,235,238,239,241] or chalcogenide glasses.[242]

The interaction mechanism in all cases is an electrochemical charge-transfer reaction [Eqs. (12) through (15)].

11.6.2.2 Field-Effect-Based Devices

The field-effect-based microfabricated potentiometric sensors, like metal-oxide semiconductor (MOS) devices in electronics, use variations in the charge distribution within the semiconductor surface space-charge region. The three field-effect device structures commonly used include the MOS capacitor (MOSCAP), the MOS diode (MOS-diode), and the MOSFET.[4,8,11,34] These structures are displayed in Fig. 11.26. In this con-

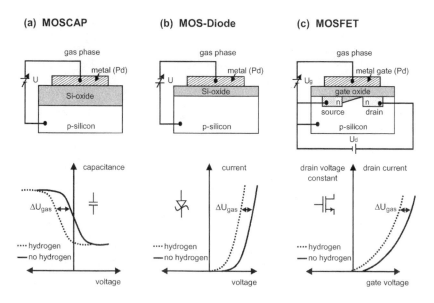

Figure 11.26 Schematic representation of the different MOS field-effect devices: (a) capacitor, (b) diode, (c) transistor. U_g denotes the gate voltage, U_d the source-drain voltage. Characteristic sensor responses (voltage shift due to exposure to hydrogen) are given at the bottom.

text, it is important to point out "that one of the most critical thrusts of the electronics industry's efforts to develop commercial field-effect devices has been predicated by the need to isolate these devices from *any* variation in their chemical environment."[34]

By replacing the metal gate with an ionic solution and a reference electrode immersed into this solution, the respective ion-sensitive device structures have been developed: the ion-sensitive capacitor (ISCAP), ion-controlled diode (ICD), and ion-sensitive field-effect transistor (ISFET) (Fig. 11.27). The gate region is now exposed to any ion present in the solution.[4,8,11,34]

Device applications are discussed in the individual device-related sections; the fabrication for the family of field-effect devices is summarized at the end in Section 11.6.2.2.6.

11.6.2.2.1 MOS Field-Effect Transistors and Ion-Selective Field-Effect Transistors (Chemotransistors)

A MOSFET in electronics is a transistor, the source-drain current (conductance) of which is modulated through an electric field perpendicular to the device surface. This electric field is generated by an isolated gate electrode and influences the charge carrier density in the conductance path between source and drain (semiconductor field effect).

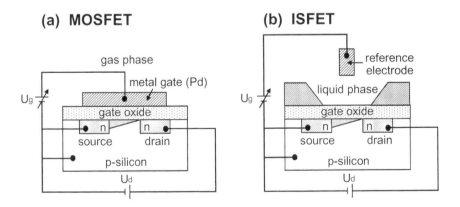

Figure 11.27 Schematic representation of (a) a MOSFET and (b) an ISFET structure. U_g denotes the gate voltage, U_d the source-drain voltage. By replacing the metal gate of the MOSFET with an ionic solution and a reference electrode immersed into this solution, the ISFET was developed.

MOSFET Transduction Principle and Sensing Characteristics

The MOSFET as used for chemical sensing has a p-type silicon substrate (bulk) with two n-type diffusion regions (source and drain). The structure is covered with a silicon dioxide insulating layer, on top of which a metal-gate electrode (originally palladium,[22] later other platinide metals[243-247]) is deposited.

When a positive voltage (with respect to the silicon) is applied to the gate electrode, holes, which are the majority carriers in the p-substrate, are depleted near the semiconductor surface. Upon applying a voltage between source and drain (U_d), the electrons from the n-doped source can pass through this depleted surface region to the drain so that a conducting n-channel between source and drain is generated in the p-substrate near the silicon/silicon dioxide interface. The conductivity of this n-channel, i.e., the magnitude of the source-drain current (I_d), can be modulated by adjusting the strength of the electrical field perpendicular to the substrate surface between the gate electrode and the silicon.[22,243,244]

Applications

Palladium-gate FET structures were demonstrated to function as hydrogen sensors by Lundström[22] and others.[245,248,249] Hydrogen molecules readily absorb on the gate metal (platinum, iridium, palladium) and dissociate into hydrogen atoms. These H-atoms can diffuse rapidly through the Pd and absorb at the metal/silicon-oxide interface partly on the metal, partly on the oxide side of the interface.[243,244] Due to the absorbed species and the resulting polarization phenomena at the interface, the drain current (I_d) is altered and the threshold voltage (U_d) is shifted. The voltage shift is proportional to the concentration or coverage of hydrogen at the oxide/metal interface. The presence of oxygen promotes the formation of water at the gas phase/metal interface due to a catalytic reaction of atomic hydrogen with atomic oxygen. With thin catalytic metal gates (containing holes and cracks), ammonia,[250,251] amines, and any kind of molecule that gives rise to polarization in a thin metal film (hydrogen sulfide, ethene, etc.) or causes charges/dipoles on the insulator surface can be detected.[243,244] However, detailed models for these processes do not yet exist. Sensitivity and selectivity patterns of gas-sensitive FET devices depend on the type and thickness of the catalytic metal used, the chemical reactions at the metal surface, and the device operation temperature. For extremely high temperatures (600 to 800°C), silicon carbide devices have been developed.[197-200] A paramount problem with MOSFET sensors has been long-term drift, which seems to be mitigated to some extent by the deposition of a thin alumina layer between the Pd gate and the silicon oxide.[252] The temperature should be kept constant. Alternative gate materials include polyaniline for the detection of water and ammonia,[253,254] and high-temperature superconducting cuprate to monitor ammonia and nitrogen oxides.[255]

ISFET Transduction Principle and Sensing Characteristics

In the case of the ISFET, the gate-metal electrode of the MOSFET is replaced by an electrolyte solution that is in contact with the reference electrode , i.e., the silicon gate oxide is directly exposed to an aqueous electrolyte solution (Fig. 11.27).[21] An external reference electrode is required for stable operation of an ISFET.[4,8,11,34,256-259] Unfortunately, including such a reference is nontrivial and subject to dedicated research.[34,260-264] An electric current (I_d) flows from the source to the drain via the channel, and, like in MOSFETs, the channel resistance depends on the electric field perpendicular to the direction of the current. It also depends on the potential difference across the gate oxide. Therefore, the source-drain current, I_d, is influenced by the interface potential at the oxide/aqueous solution. Although the electric resistance of the channel provides a measure for the gate-oxide potential, the direct measurement of this resistance gives no indication of the absolute value of this potential. However, at a defined source-drain potential (U_d), changes in the gate potential can be compensated by a modulation of U_g. This adjustment can be carried out in such a way that the changes in U_g applied to the reference electrode exactly compensate for the changes in the gate-oxide potential. This is automatically performed by ISFET amplifiers with feedback, which maintains a constant source-drain current. In this particular case, the gate-source potential is determined by the surface potential at the insulator/electrolyte interface. Mechanistic studies of the processes occurring at the solution/gate-oxide interface (site binding model[265]) and the oxide/semiconductor interface can be found in the literature.[4,8,11,34,258,259,265-268] The insulator solution interface is assumed to represent, in most cases, a polarizable interface, i.e., there will be charge accumulation across the structure but no net charge passing through. Interfaces with net charge passing through are called faradaic interfaces and are discussed in the literature.[4,8,11,34,268]

Applications

The silicon-gate-oxide surface contains reactive Si-OH groups that, besides providing pH sensitivity, can be used for covalent attachment of a variety of organic molecules and polymers.

pH-FET

The basic ISFET is an exposed-gate-oxide FET that functions as a pH sensor.[21] The surface of the gate oxide contains OH functionalities that are in electrochemical equilibrium with ions in the sample solutions (H^+ and OH^-). The hydroxyl groups at the gate-oxide surface can be protonated and deprotonated; therefore, when the gate oxide contacts an aqueous solution, a change of pH will change the silicon-oxide surface potential. A site-binding model describes the signal transduction as a function of the state of ionization of the amphoteric surface Si-OH groups.[269,270] The change in the charge state of these sites leads to a variation of the dipole layer and, consequently, to a change in the semiconductor space charge region.[4,34,267,268] Typical pH sensitivities measured with silicon-oxide ISFETs are 37 to 40 mV/ pH unit.[270] Inorganic gate materials for pH sensors such as silicon nitride (CMOS process material),[271-277] oxynitride,[278] alumina,[279] tantalum oxide,[279,280] and iridium oxide[281] have better properties than silicon oxide with regard to pH response, hysteresis, and drift. In practice, these layers are deposited on top of the first layer of silicon oxide by means of chemical vapor deposition (CVD). Sensor signals achieved with a silicon nitride membrane are given in Fig. 11.28.[273]

CHEMFET

The CHEMFET, or chemically sensitive FET,[282] is a modification of an ISFET with the original inorganic gate material covered by organic ion-selective membranes such as polyurethane, silicone rubber, polystyrene, polyamide, and polyacrylates containing ionophores [Fig. 11.29(a)].[283] A critical point of the CHEMFET is the attachment of the sensitive membrane, which can be improved by mechanical[284] or chemical[283,285,286] anchoring to the surface of the gate oxide. CHEMFETs selective to K^+,[283,287-290] Na^+,[291-293] Ag^+,[294] transition metal cations [Pb^{2+}, see Fig. 11.29(b); Cd^{2+}],[295-297] and some anions (NO^{3-})[298-300] have been developed.

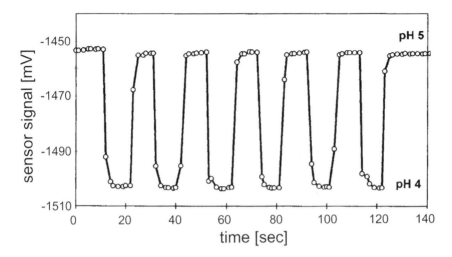

Figure 11.28 Dynamic response of a pH-ISFET in a flowthrough configuration upon repeated pH changes by one unit (4 to 5). The signals exhibit a stable baseline and a good reproducibility. According to the Nernst equation [Eq. (15)], a pH change of one unit causes a voltage change of 59 mV. Reprinted from Hoffmann and Rapp[273] with permission.

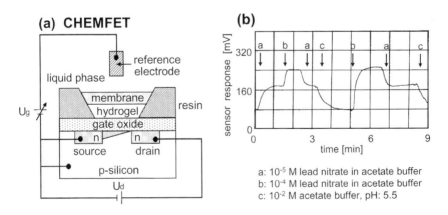

Figure 11.29 (a) Schematic representation of a CHEMFET, a modification of an ISFET with the original inorganic gate material covered by an organic ion-selective membrane and a hydrogel. (b) Sensor response of a CHEMFET with a lead-selective membrane upon different target analyte concentrations in an acetate buffer solution: *a* represents a 10^{-5} molar lead nitrate solution, *b* a 10^{-4} molar lead nitrate solution, and *c* the pure buffer solution. Adapted from Battilotti et al.[297] with permission.

Highly specific organic or biological compounds can be incorporated in the membrane as well: enzymes (ENFET)[301,302] such as glucose oxidase for the detection of glucose[301,303-305] and penicillinase for the detection of penicillin.[301] Other enzymes can be used to detect pesticides and organophosphorous compounds[306-309] or aldehydes.[310] Antibodies can be immobilized for immunoreactions (IMFET), and biological entities like whole cells[311-313] or insect antennae[314] have been used (BIOFET).

Differential CHEMFET Configuration

CHEMFETs can be applied in a differential measurement configuration[315-322] [Fig. 11.30(a)]. This device configuration offers the advantage that a variety of experimental parameters and external disturbances, e.g., light and temperature, which affect both FET structures, cancel out. The inorganic gate of one of the CHEMFETs has to be rendered insensitive to pH or other target analytes by chemical surface treatment. This "insensitive" FET [sometimes called reference FET (REFET)] should ideally show no response to the target species present in the sample phase [Fig. 11.30(b)].[323] pH sensitivity, e.g., can be suppressed by plasma-deposition of a hydrophobic polymer on the gate[324-328] or by realizing a reservoir of buffered solution with constant pH separated by a membrane from the sample phase [Fig. 11.30(a)].[323] The applications comprise target analytes similar to those for the CHEMFETs.

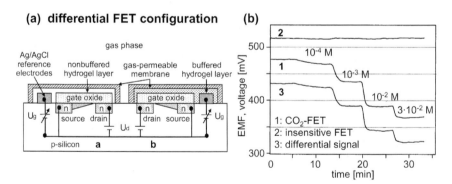

Figure 11.30 (a) Schematic of a differential FET configuration for sensing carbon dioxide. The carbon dioxide FET (formation of "carbonic acid," pH-FET) exhibits a nonbuffered hydrogel layer; the insensitive FET has a buffered one. The pH of the latter will not change due to carbon dioxide absorption, as shown in the sensor responses (b). The differential configuration thus eliminates effects common to both devices (temperature fluctuations, flow disturbances, etc.). Adapted from Shin et al.[323] with permission.

11.6.2.2.2 *MOS Diode and Ion-Controlled Diode (Chemodiodes)*

Transduction Principle and Sensing Characteristics

The chemical sensing mechanism is identical to that of the MOSFET, except for the transduction method, which is slightly different. When the gas molecules (hydrogen) diffuse to the metal/oxide interface to form a polarization layer at the interface, the height of the energy barrier of the diode is altered. This leads to a change in either the forward voltage or the reverse current of the diode. The diode characteristics are therefore shifted along the voltage axis (see Fig. 11.26), as they are with the MOSFET.[22,243,244,329-332] The response time and sensitivity of the sensor can be improved by operation at elevated temperatures. Therefore, sensing structures have been placed on thermally isolated membranes.[333,334]

The ion-controlled diode was first described by Zemel.[335] Its operation is not too different from that of an ISFET and is analogous to that of the MOS diode.[8,34,268,335,336] Polarization resulting from ion adsorption on the insulator (silicon oxide) causes changes in the effective forward voltage or the reverse current, which can be measured. A more sophisticated structure with a through-the-chip p-n junction, which enables the contacts on the backside of the chip to be isolated from the aqueous electrochemistry occurring at the front ("gate") side, is described in the literature.[34,335,336]

Applications

The applications are the same as for the FETs.

11.6.2.2.3 *MOS Capacitor and Ion-Selective Capacitor (Chemocapacitors)*

Transduction Principle and Sensing Characteristics

The sensor element is a standard MOS or electrolyte insulator semiconductor (EIS) structure, as shown in Fig. 11.26. Again, the target analyte species changes the polarization at the metal/oxide or electrolyte/oxide interface, thus affecting the flat-band voltage of the capacitor. The capacitance-voltage curve of the capacitor is shifted by a certain amount that is proportional to the target analyte concentration in the gas or liquid phase. The capacitance can be recorded as a measure of the analyte concentration or some circuitry can be used to keep the capacitance constant by varying the necessary bias voltage. Capacitor-type structures are straightforward to realize.

Applications

The applications (MOSCAP[4,243,244,248,337,338], ISCAP/EIS[339-342]) are similar to those for the FETs. A set of capacitance/voltage curves is shown in Fig. 11.31.[341]

11.6.2.2.4 Measuring Work Functions: Kelvin Probe and Suspended-Gate FET

Transduction Principle and Sensing Characteristics

The work function is defined as the minimum work required to extract an electron in vacuum from the Fermi level of a conducting phase through

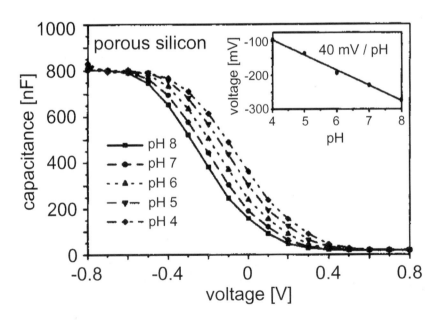

Figure 11.31 Set of capacitance/voltage curves for a porous electrolyte-insulator-semiconductor (EIS) structure exposed to solutions with different pHs. The insert shows the calibration curve (pH versus voltage). Adapted from Schöning et al.[341] with permission.

a surface and place it outside the reach of electrostatic forces at the so-called vacuum level.[343,344] When two different electronic conductors are in contact, electrons flow from the material with the lower electron affinity to that with the higher electron affinity according to the difference in the (electro-)chemical potential [Eqs. (5), (6), and (13)] of the electrons until an equilibrium is reached. A contact or Galvani potential arises, which represents the bulk-to-bulk inner contact potential of the two materials or the difference of the Fermi levels of the two materials. If the surfaces of two different materials are parallel and separated by a very thin insulator or air gap, a potential across the gap is formed, the Volta potential or outer potential, which represents the difference of the work functions of both materials. A palladium plate separated by a thin air gap (of a few µm) from a copper plate will become positively charged, and correspondingly, the copper plate will be negatively charged. The work function of a certain material, which includes the chemical potential of the electrons (changes in electron affinity, eventual band bending due to electron transfer) and the surface dipole field (which exists even at absolutely clean surfaces) is changed upon formation of surface adsorbates, e.g., from the gas phase. Therefore, work function measurements can be used to advantage in chemical or gas sensing.

The Kelvin probe relies on the displacement of one of the surfaces in a periodic oscillation. This oscillation induces charges across the surfaces and generates a sinusoidal current in the sensing plate that is proportional to the work function difference between the sensing plate and the reference plate. This current thus directly depends on the surface chemistry of the plates. Micromachined Kelvin probes with a metal sensing film supported by a dielectric membrane and a 2.5-µm-thick silicon reference plate that is electrostatically deflected by a drive electrode [schematically depicted in Fig. 11.32(a)] have been fabricated and used to detect the surface adsorption of oxygen.[345]

The SGFET (or even the standard MOSFET) is very similar to the Kelvin probe.[343,346] Although FET transducers are sometimes called work function devices, the notion of work functions has not yet been discussed in the context of FETs. A metal plate (suspended gate) is separated by an air gap (or in the case of the MOSFET by a silicon oxide layer) from a silicon plate [Fig. 11.32(b)].[343] The back of the silicon is ohmically connected to the gate metal through a variable voltage source (U_g). The plate distance, however, cannot be varied. Therefore, the drain-source current is used to interrogate the Volta potential. The magnitude of the drain-source current at a certain applied voltage depends on the work function difference between the metal gate and the silicon. It therefore directly depends on surface adsorption or absorption chemistry in sensitive layers applied between the suspended gate and the silicon.

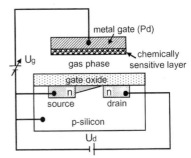

Figure 11.32 Schematic representation of (a) a micromachined Kelvin probe and (b) a suspended-gate FET. Details are given in the text. (a) and (b) are adapted from Bergstrom et al.[345] and Janata and Josowicz,[343] respectively, with permission.

Applications

Sensitive layers applied in the gap include metal oxides to detect ammonia, carbon monoxide, or nitrogen oxides;[347-351] potassium iodide to detect ozone;[352] palladium/polyaniline to detect hydrogen and ammonia;[353] and polypyrrole to detect alcohols and volatile organics.[354]

11.6.2.2.5 Light-Addressable Potentiometric Sensor

Transduction Principle and Sensing Characteristics

The light-addressable potentiometric sensor is also based on the field effect,[146] and its working principle is very closely related to that of FET devices (Fig. 11.33).[146,355-358] The LAPS device is a thin silicon plate (thinned down to a few μm thickness) with an approximately 100-nm-thick oxynitride layer in contact with an electrolyte solution. A potential is applied between the silicon plate and, e.g., a silver/silver chloride controlling electrode immersed in the electrolyte solution. The controlling electrode simultaneously serves as a reference electrode. The sign and magnitude of the applied potential are adjusted so that they deplete the semiconductor of majority carriers at the insulator interface. Upon illumination of the plate with LEDs, hole-electron pairs are created that can reach the depletion area. Due to the charge separation in the depletion area, a photocurrent flows through the device. In this way, a sinusoidally modulated light beam causes a sinusoidal photocurrent, the amplitude of

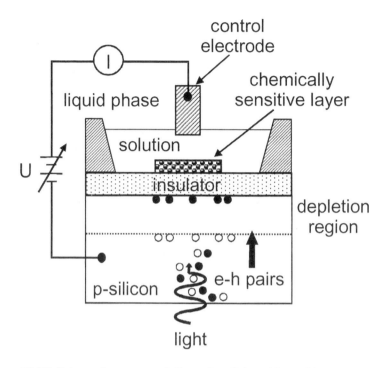

Figure 11.33 Schematic representation of a light-addressable potentiometric (LAPS) device. Details are given in the text. Adapted from Fanigliulo et al.[361] and Tanaka et al.[362] with permission.

which depends on the width of the depletion charge region: the larger the depletion region, the larger the photocurrent. The photocurrent amplitude thus depends on the absorbed species in the sensitive layer or the pH of the solution in contact with the insulator. The current-voltage curve is of sigmoidal shape and is shifted along the voltage axis due to chemical changes (e.g., pH changes) at the solution/silicon-oxide interface in a way similar to other field-effect devices.

Applications

The silicon plate is straightforward to manufacture, and arrays of LEDs and different selective layers can be used on the same chip.[146] Assessments of the lateral resolution[359,360] and benchmarking against ISFETs[361] have been performed. LAPS methods have also been applied to the gas phase.[355] Applications include monitoring cell activity via pH changes resulting from cell activity and metabolism,[362-365] and biosensing in liquids.[356,358,366] LAPS devices have been developed by Molecular Devices Inc.[367]

11.6.2.2.6 Field-Effect Device Fabrication

Additional field-effect device fabrication steps after semiconductor processing include the following:

- Patterning of metal electrodes (liftoff, thermal evaporation, sputtering) as described for MOSFETs in the literature[242-251]
- Deposition of additional metal oxides/nitrides by LPCVD (tantalum oxide, alumina, silicon nitride) as described for MOSFETs and ISFETs in the literature[271-281]
- Optional membrane formation for temperature stabilization by back-etching[333,334]
- Surface micromachining (HF etching, Al-etching) for the SGFET[349,368]
- Deposition of electro-active polymers, membrane materials, hydrogels by spin-casting, spraying, screen printing, photolithography for CHEMFETs[253-255,282-310,324-328]

The fabrication of field-effect electrochemical sensors integrated with CMOS circuitry is described in various publications.[11,271,272,275,277,278,293,337,369] A modular chip system based on a sensing and a "service" chip has been described in Domansky et al.[254] and a flow-through ISFET by Sibbald and Shaw.[370] Backside-contact ISFETs are detailed in the literature.[371,372] pH-ISFETS are commercially available from companies such as Honeywell.[373]

11.6.3 Conductometric Sensors

Conductometric techniques are a special case of AC impedance techniques. Instead of the real and imaginary components of the electrode impedance at different frequencies, only the real-valued resistive component, related to the sample (sensing material) resistance, is of interest. Since complex impedances include capacitive and inductive contributions, chemocapacitors that do not rely on the field effect are included here, in the conductometric section. The section on conductometric sensors is therefore organized in two parts, one on resistance measurements at room temperature and elevated temperatures (chemoresistors), the other on chemocapacitors.

11.6.3.1 Chemoresistors

Transduction Principle and Sensing Characteristics

Chemoresistors rely on changes in the electric conductivity of a film or bulk material upon interaction with an analyte. Conductance, G, is defined as the current, I [A], divided by the applied potential, U [V]. The unit of conductance is Ω^{-1} or S (Siemens). The reciprocal of conductance is the resistance, R [Ω]. The resistance of a sample increases with its length, l, and decreases with its cross-sectional area, A:

$$R = \frac{1}{G} = \frac{U}{I} = \frac{1}{\kappa} \cdot \frac{l}{A}. \qquad (23)$$

Conductivity or specific conductance, κ [1/Ωm], is therefore defined as the current density [A/m^2] divided by the electrical field strength [V/m]. The reciprocal of conductivity is resistivity, ρ [Ωm]. The conductivity can be thought of as the conductance of a cube of the probed material with unit dimensions.[11]

Conductometric sensors are usually arranged in a metal-electrode-1/sensitive-layer/metal-electrode-2 configuration.[4] The conductance is measured either via a Wheatstone bridge arrangement or by recording the current at an applied voltage in a DC mode or in a low-amplitude, low-frequency AC mode to avoid electrode polarization. In Fig. 11.34(a),[374] a conductance cell (in this case with metal oxides as the conductive layer) and the respective equivalent electric circuits are depicted. The goal of conductometry is to determine the sample resistance (c). The lead wire resistances can normally be neglected. The electrode impedance (a) consists of two elements: the contact capacitance and the contact resistance. By applying an AC potential, an AC will flow through the resistor cell. If the contact capacitance is sufficiently large, no potential will build up across the corresponding contact resistance. The contact resistance should be much lower than the sample resistance and should be minimized, so that the bulk contribution dominates the measured overall conductance. If surface conductivity mechanisms occur that differ from those in the bulk, this can be modeled by introducing an additional surface resistor (b) to the

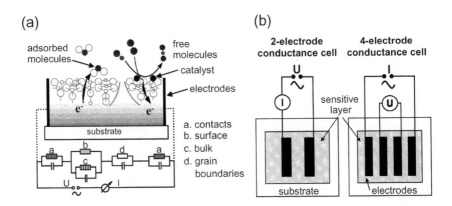

Figure 11.34 (a) Schematic representation of a conductance cell (in this case, a semiconductor sensor with tin dioxide as the sensitive layer) and of the different contributions (contacts, surface, bulk, and grains) to the overall conductivity and the respective equivalent circuits. Adapted from Göpel et al.[374] with permission. (b) Schematic representation of a two-electrode and four-electrode conductance cell. Details are given in the text.

equivalent circuit. A grain boundary in the sensing material constitutes a resistance-capacitance unit (d). The conductivity depends on the concentration of charge carriers and their mobility, either of which can be modulated by analyte exposure. In contrast to potentiometry and voltammetry, conductometric measurements monitor processes in the bulk or at the surface of the sample. Any contribution of electrode processes has to be avoided.

Therefore, in most cases, a four-electrode configuration is preferred over a simple two-electrode configuration.[Fig. 11.34(b)] The outer pair of electrodes is used for injecting an AC into the sample; the potential difference is then measured at the inner pair of electrodes. The interference of electrode impedances on the measurement results is thus excluded.

11.6.3.1.1 Low-Temperature Chemoresistors

Several classes of predominantly organic materials are used for applications with chemoresistors at room temperature. The chemically sensitive layer is applied over interdigitated electrodes on an insulating substrate. Electrode spacing is typically 5 to 100 μm, and the total electrode area is a few mm^2. The applied voltage ranges between 1 and 5 V.

Metal phthalocyanines constitute organic p-type semiconductors. The adsorption of oxidizing agents such as nitrogen oxide or ozone therefore decreases the resistance by increasing the number of holes in the conduction band.[8] Metal phthalocyanines at elevated operational temperatures (approximately 450 K) have been used to monitor nitrogen oxide,[375,376] ozone,[377] hydrogen chloride,[376] and even ammonia.[378,379]

Conducting polymers such as polypyrroles, polyaniline, and polythiophene exhibit a large conjugated π-electron system that extends over the whole polymer backbone. Partial oxidation of the polymer chain then leads to electrical conductivity because the resulting positive charge carriers (called polarons or bipolarons) are mobile along the chains.[9] Counteranions must be incorporated into the polymer upon oxidation to balance the charge on the polymer backbone.

However, the conducting polymers do not only react with oxidizing agents; they also respond to a wide range of organic vapors.[380-383] The underlying principle of this response is still unclear. Possibilities include: (1) vapor molecules could affect the charge transfer between the polymer and the electrode contact; (2) analyte molecules could interact with the mobile charge carriers on the polymer chains; (3) analyte molecules could interact with the counterions, thus modulating the mobility of the charge carriers; or (4) analyte molecules could alter the rate of interchain hopping in the conducting polymer.[354,384,385]

Fabrication

Additional fabrication steps after semiconductor processing include the following:

- Patterning of metal electrodes (liftoff, thermal evaporation, sputtering)[380-383,386-400]
- Optional membrane formation by back-etching for temperature stabilization[389,401]
- Deposition of carbon-loaded polymers, membrane materials, or hydrogels by spin-casting, spraying, screen printing, and photolithography[393-396,398-399]
- Deposition of conducting polymers by electrochemical deposition; sensor selectivity modified by changing the counterion used in the polymerization process[380-383]

The fabrication of an impedance device[402] and microbridges[390] integrated with CMOS circuitry components have been described. Complete processing sequences are detailed by Sheppard *et al.*[398] and Varlan and Sansen.[400]

Applications

Applications include a variety of polar organic volatiles such as ethanol, methanol, and components of aromas.[380-383,386-395] Conducting polymers show a high cross-sensitivity to water. Sensors are commercially available from Osmetech and Marconi.[391]

In *carbon-black-loaded polymers*, conducting carbon black is dispersed in nonconducting polymers deposited onto an electrode structure. The conductivity is by particle-to-particle charge percolation so that if the polymer absorbs vapor molecules and swells, the particles are, on average, further apart, and the conductivity of the film is reduced.[392] Carbon-black-loaded polymers are used to detect a variety of organic solvents such as hydrocarbons, chlorinated compounds, and alcohols.[393-396] Sensors are commercially available from Cyrano Sciences.[397]

Hydrogels responsive to pH changes have been applied to interdigitated electrode arrays. The hydrogel swells or shrinks to a hydration determined by the pH of the analyte solution. This leads to a corresponding increase or decrease in the mobility of the ions partitioned by the gel. The sensitivity of ion mobility to small changes in hydration causes large resistance changes.[398,399] A conductometric variant of a Severinghaus electrode (detection of carbon dioxide via dissolution in water, formation of carbonic acid, and monitoring of the pH change [4]) with a liquid reservoir has been described by Varlan and Sansen.[400]

11.6.3.1.2 High-Temperature Chemoresistors (Hotplate Sensors)

There is a wealth of literature on semiconducting-metal-oxides-based chemical sensors; this section focuses exclusively on the silicon-based, micromachined hotplates. A typical high-temperature chemoresistor includes an integrated heater, a thermometer, and a sensing film on a thermally isolated stage such as a membrane (Fig. 11.35).[403] Isolated micromachined structures (hotplates) exhibit very short thermal time constants on the order of milliseconds.

The sensitive materials used with the hotplate sensors include wide-bandgap semiconducting oxides such as tin oxide, gallium oxide, indium oxide, or zinc oxide. In general, gaseous electron donors (hydrogen) or acceptors (nitrogen oxide) adsorb on the metal oxides and form surface states, which can exchange electrons with the semiconductor. An acceptor molecule will extract electrons from the semiconductor and thus decrease its conductivity. The opposite holds true for an electron-donating surface state. A space-charge layer will thus be formed. By changing the surface concentration of donors/acceptors, the conductivity of the space-charge region is modulated.[4,374,404-406] In addition to the interaction of surface

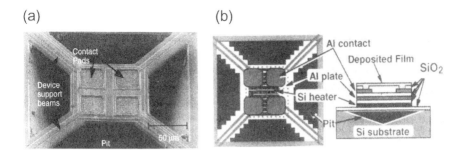

Figure 11.35 (a) Scanning electron micrograph of a microhotplate and (b) a schematic top view and side view of the device. The suspended plate exhibits a polysilicon heater, an aluminum plane for homogenous heat distribution, and aluminum electrodes for measuring the resistance of a semiconductor metal oxide. Reprinted from Suehle et al.[403] and Semancik and Cavicchi[418] with permission.

adsorbates based on electronic effects, mentioned above, diffusion of lattice defects from the bulk of the metal-oxide crystal also occurs (ionic conduction) at elevated temperatures (>600°C). The defects can act as donors or acceptors. Oxygen vacancies, for example, act as intrinsic donors. The overall conductivity in polycrystalline samples includes contributions from the individual crystallites, the grain boundaries, insulating components such as pores, and the contacts [Fig. 11.34(a)]. Thus the conduction mechanism in ceramic polycrystalline samples is difficult to analyze, and a variety of empirical data has been published.[4,374,404-406]

The most extensively investigated material, tin oxide, is oxygen-deficient and, therefore, is an n-type semiconductor since oxygen vacancies act as electron donors. In clean air, oxygen, which traps free electrons by its electron affinity, and water are absorbed on the tin-oxide particle surface, forming a potential barrier in the grain boundaries. This potential barrier restricts the flow of electrons and thus increases the resistance. When tin oxide is exposed to reducing gases such as carbon monoxide, the surface adsorbs the gases, and some of the oxygen is removed by reaction with water and oxygen at the surface. This lowers the potential barrier, thereby reducing the electric resistance. The reaction between gases and surface oxygen depends on the sensor temperature, the gas involved, and the sensor material.[4,374,404-407]

Semiconductor metal-oxide sensors usually are not very selective but respond to almost any analyte (carbon monoxide, nitrogen oxide, hydrogen, hydrocarbons). One method for modifying the selectivity pattern includes surface doping of the metal oxide with catalytic metals such as platinum, palladium, gold, or iridium.[374,405,408] Surface doping improves the sensitivity to reducing gases, reduces the response time and operational temperature, and changes the selectivity pattern.[374,405,408]

Most modern sensors operate in a regime in which the overall conductivity is determined by nanocrystalline sensing materials. As a consequence of the small grain size (better surface-to-volume ratio), the relative interactive surface area is larger, and the density of charge carriers per volume is higher. This leads to more drastic conductivity changes and hence larger sensor responses when compared to larger grains.[374,405]

Since microhotplates have a very low thermal mass, they allow for applying temperature-programmed operational modes.[409-412] By operating the device, e.g., in a cyclic thermal mode, reaction kinetics on the sensing surface are altered, producing a time-varying response signature that is characteristic for the respective analyte gas.[409-412] An example is given in Fig. 11.36.[411]

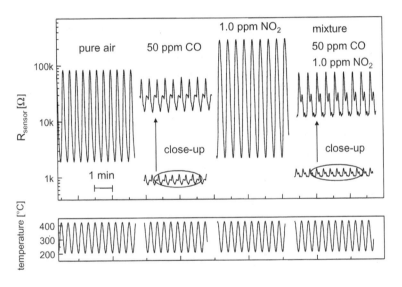

Figure 11.36 Sinusoidal modulation of the operation temperature of a tin dioxide sensor between 200 and 400°C (bottom) leads to characteristic frequency-dependent resistance features (upper part). Changes of the resistance (R sensor) of the micromachined sensor upon exposure to 50 ppm CO, 1 ppm NO_2, and a mixture of 50 ppm CO and 1 ppm NO_2 in synthetic air (50% relative humidity). Reprinted from Heilig et al.[411] with permission.

Fabrication

Additional fabrication steps after semiconductor processing include the following:

- Additional deposition/patterning of metal electrodes using liftoff, thermal evaporation, and sputtering[413-416]
- Backside (KOH)[401,413-417] or frontside (RIE, EDP)[403,418] etching for membrane formation
- Deposition of metal-oxide materials by LPCVD, sol-gel processes, sputtering, and screen printing[419-424]
- Sintering of the metal oxides (annealing) at elevated temperatures[425]

The fabrication of hotplates on a CMOS-substrate is described by Suehle et al.[403] and Semancik et al.[418] Complete processing sequences are detailed by Majoo et al.[426] and Chung et al.[427] The fabrication of a 39-electrode array with a tin-oxide gradient coating has been reported by Althainz et al.[428,429] Extensive reliability studies are reported on in the literature.[430-432] Investigations and simulations to optimize the power consumption are under way.[433-435] Devices are commercially available from Figaro, Marconi, Capteur, and Microchemical Systems.[436]

Applications

Typical applications include the detection of hydrogen,[403] oxygen,[403] nitrogen oxide,[413,437] CO,[414,415] and a variety of organic volatiles,[416,417,428,429,438] using tin oxide as the sensitive layer. Thermal cycling of the hotplate structures allows for detection of gas mixtures (CO and nitrogen oxide) with a single sensor[411] (Fig. 11.36). In other applications, temperature profiles were optimized to detect specific organic volatiles.[418,438,439]

Additional sensitive materials on hotplates include, for example, niobium and titanium oxide to detect oxygen,[440] as well as metal films covered with surface oxides. Reducing gases like hydrogen donate electrons to the metal and increase the conductance, whereas electron acceptor molecules like oxygen decrease the metal conductance.[426,441,442] An unheated tin-oxide oxygen sensor (slow response) has been reported by Atkinson et al.[443]

11.6.3.2 Chemocapacitors

Transduction Principle and Sensing Characteristics

Chemocapacitors (dielectrometers) rely on changes in the dielectric properties of a sensing material upon analyte exposure (chemical modulation of equivalent circuit capacitors in Fig. 11.34 by changes in the dielectric constant of the sensitive layer). Interdigitated structures analogous to the room-temperature chemoresistors are predominantly used.[444-447] In some cases, plate-capacitor-type structures with the sensitive layer sandwiched between a porous thin metal film (permeable to the analyte) and an electrode patterned on a silicon support are used to increase sensitivity by trapping the electric field.[448,449] The capacitances are usually measured at an AC frequency from a few up to 500 kHz.

Two effects change the capacitance of a polymeric sensitive layer upon absorption of an analyte: (1) swelling and (2) change of the dielectric constant due to incorporation of the analyte molecules into the polymer matrix.[450,451] For a simple interdigitated structure, the space containing 95% of the field lines includes the polymer volume within a distance of half of the periodicity from the electrodes.[452] For a layer thickness less than half the periodicity, swelling of the polymer upon analyte absorption always results in an increase of the measured capacitance, regardless of the dielectric constant of the analyte [Eq. (25)]. This results from the increased polymer/analyte volume within the field line region exhibiting a larger dielectric constant than that of the substituted air. The capacitance change for a polymer layer thicker than half the periodicity of the electrodes is determined by the ratio of the dielectric constants of the analyte, ε_A (the analyte is assumed to be in the liquid state), and the polymer, ε_{poly} [Eq. (24)]. If the dielectric constant of the polymer is lower than that of the analyte, the capacitance will be increased. Conversely, if the polymer dielectric constant is larger, the capacitance will be decreased (Fig. 11.37). These effects have been discussed and supported by the simulations of Steiner et al.,[451] where the following formulae have been used to describe the change of the sensor capacitance:

$$\varepsilon_{eff} = \varepsilon_{poly}(1 - VF_A c_A) + \varepsilon_A \cdot VF_A c_A \qquad (24)$$

$$h_{eff} = h(1 + S_A c_A). \qquad (25)$$

Here ε_{eff} denotes the resulting effective dielectric constant of the polymer/analyte system. ε_{poly} is the dielectric constant of the polymer, c_A the concentration of the analyte in the gas phase, VF_A the volume fraction of the analyte in the polymer per unit gas phase concentration, h_{eff} the resulting effective polymer thickness after analyte absorption, and S_A the experimental swelling coefficient of the polymer per unit gas-phase concentration for the respective analyte. VF_A and S_A are constants (Henry's law is assumed to be valid) and have to be determined experimentally for every polymer/analyte combination by mass-sensitive or

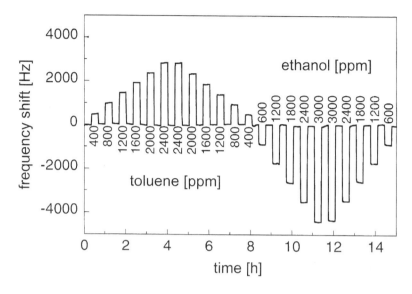

Figure 11.37 Frequency responses of a switched-capacitor device upon exposure to different analytes at 301 K. For a thick layer of PEUT (4.3 μm), toluene (dielectric constant lower than that of PEUT) causes positive frequency shifts; ethanol (dielectric constant higher than that of PEUT) causes negative frequency shifts.

optical measurements. Typical sensor signals for a polymer layer thicker than the surface-normal extension of the field lines are shown in Fig. 11.37. The ratio of the dielectric constants of polymer (2.9) and analytes (toluene: 2.4, ethanol: 24.5) controls the signs of the signals. The capacitance changes are recorded as frequency shifts using a reference/sensing capacitor configuration and on-chip circuitry (switched capacitors, $\Sigma\Delta$ conversion). Details are given in the literature.[450,453,454]

For conducting measurements at defined temperatures, sensor and reference capacitors can be placed on thermally isolated membrane structures (microhotplates).[453-455]

Fabrication

Additional fabrication steps after semiconductor processing include the following:

- Deposition/patterning of metal electrodes (liftoff, thermal evaporation, sputtering)[444-449]

- Optional back-etching, membrane formation for temperature stabilization[453-456]
- Deposition of polymers (spin-casting, spraying, photolithography)[444-454]

The fabrication of capacitors integrated with CMOS circuitry components is described in the literature.[445,447,450,451,453,454,457,458] Temperature-stabilized membranes have also been reported on in the literature.[453-455] Capacitive humidity sensors are commercially available from, e.g., Sensirion, Vaisala, and Humirel.[459]

Applications

Typical applications areas are humidity sensing using polyimide films,[444-448,457-459] since water has a relatively high dielectric constant of 78.5 (liquid state) at 298 K, leading to large capacitance changes. A variant exhibits polyimide columns sandwiched between metal electrodes.[460] More recent applications include the detection of different kinds of organic volatiles in the gas phase (hydrocarbons, chlorinated hydrocarbons, alcohols, etc.) using polymeric layers[450,451,453,454,461,462] or liquid crystals,[463] and the detection of nitrogen oxide,[464] sulphur dioxide,[465] and carbon dioxide[466,467] using ceramic materials.

The interaction mechanisms involve reversible physisorption and bulk/gas phase partitioning [see Eq. (10)].

11.6.4 Combinations of Electrochemical Principles

Combinations of different electrochemical principles are beneficial for several specific applications. For example, the concentration range can be extended by applying different electrochemical sensor types simultaneously. An integrated hydrogen system consisting of aluminum-gate FETs for low concentrations, a hydrogen-sensing chemoresistor made of palladium/nickel for high concentrations, and the necessary circuitry has been reported by Rodriguez et al.[468] and Hughes et al.[469]

Combinations of potentiometric and amperometric sensors have been used to monitor the content of oxygen, carbon dioxide, and the pH of blood.[470-473] An integrated sensor chip bearing an amperometric oxygen sensor (Clark-type: two-step-reduction of oxygen in aqueous solution via hydrogen peroxide to hydroxyl ions), a pH-FET, and a Severinghaus type pH-FET for carbon dioxide (detection of carbon dioxide via dissolution in water, formation of carbonic acid, and monitoring of the pH change) is shown in Fig. 11.38.[470,472] A sensor system that includes several ISFETs and an amperometric free-chlorine sensor for operation in water has been reported by Van den Berg et al.[474]

Disposable electrochemical multisensor systems for fast blood analysis are marketed by, e.g., I-STAT.[475] Sodium, potassium, chloride, ionized calcium, pH, and carbon dioxide are measured by ion-selective electrode potentiometry. Concentrations are calculated from the measured potential through the Nernst equation [Eq. (15)]. Urea is first hydrolyzed to ammonium ions in a reaction catalyzed by the enzyme urease. The ammonium ions are then measured by an ion-selective electrode, and the concentration is calculated through the Nernst equation [Eq. (15)]. Glucose is measured amperometrically. Oxidation of glucose, catalyzed by the enzyme glucose oxidase, produces hydrogen peroxide. The liberated hydrogen peroxide is oxidized at an electrode to produce an electric current, the intensity of which is proportional to the glucose concentration. Oxygen is measured amperometrically. The oxygen sensor is similar to a conventional Clark electrode. Oxygen permeates a gas-permeable membrane from the blood sample into an internal electrolyte solution, where it is reduced at the cathode. The oxygen-reduction current is proportional to the dissolved-oxygen concentration. Hematocrit is determined conductometrically. The measured conductivity, after correction for electrolyte concentration, is related to the hematocrit.

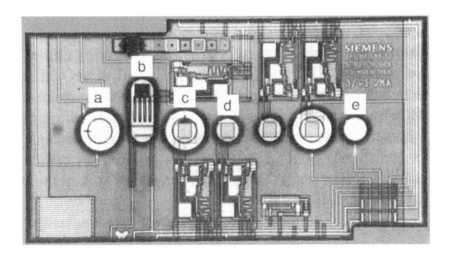

Figure 11.38 Micrograph of an integrated sensor chip combining different electrochemical principles: (a) amperometric oxygen sensor, (b) reference electrode, (c) carbon dioxide sensor (Severinghaus type pH-FET), (d) pH-FET, and (e) "pseudo" reference electrode. Reprinted from Gumbrecht et al.[470] with permission.

Acknowledgments

The authors are greatly indebted to current and former staff of the Physical Electronics Laboratory at ETH Zurich involved in chemical microsensor development. The authors would like to thank Professor Art Janata, Georgia Institute of Technology, Atlanta, Georgia, for discussions and for reading the manuscript. The authors also express their gratitude to Dr. Stephen J. Martin, Sandia National Laboratories, Albuquerque, New Mexico, and Prof. Antonio J. Ricco, Stanford University, for providing pictures and scientific material.

References

1. Göpel, W., Hesse, J., Zemel, J.N. (eds.), *Sensors: A Comprehensive Survey*, vol. 2/3, *Chemical and Biochemical Sensors*, VCH-Verlagsgesellschaft, Weinheim, 1991; Baltes, H., Göpel, W., Hesse, J. (eds.), *Sensors Updates*, VCH-Wiley, Weinheim
2. Hulanicki, A., Glab, S., Ingman, F., *Pure Appl. Chem.*, 1991, 63:1247
3. Durst, R.A., Murray, R.W., Izutsu, K., Kadish, K.M., Faulkner, L.R:, Draft IUPAC Report Commission V.5
4. Janata, J., *Principles of Chemical Sensors*, Plenum, NY, 1989; Janata, J., Huber, R.J. (eds.), *Solid State Chemical Sensors*, Academic Press, San Diego, 1985
5. Janata, J., Josowicz, M., Vanysek, P., DeVaney, M.D., *Anal. Chem.*, 1998, 179R–208R
6. Edmonds, T.E. (ed.), *Chemical Sensors*, Chapman and Hall, NY, 1988
7. Kress-Rodgers, E. (ed.), *Handbook of Biosensors and Electronic Noses*, CRC Press Inc., Boca Raton, FL, 1997
8. Madou, M.J., Morrison, S.R., *Chemical Sensing with Solid State Devices*, Academic Press, Boston, 1989
9. Gardner, J.W., Bartlett, P.N., *Electronic Noses*, Oxford University Press, 1999; Gardner, J.W., *Microsensors*, Wiley, Chichester, 1994
10. Hall, E.A., *Biosensors*, Open University Press, 1990
11. Lambrechts, M., Sansen, W., *Biosensors: Microelectrochemical Devices*, Institute of Physics Publishing, Bristol, 1992
12. Carey, W. P., Beebe, K. R., Kowalski, B. R., *Anal. Chem.*, 1986, 58:149–153
13. Ballantine, D. S., Rose, S. L., Grate, J. W., Wohltjen, H., *Anal. Chem.*, 1986, 58:3058–3066
14. Osbourn, G. C., Bartholomew, J. W., Ricco, A. J., Crooks, R. M., *Acc. Chem. Res.*, 1998, 31:297; Ricco, A. J., Crooks, R. M., Osbourn, G. C., *Acc. Chem. Res.*, 1998, 31:289
15. Hierlemann, A., Schweizer-Berberich, M., Weimar, U., Kraus, G., Pfau, A., Göpel, W., *Pattern Recognition and Multicomponent Analysis in Sensors Update* (Baltes, H., Göpel, W., Hesse, J., eds.), VCH, Weinheim, Germany, 1996

16. Massart, D.L., Vandeginste, B.G.M., Deming, S.N., Michotte, Y., Kaufman, L., *Data Handling in Science and Technology 2: Chemometrics: A Textbook*, Elsevier, Amsterdam, 1988
17. Brereton, R.G. (ed.), *Multivariate Pattern Recognition*, vol. 9 in the series *Chemometrics, Data Handling in Science and Technology*, Elsevier, Amsterdam, NY, 1992
18. Weetall, H.H., *Biosens. Bioelectron.*, 1999, 14:237–242
19. Van den Berg, A., van der Waal, P.D., van der Schoot, B.B., de Rooij, N.F., *Sens. Mater.*, 1994, 6:23–43
20. Müller, G., Deimel, P.P., Hellmich, W., Wagner, C., *Thin Solid Films*, 1997, 296:157–163
21. Bergveld, P., *IEEE Trans. Biomed. Eng.*, 1970, BME-17:70–71
22. Lundström, I., Shivaraman, S., Svensson, C., Lundkvist, L., *Appl. Phys. Lett.*, 1975, 26:55–57
23. Middelhoek, S., Audet, S.A., *Silicon Sensors*, Academic Press Inc., London, 1989; Middelhoek, S., *Sens. Actuators A*, 1994, 41–42:1–8
24. Sze, S.M. (ed.), *Semiconductor Sensors*, Wiley, NY, 1994
25. Gad-el-Hak, M., *The MEMS Handbook*, CRC Press, Boca Raton, FL, 2002
26. Kovacs, G.T.A., *Micromachined Transducers*, WCB/McGraw-Hill, NY, 1998
27. Elwenspoek, M., Hansen, H., *Silicon Micromachining*, Cambridge University Press, Cambridge, 1998
28. Madou, M., *Fundamentals of Microfabrication*, CRC Press, Boca Raton, FL, 1997
29. Muller, R.S., Howe, R.T., Senturia, S.D., Smith, R.L., White, R.M. (eds.), *Microsensors*, IEEE Press, NY, 1991
30. Tuller, H.L., Mlcak, R., *Sens. Actuators B*, 1996, 35–36:255–261
31. Picraux, S.T., McWorther, P.J., *IEEE Spectrum*, 1998, 12:24–33
32. Gardner, J.W., Varadan, V.K., Awadelkin, O.O., *Microsensor, MEMS and Smart Devices*, Wiley, NY, 2001
33. Baltes, H., Brand, O., *IEEE AES Syst. Mag.*, 1999, 14:29–34; *Proc. SPIE*, 1999, 3673:2–10
34. Zemel, J.N., *Rev. Sci. Instrum.*, 1990, 61(6):1579–1606
35. Atkins, P.W., *Physical Chemistry*, 5th edition, Oxford University Press, Oxford, 1994
36. Grate, J. W., Frye, G. C., in *Sensors Update*, vol. 2 (Baltes, H., Göpel, W., Hesse, J., eds.), VCH, Weinheim, FRG, 1996, pp. 37–83
37. Grate, J. W., Martin, S. J., White, R. M., *Anal. Chem.*, 1993, 65:940A–948A, 987A–996A
38. Ballantine, D. S., White, R. M., Martin, S. J., Ricco, A. J., Frye, G. C., Zellers, E. T., Wohltjen, H., *Acoustic Wave Sensors: Theory, Design, and Physico-Chemical Applications*, Academic Press, San Diego, 1997
39. Sauerbrey, G. Z., *Z. Phys.*, 1959, 155:206–222
40. Martin, S. J., Frye, G. C., Senturia, S. D., *Anal. Chem.*, 1994, 66:2201
41. King, W.H., *Anal. Chem.*, 1964, 36:1735
42. White, R.M., *Proc. IEEE*, 1970, 58:1238–1276
43. Nieuwenhuizen, M. S., Venema, A., *Sens. Mater.*, 1989, 5:261–300

44. Bodenhöfer, K., Hierlemann, A., Noetzel, G., Weimar, U., Göpel, W., *Anal. Chem.*, 1996, 68:2210
45. Martin, S.J., Frye, G.C., Spates, J.J., Butler, M.A., *Proc. IEEE Ultrasonics Symposium*, 1996, 1:423–434 (cat. no. 96CH35993)
46. Venema, A., *Thin Solid Films*, 1974, 22(2):S27–S29; Venema, A., Dekkers, J.M., *IEEE Trans. Microwave Theory Tech.*, 1975, MTT23(9):765–767
47. Lakin, K.M., Liu, J., Wang, K., *Proc. IEEE Ultrasonics Symposium*, 1974, pp. 302–306
48. Xia, J., Burns, S., Porter, M., Xue, T., Liu, G., Wyse, R., Thielen, C., *Proc. IEEE International Frequency Control Symposium*, 1995, pp. 879–884 (cat. no. 95CH35752)
49. Chuang, C.T., White, R.M., Bernstein, J.J., IEE*E Electron Devices Lett.*, 1982, 6:145–148
50. Martin, S.J., Schweizer, K.S., Ricco, A.J., Zipperian, T.E., *Proc. Transducers 1985*, 1985, p. 445 (cat. no. 85CH2127-9)
51. Nieuwenhuizen, M.S., Nederlof, A.J., *Sens. Actuators B*, 1992, 9:171–176
52. Brand, O., Baltes, H., "Micromachined resonant sensors," in *Sensors Update*, vol. 4 (Baltes, H., Göpel, W., Hesse, J., eds.), VCH, Weinheim, Germany, 1998
53. Polla, D.L., Muller, R.S., *IEEE Electron. Devices Lett.*, 1986, 7:254–256
54. Caliendo, C., Verardi, P., Verona, E., D'Amico, A., DiNatale, C., Saggio, G., Serafini, M., Paolesse, R., Huq, S.E., *Smart Mater. Struct.*, 1997, 6:689–699
55. Casalnuovo, S.A., Hietala, V.M., Heller, E.J., Frye-Mason, G.C., Baca, A.G., Wendt, J.R., *Technical Digest Solid-State Sensor and Actuator Workshop, Hilton Head Island*, 2000, pp. 154–157
56. Wenzel, S.W., and Hite, R.M., *IEEE Trans. Electron. Devices*, 1988, 35:735–743
57. Wenzel, S.W., White, R.M., *Appl. Phys. Lett.*, 1989, 54:1976
58. Vellekoop, M.J., Lubking, G.W., Sarro, P.M., Venema, A., *Sens. Actuators A*, 1994, 44(3):249–263
59. Wenzel, S.W., White, R.M., *Sens. Actuators A*, 1990, 21–23:700–703
60. Martin, S.J., Butler, M.A., Spates, J.J., Mitchell, M.A., Schubert, W.K., *J. Appl. Phys.*, 1998, 83:4589–4601
61. Luginbuhl, P., Collins, S.D., Racine, G.A., Grétillat, M.A., de Rooij, N.F., Brooks, K.G., Setter, N., *J. MEMS*, 1997, 6:337–346
62. Vellekop, M.J., Van Rhijn, A.J., Lubking, G.W., Venema, A., *Sens. Actuators A*, 1991, 25–27:699–703
63. Martin, B.A., Wenzel, S.W., White, R.M., *Sens. Actuators A*, 1990, 21–23:704–708
64. Schubert, W.K., Adkins, D.R., Butler, M.A., Martin, S.J., Mitchell, M.A., Kottenstette, R., Wessendorf, K.O., *Proc. ECS Meeting Honolulu, Hawaii*, 1999, 99–23:332–335
65. Grate, J.W., Wenzel, S.W., White, R.M., *Anal. Chem.*, 1991, 63:1552–1561
66. Grate, J.W., Wenzel, S.W., White, R.M., *Anal. Chem.*, 1992, 64:413–423
67. Wang, A.W., Kiwan, R., White, R.M., Ceriani, R.L., *Sens. Actuators B*, 1998, 49:13–21

68. Lange, D., Hagleitner, C., Hierlemann, A., Brand, O., Baltes, H., *Anal. Chem.*, 2002, 74:3084–3095
69. Baltes, H., Lange, D., Koll A., *IEEE Spectrum*, 1998, 9:35–38
70. Hierlemann A., Lange, D., Hagleitner, C., Kerness, N., Koll A., Brand O., Baltes, H., *Sens. Actuators B*, 2000, 70:2–11
71. Thundat, T., Chen, G.Y., Warmack, R.J., Allison, D.P., Wachter, E.A., *Anal. Chem.*, 1995, 67:519–521
72. Wachter, E.A., Thundat, T., *Rev. Sci. Instrum.*, 1995, 66:3662–3667
73. Lang, H.P., Berger, R., Battiston, F., Ramseyer J.P., Meyer E., Andreoli C., Brugger J., Vettiger P., Despont M., Mezzacasa T., Scandella L., Güntherodt H.J., Gerber C., Gimzewski J.K., *Appl. Phys. A*, 1998, 66:161–64
74. Maute M., Raible S., Prins F.E., Kern D.P., Ulmer H., Weimar U., Göpel W., *Sens. Actuators B*, 1999, 58:505–511
75. Lang, H.P., Baller, M.K., Berger, R., Gerber C., Gimzewski J.K., Battiston, F., Fornaro, P., Ramseyer J.P., Meyer E., Güntherodt H.J., *Anal. Chim. Acta.*, 1999, 393:59–65
76. Boisen, A., Thaysen, J., Jesenius, H., Hansen, O., *Ultramicroscopy*, 2000, 82:11–16
77. Jesenius, H., Thaysen, J., Rasmussen, A.A., Veje, L.H., Hansen, O., Boisen, A., *Appl. Phys. Lett.*, 2000, 76:2615–2617
78. Gimzewski, J.K., Gerber, C., Meyer, E., Schlittler, E.E., *Chem. Phys. Lett.*, 1994, 217:589–594
79. Fritz, J., Baller, M.K., Lang, H.P., Rothuizen, H., Vettiger, P., Meyer, E., Güntherodt, H.J., Gerber, C., Gimzewski, J.K., *Science*, 2000, 288:316–318
80. Berger, R., Delamarche, E., Lang, H.P., Gerber, C., Gimzewski, J.K., Meyer E., Güntherodt, H.J., *Science*, 1997, 276:2021
81. Lee, S.S., White, R.M., *Sens. Actuators A*, 1996, 52:41–45
82. Britton, C.L., Warmack, R.J., Smith, S.F., Oden, P.I., Jones, R.L., Thundat, T., Brown, G.M., Bryan, W.L., DePriest, J.C., Ericson M.N., Emery, M.S., Moore, M.R., Turner, G.W., Wintenberg, A.L., Threatt, T.D., Hu, Z., Clonts, G., Rochelle, J.M., *Proceedings 20th Anniversary Conference on Advanced Research in VLSI*, IEEE Comput. Soc, Los Alamitos, CA, 1999, pp. 359–68
83. Barnes, J.R., Stephenson, R.J., Welland, M.E., Gerber, C., Gimzewski, J.K., *Nature*, 1994, 372:79–81
84. Meijer, G.C.M., van Herwaarden, A.W., *Thermal Sensors*, Institute of Physics Publishing, Bristol, UK, 1994
85. Baker, A.R., Combustible gas-detecting, electrically heatable element, UK patent no. 892530, 1962
86. Zanini, M., Visser, J.H., Rimai, L., Soltis, R.E., Kovalchuk, A., Hoffman, D.W., Logothetis, E.M., Bonne, U., Brewer, L., Bynum, O.W., Richard, M.A., *Sens. Actuators A*, 1995, 48:187–192
87. Manginell, R.P., Smith, J.H., Ricco, A.J., Moreno, D.J., Hughes, R.C., Huber, R.J., Senturia, S.D., *Technical Digest,Solid State Sensor and Actuator Workshop, Hilton Head Island,* 1996, pp. 23–27
88. Manginell, R.P., Smith, J.H., Ricco, A.J., *Proc. 4th Annual Symposium on Smart Structures and Materials*, SPIE, 1997, pp. 273–284

89. Dabill, D.W., Gentry, S.J., Walsh, P.T., *Sens. Actuators*, 1987, 11:135–143
90. Jones, M.G., Nevell, T.G., *Sens. Actuators*, 1989, 16:215–224
91. Aigner, R., Dietl, M., Katterloher, R., Klee, V., *Sens. Actuators B*, 1996, 33:151–155
92. Aigner, R., Auerbach, F., Huber, P., Mueller, R., Scheller, G., *Sens. Actuators B*, 1994, 18–19:143–147
93. Krebs, P., Grisel, A., *Sens. Actuators B*, 1993, 13–14:155–158
94. Gall, M., *Sens. Actuators B*, 1991, 4:533–538
95. Gall, M., *Sens. Actuators B*, 1993, 15–16:260–264
96. Mastrangelo, C.H., "Thermal application of microbridges," Ph.D. thesis, UC Berkeley, 1991
97. Accorsi, A., Delapierre, G., Vauchier, C., Charlot, D., *Sens. Actuators B*, 1991, 4:539–543
98. Vandrish, G., *Key Eng. Mater.*, 1996, 122–124:225–232
99. Xie, B., Danielson, B., Norberg, P., Winquist, F., Lundstroem, I., *Sens. Actuators B*, 1992, 6(1–3):127–130
100. Danielsson, B., Mattiasson, B., Mosbach, K., *Appl. Biochem. Bioeng.*, 1981, 3:97–143
101. Xie, B., Danielson, B., *Technical Digest, Transducers 1995, Stockholm, Sweden*, pp. 501–503
102. Lysenko, V., Delhomme, G., Soldatkin, A., Strikha, V., Dittmar, A., Jaffrezic-Renault, N., Martelet, C., *Talanta*, 1996, 43:1163–1169
103. Xie, B., Danielson, B., Winquist, F., *Sens. Actuators B*, 1993, 15–16:443–447
104. Shimohgoshi, M., Karube, I., *Sens. Actuators B*, 1996, 30:17–21
105. Polla, D.L., White, R.M., Muller, R.S., *IEEE Digest of Technical Papers, Transducers 1985*, pp. 33–36
106. Dorojkine, L.M., Mandelis, A., *Opt. Eng.*, 1997, 36(2):473–481
107. Van Herwaarden, A.W., Sarro, P.M., *Sens. Actuators*, 1986, 10:321–346
108. Sarro, P.M., van Herwaarden, A.W., van der Vlist, W., *Sens. Actuators A*, 1994, 41–42:666–671
109. Van Herwaarden, A.W., Sarro, P.M., Gardner, J.W., Bataillard, P., *Sens. Actuators A*, 1994, 43:24–30
110. Kerness, N., Koll, A., Schaufelbuehl, A., Hagleitner, C., Hierlemann, A., Brand, O., Baltes, H., *Proc. IEEE Workshop on Micro Electro Mechanical Systems MEMS 2000, Myazaki, Japan*, 2000, pp. 96–101, ISBN 0-7803-5273-4
111. Koll, A., Schaufelbühl, A., Brand, O., Baltes, H., Menolfi, C., Huang, H., *Proc. IEEE Workshop on Micro Electro Mechanical Systems MEMS 99, Orlando, FL*, 1999, pp. 547–551, ISBN 0-7803-5194-0
112. Lerchner, J., Seidel, J., Wolf, G., Weber, E., *Sens. Actuators B*, 1996, 32:71–75
113. Caspary, D., Schröpfer, M., Lerchner, J., Wolf, G., *Thermochim. Acta*, 1999, 337:19–26
114. Lerchner, J., Wolf, A., Wolf, G., J., *Thermal. Anal.*, 1999, 55:212–223

115. Bataillard, P., Steffgen, E., Haemmerli, S., Manz, A., Widmer, H.M., *Biosens., Bioelectron.*, 1993, 8:89–98
116. Köhler, J.M., Kessler, E., Steinhage, G., Gründig, B., Cammann, K., *Mikrochim. Acta*, 1995, 120:309–319
117. Sarro, P.M., Yashiro, H., van Herwaarden, A.M., Middelhoek, S., *Sens. Actuators*, 1988, 14:191–201
118. Van Herwaarden, A.W., *Meas. Sci.Technol.*, 1992, 3:935–937
119. Papadopoulos, C.A., Vlachos, D.S., Avaritsiotis, J.N., *Sens. Actuators B*, 1996, 34:524–527
120. Swart, N.R., Stevens, M., Nathan, A., Karanassios, V., *Sens. Mater.*, 1997, 9(6):387–394
121. Kimura, M., Manaka, J., Satoh, S., Takano, S., Igarashi, N., Kazutoshi, N., *Sens. Actuators B*, 1995, 24–25:857–860
122. Wolfbeis, O., Boisde, G.E., Gauglitz, G., "Optochemical sensors," in *Sensors: A Comprehensive Survey*, vol. 2 (Göpel, W., Hesse, J., Zemel, J.N., eds.), VCH-Verlagsgesellschaft, Weinheim, 1991, pp. 573–646
123. Gauglitz, G., "Opto-chemical and opto-immuno sensors," in *Sensors: A Comprehensive Survey, Update*, vol. 1 (Göpel, W., Hesse, J., Zemel, J.N., eds.), VCH-Verlagsgesellschaft, Weinheim, 1996, pp. 1–49
124. Brecht, A., Gauglitz, G., Göpel, W., "Sensors in Biomolecular Interaction Analysis and Pharmaceutical Drug Screening," in *Sensors: A Comprehensive Survey, Update*, vol. 3 (Göpel, W., Hesse, J., Zemel, J.N., eds.), VCH-Verlagsgesellschaft, Weinheim, 1998, pp. 573–646
125. Mulvaney, S.P., Keating, C.D., *Anal. Chem.*, 2000, 72:145R–158R
126. http://www.affymetrix.com, http://www.apb.com, http://www.cepheid.com
127. Blum, L.J., ed., *Bio- and Chemoluminescent Sensors*, World Scientific, Singapore, 1997
128. Azzam, R.M.A., Bashara, N.M., *Ellipsometry and Polarized Light*, North Holland, NY, 1988
129. Jin, G., Tengvall, P., Lundström, I., Arwin, H., *Anal. Biochem.*, 1995, 232:69–72
130. Gauglitz, G., Brecht, A., Kraus, G., Nahm, W., *Sens. Actuators B*, 1993, 11:21
131. Piehler, J., Brecht, A., Gauglitz, G., *Anal. Chem.*, 1996, 68:139
132. Klotz, A., Brecht, A., Barzen, C., Gauglitz, G., Harris, R.D., Quigley, Q. R., Wilkinson, J. S., *Sens. Actuators B*, 1998, 51:181–187
133. Lukosz, W., *Sens. Actuators B*, 1995, 29:37–50
134. Kretschmann, E., Raether, H., *Z. Naturf.*, 1968, 23a:2135
135. *Sens. and Actuators B, Special Issue Devoted to SPR Sensors* (Yee, S.S., ed.), 1999, vol. 54 (1–2), and articles therein
136. Homola, J., Yee, S.S., Gauglitz, G., *Sens. Actuators B*, 1999, 54:3–15
137. Ehrfeld, W., Bauer, H.D., *Proc. SPIE*, 1998, 3276:2–14
138. Spichiger-Keller, U.E., *Chemical Sensors and Biosensors for Medical and Biological Application*, Wiley-VCH, Weinheim, 1998
139. Klimant, I., Kühl, M., Glud, R.N., Holst, G., *Sens. Actuators B*, 1997, 38–39:29–37

140. Holst, G., Glud, R.N., Kühl, M., Klimant, I., *Sens. Actuators B*, 1997, 38–39:122–129
141. Wolfbeis, O.S., *Anal. Chem.*, 2000, 72:81R–90R
142. Dakin, J., Culshaw, B., eds., *Optical Fiber Sensors*, vols. 3 and 4, Artech House, Norwood, MA, 1997
143. Boisde, G.E., Harmer, A., eds., *Chemical and Biochemical Sensing with Optical Fibers and Waveguides*, Artech House, Norwood, MA, 1996
144. Dickinson, T.A., White, J., Kauer, J.S., Walt, D., *Nature*, 1996, 382:697–700
145. White, J., Kauer, J.S., Dickinson, T.A., Walt, D., *Anal. Chem.*, 1996, 68:2191–2202
146. Hafeman, D.G., Parce, J.W., McConell, H.M., *Science*, 1988, 240:1182–1185
147. Sze, S.M., *Physics of Semiconductor Devices*, Wiley, NY, 2nd edition, 1981
148. Lukosz, W., Stamm, C., Moser, H.R., Ryf, R., Dübendorfer, J., *Sens. Actuators B*, 1997, 38–39:316–323
149. Tiefenthaler, K., Lukosz, W., *J. Opt. Soc. Am. B*, 1989, 6:209–220; *Thin Solid Films*, 1985, 126:205–211
150. Lukosz, W., Nellen, P.M., Stamm, C., Weiss, P., *Sens. Actuators B*, 1990, 1:585–588
151. Nellen, P.M., Lukosz, W., *Sens. Actuators B*, 1990, 1:592–596
152. Brosinger, F., Freimuth, H., Lacher, M., Ehrfeld, W., Gedig, E., Katerkamp, A., Spener, F., Cammann, K., *Sens. Actuators B*, 1997, 44:350–355
153. Busse, S., Käshammer, J., Krämer, S., Mittler, S., *Sens. Actuators B*, 1999, 60:148–154
154. Schipper, E.F., Brugman, A.M., Dominguez, C., Lechuga, L.M., Kooyman, R.P., Greve, J., *Sens. Actuators B*, 1997, 40:147–153
155. Tiefenthaler, K., Lukosz, W., *Sens. Actuators*, 1988, 15:273–284
156. Lukosz, W., Stamm, C., *Sens. Actuators A*, 1991, 25:185–188
157. Stamm, C., Lukosz, W., *Sens. Actuators B*, 1993, 11:177–181
158. Heideman, R., Kooyman, R.P., Greve, J., *Sens. Actuators B*, 1993, 10:209–217
159. Äyräs, P., Honkanen, S., Grace, K.M., Shrouf, K., Katila, P., Leppihalme, M., Tervonen, A., Yang, X., Swanson, B., Peyghambarian, N., *Pure Appl. Opt.*, 1998, 7:1261–1271
160. Grace, K.M., Shrouf, K., Honkanen, S., Äyräs, P., Katila, P., Leppihalme, M., Johnston, R.G., Yang, X., Swanson, B., *Electron. Lett.*, 1997, 33:1651–1653
161. Gauglitz, G., Ingenhoff, J., Fres. J., *Anal. Chem.*, 1994, 349:355–359
162. Puyol, M., del Valle, M., Garces, I., Villuendas, F., Dominguez, C., Alonso, J., *Anal. Chem.*, 1999, 71:5037–5044
163. Heideman, R.G., Veldhuis, G.J., Jager, E.W.H., Lambeck, P.V., *Sens. Actuators B*, 1996, 35–36:234–240
164. Heideman, R.G., Lambeck, P.V., *Sens. Actuators B*, 1999, 61:100–127
165. Maisenhölder, B., Zappe, H.P., Kunz, R.E., Moser, M., Riel, P., *Technical Digest Transducers Chicago*, 1997, pp. 79–80

166. Maisenhölder, B., Zappe, H.P., Kunz, R.E., Riel, P., Moser, M., Edlinger, J., *Sens. Actuators B*, 1997, 38–39:324–329
167. Maisenhölder, B., Zappe, H.P., Moser, M., Riel, P., Kunz, R.E., Edlinger, J., *Electron. Lett.*, 1997, 33:986–988
168. Hofstetter, D., Zappe, H.P., Epler, J.E., Riel, P., *IEEE Photonics Technol. Lett.*, 1995, 7:1022–1024
169. Hofstetter, D., Zappe, H.P., Epler, J.E., Söchtig, J., *Electron. Lett.*, 1994, 30:1858–1859
170. Zappe, H.P., Hess, M., Moser, M., Hövel, R., Gulden, K., Gauggel, H.P., Monti di Sopra, F., *Appl. Opt.*, 2000, 39:2475–2479
171. Zappe, H.P., *Sensors*, 2000, 16(9):114–117
172. Jerman, J.H., Clift, D.J., *Technical Digest Transducers*, 1991, pp. 372–375
173. Grasdepot, F., Alause, H., Knap, W., Malzac, J.P., Suski, J., *Sens. Actuators B*, 1996, 35–36:377–380
174. Alause, H., Grasdepot, F., Malzac, J.P., Knap, W., Hermann, J., *Sens. Actuators B*, 1997, 43:18–23
175. Silveira, J.P., Anguita, J., Briones, F., Grasdepot, F., Bazin, A., *Sens. Actuators B*, 1998, 48:305–307
176. Correia, J.H., de Graaf, G., Kong, S.H., Bartek, M., Wolffenbuttel, R.F., *Sens. Actuators A*, 2000, 82:191–197
177. Aratani, K., French, P.J., Sarro, P.M., Poenar, D., Wolffenbuttel, R.F., Middelhoek, S., *Sens. Actuators A*, 1994, 43:17–23
178. De Graaf, G., Wolffenbuttel, R.F., *Sens. Actuators A*, 1998, 67:115–119
179. Han, J., *Appl. Phys. Lett.*, 1999, 74:445–447
180. Han, J., Neikirk, D.P., Clevenger, M., McDevitt, J.T., *SPIE Proceedings 1996*, 2881:171–179
181. Chen, T., *Sens. Actuators B*, 1993, 13–14:284–287
182. Carson, H., *Sens. Rev.*, 1997, 17:304–306; http://www.vaisala.com
183. Yee, G.M., Maluf, N.I., Hing, P.A., Albin, M., Kovacs, G.T., *Sens. Actuators A*, 1997, 58:61–66
184. Lee, S.S., Lin, L.Y., Wu, M.C., *Appl. Phys. Lett.*, 1995, 67:2135–2137
185. Yee, G.M., Maluf, N.I., Kovacs, G.T., *Technical Digest Transducers Sendai*, 1999, pp. 1882–1883
186. Kwa, T.A., Wolffenbuttel, R.F., *Sens. Actuators A*, 1992, 31:259–266
187. Sinclair, M.B., Butler, M.A., Kravitz, S.H., Zubrizycki, W.J., Ricco, A.J., *Opt. Lett.*, 1997, 22:1036–1038
188. Simpson, M., Sayler, G., Nivens, D., Ripp, S., Paulus, M., Jellison, G., *Trends Biotechnol.*, 1998, 16:332–338
189. Simpson, M., Paulus, M., Jellison, G., Sayler, G., Applegate, B., Ripp, S., Nivens, D., *Technical Digest Solid-State Sensor and Actuator Workshop, Hilton Head*, 1998, pp. 354–357
190. Simpson, M., Sayler, G., Nivens, D., Ripp, S., Paulus, M., Jellison, G., SPIE Proc., 1998, 3328:202–212
191. Eres, G., Lowndes, D.H., U.S. patent no. 5 110 790, 1992
192. Liedberg, B., Nylander, C., Lundström, I., *Sens. Actuators*, 1983, 4:299–304

193. Garabedian, R., Gonzalez, C., Richards, J., Knoesen, A., Spencer, R., Collins, S.D., Smith, R.L., *Sens. Actuators A*, 1994, 43:202–207
194. Richards, J.D., Garabedian, R., Gonzalez, C., Knoesen, A., Smith, R.L., Spencer, R., Collins, S.D., *Appl. Opt.*, 1993, 32:2901–06
195. http://www.ti.com/sc/docs/products/analog/tspr1a150100.html; http://www.ti.com/spreeta
196. Widmer, H., *Anal. Methods Instrum.*, 1993, 1:60–72
197. Vasiliev, A., Moritz, W., Fillipov, V., Bartholomäus, L., Terentjev, A., Gabusjan, T., *Sens. Actuators B*, 1998, 49:133–138
198. Baranzahi, A., Spetz, A., Lundström, I., *Sens. Actuators B*, 1995, 26–27:165–169
199. Svennigstorp, H., Tobias, P., Lundström, I., Salomonsson, P., Martensson, P., Ekedahl, L.G., Spetz, L.A, *Sens. Actuators B*, 1999, 57:159–165
200. Samman, A., Gebremariam, S., Rimai, L., Zhang, X., Hangas, J., Auner, G.W., *Sens. Actuators B*, 2000, 63:91–102
201. Fabry, P., Siebert, E., "Electrochemical sensors," in *The GRC Handbook of Solid State Electrochemistry* (Gellings, P.J., Bouwmeester, H.J.M., eds.), CRC Press, Boca Raton, FL, 1997
202. Wollenberger, U., "Electrochemical biosensors – ways to improve sensor performance," in *Biotechnology and Engineering Reviews*, vol. 13, Artech House, Norwood, UK, 1996, pp. 237–266
203. Kas, J., Marek, M., Stastny, M., Volf, R., *Bioelectrochem. Princ. Pract.*, 1996, 3:361–453
204. Kauffman, J.M., ed., *Special Issue Bioelectrochem. Bioeng.*, 1997, p. 42
205. Schultze, J.W., Tsakova, V., *Electrochim. Acta*, 1999, 44:3605–3627
206. Neuman, M.R., Buck, R.P., Cosofret, V., Lindner, E., Liu, C.C., *IEEE Eng. Medicine Biol.*, 1994, 13:409–419
207. Buss, G., Schöning, M.J., Lüth, H., Schultze, J.W., *Electrochim. Acta*, 1999, 44:3899–3910
208. Schmitt, G., Schulze, J.W., Fassbender, F., Buss, G., Lüth, H., Schönung, M.J., *Electrochim. Acta*, 1999, 44:3865–3883
209. Sreenivas, G., Ang, S.S., Fritsch, I., Brown, W.D., Gerhadt, G.A., Woodward, D.J., *Anal. Chem.*, 1996, 68:1858–1864
210. Wittkampf, M., Chemnitius, G.C., Cammann, K., Rospert, M., Mokwa, W., *Sens. Actuators B*, 1997, 43:40–44
211. Desmond, D., Lane, B., Alderman, J., Hall, G., Alvarez-Icaza, M., Garde, A., Ryan, J., Barry, L., Svehla, G., Arrigan, D.W., Schniffner, L., *Technical Digest Transducers Stockholm*, 1995, 2:948–951
212. Kovacs, G.T.A., Storment, C.W., Kounaves, S.P., *Sens. Actuators B*, 1995, 23:41–47
213. Belmont, C., Tercier, M.L., Buffle, J., Fiaccabrino, G.C., Koudelka-Hep, M., *Anal. Chim. Acta*, 1996, 329:203–214
214. Herdan, J., Feeney, R., Kounaves, S.P., Flannery, A.F., Storment, C.W., Kovacs, G.T.A., *Environ. Sci. Technol.*, 1998, 32:131–136
215. Zhou, Z.B., Wu, Q.H., Liu, C.C., *Sens. Actuators B*, 1994, 21:101–108
216. Maseeh, F., Tierney, M.J., Chu, W.S., Joseph, J., Kim, H.O., Otagawa, T., *Technical Digest Transducers*, 1991, pp. 359–362

217. Buttner, W.J., Maclay, G.J., Stetter, J.R., *Sens. Actuators B*, 1990, 1:303–307
218. Stetter, J.R., Hesketh, P.J., Maclay, G.J., *Proc. Symposium on Chemical and Biological Sensors and Analytical Methods, 1997, ECS-Proceedings*, 97–19:70–77
219. Van den Berg, A., van der Waal, P.D., van der Schoot, B.H., de Rooij, N.F., *Sens. Mater.*, 1994, 6:23–43
220. Steinkuhl, R., Sundermeier, C., Hinkers, H., Dumschat, C., Cammann, K., Knoll, M., *Sens. Actuators B*, 1996, 33:19–24
221. Yon Hin, B.F.Y., Sethi, R.S., Lowe, C.R., *Sens. Actuators B*, 1990, 1:550–554
222. Schwake, A., Ross, B., Cammann, K., *Sens. Actuators B*, 1998, 46:242–248
223. Dilhan, M., Estève, D., Gué, A.M., Mauvais, O., Mercier, L., *Sens. Actuators B*, 1995:26–27, 401–403
224. Van den Berg, A., Grisel, A., Verney-Norberg, E., van der Schoot, B.H., Koudelka-Hep, M., de Rooij, N.F., *Sens. Actuators B*, 1993, 13–14:396–399
225. Wittkampf, M., Cammann, K., Amrein, M., Reichelt, R., *Sens. Actuators B*, 1997, 40:79–84
226. Sansen, W., de Wachter, D., Callewaert, L., Lambrechts, M., Claes, A., *Sens. Actuators B*, 1990, 1:298–302
227. Koudelka, M., *Sens. Actuators*, 1986, 9:249
228. Hinkers, H., Sundermeier, C., Lürick, R., Walfort, F., Cammann, K., Knoll, M., *Sens. Actuators B*, 1995, 26–27:398–400
229. Hinkers, H., Hermes, T., Sundermeier, C., Borchardt, M., Dumschat, C., Bücher, S., Bühner, M., Cammann, K., Knoll, M., *Sens. Actuators B*, 1995, 24–25:300–303
230. Hermes, T., Bühner, M., Bücher, S., Sundermeier, C., Dumschat, C., Borchardt, M., Cammann, K., Knoll, M., *Sens. Actuators B*, 1994, 21:33–37
231. Zhu, J., Wu, J., Tian, C., Wu, W., Zhang, H., Lu, D., Zhang, G., *Sens. Actuators B*, 1994, 20:17–22
232. Van den Berg, A., Grisel, A., Koudelka, M., van der Schoot, B.H., *Sens. Actuators B*, 1991, 5:71
233. Clark, L.C., *Trans. Am. Soc. Artif. Intern. Organs*, 1956, 2:41–48
234. Schnakenberg, U., Lisec, T., Hintsche, R., Kuna, I., Uhlig, A., Wagner, B., *Sens. Actuators B*, 1996, 34:476–480
235. Hintsche, R., Kruse, C., Uhlig, A., Paeschke, M., Lisec, T., Schnakenberg, U., Wagner, B., *Sens. Actuators B*, 1995, 26–27:471–473
236. Nernst, W., *Z. Phys. Chem.*, 1904, 47:52
237. Currie, J.F., Essalik, A., Marusic, J.C., *Sens. Actuators B*, 1999, 59:235–241
238. Goldberg, H.D., Brown, R.E., Liu, D.P., Meyerhoff, M.E., *Sens. Actuators B*, 1994, 21:171–183
239. Lauks, I., van der Spiegel, J. Sansen, W., Steyaert, M., *Technical Digest Transducers, Philadelphia*, 1985, pp. 122–124
240. Lang, T., Caron, M., Izquierdo, R., Ivanov, D., Currie, J.F., Yelon, A., *Sens. Actuators B*, 1996, 31:9–12
241. Knoll, M., Cammann, K., Dumschat, C., Sundermeier, C., Eshold, J., *Sens. Actuators B*, 1994, 18–19:51–55
242. Taillades, G., Valls, O., Bratov, A., Dominguez, C., Pradel, A., Ribes, M., *Sens. Actuators B*, 1999, 59:123–127

243. Lundström, I., *Sens. Actuators B*, 1996, 56:75–82
244. Ekedahl, L.G., Eriksson, M., Lundström, I., *Acc. Chem. Res.*, 1998, 31:249–256
245. Hughes, R.C., Schubert, W.K., Zipperian, T.E., Rodriguez, J.L., Plut, T.A., *J. Appl. Phys.*, 1987, 62:1074–1083
246. Spetz, A., Armgarth, M., Lundström, I., *J. Appl. Phys.*, 1988, 64:1274–1283
247. Spetz, A., Winquist, F., Sundgren, H., Lundström, I., in *Gas Sensors* (Sberveglieri, G., ed.), Kluwer, Dordrecht, 1992, pp. 219–279
248. Thomas, R.C., Hughes, R.C., *J. Electrochem. Soc.*, 1997, 144:3245–3248
249. Morita, Y., Nakamura, K., Kim, C., *Sens. Actuators B*, 1996, 33:96–99
250. Winquist, F., Spetz, A., Armgarth, M., Nylander, C., Lundström, I., *Appl. Phys. Lett.*, 1983, 43:839–841
251. Winquist, F., Lundström, I., Danielson, B., *Anal. Chem.*, 1986, 58:45–148
252. Armgarth, M., Nylander, C., *Appl. Phys. Lett.*, 1981, 39:91–92
253. Barker, P.S., Monkman, A.P., Petty, M.C., Pride, R., *Synthetic Metals*, 1997, 85:1365–1366; *Sens. Actuators B*, 1995, 24, 451
254. Domansky, K., Baldwin, D.L., Grate, J.W., Hall, T.B., Li, J., Josowicz, M., Janata, J., *Anal. Chem.*, 1998, 70:473–481; Polk, P.J., Smith, A.J., De Weerth, S.P., Zhou, Z., Janata, J., Domansky, K., *Electroanalysis*, 1999, 11:707–711
255. Gupta, R.P., Gergintschew, Z., Schipanski, D., Vyas, P.D., *Sens. Actuators B*, 1999, 56:65–72; *Sens. Actuators B*, 2000, 63:35–41
256. Moss, S.D., Janata, J., Johnson, C.C., *Anal. Chem.*, 1975, 47:2238–2234
257. Kelly, R.G., *Electrochim. Acta*, 1977, 22:1–8
258. Bergveld, P., de Rooij, N.F., *Sens. Actuators*, 1981, 1:5–15
259. Kelly, R.G., Owen, A.E., *IEE Proc.*, 1985, 132:227–236
260. Katsube, T. Lauks, I., Zemel, J.N., *Sens. Actuators*, 1982, 2:399
261. Desmond, D., Lane, B., Alderman, J., Glennon, J.D., Diamond, D. Arrigan, D.W., *Sens. Actuators B*, 1997, 44:389–396
262. Dumschat, C., Felcmann, C., Pötter, W., Cammann, K., *Technical Digest, Transducers, Stockholm*, vol. 2, 1995, pp. 970–972
263. Van den Berg, A., Grisel, A., van den Vlekkert, H.H., de Rooij, N.F., *Sens. Actuators B*, 1990, 1:425–432
264. Bousse, L.J., Bergveld, P., Geeraedts, H.J., *Sens. Actuators*, 1986, 9:179–197
265. Siu, W.M., Cobbold, R.S., *IEEE Trans. Electron. Devices*, 1979, ED-26:1805–1815
266. Buck, R.P., Hackleman, D.E., *Anal. Chem.*, 1977, 49:315–2321
267. Janata, J., *Sens. Actuators*, 1983, 4:255–265
268. Lauks, I., *Sens. Actuators*, 1981, 1:261–288
269. Bousse L., Bergveld P., S*ens. Actuators*, 1984, 6:65
270. Van den Berg A., Bergveld P., Reinhoudt D.N., Sudholter E.J.R., *Sens. Actuators*, 1985, 8:129
271. Martinoia, S., Lorenzeli, L., Massobrio, G., Margesin, B., Lui, A., *Sens. Mater.*, 1999, 11:279–295

272. Yeow, T.C., Haskard, M.R., Mulcahy, D.E., Seo, H.I., Kwon. D.H., *Sens. Actuators B*, 1997, 44:434–440
273. Hoffmann, W., Rapp, R., *Sens. Actuators B*, 1996, 34:471–475
274. Rapp, R., Hoffmann, W., Süss, W., Ache, H.J., Gölz, H., *Electrochim. Acta*, 1997, 42:3391–3398
275. Cané, C., Götz, A., Merlos, A., Gràcia, I., Errachid, A., Losantos, P., Lora-Tamayo, E., *Sens. Actuators B*, 1996, 35–36:136–140
276. Sakai, T., Amemiya, I., Uno, S., Katsura, M., *Sens. Actuators B*, 1990, 1:341–344
277. Cambiaso, A., Chiarugi, S., Grattarola, M., Lorenzelli, L., Lui, A., Margesin, B., Martinoia, S., Zanini, V., Zen, M., *Sens. Actuators B*, 1996, 34:245–251
278. Bausells, J., Carrabina, J., Errachid, A., Merlos, A., *Sens. Actuators B*, 1999, 57:56–62
279. Gimmel, P. Schierbaum, K.D., Göpel, W., van den Vlekkert, H.H., de Rooij, N.F., *Sens. Actuators B*, 1990, 1:345–349
280. Olthuis, W., Luo, J., van der Schoot, B., Bomer, J.G., Bergveld, P., *Sens. Actuators B*, 1990, 1:416–420
281. Hendrikse, J., Olthuis, W., Bergveld, P., *Sens. Actuators B*, 1998, 53:97–103
282. Johnson, C.C., Moss, D., Janata, J., U.S. patent no. 4020830, 1977
283. Van der Wal, P., Skowrońska-Ptasińska, M., van den Berg, A., Bergveld, P., Sudholter, E.J.R., Reinhoudt, D.N., *Anal. Chim. Acta*, 1990, 231:41
284. Blackburn, G.F., Janata , J., *J. Electrochem. Soc.*, 1982, 129:2580
285. Harrison, D.J., Teclemariam, A., Cunningham, L., *Anal. Chem*, 1989, 61:246
286. Sudholter, E.J.R., van der Wal, P., Skowrońska-Ptasińska, M., van den Berg, A., Bergveld, P., Reinhoudt, D.N., *Anal. Chim. Acta*, 1990, 230:59
287. Brzózka, Z., Holterman, H.A.J., Honig, G.W.N., Verkerk, U.H., van den Vlekkert, H.H., Engbersen, J.F.J., Reinhoudt, D.N., *Sens. Actuators B*, 1994, 18–19:38
288. Reinhoudt, D.N., Engbersen, J.F.J., Brzózka, Z., van den Vlekkert, H.H., Honig, G.W.N., Holterman, H.A.J., Verkerk, U.H., *Anal. Chem.*, 1994, 66:3618
289. Van der Wal, P.D., Sudholter, E.J.R., Reinhoudt, D.N., *Anal. Chim. Acta*, 1991, 245:159
290. Park, L.S., Hur, Y.J., Sohn, B.K., *Sens. Actuators B*, 1996, 57:239–243
291. Brunink, J.A.J., Haak, J.R.,Bomer, J.G., Reinhoudt, D.N., McKervey, M.A., Harris, S.J., *Anal. Chim. Acta*, 1991, 254:75
292. Brunink J.A.J., Bomer J.G., Engbersen J.F.J., Verboom W., Reinhoudt D.N., *Sens. Actuators B*, 1993, 15–16:195
293. Tsukuda, K., Miyahara, Y., Shibata, Y., Miyagi, H., *Technical Digest Transducers*, 1991, pp. 218–221
294. Brzózka, Z., Cobben, P.L.H.M., Reinhoudt, D.N., Edema, J.J.H., Buter, J., Kellogg, R.M., *Anal. Chim. Acta*, 1993, 273:139
295. Cobben, P.L.H.M., Egberink, R.J.M., Bomer, J.G., Bergveld, P., Verboom, W., Reinhoudt, D.N., *J. Am. Chem. Soc.*, 1992, 114:10573

296. Cobben, P.L.H.M., Egberink, R.J.M., Bomer, J.G., Haak, J.R., Bergveld, P., Reinhoudt, D.N., *Sens. Actuators B*, 1992, 6:304
297. Battilotti, M., Mercuri, R., Mazzamuro, G., Giannini, I., Giongo, M., *Sens. Actuators B*, 1990, 1:438–440
298. Dumschat, C., Fromer, R., Rautschek, H., Müller, H., Timpe, H.-J., *Anal. Chim. Acta*, 1991, 243:179; *Sens. Actuators B*, 1990, 2:271–276
299. Rocher, V., Jaffrezic-Renault, N., Perrot, H., Chevalier, Y., Le Perchec, P., *Anal. Chim. Acta*, 1992, 256:251
300. Antonisse, M.G., Snellink-Ruël, B.H., Lugtenberg,R.J., Engbersen,J.F., van den Berg, A., Reinhoudt, D.N., *Anal. Chem.*, 2000, 72:343–348
301. Caras, S.D., Janata, J., *Technical Digest Transducers*, 1985, pp. 158–161
302. Colapicchioni,, C., Barbaro, A., Porcelli, F., *Sens. Actuators B*, 1992, 13:202–207
303. Kuriyama, T., *Technical Digest Transducers, Stockholm*, 1995, vol. 1, pp. 447–449
304. Katsube, T., Katoh, M., Maekawa, H., Hara, M., Yamaguchi, S., Uchida, N., Shiomura, T., *Sens. Actuators B*, 1990, 1:504–507
305. Kimura, J., Murakami, T., Kuriyama, T., Karube, I., *Sens. Actuators*, 1988, 15:435–443
306. Janata, J., Huber, R.J., Cohen, R., Kolesar, E.J., *Aviat. Space Environ. Med.*, 1981, 52:666–671
307. Campanella, L., Colapicchioni, C., Favero, G., Sammartino, M.P., Tomassetti, M., *Sens. Actuators B*, 1996, 33:25–33
308. Wan, K., Chovelon, J.M., Jaffrezic-Renault, N., Soldatkin, A.P., *Sens. Actuators B*, 1999, 58:399–408
309. Dumschat, C., Müller, H., Stein, K., Schwedt, G., *Anal. Chim. Acta*, 1991, 252:7–9
310. Vianello, F., Stefani, A., DiPaolo, M.L., Rigo, A., Lui, A., Margesin, B., Zen, M., Scarpa, M., Soncini, G., *Sens. Actuators B*, 1996, 37:49–54
311. Baumann, W.H., Lehmann, M., Schwinde, A., Ehret, R., Brischwein, M., Wolf, B., *Sens. Actuators B*, 1999, 55:77–89
312. Fromherz, P., Offenhäuser, A., Vetter, T., Weis, J., *Science*, 1991, 252:1290
313. Offenhäuser, A., Sprössler, C., Matsuszawa, M., Knoll, W., *Biosens. Bioelectron.*, 1997, 12:819–826
314. Schöning, M.J., Schütz, S., Schroth, P., Weissbecker, B., Steffen, A., Kordos, P., Hummel, H.E., Lüth, H., *Sens. Actuators B*, 1998, 47:235–238
315. Comte, P.A., Janata, J., *Anal. Chim. Acta*, 1978, 101:247–252
316. Van den Berg, A., Grisel, A., van den Vlekkert, H.H., de Rooij, N.F., *Sens. Actuators B*, 1990, 1:395–400, 425–252
317. Bergveld, P., van den Berg, A., van der Wal, P.D., Skowronska-Ptasinska, M., Sudhölter, E.J., *Sens. Actuators*, 1989, 18:309–327
318. Matsuo, T., Nakajiama, H., *Sens. Actuators,* 1984, 5:425–435
319. Cané, C., Gràcia, I., Merlos, A., Lozano, M., Lora-Tamayo, E., Esteve, J., *Technical Digest Transducers*, 1991, pp. 225–228
320. Skowroñska-Ptasiñska M., van der Wal P., van den Berg A., Bergveld P., Sudholter E.J.R., Reinhoudt, D.N., *Anal. Chim. Acta*, 1990, 230:67

321. Baccar, Z.M., Jaffrezic-Renault, N., Martelet, C., Jaffrezic, H., Marest, G., Plantier, A., *Sens. Actuators B*, 1996, 32:101–105
322. Wilhelm, D., Voigt, H., Treichel, W., Ferretti, R., Prasad, S., *Sens. Actuators B*, 1991, 4:145–149
323. Shin, J.H., Lee, H.J., Kim, C.Y., Oh, B.K., Rho, K.L., Nam, H., Cha, G.S., *Anal. Chem.*, 1996, 68:3166–3172
324. Nakajima, H., Esashi, M., Matsuo, T., *J. Electroanal. Chem.*, 1982, 129:141
325. Fujihira, M., Fukui, M., Osa, T., *J. Electroanal. Chem.*, 1980, 106:413
326. Tahara, S., Yoshii, M., Oka, S., *Chem. Lett.*, 1982, 3:307–310
327. Matsuo, T., Nakajima, H., *Sens. Actuators*, 1984, 5:293
328. Kimura, J., Kuriyama, T., Kawana, Y., *Sens. Actuators*, 1980, 9:373
329. Steele, M.C., MacIver, B.A., *Appl. Phys. Lett.*, 1976, 28:687–788
330. Shivamaran, M.S., Lundström, I., Svensson, C., Hammarsten, H., *Electron. Lett.*, 1976, 12:483–484
331. Ito, K., *Surface Sci.*, 1979, 86:345–352
332. Yamamoto, M., Tonomura, S., Matsuoka, T., Tsubomura, H., *Surface Sci.*, 1980, 92:400–406
333. Liu, C.C., Materials, *Chem. Phys.*, 1995, 42:87–90
334. Wu, Q.H., Lee, K.M., Liu, C.C., *Sens. Actuators B*, 1993, 13–14:1
335. Zemel, J., U.S. patent no. 4103227, 1978
336. Wen, C.C., Lauks, I., Zemel, J.N., *Thin Solid Films*, 1980, 70:333
337. Dura, H.G., Schöneberg, U., Mokwa, W., Hostika, B.J., Vogt, H., *Sens. Actuators B*, 1992, 6:162–164
338. Westcott, L., Rogers, G., *J. Phys. E: Sci. Intrum.*, 1985, 18:577–581
339. Schöning, M.J., Thust, M., Müller-Veggian, M., Kordos, P. Lüth, H., *Sens. Actuators B*, 1998, 47:225–230
340. Vogel, A., Hoffmann, B., Sauer, t., Wegner, G., *Sens. Actuators B*, 1990, 1:408–411
341. Schöning, M.J., Ronkel, F., Crott, M., Thust, M., Schultze, J.W., Kordos, P., Lüth, H., *Electrochim. Acta*, 1997, 42:3185–3193
342. Gabusjan, T., Bartolomäus, L., Moritz, W., *Sens. Mater.*, 1998, 10:263–273
343. Janata, J., Josowicz, M., *Anal. Chem.*, 1997, 69:293A–296A
344. Trasatti, S., Parsons, R., *Pure Appl. Chem.*, 1986, 58:37
345. Bergstrom, P.L., Patel, S.V., Schwank, J.W., Wise, K.D., *Sens. Actuators B*, 1997, 42:195–204; *Technical Digest, Transducers, Stockholm*, 1995, vol. 2, pp. 993–996
346. Josowicz, M., Janata, J., in *Chemical Sensor Technology* (Seiyama, T., ed.), Elsevier, Amsterdam, 1988, pp. 153–177
347. Scharnagl, K., Bögner, M., Fuchs, A., Winter, R., Doll, T., Eisele, I., *Sens. Actuators B*, 1999, 57:35–38
348. Bögner, M., Fuchs, A., Scharnagl, K., Winter, R., Doll, T., Eisele, I., *Sens. Actuators B*, 1998, 47:145–152
349. Lorenz, H., Peschke, M., Riess, H., Janata, J., Eisele, I., *Sens. Actuators A*, 1990, 21–23:1023–1026
350. Peschke, M., Lorenz, H., Riess, H., Eisele, I., *Sens. Actuators B*, 1990, 1:21–24

351. Gergintschew, Z., Kornetzky, P., Schipanski, D., *Sens. Actuators B*, 1996, 35–36:285–289
352. Fuchs, A., Bögner, M., Doll, T., Eisele, I., Sens. *Actuators B*, 1998, 48:296–299
353. Domansky, K., Baldwin, D.L., Grate, J.W., Hall, T.B., Li, J., Josowicz, M., Janata, J., *Anal. Chem.*, 1998, 70:473–481
354. Josowicz, M., Blackwood, D.J., *J. Phys. Chem.*, 1991, 95:493
355. Ito, Y., Morimoto, K., Tsunoda, Y., *Sens. Actuators B*, 1993, 13–14:348–350
356. Adami, M., Sartore, M., Rapallo, A., Nicolini, C., *Sens. Actuators B*, 1992, 7:343–346
357. Adami, M., Sartore, M., Baldini, E., Rossi, A., Nicolini, C., *Sens. Actuators B*, 1992, 9:25–31
358. Owicki, J.C., Bousse, L.J., Hafeman, D.G., Kirk, G.L., Olson, J.D., Wada, H.G., Parce, J.W., *Ann. Rev. Biophys. Biomol. Struct.*, 1994, 23:87–113
359. Parak, W.J., Hofmann, U.G., Gaub, H.E., Owicki, J.C., *Sens. Actuators B*, 1997, 63:45–57
360. Ito, Y., *Sens. Actuators B*, 1998, 52:107–111
361. Fanigliulo, A., Accossato, P., Adami, M., Lanzi, M., Marinoia, S., Paddeu, S., Parodi, M.T., Rossi, A., Sartore, M., Grattarola, M., Nicolini, C., *Sens. Actuators B*, 1996, 32:41–48
362. Tanaka, H., Yoshinobu, T., Iwasaki, H., *Sens. Actuators B*, 1999, 59:21–25
363. Gavazzo, P., Paddeu, S., Sartore, M., Nicolini, C., *Sens. Actuators B*, 1994, 18–19:368–372
364. Bousse, L., *Sens. Actuators B*, 1996, 34:270–275
365. Owicki, J.C., Parce, W., *Biosens. Bioelectron.*, 1992, 7:255–272
366. Lee, W.E., Thompson, H.G., Hall, J.G., Bader, D.E., *Biosens. Bioelectron.*, 2000, 14:795–804
367. http://www.moleculardevices.com/
368. Zhang, T.H., Petelenz, D., Janata, J., *Sens. Actuators B*, 1993, 12:175–180
369. Wong, H.S., White, M.H., *IEEE Trans. Electron. Devices*, 1989, 36:479–487
370. Sibbald, A., Shaw, J.E., Sens. *Actuators*, 1987, 12:297–300
371. Osorio-Saucedo, R., Luna-Arredondo, E.J., Calleja-Arriaga, W., Reyes-Barranca, M.A., *Sens. Actuators B*, 1996, 37:123–129
372. Sakai, T., Amemiya, I., Uno, S., Katsura, M., *Sens. Actuators B*, 1990, 1:341–344
373. "DuraFET," see http://www.honeywell.com
374. Göpel, W., Reinhardt, G., *Metal Oxide Sensors in Sensors Update*, vol. 1 (Baltes, H., Göpel, W., Hesse, J., eds.), VCH, Weinheim, 1996, pp. 49–120
375. Jones, T.A., Bott, B., *Sens. Actuators*, 1984, 5:43
376. Bott, B. Jones, T.A., *Sens. Actuators*, 1986, 9:27
377. Jones, T.A., Bott, B., Hurst, N.W., Mann, B., *Surf. Sci.*, 1980, 92:90
378. Wohltjen, H., Barger, W.R., Snow, A.W., Jarvis, N.L., *IEEE Trans. Electron. Devices*, 1985, ED-32:1170
379. Nylander, C., Armgarth, M., Lundström, I., *Surf. Sci.*, 1980, 92:400
380. Bartlett, P.N., Ling-Chung, S.K., *Sens. Actuators*, 1989, 19:125–140, 141–150; *Sens. Actuators*, 1989, 20:287–292

381. Persaud, K., Dodd, G.H., *Nature*, 1982, 299:352–355
382. Persaud, K.C., Pelosi, P., *Trans. Am. Soc. Artif. Intern. Organs*, 1985, 31:297–300
383. Persaud, K.C., Pelosi, P., "Sensor arrays using conducting polymers, in *Sensors and Sensory Systems for an Electronic Nose* (Gardner, J.W., Bartlett, P.N., eds.), Kluwer, Dordrecht, 1992
384. Bartlett, P.N., Gardner, J.W., "Odor sensor for an electronic nose," in *Sensors and Sensory Systems for an Electronic Nose* (Gardner, J.W., Bartlett, P.N., eds.), Kluwer, Dordrecht, 1992
385. Topart, P., Josowicz, M., *J. Phys. Chem.*, 1992, 96:7824–30
386. Gardner, J.W., Pearce, T.C., Friel, S., Bartlett, P.N., Blair, N., *Sens. Actuators B*, 1994, 18–19:240–243
387. Gardner, J.W., Bartlett, P.N., *Synth. Met.*, 1993, 55–57:3665–70
388. Cole, M., Gardner, J.W., Lim, A.W.Y., Scivier, P.K., Brignell, J.E., *Sens. Actuators B*, 1999, 58:518–525
389. Bartlett, P.N., Eliott, J.M., Gardner, J.W., *Measurement and Control*, 1997, 30:273–278
390. Gardner, J.W., Vidic, M., Ingleby, P., Pike, A.C., Brignell, J.E., Scivier, P., Bartlett, P.N., Duke, A.J., Elliott, J.M., *Sens. Actuators B*, 1998, 48:289–295
391. http://www.osmetech.plc.uk/, http://www.marconitech.com/
392. Lonergan, M.C., Severin, E.J., Doleman, B.J., Beaber, S.A., Grubbs, R.H., Lewis, N.S., *Chem. Mater.*, 1996, 8:2298–2312
393. Severin, E.J., Doleman, B.J., Lewis, N.S., *Anal. Chem.*, 2000, 72:658–668
394. Doleman, B.J., Lonergan, M., Severin, E.J., Vaid, T.P., Lewis, N.S., *Anal. Chem.*, 1998, 70:4177–4190
395. Patel, S.V., Jenkins, M.W., Hughes, R.C., Yelton, W.G., Ricco, A.J., *Anal. Chem.*, 2000, 72:1532–1542
396. Eastman, M.P., Hughes, R.C., Yelton, W.G., Ricco, A.J., Patel, S.V., Jenkins, M.W., *J. Electrochem. Soc.*, 1999, 146:3907–3913
397. http://cyranosciences.com/
398. Sheppard, N.F., Lesho, M.J., McNally, P., Francomarco, A.S., *Sens. Actuators B*, 1995, 28, 95–102; *Sens. Actuators B*, 1996, 37:61–66
399. Sheppard, N.F., Tucker, R.C., Salehi-Had, S., *Sens. Actuators B*, 1993, 10:73–77
400. Varlan, A.R., Sansen, W., *Sens. Actuators B*, 1997, 44:309–315
401. Gardner, J.W., Pike, A., de Rooij, N.F., Koudelka-Hep, M., Clerc, P.A., Hierlemann, A., Göpel, W., *Sens. Actuators B*, 1995, 26:35–139
402. Beckmann, F., Marschner, J., Hofmann, T., Laur, R., *Sens. Actuators B*, 1997, 62:734–738
403. Suehle, J.S., Cavicchi, R.E., Gaitan, M., Semancik, S., *IEEE Electron. Devices Lett.*, 1993, 14:118–120
404. Vandrish, G., *Key Engineering Materials*, 1996, 122:185–224
405. Barsan, N., Schweizer-Berberich, M., Göpel, W., *Fresenius J., Anal. Chem.*, 1999, 365:287–304; Simon, I., Barsan, N., Bauer, M., Weimar, U., *Sens. Actuators B*, 2001, 73:1–26
406. Geistlinger, H., *Sens. Actuators B*, 1993, 17:47–60

407. Heiland, G., Kohl, D., in *Chemical Sensor Technology*, vol. 1 (Seiyama, T., ed.), Elsevier, Amsterdam, 1988
408. Kohl, D., *Sens. Actuators B*, 1990, 1:158–165
409. Lee, A.P., Reedy, B.J., *Sens. Actuators B*, 1999, 60:35–42
410. Yea, B., Osaki, T., Sugahara, K., Konishi, R., *Sens. Actuators B*, 1997, 41:121–129
411. Heilig, A., Barsan, N., Weimar, U., Schweizer-Berberich, M., Gardner, J.W., Göpel, W., *Sens. Actuators B*, 1997, 43:45–51
412. Corcoran, P., Lowery, P., Anglesa, J., *Sens. Actuators B*, 1998, 48:448–455
413. Sberveglieri, G., Hellmich, W., Müller, G., *Microsyst. Technol.*, 1997, 3:183–190
414. Demarne, V., Grisel, A., *Sens. Actuators*, 1988, 13:301–313
415. Faglia, G., Comini, E., Pardo, M., Taroni, A., Cardinali, G., Nicoletti, S., Sberveglieri, G., *Microsyst. Technol.*, 1999, 6:54–59
416. Park, H.S., Shin, H.W., Yun, D.H., Hong, H-K., Kwon, C.H., Lee, K., Kim, S-T., *Sens. Actuators B*, 1995, 24–25:478–481
417. Götz, A., Gràcia, I., Cané, C., Lora-Tamayo, E., Horrillo, M.C., Getino, G., García, C., Gutiérrez, J., *Sens. Actuators B*, 1997, 44:483–487
418. Semancik, S., Cavicchi, R.E, *Acct. Chem. Res.*, 1998, 31:279–287
419. Poirier, G.E., Cavicchi, R.E., Semancik, S., *J. Vac. Sci. Technol. A*, 1993, 11:1392–1395; *J. Vac. Sci. Technol. A*, 1994, 12:2149–2152
420. Frietsch, Dimitrakopoulos, L.T., Schneider, T., Goschnick, J., *Surface Coatings Technology*, 1999, 120–121:265–271
421. Semancik, S., Cavicchi, R.E., Kreider, K.G., Suehle, J.S., Chaparala, P., *Sens. Actuators B*, 1996, 34:209–12
422. Sberveglieri, G. Faglia, G., Gropelli, S., Nelli, P., Camanzi, A., *Semicond. Sci. Technol.*, 1990, 5:1231
423. DiMeo, F., Cavicchi, R.E., Semancik, S., Suehle, J.S., Tea, N.H., Small, J., Armstrong, J.T., Kelliher, J.T., *J. Vac. Sci. Technol. A*, 1998, 16:131–138
424. Schweizer-Berberich, M., Zheng, J.G., Weimar, U., Göpel, W., Barsan, N., Pentia, E., Tomescu, A., *Sens. Actuators B*, 1996, 31:1–5
425. Briand, D., Krauss, A., van der Schoot, B., Weimar, U., Barsan, N., Göpel, W., de Rooij, N.F., *Sens. Actuators B*, 2000, 68:223–233
426. Majoo, S., Gland, J.L., Wise, K.D., Schwank, J.W., *Sens. Actuators B*, 1996, 35–36:312–319
427. Chung, W.Y., Shim, C.H., Choi, S.D., Lee, D.D., *Sens. Actuators B*, 1994, 20:139–143
428. Althainz, P., Goschnick, J., Ehrmann, S., Ache, H.J., *Sens. Actuators B*, 1996, 33:72–76
429. Althainz, P., Dahlke, A., Frietsch-Klarhof, M., Goschnick, J., Ache, H.J., *Sens. Actuators B*, 1995, 24–25:366–369
430. Swart, N.R., Nathan, A., *IEEE Proc. Solid State Sensor and Actuator Workshop, Hilton Head*, 1994, pp. 119–122
431. Bosc, J.M., Guo, Y., Sarihan, V., Lee, T., *IEEE Trans. Reliab.*, 1998, 47:135–141
432. Bosc, J.M., Odile, J.P., *Microelectron. Reliab.*, 1997, 37:1791–1794

433. Pike, A., Gardner, J.W., *Sens. Actuators B*, 1997, 45:19–26
434. Lee, D-D., Chung, W-Y., Choi, M-S., Baek, J-M., *Sens. Actuators B*, 1996, 33:147–150
435. Dumitescu, M., Cobianu, C., Lungu, D., Dascalu, D., Pascu, A., Kolev, S., van den Berg, A., *Sens. Actuators A*, 1999, 76:51–56
436. http://www.figarosensor.com/, http://www.microchemical.com/, http://www.marconitech.com/, http://www.capteur.co.uk/
437. Horrillo, M.C., Sayago, I., Arés, L., Rodrigo, J., Gutiérrez, J., Götz, A., Gràcia, I., Fonseca, L., Cané, C., Lora–Tamayo, E., *Sens. Actuators B*, 1999, 58:325–329
438. Cavicchi, R.E., Suehle, J.S., Kreider, K.G., Gaitan, M., Chaparala, P., *Sens. Actuators B*, 1996, 33:142–146
439. Kunt, T.A., McAvoy, T.J., Cavicchi, R.E., Semancik, S., *Sens. Actuators B*, 1998, 53:24–43
440. Demarne, V., Balkanova, S., Grisel. A., Rosenfeld, D., Lévy, F., *Sens. Actuators B*, 1993, 14–15:497–498
441. Majoo, S., Schwank, J.W., Gland, J.L., *AIChE J. Ceramics Proc.*, 1997, 43:2760–2765
442. Patel, S.V., Wise, K.D., Gland, J.L., Zanini-Fisher, M., Schwank, J.W., *Sens. Actuators B*, 1997, 42:205–215
443. Atkinson, J., Cranny, A., Simonis de Cloke, C., *Sens. Actuators B*, 1998, 47:171–180
444. Sheppard, N.F., Day, D.R., Lee, H.L., Senturia, S.D., *Sens. Actuators*, 1982, 2:263–274
445. Senturia, S.D., *Technical Digest Transducers*, 1985, pp. 198–201
446. Glenn, M.C., Schuetz, J.A., *Technical Digest Transducers*, 1985, pp. 217–219
447. Denton, D.D., Senturia, S.D., Anolick, E.S., Scheider, D., *Technical Digest Transducers*, 1985, pp. 202–205
448. Delapierre, G., Grange, H., Chambaz, B., Destannes, L., *Sens. Actuators*, 1983, 4:97–104
449. Shibata, H., Ito, M., Asakursa, M., Watanabe, K., *IEEE Trans. Instrumen. Meas.*, 1996, 45:564–569
450. Cornila, C., Hierlemann, A., Lenggenhager, R., Malcovati, P., Baltes, H., Noetzel, G., Weimar, U., Göpel, W., *Sens. Actuators B*, 1995, 24–25:357–361
451. Steiner, F.P., Hierlemann, A., Cornila, C., Noetzel, G., Bächtold, M., Korvink, J.G., Göpel, W., Baltes, H., *Technical Digest Transducers*, 1995, vol. 2, pp. 814–817
452. Van Gerwen, P., Laureys, W., Huyberechts, G., Op De Beeck, M., Baert, K., Suls, J., Varlan, A., Sansen, W., Hermans, L., Mertens, R., *Technical Digest Transducers*, 1997, vol. 2, pp. 907–910
453. Hagleitner, C., Koll, A., Vogt, R., Brand, O., Baltes, H., *Technical Digest Transducers Sendai, Japan*, 1999, vol. 2, pp. 1012–1015
454. Koll, A., Kummer, A., Brand, O., Baltes, H. *Proc. SPIE Smart Structures and Materials, Newport*, 1999, vol. 3673, pp. 308–317

455. Hille, P., Strack, H., *Sens. Actuators A*, 1992, 32:321–325
456. Mutschall, D., Obermeier, E., *Sens. Actuators B*, 1995, 24–25:412–414
457. Boltshauser, T., Baltes, H., *Sens. Actuators A*, 1991, 25–27:509–512
458. Boltshauser, T., Chandran, L., Baltes, H., Bose, F., Steiner, D., *Sens. Actuators B*, 1991, 5:161–164
459. *http://www.sensirion.com/, http://www.vaisala.com/, http://www.humirel.com/*
460. Kang, U., Wise, K.D., *Technical Digest Solid State Sensor and Actuator Workshop, Hilton Head*, 1998, pp. 183–186
461. Endres, H.E., Drost, S., *Sens. Actuators B*, 1991, 4:95–98
462. Casalini, R., Kilitziraki, M., Wood, D., Petty, M.C., *Sens. Actuators B*, 1999, 56:37–44
463. Dickert, F.L., Zwissler, G.K., Obermeier, E., *Ber. Bunsenges, Phys. Chem.*, 1993, 97:184–188
464. Ishihara, T., Fujita, H., Takita, Y., *Sens. Actuators B*, 1998, 52:100–106
465. Lin, J., Möller, S., Obermeier, E., *Sens. Actuators B*, 1991, 5:219–221
466. Balkus, K.J., Ball, L.J., Gimon-Kinsel, M.E., Anthony, J.M., Gnade, B.E., *Sens. Actuators B*, 1997, 42:67–79
467. Endres, H.E., Hartinger, R., Schwaiger, M., Gmelch, G., Roth, M., *Sens. Actuators B*, 1999, 57:83–87
468. Rodriguez, J.L., Hughes, R.C., Corbett, W.T., McWorther, P.J., *Technical Digest IEEE Intl. Electron Devices Meeting, San Francisco*, 1992, pp. 521–524
469. Hughes, K.L., Miller, S.L., Rodriguez, J.L., McWorther, P.J., *Sens. Actuators B*, 1996, 37:75–81
470. Gumbrecht, W., Peters, D., Schelter, W., Erhardt, W., Henke, J., Steil, J., Sykora, U., *Sens. Actuators B*, 1994, 18–19:704–708
471. Schelter, W., Gumbrecht, W., Montag, B., Sykora, U., Erhardt, W., *Sens. Actuators B*, 1992, 6:91–95
472. Lauwers, E., Suls, J., Gumbrecht, W., Maes, D., Gielen, G., Sansen, W., *IEEE J. Solid-State Circuits*, 2001, 36:2030–2038
473. Arquint, P., van den Berg, A., van der Schoot, B., de Rooij, N.F., Bühler, H., Morf, W.E., Dürselen, L.F., *Sens. Actuators B*, 1993, 13–14:340–344
474. Van den Berg, A., van der Wal, P.D., van der Schoot, B., de Rooij, N.F., *Sens. Mater.*, 1994, 6:23–43
475. *http://www.i-stat.com/*

12 Microfluidics

Jens Ducrée[1], Peter Koltay[2], and Roland Zengerle[1,2]

[1]HSG-IMIT—Institute of Micro and Information Technology,
Villingen-Schwenningen, Germany
[2]Institute of Microsystem Technology IMTEK
University of Freiburg, Germany

12.1 Introduction

The field of microfluidics has become one of the most dynamic disciplines of microtechnology. On the one hand, microfluidics offers the mere benefits of miniaturization, enabling many fields of application, in particular where small liquid volumes, transportable and cheap devices, or integrated process control are beneficial. On the other hand, microfluidics provides an elegant and often exclusive access to the nanoworld of biomolecular chemistry and cell handling, leveraging many novel biotechnological applications. This chapter first outlines the fluidic properties and working principles underlying microfluidic devices, such as diffusion, heat transport, interfacial surface tension, and electrokinetic effects. It then introduces fabrication techniques and sketches microfluidic components for flow control, pumping, physical sensing, and dispensing and their applications in (bio-)analytical chemistry, drug discovery, and chemical process engineering.

Microfluidic devices are one of the earliest success stories in the commercialization of microelectromechanical systems (MEMS). Efforts to dispense minute amounts of liquid at high precision date back to the early 1950s and constitute the basics of contemporary inkjet technology. Since then, enterprises have continuously improved and diversified this technology, the worldwide annual revenues of which currently approach $10 billion.

With the groundbreaking progress in microtechnology, starting in the 1970s, other types of microdevices were developed in academic and industrial labs. "Killer" applications such as microelectronic memory chips and processors as well as microelectromechanical read/write heads for hard disks shine, with staggering growth rates in performance and revenues. When MEMS were still chiefly an academic topic in the 1980s, microfluidic research focused on miniaturized conventional components. The first microfabricated pumps and valves were presented; they relied basically on the same principles as their macroscopic counterparts.

The transition from these prototypes to commercial products was often much more tedious than expected, or even failed completely. Initial hopes arose from the apparent analogies to the microelectronics industry, where similar manufacturing processes have been used. However, microelectronic components work in a well-defined, encapsulated environment designed for the relatively straightforward task of moving electrons in a controlled manner.

Apart from the inkjet industry, the first attempts to commercialize microfluidic devices were micropumps and valves, in the late 1980s. The expectations of wide commercial proliferation were again disappointed by the lack of economically viable new markets and the unwillingness of the potential customers to switch from proven conventional technology. With the growing maturity of the technology, the attitude of the industry has gradually changed. There is now an increasing commercial demand for liquid handling of miniaturized volumes. Production numbers are still low because the devices are still too expensive to conquer high-volume markets.[1,2]

By the beginning of the 1990s, the paradigms in microfluidics began to shift. Instead of miniaturizing conventional components, novel application types emerged in chemistry and biotechnology that could not be realized by simply downscaling traditional solutions. Chemists sought compact onsite analysis systems known as μTAS that are capable of taking, processing, and analyzing minute amounts of sample in an integrated manner.[3-6] These "labs-on-a-chip" have also demonstrated commercial feasibility for medical and military applications such as point-of-care analysis and portable chemical warfare equipment.[7]

In the same period, commercialization of biotechnology advanced rapidly. Fields such as drug discovery and genetic research came up with an ever-increasing number of substances to be synthesized, analyzed, and tested. Due to the high costs associated with personnel and material, automated "rationalized" procedures handling minute volumes of precious material had to be implemented, avoiding manual intervention by operators and experts. In the wake of these innovations, the well-established microtiter plate (MTP) technology was progressively miniaturized, requiring novel design principles in liquid handling, processing, and detection.

Microtechnology proved to be ideally suited for solving these technological challenges. Microfabricated pipettes and dispensers successfully interfaced between macroscopic reagent or sample volumes on the inlet side and the microworld on the outlet side. The devices benefitted from the intrinsic precision and cost scaling of batched microfabrication as well as their potential to operate in a highly parallel manner. In the quest for new drugs, these pipetting devices help the pharmaceutical industry to generate comprehensive libraries of drug candidates in MTPs and to test them against targets by high-throughput screening (HTS) procedures.

While the miniaturization of the MTP world has somewhat reached its limits in the eyes of many experts, microarrays, a new genuinely microfluidic technology, has appeared on the horizon.[8-16] As liquid volumes shrink beyond the microliter range, surface tension forces prevail over gravity to confine liquid volumes to small droplets, thus obviating container vessels. Nanoliter and sub-nanoliter droplets are dispensed in an equidistant lattice arrangement on flat slides, each of them featuring a distinct probe that is "marked" by its planar coordinates. With grid spacings of a few hundred microns only, densities can easily reach several thousand probe sites per square centimeter. With a sophisticated technology, microarrays displaying the full yeast genome, each spot representing a certain oligomer DNA sequence, have been presented.[17] It has further been demonstrated that massively parallel, multiplexed assays can be conducted by exposing the microarray to a sample containing different types of target molecules.

Stocks of many startup companies involved in microarray and lab-on-a-chip technology have recently gone through some critical turbulence. While many analysts forecast a tremendous market potential, they simultaneously shed doubts about when commercial maturity will have been achieved. With the technology still in its infancy, technologies presently serve mainly as enabling technologies for the pharmaceutical industry and academic institutions. Experience from other technologies teaches that the path to adulthood, at which point these microfluidic products are directly accessible by the consumer, is commonly accompanied by several cycles of drawbacks and progress lasting at least a decade. However, the widely predicted biotechnological revolution in the 21st century will unquestionably turn medicine and biology into information sciences entailing the retrieval of enormous amounts of data.[2,18] One day, for example, a cost-effective, extensive genetic or immunological analysis will be required to prescribe a patient-optimized drug cocktail, a scenario that is unimaginable without microfluidics.

Another hot topic in microfluidics is chemical reaction technology.[19] The fundamental steps of chemical process engineering are mixing and temperature control. Microreactors offer unique performance advantages in this area, such as tight thermal control and rapid mixing, to improve the efficiency and quality of existing synthesis protocols and products significantly. They clearly bear the potential for exploring yet-untouched reaction regimes. High throughput may be reached by running arrays of microreactors in parallel. A whole range of other device concepts has been suggested and realized. Application-oriented MEMS developers engaged in these areas generally experience strong and still-increasing commercial interest in their development.

From physics and engineering points of view, microfluidics can be distinguished from traditional fluidics in the macroworld by the small masses and high surface-to-volume ratios involved. Effects such as interfacial surface tension, evaporation, stiction in the wall region, and electrical double layers have a massive impact on the fluid dynamics. The short distances of microstructures also boost the rapid completion of transport phenomena such as diffusion and the conduction of heat. Furthermore, the presence of mesoscopic particles can no longer be neglected, and even molecular effects such as the spatial extension of molecules or rarefaction have to be considered in certain situations. Although these phenomena often allow for new design principles, they can at the same time severely interfere with the device operation. Getting rid of these "microfluidic" effects is often cumbersome or even impossible because they are in most cases intrinsic properties of the fluidic system and thus cannot be "switched off."[20,21]

This chapter begins with the fundamental principles and technologies underlying microfluidic devices: the properties of fluids and their embedding into microfluidic structures and actuators. It introduces typical manufacturing processes. Basic system components are outlined, and most of the applications of commercial interest are investigated. Comprehensive textbooks and the reference database provide a more thorough insight into the field of microfluidics.[2,18,22-25]

12.2 Properties of Fluids

Many reviews of microfluidic systems do not discuss the properties of fluids, i.e., liquids and gases, in detail. Fluids are instead characterized by a few macroscopic parameters such as density and viscosity, much as they are in macrosystems. This is certainly justified when coping with simple substances such as water that are more or less treated as technical working fluids, for example, to exert pressure. However, most microfluidic devices make use of the physical and chemical properties of fluids. A short discussion of the properties of fluids is therefore given in this section.

12.2.1 Volumes and Length Scales

First picture the typical dimensions encountered in microfluidics. The largest cube in Fig. 12.1 possesses an edge length of $l = 1$ mm, which cor-

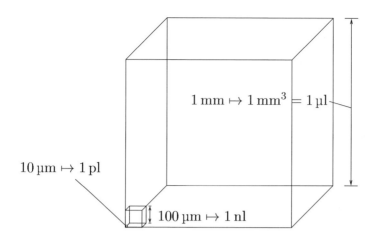

Figure 12.1 Illustration of length scales and related volumes in microfluidics. A cube with an edge length of 1 mm corresponds to l mm^3 = 1 µl; each order of magnitude in linear dimension translates into three orders of magnitude in the volume.

responds to a volume of 1 µl. Since the volume V scales with l^3, each order of magnitude on the linear scale yields three orders of magnitude in volume, and the surface-to-volume ratio is proportional to l^{-1}. Accordingly, a microfluidic structure with a typical linear feature size of 100 µm hosts a volume in the nanoliter range, and a linear dimension of 10 µm translates into picoliters.

12.2.2 Mixtures

Pure substances are composed of one given atomic or molecular constituent. The ratio between different types of atoms is strictly fixed by stoichiometry. Pure substances such as water are rarely processed in microfluidic devices, although sometimes they serve as working fluids. Mixtures are matter that consists of more than one substance or phase that are physically combined by no particular proportion of mass. Mixtures either appear as homogeneous solutions or as heterogeneous dispersions. Distinguishing between colloids or suspensions depends on the size of the dispersed phase. The combination of the carrier matrix (gas or liquid) and dispersed phase (fluid or solid) defines the type of a colloid, e.g., an emulsion (dispersion of immiscible liquids) or an aerosol (solid or liquid dispersed through a gas).

12.2.3 Physical Properties

Apart from their chemical or biochemical properties, the physical properties of the fluid are often decisive for processing and detection technology. Here we compile the properties that are important in microfluidic devices.

Compressibility

The response of a fluid volume V to a change in the external pressure p is characterized by the coefficient

$$\kappa = -\frac{1}{V}\frac{dV}{dp} = p^{-1}, \qquad (1)$$

known as the compressibility. While the (isothermal) compressibility of ideal gases is high, the compressibility of liquids ranges around 10^{-9} Pa^{-1}, and liquids can therefore, in many cases, be regarded as incompressible.

Thermal Expansion

The thermal volume expansion coefficient α_V is connected to the thermal motion and oscillations of molecules increasing with the temperature. It is defined as

$$\alpha_V = \frac{1}{V}\frac{\partial V}{\partial T}\bigg|_p = \frac{1}{V}\frac{nR_g}{p}, \qquad (2)$$

at constant pressure. In general, the expansion coefficient increases from solids to liquids and gases. A cube expands according to

$$V(T) = V_0(1 + \alpha_v T) \qquad (3)$$

and α_V corresponds to three times the linear expansion coefficient (in linear approximation). Typical values of α_V are in the vicinity of 10^{-3} K^{-1} for liquids, i.e., about 1%/10°C.

12.2.4 Vapor Pressure

In a certain region of the parameter space (p,T), the liquid and gaseous phase coexist in a thermodynamic equilibrium. In a closed vessel initially filled with a liquid, a vapor forms to create a pressure that increases with the temperature. Above the boiling point, gas bubbles form within the bulk liquid as the vapor pressure exceeds the environmental pressure. If the vessel is sufficiently large or open, the liquid evaporates. Due to their large surface-to-volume ratio, microdroplets quickly evaporate, fostered by convection and low humidity.

12.2.5 Surface Tension

For microvolumes, the force related to the surface tension is often of the same magnitude or greater than other effects such as gravity, inertia, or friction. In the static case, for example, the surface tension shapes the fluid volume, a phenomenon unknown to the macroworld. By definition, the surface tension

$$\sigma = \frac{F_\sigma}{l} \tag{4}$$

is given by the quotient of the force F_σ acting on a movable wire along the edge length l of a liquid membrane (see Fig. 12.2). The surface tension of a solution may vary severely with the solute concentration, in particular if surface active agents (surfactants) are involved.

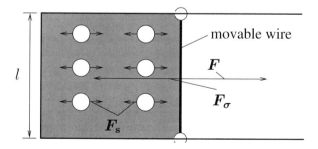

Figure 12.2 Surface tension a can be measured by pulling a liquid membrane by a wire of edge length *l* with the force **F**, which counteracts the overall forces due to surface tension $\boldsymbol{F}_\sigma = \Sigma \boldsymbol{F}_s$

12.2.6 Electrical Properties

The electrical characteristics of a fluid are determined by the type of charge carriers. In the case of gases, radiation or electron impact ionization generates ions as well as electrons as free charge carriers that can move in an external field to generate a current. In the absence of artificial ionization sources, natural radioactivity establishes a permanent concentration of about 10^9 ions/m^3, which is usually too low to allow a transmission of larger currents. Exposed to high electric fields, however, the energy of the electrons and ions is large enough to generate avalanches of secondary charge carriers, possibly igniting self-sustained gas discharges. In this case, violent phenomena such as sparks and electrical breakthroughs are observed, which often harm elements such as electrostatic actuators. It is therefore important to leave the field strengths below the so-called Paschen curve.[22]

The situation is different for electrolyte solutions where no free electrons are available and the charge carriers are solvated ions of either charge. In Ohm's law for the current density $j_q = \sigma_E E$, the electrolytic conductivity

$$\sigma_E = \frac{|j_q|}{|E|} = e(z_+\mu_+c_+ + z_-\mu_-c_-) \tag{5}$$

results from the concentration c_i, the number of unit electron charges e per molecule z_i, and the ionic mobilities μ_i of anions and cations.

Insulating fluids can only interact with alternating external fields if they are composed of polar or electrically polarizable particles. Large particles such as macromolecules and cells may assume great dipole moments $p_q = qd$ because they can displace a charge q by a large vector d. This can be very useful for dielectric phenomena where polarizable particles can be separated in inhomogeneous alternating fields according to their polarizability.

12.2.7 Optical Properties

Detection in analytical devices is often based on optical effects such as refractive index, absorption, optical activity, or light scattering. However, these approaches are often not feasible because of the short optical path lengths typical for microfluidic systems. When high sensitivity is required in biological assays, technologies based on luminescence are the method of choice. This can be chemiluminescence, bioluminescence, or laser-induced fluorescence of fluorescently tagged target molecules (see Fig. 12.3). To meet the economic requirements of mass markets, companies increasingly try to replace costly optical detection by electronic detection (amperometric, capacitive).

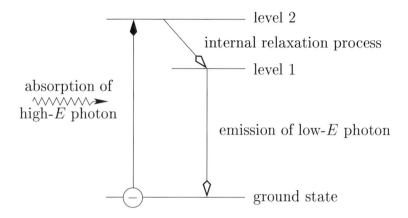

Figure 12.3 The principle of laser-induced fluorescence. In a three-level atomic system, an electron is elevated by absorbing a high-energy photon from the coherent laser field to the upper level 2. Part of the energy is released by an internal process. The final relaxation step proceeds under emission of a low-energy photon, which is used as the fluorescence signal.

12.2.8 Transport Phenomena

The random thermal motion of molecules is the physical origin of a class of processes referred to as transport phenomena. They obey the general pattern

$$\text{flow} = \text{coefficient} \times \text{force}, \tag{2.6}$$

being of the form of Ohm's law. Other than in Ohm's law, its origin does not have to be a (macroscopic) force field because statistical Brownian motion can also lead the system to abandon a gradient. Important transport phenomena in microfluidics, are diffusion, viscosity, and heat diffusion (conduction), as summarized in Table 12.1. The laws governing these effects are discussed below.

Diffusion

The mechanism of diffusion counteracts the formation of nonuniform particle density distributions $\rho_N(r)$ [or concentrations $c(r)$]. In the absence of other driving forces such as potential gradients, systems tend to assume a state with $\rho_N(r) = \text{const}$. In the presence of sources, drains, or external fields that do not vary in time, a stationary profile of the concentration is built up. Diffusive transport can be initiated by a gradient in the concentration of the solute or an inhomogeneous distribution of partial pressures within a gas volume.

Table 12.1 Summary of Phenomenological Laws of Transport and Coefficients Calculated for Ideal Gases. The transported quantities are particle number N (with particle density/concentration ρ_N), momentum p_z, heat Q, and change q. The thermodynamic state of the system is represented by the average thermal velocity v_T and the mean free path l_{mfp}. For the viscosity, the z-direction delineates the direction of flow and x the transversal axis.

Effect	Transported	Gradient	Coefficient	Law
Diffusion	N	$\dfrac{d\rho_N}{dz}$	$D \simeq \dfrac{1}{3} v_T l_{\text{mfp}}$ (diffusion coefficient)	$\boldsymbol{j}_N = -D\nabla\rho_N$ (Fick)
Viscosity	mv_z	$m\dfrac{dv_z}{dz}$	$\eta \simeq \dfrac{1}{3}\rho v_T l_{\text{mfp}}$ (viscosity)	$j_{p,x} = -\eta\dfrac{dv_z}{dx}$ (Newton)
Conduction of Heat	Q	$\rho C_m \dfrac{dT}{dz}$	$\lambda \simeq \dfrac{1}{3}\rho C_m v_T l_{\text{mfp}}$ (thermal conductivity)	$\boldsymbol{j}_q = -\sigma_E \nabla T$ (Fourier)
Electrical Conductivity	q	$-\dfrac{d\phi}{dz} = E_z$	$\sigma_E \simeq \dfrac{\rho_q q^2 l_{\text{mfp}}}{mv_T}$ (electrical conductivity)	$\boldsymbol{j}_q = -\sigma_E \nabla \phi$ (Ohm)

Fick's laws describe the evolution of an initially nonuniform particle density toward ρ_N via restoring particle currents $\boldsymbol{j}_N$. The first law

$$\boldsymbol{j}_N = -D\nabla\rho_N \qquad (7)$$

expresses that $\boldsymbol{j}_N$ always points antiparallel to the direction of the largest gradient. The diffusion coefficient D typically amounts to some $10^{-9}\,\text{m}^2\,\text{s}^{-1}$ for solvated ions. Fick's second law emerges by combining the equation of continuity with Eq. (7), yielding

$$\frac{\partial \rho_N}{\partial t} = D\Delta\rho_N, \qquad (8)$$

with the Laplacian Δ of ρ_N matching the partial time derivative of ρ_N.

The bilateral diffusion from a two-dimensional layer of molecules within pure solvent is illustrated in Fig. 12.4. The spreading over a certain distance l by mere diffusion affords a typical time $t = l^2/D$.

Viscosity

The internal friction of a fluid is referred to as viscosity. It is related to the transfer of momentum $p = mv$ from one fluid plane sliding parallel to another. For a pressure-driven flow in the z-direction with a velocity profile $v_z(x)$ in the transverse x-direction, for example, the viscosity seeks to minimize the gradient dv_z/dx, which originates from the adhesion of fluid at the walls of the channel. The coefficient of viscosity η is defined by the ratio of the force component F_z parallel to a surface A:

$$\frac{F_z}{A} = j_{p,x} = -\eta \frac{dv_z}{dx}, \tag{9}$$

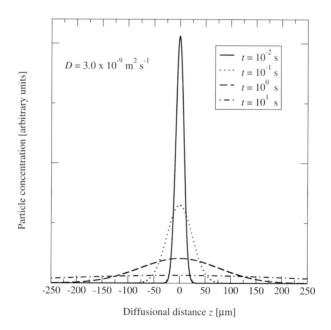

Figure 12.4 Diffusion of a thin layer of solved molecules centered at $z = 0$ at the time $t = 0$ in the surrounding pure solvent at times t for $D = 3.0 \times 10^{-9}$ m²sec⁻¹.

which is equivalent to the flow density $j_{p,x}$ of momentum $mv_z(x)$ in the x-direction. Values of η roughly amount to 10^{-5} Pa s for gases and 10^{-3} Pa s for liquids.

Thermal Conductivity

There are two dominant modes of heat transport in a fluid: diffusion and convection. The convective contribution connected to turbulences is very complicated to describe; it is, however, widely suppressed in microdevices, where laminar flow prevails. Flow phenomena are treated in Section 12.3; we focus on heat diffusion here. The statistical phenomenon of diffusion can be described by Fourier's law

$$j_Q = -\lambda \nabla T$$
$$\lambda \stackrel{gas}{=} DC_V \tag{10}$$

via the thermal conductivity λ. For an ideal gas, λ is represented by the product of the diffusion constant D [Eq. (7)] and the heat capacity per unit volume $C_V = \rho C_m$. The thermal conductivity λ varies with the temperature and displays a wide range of values depending on the material and its state. Gases exhibit values near 10^{-2} to 10^{-1} W m^{-1} K^{-1}; λ is usually one order of magnitude higher for liquids.

12.3 Physics of Microfluidic Systems

12.3.1 Navier-Stokes Equations

The Navier-Stokes equations

$$\rho\left[\frac{\partial}{\partial t}\mathbf{u} + (\mathbf{u}\cdot\nabla)\mathbf{u}\right] = -\nabla p + \eta\nabla^2\mathbf{u} + \rho\mathbf{g} \tag{11}$$

$$\nabla \cdot \mathbf{u} = 0 \tag{12}$$

reflect the fluid mechanical equivalent of the conservation of momentum [Eq. (11)] and mass [Eq. (12)], here expressed for incompressible fluids. The two terms on the left-hand side of Eq. (11) delineate the total time derivative of the momentum (density) field $\rho\mathbf{u}(\mathbf{r},t)$.

The three terms on the right-hand side of Eq. (11) represent the force densities acting on a given volume element. These are the pressure drop ∇p, the viscous force varying with the coefficient η [Eq. (9)] and with the curvature of the velocity profile $\nabla^2 u$, as well as volume forces such as gravity $\propto g$, centrifugal forces $\propto v^2 r$, or buoyancy $\propto \Delta\rho$. Due to the force scaling in microspace, effects related to gravity and buoyancy can overwhelmingly be neglected.

12.3.2 Laminar Flow

Reynolds Number

It can be shown that in the absence of volume forces, the momentum equation (11) determining the dynamics of a fluidic system is basically determined by the Reynolds number

$$Re = \frac{\rho v l}{\eta}, \qquad (13)$$

incorporating a characteristic velocity v and a characteristic dimension l in the system under investigation. Re relates to the ratio between the mechanical work spent on acceleration and viscous effects.

The Reynolds number also determines the character of flow. Below a geometry-dependent critical value, e.g., 2300 for round tubes with smooth walls, the flow remains laminar because fluid layers slide along each other without transverse mixing. This is the typical flow regime for microdevices, as can be seen from inserting some typical values in Eq. (13). Above the critical value, unsteady turbulent flow patterns are observed, which are much harder to predict.

Laminar Flow Profile

In a pressure-driven laminar flow through a tube of radius r_0, a parabolic velocity profile

$$v_z(r) = \frac{\Delta_p}{4\eta l}(r_0^2 - r^2) \qquad (14)$$

is observed, culminating in the center and vanishing toward the wall. It clearly contrasts a turbulent flow profile that flattens out with increasing flow velocity (see Fig. 12.5). The law of Hagen-Poiseuille,

$$I_V = \frac{\pi}{8\eta} \frac{\Delta p}{l} r_0^4, \qquad (15)$$

reveals that the volume flow I_V is proportional to the pressure gradient $\Delta p/l$. The scaling of I_V with r_0^4, i.e., the square of the cross-sectional area A, shows that the throughput of pressure-driven flows is limited: Applying the same $\Delta p/l$ to N tubes of cross section A/N in parallel instead of a single macro-tube ($N = 1$) of cross section A results in a decrease of I_V by a factor of $1/N$. Microfluidic devices are therefore rarely implemented if throughput is the primary objective. Instead, microfluidics excels where quality, precious reagents, and speed are issues, e.g., for analytics or high-quality synthesis.

The laminar flow regime is characterized by smooth streamlines that are predictable in time. Material transport other than in the direction of flow can only proceed via diffusion. Laminar flows are therefore well controllable, and characteristic effects such as hydrodynamic focusing and jet contraction appear (see Fig. 12.6).

12.3.3 Dynamic Pressure

For a frictionless ($\eta = 0$), stationary ($\partial u/\partial t = 0$) laminar flow, the Navier-Stokes equation (11) reduces to

$$p + \frac{\rho}{2} v^2 = p_{\text{tot}} = \text{const.}, \qquad (16)$$

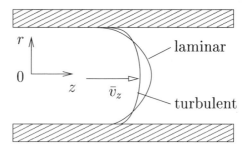

Figure 12.5 Schematic of laminar and turbulent flow profiles.

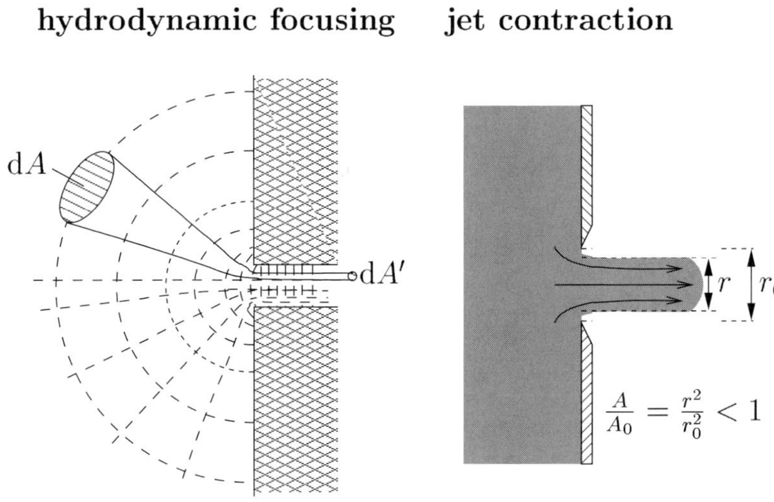

hydrodynamic focusing   **jet contraction**

$$\frac{A}{A_0} = \frac{r^2}{r_0^2} < 1$$

Figure 12.6 Effects in laminar flow. In hydrodynamic focusing (left), the full solid angle is projected on the opening of the capillary as fluid is sucked in. The vertical position of an active surface element dA therefore corresponds to a tiny section in the entrance region. Jet contraction (right) is observed when fluid escapes from the orifice of physical radius r_0. Due to the continuity of stream lines within the sharp exit contours, the jet radius r falls short of r_0.

also known as the Bernoulli equation. The relationship [Eq. (16)] reveals that the total pressure p_{tot} corresponding to the sum of kinetic and potential energy is preserved along the flow. The static pressure p complements the dynamic pressure $\propto v^2$ (see Fig. 12.7).

In regions of high flow velocity v, p may fall below the vapor pressure (Section 12.2.4) of a liquid, and gas bubble formation sets in. This so-called cavitation may significantly impair the functionality of microfluidic devices.

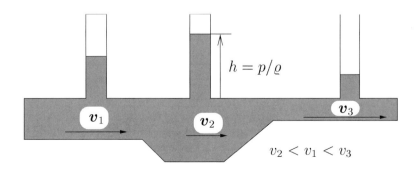

Figure 12.7 Static pressure p at different cross sections of a fluidic duct.

12.3.4 Fluidic Networks

In most situations of practical relevance, the Navier-Stokes equations can only be solved numerically. Several powerful commercial software packages are available that carry out full-fledged three-dimensional simulations.[26-31] As an alternative approach, the complexity of fluidic devices can be significantly reduced, in many cases, before starting a numerical solver by using so-called lumped-element or network models. This very instructive approach resembles the circuit modeling of electronics, where sections of the hardware are replaced by abstract elements "condensed" to discrete locations. One of the trends in microfluidic simulations points in this direction.

The fluidic analogs of voltage and current are usually identified with the pressure drop Δp and the mass flow rate $I_m = dm/dt$. The impedance functions

$$R_{hd} = \frac{\Delta_p}{I_m} = 8\pi \frac{\eta l}{\rho A^2} \quad (17)$$

$$L_{hd} = \frac{\Delta_p}{\dot{I}} = \frac{l}{A} \quad (18)$$

$$C_{hd} = \frac{dm}{dp} = \rho k_{elast} \quad (19)$$

as evaluated for round tubes of length l and cross section A are the fluidic analogies of resistors, inductors, and capacitors. In the static case ($\dot{I}_m = 0$ and $\dot{p} = 0$), only R_{hd} [Eq. (17)] governs the flow behavior, and equations such as the law of Hagen-Poiseuille [Eq. (15)] are obtained from Eq. (17).

The inertia of the fluid is represented by the hydraulic inductance L_{hd} [Eq. (18)]. Compressible fluids $\kappa \neq 0$ [Eq. (1)], elastic tubing, or membranes exhibiting a nonvanishing elastic constant k_{elast} introduce a fluidic capacitance C_{hd} [Eq. (19)] into the system. The combined action of R_{hd}, L_{hd}, and C_{hd} in Fig. 12.8 constitutes a fluidic low-pass filter suppressing high-frequency current oscillations. The network paradigm, which also holds for other physical domains, is supported by well-known simulation tools.[32,33]

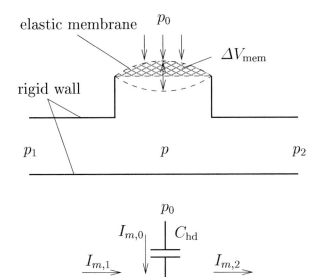

Figure 12.8 Cavity with an elastic membrane acting as a fluidic capacitance. (top) The membrane displacement volume ΔV_{mem} adjusts to the difference Δp between the external and the internal pressure p_0 and p_2, respectively. (bottom) Electric circuit equivalent of the elastic element with hydraulic resistances $R_{hd,i}$, inductances $L_{hd,i}$ and capacitance $C_{hd,i}$ governing flows $I_{m,ij}$ in the corresponding channel sections.

12.3.5 Heat Transfer

Thermodynamics teaches that each (isolated) system seeks a uniform temperature in the absence of external heat sources. Temperature gradients are counteracted by an energy flow

$$j_Q = h_Q \Delta T, \qquad (20)$$

with a material- and geometry-dependent heat transfer coefficient h_Q. Heat is typically interchanged between two vessels separated by an intermediate wall (see Fig. 12.9). While the thermal conductivity λ [Eq. (10)] equilibrates the temperature profile within each vessel and across the wall, the power of the (radiative/conductive) heat transmission

$$P = \alpha A \Delta T \qquad (21)$$

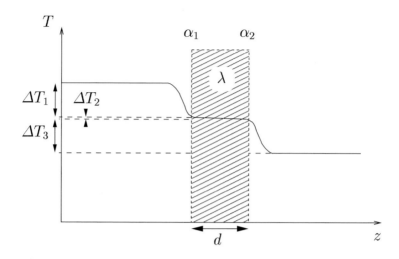

Figure 12.9 Transition of heat between two vessels at T_1 and T_3 separated by a wall.

is characterized by the heat transmission coefficient α amounting to roughly 6 W m^{-2} K^{-1} at room temperature. The overall heat transfer in Fig. 12.9 is obtained by inserting the heat transmission coefficient

$$\frac{1}{h_Q} = \frac{1}{\alpha_1} + \frac{d}{\lambda} + \frac{1}{\alpha_2} \qquad (22)$$

in Eq. (20), corresponding to the rule for series-connected conductors in electrical circuits. Microdevices display small wall thicknesses d and tiny vessel dimensions, such that efficient heat transfer and swift thermalization can be achieved. This has important implications for thermal control in process engineering or thermocycling in life science applications. Also, convection and turbulences enhance the heat transport across the fluid. Therefore, the magnitude of the often-cited Nusselt number

$$Nu = \frac{Q_{conv}}{Q_{diff}}, \qquad (23)$$

as defined by the ratio of convective Q_{conv} and diffusive heat transfer Q_{diff}, embodies the efficiency of the overall heat transfer mechanism. High Nusselt numbers are, for example, achieved in micro heat exchangers by artificially roughening the wall surface to induce turbulence.[34,35]

12.3.6 Interfacial Surface Tension

The surface tension σ, known from Section 12.2.5 describes the interaction between a liquid and its own vapor. In microfluidic systems, the interaction between solid structures and the fluid phases also plays an important role. The impact of this interfacial surface tension on effects such as wetting and capillarity is summarized by the contact angle Θ resulting from the equilibrium of forces

$$F_{SL} + F_{LV} \cos\Theta + F_{SV} = 0 \qquad (24)$$

at the three-phase contact point of a droplet situated on a solid surface (see Fig. 12.10).

The conversion of interfacial energy into liquid motion is represented by the capillary pressure

$$p_\Theta = \frac{2\sigma}{r} \cos\Theta, \qquad (25)$$

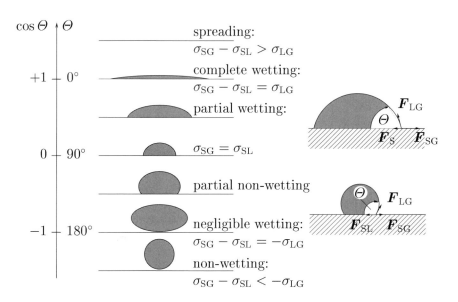

Figure 12.10 Interfacial surface tension. The contact angle Θ is defined by Eq. (24) via the equilibrium of forces attributed to the interfacial surface tensions σ_{ij} between the solid (S), liquid (L) and vapor (V) phases. For $\Theta > 90°$, the liquid wets the solid. At $\Theta > 90°$, the term $\cos\Theta$ switches its sign and the droplet tends to detach from the surface.

pointing either into or out of the capillary, depending on the sign of cos Θ (see Fig. 12.11). Note that p_Θ is only active in partially filled capillaries where a liquid meniscus continuously advances to still-unwetted regions. In contrast to macrofluidics, interfacial surface tensions and the associated capillarity are important—and unavoidable—effects in microfluidics, often prevailing over other forces. Usually the hydrophilicity of surfaces has to be well controlled to achieve reproducible liquid motion.

12.3.7 Electrokinetics

Electrical Double Layers

Many molecules situated on the surface of a wall tend to dissociate as they are brought into contact with an appropriate solvent. This occurs because the Gibbs enthalpy is governed by the energy and the entropy of the dissolution process for surface-bound molecules. The dynamic equilibrium

$$\text{SiOH(s)} \rightleftharpoons \text{SiO}^-(\text{s}) + \text{H}^+(\text{aq}) \tag{26}$$

of silanol groups on the surface of a glass tube in contact with a solvent, for example, severely shifts to the dissociated state as the pH value of the solution increases. A net negative charge of the SiO$^-$ molecules attached to the wall is compensated by solvated cations from the liquid phase (see Fig. 12.12). This charge distribution, starting with the strongly bound, nanometer-size Stern layer close to the wall, is associated with the so-called ζ-potential ("zeta") quantifying the potential drop from the surface toward the liquid bulk.

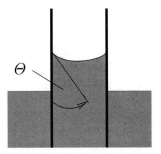
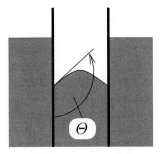

Figure 12.11 Capillary pressure visualized in a tube the inner surface of which is coated with a hydrophilic (cos Θ > 0, left) and hydrophobic (cos Θ < 0, right) layer.

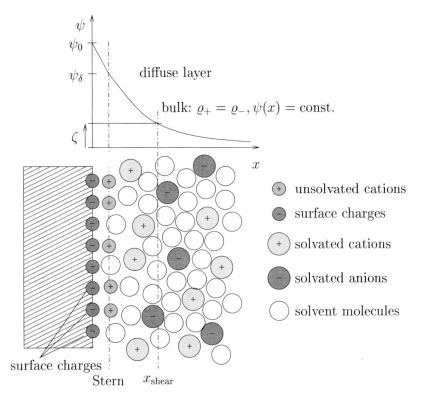

Figure 12.12 Structure of the first molecular layers of the fluid next to a negatively charged surface. The curve of the surface potential $\psi(x)$ reflects the transition from the surface over the immobilized Stern plane at $x = \delta$ and the diffuse layer to the bulk solution with $\psi(x \mapsto \infty) = 0$.

Streaming Potential and Electroosmotic Flow

As a pressure gradient is applied across the tube in Fig. 12.13 (left), the transport of mobile ions in the boundary layer leads to a so-called streaming potential counteracting the fluid motion. On the other hand, an axial external electrical field induces a motion of the boundary layer and—by virtue of the viscosity—a collective motion of the liquid bulk known as the electroosmotic flow (EOF). This EOF is characterized by a very flat velocity profile (Fig. 12.13, right) such that, for example, the hydrodynamic dispersion of sample plugs, as observed in a pressure-driven flow (Fig. 12.5), is widely suppressed.

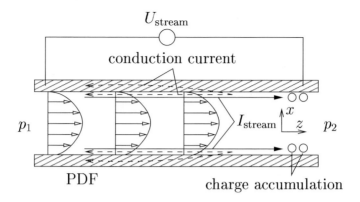

Figure 12.13 Streaming potential U_{stream} and current I_{stream} in a pressure-driven flow (PDF). The mobile charges from the electrical double layer are dragged in the direction of the PDF constituting a charge flow I_{stream}. Due to charge accumulation and depletion at the ends of the capillary, a streaming potential U_{stream} forms. U_{stream} turn initiates a conduction current through the wall or the liquid, depending on the conductivities of the media. (Exaggerated; not to scale.)

Electrophoresis

Apart from the collective EOF, all ions in the liquid phase also interact with the external electrical field E

$$v_{ep,i} = \mu |E| \tag{27}$$

$$\mu = \frac{v_i}{|E|} = \frac{q_i}{6\pi\eta r_i} \tag{28}$$

with individual drift velocities $v_{ep,i}$. Their individual electrophoretic mobilities μ_i are governed by the radius r_i and charge q_i of the solvated ions. Cations and anions travel in opposite directions, but usually $v_{ep,i} \ll v_{eof}$, and all ions arrive (at different times) at the same electrode. (See also Chapter 11.)

Electrophoresis has become a very popular technique in analytical chemistry, in particular for fast and high-quality chip-based separations, because the speed of electrokinetic motion follows the electric field. Note, however, that the liquid between the electrodes acts as an ohmic resistor. In the course of time, the Joule heating elevates the temperature of the liquid. Because diffusive heat transport to the wall is fastest in the outer layers, a nonuniform radial temperature distribution establishes. The hot liquid in the center displays the lowest viscosity and easily slips past the adjacent layers. On the other hand, the ionic mobility increases with the temperature, imposing a radial velocity gradient for each ion type. At high electric field strength, an initially well-defined plug distorts to deteriorate the resolution of analytical separations.

Dielectrophoresis

In an alternating and inhomogeneous electrical field E, the so-called dielectrophoretic (DEP) force

$$F_{\text{DEP}} = \frac{\text{Re}[p_q]}{2|E|} \nabla(|E|^2)$$

$$p_q(v) \stackrel{\text{sphere}}{=} \alpha_\varepsilon(v)E$$

(29)

acts on particles exhibiting a certain frequency-dependent polarizability $\alpha_\varepsilon(v)$. DEP is applied, for example, to particle trapping in fluids or to distinguish different types and states of cells according to their dielectric susceptibility.[36,37] (See also Chapter 11.)

12.4 Fabrication Technologies

The manufacturing of microfluidic components is guided by a set of boundary conditions that in some ways coincide, and in other aspects severely deviate, from conventional micromachining. Whereas many pioneering microfluidic projects relied on silicon microfabrication, the quest for alternate technologies has grown with increasing commercial engagement and the scope of potential applications.

Economic constraints are rooted in the strong competition with existing technologies.[38] There are few areas of microfluidic applications where miniaturization constitutes an exclusive pathway toward a solution. The vast majority of problems have been solved by conventional means that have already run through multiple cycles of cost reduction. Since microfluidic devices usually require a certain minimum wafer area, material expenses are an issue, often eliminating silicon from the choice of base materials.

The strongest and commercially most promising branches of microfluidics lie in the fields of analytical chemistry and the life sciences.

These broad fields are characterized by the close interaction of sample liquids with the walls, which considerably limits the scope of base materials. Here again, silicon exhibits properties that are unfavorable for many application areas. However, some technical applications, e.g., pumping and flow control of rather inert working fluids, have been successfully realized in silicon.

Another important aspect is that typical features in microfluidic devices often do not require cutting-edge technology regarding dimensions. For example, channels usually measure between 10 and 100 µm in diameter, which is easily realized even with old-fashioned micromachining equipment. On the other hand, microchannels have to be reliably sealed to withstand high pressures and stress and to encapsulate internally stored biofluids. The cover sometimes has to exhibit distinct properties such as optical transparency and low background fluorescence.

The point is that the choice of technology is guided not by the manufacturing of channels but by sealing channels, alignment of hybrid structures, and fulfilling economic and application-specific boundary conditions for the material. Once the material has been chosen, a manufacturing process has to be found. Typical questions concern the initial investment costs, the external manufacturing facilities, and the cost for prototyping, as well as the suitability for batch fabrication and mass production. This section summarizes the basic properties of the most prominent materials, it does not provide a comprehensive outline of manufacturing technologies. Extensive literature describes the structuring of other materials such as metals and ceramics as well as alternative structuring methods such as laminating, LIGA, and electrodischarge machining (EDM).[39-41]

12.4.1 Silicon

Silicon technology is undoubtedly the most advanced manufacturing technology. Structuring by lithography and subsequent etching allow structures to be carved into wafers with high precision (in the micron range).

Wet Etching

Conventional KOH wet etching is a well-explored batch process. The etch rate is tightly linked to the crystalline axis, peaking at about 1 µm min^{-1}. This anisotropy implies a limited degree of freedom for capillary pathways in the wafer plane as well as for the channel profiles and aspect ratios.

Dry Etching

To remove the restrictions on channel pathways, a single wafer process referred to as dry etching has been elaborated. Material removal by directed ion bombardment onto the surface proceeds at rates up to some 10 μm/min. Dry etching produces rectangular channel cross sections and allows much higher aspect ratios than standard wet etching.

Bonding

Silicon channels are often covered by wafers of Pyrex, a transparent glass whose thermal expansion coefficient [Eq. (2)] is adapted to silicon. In anodic bonding, the cleaned surfaces are pressed against each other and exposed to an external electrical potential measuring up to about 1 kV. At a process temperature below 500°C, ion migration sets in and a remnant ion distribution establishes. The silicon-Pyrex bond is robust and, to a certain extent, tolerant with respect to the surface quality.

So-called silicon fusion bonding can be used to attach two silicon wafers to each other. Prebonding is realized between hydrophilic Si surfaces via van der Waals interactions. These physical bonds are reinforced at temperatures around 1000°C when chemical Si-O-Si bonds build up. Special equipment is needed to achieve alignment errors below 20 μm with the opaque material.

Production Costs

The costs of silicon micromachining given here are rough estimates. The cost of raw material for a four-inch wafer ranges from \$25 to \$50. Each structuring step accounts for \$100 to \$200. For example, a simple flap valve as depicted in Section 12.5.1 consists of two high-bond quality wafers and two structuring steps for the flap and the valve seat, respectively. The total wafer cost therefore amounts to roughly \$700. With a chip area of 3.5×3.5 mm^2, a four-inch wafer roughly accommodates 480 chips, leaving about 336 valves if a typical yield of 70% is assumed. The cost per chip therefore settles around \$700/336 ≈ \$2 per valve (plus sizeable costs for packaging and testing).

The fairy tale of traditional silicon technology tells that by ramping up production numbers, silicon chips always become cheap. The calculation above reveals, however, that this statement does not hold for microfluidic devices because the surface area per chip dominates the production costs. The degree of functionality, e.g., memory bits or functional units, versus the occupied surface area scales differently than in microfluidics, where the potential for further miniaturization is also often limited by restrictions such as minimum flow rates. These are linked to the application and not to the fabrication technology.

12.4.2 Plastics

In seeking a more competitive cost structure for mass products, MEMS engineers looked for cheap replication processes, which they found in plastics technology. The huge variety of plastics materials, displaying a wide range of properties, also opened a new parameter space with a possibility for "tuning" the liquid-solid interaction via the material properties.

The ideal manufacturing technology fulfills the needs of the application while meeting the requirements on costs per unit. These effective costs comprise the initial capital investment and the running and maintenance costs, as well as the desired or maximum throughput. The low weight of microfluidic components to a large extent eliminates the material expenses per chip from the cost sheet in plastics technology. There is a set of common plastic replication technologies that cannot be discussed in detail here. We focus on the two most prominent replication technologies: hot embossing and microinjection molding (µIM). Techniques such as lamination,[42] laser ablation,[43,44] cast molding, and soft lithography[45] are not covered. This section concludes with remarks on master fabrication and sealing plastic channels.

Hot Embossing

A thermoplastic polymer film is brought beyond its glass transition temperature and pressed by a large force (several kN) into a microstructured mold insert situated in an evacuated chamber (see Fig. 12.14). As the polymer occupies the complete space, it assumes the negative image

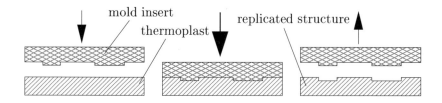

Figure 12.14 Schematic of hot embossing. In an evacuated chamber, the heated mold insert is pressed against a thermoplastic substrate to assume its inverse shape. Upon cooling, the replicated structured is released.

of the master. Upon cooling, the replicated plastic structure is de-embossed from the mold insert. The short flow length and the use of slow flow speeds ($\approx \mu m sec^{-1}$) allows the production of very fine and widely stress-free microstructures that exhibit high aspect ratios (up to some hundreds). The time scale of this variotherm (force-temperature-time) process ranges on the order of several minutes per cycle. The thickness of the residual layer in contact with the press can be kept significantly below 100 µm to avoid extensive shrinkage and stresses. Defects often arise due to trapped air bubbles; cracks occur due to inner mechanical stress or shrinkage. A cost structure analysis shows that hot embossing is currently most suitable for small-series production.

Microinjection Molding

Injection molding (IM) is a well-established, cheap, and high-throughput replication technology on an industrial scale; it is mostly used with thermoplastics. A structured mold insert is filled with a flow of liquidized resin; the molded piece is subsequently released by opening the tool. Since the flow in microdimensions is governed by different effects, the process has been adapted for the production of microstructures. Effects to be considered for microinjection molding are, for example, rapid, inhomogeneous cooling due to the high surface-to-volume ratios, excessive hydrodynamic shear stresses, and large pressure drops in tiny and long channels. It is also difficult to incorporate escape channels for trapped air, which would be the same size as the microstructures. At least for high-end manufacturing, vacuum conditions are therefore mandatory.

Dedicated simulation tools (e.g., POLYFLOW,[29] FLUENT,[30] and C-MOLD[46]) have been developed, helping to elaborate a reliable µIM protocol and to minimize cycle times down to a few minutes or even seconds, depending on the geometry of the molded part. Simple microstructures such as compact disks can be produced within a few seconds; high-aspect-ratio structures displaying a small wall thickness last several minutes because they usually require a variotherm process. In powder-injection molding (PIM), submicron metal or ceramic particles are dispersed through a moldable binder. After debinding, e.g., in a thermal or catalytic process step, and sintering, metal or ceramic parts are obtained.

Mold Inserts

A replication tool hosts a mold insert that can be fabricated by various fabrication methods, e.g., silicon micromachining, micromilling, electrodischarge machining, LIGA, or laser ablation. Apart from quality issues such as low thermal loads for rapid cycling, dimensional accuracy, and high mechanical and thermal stabilities, typical issues of prototyping apply to the fabrication of the mold master. Cost issues usually compete with the lifetime and dimensional requirements. Tough and highly accurate mold inserts fabricated by LIGA are much more expensive than "soft" tools such as silicon structures, which are only acceptable if certain tolerances and degradation are acceptable.

Scaling Strategies

As the previously described sealing mechanisms for silicon obviously do not apply, dedicated sealing techniques have to be used for plastic structures. Conventional gluing, e.g., by self-adhesive foils, is certainly not a good idea because the interfacial surface tension of the glue is high and capillary forces make the glue flow into the microcavities. Special fast-setting glues have therefore been used. Other strategies involve physical adsorption, which, for example, takes place between PDMS and plain surfaces. Other means of sealing are ultrasonic or laser welding. The bonding method also depends on technical issues such as the maximum mechanical or thermal stress. Criteria such as transparency and background fluorescence are often of pivotal significance for detection.

12.4.3 Quartz

Wafers of grown crystalline quartz (SiO_2) are frequently used in MEMS technology, e.g., as supporting membranes of photolithographic masks (blanks). These blanks are comparable to silicon with regard to both pricing and their anisotropic chemical-etching behavior. Quartz, however, is piezoelectric, is well-amenable for metallization, and possesses a far lower thermal conductivity and a nearly temperature-independent thermal expansion coefficient. Quartz has been used, for example, for chip-based electrophoresis applications.[47,48]

Structuring

The photolithography and the wet-etching protocol of quartz proceed similar to silicon. An additional thin metal (e.g., Cr-Au) layer is required between the substrate and the photoresist to reinforce the bond. The thin metal layer and quartz are subsequently etched by an HF solution; finally, the remnant metal layer is removed. Mechanical machining of quartz can be performed, for example, by diamond saw cutting, grinding, lapping and polishing to produce blanks as thin as 100 µm. Used as a cover plate, through-holes can be realized by ultrasonic drilling.

Bonding

Quartz surfaces have to be carefully cleaned before bonding, for example, by a sequence of ultrasonically assisted removal of organic contamination in acetone and methanol, rinsing with DI water, and etching with $H_2SO_4 + H_2O_2$ and HF. Then the base and cover plate are stacked, and a dilute HF solution fills the intermediate gap by capillary action. A pressure is exerted at room temperature to establish the bond.

12.4.4 Glass

Glass possesses a three-dimensional microstructure in the form of a network, but it lacks the long-range periodicity of a crystalline material. It displays the atomic structure of a liquid and is therefore amorphous, exhibiting isotropic behavior. Silica glass is solely composed of pure molten quartz (SiO_2). Silicate glasses are based on an Si^{4+} network-forming (NWF) atom polyhedrally bonded to O^{2-}, which can either act as a bridge to a neighboring NWF or just saturate the excess charge. In alkali silicate glasses, some interstitial positions are occupied by the network-modifying (NWM) alkali cations such as Li^+, Na^+, K^+, Rb^+, or Cs^+. Non-silicate glasses feature boron $B3^+$, germanium Ge^{4+}, or phosphorus P^{5+} as the NWF cation.

Glass is frequently used for properties such as great durability under exposure to harsh (natural) environments, transparency, and biocompatibility. Many of its properties can be tuned by varying the constituents or their respective ratios. Commercial suppliers offer a broad range of glasses with properties optimized for distinct application areas. Prominent examples in MEMS technology are borosilicate glasses such as Pyrex[TM], Tempax[TM], and Corning 7740[TM], whose thermal expansion coefficient is adapted to silicon for an optimized bonding, or Borofloat 33[TM], AF45[TM], and D263[TM] wafers from Schott-Desag. Microfluidic applications realized in glass technology include microreactors, filters, implants, and well plates.

Structuring

Numerous glass-etching techniques have been developed for high-aspect-ratio features or precision engineering; for example, wet (isotropic) etching and reactive ion beam etching (RIBE). We focus here on a very interesting anisotropic etching method that has been elaborated for Foturan™ glass as sketched in Fig. 12.15.[49] Upon UV exposure, an additional electron is ejected from Ce^{3+} ions to reduce Ag^+ ions. At an appropriate temperature, small micron-scale clusters crystallize around the Ag atoms, which can subsequently be removed with an HF solution at etch angles of a few degrees only. Feature sizes well below 100 µm have been demonstrated.

Bonding

Two polished glass wafers can, for example, be bonded by thermal diffusion processes at sufficiently high temperatures. The bond becomes permanent after cooling. An interesting alternative, in particular if thermal processes are harmful due to mechanical stress, is glass soldering. In these cases, a low-melting-point solder can be used to connect two glass wafers, as well as connecting glass with other materials.

12.5 Flow Control

Flow control is one of the most important topics in commercial microfluidics. There are two branches: The first is represented by integrated microfluidic systems where minute amounts of liquid have to be delivered via microchannels between a source and a destination. Except for the pumping mechanism, which can sometimes be delegated to exter-

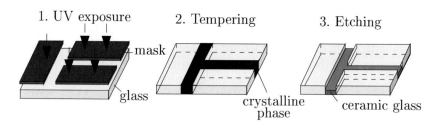

Figure 12.15 Structuring of photosensitive glass (Foturan™) by UV lithography, tempering, and etching.[49]

nal macrodevices, simple passive components are required. In this field, the right choice has to be made between different microtechnological pathways. The choice is frequently guided by cost issues and the strong coupling between the fluid and the device.

The other branch is associated with technical applications such as pneumatic control of macroscopic equipment where, typically, high-pressure gas flows have to be controlled. In this arena, MEMS offers the advantage of installing decentralized active microvalves with very low power consumption, or passive flow rectifiers, while size issues are often not very challenging. Problems are usually attributed to flow rates, maximum pressure heads, and coping with particles in the fluid.

12.5.1 Check Valves

A check valve is a flow rectifier with a direct ion-dependent flow resistance. Ideally, the flow in the forward direction proceeds with a negligible flow resistance while the reverse flow should be completely blocked, even at high back pressures. To meet these requirements, different mechanical valve concepts have been devised.

Mechanical Check Valves

Figure 12.16 shows the two popular concepts of membrane and flap valves. Each displays unique performance characteristics. These mechanical valves withstand comparatively high back pressures. Note, however, that apart from performance, technological and economic issues such as the number of process steps and the assembly costs often have to be considered.[50-53]

Dynamic Check Valves

A direction-dependent flow resistance can also be introduced by the channel geometry manufactured in a single-wafer process, which is usually much easier to handle than the rather complex configurations of mechanical check valves. Also, no moving parts are involved in this kind of valve. However, backward flow can only be reduced relative to the forward direction; it cannot be completely suppressed with these fixed-geometry valves.

Streamlines of high-velocity flow through geometric expansions (diffusers) or contractions (nozzles) are asymmetric ($d_{diff} \neq d_{nozz}$) (see

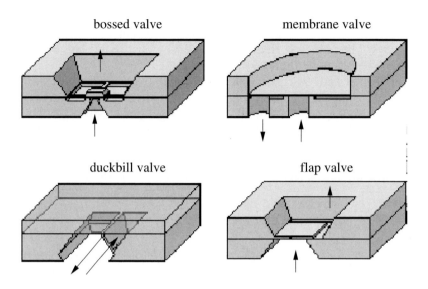

Figure 12.16 Types of passive valves. Each type possesses a distinct working surface, resulting in different hydraulic pressures for switching the valve. A typical edge length measures 1 to 2 mm.

Fig. 12.17). For opening angles larger than about 3 degrees, additional turbulent motion accounts for an increased hydrodynamic resistance [Eq. (17)] in the nozzle direction. A diffuser-nozzle valve therefore displays asymmetric flow rates throttling backward with respect to forward flow.[54-57]

Another variant of a dynamical geometrical valve is a bypass structure known as a Tesla valve, depicted in Fig. 12.18.[58,59] In an asymmetric geometry, a channel is split off and then rejoined with the main flow. Again, deviating flow patterns and turbulences in forward and backward directions induce an asymmetric hydrodynamic flow resistance [Eq. (17)].

12.5.2 Capillary Breaks

In hydrophobic microfluidics,[60-63] the capillary pressure [Eq. (25)] is used to stop the flow up to a certain threshold if a meniscus is present and the hydrophobic part of the surface is not wetted yet (see Fig. 12.19). The required pressure drop arises from the interplay between the change in hydrophilicity within the narrow channel and the abrupt shift in channel radius. The hydrophobic break interrupts the fluid flow until a certain

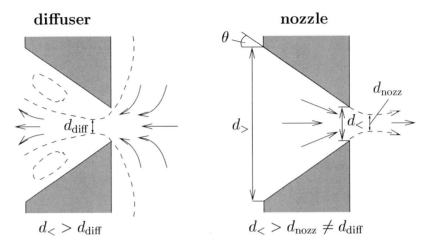

Figure 12.17 Asymmetry of flow ($d_{diff} \neq d_{nozz}$) and jet contraction ($d_< > d_{diff}$ and $d_< > d_{nozz}$) in diffuser and nozzle direction with diameters $d_<$ and $d_>$.

pressure threshold is exceeded.[60] Once the surface is completely wetted, the law of Hagen-Poiseuille [Eq. (15)] solely governs the flow.

12.5.3 Active Microvalves

Whereas the capabilities of passive microvalves are mostly used for flow rectification in micropumps, actively controlled valves target other markets such as industrial pneumatics or drug delivery systems. For most applications, outer dimensions in the centimeter range are acceptable. However, there are already many off-the-shelf conventional technologies on the market, in particular in the field of industrial pneumatics. Designated precision-engineered, piezoelectrically or electromagnetically actuated two- and three-way valves feature fast response times of about 10 msec and an almost negligible power consumption of less than 0.1 W.

Figure 12.18 Flow through a bypass in forward and reverse directions.[58-59]

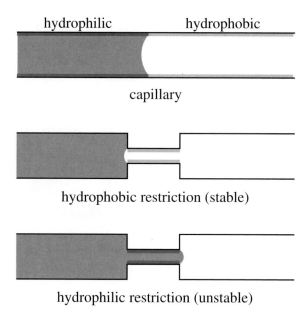

Figure 12.19 Liquid meniscus at transitions between hydrophilic and hydrophobic capillary coatings. The constrained hydrophobic channel acts as a passive capillary pressure barrier.[60-64]

These conventionally manufactured valves are thus amenable to direct control via TTL logical signals.

Market Situation

MEMS engineers need to add a competitive edge to microfabricated valves in order to gain a significant share of the market. Commercial devices were initially presented by Redwood Microsystems,[65] IC Sensors,[66] Hewlett Packard,[67] and later HSG-IMIT.[68] Two strategies for the commercial success of microvalves have been pursued.[69] The first strategy tries to outperform traditional valves, in terms of an integrated manufacturing and reduced costs, while supplying an equivalent performance. In our view, the economics of MEMS, in particular the high costs for assembly and testing, rule out this approach. The second, more realistic strategy seeks to improve the performance of the valve regarding temperature range, power consumption, size, weight, and amenability for integration at the same price level as conventional solutions. Also, compact, integrated pressure control units incorporating an active valve, pressure sensors, and flow resistors seem to have a certain potential as they face an as-yet unexplored market territory.[70]

Micromachined Valves

Many MEMS engineers have focused on thermal principles, such as volume expansion or bimetallic beams, to switch their microvalves. Thermal actuation is attractive due to its low fabrication costs and the potential for system integration. In terms of performance, however, these valves exhibit a limited operating range, typically between 0 and 60°C, and fast response times have to be paid for by large power consumption. Alternatively, if actuation time is not critical, for example, in drug delivery systems, the feasibility of electrochemical or hygrogel-based, pH-sensitive actors has been demonstrated.[71-73]

For industrial pneumatics, the cost issue is critical. The main ingredients for manufacturing competitive devices are a three-way functionality, small size, a cheap and reliable manufacturing technology, and, again, low power consumption combined with fast switching times. In our eyes, the best candidates in this arena are piezoelectric and electrostatic actuation principles (Fig. 12.20).[74]

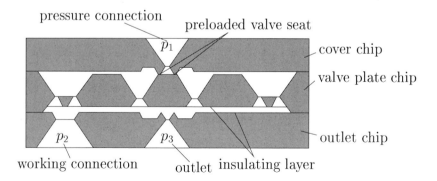

Figure 12.20 Schematic cross section of a normally closed three-way valve.[74] The preloaded valve plate chip shuts off the pressure connection at p_1. Upon applying a voltage, the valve plate closes the outlet (p_3), thus guiding the gas flow from the pressure connection to the working connection at p_2.

12.6 Micropumps

The integration of pumping functionality into microdevices has been one of the pioneering applications of microfluidics.[50,51,53,54,75,76] The capability to pump implies a pressure-generating mechanism. This can either be a kind of scaled-down macromechanism such as volume displacement or a unique microfluidic actuation principle such as electroosmosis.[77-79] Sometimes the entire setup does not need to be "micro." For example, a conventional actuation unit such as a syringe pump may be combined with a microstructure.

12.6.1 Microdisplacement Pumps

A periodic volume displacement can be used to propel a fluid flow.[51,76,80-82] In such a microdisplacement pump, an actuator compresses the fluid in a working chamber whose entrance and exit flows are rectified by check valves (see Fig. 12.21). The previously discussed passive valve types, such as the mechanical check valves or fixed-geometry valves, have been used, and mechanical, pneumatic, piezoelectric, electrostatic, electromagnetic, or thermal actuation has been applied. As an alternative to incorporating separate check valves, pumping can also be effected by a series of membranes operating in a peristaltic sequence.[83]

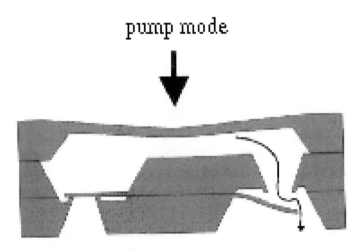

Figure 12.21 Working principle of displacement pumps, in this case with two check valve rectifiers.

Typical microdisplacement pump rates amount to some 100 μm/min. for liquids and a few milliliters per minute for gases. Maximum pressure heads range between a few 100 hPa for gases and a few atmospheres for liquids. Microdisplacement pumps for liquids are usually sensitive to gas bubbles because they severely enhance the average compressibility within the pump chamber. Gas bubbles result, for example, from incomplete priming or cavitation. The presence of particles or impurities can also lead to failures due to clogging. Intrinsic limitations of the performance, in terms of pressure head and throughput, as well as reliability problems and costs per unit, have so far prevented microdisplacement pumps from conquering high-volume markets.

12.6.2 Charge-Induced Pumping Mechanisms

A net charge in the fluid can be moved by an electric field. The bulk fluid is dragged along by viscous forces. The charge surplus can, for example, be associated with the electrical double layer (EDL) near the wall leading to electroosmotic flow (Section 12.7), which is one of the most popular pumping mechanisms in microanalytical systems.[77-79] Also, the injection of electric charge carriers (in an insulating liquid) or magnetic particles promotes flow in appropriate external fields.[84-87] The latter techniques are referred to as electrohydrodynamic (EHD) and magnetohydrodynamic (MHD) pumping, respectively.

12.6.3 Other Pumping Mechanisms

For special applications, other, more exotic pumping mechanisms have been proposed; they cannot be discussed here in detail. Centrifugal forces generated in a compact disk player can be used.[88,89] Mechanical annular gear pumps are another alternative.[90] Finally, we mention the pumping imposed by a flexural plate wave (FPW) traveling along the active surface to drag the bulk fluid.[91,92]

12.7 Sensors

Physical or chemical sensors are often required to monitor certain parameters of the fluid, either for precise dosage, for analytical purposes, or to supply feedback for actuation control.[93,94] Microtechnology helps to build small sensing units that can be deployed very close to the point of interest since their interaction with the regular flow can in many cases be minimized. For certain applications, the sensors themselves do not have to be small; they just have to be capable of measuring within microstructures. External detectors can then be used, e.g., based on optical principles such as light absorbance or laser fluorescence in applications such as particle image velocimetry (PIV).

We focus on the first genuine MEMS-type sensors and further distinguish between physical and chemical microsensors. Various physical parameters have been determined by microsensors; the most important ones are pressure, temperature, and flow rate. Pressure is often measured by the deflection of a membrane that can be detected by the changing resistance of a piezoresistive element. Temperature can, for example, be sensed by the variation of a defined resistor with the temperature. Details on the realization of such sensor types are found in abundance throughout the literature.[95] We point out here how a flow sensor can be constructed.

12.7.1 Flow Sensors

Pitot Tubes

The Bernoulli equation [Eq. (16)] reveals that a measurement of the difference between the total p_{tot} and the static pressure p allows the flow velocity v (in a stationary laminar flow, and neglecting viscous forces) to be determined.[96] Due to its compact size, such an inflow microsensor leaves the flow resistance of a macroscopic tube nearly unchanged. A design elaborated at HSG-IMIT is shown in Fig. 12.22.[97,98] The hexagonal shape of the injection-molded housing guarantees laminar flow conditions, and the opening for the membrane avoids clogging due to suspended particles. A parylene C layer protects the membrane from corrosion.

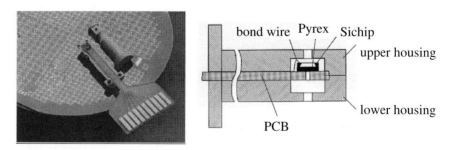

Figure 12.22 Flow measurement in macroscopic channels by the Pitot tube principle.[97,98]

Thermal Flow Sensors

In terms of energy transport, a flow is a convective mechanism that transports heat in the direction of flow. Thermal flow sensors consist of a heating element that dissipates energy to the fluid.[99-101] Different modes of flow rate measurement are possible; they differ in the optimum measurement range and the signal generation. In the constant-temperature mode, the power of the heater P is adjusted to keep the temperature of the heating element unchanged, and P is a measure for the flow velocity. Vice versa, at constant P, the temperature measured near or at the heating element can be converted into a flow rate after calibration.

The time-of-flight method records the delay between the generation of a short thermal pulse at the heater and its arrival at a downstream temperature sensor. An additional upstream sensor can be used for calibration purposes or to enable bidirectional flow rate measurements (see Fig. 12.23). The measurement setup takes advantage of the small masses involved, which enable short thermal response times. Micromachining also allows the cross section of the membrane supporting the heater and the temperature sensor to be kept small to widely suppress conduction of heat through the substrate. A time resolution of the microfabricated flow sensors within the millisecond range is possible.

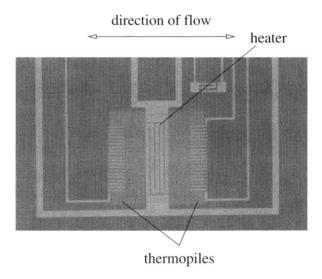

Figure 12.23 Time-of-flight anemometer developed at HSG-IMIT. The central heating element is enclosed by two thermopiles acting as upstream and downstream temperature sensors.[98]

12.7.2 Chemical Sensors

Many commercial projects in microfluidics are targeted at the development of transportable, point-of-care analysis systems. Therefore, the mere microfluidic system has to be extended by a chemical or biomolecular sensing capability. Such a setup, in turn, often depends on microfluidics, for example, by implementing reference sensors or by embedding the sample plug into a working fluid. In this way, problems that affect the long-term stability of sensitive surfaces or calibration can often be avoided. (See also Chapter 11.)

ISFETs and ChemFETs

Chemical sensing requires the transduction of chemical properties into an electrical, thermal, optical, or other signal that can be detected by a microsensor or an external macrosensor. We highlight the popular direct coupling of liquid solutions with semiconductor devices based on the MOSFET (metal-oxide field effect transistor) principle in Fig. 12.24. For a given voltage U_{ds}, the gate voltage U_{gate} regulates the current between the source and the drain. In an ion-selective FET (ISFET), the gate is replaced by an electrolyte solution. Now the electrochemical potential of the dissolved ions governs the current. A chemical sensor (ChemFET) is obtained when the gate zone is covered by a layer that is permeable to a selected set of ions only.[102-104] (See also Chapter 11.)

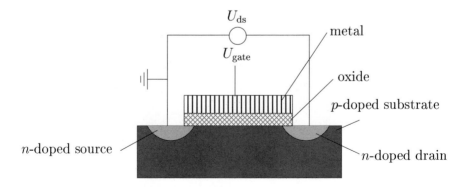

Figure 12.24 Schematic of a MOSFET.

Biosensors

The high specificity of biomolecular binding to complementary structures, also known as molecular recognition, is one of the key features of living organisms. Famous examples of such pairing activity are the two DNA strands, as well as antigen-antibody and protein-enzyme complexes. By making use of this high specificity, artificial biosensors can be fabricated by immobilizing biomolecules on a surface that capture their complementary partners from the sample solution. Immobilization schemes comprise crosslinking, adsorption, entrapment in polymer matrices, and covalent binding.

Signal transduction picks up the change of the surface properties induced by the specific attachment of target molecules from the solution. Such surface-sensitive detection can, for example, be realized in a ChemFET setup by replacing the ion-selective membrane with the layer of biornolecules. Conductometric, amperometric, and optical methods have also been used. Problems in sensing arise from the low concentrations typically encountered in biological samples and from the long-term stability of the sensitive layers.[105]

12.8 Pipettes and Dispensers

Automated liquid handling is one of the key technologies in biotechnology, e.g., for combinatorial chemistry. Microfluidic liquid handling devices are currently used to allow for fast and precise delivery of minute amounts of often precious substances such as enzymes.[106]

12.8.1 Pipettes

Pipettes exhibit a built-in capability to aspirate and dispense liquid through their tips. Pipettes are therefore very flexible and are often used in well-plate technology to deliver liquids from one reservoir well to a set of target wells. However, sample loading involves contact between the tip and the liquid to be aspirated and thus bears the threat of cross-contamination. Furthermore, as the tip dips into the reservoir, the interfacial surface tension (see Section 12.3.6) makes the liquid adhere to its outer surface, which can significantly impair the accuracy of the following dispense processes. Each sample aspiration process thus has to be accompanied by careful washing and drying steps. The dispense cycle then takes place in a similar manner, as outlined in Section 12.8.2.[107-110]

12.8.2 Dispensers

Sample Loading

Mere dispensers rely on an external sample loading mechanism in the sense that the liquid is not loaded through the same orifice as the mechanism subsequently used for dispensing. Because it is often the case that only one reservoir is attached to a dispenser, flexibility is greatly reduced. On the other hand, contact between the liquid in the target well and the tip is not necessary, so intermediate washing steps can be omitted. This makes mere dispensers ideally suited for high-precision dosage or high-throughput dispense sequences for a given substance.

Drop-on-Demand Technology

To eliminate widely the complex interaction between the solid tip, the liquid to be dispensed, and (for contact principles) the solid or liquid target, the so-called drop-on-demand technology has been developed. These dispensers use a highly dynamic actuation principle to overcome the surface tension effects at the tip and eject a defined liquid volume from the tip. The liquid droplets travel a short distance through the air before they contact the target.

The earliest drop-on-demand dispensers were inkjet print heads, which have conquered a multibillion-dollar market with their capability to deliver individual, well-defined picoliter droplets of ink at a desired location on a solid substrate accurately. The droplets are released on demand by electronic control of a piezoactuator or thermal actuator.

Over several decades of development, commercial inkjet print heads have been highly optimized for operating with a distinct, brand-specific ink. However, their accuracy and reliability suffers significantly from exchanging the ink with the liquids common in modern biotechnology. Also, the popular thermal actuation may degrade the sample. In complying with the requirements of biotechnology and specializing for the specific application, the inkjet process has been adapted.

In many applications, the overall amount of dispensed liquids matters while the droplet size and shape as well as the angular divergence of the ejected jet are of lesser importance. These dispensers typically operate by displacing a well-defined volume of liquid in a compression chamber with a piezoactuator.[108,109,111,112] In other applications, such as pulmonary drug delivery, atomizers that deliver a uniform distribution of droplets with a well-defined size are needed. In this arena, microfabrication can supply highly precise and massively parallel nozzle arrays as well as low-power actuation.

Flow-on-Demand Dispensers

Compact size and low power consumption make microfluidic dispensers also suitable for application areas such as implantable drug delivery systems used[113] in pain therapy, where a minute flow rate has to be maintained over a long time. If the tip injects the drug directly into the bloodstream, surface tension effects are not important. Instead, the stability of the flow rate and a reliable on-off mechanism are issues. The constant flow rate is established by pressurizing the drug reservoir, for example, with a micropump or a separate chamber filled with a high-vapor-pressure liquid that is connected to a throttle, such as a high-flow-resistance capillary. The flow can be interrupted by disabling the pump or switching an integrated valve.

12.9 Microarrays

12.9.1 Concept

In so-called ligand assays,[114] target molecules in a given sample are identified by their binding ("hybridization" for DNA assays) to a specific site hosting a complementary probe molecule. Multiple assays can be conducted simultaneously if each probe site and each target molecule under investigation are marked. Reactions between targets and probes thus yield information on the molecular content of the sample. The most popular approach for building such a parallel assay is to immobilize probe sites in a two-dimensional array on a flat substrate and to label the target molecules by a fluorescent dye for subsequent detection by laser light.[8-10,13,16,115,116]

A droplet of the liquid sample are then applied to the array. Molecular reactions, i.e., contact between targets and probes on the molecular level, are achieved by diffusion, which obviously proceeds very fast if the outer array dimension is kept small. An efficient, highly parallel assay can therefore be realized on a densely packed spot array. In this sense, microarrays resemble arrays of (disposable) biosensors. Detection, however, prescribes a certain level of signal-to-noise ratio restricting the minimum spot size for a given surface density of immobilized probe molecules. Other size limitations arise from manufacturing issues.

12.9.2 Fabrication

Production methods for microarrays can be categorized into spotting of premade probes and on-site synthesis of probe molecules. In contrast to the dispensers outlined in Section 12.8, liquids are transferred onto a solid microarray substrate instead of being dispensed into another fluid. For this application, issues such as the final shape, alignment precision, and homogeneity of the individual droplets matter, as well as the overall array after the evaporation process.

There is no universal fabrication method; each technique performs differently on important issues such as costs, speed, spot density, reproducibility, spot quality, scope of probe molecules, cross-contamination, and intellectual property rights. We further distinguish the overall number of spots required, which is tightly linked to the application. Low-density microarrays feature fewer than 100 spots, medium-density arrays up to about 1000, and high-density microarrays to about 10,000 spots.

Pin Printing

The underlying principle of pin printing is very simple, as illustrated in Fig. 12.25.[8,117] A pin dips into a reservoir of probe molecule solution. Interfacial tension makes a certain amount of liquid adhere to the tip. Upon transient contact, a portion of the liquid is transferred to the substrate. The maximum number of subsequent dispense cycles per load depends on the liquid storage capability of the pin, which can be extended, for example, by an inner capillary channel. The geometry and movement of the pin tip decisively determine the size, shape, and quality of the spots.

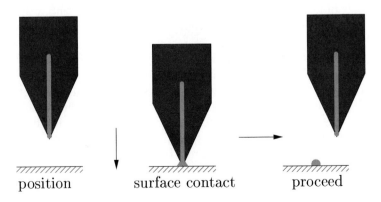

Figure 12.25 Principle of contact pin printing.

Pin printing is a very popular, straightforward arraying technique that has been integrated into workstations that incorporate washing stations, automated well-plate handling, and computer-controlled motion of multiple-pin print heads. These workstations are very flexible and well-suited to deliver small numbers of low-density arrays. The grid distance can be reduced with respect to the well plates by an interlaced printing pattern. Intrinsic mechanical tolerances and wear of the pins, however, limit the maximum density and throughput.

Inkjet Spotting

Inspired by inkjet dispensers, several drop-on-demand methods for microarray fabrication have been developed.[112,117,118,120118] Liquid is stored in a capillary that ejects individual droplets upon actuation. Various actuation principles, such as electronically controlled squeezing with a piezoactuator, have been pursued. Equipped with a pipetting capability, this drop-on-demand method can be automated in a way similar to pin printing. Due to the missing contact step, a greater liquid storage capacity, and reduced mechanics, the typically more expensive drop-on-demand workstations operate faster and at higher quality, reproducibility, and throughput.

Often the addressability of individual nozzles is not necessary. Instead, fast repetitive printing of the same set of analytes is required; for example, in a high-throughput microarray production line. With this in mind, the so-called TopSpot principle[118-120] has been developed, which allows massively parallel printing on an industrial scale (see Fig. 12.26).

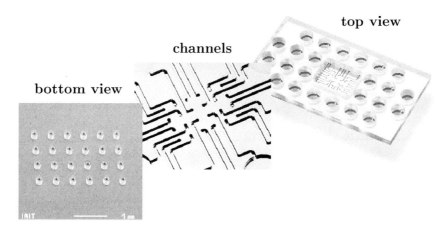

Figure 12.26 The TopSpot print head offers parallelism on the input as well as the output side.[118-120]

The top side of the micromachined print chip features liquid reservoirs that are sufficiently large to allow several thousand dispense cycles. The reservoirs are spaced to align with standard well plates ("microtiter plates") to facilitate interfacing with standard pipetting robots. Each reservoir is connected via a distinct horizontal channel to a nozzle in a central array. A pneumatic pulse applied to the upper side of the central window simultaneously releases nanoliter droplets from the orifices on the bottom side of the chip. Up to 384 nanoliter droplets can be printed simultaneously in a pitch of a few 100 μm onto the substrate underneath the print head. Extending the number of reservoirs on the chip or offset printing with an automated xy-table guiding the print head even allow printing of medium- and high-density microarrays.

Onchip Synthesis

In the field of high-density DNA arrays, photochemical protocols have been elaborated to successively concatenate nucleotides to short DNA strands on the surface of the substrate.[17,121,122] This onsite synthesis is guided by photolithographic masks or digital light processors and can thus be conducted at high precision. Whole genomes have already been integrated on a single chip. This fabrication technique is, however, more related to photochemistry and lithography, and is therefore not covered here in detail.

12.9.3 Particle-Based Microarray Concepts

The identification of probe spots by their lattice position implies precise fabrication techniques and possibly long diffusion times for the target molecules in the sample droplet. Faster hybridization and easy fabrication can be achieved using the surfaces of labeled particles, such as microbeads, for probe immobilization. Encoding the particles can, for example, be realized by color schemes,[123] microtransponders,[124] or fiberoptical readout.[125] The particles can then simply be dispersed in the analyte solution. In such particle-based microarray concepts, the alignment problem is transferred from the arraying of the probe droplet to the readout, where particle *and* target molecule labels have to be screened in parallel. Readout can, for example, be made by sequential methods, forcing the particles to pass a detector in single file. The setup resembles conventional flow cytometers.

12.10 Microreactors

Chemical process engineering is regarded as one of the key application areas for microfluidics. The fundamental process steps are heat exchange to manage reaction kinetics and mixing to establish the contact of reagents on a molecular scale. Both steps can be performed in a very tightly controllable manner while guaranteeing a high degree of homogeneity across the reaction volume. These aspects sometimes allow for new, more aggressive reaction pathways that are not accessible in macrodevices, e.g., for physical, quality, or security reasons. Furthermore, reactors may be integrated into microanalytical devices, for example, in sample preparation or amplification (such as PCR) steps, before analysis. The throughput in microreactors is limited, however, because the flow rate scales with the *square* of the channel cross section. The resulting high degree of parallelism thus makes numbering-up strategies very costly and prone to technical failure.

The term microreactor defines a very broad class of devices that can be further differentiated regarding complexity, integration, and physical and chemical working principles as well as application areas.[19,126] Single-step reactors often conduct a fundamental process such as mass or heat exchange, phase mixing, and catalytic conversions before entering a microreaction or macroreaction chamber. More complex reactors integrate multiple processes in a single device, for example, a PCR synthesis, including reagent mixing and temperature control.

12.10.1 Micromixers

One fundamental step in process engineering is mixing. Within the same phase, the Brownian motion establishes homogeneous particle concentrations (see Section 12.2.8). Macroscopic devices usually enhance this diffusional mixing activity by turbulent stirring or shaking, which is obviously not feasible for the laminar flow conditions in microstructures. Micromixers instead use an appropriate channel geometry to establish large contact-surface-to-volume ratios between streams, for example, by following a multilamination strategy, as shown in Fig. 12.27.[127-132]

Sometimes different phases have to be mixed, for example, in emulsions where the interfacial surface tension keeps the phases apart. For long-term stability, suspended particles such as fat droplets in water have to be small enough to avoid segregation. Also, large surface-to-volume ratios can be important in multiphase reactions to accelerate the dissolu-

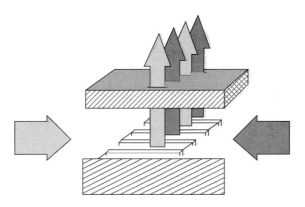

Figure 12.27 Mixing by multilamination of fluid layers.[132]

tion process. In other situations, e.g., for the industrial production of multiphase dispersions, only slight variations in droplet diameters are tolerable to assure sufficient product quality. The strength of micromixing is particularly pronounced in these multiphase applications. The capability to produce very uniform emulsions on an industrial scale has, for example, already been demonstrated.[133]

12.10.2 Heat Exchangers

Thermal energy migrates from hot to cold regions to establish a uniform temperature profile. Diffusional heat transport (Section 12.2.8) is linked to mass diffusion and lets a temperature gradient across a closed fluid volume vanish in the same way as a concentration gradient in a mixer. In many situations, diffusive mixing has to be avoided, implying that the fluids have to be separated by an impenetrable solid wall. In a separate loop, a working liquid is heated or cooled to control thermally a neighboring chemical reaction chamber. The speed of thermalization within the vessels improves with shrinking diameters of the vessels, and—most importantly—the heat flow through the separating wall scales with the temperature *gradient,* thus increasing with shrinking wall thickness [see Eq. (22)]. Apart from their amenability to integration, microstructured heat exchangers thus excel with very large heat exchange rates per unit reaction volume.[35,134-136] Absolute rates up to 20 kW have been reported.[137]

12.10.3 Chemical Reactors

Typical processes run by chemical reactors are heterogeneous catalysis, oxidations, hydrogenations, biomolecular synthesis and other single or multiphase reactions.[135,138-144] In areas such as combinatorial. Chemistry, a tremendous number of different products are generated in a batch mode, using, for example, well plates in combination with lab automation schemes. Flowthrough systems, which are commonly referred to as microreactors, queue several reaction steps in space and time, taking reagents and energy supply as input and the reaction product as output. Their layout is tightly connected to the particular reaction protocol. In this field, miniaturization and integration offer great advantages over macrosystems, ideally allowing onsite, on-demand, and flexible production of small amounts of chemicals that are either not stable in long-term storage, too dangerous for transportation, dependent on the local conditions, or not economically amenable for large-scale production.

12.11 Microanalytical Chips

Other than for microreactors, analytical systems eventually create information instead of reaction product volumes. Miniaturized analytical systems, with their potential for system integration, automation, parallelism, and onsite deployment, as well as low sample, reagent, and waste volumes, do evidently not have to deal with the problem of material throughput. Microanalytical systems can thus focus entirely on generating high-quality information in a fast and possibly parallel manner. In addition, analysis techniques can be used that may not be accessible to macrodevices.

12.11.1 Lab-on-a-Chip Systems

Devices incorporating the functionality of sample taking, sample preparation, sensing, and detection on a single microfluidic chip [3,4,6,7,145-155] are commonly termed lab-on-a-chip or miniaturized total analysis systems (µTAS). At present, there is no system on the market that completely integrates the full scope of these tasks. However, lab-on-a-chip systems have been developed that partially use macroperiphery to perform crucial steps, for example, sample preparation and detection. The main market drivers for these chips are shortened analysis times, reduced human intervention, and smaller volumes of precious or hazardous analyte, reagent, and waste. At the same time, important performance characteristics such as sensitivity and resolution should at least be preserved with respect to macrosystems.

In flow-injection analysis (FIA), a fluid stream is monitored on a continuous basis, e.g., by direct chemoelectrical transduction or by adding reagents that induce a detectable reaction product further downstream.[156,157] If a sample is to be analyzed at a fixed point in time or at least only in certain discrete time steps, an analytical separation can be carried out. To this end, a spatially well-defined volume is formed from the sample and introduced at the entrance to a separation channel. A force, typically electroosmosis for electrophoresis or pressure for chromatography, drives the sample plug whose constituents interact with the carrier liquid or the material within the column according to their physical or chemical properties. In combination, for example, with laser-induced fluorescence of labeled sample molecules, a robust detection is possible at the end of the channel where a time signal, peaking for each species passing by, is recorded. Separations therefore offer an automated sample pretreatment step as each molecular species passes the detector one at a time, thus widely eliminating interference of the different signals.

Capillary electrophoresis (CE) and high-performance liquid chromatography (HPLC) are commonly used separation schemes. These methods differentiate molecular species according to their passage times under the impact of an external electrical field or a hydraulic pressure gradient, respectively (see Chapter 11). Although the principle of chip-based gas chromatography was demonstrated in the 1970s, it is so far seldom used in chip-based separations. Pressure generators are hard to integrate and, more importantly, chromatography relies on the interaction of the sample molecules with a stationary phase, which is hard to introduce in a microchannel and which also increases the already severe pressure drops in microchannels.

12.11.2 Chip-Based Capillary Electrophoresis

Electronic signals and high voltages can conveniently be controlled on microdevices. Another charming aspect of chip-based CE is the electrically controllable plug formation, which takes advantage of the precision of the micromachined channels (see Figure 12.28). Electrophoretic separation techniques have been massively pursued by industrial and academic developers. Several commercial devices have already been presented, primarily targeting the bioanalytical market.[153,154,158]

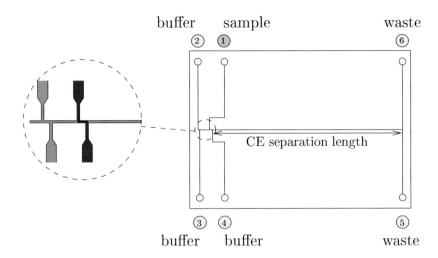

Figure 12.28 Schematic diagram of chip-based CE. After filling the chip with buffer solution, a sample plug is defined by initiating EOF between electrodes 1 and 4 and then 2 and 5. The chip length amounts to 78 mm; the distance between electrodes 1 and 2 at the intersection with the central channel measures only a few hundred µm.

References

1. J. Ducrée, H. Sandmaier, and W. Lang, "MEMS: from development to production," *Seisan-Kenkyu*, 52(6):22–25 (2000)
2. Microfluidics Roadmap for the Life Sciences, J. Ducrée and R. Zengerls (eds.), Books-on-Demand GmbH, Norderstedt, ISBN 3-8334-0744-1, www.microfluidics-roadmap.com (2004)
3. A. Manz, N. Graber, and H. M. Widmer, "Miniaturized total chemical analysis systems: a novel concept for chemical sensing," *Sensors & Actuators B*, 1(1–6):244–248 (1990)
4. A. van den Berg and T. S. J. Lammerink, "Micro total analysis systems: microfluidic aspects, integration concept and applications," in *Microsystem Technology in Chemistry and Life Science*, vol. 194, Springer, Berlin, Heidelberg, New York (1997), pp. 21–49
5. S. Shoji, "Micro total analysis system (µTAS)," *Electronics & Communications in Japan—Part 2: Electronics*, 82(2):21–29 (1999)
6. D. Figeys and D. Pinto, "Lab-on-a-chip: a revolution in biological and medical sciences," *Anal. Chem.*, 72(9):330–335A (2000)
7. Cepheid, *www.cepheid.com,* August 2002

8. M. Schena, D. Shalon, R. W. Davis, and P. O. Brown, "Quantitative monitoring of gene expression patterns with a complementary DNA microarray," *Science*, 270(5235):467–470 (1995)
9. B. Lemieux, A. Aharoni, and M. Schena, "Overview of DNA chip technology," *Molecular Breeding*, 4(4):277–289 (1998)
10. P. K. Gupta, J. K. Roy, and M. Prasad, "DNA chips, microarrays and genomics," *Current Sci.*, 77(7):875–884 (1999)
11. J. G. Hacia, "Resequencing and mutational analysis using oligonucleotide microarrays," *Nature Genetics*, 21:42–47 (1999)
12. E. S. Lander, Array of hope," *Nature Genetics*, 21:3–4 (1999)
13. V. G. Cheung, M. Morley, F. Aguilar, A. Massimi, R. Kucherlapati, and G. Childs, "Making and reading microarrays," *Nature Genetics*, 21:15–19 (1999)
14. D. D. L. Bowtell, "Options available—from start to finish—for obtaining expression data by microarray," *Nature Genetics*, 21:25–32 (1999)
15. C. M. Roth and M. L. Yarmush," Nucleic acid biotechnology," *Annu. Rev. Biomed. Eng.*, 1:265–297 (1999)
16. *Microarray Biochip Technology* (M. Schena, ed.), Eaton Publishing (2000)
17. Affymetrix, *www.affymetrix.com*, August 2002
18. J. Ducrée and R. Zengerle, "Microfluidics—markets and technologies," *MST News*, 05(02):8–10 (2002)
19. W. Ehrfeld, V. Hessel, and H. Löwe, *Microreactors, New Technology for Modern Chemistry*, Wiley/VCH, Weinheim, Germany (2000)
20. P. Gravesen, J. Branebjerg, and O. S. Jensen, "Microfluidics—a review," *J. Micromech. Microeng.*, 3(4):168–182 (1993)
21. C. M. Ho, "Fluidics—the link between micro and nano sciences and technologies," in *Proceedings of the 14th IEEE International Conference on Micro Electro Mechanical Systems (MEMS'01), January 21–25, Interlaken, Switzerland* (2001), pp. 375–384
22. J. Ducrée and R. Zengerle, *Microfluidics*, Springer, Berlin, Heidelberg, New York (2005)
23. M. Koch, A. Evans, and A. Brunnschweiler, *Microfluidic Technology and Applications,* Microtechnologies and Microsystems (series), RSP Research Studies Press Ltd., Baldock, Hertfordshire, England (2000)
24. myFluidix.com—Free Online Resources in Microfluidics, *www.myfluidix.com*, June 2003
25. Microfluidics Worldwide Information Service, *www.microfluidics.de*, August 2002
26. CFD Research Corporation, *www.cfdrc.com,* 2000
27. Coventor (formerly Microcosm Technologies), *www.coventor.com*, August 2002
28. FIDAP (CFDRC), *www.cfdrc.com/software/fidap*, October 2002
29. POLYFLOW for Laminar Flows, *www.cfdrc.com/software/polyflow*, CFDRC, October 2002
30. FLUENT, *www.fluent. com*, 2002
31. FEMLAB—Multiphysics in Matlab (comsol), *www.femlab.com*, 2002

32. Saber—Mixed-Signal Simulator, *www.analogy. com*, 2000
33. SPICE, bwrc.eecs.berkeley.edu/Classes/IcBook/SPICE, August 2002
34. P. Wu and W. A. Little, " Measurement of friction factors for the flow of gases in very fine channels for microminiature Joule-Thomson refrigerators," *Cryogenics*, 23:273–277 (1983)
35. P. Wu and W. A. Little, "Measurement of the heat transfer characteristics of gas flow in fine heat exchangers used for microminiature refrigerators," *Cryogenics*, 24:415–420 (1984)
36. G. Fuhr, T. Müller, T. Schnelle, R. Hagedorn, A. Voigt, S. Fiedler, W. M. Arnold, U. Zimmermann, B. Wagner, and A. Heuberger, "Radio-frequency microtools for particle and live cell manipulation, "*Naturwissenschaften,* 81(12):528–535 (1994)
37. T. Schnelle, T. Müller, G. Gradl, S. G. Shirley, and G. Fuhr, "Dielectrophoretic manipulation of suspended submicron particles," *Electrophoresis*, 21(1):66–73 (2000)
38. J. M. Wilkinson, "Cost models for microsystem technology," *Micromachine Devices*, 4(3):1–3 (1999)
39. M. Madou, *Fundamentals of Microfabrication.* CRC Press (1997)
40. M. Madou and J. Florkey, "From batch to continuous manufacturing of microbiomedical devices," *Chem. Rev.*, 100(7):2679–2692 (2000)
41. W. Menz, *Microsystem Technology*, Wiley/VCH, Weinheim (2000)
42. M. Bachman, Y.-M. Chiang, C. Y. Chu, and G. Li, "Laminated microfluidic structures using a micromolding technique," in *Proceedings of SPIE— Microfluidic Devices and Systems II*, vol. 3877 (C. H. Ahn and A. B. Frazier, eds.), pp. 139–146 (1999)
43. J. Arnold, U. Dasbach, W. Ehrfeld, K. Hesch, and H. Löwe, "Combination of excimer laser micromachining and replication processes suited for large scale production," *Appl. Surface Sci.,* 86:251–258 (1995)
44. H. Becker and C. Gärtner, "Polymer microfabrication methods for microfluidic analytical applications," *Electrophoresis,* 21(1):12–26 (2000)
45. J. C. McDonald, D. C. Duffy, J. R. Anderson, D. T. Chiu, H. Wu, O. J. A. Schueller, and G. M. Whitesides, "Fabrication of microfluidic systems in poly (dimethylsiloxane)," *Electrophoresis*, 21(1):27–40 (2000)
46. CMOLD, *www.cmold. com*, November 2002
47. Y. Fintschenko and A. van den Berg, "Silicon microtechnology and microstructures," *J. Chromatography A*, 819(1–2):3–12 (1998)
48. J. H. Chan, A. T. Timperman, D. Qin, and R. Aebersold, "Microfabricated polymer devices for automated sample delivery of peptides for analysis by electrospray ionization tandem mass spectrometry," *Anal. Chem.*, 71:4437–4444 (1999)
49. MGT Mikroglas Technik AG, Mainz, *www.mikroglas.de*, August 2002
50. S. Shoji, S. Nakagawa, and M. Esashi, "Micropump and sample-injector for integrated chemical analyzing systems," *Sensors & Actuators A*, 21(1–3):189–192 (1990)
51. H. T. G. van Lintel, F. C. M. van de Pol, and S. Bouwstra, "A piezoelectric micropump, based on micromachining of silicon," *Sensors & Actuators*, 15(2):153–167 (1988)

52. X.-Q Wang and Y.-C Tai, "A normally closed in-channel micro check valve," in *Proceedings of the 13th Annual International Conference on Micro Electro Mechanical Systems (MEMS'2000), January 23–27, Miyazaki, Japan*, IEEE, Piscataway, NJ (2000), pp. 68–71
53. R. Zengerle, A. Richter, and H. Sandmaier, "A micro membrane pump with electrostatic actuation," in *Proceedings of IEEE Micro Electro Mechanical Systems (MEMS'92), February 4–7, Travemünde, Germany* (W. Benecke and H.-C. Petzold, eds.), IEEE (1992), pp. 19–24
54. E. Stemme and G. Stemme, "A valveless diffuser/nozzle-based fluid pump," *Sensors & Actuators A*, 39(2):159–167 (1993)
55. T. Gerlach and H. Wurmus, "Working principle and performance of the dynamic micropump," *Sensors & Actuators A*, 50(1–2):135–140 (1995)
56. A. Olsson, O. Larsson, J. Holm, L. Lundbladh, O. Ohman, and G. Stemme, "Valve-less diffuser micropumps fabricated using thermoplastic replication," in *Proceedings of the 10th IEEE Annual International Workshop on Micro Electro Mechanical Systems (MEMS'97), January 26–30, Nagoya, Japan*, IEEE, New York (1997), pp. 305–310
57. W. van der Wijngaart, H. Andersson, P. Enoksson, K. Noren, and G. Stemme, "The first self-priming and bi-directional valve-less diffuser micropump for both liquid and gas," in *Proceedings of the 13th IEEE Annual International Conference on Micro Electro Mechanical Systems (MEMS'2000), January 23–27, Miyazaki, Japan*, IEEE, Piscataway, NJ (2000), pp. 674–679
58. Nikola Tesla, valvular conduit, U.S. patent no. 1,329,559, 1920
59. Fred Forster, *lettuce.me.washington.edu/micropump*, University of Washington, August 2002
60. M. R. McNeely, M. K. Spute, N. A. Tusneem, and A. R. Oliphant, "Hydrophobic microfluidics," in *Proceedings of SPIE—Microfluidic Devices and Systems II*, vol. 3877 (C. H. Ahn and A. B. Frazier, eds.), pp. 210–220, (1999)
61. K. Handique, D. T. Burke, C. H. Mastrangelo, and M. A. Burns, "Nanoliter-volume discrete drop injection and pumping in microfabricated chemical analysis systems," in *Proceedings of the 8th Solid-State Sensor & Actuator Workshop, June 8–11, Hilton Head Island, SC* (K. H. Chau and P. J. French, eds.), pp. 346–349 (1998)
62. K. Hosokawa, T. Fujii, and T. Endo, "Handling of small volume droplets using hydrophobic microcapillary vent," *Trans. Inst. Elect. Eng. Japan*, A119–EE(10):470–475 (1999)
63. E. T. Lagally, P. C. Simpson, and R. A. Mathies, "Monolithic integrated microfluidic DNA amplification and capillary electrophoresis analysis system," *Sensors &Actuators B*, 63:138–146 (2000)
64. D. K. Jones, C. H. Mastrangelo, M. A. Burns, and D. T. Burke, "Selective hydrophobic and hydrophilic texturing of surfaces using photolithographic photodeposition of polymers [for microfluidic sample control]," in *Proceedings of SPIE—Microfluidic Devices and Systems*, vol. 3515 (A. B. Frazier and C. H. Ahn, eds.), pp.136–143 (1998)

65. M. J. Zdeblick, R. Anderson, J. Jankowski, B. Kline-Schoder, L. Christel, R. Miles, and W. Weber, "Thermopneumatically actuated microvalves and integrated electro-fluidic circuits," in *Proceedings of the 6th Solid-State Sensor & Actuator Workshop, June 13–16, Hilton Head Island, SC*, Transducer Res. Found., Cleveland Heights, OH (1994), pp. 251–255
66. H. Jerman. "Electrically activated normally closed diaphragm valves," *J. Micromech. Microeng.*, 4(4):210–216 (1994)
67. P. W. Barth, C. C. Beatty, L. A. Field, J. W. Baker, and G. B. Gordon, "A robust normally closed silicon microvalve," in *Proceedings of the 6th Solid-State Sensor & Actuator Workshop, June 13–16, Hilton Head Island, SC* (1994), pp. 248–250
68. S. Messner, M. Müller, V. Burger, J. Schaible, H. Sandmaier, and R. Zengerle, "A normally-closed, bimetallically actuated three-way microvalve for pneumatic applications," in *Proceedings of the 11th IEEE Annual International Workshop on Micro Electro Mechanical Systems (MEMS'98), January 25–29, Heidelberg, Germany*, IEEE, New York (1998), pp. 40–44
69. R. Zengerle, "Microfluidics," in Proceedings of MME'98—Micromechanics Europe, Ulvik, Norway (1998), pp.111–122
70. M. Dunbar, "Electronic pressure regulators using micromachined silicon pressure sensors and silicon microvalves," in *Proceedings 7th International Conference for Sensors Transducers and Systems (SENSOR 95)*, pp. 381–385, (1995)
71. M. W. Hamberg, C. Neagu, J. G. E. Gardeniers, D. J. Ijntema, and M. Elwenspoek, "An electrochemical micro actuator," in *Proceedings of IEEE Micro Electro Mechanical Systems (MEMS '95), January 29–February 2, Amsterdam, The Netherlands* (1995), pp. 106–110
72. S. Böhm, W. Olthuis, and P. Bergveld, "A bi-directional electrochemically driven micro liquid dosing system with integrated sensor/actuator electrodes," in *Proceedings of the 13th IEEE Annual International Conference on Micro Electro Mechanical Systems (MEMS'2000), January 23–27, Miyazaki, Japan,* IEEE, Piscataway, NJ (2000), pp. 92–95
73. D. Eddington, R. Lui, J. Moore, and D. Beebe, "An organic self-regulating microfluidic system," *Lab on a Chip*, 1(2):96–99 (2001)
74. S. Messner, M. Müller, J. Schaible, and R. Zengerle, "A modular housing concept for pneumatic microsystems," in Proceedings of the 11th IEEE Annual International Workshop on Micro Electro Mechanical Systems (MEMS'98), January 25–29, Heidelberg, Germany, VDE Verlag, Berlin, Germany (1998), pp. 615–620
75. J. G. Smits and N. V. Vitafin, Piezo-electrical micropump, patents EP0134614 and NL8302860, 1985
76. R. Zengerle, S. Kluge, J. Richter, and A. Richter, "A bidirectional silicon micropump," in *Proceedings of IEEE Micro Electro Mechanical Systems (MEMS'95), January 29–February 2, Amsterdam, The Netherlands*, IEEE, New York (1995), pp. 19–24

77. D. J. Harrison, K. Fluri, K. Seiler, F. Zhonghui, C. S. Effenhauser, and A. Manz, "Micromachining a miniaturized capillary electrophoresis-based chemical analysis system on a chip," *Science*, 261(5123):895–897 (1993)
78. A. Manz, C. S. Effenhauser, N. Burggraf, D. J. Harrison, K. Seiler, and K. Fluri, "Electroosmotic pumping and electrophoretic separations for miniaturized chemical analysis systems," *J. Micromech. Microeng.*, 4(4):257–265 (1994)
79. C. T. Culbertson, R. S. Ramsey, and J. M. Ramsey, "Electroosmotically induced hydraulic pumping on microchips: differential ion transport," *Anal. Chem.*, 72:2285–2291 (2000)
80. F. C. M. van de Pol, H. T. G. van Lintel, M. Elwenspoek, and J. H. J. Fluitman, "A thermopneumatic micropump based on micro-engineering techniques," *Sensors & Actuators A*, 21(1–3):198–202 (1990)
81. K. P. Kämper, J. Döpper, W. Ehrfeld, and S. Oberbeck, "A self-filling low-cost membrane micropump," in *Proceedings of the 11th IEEE Annual International Workshop on Micro Electro Mechanical Systems (MEMS'98), January 25–29, Heidelberg, Germany*, IEEE, New York (1998), pp. 432–437
82. R. Linnemann, P. Woias, C. D. Senfft, and J. A. Ditterich, "A self-priming and bubble-tolerant piezoelectric silicon micropump for liquids and gases, in *Proceedings of the 11th IEEE Annual International Workshop on Micro Electro Mechanical Systems (MEMS'98), January 25–29, Heidelberg, Germany*, IEEE, New York (1998), pp. 532–537
83. J. G. Smits, "Piezoelectric micropump with three valves working peristaltically (for insulin delivery)," *Sensors & Actuators A*, 21(1–3):203–206 (1990)
84. J. M. Crowley, G. Wright, and J. C. Chato, "Selecting a working fluid to increase the efficiency and flow rate of an EHD pump," in *Conference Record of the 1987 IEEE Industry Applications Society Annual Meeting*, vol. 2, IEEE, New York (1987), pp. 1433–1438
85. S. F. Bart, L. S. Tavrow, M. Mehregany, and J. H. Lang, "Microfabricated electrohydrodynamic pumps," *Sensors & Actuators A*, 21(1–3):193–197 (1990)
86. A. Richter, "An electrohydrodynamic micropump," in *Actuator 90, Proceedings of 2nd International Technology-Transfer Congress, Bremen, Germany* (1990), pp. 172–177
87. G. Fuhr, R. Hagedorn, T. Müller, W. Benecke, and B. Wagner, "Pumping of water solutions in microfabricated electrohydrodynamic systems, in *Proceedings of IEEE Micro Electro Mechanical Systems (MEMS'92), February 4–7, Travemünde, Germany*, IEEE, New York (1992), pp. 25–30
88. D. C. Duffy, H. L. Gillis, J. Lin, N. F. Sheppard, and G. J. Kellogg, "Microfabricated centrifugal microfluidic systems: characterization and multiple enzymatic assays," *Anal. Chem.*, 71(20):4669–4678 (1999)
89. G. Ekstrand, C. Holmquist, A. Edman Örlefors, B. Hellman, A. Larsson, and P. Andersson, "Microfluidics in a rotating CD," in *Proceedings of the Micro Total Analysis Systems Symposium (μTAS 2000), May 14–18, Enschede, The Netherlands* (A. van den Berg, W. Olthuis, and P. Bergveld, eds.), Kluwer Academic, The Netherlands (2000), pp. 311–314

90. HNP Mikrosysteme GmbH, *www.hnp-mikrosysteme.de*. Parchim (Germany), August 2002
91. R. M. Moroney, R. M. White, and R. T. Howe, "Ultrasonically induced microtransport," in *Proceedings of IEEE Micro Electro Mechanical Systems (MEMS'91), January 30–February 2, Nara, Japan*, IEEE (1991), pp. 277–282
92. K. Ikekame, K.-H. Hashimoto, and M. Yamaguchi, "Micro-actuators using acoustic streaming induced by high-frequency ultrasonic waves," *Trans. Inst. Elect. Eng. Japan,* A118-E(7–8):347–351 (1998)
93. S. Shoji, "Fluids for sensor systems," in *Microsystem Technology in Chemistry and Life Sciences* (A. Manz and H. Becker, eds.), Springer, Berlin, Heidelberg, New York (1997), pp. 163–188
94. G. A. Urban and G. Jobst, "Sensor systems," in *Microsystem Technology in Chemistry and Life Sciences* (A. Manz and H. Becker, eds.), Springer, Berlin, Heidelberg, New York (1998), pp. 189–213
95. S. Middelhoek and S. A. Audet, *Silicon Sensors*, Academic Press Ltd., London, UK (1989)
96. M. A. Boillat, B. van der Schoot, P. Arquint, and N. F. de Rooij, "Controlled liquid dosing in microinstruments," in *Proceedings of SPIE—Microfluidic Devices & Systems II, August 1999*, vol. 3877 (C. H. Ahn and A. B. Frazier, eds.), pp. 20–27
97. M. Ashauer, H. Glosch, H. Ashauer, H. Sandmaier, and W. Lang, "Liquid mass flow sensor using dynamic pressure detection," in *Micro-Electro-Mechanical Systems (MEMS)—1998, ASME International Mechanical Engineering Congress and Exposition* (L. Lin, F. K. Forster, N. R. Aluru, and X. Zhang, eds.), New York (1998), pp. 421–425
98. HSG–IMIT, *www.hsg-imit.de,* August 2002
99. M. Elwenspoek, T. S. J. Lammerink, R. Miyake, and J. H. J. Fluitman, "Towards integrated microliquid handling systems," *J. Micromech. Microeng.*, 4(4):227–245 (1994)
100. T. Ebefors, E. Kalvesten, and G. Stemme, "Three dimensional silicon triple-hot-wire anemometer based on polyimide joints," in *Proceedings of the 11th IEEE Annual International Workshop on Micro Electro Mechanical Systems (MEMS'98), January 25–29, Heidelberg, Germany*, IEEE, New York (1998), pp. 93–98
101. M. Ashauer, H. Glosch, F. Hedrich, N. Hey, H. Sandmaier, and W. Lang, "Thermal flow sensor for liquids and gases based on combinations of two principles," *Sensors & Actuators A,* 73(1–2):7–13 (1999)
102. S. Shoji, M. Esashi, and T. Matsuo, "Prototype miniature blood gas analyser fabricated on a silicon wafer," *Sensors & Actuators*, 14(2):101–107 (1988)
103. P. Bergveld, "Exploiting the dynamic properties of FET-based chemical sensors," *J. Phys. E*, 22(9):678–683 (1989)
104. A. van den Berg, P. Bergveld, D. N. Reinhoudt, M. Elwenspoek, and J. H. J. Fluitman, "Miniaturized chemical analysis systems," in *5th International Symposium on Micro Machine and Human Science*, IEEE, New York (1994), pp. 181–184

105. P. Bergveld, "The future of biosensors," *Sensors & Actuators A,* 56(1–2):65–73 (1996)
106. P. Koltay, J. Ducrée, and R. Zengerle, "Nanoliter and picoliter liquid handling," in *Micro Total Analysis Systems* (A. van den Berg and E. Oosterbroek, eds.), Elsevier Science, Amsterdam, The Netherlands (2003), pp. 151–171
107. J. Driscoll, R. Delmendo, R. Papen, and D. Sawutz, "Multi PROBE nL complements drug discovery assay miniaturization," *J. Biomolecular Screening,* 3(3):237–239 (1998)
108. GeSiM, Gesellschaft für Silizium-Mikrosysteme mbH, w*ww.gesim.de,* August 2002
109. Microdrop—Gesellschaft für Mikrodosiersysteme mbH, *www.microdrop.de,* August 2002
110. J. Comley, "Nanoliter dispensing—on the point of delivery," *Drug Discovery World,* 3(3):33–44 (2002)
111. P. Koltay, G. Birkle, R. Steger, H. Sandmaier, and R. Zengerle, "Highly parallel and accurate nanoliter dispenser for high-throughput synthesis of chemical compounds," in *Proceedings of iMEMS* (2000)
112. Packard Instrument Company, *www.lifesciences.perkinelmer.com,* August 2002
113. D. Maillefer, H. van Lintel, G. Rey-Mermet, and R. Hirschi, "A high-performance silicon micropump for an implantable drug delivery system," in *Proceedings of the 12th IEEE Annual International Conference on Micro Electro Mechanical Systems (MEMS'99), January 17–21, Orlando, Florida,* IEEE (1999), pp. 541–546
114. R. P. Ekins, "Ligand assays: from electrophoresis to miniaturized microarrays," *Clinical Chem.,* 44(9):2015–2030 (1998)
115. R. Ekins and F. W. Chu, "Microarrays: their origins and applications," *Trends in biotechnology,* 17(6):217–218 (1999)
116. D. Shalon, S. J. Smith, and P. O. Brown," A DNA microarray system for analyzing complex DNA samples using two-color fluorescent probe hybridization," *Genome Research,* 6:639–645 (1996)
117. Stanford Tip Gallery, *cmgm.stanford.edu/pbrown/mguide/tips.html,* The Brown Lab, August 2002
118. J. Ducrée, H. Gruhler, N. Hey, M. Müller, S. Békési, M. Freygang, H. Sandmaier, and R. Zengerle, "TopSpot—a new method for the fabrication of microarrays," in *Proceedings of the 13th IEEE Annual International Conference on Micro Electro Mechanical Systems (MEMS'2000), January 23–27, Miyazaki, Japan,* IEEE (2000), pp. 317–322
119. J. Ducrée, B. de Heij, H. Sandmaier, and R. Zengerle, "Fabrication of microarrays on an industrial scale with TopSpot," *MST News,* 4:22–23 (2000)
120. Lab for MEMS Applications, *www.imtek.de/anwendungen/english/,* IMTEK, August 2002
121. R. C. Anderson, G. McGall, and R. J. Lipshutz, "Polynucleotide arrays for genetic sequence analysis," in *Microsystem Technology in Chemistry and Life Science,* vol. 194, Springer, Berlin, Heidelberg (1997), pp. 117–129

122. R. J. Lipshutz, S. P. A. Fodor, T. R. Gingeras, and D. J. Lockhart, "High density synthetic oligonucleotide arrays," *Nature Genetics*, 21:20–24 (1999)
123. Luminex Corporation, *www.luminexcorp.com*, August 2002
124. PharmaSeq, *www.pharmaseq.com*, August 2002
125. Illumina, Inc., *www.illumina.com*, August 2002
126. *Proceedings of the 3rd International Conference on Microreaction Technology (IMRET 3), April 18–21, Frankfurt, Germany* (W. Ehrfeld, ed.), Springer (1999)
127. C. Erbacher, F. G. Bessoth, M. Busch, E. Verpoorte, and A. Manz, "Towards integrated continuous-flow chemical reactors," *Microchimica Acta*, 131(1/2):19–24 (1999)
128. F. G. Bessoth, A. J. de Mello, and A. Manz, "Microstructure for efficient continuous flow mixing," *Analytical Commun.*, 36(6):213–215 (1999)
129. W. Ehrfeld, K. Golbig, V. Hessel, H. Löwe, and T. Richter, "Characterization of mixing in micromixers by a test reaction: single mixing units and mixer arrays," *Ind. Eng. Chem. Res.*, 38(3):1075–1082 (1999)
130. J. Evans, D. Liepmann, and A. P. Pisano, "Planar laminar mixer (MEMS fluid processing system)," in *Proceedings of the 10th IEEE Annual International Workshop on Micro Electro Mechanical Systems (MEMS'97), January 26–30, Najoya, Japan*, IEEE, New York (1997), pp. 96–101
131. M. Koch, D. Chatelain, A. G. R. Evans, and A. Brunnschweiler, "Two simple micromixers, based on silicon," *J. Micromech. Microeng.*, 8:123–126 (1998)
132. IMM—Institut für Mikrostrukturtechnik Mainz, *www.imm-mainz.de*, August 2002
133. N. Schwesinger, T. Frank, and H. Wurmus, "A modular microfluid system with an integrated micromixer," *J. Micromech. Microeng.*, 6(1):99–102 (1996)
134. T. S. Ravigururajan, "Impact of channel geometry on two-phase flow heat transfer characteristics of refrigerants in microchannel heat exchangers," *Journal of Heat Transfer—Transactions of the ASME*, 120(2):485–504 (1998)
135. D. Hönicke, "Microchemical reactors for heterogeneously catalyzed reactions," *Studies in Surface Science & Catalysis*, 122:47–62 (1999)
136. FZ Karlsruhe, Projekt Mikrosystemtechnik, *www.fzk.de/pmt*, August 2002
137. W. Bier, W. Keller, W. Linder, D. Seidel, and K. Schubert, "Manufacturing and testing of compact micro heat exchangers with high volumetric heat transfer coefficients," *ASME DSC*, 19:189–197)(1990)
138. F. Jones, A. Kuppusamy, and A.-P. Zheng, "Dehydrogenation of cyclohexane to benzene in microreactors," in *Proceedings of SPIE—Microfluidic Devices and Systems II*, vol. 3877 (C. H. Ahn and A. B. Frazier, eds.), pp. 160–168 (1999)
139. E. L'Hostis, P. E. Michel, G. C. Fiaccabrino, D. J. Strike, N. F. de Rooij, and M. Koudelka-Hep, "Microreactor and electrochemical detectors fabricated using Si and EPON SU-8, *Sensors & Actuators B*, 64(1–3):156–162 (2000)
140. H. Löwe and W. Ehrfeld, "State-of-the-art in microreaction technology: concepts, manufacturing and applications," *Electrochimica Acta*, 44(21–22):3679–3689 (1999)

141. P. M. Martin, D. W. Matson, W. D. Bennettan, D. C. Stewart, and C. C. Bonham, "Laminated ceramic microfluidic components for microreactor applications, in *AIChE 2000 Spring Meeting, Atlanta, GA* (2000)
142. P. L. Mills and R. V. Chaudhari, "Multiphase catalytic reactor engineering and design for pharmaceuticals and fine chemicals," *Catalysis Today*, 37(4):367–404 (1997)
143. R. Srinivasan, I. M. Hsing, P. E. Berger, K. F. Jense, S. L. Firebaugh, M. A. Schmidt, M. P. Harold, J. J. Lerou, and J. F. Ryley, "Micromachined reactors for catalytic partial oxidation reactions," *AIChE Journal*, 43(11):3059–3069 (1997)
144. A. Y. Tonkovich, J. L. Zilka, D. M. Jimenez, M. J. Lamont, and R. S. Wegeng, "Microchannel chemical reactors for fuel processing," in *AIChE 1998 Spring National Meeting, New Orleans* (1998)
145. N. Burggraf, A. Manz, E. Verpoorte, C. S. Effenhauser, H. M. Widmer, and N. F. de Rooij, "A novel approach to ion separations in solution: synchronized cyclic capillary electrophoresis (SCCE)," *Sensors & Actuators B*, 20(2–3):103–110 (1994)
146. A. G. Hadd, D. E. Raymond, J. W. Halliwell, S. C. Jacobson, and J. M. Ramsey, "Microchip device for performing enzyme assays," *Anal. Chem.*, 69(17):3407–3412 (1997)
147. S. Shoji, "Micro total analysis system (μTAS) chemical analysis," *Transactions of the Institute of Electronics, Information & Communication Engineers of Japan*, J81C-1(7):385–393 (1998)
148. H. Becker and W. Dietz, "Microfluidic devices for μ-TAS applications fabricated by polymer hot embossing," in *Proceedings of SPIE—Microfluidic Devices & Systems, Santa Clara, 21–22 September 1998*, vol. 3515 (A. B. Frazier and C. H. Ahn, eds.), pp. 177–182
149. G. Dutton, "Hewlett Packard and Caliper launch microfluidics lab-on-chip—HP 2100 bioanalyzer debuts for biotech and pharmaceutical industries," *Genetic Eng. News*, 19(16):1 (1999)
150. D. J. Harrison, C. Wang, P. Thibeault, F. Ouchen, and S. B. Cheng, "The decade search for the killer ap in μ-TAS, in *Proceedings of the Micro Total Analysis Systems Symposium (μTAS 2000), May 14–18, Enschede, The Netherlands* (A. van den Berg, W. Olthuis, and P. Bergveld, eds.), Kluwer Academic Publishers (2000), pp. 195–204
151. J. P. Kutter, "Current developments in electrophoretic and chromatographic separation methods on microfabricated devices," *TrAC—Trends in Anal. Chem.*, 19(6):352–363 (2000)
152. S. Verpoorte, "Nanoscale chemical analysis," *TrAC—Trends in Analytical Chemistry*, 19(6):350–351 (2000)
153. Aclara BioSciences, Inc., *www.aclara.com*, August 2002
154. Caliper Technologies, *www.calipertech.com*, August 2002
155. *Proceedings of the Micro Total Analysis Systems Symposium (μTAS 2002), November 3–7, Nara, Japan* (Y. Baba, S. Shoji, and A. van den Berg, eds.), Kluwer Academic Publishers (2002)

156. T. T. Veenstra, T. S. J. Lammerink, M. C. Elwenspoek, and A. van den Berg, "Characterization method for a new diffusion mixer applicable in micro flow injection analysis systems," *J. Micromech. Microeng.*, 9:199–202 (1999)
157. A. G. Hadd, S. C. Jacobson, and J. M. Ramsey, "Microfluidic assays of acetylcholinesterase inhibitors," *Anal. Chem.*, 71(22):5206–5212 (1999)
158. B. Stanislawski, D. Kaniansky, M. Masár, and M. Jöhnck, "Design principles, performance and perspectives of a complete miniaturized electrophoretic system, in *Proceedings of the Micro Total Analysis Systems Symposium (μTAS 2002), November 3–7, Nara, Japan*, vol. 1 (Y. Baba, S. Shoji, and A. van den Berg, eds.), Kluwer Academic Publishers (2002), pp. 350–352

13 Biomedical Systems

Whye-Kei Lye and Michael Reed

Department of Electrical and Computer Engineering
University of Virginia, Charlottesville, Virginia

13.1 Introduction and Overview

The phenomenal success and widespread proliferation of microelectronic devices can be attributed in large part to the mass production technologies that have been developed for their manufacture. These microfabrication techniques have been honed to maximize device yield and microelectromechanical systems (MEMS) capitalize on these semiconductor microfabrication tools to create miniature systems that meld electronic and mechanical functions. Miniaturization brings with it the benefits of smaller devices that cost less, require less power, and incorporate greater functionality. The use of microsystems in biomedical applications holds the promise of improving patient care in a minimally invasive manner while simultaneously reducing health care costs.

The biomedical applications of microsystems fall broadly into the two categories of diagnostic and therapeutic systems. Diagnostic applications include DNA diagnostics,[1-3] systems on a chip,[4-6] and cell[7] and molecule[8] sorting. Therapeutic systems include drug and gene delivery,[9-15] tissue augmentation/repair,[16] micro/minimally invasive surgical systems,[17,18] and biocapsules.[19,20]

These categories are necessarily broad, and the applications listed are by no means exhaustive: A tremendous number of biomedical microsystems have been proposed and are under development. Further, with increased opportunities and capabilities to integrate functionality on microsystems, it is likely that we will see the advent of devices that bridge these categories.

Significant opportunities exist to leverage the capabilities of microsystems in therapeutic applications. This chapter discusses specific requirements and challenges unique to the development of biomedical devices in general, with particular focus on therapeutic systems.

13.2 Materials and Fabrication Techniques

A considerable impediment facing the development of biomedical microsystems is the stringent regulatory controls to which they are subject. Before a medical device is approved for market, it must pass through an obstacle course of legislation designed to protect the patient. Medical devices must be clinically tested, shown to be appropriate for treating the indications targeted, and demonstrated to be safe. The manufacture of these devices is further subject to extensive legislation with an eye to maximizing safety. These regulations need to be considered throughout the development process, and devices must be engineered to meet these requirements. Critical to meeting those requirements is the consideration of biocompatibility: the manner in which the materials comprising a microsystem are affected by, and in turn affect, the patient.

This section discusses the material requirements for biomedical microsystems, presents some information on biocompatible materials, and explores some of the methods used for micromachining these materials.

13.2.1 Material Requirements

The choice of materials for a microsystem is dictated by its functional specifications and operating environment. In the case of biomedical applications, the environment presented by the patient's physiology is profoundly diverse and challenging. Microsystems may be subject to peak mechanical stresses that can vary from 4 MPa for muscle contractions to 40 and 80 MPa in tendons and ligaments, respectively, with the annual number of stress cycles ranging from 10^5 for peristalsis to 5×10^7 for heart contractions. Local acidity varies from pH 1 for gastric contents to pH 7.2 for blood, with some fluids such as urine spanning a large range from pH 4.5 to pH 6.[21] When exposed to these conditions, materials can undergo deformation and failure, friction and wear, corrosion and dissolution, swelling and leaching, and reactions with biological molecules. To further complicate matters, all medical devices must be sterilized prior to use in order to prevent the transmission of disease and minimize the risk of infection. Several methods are used for sterilization, and the microsystems must survive this process. The techniques used include immersion in commercial solutions of formaldehyde, heat (dry, 160 to 175°C; wet, 120 to 130°C), gas (ethylene oxide, 400 to 1200 mg/liter), and irradiation (^{60}Co gamma; 10 to 40 kGy dose). Reusable devices must be able to survive repeated sterilization processes without degradation.

The mechanical characteristics necessary for microsystem operation must be tempered by the requirement for biocompatibility. Microsystems are, necessarily, objects foreign to the human body, and on introduction will stimulate a physiological response in the patient. Some possible reactions are inflammation, coagulation and hemolysis, immune and allergenic response, tissue growth, cytotoxicity and carcinogenesis, organ failure, and death. While some sort of physiological response is inevitable, the materials used should not exacerbate these reactions. Most materials in clinical use elicit specific, controlled, and beneficial responses in the host.

Given the complexity of biological systems, the assessment of biocompatibility is best done *in vivo*. One method is to evaluate tissue response using light microscopy. After being implanted with a foreign body, tissue undergoes a series of well-characterized physiological reactions. Acute inflammation, which occurs over the first two days, manifests itself in terms of rubor (local vascular dilation causing redness), calor (increased blood flow causing a rise in temperature), tumor (increased vascular permeability resulting in swelling), and dolor (the accumulation of chemical mediators causing pain).

The severity of the inflammation is related in part to the types of proteins absorbed onto the surface of the materials in contact with the tissue. Common blood proteins like albumin and fibrinogen elicit very different responses: Albumin does not appear to contribute to inflammation, while fibrinogen antagonizes inflammation by causing the accumulation of phagocytes. Absorption studies with these proteins can provide useful *in vitro* information in the early stages of material evaluation.

Acute inflammatory responses usually subside after one week. Over a longer period, the implant becomes covered in a capsule of fibrotic tissue. Capsules around biocompatible materials tend to be thin, minimally fibrotic, and have few leukocytes. Antagonistic materials will cause a continued tissue reaction, resulting in thick, more vascularized capsules with elevated numbers of white blood cells. Measurement of systemic levels of phagocytes is often used to evaluate a patient's response to surgical implants.

A summary of materials in clinical use is presented in Table 13.1. These materials include metals and alloys, ceramics, glasses, and plastics. The compatibility of a material for a given application is highly specific because histologic responses vary: Different tissues react differently to a given material. Materials that are suitable in one application may fail in another.

Biocompatible polyurethane elastomers, for example, have been used successfully to fabricate catheters and tubing and to insulate pacemaker leads. When applied to the vascular system as a drug delivery coating on coronary stents, however, polyurethane stimulates post-implant cell proliferation, aggravating the restenotic response in patients, with possibly severe consequences.

Table 13.1. Biocompatible Materials and Applications

Material	Application
Stainless steels	Surgical instruments
	Screw implants
	Hip endoprostheses
	Coronary stents
	Carotid stents
Cobalt-chromium alloys	Joint replacement
Titanium alloys	Orthopedic implants
	Coronary stents
Shape memory alloys (NiTi)	Self-expanding stents
	Actuators
Ceramics: alumina and zirconia	Dental implants
	Joint replacement
	Bone screws
Bioactive glasses: compositions of Na_2O, CaO, P_2O_5, and SiO_2	Bone implants
	Vertebral prostheses
Carbon composites	Hip endoprostheses
	Heart valves
	Sutures
Polymers	Balloon catheters
	Tubing
	Bioactive membranes
	Controlled-release delivery systems
	Artificial cells
	Vascular grafts
	Reconstructive implants
	Artificial corneas

Rapoport et al.[22] reported on the development of a polysilicon rotor for blood flow measurements. The device worked successfully with gas and liquid flows, but gave unexpected results when tested with canine blood. Figure 13.1 shows the interaction of this microsystem with blood. The round features adhering to the device are erythrocytes (red blood cells), which tend to stick preferentially to rough polysilicon surfaces. The dimensions of these cells, 5 to 7 µm in diameter, are comparable to the gaps between the polysilicon layers, with the resulting possibility of adherent cellular material impeding rotor function. In retrospect, a thrombogenic study of polysilicon might have brought these issues to light at an earlier stage of microsystem development.

Human physiology presents an extremely hostile environment with highly complex responses that can be difficult to anticipate. Regardless of

Figure 13.1 The impact of canine erythrocytes on polysilicon microsystem function.[19]

the materials used, all medical implants are artificial constructs and, as such, are foreign bodies that elicit physiological response in the patient. Clinical approval of these devices is dependent on understanding the effect of these responses on patient health and device performance. It is beyond our purview to address all the salient issues in this matter; several comprehensive works have been written on biocompatible materials and biocompatibility testing.[21]

13.2.2 Fabrication Techniques

The fabrication of microstructures and micromachines for biomedical applications requires the use of tools from semiconductor microfabrication and conventional machining. There are two reasons for this. First, the sizes of biomedical microsystems and their components range from a few microns to a few millimeters. This range sits at the nexus of conventional machining and microfabrication. Second, a review of Table 13.1 reveals a dearth of biocompatible materials that can be processed using semiconductor microfabrication techniques. It is therefore imperative to look beyond the semiconductor arena to identify and adopt the tools necessary for working with these materials on the micron scale.

The semiconductor microfabrication tools that were initially developed for manufacturing microelectronic devices have provided the foundation for the development of microsystems technology. These microelectronic devices continue to play an integral role in medical microsystems, providing functionality such as computation, control, sensing, and actuation that cannot otherwise be realized. Chapter 15 and other salient references[23-26] provide a more complete discussion of semiconductor and microsystem fabrication.

While the techniques described above have generally been applied to semiconductor materials, they have also been adapted to meet the specialized requirements of microsystem fabrication. Thick, photosensitive polymeric films, such as the epoxy-based SU-8, have been developed. These polymeric films, which can be in excess of 100 μm thick, are lithographically patterned and used as structural materials for microsystems.[27] SU-8 has been used to fabricate electrophoretic channels,[28] as well as prototype seat belt actuator restraint mechanisms. Another thick film polymeric process is LIGA (German for lithographie, galvanoformung, abformung). This process uses poly(methyl methacrylate) (PMMA) layers as thick as 200 μm that are lithographically patterned by X-ray sources. The developed PMMA material forms a template into which thick metal films are electrodeposited. When the original PMMA layer is removed, the resultant metal form serves as a template for injection molding of polymers; this makes possible high-volume production of microstructured polymeric parts.

Additionally, Lee *et al.* recently reported on surface modification methods that allow silicon microfabrication tools to be applied to inert fluoropolymers such as polytetrafluoroethylene (PTFE).[29] The low surface energies of these materials make them difficult to coat with photoresists, preventing the use of lithography. Lee *et al.* demonstrated that low-energy ion bombardment improves the wetting characteristics of these materials, allowing photoresist layers to be applied and patterned. Ion beam etching is then used to transfer the masked patterns into the substrate. Submicron-scale structures including undercut cantilevers were fabricated with this technique.

Fabrication techniques based on lithographic pattern transfer, such as those described above, are limited to planar substrates. The utility of the processes is also restricted to a specific set of materials. To realize fully the potential of biomedical microsystems, it is necessary to explore the use of nontraditional microfabrication technologies. This is motivated by the desire to build microsystems with nonplanar geometries and to use a wider range of materials.

High-precision machining tools such as laser micromachining and electrical discharge machining (EDM) have been successfully used in the fabrication of microsystems. Used in conjunction with computer numeric control (CNC) systems, these techniques allow for precision machining of a wide variety of materials.

Most of the materials in Table 13.1 can be laser-machined, provided that the beam characteristics are properly tailored to the material. Metals and alloys such as stainless steel, shape-memory alloys (NiTi), and copper can be laser-machined, as can ceramics like alumina (Al_2O_3) and aluminum nitride (AlN).[30] Laser cutting is routinely used in the manufacture of coronary stents made of stainless steel and titanium. Excimer lasers can also be used to photo-ablate a wide range of polymers, including polystyrene, polycarbonate, poly(tetrafluoroethylene) (Teflon), poly(ethylene terephthalate) (PET), and PMMA.[31]

EDM, which is primarily used to machine conductive materials, including titanium, stainless steel, and nitinol,[32] can also be used on certain semiconductors (including silicon)[33] and a variety of ceramic materials. The electrodes can also be immersed in a conducting electrolyte such as a sodium hydroxide solution. This modified technique, known as electrochemical discharge machining (ECDM), has been used to microstructure glass.[34]

The combination of semiconductor microfabrication and precision machining is key to building viable biomedical microsystems. The following sections examine some medical microsystems and the ways in which these tools are important in their fabrication.

13.3 Surgical Systems

One of the most promising applications for microsystems is in the field of surgery. Conventional surgical procedures are generally intrusive and traumatic for the patient. Large incisions are typically required to gain entry into the body. Additional trauma may be sustained as organs and bone are moved to provide access to the region of interest for the surgeon's hands and tools.

The massive physiological insult to which the patient is subjected during conventional surgery places extreme stress on the body. This causes the patient pain and increases the risk of post-operative complications. Long convalescent times associated with invasive surgical techniques also increase health care costs. Scarring resulting from surgery is a further problem. The highly invasive nature of conventional operations also limits its efficacy in treating patients who have weak immune systems or who are in generally poor health and are unlikely to recover from the procedures.

The use of microsystems in minimally invasive surgery holds the promise of reducing surgical trauma and shortening recovery times and hospital stays. By using small incisions or natural orifices to access the region of interest, these procedures are less invasive, causing the patient less pain and minimizing scarring. It may also be possible for some procedures to be performed on an outpatient basis, further lowering health care costs. The initial capital costs of such minimally invasive microsurgical systems are more than paid for by savings.

Microsurgical tools, catheters, and endoscopes allow for the development of new procedures and approaches to surgery. Surgical microsystems hold the promise of extending the reach of surgery into previously inaccessible areas and are key to the future advancement of medicine. Areas of application of minimally invasive surgery (Fig. 13.2) include the brain, heart and blood vessels (vascular), lungs (thoracoscopy), joints (arthroscopy), gallstones and kidney stones, esophagus and stomach (endoscopy), and abdomen (laparoscopy). In fact, most of the gall bladder removals, hysterectomies, and other urinogenitary tract procedures performed in the United States are minimally invasive in nature. It has been suggested that 75% of thoracic and abdominal operations can be replaced by minimally invasive procedures.

Active medical tools with intelligent control systems have been proposed. Martelli *et al.* have reported that robot-assisted knee replacement surgery improves on implant accuracy and reduces operating time and surgical errors.[35] While not minimally invasive, the use of robotic systems to assist in surgery clearly points the way for future technological development in this arena, e.g., remote telesurgery, which enables complex operations to be performed on patients in rural or hazardous locations who would otherwise not have access to appropriate medical care.

While there are many reasons to develop microsystems for minimally invasive surgery, significant challenges need to be addressed. Foremost among these is that minimally invasive surgical systems severely restrict the visual and tactile information available to surgeons. Surgical technique relies critically on the sense of touch, and much of the training a surgeon receives serves to hone this sense. Visual details, while important, only convey part of the anatomic information needed. Hidden tissue planes can be identified by palpation, as can regions of tissue thickening or hardening associated with infection or cancer. In particular, tactile feedback is essential in the accurate targeting of cancerous tissues and in delineating tissue boundaries for surgical resection.[36] Another technological challenge is the development of sufficiently maneuverable systems that will not hamper the surgeon. Passive instruments do not provide the requisite degree of control necessary for doctors to navigate the tortuous paths within the body in order to complete operations successfully. These limitations have

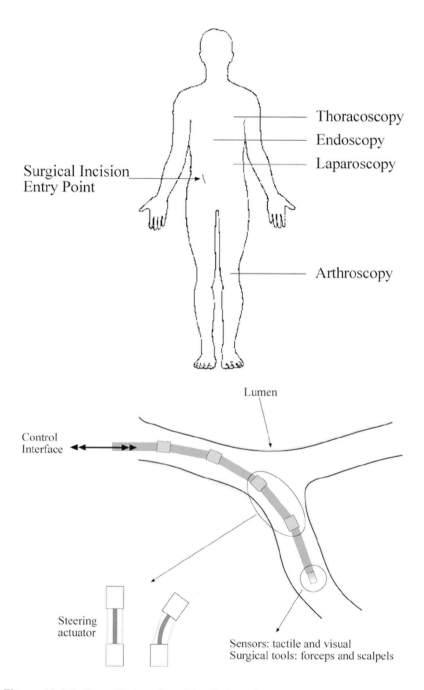

Figure 13.2 Active catheters for minimally invasive surgery.

prompted the development of active catheter systems with integrated sensors and actuators that will enhance dexterity and perception.

The development of surgical microsystems is a multidisciplinary effort involving ergonomics, control systems, and microdevices. Three areas in which microsystems have applications in advanced surgical instruments are sensor systems (tactile/visual), motion control (active steering of catheters), and surgical tooling (forceps, grippers, and scalpels).

13.3.1 Sensors

Takizawa et al. reported on the development of tactile sensors on a catheter to aid in navigation through vessels.[37] The multifunctional integrated film (MIF) technology on which these sensors are based allows for their monolithic fabrication on pliant films. Piezoelectric tactile sensors for endoscopic graspers have been demonstrated using polyvinylidene fluoride (PVDF) films as a sensor material.[38]

Catheter-mounted silicon sensors fabricated by combining bipolar electronics with surface micromachining have also been reported.[39] These devices allow for the measurement of local blood pressure, flow velocity, and oxygen saturation.

Coherent fiberoptic bundles are currently used as a means for providing visual information endoscopically. However, this adds to catheter size, provides limited resolution, and does not allow for stereopsis. Microcamera systems have been proposed to address these issues, although the optical systems remain a significant hurdle.

13.3.2 Motion Control

Manipulation and steering of passive catheters is a challenge at best and a threat to the patient at worst. Active control of catheter actuation increases dexterity and surgical precision. A multilink active catheter using silicon-based electronics with flexible circuitry has been demonstrated by Park et al.[40] The integrated CMOS circuits function as communications and control systems, allowing the selective actuation of shape-memory alloy (SMA) actuators along the catheter. This provides six degrees of freedom at each joint. Polyimide interconnects are used as a signal bus between the CMOS interface devices, minimizing the number of leads needed to control the catheter. The hybrid circuitry was fabricated monolithically, using standard CMOS processes in conjunction with bulk micromachining to create isolated circuits connected by a flexible polyimide membrane.

Nickel-titanium shape memory alloys provide the most common form of actuation for microsurgical catheters. These actuators provide a high power-to-weight ratio and are easy to drive in comparison with piezo-electric, electrostatic, or bimetallic actuators. To increase actuation distance, flat springs are fabricated from SMA. Specialized methods have been developed allowing for lithographic patterning and batch fabrication of planar-shape-memory alloy actuators.[41] Serpentine planar SMA springs with widths of 30 µm have been demonstrated with this process.

13.3.3 Microinstruments

A variety of catheter-based surgical microinstruments have been developed, including rotary cutting blades for atherectomy[42,43] and forceps.[18,37] Guber et al. reported on the fabrication of passive microforceps from NiTi wires using EDM and laser welding.[18] Carrozza et al. have also proposed piezoelectric-actuated microgrippers fabricated using the LIGA process and implemented on catheters.[44]

An interesting concept for a microsurgical instrument is an ultrasonic microscalpel tool.[45] The device is a microsurgical system that is bulk-micromachined from silicon and incorporates integrated force sensors (piezoresistive sensors in a Wheatstone configuration) and microfabricated cutting blades. The blades are made of silicon nitride coated with a sputtered titanium/platinum layer. Cutting action is driven by an ultrasonic transducer that is mounted on the silicon substrate.

Minimally invasive surgical procedures have numerous advantages over conventional techniques. The successful development and adoption of microsurgical tools is critical to advancing minimally invasive medicine. While these tools have had a promising start, significant challenges still need to be met. At a systems level, the development of feedback systems (haptic and stereopsis) and active control systems (facilitating motion with multiple degrees of freedom) are key to wider acceptance of these tools by the medical community. From an engineering standpoint, the complexity of these surgical tools makes reliability a critical concern, and the wide range of materials used in their manufacture raises issues such as biocompatibility and sterilization. Nevertheless, the arena of surgical microsystems provides fertile ground for exploring and developing new fabrication techniques in the creation of microstructures from a wide variety of materials.

13.4 Tissue Repair

The application of devices to help seal surgical incisions or repair damaged tissue is not new. Devices like tissue staples and coupling rings have existed in various forms for more than a century. The use of microfabrication techniques has allowed for their refinement, as well as for the development of novel devices and applications.

Conventional vascular anastomosis, or the joining of blood vessels, is a difficult and time-consuming procedure. Even in experienced hands, suturing small blood vessels together is a challenging proposition that requires the utmost care and patience. The suturing process itself can cause significant damage to blood vessel walls, leading to thrombus formation and possibly causing obstruction of the complete anastomosis. Given these considerations, and the move towards minimally invasive or endoscopic procedures, it is not surprising that there is a demand for a simple, timesaving, facilitated method of anastomosis. Some of the methods of facilitated anastomosis that have been described include the use of tissue glues,[46] laser tissue welding,[46] and mechanical fastening devices.

Dizon et al. proposed an anastomosis device fabricated by bulk micromachining of silicon[16] (Fig. 13.3). Reentrant barb structures were designed to ensure that the devices remain fastened to tissue. To facilitate anastomosis with this technology, a connecting annulus covered with

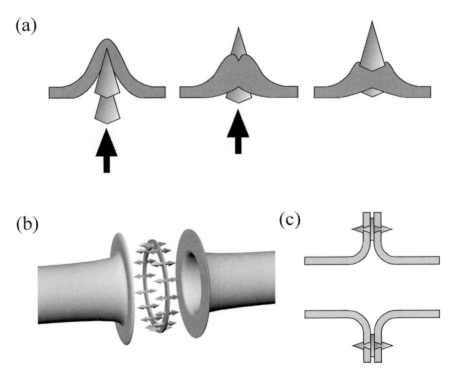

Figure 13.3 Microfabricated device for anastomosis of blood vessels: (a) reentrant barb for joining tissue; (b) and (c) rendition of device in use.[13]

these reentrant barbs on both sides could be used to join the tissues together, as shown in Fig. 13.3.

To adhere to vascular tissues, the barbs must be tall enough to pierce through the vessel wall such that the entire upper pyramid protrudes and locks in place (Fig. 13.3). Fabrication of these structures presents a difficult challenge, which was met by using the bulk micromachining properties of single-crystal silicon combined with ion beam milling for pattern transfer. The barbs fabricated with this process (Fig. 13.4) were 140 μm tall. Compensation structures were defined on the mask to reduce mask undercut resulting from convex corners. This served to increase the den-

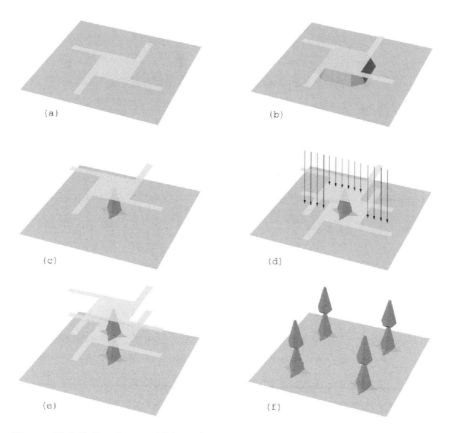

Figure 13.4 Bulk micromachining of reentrant barbs: (a) Mask pattern with compensation structures. (b) First anisotropic etch in KOH; the long arms are convex corner compensation structures that retard mask undercut and result in an increased height of the final structure. (c) Completion of the first etch step. (d) A second oxide layer is grown and patterned by ion-beam milling. (e) A second anisotropic etch to define the lower pyramidal structure. (f) Reentrant barbs after final oxide removal.[13]

sity of barbs for a defined height. Figure 13.5 is an electron micrograph of a reentrant barb fabricated by this process.

13.5 Therapeutic Systems

A major challenge in the administration of an effective pharmacological therapy is the maintenance of a steady therapeutic drug concentration

Figure 13.5 Electron micrograph of a reentrant silicon microbarb.

level in the patient. Conventional methods of dosing by intravenous or oral administration have potential disadvantages, including wide fluctuations in drug concentrations. At high concentrations, drugs can be potentially toxic, while at low concentrations they are ineffective. In the specific case of antibiotics, sustained subtherapeutic levels can also result in the development of resistant bacteria. Elimination of such fluctuations requires that the therapeutic agent be delivered to the patient at a rate based on the pharmokinetics of the specific drug. This requires constant monitoring of drug levels, a process that is usually not practical outside of medical facilities. Additionally, if the drug is to be targeted at a specific tissue or site in the body, systemic delivery only serves to raise the requisite dosage levels for efficacy and unnecessarily exposes healthy tissue to possible toxicity.

13.5.1 Implantable Delivery Systems

Implantable microsystems for drug delivery are suitably applied in therapies for chronic indications requiring frequent injections. Microsystems placed under the skin in a suitable position can be refilled by injection, on a less frequent basis, lowering the risk of infection.

Therapeutic delivery is facilitated by micropumps that deliver precise doses of drugs as needed. Micropumps based on shape-memory alloy actuation have been developed by Benard et al.[47] They fabricated micropumps using anisotropically etched silicon substrates with NiTi films monolithically applied by sputter deposition. Piezoelectric actuators have also been used to drive pumps in microsystems.[43]

Santini et al. reported on the development of a novel solid-state pulsatile-release drug delivery system based on the electrochemical dissolution of a gold membrane covering therapeutic reservoirs.[9] The reservoirs are anisotropically etched into silicon, filled with the drug, and capped with an adhesive plastic layer. The release orifice is covered with a gold membrane that can be selectively addressed for pulsatile drug delivery. When a positive bias is applied to the gold membrane, it dissolves, releasing the stored therapeutic.

One advantage of active drug delivery systems is that they allow the regulation of drug doses to be adapted based on physical activity, food ingestion, and circadian rhythms. The optimal treatment of diabetes, in particular, will require closed-loop control of insulin delivery based on blood glucose levels. While the concept of such systems dates back to the mid-1960s, the functional realization of such a device has yet to be demonstrated. The critical challenge here is the development of reliable glucose sensors that are functionally stable in the complex *in vivo* environment.

Tangential to the approach of artificial control systems is the use of implantable biocapsules to create an artificial pancreas (Fig. 13.6).[19,20] Here, microfabricated membranes with well-controlled nanometer-sized pores serve as immuno-isolation barriers. The pores allow for the free diffusion of glucose, insulin, and cellular nutrients but physically inhibit the access of antibodies and immunologic cells. This allows for the proper physiological function of the cells encapsulated within the device. Given proper immuno-isolation, pancreatic islet cells respond physiologically to appropriate stimuli, secreting bioactive compounds such as insulin. Immuno-isolation also allows for the use of allografts and xenografts in the treatment of a wide variety of indications, including diabetes, hemophilia, and Parkinson's disease. The fabrication process of these devices uses a combination of bulk and surface micromachining. The cellular cavity is formed by bulk micromachining of a silicon substrate, while surface micromachining is used to form the nanoporous membrane.

13.5.2 Mechanical Delivery Systems

The transdermal delivery of therapeutic agents overcomes limitations related to drug degradation in the gastrointestinal tract. Microfabricated needles have been used to increase epidermal permeability for this purpose.[10,13,14] Microneedles between 50 and 100 µm in length were fabricated, using reactive ion etching, and shown to effect painless transdermal drug delivery. The transport rate of therapeutics across the skin can further be enhanced by ultrasound, electroporation, and iontophoresis. The

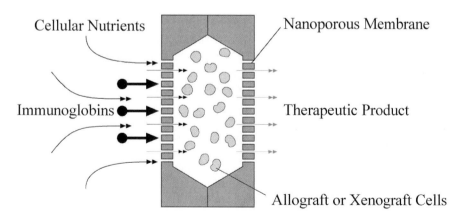

Figure 13.6 Microstructures facilitating immuno-isolation of xenografts.[16]

use of microfabrication also allows for the monolithic integration of microneedle structures with pumps, sensors, and transducers.

Similar structures have been demonstrated to facilitate gene transfection. Arrays of micropiercing structures were used to transfect tobacco leaves,[48] nematodes,[11] and smooth muscle cells in culture (Fig. 13.7).[12] These structures penetrate and allow a path for genetic material to enter the cell. Transfection was effected with naked DNA, without the use of viral vectors.

13.6 Summary

It is clear that biomedical microsystems hold great promise for improving the way in which ailments are diagnosed and treated. The key to achieving widespread adoption of these tools while keeping health care costs down is to improve fabrication efficiency.

Figure 13.7 Gene transfection of the β-galactosidase marker gene into cells in culture by mechanical injection. The small circular features are the entry points of the microbarbs.[12]

The materials used in biomedical microsystems, and their physical dimensions, necessitate the use of tools from both semiconductor microfabrication and conventional machining in their manufacture. Semiconductor fabrication techniques for the most part require little further development, as the minimum dimensions achievable by lithographic means are more than sufficient for biomedical applications.

The critical challenge that remains is to develop tools for three-dimensional shaping of a great variety of materials on the micron scale. The tools described herein, such as laser machining and electrodischarge machining, have made significant inroads, but they lack the massive batch process capability of lithography, which is key to lowering costs.

References

1. C.H. Mastrangelo, M.A. Burns, and D.T. Burke, "Microfabricated devices for genetic diagnostics," *Proc. IEEE,* 86(8):1769–1787 (1998)
2. M.U. Kopp, A.J. de Mello, and A. Manz, "Chemical amplification: continuous-flow PCR on a chip," *Science*, 280:1046–1048 (1998)
3. M.A. Northrup, M.T. Ching, R.M. White, and R.T. Watson, "DNA amplification with a microfabricated reaction chamber," *Proc. Int. Conf. Solid-State Sensors and Actuators (Transducers 93),* June 1993, pp. 924–926
4. E.T. Lagally, P.C. Simpson, and R.A. Mathies, "Monolithic integrated microfluidic DNA amplification and capillary electrophoresis analysis system," *Sensors and Actuators B*, 63:138–146 (2000)
5. J.R. Webster, M.A. Burns, D.T. Burke, and C.H. Mastrangelo, "Monolithic capillary electrophoresis device with integrated fluorescence detector," *Anal. Chem.*, 73:1622–1626 (2001)
6. M.A. Burns, B.N. Johnson, S.N. Brahmasandra, K. Handique, J.R. Webster, M. Krishnan, T.S. Sammarco, P.M. Man, D. Jones, D. Heldsinger, C.H. Mastrangelo, and D.T. Burke, "An integrated nanoliter DNA analysis device," *Science*, 282:484–487 (1998)
7. R.H. Carlson, C.V. Gabel, S.S Chan, R.H. Austin, J.P. Brody, and J.W. Winkelman, "Self-sorting of white blood cells in a lattice," *Phys. Rev. Lett.*, 79(11):2149–2152 (1997)
8. T.A.J. Duke, and R.H. Austin, "Microfabricated sieve for the continuous sorting of macromolecules," *Phys. Rev. Lett.*, 80(7):1552–1555 (1998)
9. J.T. Santini, Jr., M.J. Cima, and R. Langer, "A controlled-release microchip," *Nature*, 397:335–338 (1999)
10. S. Henry, D.V. McAllister, M.G. Allen, and M.R. Prausnitz, "Micromachined needles for the transdermal delivery of drugs," *Proc. IEEE MEMS 1998*, pp. 494–498
11. S. Hashmi, P. Ling, G. Hashmi, M. Reed, R. Gaugler, and W. Trimmer, "Genetic transformation of nematodes using arrays of micromechanical piercing structures," *BioTechniques*, 19:766–770 (1995)

12. M.L. Reed, C. Wu, J. Kneller, S. Watkins, D.A. Vorp, A. Nadeem, L.E. Weiss, K. Rebello, M. Mescher, A.J. Conrad Smith, W. Rosenblum, and M.D. Feldman, "Micromechanical devices for intravascular drug delivery," *J. Pharma. Sci.*, 8(11):1387–1394 (1998)
13. A. Trautman, P. Ruther, and O. Paul, "Microneedle arrays fabricated using suspended etch mask technology combined with fluidic through wafer vias", *Proc. MEMS 2003*, Kyoto, Japan, Jan. 2003, pp. 682–691
14. P. Griss and G. Stemme, "Novel, side opened out-of-plane microneedles for microfluidic transdermal interfacing", *Proc. MEMS 2002*, Las Vegas, USA, Jan. 2002, pp. 467–470
15. J.G.E. Gardeniers, J.W. Berenschot, M.J. de Boer, Y. Yeshurun, M. Hefetz, R. van't Oever, and A. van den Berg, "Silicon micromachined hollow microneedles for transdermal liquid transfer", *Proc. MEMS 2002*, Las Vegas, USA, Jan. 2002, pp. 141–144
16. R. Dizon, H. Han, A. G. Russell, and M.L. Reed, "An ion milling pattern transfer technique for fabrication of three-dimensional micromechanical structures," *J. Microelectromechanical Systems*, 2(4):151–159 (1993)
17. M. Esashi, "Micro electro mechanical systems by bulk silicon micromachining," *Proc. 2nd Int. Symp. Microstructures and Microfabricated Systems*, Electrochem. Soc. (1995), pp. 11–23
18. A.E. Guber, N. Giordano, A. Schüssler, O. Baldinus, M. Loser, and P. Wieneke, "Nitinol-based microinstruments for endoscopic neurosurgery," *Actuator 96, 5th Int. Conf. New Actuators* (1996), pp. 375–378
19. T.A. Desai, D.J. Hansford, L. Kulinsky, A.H. Nashat, G. Rasi, J. Tu, Y. Wang, M. Zhang, and M. Ferrari, "Nanopore technology for biomedical applications," *J. Biomed. Microdevices*, 2(1):11–40 (1999)
20. L. Leoni and T.A. Desai, "Biotransport and biocompatibility of nanoporous biocapsules," *1st Ann. Int. IEEE-EMBS Special Topic Conf. on Microtechnologies in Medicine & Biology* (2000), pp. 113–117
21. J. Black, *Biological Performance of Materials*, 3rd ed., Marcel Dekker (1999)
22. S.D. Rapoport, M.L. Reed, and L.E. Weiss, "Fabrication and testing of a microdynamic rotor for blood flow measurements," *J. Micromech. Microeng.*, 1:60–65 (1991)
23. R.C. Jaeger, *Introduction to Microelectronic Fabrication*, Addison-Wesley (1993)
24. S. Wolf and R. N. Tauber, *Silicon Processing for the VLSI Era*, vol. 1, *Process Technology*, Lattice Press (1986)
25. H.C. Hoch, L.W. Jelinski, and H.G. Craighead (eds.), *Nanofabrication and Biosystems: Integrating Materials Science, Engineering, and Biology*, Cambridge (1996)
26. S.D. Senturia, *Microsystem Design*, Kluwer Academic (2001)
27. H. Lorenz, M. Despont, N. Fahrni, N. LaBianca, P. Renaud, and P. Vettiger, "SU-8: A low-cost negative resist for MEMS," *J. Micromech. Microeng.*, 7:121–124 (1997)

28. B.-H. Jo, L.M. Lerberghe, K.M. Motsegood, and D.J. Beebe, "Three-dimensional micro-channel fabrication in polydimethylsiloxane (PDMS) elastomer," *J. Microelectromech. Sys.*, 9(1):76–81 (2000)
29. L.P. Lee, S.A. Berger, D. Liepmann, and L. Pruitt, "High aspect ratio polymer microstructures and cantilevers for bioMEMS using low energy ion beam and photolithography," *Sensors and Actuators A*, 71:144–149 (1998)
30. H.K. Tönshoff, F. von Alvensleben, A. Ostendorf, G. Willmann, and T. Wagner, "Precision machining using UV and ultrashort pulse lasers," *SPIE – Int. Soc. Opt. Eng. Proc. SPIE*, 3680:536–545 (1999)
31. R. Srinivasan and B. Braren, "Ultraviolet laser ablation of organic polymers," *Chem. Rev.*, 89(6):1303–1316 (1989)
32. D. Petrovic, G. Popovic, W. Brenner, E. Chatzitheodoridis, P. Herbst, and A. Vujanic, "Microtools and microactuators processed with μ-EDM," *Ninth Micromechanics Europe Workshop MME'98 Proc.*, pp. 244–247
33. D. Reynaerts, W. Meeusen, X. Song, H. Van Brussel, S. Reyntjens, D. De Bruyker, and R. Puers, "Integrating electro-discharge machining and photolithography: work in progress," *J. Micromech. Microeng.*, 10:189–195 (2000)
34. V. Fascio, R. Wüthrich, D. Viquerat, and H. Langen, "3D microstructuring of glass using electrochemical discharge machining (ECDM)," *Proc. 1999 Int. Symp. Micromechatronics and Human Science*, pp. 179–183
35. M. Martelli, M. Marcacci, L. Nofrini, F. La Palombara, A. Malvisi, F. Iacono, P. Vendruscolo, and M. Pierantoni, "Computer- and robot-assisted total knee replacement: analysis of a new surgical procedure," *Ann. Biomed. Eng.*, 2:1146–1153 (2000)
36. C. Nies, R. Leppek, H. Sitter, H.J. Klotter, J. Riera, K.J. Klose, W.B. Schwerk, and M. Rothmund, "Prospective evaluation of different diagnostic techniques for the detection of liver metastases at the time of primary resection of colorectal carcinoma," *Eur. J. Surg.*, 162(10: 811–816 (1996)
37. H. Takizawa, H. Tosaka, R. Ohta, S. Kaneko, and Y. Ueda, "Development of a microfine active bending catheter equipped with MIF tactile sensors," *Proc. IEEE MEMS* (1999), pp. 412–417
38. J. Dargahi, M. Parameswaran, and S. Payandeh, "A micromachined piezoelectric tactile sensor for an endoscopic grasper—theory, fabrication and ," *J. Microelectromech. Sys.*, 9(3):329–335 (2000)
39. J.F.L. Goosen, D. Tanase, and P.J. French, "Silicon sensors for use in catheters. *1st Annual International IEEE-EMBS Special Topic Conference on Microtechnologies in Medicine & Biology* (2000), pp. 152–155
40. K.-T. Park and M. Esashi, "A multilink active catheter with polyimide-based integrated CMOS interface circuits," *Microelectromech. Sys.*, 8(4):349–357 (1999)
41. T. Mineta, T. Mitsui, Y. Watanabe, S. Kobayashi, Y. Haga, and M. Esashi, "Batch fabricated flat meandering shape memory alloy actuator for active catheter," *Sensors and Actuators A*, 88:112–120 (2001)
42. A. Ruzzu, J. Fahrenberg, M. Müller, C. Rembe, and U. Wallrabe, "A cutter with rotational-speed dependent diameter for interventional catheter systems," *Proc. MEMS 98*, pp. 499–503

43. D. Maillefer, H. van Lintel, G. Rey-Mermet, and R. Hirschi, "A high-performance silicon micropump for an implantable drug delivery system," *Proc. IEEE MEMS 1999*, pp. 541–546
44. M.C. Carrozza, P. Dario, A. Menciassi, and A. Fenu, "Manipulating biological and mechanical micro-objects using LIGA-microfabricated end-effectors," *Proc. 1998 IEEE Int. Conf. on Robotics and Automation* (1998), pp. 1811–1816
45. I.-S. Son, A. Lal, B. Hubbard, and T. Olsen, "A multifunctional silicon-based microscale surgical system," *Sensors and Actuators A*, 91:351–356 (2001)
46. D. Barrieras, P.P. Reddy, G.A. McLorie, D. Bagli, A.E. Khoury, W. Farhat, L. Lilge, and P.A. Merguerian, "Lessons learned from laser tissue soldering and fibrin glue pyeloplasty in an *in vivo* porcine model," *J. Urol.*, 164(3):1106–1110 (2000)
47. W.L. Benard, H. Kahn, A.H. Heuer, and M.A. Huff, "Thin-film shape-memory alloy actuated micropumps," *J. Microelectromech. Sys.*, 7(2):245–251 (1998)
48. W. Trimmer, P. Ling, C.-K. Chin, P. Orton, R. Gaugler, S. Hashmi, G. Hashmi, B. Brunett, and M. Reed, "Injection of DNA into plant and animal tissues with micromechanical piercing structures," *Proc. IEEE Micro Electro Mechanical Systems 1995*, pp.111–115

14 Microactuators

Jack W. Judy

University of California at Los Angeles

14.1 Introduction

This chapter provides the information necessary to understand the capabilities and limitations of existing electrically operated microactuators so that a microactuator can be selected or designed for a specific application and set of design constraints. It begins by providing introductory information about the field of microelectromechanical systems (MEMS), with the specific perspective of microactuator development. A brief but broad discussion of the transduction mechanisms used by microactuators is presented. Microactuators based on electromechanical transduction mechanisms are then described and analyzed, and specific examples are cited. References to published work are provided as existence proofs and for in-depth study of the individual cases.

Microsensors have enjoyed much commercial success, e.g., pressure sensors and inertial devices in automotive and consumer applications. In contrast, the development and successful commercialization of microactuators has been more troublesome. Early commercial microactuators were designed for applications that required operation in an uncontrolled environment (such as microvalves) or reliable repetitive physical contact (such as microrelays). In addition, it is often difficult to interface a *micro*actuator with the *macro*scopic world. It was not until applications were developed that did not necessarily require physical contact and where use of a controlled environment could be made, e.g., radio-frequency MEMS and micro-opto-electromechanical systems (MOEMS), that microactuator development expanded tremendously.

To limit the scope of this chapter, we will discuss microactuators that are physically smaller than 1 mm^3 and made of critical components less than 1 mm in size, or that are produced with micromachining process technologies.

14.2 Actuators: Transducers with Mechanical Output

An actuator is a device that converts energy from one form, such as electrical, mechanical, thermal, magnetic, chemical, and radiation energy, into the mechanical form. (For example, resistive heating elements convert electrical energy into bending of a bimorph microstructure.) In some cases, a microactuator may convert energy into intermediate forms before resulting in the final mechanical output (such as inductively coupled heating elements, which convert electrical energy first into magnetic energy before finally resulting in thermal energy serving to deflect a microelement). Such devices are said to be based on tandem transduction.

In some actuators, the mechanical output signal is in turn used to direct or modulate a nonmechanical signal. This situation is encountered in numerous optical applications, many of which are described in detail in Chapter 7.

14.2.1 Transduction Mechanisms

Prominent actuation mechanisms used in MEMS are listed in Table 14.1. Specific examples of MEMS actuators that make use of these transduction mechanisms are described, analyzed, compared, and contrasted throughout the rest of this chapter.

Each mechanism has its own advantages, disadvantages, and appropriate set of applications. Electromechanical transduction mainly relies on two mechanisms. The first makes use of electrostatic forces to attract or

Table 14.1 Common Microactuation Mechanisms

Input	Output: Mechanical
Electrical	Electrostatics
	Piezoelectricity
Magnetic	Magnetostatics
	Magnetostriction
Mechanical	Pneumatics
	Hydraulics
Thermal	Expansion
	Shape memory effect
	Phase change
Chemical	Combustion

repel micromechanical structures. The second uses the piezoelectric effect in appropriate materials to induce a mechanical strain in a stack of materials that contains a piezoelectric component or thin film. Electromechanical transduction dominates in the MEMS field, particularly in the applications. The basics and applications of electrostatic and piezoelectric microactuation are described in detail in Sections 14.3 to 14.4 and in Section 14.5, respectively. Other actuation mechanisms are briefly reviewed in the following, support by citations from recent MEMS Conferences and Solid-State Sensor and Actuator Conferences. Reviews of less recent work are found in Fujita.[1,2]

Mechanical-mechanical microactuation has tried to mimic macroscopic hydraulic and pneumatic systems. Examples include a LIGA-based microsystem able to move an optical component in a microtriangulation system for distance measurements[3] and a thermopneumatic valve.[4]

Similar to electromechanical transduction, magnetomechanical microactuation exploits mainly two effects: the action of electromagnetic (magnetostatic) forces and magnetostriction (the strain magnetically inducible in ferromagnetic materials). Examples of electromagnetic actuators and their applications are found in several recent works.[5-12] Magnetostriction has been used mostly in the context of micromirror actuation. [13-15]

Thermomechanical microactuation exploits the thermal expansion of materials and material sandwiches.[16-30] The thermal strain is usually amplified in the out-of-plane direction of planar microstructures by the bimorph effect. In-plane amplification has been achieved using various lever mechanisms.[16,26,29] A second thermomechanical actuation mechanism, termed shape memory effect, exploits the martensitic phase change of appropriate shape memory alloys (SMAs) belonging, for example, to the NiTi systems.[31-35]

In the area of chemical-mechanical microactuation, osmosis is probably one of the more exotic phenomena to be exploited. In view of their rather long time constant,[36] reported devices are likely of more academic than commercial interest. Actuation by self-oscillating reactions has been applied in the actuation of microcilia actuation.[37] On the other hand, chemical-mechanical transduction has recently attracted much interest, in such spectacular manifestations as microturbomachinery[38] and microcombustion engines for space applications.[39]

14.2.2 Scaling Advantages and Issues

It is critical to have a good understanding of the scaling properties of the transduction mechanisms, the overall design, the materials, and the fabrication processes involved when miniaturizing any actuator. The scaling properties of any one of these components could present a formidable barrier to adequate performance or economic feasibility. Due to powerful scaling functions and the shear magnitude of the scaling involved, our experience and intuition of macroscale phenomena and designs will not transfer directly to the microscale.

When designing microactuators, it is important to be aware that the properties of thin-film materials are often significantly different than those of their bulk or macroscale form. Much of this disparity arises from the difference in the processes used to produce thin-film materials and bulk materials. Another source of variation is the fact that the assumption of homogeneity, commonly used with accuracy for bulk materials and devices, becomes unreliable when used to model devices that have dimensions of the same scale as grains and other microscopic fluctuations in material properties. Thus local changes in grain size and other characteristics can significantly alter the performance of microactuators produced either together in one batch or in different runs. One potential advantage of scaling microactuators or components critical for their operation (such as micromechanical flexures) to dimensions approaching the defect density of the material is that devices or components can be produced with a very low total defect count. This is one reason the reliability of some microactuators, particularly those of simple mechanical design, such as cantilevers, can have a higher reliability than macroscopic versions.[40] However, due to the high surface-to-volume ratio of MEMS, more attention must be paid to controlling surface properties.

In addition to the material properties directly related to the transduction mechanisms, other important material properties to characterize for microactuators include elastic modulus, Poisson's ratio, coefficient of friction, wear resistance, fracture stress, yield stress, residual in-plane stress, and vertical stress gradient. Due to the flexibility of microfabrication, it is typically convenient to integrate microstructures that can be used to provide *in situ* measurements of material properties.[41-43] Many such structures have been designed and used to reveal that the characteristics of thin-film materials can vary tremendously from film to film without careful process control.[44-46] In fact, any high-precision and high-reliability MEMS application requires that significant effort be directed toward quantifying the precise material properties of the films being used. Chapter 2 gives a detailed account of material characterization efforts in the field of MEMS and measured data.

14.2.3 Electrical Microactuators

The remainder of this chapter describes electrical microactuators. The physics involved in both electrostatic and piezoelectric actuation are briefly described and the corresponding scaling properties are analyzed. Design strategies and microactuator configurations are discussed, and practical issues involved with their implementation and common or potential solutions are presented.

Electrical microactuators are by far the most common and diverse type of microactuator. This is primarily because of the ease with which most electrical microactuators can be produced using conventional microfabrication processes and materials. Examples of electric microactuators include static, resonant, rotary, and stepper-motor configurations. Electrical microactuators can be driven by an electrical-to-mechanical conversion that makes use either of the direct electrostatic forces between charged objects or a piezoelectric material that can mediate the energy transformation. Electrostatic and piezoelectric transduction mechanisms, the physical relationships involved, the material properties that govern their operation, and examples of microactuators realized with each technology are described in the following sections.

14.3 Electrostatic Forces

Electrostatic forces are commonly observed even on the macroscale during everyday activities, such as combing dry hair or drying clothes. The discovery of elemental charge and the concept of electric fields have led to an understanding of electrostatic energy and the forces acting between objects charged to a different extent.[47,48] To determine the magnitude of the driving forces, an understanding of the dependence of electrostatic energies on physical parameters is needed. We must also realize that forces exist simply to drive systems to lower energy. Force is defined as being equal to the negative derivative of energy with respect to the change in the parameter of interest.

14.3.1 Electrostatic Systems

14.3.1.1 Constant-Charge System

If two isolated conducting bodies each have a fixed amount of charge, energy conservation demands that changing their relative separation will require mechanical work ∂W_m at the expense of electrostatic energy ∂W_e (i.e., $\partial W_e + \partial W_m = 0$). Therefore, in constant-charge electrostatic systems, force is expressed as $F = \nabla W_m = -\nabla W_e$ and torque as $T_\varphi = \partial W_m / \partial \varphi = -\partial W_e / \partial \varphi$ for a given angular variable φ.[49]

The electric field E is proportional to the spatial gradient of potential V (e.g., $E = -\nabla V$, with the negative sign indicating that in accordance with convention, V drops in the direction of the electric field) and the potential energy of an electrical charge q is proportional to the potential V at its current position (i.e., $W_e = qV$). The electrostatic force F_e on a charge q in an electrical field E is therefore given by:

$$F_e = -\nabla W_e = -\nabla(qV) = qE, \tag{1}$$

such that the force on a positive charge is directed toward lower potentials 49.

It is also known that in free space, the divergence of the electric field is given by $\nabla E = \rho / \varepsilon_0$, with volume charge density ρ and permittivity of free space ε_0. The integral form of this equation is Gauss's law, $\oint_c E ds = Q / \varepsilon_0$, with total charge Q contained within a volume bounded by the surface s. For a point charge q_1 placed at the center of free space, the magnitude of the electric field E_1 at a distance r in any direction from the central charge can be found with Gauss's law to be:

$$E_1 = \frac{q_1}{4\pi\varepsilon_0 r^2}, \tag{2}$$

which is directed away from the central charge q_1.[49] If another point charge q_2 is placed within the electric field of point charge q_1, the force F_{12} it will experience can be computed with Eqs. (1) and (2). The result is:

$$F_{12} = \frac{q_1 q_2}{4\pi\varepsilon_0 r^2}, \tag{3}$$

otherwise known as Coulomb's law.[49]

14.3.1.2 Constant-Voltage System

If two conducting bodies are held at fixed potentials by an external source, such as batteries, a relative change in spatial separation will lead to a change in total energy. Specifically, the electrostatic energy ∂W_e plus the mechanical work ∂W_m required to change the conductor separation must equal the energy supplied by the source ∂W_{source}; thus $\partial W_e + \partial W_m = \partial W_{source}$.

Note that energy equivalent to the change in the electrostatic energy ∂W_e is lost in the charging process. Therefore, the total energy supplied by the source is twice the change in the electrostatic energy ∂W_e. As a result, the force in constant-voltage electrostatic systems is given by $F = \nabla W_e$ and torque is given by $T_\varphi = \partial W_e / \partial \varphi$ for a given angular variable φ.[49]

14.3.1.3 Electrostatic Energies and Energy Densities

The electrostatic energy W_e in an electric field E of a volume υ is expressed as:

$$W_e = \frac{1}{2} \int_\upsilon (DE) d\upsilon, \tag{4}$$

with the electric flux density D.[49] Assuming D is related to E by $D = \varepsilon_r \varepsilon_0 E$, which is true for a linear dielectric material with relative permittivity ε_r, Eq. (4) can be reduced to:

$$W_e = \frac{1}{2} \int_\upsilon \varepsilon_r \varepsilon_0 E^2 d\upsilon. \tag{5}$$

The energy stored within a volume that can be mechanically expanded or contracted is energy that can be converted into mechanical energy. Dividing this total energy by the volume yields the energy density, which is the parameter to use when comparing actuators of different sizes. Since the energy density can be scale-dependent, as discussed below, its scaling properties must be studied carefully to ensure that the miniaturized device functions as expected. The energy density of electrostatic devices w_e is defined as:

$$w_e = \frac{W_e}{\upsilon} = \frac{1}{2} \varepsilon_r \varepsilon_0 E^2. \tag{6}$$

Given the generic structure shown in Fig. 14.1, with air gap g and electrode area A, the air gap volume is $\upsilon = gA$. The structure is assumed to be mechanically rigid, and the fringing fields near the air gap are ignored. In

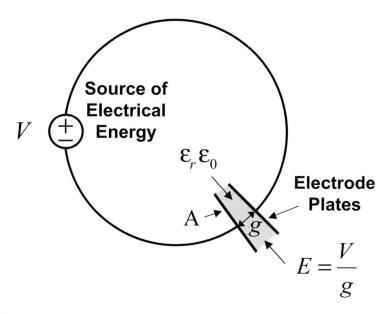

Figure 14.1 Generic electrostatic actuator.

this equilibrium electrical system, it is assumed that no dc currents flow and that no electric field exists in the conductors connecting the voltage source to the plates at the gap. These assumptions are valid because the ratio of the conductivity of the conductor ($\sigma \sim 10^7\ \Omega^{-1}\mathrm{m}^{-1}$) to that of the air gap ($\sigma \sim 10^{-15}\ \Omega^{-1}\mathrm{m}^{-1}$) is easily greater than 10^{20}.[50] Such a large difference in conductivity allows electric fields to be tightly confined and controlled.

Since the electric field inside the conductors linking the voltage source to the plates is assumed to be zero, the integral in Eq. (5) is nonzero only over the volume of the air gap. The electrical potential V generated by the voltage source is completely transferred to the plates of the air gap. The electric field in the air gap can be expressed as:

$$E = V/g, \tag{7}$$

and Eq. (5) can be reduced to:

$$W_e = \frac{1}{2}\int_v \varepsilon_r \varepsilon_0 E^2 dv = \int_v \frac{\varepsilon_r \varepsilon_0 V^2}{2g^2} dv = \frac{\varepsilon_r \varepsilon_0 A V^2}{2g}. \tag{8}$$

Thus the electrostatic energy density is proportional to the square of the voltage applied and inversely proportional to the gap. The substantial miniaturization possible with micromachining exploits the inverse dependence on gap to generate higher energy densities for a given applied voltage. The ultimate limits of electrostatic energy density are investigated in Section 14.3.3.

14.3.2 Forces in Electrostatic Systems

14.3.2.1 Constant-Voltage System

If the voltage on each plate is constant as the air gap g varies, the constant-voltage electrostatic energy $W_{e(V)}$ is given by:

$$W_{e(V)} = -\left(\frac{\varepsilon_r \varepsilon_0 A V^2}{2g}\right). \tag{9}$$

It is assumed that the energy of the system is zero when the plates are infinitely far apart and becomes negative as the plates are brought closer together.[49] Taking the derivative of the electrostatic energy with respect to the gap g yields the constant-voltage electrostatic force $F_{e(V)}$:

$$F_{e(V)} = \frac{\partial W_{e(V)}}{\partial g} = -\left(\frac{\varepsilon_r \varepsilon_0 A V^2}{2g^2}\right). \tag{10}$$

Thus the electrostatic forces in constant-voltage electrostatic systems are inversely proportional to the gap squared. This result illustrates the opportunities available to increase the forces in constant-voltage electrostatic systems through miniaturization.

14.3.2.2 Constant-Charge Systems

The capacitance of the generic structure shown in Fig. 14.1 is given by:

$$C = \frac{\varepsilon_r \varepsilon_0 A}{g}. \tag{11}$$

The charge Q on each plate is proportional to the capacitance C and the voltage V:

$$Q = CV. \qquad (12)$$

If the charge on each plate is held constant, the constant-charge electrostatic energy $W_{e(Q)}$ is obtained by substituting Eq. (12) into Eq. (8) to yield:

$$W_{e(Q)} = -\left(\frac{\varepsilon_r \varepsilon_0 A Q^2}{2gC^2}\right) = -\left(\frac{gQ^2}{2\varepsilon_r \varepsilon_0 A}\right). \qquad (13)$$

In this case, we have defined the zero energy point to be when the separation is zero.[49] If the charges have the same polarity, the energy of a constant-charge electrostatic system will have an energy that becomes increasingly negative (i.e., lower energy). If, however, the charges have opposite polarity, the energy will actually increase with their separation. Taking the derivative of the electrostatic energy with respect to the gap g yields the constant-charge electrostatic force $F_{e(Q)}$:

$$F_{e(Q)} = -\left(\frac{\partial W_{e(Q)}}{\partial g}\right) = \frac{Q^2}{2\varepsilon_r \varepsilon_0 A}. \qquad (14)$$

Therefore, charges of the same sign will experience a repulsive force, and charges of the opposite sign will experience an attractive force. Note that the forces in constant-charge electrostatic systems are *independent* of the gap size, which is particularly useful if actuation over large gaps is needed. However, the effective implementation of a constant-charge electrostatic system is substantially less straightforward than a constant-voltage electrostatic system. As a result, nearly all electrostatic microactuators operate as constant-voltage systems.

14.3.3 Scaling Properties

As with everything else in MEMS, the scaling properties of electrostatic forces must be carefully scrutinized to look for opportunities to exploit and problems to avoid. In electrostatics, the ultimate limit in performance is determined by electrostatic breakdown.[51] Paschen's curve (Fig. 14.2) is a graphical representation of the electrostatic breakdown

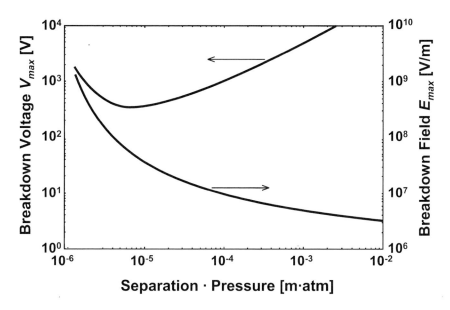

Figure 14.2 The Paschen curve of the breakdown voltage V_{max} vs. the product of pressure P and gap width g, and the breakdown field E_{max} vs. Pg at an ambient pressure of 1 atm.[51, 52]

limit. This section investigates the Paschen curve and the origin of electrostatic breakdown in detail to determine the ultimate performance limitations of electrostatic microactuators.

The magnitude of electrostatic force is ultimately limited by electrostatic breakdown (i.e., the voltage and field at which the dielectric no longer electrically isolates the charged bodies). The maximum voltage V_{max} that can be sustained before breakdown occurs decreases less than linearly with a reduction in electrode separation, thereby increasing the breakdown field (Fig. 14.2). Paschen discovered that the breakdown voltage reaches a minimum value at a separation of several micrometers at atmospheric pressure.[51] When the electrode separation is reduced below several micrometers, the breakdown voltage and field increase rapidly. A plot of the breakdown voltage as a function of the product of electrode separation and pressure, known as the Paschen curve, is given in Fig. 14.2.[52] To the left of the minimum breakdown voltage there are too few impacts of ionized gas molecules to achieve a regenerative avalanche breakdown unless the applied voltage increases. At the minimum breakdown voltage, the electrostatic energy is most efficiently linked to the avalanche process. To the right of the minimum breakdown voltage, energy is divided among ionizing and nonionizing excitations of gas molecules. For a more complete derivation of the Paschen curve than will be given below, see Von Hippel[52] and Judy.[53] The importance of Paschen's curve to the design of electrostatic microactuators was first reported by Bart et al.[54] and is further illustrated in Sections 14.3.3.2 and 14.3.3.3.

14.3.3.1 Paschen Curve

Electrostatic breakdown occurs because of electron impact ionization. When electrons fall a distance down an electric field E, they generate $dn = n\alpha dx$ new electrons, with α defined as the ionization coefficient (i.e., the number of electron-ion pairs generated per unit distance traveled by the electron). Therefore, a single electron falling a distance g, the full gap between the electrodes, will be amplified into $e^{\alpha g}$ electrons and $e^{\alpha g} - 1$ positive ions.[52] To achieve a steady-state constant breakdown current, a constant supply of avalanche-forming electrons near the negatively biased electrode, the cathode, is needed. If we hypothesize that the positive ions create avalanche-forming electrons at the cathode, with a probability γ per ion that strikes the cathode, then a self-regenerative breakdown condition is satisfied when:

$$\gamma\left(e^{\alpha g} - 1\right) = 1. \tag{15}$$

Paschen's curve can be mathematically derived with two simplifying assumptions. First, assume each electron falling a distance l though an electric field of strength E between the electrodes loses all of its kinetic energy on impact. Second, assume that each electron ionizes as soon as its kinetic energy El equals the ionization potential of the gas molecules V_i.[52] The probability that an electron will travel the distance x_i required to accumulate an energy V_i is $e^{-x_i/l} = e^{-V_i/El}$, with the number of impacts per unit distance being $1/l$. Therefore, the number of electron-ion pairs formed per unit distance can be expressed as:

$$\alpha = \frac{1}{l} e^{-V_i/El}. \tag{16}$$

At breakdown, the electric field is $E = V_{max}/g$, and is determined by Eq. (15), which can be rewritten as:

$$\alpha = \frac{1}{g}\ln\left(1+\frac{1}{\gamma}\right). \tag{17}$$

Combining Eqs. (16) and (17), the breakdown voltage can be expressed as:

$$V_{max} = \frac{gV_i}{-l\ln\left(\dfrac{l}{g}\ln\left(1+\dfrac{1}{\gamma}\right)\right)}. \tag{18}$$

The mean free path l can be expressed in terms of the pressure P

$$l = l_0\frac{P_0}{P}, \tag{19}$$

with the mean free path at a pressure of $P_0 = 1$ atm being l_0 (1 atm = 1.013 × 10^5 Pa). Equation (19) can be substituted into Eq. (18), with the result:

$$V_{max} = \frac{APg}{\ln(Pg)+B}, \tag{20}$$

where A and B are expressed as:

$$A = \frac{V_i}{P_0 l_0} \tag{21}$$

and

$$B = -\ln\left(P_0 l_0 \ln\left(1+\frac{1}{\gamma}\right)\right), \tag{22}$$

respectively.[52] A plot of Eq. (20) reveals the features shown in Paschen's curve (Fig. 14.2). To the left of the minimum breakdown voltage, there are too few impacts to achieve the regenerative breakdown condition given in Eq. (15). The applied voltage must increase by an extreme amount to generate the necessary avalanche by creating enough total impacts or by increasing the probability that positive ions create avalanche-forming electrons at the cathode (i.e., higher γ). To the right of the minimum breakdown potential, the number of impacts is too large, and much of the necessary energy is wasted in minor electronic excitation.

14.3.3.2 Field-Limited Maximum Electrostatic Energy Densities

By using the curve for the breakdown voltage shown in the Paschen curve (Fig. 14.2), the maximum electrostatic-energy density $w_{e(max)}$ can be plotted (Fig. 14.3).[55] In an ideal device, the electrode surfaces are perfectly smooth and fringing fields can be neglected. However, in practice it is dangerous to operate an electrostatic device near the Paschen limit, since nonuniformities on the electrode surfaces (e.g., spikes, dimples, edges) can increase the electric field locally and cause premature electrostatic breakdown. A gray region is plotted below the maximum electrostatic-energy density curve in Fig. 14.3, to indicate a region of operation that requires caution (i.e., within 10X of the ultimate limit).

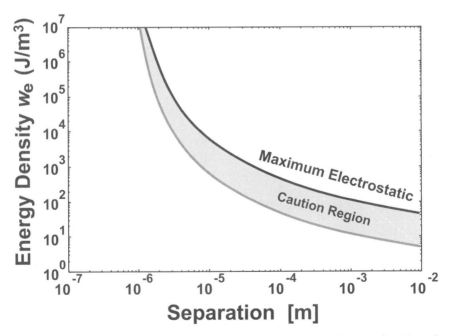

Figure 14.3 Maximum field-limited electrostatic energy density as a function of electrode separation g at an ambient pressure of 1 atm.[55]

14.3.3.3 Voltage-Limited Electrostatic Energy Densities

The energy density of an electrostatic system is often limited by the amount of voltage and the sizes of electrode gap that are practical to achieve. Figure 14.4 shows plots of the electrostatic-energy density of a device at atmospheric pressure as a function of electrode separation and for applied voltages of 5 to 100 V.

14.4 Electrostatic Microactuator Configurations

The configuration and design of electrostatic microactuators can be classified into two general groups. The first are the gap-closing actuators, where the forces act to reduce the separation between the electrodes. The second group is the constant-gap actuators, where the forces act to increase the degree to which the electrodes overlap for a fixed gap. Each class of electrostatic microactuators will be treated individually, with gap-closing actuators discussed in Sections 14.4.1 and 14.4.2, and constant-gap actuators treated in Section 14.4.3. After the important operational aspects are identified and analyzed, examples are given and additional design considerations are discussed.

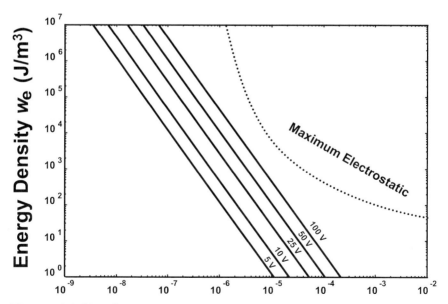

Figure 14.4 Plot of voltage-limited electrostatic energy density as a function of electrode separation at an ambient pressure of 1 atm.[55]

14.4.1 Gap-Closing Electrostatic Microactuators

The design and operation of gap-closing microactuators shown in Fig. 14.5 are straightforward. As shown in Section 14.3.2.1, two adjacent electrodes of area A separated by a gap g and biased with a potential difference V will experience an attractive force given by Eq. (10), when the surrounding medium is free space (e.g., air). If we fix one electrode and adjust the position of the other, the force acting on the plates will be inversely proportional to the square of the gap. The force on the electrodes will become very high as the gap is reduced and drops off rapidly as the gap is increased. Ultimately, the forces are limited by the maximum electric field that can be sustained before electrostatic breakdown occurs.

Although the use of small gaps can enable high forces to be generated with relatively small voltages, the range of available movement will also be reduced. Major design considerations for gap-closing microactuators involve the tradeoff that exists among level of force, range of motion, and applied voltage.

Consider the modified structure shown in Fig. 14.6, in which the movable electrode is now suspended by a spring of stiffness k_z. When no voltage is applied, the electrodes are separated by a gap g_o. As the voltage increases, an electrostatic force is generated that will pull against the spring and force the movable electrode down toward the stationary electrode. As the electrode is pulled downward, the linear spring will generate a mechanical restoring force that will counteract the electrostatic forces until the plate stops moving down, as shown in Fig. 14.6(a). (That is, in equilibrium, the electrostatic force has been balanced by the mechanical restoring force of the spring.) Although this is true for small voltages and correspondingly small deflections, it is not the case when the resulting deflection is large, as will be shown in the next section.

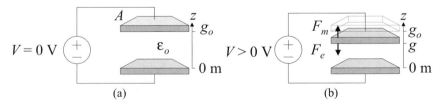

Figure 14.5 Schematic view of a gap-closing microactuator (a) with no voltage applied and in the initial position, and (b) with a voltage applied and a gap that is rigidly varied.

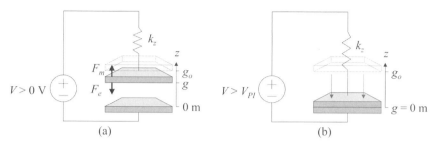

Figure 14.6 Schematic view of a gap-closing microactuator (a) held open by a spring with stiffness k_z, and (b) with the top electrode pulled completely down to the lower electrode due to $V > V_{PI}$.

14.4.1.1 Electrostatic Pull-In

To analyze carefully electrostatic gap-closing actuators being driven through large displacements, we need to consider the energies involved in the system. The electrostatic energy Eq. (9) is given by:

$$W_e = -\left(\frac{\varepsilon_0 A V^2}{2z}\right), \qquad (23)$$

with electrode separation $z = g$; the energy stored in the spring is:

$$W_m = \frac{k_z(g_0 - z)^2}{2}, \qquad (24)$$

with mechanical stiffness of the spring in the z direction k_z. The initial position of the movable electrode is $z = g_o$, which will gradually reduce as an increasing voltage is applied. The total energy of the system is given by:

$$W_{tot} = W_m + W_e = \frac{k_z(g_0 - z)^2}{2} - \left(\frac{\varepsilon_0 A V^2}{2z}\right). \qquad (25)$$

To find the forces involved, we need to take the derivative of the total energy:

$$\frac{\partial W_{tot}}{\partial z} = -k_z(g_0 - z) + \frac{\varepsilon_0 A V^2}{2z^2}, \qquad (26)$$

where clearly the first term is the restoring force of the spring and a second term is the attractive electrostatic force. In order to have mechanical equilibrium, we set the forces equal to each other:

$$k_z(g_0 - z) = \frac{\varepsilon_0 A V^2}{2z^2}. \qquad (27)$$

To determine the stability of the solutions, we need to take the second-order derivative. Stable solutions correspond to values greater than or equal to 0:

$$\frac{\partial^2 W_{tot}}{\partial z^2} = k_z - \left(\frac{\varepsilon_0 A V^2}{z^3}\right) \geq 0. \qquad (28)$$

Substituting Eq. (28) back into Eq. (27) by replacing k_z yields an important condition for electrostatic microactuator design. Specifically, the position z of the movable electrodes must always be:

$$z \geq \frac{2}{3} g_0. \qquad (29)$$

The conditions for pull-in and the regions of stable and unstable equilibrium are graphically shown in Fig. 14.7. In practical terms, Eq. (29) means that only one-third of the gap can be traversed with stability and hence used for stable microactuation. This result can be seen graphically on the plot of total system energy. The plot shows an energy well that provides a stable solution near $z = g_0$. As the applied voltage is increased, the energy barrier is reduced. Increasing the voltage further can result in an energy well that forces the moving electrode down to the stationary electrode.

Substituting Eq. (28) into Eq. (27) and solving for the voltage required to traverse one-third of the gap yields the voltage V_{PI} required to bring the movable electrode to the verge of being pulled down, namely:

$$V_{PI} = \sqrt{\frac{8}{27} \cdot \frac{k_z g_0^3}{\varepsilon_0 A}}. \qquad (30)$$

This is usually termed the pull-in voltage. Voltages larger than or equal to V_{PI} will result in the movable electrode being immediately pulled all the way down to the stationary electrode [Fig. 14.6(b)]. If the bias

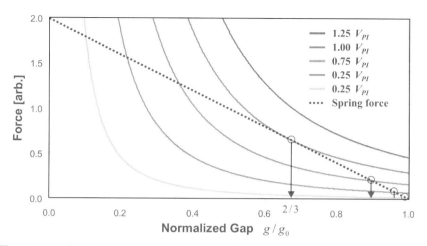

Figure 14.7 Plot of the electrostatic and mechanical restoring force in spring-suspended electrostatic microactuators as a function of normalized gap size ($g = g_0$) for different voltages. Intersections with linear spring-force lines indicate where forces are equal; intersections on the right-hand side of $(2/3)g/g_0$ are stable, whereas those on the left are unstable.

remains when the conductors touch, sparking, shorting, welding, and other disastrous effects can occur that usually result in the destruction of the device. Thus, electrostatic gap-closing microactuators made with bare conductors are typically designed to avoid this voltage mode of operation.

14.4.1.2 Solutions to the Pull-In Issues

There are few methods for circumventing the pull-in issue so that devices are not destroyed and a larger range of motion can be achieved. The most straightforward solution is to use a limit stop to prevent the biased electrodes from smashing together. The key design characteristics of limit stops are (1) that they need to be biased to the same voltage as the movable electrode to prevent any forces acting between the limit stop and the movable electrode, and (2) that they need to provide a sufficient gap to ensure the electrodes will not touch. A concern of repeated use of the limit stop is mechanical wear and the possible adhesion of the movable electrode to the limit stop.

One method of achieving controlled actuation over more than one-third of the gap is to add a capacitor in series with the electrostatic microactuators.[56] The added capacitor, which forms a voltage divider with the microactuators, stabilizes the microactuator by providing negative feedback. If the movable electrode is pulled down, the capacitance of the microactuator will increase since it is inversely proportional to the gap (i.e., $C_{microactuator} \, \varepsilon_0 \, A/g$). Thus as the microactuator plate is pulled toward the stationary electrode, the voltage divider will drop more voltage across the fixed supplementary capacitor instead of the microactuator, thereby protecting it. As the voltage across the microactuator drops, the electrostatic force will also be reduced, and eventually the downward motion will be stopped by the restoring force of the spring. A complete treatment of this concept is provided by Seeger and Crary.[56] A conclusion from their analysis is that as long as the supplemental fixed capacitor C_f has a magnitude that is less than one-half of the capacitance of the microactuator in its rest state (e.g., $C_f \leq 0.5 \, C_{microactuator} = 0.5\varepsilon_0 A/g_0$), the microactuator can be driven to any point position in the gap with stability. One caveat of this approach is that often the capacitance of the microactuators is very low (e.g., for an area $A = 10 \times 100$ µm, $g_o = 1$ µm, then $C_{microactuator} = 8.8$ fF, and thus $C_f \leq 4.4$ fF). Consequentially, it can be a challenge to fabricate and integrate precise supplemental fixed capacitors reliably at this level of capacitance while avoiding stray parasitic capacitance. In addition, in order to get the full range of motion, a drive voltage greater than $3\sqrt{3}V_{PI}$ is required. Thus, this high-voltage requirement can also present a practical barrier to the use of this technique. An alternative approach is to use a MOS capacitor biased into depletion. It has been shown that a MOS capacitor with a thin gate oxide, light substrate doping, and gate area

$$A_{gate} \leq \sqrt{\frac{\sqrt{\frac{2k_z g_0^2}{\varepsilon A}}}{2q\varepsilon_{Si}\varepsilon_0 N_A}} \tag{31}$$

will stabilize the device over the full range, and a voltage of only $(3/2)\sqrt{3}V_{PI}$ is needed to operate the device over the entire range of the air gap.[56]

It is also possible to achieve closed-loop control of electrode motion over more than one-third of the gap with a method that uses high-speed electronics and a Σ–Δ control strategy. Such a system was designed, fabricated, and tested by Fedder and Howe.[57]

14.4.2 Examples of Gap-Closing Electrostatic Microactuators

This section describes the leading examples of gap-closing electrostatic microactuators.

14.4.2.1 Resonant-Gate Transistor

The first gap-closing electrostatic microactuator, known as the resonant-gate transistor, developed by Nathanson at RCA in the 1960s, was the first engineered microactuator ever made.[58] The resonant-gate transistor, shown in Fig. 14.8, is nothing more than a single-end supported cantilever beam. Voltages applied to the cantilever caused it to move up and down and thereby to modulate the flow of charge carriers in the substrate below the cantilever (i.e., the cantilever acted as the gate electrode in this micromechanical MOSFET without the oxide layer). The voltages were typically applied at the resonant frequency of the cantilever so that a maximum deflection and hence current modulation were achieved. This earliest of all microactuators did not have a limit stop and thus would be susceptible to pull-in voltage-induced device failure.

14.4.2.2 Digital Mirror Device

A prominent example of a gap-closing microactuator with a limit stop is the digital mirror device (DMD) developed by Texas Instruments, Inc.[59] Voltages applied to lower electrodes pull down the micromirror. However, instead of landing on powered electrodes, the mirror plate touches down on a neutrally biased landing electrode. The use of a pro-

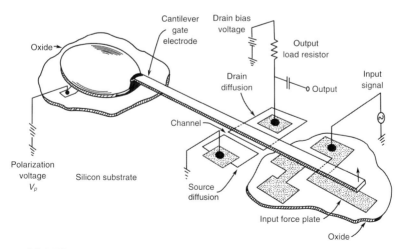

Figure 14.8 The resonant-gate transistor, which was the first engineered microactuator, is an electrostatic gap-closing microactuator.[58]

prietary surface treatment prevents the parts from sticking after repeated use. The reliability of the devices is excellent. (For example, millions of mirrors in a projection display operate for billions of cycles each without any failing.) Examples of DMD structures are shown in Fig. 14.9.

14.4.2.3 Curved Electrode

An interesting design that takes the case of limit stops to a more complex extreme is that of the curved electrostatic microactuators (Fig. 14.10).[60] The well-positioned placement of many limit stops allows a normally high-force actuation technique, which has a pull-in voltage-limited displacement, to be converted into a relatively longer range, lower force microactuation technique.

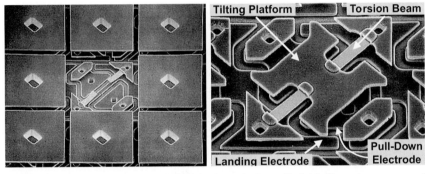

Figure 14.9 Texas Instruments' digital mirror device with (at left) a mirror removed to reveal the supporting micromechanisms and (at right) a closeup of the supporting micromechanisms, which include pull-down and landing or limit-stop electrodes.

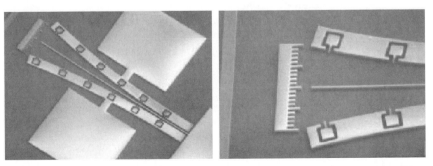

Figure 14.10 Curved electrostatic microactuators with integrated limit stops.[60]

14.4.2.4 S-Shaped Actuator

A similar approach that makes use of a continuous dielectric layer instead of many limit stops is called an S-shaped actuator (Fig. 14.11).[61] The application of a sufficient voltage to the lower electrode causes the S-shaped electrode to zip down and along it in a manner similar to a zipper. The application of a voltage on the upper electrode causes the S-shaped actuator to zip back up and along it.

14.4.2.5 Scratch-Drive Electrostatic Stepper Motor

A method of converting the high-force short-range microactuation of gap-closing microactuators into one with a nearly limitless range is to use a stepper motor configuration. The best example of this approach is the scratch-drive microactuators developed by Akiyama *et al.* (Fig. 14.12).[62] This device operates by electrostatically collapsing a microstructure, which has a bushing or standoff along one edge, down onto a lower electrode that is coated with a dielectric layer. As the electrostatic forces increase, the plate is deformed and held firmly in place. The plate bending that occurs near the bushing results in the advancement of the edge of the bushing. When the voltages are turned off, the mechanical restoring force of the bent plate propels the microactuators forward by a small step. Indeed, for the typical device geometry of plate length 80 µm and plate thickness 1.5 µm, the step size ranges from 10 nm for 50 V pulses and 1-µm-tall bushings to 80 nm for 200 V pulses and 2-µm-tall bushings. Cycling the voltage pulses rapidly can result in a substantial device velocity (e.g., 5 µm/sec for 1-µm-tall bushings and a cycle frequency of 100 Hz to 80 µm/sec for

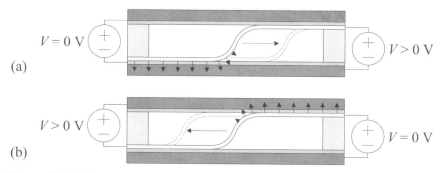

Figure 14.11 S-shaped electrostatic microactuator being (a) pulled down to the lower electrode, and (b) pulled up to the upper electrode; after Shikida *et al.*[61]

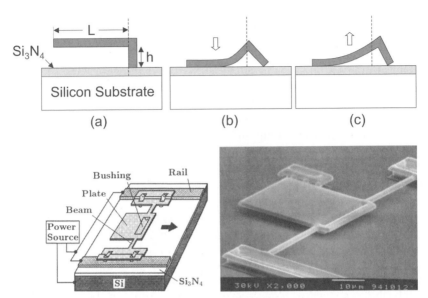

Figure 14.12 Scratch-drive actuator: schematic of device operation, isometric schematic, and SEM micrograph of microfabricated device.[62]

2-μm-tall bushings and a cycle frequency of 1 kHz). This speed has been shown to be linear with frequency from near dc to 1 kHz.

A possible disadvantage of this device is that it requires tethers enabling the plate to be biased appropriately. This issue was successfully addressed by a recent design with an untethered and electrically floating plate attracted toward the substrate by an arrangement of parallel interdigitated electrodes covering the substrate surface and protected by silicon nitride.[63] A voltage difference between the electrodes generates a field between the electrodes and floating plate, with resulting attractive force.

14.4.2.6 Grating Light Valve with Electromechanical Memory

In some instances, devices are designed to take particular advantage of the pull-in characteristic of electrostatic microactuators. The best example of this approach is the grating light valve (GLV) or modulator (GLM),[64,65] which was commercialized by Silicon Light Machines, Inc. (Fig. 14.13).[64]

In this device, the movable electrode is made of two materials sandwiched together: an upper conducting layer and a lower dielectric layer, as schematically shown in Fig. 14.14. Assume that the dielectric layer,

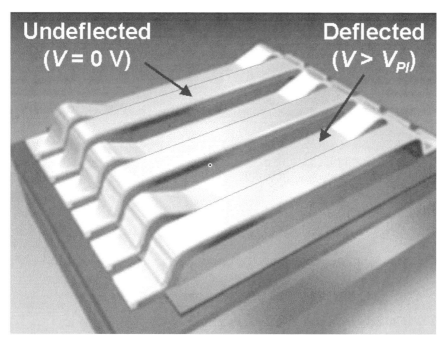

Figure 14.13 Schematic drawing of an electrostatic microactuator that takes advantage of the electromechanical hysteresis to create an addressable and programmable microactuator.[64]

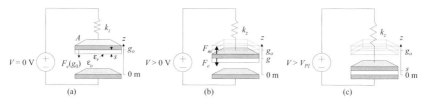

Figure 14.14 Schematic diagram of a gap-closing electrostatic microactuator with upper electrode coated with a dielectric layer (a) with no voltage applied, (b) with a voltage applied that causes the upper electrode to move, and (c) with voltage higher than the pull-in voltage.

which acts as a limit stop, has a thickness s and a relative permittivity constant ε_r. The total capacitance of the overall structure is the series combination of the air capacitance and the capacitance of the dielectric layer.

$$C_{total} = \frac{1}{\dfrac{1}{C_{dielectric}} + \dfrac{1}{C_{air}}} = \frac{1}{\dfrac{g_0 - s}{\varepsilon_0 A} + \dfrac{s}{\varepsilon_r \varepsilon_0 A}}. \tag{32}$$

When the movable electrode moves downward to a position g, Eq. (33) can be reduced to:

$$C_{total} = \frac{\varepsilon_0 A}{g - \left(1 - \frac{1}{\varepsilon_r}\right)s}, \qquad (33)$$

with g_o being replaced with g. Calculating the pull-in voltage of this structure follows the same general procedure used before and yields a pull-in voltage V_{PI} of

$$V_{PI} = \sqrt{\frac{8}{27} \cdot \frac{k_z \left(g_0 - \left(1 - \frac{1}{\varepsilon_r}\right)s\right)^3}{\varepsilon_0 A}}. \qquad (34)$$

The pull-in occurs when the electrode has moved a distance

$$\Delta g = -\frac{1}{3}\left[g_0 - \left(1 - \frac{1}{\varepsilon_r}\right)s\right]; \qquad (35)$$

thus the dielectric layer actually reduces the pull-in voltage and reduces the range of motion before pull-in occurs. After the movable electrode has been pulled down, it will remain clamped to the stationary electrode by electrostatic forces. The applied voltage can then be reduced substantially until the electrostatic clamping force is eventually exceeded by the mechanical restoring force of the spring. The voltage at which the movable electrode is pulled back up is called the pull-up voltage, V_{PU}. The pull-up voltage for this configuration can be shown to be

$$V_{PU} = \sqrt{\frac{k_z (g_0 - s) 2 s^2}{\varepsilon_r \varepsilon_0 A}}. \qquad (36)$$

To arrive at values for V_{PI} and V_{PU}, the nature of the mechanical stiffness of the supporting spring must be known. In the context of the grating light valve, the mechanical support is a doubly anchored beam with length >> width >> thickness. The mechanical response of the spring is often dominated either by bending forces or tensile forces.

In the case where bending forces govern the mechanics of the microstructure, the stiffness is given by $k_z = \sqrt{16wt^3 E/l^3}$. In the case where tensile forces govern the mechanics of the microstructure, the stiffness is given by $k_z = 4\sigma wt/l$.

A resulting novel characteristic of this device is that it has electromechanical hysteresis, as shown by a graph of electrode position versus applied voltage (Fig. 14.15); as a result, the device has electromechanical *memory*. Therefore, a snapped-down position can be programmed in with a voltage $V > V_{PI}$ and then held indefinitely simply by applying a much lower voltage $V_{PI} > V > V_{PU}$.

14.4.2.7 Electrostatic Muscle

One strategy for overcoming the small displacement normally associated with linear gap-closing microactuators is to stack many gap-closing actuator in series. A two-dimensional array of gap-closing actuators combined into an integrated force array has been designed and fabricated by Bobbio *et al.*[66] and is schematically shown in Fig. 14.16. By integrating a dielectric layer onto the electrodes, displacements equal to the entire initial gap can be achieved, just as described in the previous section. Thus, although each device my contribute only 1 μm of motion, by stacking many in series, a much larger next motion can be achieved. The design described by Bobbio *et al.* used metal-coated polyimide structures that

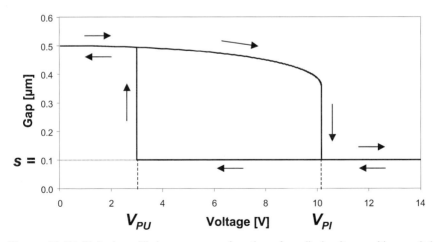

Figure 14.15 Plot of equilibrium gap as a function of applied voltage with $s = 0.1$ μm. The arrows indicate the motion of the plate as the voltage is ramped from 0 to 14 V and back to 0 V. Note the hysteresis at the center of the device response.

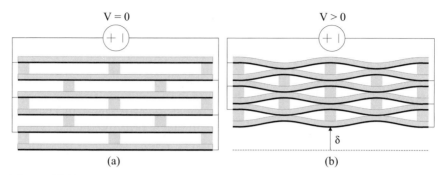

Figure 14.16 Schematic diagram of the integrated array of gap-closing microactuators with (a) no voltage applied, and (b) with enough voltage to fully close the array; after Bobbio et al.[66]

where stacked horizontally to achieve large motions in the plane of the wafer.[66] Preliminary results indicate that this approach can be used to achieve a 30% compression and will exert the sizable forces associated with gap-closing microactuators.

14.4.3 Constant-Gap Electrostatic Microactuators

Electrostatic microactuators that have a constant gap are perhaps even more frequently used than gap-closing electrostatic microactuators. Actuators with a constant gap derive a force by varying the degree of electrode overlap and thus are often called variable-capacitance or variable-reactance actuators.[67]

Consider the lateral variable-capacitance structure shown in Fig. 14.17, where the upper electrode or plate is shifted over to the right by a distance x but the gap is maintained at g_o. Clearly, the capacitance will be a function of the overlap $(L - x)$ and is given by $C(x) = \varepsilon_0 t (L-x)/g_o$, with finger thickness t and where we allow x to range from 0 to L. When the electrodes are biased to a potential V, the constant-voltage electrostatic energy $W_{e(V)}$ will be given by $W_{e(V)} = \varepsilon_0 t (L-x) V^2 / 2g_o$. Since this is a constant-voltage system, the force is given by $F_{e(V)}(x) = \partial W_{e(V)} / \partial x = -\varepsilon_0 t V^2 / 2g_o$, which attempts to forcibly realign the electrodes. Note that this force is not a function of position, displacement, or overlap (i.e., the force is constant with respect to x), but it is proportional to voltage squared and inversely proportional to the electrode separation or gap g_o. Micromachining technologies are used to reduce the gap to increase the force. The most common configuration for a constant-gap electrostatic microactuator is the comb-drive actuator shown in

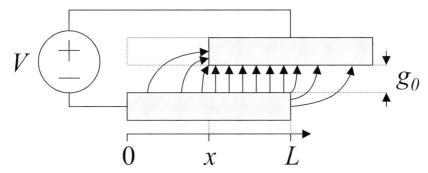

Figure 14.17 Schematic of a generic constant-gap electrostatic microactuator.[67]

Fig. 14.18.[68] The interdigitated comb fingers increase the number of electrodes that overlap, which allows the force generated by the structure to increase linearly with the number of comb fingers.

For a typical comb drive with $g = 2$ μm, $w = 2$ μm, and $V = 10$ V, the force generated per finger is 0.44 nN/finger/V^2 or 0.044 μN/finger at $V = 100$ V. With 100 fingers, the force becomes 4.4 μN, a respectable but not tremendously large value for microactuators. (For example, with a typical microflexure stiffness of 1 N/m, this force would generate a displacement of 4.4 μm.)

The primary advantage of the comb-drive actuator is its ability to generate a force that is independent of displacement and thus can act over a much longer range than the gap-closing actuator. However, there is a very clear tradeoff between force and displacement. Microactuators based on this approach were first demonstrated by W. Tang[68] but have now been very widely used for many different applications.

Figure 14.18 Scanning electron microscope image of a comb-drive microactuator (a) at rest, and (b) resonating.[68]

14.4.4 Examples of Constant-Gap Electrostatic Microactuators

This section describes some leading examples of constant-gap electrostatic microactuators. Further examples of interdigitated structures and their use in micromachined gyroscopes are provided in Chapter 10.

14.4.4.1 Electrostatic Comb Drive

Electrostatic comb-drive microactuators have become extremely popular due to their design simplicity, performance capabilities, and ease of production. In theory, comb-drive microactuators can be fabricated out of a single conductive structural layer with only one photolithography step and a timed etch of the sacrificial layer to release the device. Although in practice several masks are used to realize high-quality comb drives, the fabrication process is still quite simple and leads to a high device yield and low production costs. The examples of microsensor and microactuator applications that make use of the comb drive are too numerous to list, because the comb drive is also an effective structure for capacitive detection of motion. The most dramatic microactuator examples include Sandia National Laboratories[69] microengines, in which comb drives that have fifty 6-µm-tall fingers separated by 1-µm-wide gaps and operated with a voltage of 100 V generate a force of 13.3 µN. Sandia has used these comb drives in a reciprocating manner to drive around a microgear with tremendous speed (Fig. 14.19).[70,71]. The microgear is linked to other larger structures to result in macroscale motion.

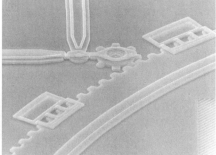

Figure 14.19 Scanning electron microscope images of electrostatic comb-drive arrays (at left) used to drive a rotary gear (at right). These impressive devices have been developed by Sandia National Laboratories and can be produced with their SUMMIT IV or SUMMIT V foundry process.[69-71]

As Sandia has demonstrated, even surface-micromachined comb fingers can also be quite thick or made of multiple layers, as shown in Fig. 14.20, to increase the overlap and hence the force generated. The use of fabrication technologies that allow the production of extremely tall high aspect ratio comb fingers (e.g., LIGA, DRIE) can allow the same in-plane geometries to realize comb drives with correspondingly much higher force.[72,73] For example, a finger-thickness-to-gap-width ratio of 10:1 yields a force that is 0.4 nN/finger/V^2, which for 100 V is 4.4 μN/finger and for 50 fingers is 221 μN).

This speed or frequency response of electrostatic microactuators is typically only limited by the electromechanical frequency response of the overall system, which is largely available to be tailored by design. Due to the relatively small deflections that can be generated with common voltages (~10 V), even in a comb-drive configuration, the devices are typically designed to operate at the resonant frequency for maximum deflection. Another limitation of this design is that as higher voltages are

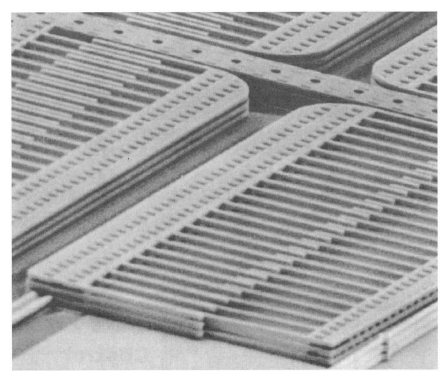

Figure 14.20 Multilevel comb fingers that have been developed by Sandia National Laboratories and can be produced with their SUMMIT IV or SUMMIT V foundry process.[69-71]

used to achieve larger deflections for a given structure, the forces generated between interdigitated comb fingers in the overlapping region will become very large. If, during use, the comb finger vibrates to a position slightly off center, then the strong forces orthogonal to the finger could cause the finger to pull in to the adjacent finger (i.e., lateral or orthogonal pull-in), which will usually cause device failure.

One comb-based electrostatic actuator, reported by Grade et al.[74] achieves a large linear static deflection by making use of a number of design innovations (Fig. 14.21). (1) Reducing the traditional full-comb design in half (i.e., one-sided design) and then in half again (i.e., pull-only design) can reduce the chip area required by a factor of 2 to 4. (2) Placing the mechanical suspension on the outside instead of the inside of the device will reduce the rotational stiffness of the suspension, which can reduce finger misalignment. (3) The finger length is linearly shortened across the comb to reduce the finger overlap and hence the orthogonal

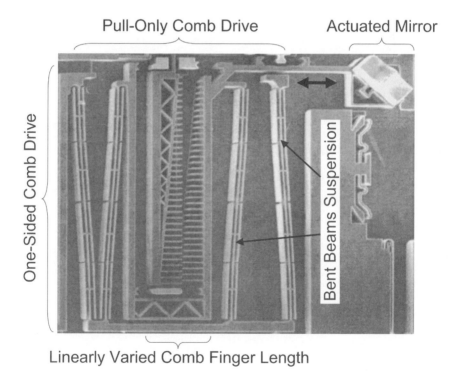

Figure 14.21 SEM image of a one-sided pull-only comb-based microactuator design with linearly varying comb finger length that can achieve large deflections that are laterally stable.[74]

force responsible for lateral pull-in by a factor of two. (4) Pre-bending the suspension can enable a larger deflection with better lateral stability. In their design, they achieved static deflection of 120 μm with 150 V. Operation at the resonant frequency could achieve the same large deflection for a much lower voltage due to the high-Q nature of the structure.

Although linear in-plane horizontal comb drives are commonly realized and used in MEMS, the development of angular and z-axis or vertical comb drives has opened up additional application opportunities of the comb drive (Fig. 14.22). The tradeoff of the higher force (10 μN to 1 mN) and shorter displacements (e.g., 0.1 to 1 μm) available with gap-closing microactuators for the far longer range (e.g., 1 to 100 μm) and lower force (e.g., 1 to 100 μN) of constant-gap actuators is commonly found to be advantageous. The use of microflexures instead of bearing surfaces, such as hubs, enables this device to have an extremely long operational lifetime.

14.4.4.2 Electrostatic Rotary Micromotors

Other constant-gap electrostatic microactuators that have been heavily investigated are those that have a rotary configuration. Specifically,

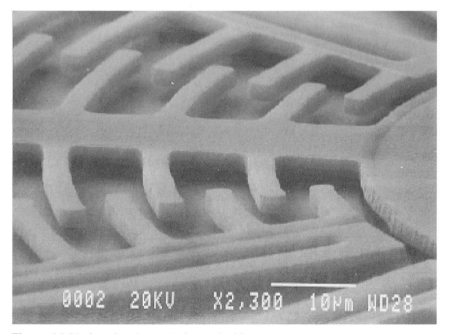

Figure 14.22 Angular electrostatic comb drive.

rotary electrostatic microactuators that make use of bearing surfaces (i.e., hubs) have been demonstrated and are often used to illustrate what is possible with MEMS technology. A very prominent and seminal surface micromachined electrostatic microactuator was the rotary micromotor developed simultaneously at the University of California, Berkeley, by Tai et al.[75,76] and at the Massachusetts Institute of Technology (MIT) by Mehregany et al.[77] The device from Berkeley is shown in Fig. 14.23(a) and (c); that from MIT is shown in Fig. 14.23(b) and (d). The electrostatic rotary micromotor was a groundbreaking device that demonstrated some of the potential of micromachining and MEMS technologies, particularly to the world at large through the lay press. Although these early demonstrations expanded the intellectual horizons for many researchers in the field of MEMS in 1988, including this author, the devices themselves did not function for a long time (e.g., minutes) due to severe wear problems.

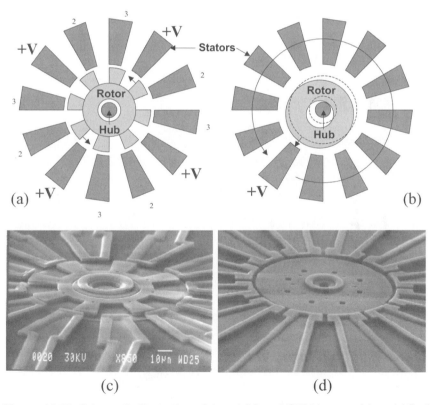

Figure 14.23 Schematic illustrations (a) and (b) and SEM images (c) and (d) of an electrostatic rotary side-drive and wobble micromotor.[75-77]

The lack of an adequate lubrication and bearing surface has plagued the development of all subsequent rotary, hub-based micromotor development. In fact, all microactuator development that required rubbing surfaces to operate has had to deal with this issue. Although there have been a number of efforts to improve the durability and wear coatings and their surface treatment, to date there remains no long-term solution to this problem in the conventional polycrystalline silicon micromotors, which has traditionally been one of the most popular materials with which to build microactuators.[75-77] The torque generated by these micromotors when surface-micromachined is quite small (~5 to 67 pN-m) but scales up with thickness and thus is larger for thick-film devices, such as LIGA or DRIE. Since the torque is dependent on the voltage squared and inversely dependent on the gap, microfabrication can be advantageous, but relatively high voltages (~10 to 100 V) are still typically necessary.

The speed, torque, and wear rate are also a function of the way in which the rotor spins around the hub. In some designs, it does indeed spin. In other designs, where only one stator is powered at a time, the rotor wobbles around the hub by being pulled tightly toward the single activated drive electrode (that is, it is not a balanced drive, in contrast with the side-drive motors).

14.4.5 Hybrid Electrostatic Microactuators

A hybrid electrostatic microactuator is one that directly translates gap-closing electrostatic forces into long-range microactuation. The leading example is the tangential drive microactuators designed, fabricated, and tested by Brennan et al.[78] This device, shown in Fig. 14.24, uses a sim-

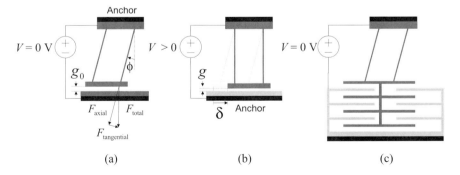

Figure 14.24 Schematic diagrams of a tangential-drive electrostatic microactuator: (a) with no voltage applied, (b) with voltage applied, and (c) in a tangential comb-drive configuration; after Brennan et al.[78]

ple micromechanical flexure (for example, a bent beam) to transform gap-closing electrostatic forces into tangential motions of electrostatic comb fingers. By carefully designing the angle of the supporting flexure and the initial gap between comb fingers and the number of fingers, the force generated and the range of motion of a tangential drive can be tailored. Brennan et al. provide a complete treatment of this device.[78]

14.4.6 Electrostatic Induction

Another class of electrostatic microactuators exploits charge induction to realize constant-charge devices. The best example is the planar three-phase electrostatic surface-drive actuator, shown in Fig. 14.25.[79] The device consists of a series of parallel electrodes fabricated into a planar substrate. The moving part is not attached to the substrate in any way; instead, it is a slider made of two layers: an insulating layer and a weakly conductive layer into which charges are induced. Actuation is achieved in a multistep process. (1) The slider is positioned on top of the substrate [Fig. 14.25(a)], biasing selected electrodes to either +V, –V, or 0 V. (2)

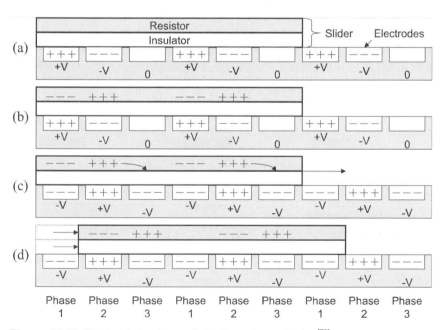

Figure 14.25 Electrostatic charge induction microactuator.[79]

After a period of time, mirror image charges are induced into the weakly conductive layer of the slider that have a sign opposing the biasing of each underlying electrode [Fig. 14.25(b)]. (3) The polarity of the lower electrodes is changed far more rapidly than the induced charges can move [Fig. 14.25(c)]. The result is that both a levitation and a translation force are generated, causing the slider to jump up and over one electrode [Fig. 14.25(d)]. This process can be repeated rapidly to cause the slider to move rapidly. The performance of reported devices yields a translation force that is proportional to surface area (e.g., 2 to 40 N/m^2 and velocities of 15 to 350 µm/sec, with electrode pitches ranging from 50 to 420 µm and drive voltages ranging from 130 to 750 V). In theory, it would be possible to stack multiple layers to realize much higher forces.

14.4.7 Issues and Challenges

The large driving voltage prevents electrostatic microactuators from being conveniently driven with typical on-chip circuits and voltages (e.g., <5 V). In addition, high voltages on small features create tremendous electric-field gradients that can attract dust particles. The attraction of a dust particle onto an electrostatic microactuator typically results in the catastrophic failure of the device. Also, electrostatic actuators will not function in conductive fluids because electric potentials and charge distributions cannot be maintained. A more recent electrostatic actuator design that can achieve very large forces (~mN) and can travel long distances (6 mm) is an electrostatic *scratch-drive* stepper motor.[62] This highly area-efficient design has been used to assemble complex, hinged 3D microstructures used for optical applications.[80]

14.5 Piezoelectric Microactuators

The word *piezo* is of Greek origin and means "pressure." Crystalline materials that (1) develop an electric polarization when stressed and conversely are (2) strained by the application of an electric field are said to be piezoelectric if, respectively, (1) the sign of the polarization depends on the sign of the stress and (2) the sign of the induced strain depends on the sign of the applied electric field. The physical origin of piezoelectricity is due to the noncentrosymmetric arrangement of atoms and their net charges in a lattice.[81-84] A good example that illustrates this concept is the crystal structure of quartz. The crystal arrangement of Si atoms (net positive charge) and O_2 (net negative charge) is not symmetric about any point of the lattice (Fig. 14.26).

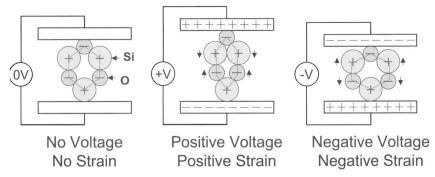

Figure 14.26 Noncentrosymmetric atomic lattice results in a voltage when compressed or pulled. Conversely, when a voltage is applied, the respective mechanical output is achieved.[81-84]

When compressing or stretching a lattice, the sign of the induced polarization will change and the magnitude will be approximately linearly dependent on the stress, which is a useful property for sensors. By applying an electric field, electrostatic forces given by Eq. (1) will act on the noncentrosymmetric arrangement of atoms and cause the lattice to compress or stretch, depending on the sign of the applied voltage and the sign of the piezoelectric coupling coefficient. The ability to apply a voltage and achieve direct mechanical output is a characteristic of piezoelectric materials that is well suited for microactuator applications.

The constitutive equations for piezoelectricity can be derived from thermodynamics and are typically expressed in a tensor form. The relationship among the strain S_{ij}, the stress T_{kl}, and the electric field E_k is given by:

$$S_{ij} = s^E_{ijkl} T_{kl} + d_{kij} E_k, \tag{37}$$

with elastic compliance at constant electric field s^E_{ijkl} [m²/N] and piezoelectric coupling coefficient d_{kij} [C/N = V/m].[82-84] The relationship among electric displacement D_i, stress, and electric field is given by:

$$D_i = d_{ikl} T_{kl} + \varepsilon^T_{ik} E_k, \tag{38}$$

with permittivity at constant stress ε^T_{ik} [F/m].[82-84] Typically, higher order tensors can be reduced to mathematical objects with fewer subscripts, by using engineering notation.[82] As an example, the elastic compliance tensors are then represented as a 6 × 6 matrix, with two instead of four subscripts.

In the field of microactuators, the most common physical device configuration is with a z-axis electric field E_3 (i.e., through the film thickness) and an x-axis or y-axis plane stress T_1 (i.e., in the plane of the thin-film). In this specific configuration, the tensor notation can be reduced to:

$$S_1 = s_{11}^E T_1 + d_{31} E_3 \tag{39}$$

and

$$D_3 = d_{31} T_1 + \varepsilon_{33}^T E_3, \tag{40}$$

with x-axis strain and stress, S_1 and T_1, respectively; x-axis mechanical compliance at constant electric field s^E_{11}; z-axis to x-axis piezoelectric coupling coefficient d_{31}; z-axis permittivity at constant stress ε^T_{33}; z-axis electric displacement D_3; and z-axis electric field E_3.[84] Table 14.2 gives the key physical parameters of piezoelectric materials.

14.5.1 Piezoelectric Energy Density

The electrical energy density stored in a piezoelectric microactuator has the same form as that for an electrostatic device, $w_e = \varepsilon_r \varepsilon_0 E^2 / 2$.[83] The ultimate maximum energy density is determined by using the relative permittivity of the piezoelectric material, $\varepsilon_r = \varepsilon_{piezo}$, and the maximum sustainable electric field before breakdown occurs, $E = E_{piezo(max)}$. For high-quality thin-film lead zirconate titanate (PZT), $\varepsilon_{PZT} \approx 1300$, $E_{PZT(max)} \approx 3 \times 10^8$ V/m, and thus $w_{e(PZT)} \approx 5.2 \times 10^7$ J/m^3. With 1-μm-thick layer,

Table 14.2 Material Properties of Common Piezoelectric Materials[87]

Material	Piezoelectric Coefficients		Compliance	Relative Permittivity
	d31 [pC/N]	d33 [pC/N]	s^E_{11} [m2/N]	$\varepsilon^T_{33}/\varepsilon_0$ [-]
ZnO	−4.7	12	6	8.2
Sol-Gel PZT	−88.7	220	—	1300
PVDF	−23	−35	—	4
PZT-5	−171	80 to 593	16.4	1700

the drive voltage needed to achieve this maximum energy density would be 300 V. For a zinc oxide (ZnO) thin film, $\varepsilon_{ZnO} = 8.2$, $E_{ZnO(max)} \approx 3 \times 10^8$ V/m, and thus $w_{e(ZnO)} \approx 3.3 \times 10^6$ J/m^3.[85] For a thin film of polyvinylidene fluoride (PVDF) $\varepsilon_{PVDF} = 4$, $E_{PVDF(max)} \approx 3 \times 10^8$ V/m, and thus $w_{e(PVDF)} \approx 1.6 \times 10^6$ J/m^3. In each case, the energy density available is quite large when compared to electrostatic energy density (Fig. 14.4).

The mechanical energy density for piezoelectric microactuators should be considered when considering the energy available for mechanical output; it is the strain energy produced in the piezoelectric material that is available for mechanical work. The strain-energy density in an elastic material is equal to one-half the product of the strain and the stress, expressed in piezoelectric parameters as $w_{piezo(max)} = S_1 T_1 / 2$.[83] Given the maximum z-axis electric field before breakdown $E_{3(max)} = E_{piezo(max)}$, the maximum x-axis strain that can be induced is given by $S_1 = d_{31} E_{3(max)}$. The resulting maximum mechanical strain energy density of a piezoelectric microactuator is $w_{piezo(max)} = \dfrac{S_{1(max)}^2}{2 s_{11}^E} = \dfrac{(d_{31} E_{3(max)})^2}{2 s_{11}^E}$.

For bulk PZT-5, the most responsive of all piezoelectric materials, this mechanical strain energy is ~1.6×10^8 J/m^3. For ZnO, which is another popular material, the maximum mechanical strain energy density is ~3×10^5 J/m^3. Both of these resulting energy densities are large when compared to most electrostatic energy densities (Fig. 14.4). Although, in practice, thin-film quality will limit the maximum sustainable electric field and hence energy density, piezoelectric materials will still offer a very large energy density.

14.5.2 Piezoelectric Microactuator Configurations

The configuration and design of piezoelectric microactuators can be classified into two general groups. The first can be referred to as linear actuators, where the forces act to expand or contract the piezoelectric element linearly. The second group can be referred to as bimorph actuators, where composite structures are formed to generated bending motion and lateral or out-of-plane forces. The two groups of piezoelectric microactuator are treated individually, with linear piezoelectric actuators discussed in Sections 14.5.2.1 and 14.5.2.2 and bimorph piezoelectric actuators in Section 14.5.2.3. After the important operational aspects are identified and analyzed, examples are given and additional design considerations are discussed.

14.5.2.1 Linear Piezoelectric Microactuators

The design and operation of linear piezoelectric microactuators, shown in Fig. 14.27, are straightforward. As discussed above, the application of a voltage across a piezoelectric film will cause a physical strain that results in a change in length. To apply the electric field, electrodes are integrated onto both the top and the bottom of the film. These electrodes are intentionally thin and compliant so as not to interfere with the mechanical deformation of the piezoelectric material.

To generate a substantial mechanical displacement or force, a high electric field in the piezoelectric material is needed. If the separation of the electrodes is large (i.e., the piezoelectric material is thick), a very high voltage will be required. Consequently, most piezoelectric microactuator designs make use of thin-film piezoelectric materials sandwiched between two conductors. Depending on the material, orientation of the crystal axis, and polarity of the voltage applied, the induced deflection could be an expansion or contraction. Yet, despite using the most responsive materials, the total motion at the end of a cantilever of length L_0 will be $\Delta L = S_1 L_0 = d_{31} E_3 L_0$. For example, a 100-µm-long and 1-µm-thick beam of PZT will have a linear tip motion of 17 nm/V. At an applied voltage of 10 V, the linear tip motion is only 0.17 µm, a very small motion even for microactuators.

However, if the device were physically prevented from expanding, the applied field would instead generate a stress in the film. The pressure exerted by a 1-µm-thick PZT microactuator at the tip of the cantilever would be $T_1 = d_{31} E_3 / s_{11}^E = (-171 \ \text{pC/N})(1 \ \text{V/µm})/(16.4 \times 10^{-12} \ \text{m}^2/\text{N}) = 10$ MPa/V. If the PZT cantilever were also 10 µm wide, the force exerted by this device would be 104 µN/V. Thus, the application of even a relatively modest voltage (10 V) will generate a high level of stress (100 MPa) and a large tip force (1 mN). The force generated is a relatively large force for such a small device (i.e., much larger electrostatic comb drives require much more chip area and generate much less force). Thus,

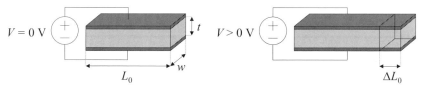

Figure 14.27 Schematic diagram of a linear piezoelectric microactuator (a) no voltage applied, and (b) with voltage applied that causes the element to extend.

although linear piezoelectric microactuators are capable of generating high forces, they do not generate large displacements.

Perhaps the most unique characteristic and capability of piezoelectric materials in the context of microactuators are their reciprocal nature (i.e., they can also sense forces exerted on them from the outside). The intrinsic ability of piezoelectric microactuators to sense externally applied motions and forces can enable these devices to be used effectively in feedback loops for control applications.

14.5.2.2 Piezoelectric Stepper Motor: Inch Worm

The speed of piezoelectric microactuators is ultimately limited by the speed with which a strain can be changed within a piezoelectric to material. This speed is related to the acoustic velocity in the material. For material PZT and ZnO, this acoustic velocity is 4.5 mm/μsec and 6.4 mm/μsec, respectively.[86] Consequently, piezoelectric microactuators can be operated at very high frequencies. The design of the overall electrochemical system limits the frequency response.

The high-frequency response of piezoelectric materials (e.g., ZnO and PZT) enables small repeated displacements to accumulate rapidly when configured in a stepper-motor design (Fig. 14.28).[87] Piezoelectric stepper

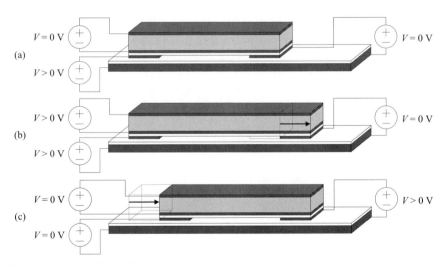

Figure 14.28 Schematic of a piezoelectric microscale stepper motor (a) clamped in position by the left electrostatic clamping capacitor, (b) extended by the application of a voltage to the piezoelectric material, and (c) activating the right electrostatic clamp, releasing the left electrostatic clamp, and relaxing the voltage on the piezoelectric material so the element shrinks and advances one step.[87]

micromotors have been used in surgical applications (e.g., a smart force-feedback knife[88] and an ultrasonic cutting tool[89]).

14.5.2.3 Piezoelectric Bimorph

The preferred microactuator configuration to translate the high forces that are easily generated with piezoelectric materials into large displacements is a bimorph structure (Fig. 14.29).[85,90-92] A bimorph is a composite structure that is typically made of two or more layers where at least one of the layers undergoes a linear expansion or contraction. The other mechanically significant layers are either passive nonpiezoelectric materials [Fig. 14.29(a)] or another layer of piezoelectric material with an oppositely induced electrical polarization which therefore generates a stress and force in the opposite direction [Fig. 14.29(b)]. Again, the electrodes are made very thin and flexible, and thus are not considered mechanically significant.

Passive

If the passive material is much thicker than the PZT, Stoney's equation can be used to relate the stress in the film to the deflection or radius of curvature.[93] Specifically, in a cantilever configuration, Stoney's equation can be used to determine the tip deflection:

$$\delta = \frac{1-v_s}{E_s} \frac{3L_0^2}{t_s^2} \sigma_{piezo} t_{piezo}, \tag{41}$$

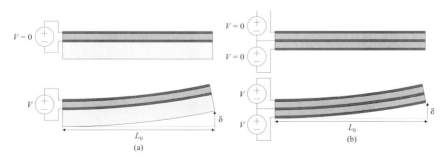

Figure 14.29 Schematic of bimorph piezoelectric microactuators with a layer of piezoelectric material and (a) a passive mechanical layer, and (b) another piezoelectric layer.[90]

with elastic modulus of the passive material E_s; Poisson's ratio of the passive material v_s, thickness of the passive material and of the piezoelectric film t_s and t_{piezo}, respectively; stress in the piezoelectric film σ_{piezo}; and cantilever length L_0. Using the stress found in Section 14.5.2.1, a film that is a 1-µm-thick layer of PZT on a 100-µm-long and 5-µm-thick passive cantilever that has an elastic modulus $E_s = 200$ GPa and Poisson's ratio $v_s = 0.3$ results in a tip deflection of $\delta = 0.2$ µm/V, a displacement that is 10 times larger than in the case of the linear microactuator of essentially the same dimensions.

Active

As in Fig. 14.29(b), if both layers in a bimorph structure are made of piezoelectric material, each layer has the same thickness ($t_1 = t_2$), and an opposite voltage is applied to each layer ($E_{layer1} = -E_{layer2}$), then the tip deflection is governed by:

$$\delta(L_0) = \frac{3d_{31}L_0^2}{8t^2}V. \tag{42}$$

For example, if both films are 1 µm thick and the bimorph cantilever is 100 µm long, the tip deflection will be 0.6 µm/V, a deflection much larger than that from a typical bimorph made with a passive material.[90]

In addition, other modes of motion of the bimorph cantilever tip can be achieved. Specifically, the tip of the beam can also experience a linear expansion or contraction and thus act like a linear piezoelectric microactuator. In theory, additional complex modes of motion can be achieved if the composite beam is made of two parallel bilayer strips (Fig. 14.30). Not only could large positive and negative curvatures and small linear extensions and contractions be achieved, but tip rotation in both directions about the axis of the beam could also be achieved.

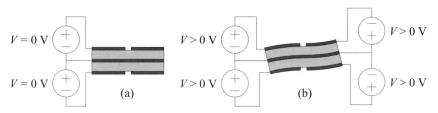

Figure 14.30 (a) Schematic diagram of a composite bimorph piezoelectric microactuator made of two parallel bi-layer strips of piezoelectric material. This device is capable of linear, curling, and (b) twisting actuation.

14.5.2.4 Membranes

By fabricating devices that consist of a composite thin plate or membrane (i.e., that are made of a mechanically passive material, a thin piezoelectric layer, and a set of interdigitated electrodes that are mechanically compliant), a physical deformation wave pattern along the interdigitated electrodes will be mechanically induced (Fig. 14.31). Switching the potentials applied to the interdigitated electrodes will shift the same physical wave pattern one-half of the spatial period of the interdigitated fingers. Switching the applied voltage at high frequencies, particularly at the resonant frequency of the electromechanical system, will yield the largest deflections. This general configuration has been used to generate devices exciting surface acoustic waves (SAWs) in plates and Lamb waves in membranes. In-depth treatments of the various types of waves in solids, including thin slabs, are found in the literature.[94-98] Resonant devices of this general configuration have been used to pump fluids, transport solid objects, and realize sensors based on the selective mass loading of the resonating surfaces, plates, or membranes.

14.5.3 Piezoelectric Microactuator Design Issues

This section discusses some of the key design issues for piezoelectric microactuators that should be taken into account when considering this transduction mechanism.

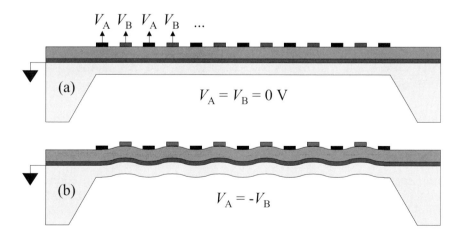

Figure 14.31 Schematic representation of a resonant piezoelectric membrane-based microactuator.

14.5.3.1 Thin-Film Piezoelectric Materials

Although it is possible to design, fabricate, and test a seemingly simple cantilever-based microactuator, the fabrication complexity of piezoelectric cantilever-based microactuators and the similarity of operational outcome usually mean that an alternative actuation technology, such as electrostatic forces, is chosen. However, piezoelectric cantilever devices should be considered if the following performance specifications are needed: push-pull bidirectional actuation, lower drive voltages, higher forces than typically available for electrostatics, and the ability to sense forces applied to the structure. In practice, piezoelectric microactuators made with thin films have been the most successful when exploiting the high forces and high operating frequencies available. Specific examples include devices that resonate passive mechanical structures, such as plates and membranes.

14.5.3.2 Bulk Piezoelectric Materials

Although most piezoelectric MEMS have been constructed with thin films of piezoelectric materials deposited by sputtering (for example, ZnO,[99,100] PZT[101]) or sol-gel deposition (for example, [102]), an alternative is to work with macroscopic or bulk-grade piezoelectric materials. Bulk-grade materials typically have excellent piezoelectric properties and are commercially available in large quantities. The challenges of this approach are (1) the machining of small (less than 1 mm^3) parts with integrated compliant electrodes and (2) integrating (for example, bonding) the bulk piezoelectric materials with the microstructure of interest. Although it is possible to macroscopically machine bulk piezoelectric materials that typically are brittle, an alternative manufacturing technology capable of high-resolution machining should be considered (such as electron-discharged machining, EDM). Although EDM is generally serial in nature, wafer-scale batch approaches have been developed. An additional alternative is to use laser machining.

14.5.3.3 Curie Temperature

The piezoelectric properties of materials are significantly temperature-dependent. In fact, above a temperature known as the Curie temperature, piezoelectric properties are completely lost. The Curie temperatures of most commonly used piezoelectric materials are only 200 to 300°C.[86]

14.6 Electrostriction, Electrets, and Electrorheological Fluids

Materials that enable other lesser-used electrical transduction mechanisms that are similar to piezoelectricity include electrostrictive materials, electrets, and electrorheological fluids. Electrostrictive materials experience a mechanical deformation when an external electric field is applied due to the rotation of small electrical domains within the material.[103] For nearly all dielectric materials, this effect is quite small. Electrets are materials that possess a permanent electric polarization. Typically, electrets are inserted between the plates of a capacitor so that movement of the plates can be detected (e.g., pressure sensors and microphones).[104,105] Currently, the use of electrets to provide electrostatic levitation in microactuators has been theorized but not demonstrated.

Electrorheological fluids are colloidal suspensions, which reversibly transform into a semi-solid when subjected to an electric field.[106] When the electric field is applied, electrorheological fluids change phase by forming fibrous structures parallel to the applied field. The viscosity can increase by a factor of up to 10^5.

References

1. H. Fujita, "Microactuators and micromachines," *Proc. IEEE*, 86(8):1721–1732 (1998)
2. H. Fujita, "Future of actuators and microsystems," *Sens. Actuators*, A56(1-2):105–111 (1996)
3. P. Ruther, W. Bacher, K. Feit, and W. Menz, "LIGA-microtesting system with integrated strain-gauges for force measurement," *Proc. IEEE MEMS 1997 Workshop,* Nagoya, Jan. 1997, pp. 541–545
4. C. Grosjean, X. Yang, and Y.-C. Tai, "A thermopneumatic microfluidic system," *Proc. IEEE MEMS 2002 Conf.,* Las Vegas, Jan. 2002, pp. 24–27
5. C. Vancura, M. Rüegg, Y. Li, D. Lange, C. Hagleitner, O. Brand, A. Hierlemann, and H. Baltes, "Magnetically actuated CMOS resonant cantilever gas sensor for volatile organic compounds," *Dig. Tech. Papers Transducers 2003 Conf.,* Boston, June 2003, pp. 1355–1358
6. K.-C Lee, Sekwang Park, and Eun Sok Kim, "Improved electromagnetic displacement transducer with large force for implantable middle ear hearing aid," *Dig. Tech. Papers Transducers 2003 Conf.,* Boston, June 2003, pp. 1217–1220
7. H. H. Gatzen, E. Obermeier, T. Kohlmeier, T. Budde, H. D. Ngo, B. Mukhopadhyay, and M. Farr, "An electromagnetically actuated bistable MEMS optical microswitch," *Dig. Tech. Papers Transducers 2003 Conf.,* Boston, June 2003, pp. 1514–1517

8. C.-H. Ji, Y. Yee, J. Choi, H.-H. Oh, and J.-U. Bu, "Latchable electromagnetic 2 x 2 MEMS optical switch," *Proc. IEEE MEMS 2003 Conf.*, Kyoto, Jan. 2003, pp. 239–242
9. J. Bernstein, W. P. Taylor, J. Brazzle, G. Kirkos, J. Odhner, A. Pareek, and M. Zai, "Two-axis-of-rotation mirror array using electromagnetic MEMS," *Proc. IEEE MEMS 2003 Conf.*, Kyoto, Jan. 2003, pp. 275–278
10. I.-J. Cho, K.-S. Yun, H.-K. Lee, J.-B. Yoon, and E. Yoon, "A low-voltage two-axis electromagnetically actuated micromirror with bulk silicon mirror plates and torsion bars," *Proc. IEEE MEMS 2002 Conf.*, Las Vegas, Jan. 2002, pp. 540–543
11. H. Rothuizen, M. Despont, U. Drechsler, G. Genolet, W. Häberle, M. Lutwyche, R. Stutz, and P. Vettiger, "Compact copper/epoxy-based electromagnetic scanner for scanning probe applications," *Proc. IEEE MEMS 2002 Conf.*, Las Vegas, Jan. 2002, pp. 582–585
12. S. Reyntjens and R. Puers, "RASTA: the real-acceleration-for-self-test accelerometer," *Dig. Tech. Papers Transducers 2001 Conf.*, Munich, June 2001, pp. 434–437
13. A. Bayrashev, A. Parker, W. P. Robbins, and B. Ziaie, "Low frequency wireless powering of microsystems using piezoelectric-magnetostrictive laminate composites," *Dig. Tech. Papers Transducers 2003 Conf.*, Boston, June 2003, pp. 1707–1710
14. T. Bourouina, E. Lebrasseur, G. Reyne, H. Fujita, T. Masuzawa, A. Ludwig, E. Quandt, H. Muro, T. Oki, and A. Asaoka, "A novel optical scanner with integrated two-dimensional magnetostrictive actuation and two-dimensional piezoresistive detection," *Dig. Tech. Papers Transducers 2001 Conf.*, Munich, June 2001, pp. 1328–1331
15. A. Garnier, T. Bourouina, and H. Fujita, "A fast, robust and simple 2-D micro-optical scanner based on contactless magnetostrictive actuation," *Proc. IEEE MEMS 2000 Conf.*, Miyazaki, Jan. 2000, pp. 715–718
16. C.-P. Hsu, Tingsin Liao, and Wensyang Hsu, "Electrothermally driven long stretch micro drive with monolithic cascaded actuation units in compact arrangement," *Dig. Tech. Papers Transducers 2003 Conf.*, Boston, June 2003, pp. 348–351
17. R. Arya, M. M. Rashid, D. Howard, S. D. Collins, and R. L. Smith, "Thermally actuated, bistable, snapping, silicon membrane," *Dig. Tech. Papers Transducers 2003 Conf.*, Boston, June 2003, pp. 1411–1414
18. E. Pichonat-Gallois and M. de Labachelerie, "Thermal actuators used for a micro-optical bench: application for a tunable Fabry-Perot filter," *Dig. Tech. Papers Transducers 2003 Conf.*, Boston, June 2003, pp. 1419–1422
19. C. S. Lee, W.-H. Jin, H.-J. Nam, S.-M. Cho, Y.-S. Kim, and J.-U. Bu, "Micro cantilevers with integrated heaters and piezoelectric detectors for low power SPM data storage application," *Proc. IEEE MEMS 2003 Conf.*, Kyoto, Jan. 2003, pp. 28–31
20. J. Qiu, J. H. Lang, A. H. Slocum, and R. Strümpler, "A high-current electrothermal bistable MEMS relay," *Proc. IEEE MEMS 2003 Conf.*, Kyoto, Jan. 2003, pp. 64–67

21. L.-A. Liew, V. M. Bright, M. L. Dunn, J. W. Daily, and R. Raj, "Development of SiCN ceramic thermal actuators," *Proc. IEEE MEMS 2002 Conf.*, Las Vegas, Jan. 2002, pp. 590–593
22. Y. Wang, Z. Li, D. T. McCormick, and N. C. Tien, "Low-voltage lateral-contact microrelays for RF applications," *Proc. IEEE MEMS 2002 Conf.*, Las Vegas, Jan. 2002, pp. 645–648
23. W.-C. Chen, J. Hsieh, and W. Fang, "A novel single-layer bi-directional out-of-plane electrothermal microactuator," *Proc. IEEE MEMS 2002 Conf.*, Las Vegas, Jan. 2002, pp. 693–697
24. M. Sinclair, "A high frequency resonant scanner using thermal actuation," *Proc. IEEE MEMS 2002 Conf.*, Las Vegas, Jan. 2002, pp. 698–701
25. P. Vettiger, G. Cross, M. Despont, U. Drechsler, U. Dürig, W. Häberle, M. I. Lutwyche, H. E. Rothuizen, R. Stutz, R. Widmer, and G. K. Binnig, "The 'Millipede'—more than 1000 tips for parallel and dense data storage," *Dig. Tech. Papers Transducers 2001 Conf.*, Munich, June 2001, pp. 1054–1057
26. L. L. Chu, J. A. Hetrick, and Y. B. Gianchandani, "Compliant micro transmissions for rectilinear electrothermal actuators," *Dig. Tech. Papers Transducers 2001 Conf.*, Munich, June 2001, pp. 714–717
27. Y. Liu, X. Li, T. Abe, Y. Haga, and M. Esashi, "A thermomechanical relay with microspring contact array," *Proc. IEEE MEMS 2001 Conf.*, Interlaken, Jan. 2001, pp. 220–223
28. G. Lammel and P. Renaud, "3D flip-chip structure of porous silicon with actuator and optical filter for microspectrometer applications," *Proc. IEEE MEMS 2000 Conf.*, Miyazaki, Jan. 2000, pp. 132–135
29. J.-S. Park, L. L. Chu, E. Siwapornsathain, A. D. Oliver, and Y. B. Gianchandani, "Long throw and rotary output electro-thermal actuators based on bent-beam suspensions," *Proc. IEEE MEMS 2000 Conf.*, Miyazaki, Jan. 2000, pp. 680–685
30. O. Brand, M. Hornung, H. Baltes, and C. Hafner, "Ultrasound barrier microsystem for object detection based on micromachined transducer elements," *J. Microelectromech. Syst.*, 6:151–160 (1997)
31. T. Mineta, T. Mitsui, Y. Watanabe, S. Kobayashi, Y. Haga, and M. Esashi, "An active wire with shape memory alloy bending actuator fabricated by room temperature process," *Dig. Tech. Papers Transducers 2001 Conf.*, Munich, June 2001, pp. 698–701
32. B. Winzek, T. Sterzl, and E. Quandt, "Bistable thin film composites with TiHfNi-shape memory alloys," *Dig. Tech. Papers Transducers 2001 Conf.*, Munich, June 2001, pp. 706–709
33. M. Kohl, B. Krevet, and E. Just, "SMA microgripper system," *Dig. Tech. Papers Transducers 2001 Conf.*, Munich, June 2001, pp. 710–713
34. C. C. Ma, R. Wang, Q. P. Sun, Y. Zohar, and M. Wong, "Frequency response of TiNi shape memory alloy thin film micro-actuator," *Proc. IEEE MEMS 2000 Conf.*, Miyazaki, Jan. 2000, pp. 370–373
35. T. Mineta, T. Mitsui, Y. Watanabe, S. Kobayashi, Y. Haga, and M. Esashi, "Batch-fabricated flat winding shape memory alloy actuator for active catheter," *Proc. IEEE MEMS 2000 Conf.*, Miyazaki, Jan. 2000, pp. 375–378

36. Y.-C. Su, L. Lin, and A. P. Pisano, "Water-powered, osmotic microactuator," *Proc. IEEE MEMS 2001 Conf.,* Interlaken, Jan. 2001, pp. 393–396
37. O. Tabata, H. Kojima, T. Kasatani, Y. Isono, and R. Yoshida, "Chemo-mechanical actuator using self-oscillating gel for artificial cilia," *Proc. IEEE MEMS 2003 Conf.,* Kyoto, Jan. 2003, pp. 12–15
38. M. A. Schmidt, "Technologies for microturbomachinery," *Dig. Tech. Papers Transducers 2001 Conf.,* Munich, June 2001, pp. 2–5
39. P. Q. Pham, D. Briand, C. Rossi, and N. F. de Rooij, "Downscaling of solid propellant pyrotechnical microsystems," *Dig. Tech. Papers Transducers 2003 Conf.,* Boston, June 2003, pp. 1423–1426
40. W. W. Van Arsdell and S. B. Brown, "Subcritical crack growth in silicon MEMS," *J. Microelectromech. Syst.,* 8(3) (Sept. 1999)
41. P. M. Osterberg and S. D. Senturia, "M-TEST: A test chip for MEMS material property measurement using electrostatically actuated test structures," *J. Microelectromech. Syst.,* 6(2) (June 1997)
42. L. L., A. P. Pisano, and R. T. Howe, "A micro strain gauge with mechanical amplifier," *J. Microelectromech. Syst.,* 6(4) (Dec. 1997)
43. H. Guckel, T. Randazzo, and D. W. Burns, "A simple technique for the determination of mechanical strain in thin films with applications to polysilicon," *J. Appl. Phys.,* 57(5) (Mar. 1985)
44. H. Guckel, D. W. Burns, H. A. C. Tilmans, D. W. DeRoo, and C. R. Rutigliano, "Mechanical properties of fine grained polysilicon – the repeatability issue,"*1988 Solid State Sensor and Actuator Workshop Technical Digest,* Hilton Head Island, SC, 6–9 June 1988, cat. no.88TH0215-4
45. M. Biebl and H. von Philipsborn, "Fracture strength of doped and undoped polysilicon,"*8th Int. Conf. on Solid-State Sensors and Actuators,* Stockholm, Sweden, 25–29 June 1995, vol. 2
46. M. Biebl, G. T. Mulhern, and R. T. Howe, "In situ phosphorus-doped polysilicon for integrated MEMS,"*8th Int. Conf. on Solid-State Sensors and Actuators,* Stockholm, Sweden, 25–29 June 1995, vol. 1
47. Sir Edmund Whittaker, *A History of Theories of Aether and Electricity,* American Institute of Physics, NY, ISBN: 0883185237 (1987)
48. E. R. Williams, J. E. Faller, and H. A. Hall, *Phys. Rev. Letters,* 26:721 (1971)
49. D. K. Cheng, *Field and Wave Electromagnetics,* 2nd ed., Addison-Wesley Publishing Company, Inc. (1983), ISBN 0201128195
50. A. E. Knowlton, *Standard Handbook for Electrical Engineers,* McGraw-Hill Book Company, Inc., NY (1949), pp. 474–475
51. F. Paschen, *Ann. Physik,* 37:69 (1889)
52. A. R. Von Hippel, *Molecular Science and MolecularEengineering,* Technology Press of MIT and J. Wiley, NY (1959)
53. J. W. Judy, "Magnetic microactuators with polysilicon flexures," M.S. thesis, University of California Berkeley Electrical Engineering and Computer Sciences Department, Aug. 1994
54. S. F. Bart, T. A. Lober, R. T. Howe, J. H. Lang, and M. F. Schlecht, "Design considerations for micromachined electric actuators," *Sens. Actuators,* 14:269–292 (July 1988)

55. J. W. Judy, "Batch-fabricated ferromagnetic microactuators with silicon flexures," Ph.D. thesis, University of California Berkeley Electrical Engineering and Computer Sciences Department, Dec. 1996
56. J. I. Seeger and S. B. Crary, "Stabilization of electrostatically actuated mechanical devices," *1997 Int. Conf. on Solid-State Sensors and Actuators, Digest of Technical Papers*, IEEE, NY, 2:1133–1136 (1997), cat. no. 97TH8267
57. G. K. Fedder and R. T. Howe, "Multimode digital control of a suspended polysilicon microstructure," *J. Microelectromech. Syst.*, 5(4):283–297 (Dec. 1996)
58. H. C. Nathanson, William E. Newell, Robert A. Wickstrom, and John R. Davis, Jr., "The resonant gate transistor," *IEEE Trans. Electron Devices*, ED-14:117 (1967)
59. L. J. Hornbeck, "Digital Light Processing™: A new MEMS-based display technology," available from Texas Instruments and from the white paper library at *www.dlp.com/dlp_technology/dlp_technology_research.asp*
60. R. Legtenberg, J. Gilbert, S. D. Senturia, and M. Elwenspoek, "Electrostatic curved electrode actuators," *J. Microelectromech. Syst.*, 6(3):257–265 (Sept. 1997)
61. M. Shikida, K. Sato, and T. Harada, "Fabrication of an S-shaped microactuator," *J. Microelectromech. Syst.*, 6(1):18–24 (March 1997)
62. T. Akiyama and K. Shono, "Controlled stepwise motion in polysilicon microstructures," *J. Microelectromech. Syst.*, 2(3):106–110 (Sept. 1993)
63. B. R. Donald, C. G. Levey, C. D. McGray, D. Rus, and M. Sinclair, "Power delivery and locomotion of untethered microactuators," *Proc. IEEE MEMS 2003 Conf.*, Kyoto, Jan. 2003, pp. 124–127
64. R. B. Apte, F. S. A. Sandejas, W. C. Banyai, and D. M. Bloom, "Deformable grating light valves for high resolution displays," *Solid-State Sensor and Actuator Workshop Technical Digest*, Hilton Head Island, SC, 13–16 June 1994, pp. 1–6
65. D. R. Pedersen and O. Solgaard, "Free-space communication link using a grating light modulator," *Sens. Actuators*, A83:6-10 (2000)
66. S. M. Bobbio, M. D. Kellam, B. W. Dudley, S. Goodwin-Johansson, S. K. Jones, J. D. Jacobson, F. M. Tranjan, and T. D. DuBois, "Integrated force arrays," *Proc. IEEE Micro Electro Mechanical Systems (MEMS 1093)*, Fort Lauderdale, FL, 7–10 February 1993, pp. 149–154
67. B. Bollée, "Electrostatic motors," *Phillips Tech. Rev.*, 30:178–194 (1969)
68. W. C. Tang, C. T.-C. Nguyen, and R. T. Howe, "Laterally driven polysilicon resonant microstructures," *Sens. Actuators*, A20(1-2):25–32 (1992)
69. Sandia SUMMiT™ and other technologies: *www.sandia.gov/mstc*
70. E. J. Garcia and J. J. Sniegowski, "Surface micromachined microengine," *Sens. Actuators*, A48(3):203–214 (30 May 1995)
71. J. J. Sniegowski and E. J. Garcia, "Surface-micro-machined gear trains driven by an on-chip electrostatic microengine," *IEEE Electron Device Lett.*, 17(7):366–368 (July 1996)
72. J. Mohr, M. Kohl, and W. Menz, "Micro optical switching by electrostatic linear actuators with large displacements," *Proc. Int. Solid-State Sensors and Actuators Conf. (Transducers 1993)*, Yokohama, Japan, June 1993, pp. 121–123

73. E. H. Klaassen, K. E. Petersen, J. Noworolski, J. Logan, N. I. Maluf, J. Brown, C. Storment, W. McCulley, and G. T. A. Kovacs, "Silicon fusion bonding and deep reactive ion etching: a new technology for microstructures," *Sens. Actuators*, A52(1-3):132–139 (March–April 1996)
74. J. D. Grade, J. Jerman, and T. W. Kenny, "A large-deflection electrostatic actuator for optical switching applications," *Solid-State Sensor and Actuator Workshop Technical Digest,* Hilton Head Island, SC, 4–8 June 2000, pp. 97–100
75. L.-S. Fan, Y.-C. Tai, and R. S. Muller, "IC-processed electrostatic micromotors," *Sens. Actuators*, 20(1-2):41–47 (Nov. 1989)
76. Y.-C. Tai and R. S. Muller, "IC-processed electrostatic synchronous micromotors," *Sens. Actuators*, 20(1-2):49–55 (Nov. 1989)
77. M. Mehregany, S. D. Senturia, J. H. Lang, and P. Nagarkar, "Micromotor fabrication," *IEEE Trans. Electron Devices*, 39(9):2060–2069 (Sept. 1992)
78. R. A. Brennan, M. G. Lim, A. P. Pisano, and A. T. Chou, "Large displacement linear actuator," *Tech. Digest, IEEE Solid-State Sensor and Actuator Workshop,* 1990, IEEE cat. no. 90CH2783-9, pp.135–139
79. S. Egawa, T. Niino, and T. Higuchi, "Film actuators: planar, electrostatic surface-drive actuators," *Proc. IEEE Micro Electro Mechanical Systems (MEMS 1993),* Nara, Japan, 30 Jan.–2 Feb. 1991, IEEE cat. no. 91CH2957-9, pp. 9–14
80. M. C. Wu, "Micromachining for optical and optoelectronic systems," *Proc. IEEE*, 85(11):1833–1856 (Nov. 1997)
81. J. Curie and P. Curie, "Development by pressure of polar electricity in hemihedral crystals with inclined faces," *Bull. Soc. Min. France*, 3:90-93 (1880)
82. J. F. Nye, *Physical Properties of Crystals – Their Representation by Tensors and Matrices*, Oxford University Press (1985)
83. W. G. Cady, *Piezoelectricity*, McGraw-Hill, NY (1946), and Dover Publishers, ASIN 0486610942 (1978)
84. T. Ikeda, *Fundamentals of Piezoelectricity*, Oxford Science Publications (1990)
85. D. L. DeVoe, *Thin-Film Zinc-Oxide Microsensors and Microactuators*, Ph.D. thesis, University of California Berkeley Mechanical Engineering Department, July 1997
86. H. Jaffe and D. A. Berlincourt, "Piezoelectric Transducer Materials," *Proc. IEEE*, 55(10): 1372–1385 (Oct. 1965)
87. J. W. Judy, D. L. Polla, and W. P. Robbins, "A linear piezoelectric stepper motor with sub-micrometer step size and centimeter travel range," *IEEE Trans. Ultrasonics, Ferroelectrics and Frequency Control*, UFFC-37(5):428–437 (1990)
88. U.S. patent 5,629,577, piezoelectric microactuator useful in a force-balanced scalpel
89. A. Lal, "Silicon-based ultrasonic surgical actuators," *Proc. 20th Annual Int. Conf. IEEE Eng. in Medicine and Biology Soc.,* Hong Kong, China, 20:2785–2790(29 Oct. 1998)

90. D. L. DeVoe and A. P. Pisano, "Modeling and optimal design of piezoelectric cantilever microactuators," *J. Microelectromech. Syst.*, 6(3):266–270 (Sept. 1997)
91. M.-N. Niu and E. S. Kim, "Bimorph piezoelectric acoustic transducer," *Dig. Tech. Papers Transducers 2001 Conf.*, Munich, June 2001, pp. 110–113
92. Y. Yee, H.-J. Nam, S.-H. Lee, J.-U. Bu, Y.-S. Jeon, S.-M. Cho, "PZT actuated micromirror for nano-tracking of laser beam for high density optical data storage," *Proc. IEEE MEMS 2000 Conf.*, Miyazaki, Jan. 2000, pp. 435–438
93. M. Ohring, *The Materials Science of Thin Films*, Academic Press (1992), p. 418
94. J. Kondoh and S. Shiokawa, "Shear-Horizontal Surface Acoustic Wave Sensors," in: *Sensors Update*, vol. 6 (H. Baltes, W. Göpel, and J. Hesse, eds.), Wiley-VCH (2000), pp. 59–78
95. J. W. Grate and G. C. Frye, "Acoustic Wave Sensors," in *Sensors Update*, vol. 2 (H. Baltes, W. Göpel, and J. Hesse, eds.), Wiley-VCH (1996), pp. 37–83
96. J. W. Grate, S. J. Martin, and R. M. White, "Acoustic wave microsensors, part I," *Anal. Chem.*, 65:940-948 (1993)
97. J. W. Grate, S. J. Martin, and R. M. White, "Acoustic wave microsensors, part II," *Anal. Chem.*, 65:987-996 (1993)
98. B. A. Auld, *Acoustic fields and waves in solids*, Krieger Publishing, Malabar, FL (1990)
99. A. Kuoni *et al.*, "A high insulating thin film ZnO piezoelectric actuator on a polyimide substrate," *Proc. Eurosensors XVI*, Prague, Sept. 2002, pp. 228–231
100. B.T. Khuri-Yakub and J.G. Smits, "Reactive magnetron sputtering of ZnO," *J. Appl. Phys.*, 52(6):4772-4774 (1981)
101. M. A. Dubois *et al.*, "Measurement of the effective transverse piezoelectric coefficient $e_{31,f}$ of AlN and $Pb(Zr_x,Ti_{1-x})O_3$ thin films, *Sens. Actuators*, A77:106–112 (1999)
102. T. G. Cooney and L. F. Francis, "Processing of sol-gel derived PZT coatings on non-planar substrates," *J. Micromech. Microeng.*, 6:291-300 (1996)
103. V. Sundar and R. E. Newnham, "Electrostriction," in *The Electrical Engineering Handbook* (R. C. Dorf, ed.), CRC Press LLC, Boca Raton, FL (2000), chap. 50
104. J. Boland, C.-H. Chao, Y. Suzuki, and Y.-C. Tai, "Micro electret power generator," *Proc. IEEE MEMS 2003 Conf.*, Kyoto, Jan. 2003, pp. 538–541
105. T. Y. Hsu, W. H. Hsieh, Y. C. Tai, and K. Furutani, "A thin film teflon electret technology for microphone applications," *Tech. Dig. Solid-State Sensor and Actuator Workshop*, Hilton Head Island, SC, June 1996, pp. 235–239
106. *Proc. 8th Intl. Conf. on Electrorheological Fluids and Magnetorheological Suspensions*, Nice, France, World Scientific (July 2001)

15 Micromachining Technology

Paddy J. French and Pasqualina M. Sarro

Delft University of Technology, The Netherlands

15.1 Introduction

The term *micromachining* usually refers to the fabrication of micromechanical structures with the aid of etching techniques to remove part of the substrate or a thin film. Silicon has excellent mechanical properties,[1] making it an ideal material for machining. An early silicon (pressure) sensor was made by Honeywell in 1962 using isotropic etching.[2] In 1966, Honeywell developed a technique to fabricate thin membranes using mechanical milling. Crystal-orientation-dependent etchants led to more precise definition of structures and increased interest.[3] Anisotropic etching was introduced in 1976. An early silicon pressure sensor, based on anisotropic etching, was made by Greenwood in 1984.[4] Surface micromachining also dates back to the 1960s. Early examples included metal mechanical layers.[5] Basically, surface micromachining involves the formation of mechanical structures from thin films on the surface of the wafer. The 1980s saw the growth of silicon-based surface micromachining using a polysilicon mechanical layer.[6,7] In recent years, a number of new technologies have been developed using both silicon and alternative materials. These include the epi-processes where the epilayer is used as a mechanical layer and a number of deep plasma etching processes. This chapter concentrates on silicon-based micromachining processes.

15.2 Bulk Micromachining

Bulk micromachining covers all techniques that remove significant amounts of the substrate (bulk) material and the bulk is part of the micromachined structure. This microstructuring of the substrate is often done to form structures that can physically move, such as floating membranes or cantilever beams. Other types of structures that can be realized by bulk micromachining are wafer-through holes, often used for through-wafer interconnects in chip stacks and very deep cavities or channels to form microwells or reservoirs for biochemical applications. Typical bulk micromachined structures are shown in Fig. 15.1.

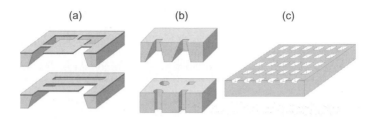

Figure 15.1 Typical bulk micromachined structures: (a) membranes and beams; (b) wafer-through holes; (c) microwells.

The substrate (silicon) removal can be done using a variety of methods and techniques. In addition to processes using wet (or liquid) etchants, techniques using etchants in vapor and plasma states (generally referred to as dry) are available. In this section, a number of currently available processes are introduced, and their potentials and limitations are discussed. Various aspects, such as etch characteristics, compatibility to conventional IC processes, complexity, and costs, are illustrated to evaluate the suitability of each technique for a specific application.

15.2.1 Wet Etching

Wet etching of silicon[8] is often used if large amounts of the silicon bulk have to be removed. Until recently, it was more widely used in micromachining than dry etching, the preferred method used in IC technology. The reasonably fast etch rates that can be achieved, the low cost of wet etching due to low-complexity equipment, and the availability of masking materials to perform the process selectively are among the major reasons for the large use of wet etching of silicon. Chemical solutions that remove the silicon anisotropically (orientation-dependent etch rates) or isotropically (etch rate is equal in all directions) are available. These two types of etching are illustrated in Fig. 15.2 and discussed below.

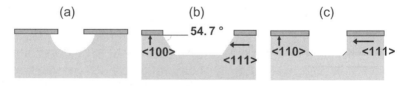

Figure 15.2 Schematic cross section of wet etching profiles: (a) isotropic; (b) anisotropic (100); (c) anisotropic (110).

15.2.1.1 Anisotropic Etching

Wet anisotropic etching of the silicon substrate is the more mature technology and the most widely used process for the fabrication of mechanical microstructures for commercially available microsensors, such as pressure sensors and accelerometers. The selective removal of the bulk silicon in an anisotropic etchant is used in combination with an etch-stop technique to determine accurately the three-dimensional (3D) microstructures. An anisotropic etchant etches silicon preferentially along given crystal planes. This results in unique structures that can be accurately predetermined, once the etching characteristics of the etchant are known. Square or rectangular cavities and pits bounded by (111) planes, V-grooves, and even holes or channels with vertical walls can be realized by properly dimensioning the size and orientation of the structures included in the layout.

The much higher etch rate in one direction with respect to another results in the exposure of the slowest etching planes over time. In silicon, the (111) planes are at an angle of 54.74 degrees to the wafer surface for the most commonly used wafer orientation, i.e., (100), and at 90 degrees for the less frequently used (110) silicon wafers. The difference in etch rates of the silicon crystal planes in several anisotropic etchants results in a degree of anisotropy that can be even higher than 1000 and is very dependent on the type of solution, the concentration, the temperature, and the presence of additives or dopants. Several data are reported in the literature,[9,10] and extensive studies have been published by Sato *et al.*[11] and Shikida *et al.*[12]

Etchants

Crystal-orientation-dependent etchants of silicon are:

- Hydroxides of alkali metal, such as KOH, NaOH, or CsOH
- A mixture of ethylene-diamine-pyrocatechol-water (EDP)
- Hydrazine
- Ammonium hydroxide (NH_4OH)
- Tetra-methyl-ammonium hydroxide (TMAH)

The most frequently used anisotropic wet etchants for MEMS are KOH,[1,3,8,13,14] EDP,[15,16] and TMAH.[17-19] For these etchants, many studies have been done and much data is available in literature.[1,3,8-19] By far the largest amount of information is on KOH. Despite its toxicity and handling problems, EDP has sometimes been preferred to KOH because of its greater selectivity to heavily p-type doped regions[1] and, in CMOS post-process bulk micromachining,[20] because of its large selectivity to oxide and aluminum. Recently, EDP use is diminishing; it is often being replaced by TMAH solutions that also have a good selectivity to oxide. Moreover, the lack of metal ions in TMAH assures full compatibility with IC processes. A few important characteristics of these three anisotropic etchants are given in Table 15.1. Often a range of values is given instead of a specific value because solution concentration, temperature, and other factors influence these values.

Limited data are available on the other less frequently used etchants. A few characteristics are summarized here.

Etch rates of Si in hydrazine[20-23] increase with solution temperature and decrease with increasing mole concentrations. Values between 1.5 and 3.5 µm/min. are reported, although selectivity to (111) planes is only about 20. Oxide is an excellent masking material (a dip etch to remove native oxide is required), as is nitride. Aluminum is etched, but passivation in saturated solutions is possible. Very high etch rates (>100 µm/min.) have been reported for a 50-50 hydrazine-water solution at 120°C,[22] with somewhat better selectivity to the (111) planes. Hydrazine is extremely toxic and tends to age rather quickly if not properly stored, and there is a risk of spontaneous ignition. For these reasons, it is seldom used.

Ammonium hydroxide is particularly interesting because it does not incorporate alkali ions that might interfere with IC compatibility. Concentrations between 1 and 18 wt-% have been used.[24] An etch rate of (100) Si is generally lower than in KOH, but rates as high as 30 µm/hr have been reported. Oxide and nitride are used as masking layers, and alu-

Table 15.1 Characteristics of Commonly Used Anisotropic Etchants

Property	KOH[1,3,8-14]	EDP[1,8,9,13,15,16,20]	TMAH[1,8,12,13,17-19]
Si-etch rate (100) [µm/h]	50 to 150	30 to 35	20 to 70
Quality etched surfaces	High	High	Medium
Selectivity (111)/(100)	1:30 to 1:400	1:20	1:10 to 1:50
Underetch rate [µm/h]	0.5 to 1.5	1.4 to 1.5	0.2 to 2.0
CMOS compatible (front)	No	Yes	Yes
Masking materials	Si_3N_4, SiO_2, Au, Pt, Cu	Si_3N_4, SiO_2, Au, Cr, Ag, Ta	Si_3N_4, SiO_2, Au, Pt
Selectivity PECVD SiO_2/Si	1:100 to 1:300	$1:10^4$	$1:10^2$ to $1:10^3$
Selectivity PECVD SiN/Si	$1:10^4$		1:150 to 200
Attack of aluminum	High	Medium	With Si low
Toxicity	Low	High	Low
Long-term stability	High	Low	Medium
Cost	Low	High	Medium

minum can be passivated in saturated solutions. A major drawback is the tendency to form extremely rough etch-front surfaces.

An extensive study of etch-rate, selectivity, and anisotropy of CsOH solutions has been performed by Clark et al.[25] A large (110)/(111) selectivity and a very low oxide etch rate are interesting features, although surface roughness is often a problem.

Etch-Stop Techniques

In the fabrication of a 3D microstructure, it is often essential to control the vertical dimension of the structures with high accuracy and uniformity. This means that the etching of the bulk silicon must stop once the predetermined membrane thickness has been reached. A few techniques are available for this purpose. For somewhat thicker membranes (10 to 50 µm), *time stop* is generally used. A reproducible and constant etch rate is necessary in this case. Another frequently used etch-stop technique is the *boron etch stop*, shown in Fig. 15.3.[1,9,26,27] Highly boron-doped silicon (concentration $>5 \times 10^{19}/cm^3$) strongly reduces the etch rate in all alkaline etchants. By selective doping, silicon regions can be made resistant to etching, while undoped or low-doped regions will be etched. In this way, not only membrane fabrication but also 3D structuring are possible. However, the high boron concentration required limits the possible application. It is difficult to create such heavily doped regions by diffusion, and the high doping can introduce stress in the material. Consequently, there is a practical limit to the structure thickness (<15 µm). Another possibility is the use of buried layers made of etch-resistant materials, such as silicon oxide, silicon nitride, and silicon carbide.[28] These buried layers can be formed by ion implantation of oxygen, nitrogen, or carbon, or by using SOI wafers. The limited availability of these techniques, the high costs often involved, and the fact that only two-dimensional structures can be fabricated in a single etch step strongly limit their application in many cases.

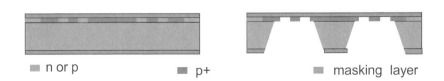

Figure 15.3 Schematic view of the p+ etch-stop technique.

An alternative way to control the membrane thickness is the use of electrochemical passivation of silicon. Among these techniques, the ones used most are the electrochemically controlled pn etch-stop (ECE), the photovoltaic etch-stop, and the galvanic etch-stop. These techniques allow the fabrication of structures with a reproducible thickness, although an external power and electrodes are needed to stop the etching process. The *ECE* technique,[9,29,30] illustrated in Fig. 15.4, has been used for several years as a post-process to fabricate microstructures with integrated devices.[1,2,20,31,32] Generally, the pn junction where the etch process stops consists of a p-type silicon wafer with an n-type epilayer on top or, if devices are fabricated in a CMOS process, the n-type well in a p-type epilayer or substrate. A positive voltage is applied to the n side of the junction with respect to a Pt counterelectrode. The p side of the junction is etched, and when the etch-front reaches the junction, the etch stops as the n side is passivated by the applied voltage. The special wafer holder required and the proper contact pattern to passivate all n regions on the wafer are often indicated as factors limiting the use of this technique. Recently, wafer holders and computer-controlled potentiostats[33] have been made available, helping to reduce some of these concerns.

A more recent technique is the *photovoltaic etch-stop*.[34,35] This technique does not require external electrodes because the external power source is a high-intensity light source. Contrary to the ECE technique, the p side of a junction is passivated and the n side is etched. Figure 15.5 (left) shows a schematic view of the sample cross-section. A platinum/titanium film is sputtered on the wafer back side and acts as a masking layer as well as a contact to the n-type silicon. The wafer is illuminated by a strong light source to ensure an etch-stop at the p-type epilayer. The n-type substrate and the platinum film interact galvanically. This method still presents some difficulties because a large platinum electrode is needed and the required high-power light source complicates the setup.

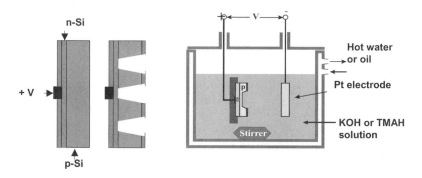

Figure 15.4 ECE-stop technique: (left) schematic cross section of the wafer; (right) etch setup.

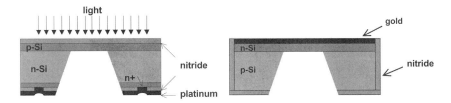

Figure 15.5 Schematic view of photovoltaic (left) and galvanic (right) etch-stop techniques.

Galvanic etch-stop, a new etch-stop technique that does not use any external power source, was introduced a few years ago.[36] It uses a structure very similar to the one used for the ECE technique, as shown in Fig. 15.5 (right). The electrical power used to stop the etching process is generated within the structure itself. The gold/silicon combination forms a galvanic cell.[37,38] The reduction of oxygen at the gold electrode generates the cell current. When the structure is immersed in the etching solution, the gold electrode and the p-type Si bulk are insulated by the reverse-biased pn junction. If the etch front reaches the junction, the insulation is destroyed and the galvanic cell is formed. When a sufficiently large gold electrode is used, etching stops and an n-type membrane is obtained. A disadvantage of this technique is that often, a relatively large gold electrode is required. Further investigation is needed to establish accurately the potential of this technique.

Table 15.2 compares the most frequently used or more promising etch-stop techniques.

Table 15.2 Comparison of Etch-Stop Techniques for Anisotropic Wet Etching of Silicon; Adapted from Ashruf *et al.*[37]

Etch-Stop Technique	Thickness Control	External Power Source	Stop on p- or n- Type	References
Time	Poor	No	Yes	38
High boron doping	Limited range	No	p	26, 27
Buried masking layers	Good	No	p and n	28
Electrochemically controlled pn	Good	Yes	n	29–31
Photovoltaic	Good	Yes	p	34
Galvanic	Good	No	n	36, 37

Convex Corner Undercutting

If the structure to be etched in any of the above-mentioned anisotropic etchants has one or more convex corners, measures must be taken. This is necessary because, in this case, the structures are not bounded by slow-etching (111) planes. Rather, the fast-etching planes are exposed to the etching solution, which will cause the convex corner to be etched away, resulting in a severe alteration of the desired structure. Several experiments have been carried out to identify the fast-etching planes that are strongly related to the etchant concentration, temperature, and substrate doping.[8,39-41] It is possible to some extent to compensate for the undercut of convex corners by modifying the mask layout. The size and shape of the corner compensation structure depends mainly on the etch rate of the fast-etching plane with respect to the etch rate of the (100) plane. Several compensating structures added to the convex corner have been proposed, and the equations that can be used to define design rules for such structures have been derived.[39-42] The etching of a convex corner is illustrated in Fig. 15.6, together with the most frequently adopted compensation pattern. Next to triangles, rectangles, and beams, more sophisticated structures such as asymmetrical folded strips, double<110> or <010>-oriented beams, and split beams have been used with encouraging results. The

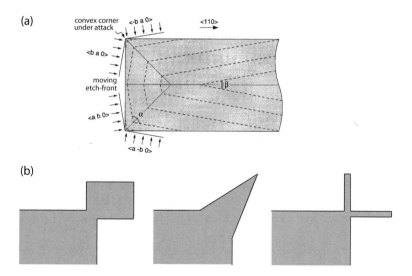

Figure 15.6 Schematic view of (a) convex corner undercutting[41] and (b) basic convex corner compensation structures.

major limitation is given by the minimum distance between compensating structures of adjacent convex corners for a given etch depth.

15.2.1.2 Isotropic Etching

Silicon can also be etched isotropically in HF-based solutions. The main dissolution mechanism is anodic, for which valence band holes are required. The reaction can be controlled by changing the surface hole concentration. This can be achieved with an electrochemical cell (anodic dissolution), by using an oxidizing agent (electroless or chemical dissolution), or by high-intensity illumination (open-circuit light-assisted dissolution).[37] The latter is generally not used for micromachining of silicon and will not be discussed here.

Etchants

In the case of *anodic dissolution*, the silicon wafer is the anode in an electrochemical cell. A Pt electrode is used as a counter-electrode. The setup used, which is very similar to the one used for anisotropic ECE, is shown in Fig. 15.7. More often, a three-electrode configuration with a Ag/AgCl reference electrode is used. The important process variables are the current density and the solution concentration. There is a critical current density for each value of HF concentration. Above this value, silicon is uniformly etched (electropolishing), while below this value, a porous silicon layer is formed. Doping and illumination determine the general type of porous film, but the morphology depends on current density and

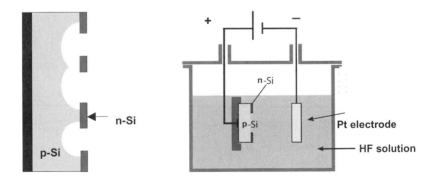

Figure 15.7 Anodic etching of silicon in HF solutions: (left) schematic cross section of etched wafer; (right) etch setup.

etchant concentration.[37,42,43] Often an additive is added to the HF solution to enhance the formation process.[44] These additives act as surfactants, reducing the surface tension of the solution and allowing the hydrogen formed as a by-product of the process to escape freely. This prevents the hydrogen from sticking to the silicon and erroneously masking it, causing a nonhomogeneous layer.

HF concentrations range from 1 to 40%, although recently, some higher concentrations (73%) have been used[45] because the etch rate of aluminum is strongly reduced at this high concentration. LPCVD silicon nitride or noble metals can be used as etch masks. Less aggressive with respect to photoresist, but with lower etch rates, are mixtures of ammonium fluoride, HF, and water, generally referred to as buffered HF solutions.[46] In addition to being used as a dielectric in humidity sensors, porous silicon is also an interesting material as a sacrificial layer in epi-micromachining (see Section 15.4.5), enabling deep trenches and 3D free-standing silicon microstructures to be fabricated using the effect of light on porous silicon formation (see Section 15.2.2).

Electroless dissolution requires a strong oxidizing agent added to the HF solution. The oxidizing agent is reduced and thereby injects holes into the valence band.[37] The holes are then consumed in the silicon dissolution reaction. The etch rate can be increased by increasing the oxidizing agent concentration responsible for an increase in the hole injection current. If the reduction reaction is mass-transport-controlled,[13] agitation of the solution generally results in an enhanced reaction rate since it has a considerable effect on the hole injection current. Two types of surface morphology are possible: a uniformly etched surface or a porous layer is formed (stain etching). If the dissolution reaction is limited by hole injection, a porous layer is formed; if it is limited by diffusion of HF species to the surface, electropolishing occurs. Commonly used etchants are HNA, i.e., a mixture of HF, nitric acid (HNO_3) as oxidizing agent, and acetic acid (CH_3COOH) to stabilize the oxidizing agent concentration,[47,48] and mixtures of HF, HNO_3, and H_2O.[8,47-49] For both systems, the etch rates of silicon and the quality of the etched surfaces strongly depend on the proportion of the acids in the mixture. A mixture of $HNO_3:H_2O:NH_4F$ (126:60:5) has also been used.[50] The use of NHF instead of HF results in a buffer action, keeping the HF and HF_2^- concentrations from changing rapidly with use. Moreover, photoresist can be used as a masking layer.

Table 15.3 lists a few important characteristics of these isotropic etchants. Once again, a range of values is given instead of a specific value because solution composition, temperature, stirring, and other factors influence the values.

Table 15.3 Characteristics of Commonly Used Isotropic Etchants

Solution	HF:H2O [43-45,54]	HF:HNO3:H2O [8,47-49]	HF:HNO3:CH3COOH [8,13,47,48]	HNO3:H2O:NH4F [50]
Etch-rate Si [μm/min.]	0.5 to 2*	1 to 800	1 to 470	1.5
Etch-rate oxide [nm/min.]	20 to 2000	60 to 1350	10 to 30	8.7 to 11
Quality etched surfaces	Medium	Rough-smooth	Rough-smooth	—
Masking materials	Si_3N_4, Au, Pt	Si_3N_4, Cr/Au, Pt, resist (ma-P1275)	Si_3N_4, Au, Pt	Si_3N_4, Cr/Au, Pt, resist (OCG820, OlinHPR)

*Anodic etching

Etch-Stop Techniques

The most commonly known etch-stop techniques in isotropic etchants are of the extrinsic type. The only intrinsic etch-stop technique is based on lightly doped silicon. The HNA system etches heavily doped silicon preferentially with respect to lightly doped silicon.[51] For some solutions, composition etch rates between 0.7 and 3 μm/min. have been reported for 10^{-2} Ωcm silicon with no appreciable etch rate for 6.8×10^{-2} Ωcm silicon. However, the HNA system shows a poor reproducibility for etch rate and selectivity, and a preferential etching of defects. To ensure a smooth finish, the low-doped layers should be defect-free, which is generally not the case because this layer must be grown on top of heavily doped substrates.

Several extrinsic etch-stop techniques are available. The *pn etch stop* is the most commonly used technique in porous-silicon micromachining applications.[37,52,53] The n-type mechanical structure is realized in a p-type bulk wafer. The p-type is anodized in the dark (otherwise, enough holes are generated to make the n-type porous as well) and made porous, while the n-region is not. Then the porous silicon is removed in a weak alkaline solution. The pn etch-stop is also used to etch uniformly the p-Si substrate using the setup and the configuration shown in Fig. 15.7. The sharp selectivity of HF anodic etching between p-type and n-type is used to realize single-crystal silicon microstructures.[54] Because low/moderately doped (up to $10^{16}/cm^3$) n-type silicon is not etched in weak HF, the substrate can be etched away, and free-standing n-type silicon microstructures are formed. Heavily doped n-type silicon ($>10^{17}/cm^3$) etches quite fast; the n/n+ selectivity can be used as well,[51,54] although the applied bias must be carefully adjusted to avoid breakdown.

Selective porous formation can also be achieved by *electrical insulation* of the mechanical structure from the substrate. This can be done using a polysilicon on oxide as a mechanical layer.[55,56] The underlying p-type bulk is contacted and made porous, and is subsequently removed. The doping of the polysilicon can be either p or n type, and rather large gaps can be achieved.[57-59] The lightly doped etch stop uses p-type layers for the mechanical structures.[60] These regions are protected from the solution by a masking layer, and the n-type bulk is positively biased. Only the n-type becomes porous because the pn junction provides a barrier for the charge carriers. Low-doped p-type structures can be fabricated with this method, which is therefore complementary to the pn etch stop.

The last two techniques, the *photovoltaic etch stop*[61] and the galvanic etch stop[62] are contactless. For the photovoltaic etch-stop, n-type regions in a p-type wafer are defined. The n-type regions will form the mechanical structures. When the wafer is immersed in the HF solution in the presence of illumination, the pn junction functions as a photovoltaic cell short-circuited by the HF solution. The photocurrent flows from the positive n side to the negative p side, resulting in its anodic dissolution. The etch rate is dependent on the light intensity, and for a high level of illumination, it might become limited by the electrochemical reactions occurring at both sides of the pn junction. The *galvanic etch stop* uses a structure similar to that of the pn etch stop. Anodization is accomplished without an external power source. A platinum/chromium or gold/chromium layer is used to make (back-side) contact to the p-type silicon wafer. The oxygen (or other strong oxidizing agent) is reduced at the metal surface. The holes generated at the metal/HF solution interface flow to the p-type Si/HF solution interface where the p-type is anodically dissolved. The n-type regions are not attacked in HF in dark conditions, resulting in intact freestanding structures. The major characteristics of these techniques are summarized in Table 15.4.

15.2.2 High-Aspect-Ratio Micromachining

Quite often, high-aspect-ratio (depth/width) microstructuring of silicon is required. Several new techniques developed in the past decade have been used to realize high-aspect-ratio structures, three-dimensional structuring of the substrate, or fast prototyping. Here we discuss only those that can be applied to silicon. These are mostly based on dry etching of silicon, either isotropically or anisotropically, with the exception of porous silicon, which is a wet etching technique.

Table 15.4 Comparison of Etch-Stop Techniques for Isotropic Wet Etching of Silicon; Adapted from Ashruf et al.[37]

Etch-Stop Technique	Thickness Control	External Power Source	Stop on p- or n-Type	References
Intrinsic	Poor	No	p and n	51
pn	Reasonable/good	Yes	n	38, 52-54
Selective porous formation by electrical insulation	Good	Yes	p and n	57-59
Resistivity gradient (n/n+)	Poor	Yes	n	51, 55, 56
Lightly doped	Good	Yes	p	60
Photovoltaic	Good	Yes (contactless)	n	52, 61
Galvanic	Good	No	n	62

15.2.2.1 Porous Silicon

The presence of light has a strong impact on the isotropy of the porous silicon formation process, both in HF or AFEM solutions. By properly using this effect, both isotropic and anisotropic modes of porous silicon formation can be achieved in a single etch step. This approach results in high-aspect-ratio microstructures, made in a single-step electrochemical etching, using only one mask.[63-65] This process is based on the formation mechanism of macro porous silicon, studied by Lehmann,[66] and the combination of both isotropic and anisotropic modes. The diameter of pores or the width of the trench can be controlled by the current density. First, vertical walls are formed. After the desired depth is obtained, the light intensity is adjusted to increase the current density. In this way, the width of the trenches under the structures is increased without affecting the width of the existing trenches.

Conventional porous silicon formation is isotropic. An anisotropic mode is obtained using n-type silicon and the setup depicted in Fig. 15.8.[67] First a pit is etched in KOH and used as a starting point for macropores. The holes generated at the back side of the sample diffuse to the front side and reach the pore tip, resulting in straight pores. Once the desired depth is obtained, a switch to an isotropic mode takes place to under-etch the trenches. The diameter of the pores is controlled by adjusting the light intensity. Increasing the light intensity generates more holes that are used to enlarge the pore diameter. However, the current density should be kept lower than the critical current density; otherwise, electropolishing occurs at the silicon surface under the structure. The formation

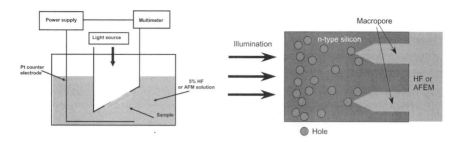

Figure 15.8 Schematic drawing of the electrochemical etching setup (left) and the macro pores formation mechanism (right).[67]

rate, on the other hand, is more or less independent of the current density. This process has been done in both HF and ammonium fluoride solutions. The best results so far for free-standing structures have been achieved in a 4.5% ammonium fluoride solution.[67]

15.2.2.2 Deep Reactive Ion Etching

Reactive ion etching (RIE) of silicon has been used for a long time in IC technology; it has been playing an increasingly important role in silicon micromachining as well. The major problems encountered in plasma etching of silicon are the low etch rate, the limited selectivity to dielectrics or photoresist, and the limited aspect ratio (depth/width). All of these aspects need to be addressed when very deep or wafer-through etching is required, as is the case with many micromachined sensors. New developments in both equipment and process for deep reactive ion etching of silicon[68-73] have made this technology a good candidate for IC-compatible bulk micromachining. In fact, a new generation of dry etchers[69-71] eliminates most of the limitations. Most deep RIE (DRIE) systems use an inductively coupled high-density plasma source and a fluorine-based non-corrosive etch chemistry, resulting in high etch rates. Two types of processes can be used. In the room-temperature process regime, developed by Bosch,[72] thin fluorocarbon polymer film is used for sidewall protection to control the mask undercut. This process, which alternates etch cycles and deposition cycles, is illustrated in Fig. 15.9. The etch cycle uses SF_6 to etch silicon. Then a fluorocarbon polymer made of a chain of CF_2 molecules is deposited using C_4F_8 as source gas. In the etch cycle that follows, the SF_x^+ ions remove the protective polymer at the bottom of the trench, while it remains intact along the sidewalls. This process results in a very directional etch at rates of several microns per minute. By increas-

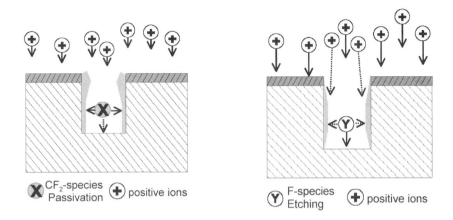

Figure 15.9 Sidewall passivation mechanism.[71]

ing the flow rate of SF_6 and the RF coil excitation power, higher etch rates are possible. In the cryogenic process regime (between −170° and −100°C), an ultra-thin layer of silicon oxide is used for sidewall protection to control the anisotropy.[73,74] The main characteristics of several DRIE anisotropic processes developed for bulk micromachining applications are reported in Table 15.5. For the etch rate of silicon and the selectivity to oxide or photoresist, typical or maximum values are given because they are strongly dependent on plasma process parameters and pattern configurations.

The high selectivity to silicon oxide and to photoresist in combination with the high etch rates that can be achieved make it possible to etch 300 to 500 µm of silicon in a reasonable time. A limitation of DRIE is the dependence of the etch rates on the aspect ratio of the trench. The etch rate is diffusion-limited and drops significantly for narrow trenches. This effect can be minimized by adjusting some process parameters at the expense of the etch rate, which will decrease. Recent studies have shown a number of differences between the cyrogenic and room-temperature processes in terms of crystal orientation dependence and thus the type of structure that can be fabricated.[73,74]

The true post-processing character of this technology is another important advantage. When merged with silicon fusion bonding processes, deep reactive ion etching can realize micromechanical devices with a large degree of versatility and flexibility, while the IC compatibility is still preserved.[75] This is a good example of a combination of pre-processing (the fusion bonding) and post-processing (deep RIE), which are compatible with either bipolar or CMOS IC processing.

Table 15.5 Characteristics of a Few Anisotropic DRIE Processes

DRIE Process	Si Etch-Rate [μm/min.]	Selectivity to SiO_2	Selectivity to Photoresist	Remarks	References
$CF_4/SF_6/O_2$ ICP system (STS)	6	300:1	200:1	Profile control Room-temperature process	72
SF_6/O_2 DECR system (Alcatel)	5	100:1		Cryogenic regime Anisotropic profile	68
SF_6/C_4F_8 ICP system (STS)	7		170	−10 to 20°C	70
SF_6/C_4F_8 ICP system (Alcatel)	10	1000:1	250:1	Room temperature or cryogenic regime	69
SF_6/O_2 ICP system METlab (Alcatel)	10	1000:1		Cryogenic regime	73
SF_6/O_2 ICP system Plasmalab100 (Oxford Plasma Technology) or METlab (Alcatel)	15	10000	1500	Cryogenic regime −90 to −110°C Fully anisotropic profile	74

15.2.2.3 Laser Micromachining

Another interesting upcoming technology for IC-compatible silicon micromachining is direct etching of silicon by laser. A schematic view of this process[76] is illustrated in Fig. 15.10. This technique is based on melting silicon in a chlorine atmosphere to etch arbitrary, 3D microstructures directly onto silicon substrates. The silicon sample is exposed to a chlorine atmosphere and locally heated above the melting point by a tightly focused laser beam incident through a quartz window. A reaction occurs between the gas and the molten silicon, forming silicon chlorides, which are subsequently removed in the vapor phase. The molten silicon etch rate is much higher than that for solid silicon. Using a 20 W Ar-ion laser, a removal rate as high as 105 mm^3/sec has been achieved and submicron resolution has been demonstrated as trench widths of less than 100 nm have been realized.[77] Another important property of the chlorine reaction is its high Si/SiO$_2$ selectivity (greater than 1000:1). Continuous etch trenches result by translating the sample during exposure using a computer-controlled xyz stage. Arbitrary patterns can be composed of line segments etched by moving the sample with the focused laser beam switched on. The width of a trench is influenced by the absorbed power and can be decreased by decreasing the incident power. The depth of the etched structures depends on chlorine pressure and scan speed. This process can be used to make via holes, channels, diffusers/nozzle elements for microfluidic devices, and rather complex structures. Commercial tools are currently available.[78] Laser micromachining is also used as a nonlinear dry cutting process for downsizing of 300 mm silicon wafers (to allow post-processing at smaller wafer size facilities); microdrilling of

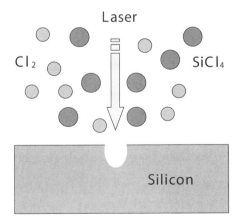

Figure 15.10 Laser micromachining of silicon.

holes, slots, or channels in glass or plastics for small and complex medical devices; and as an alternative to dicing for silicon wafers with micromachined structures.

The attractiveness of this technique is related to its truly post-processing character. Further, the crystal orientation and doping independence make laser micromachining a versatile, flexible tool, particularly suitable for prototyping and for complexly shaped patterns or molds.

15.2.2.4 Focused Ion Beam Micromachining

Recently, the possibility of focused ion beam machining in microsystem technology has been explored.[79,80] The FIB technology enables localized milling and deposition of conductors and insulators with high precision. It has been used for device modification, mask repair, process control, and failure analysis. The system is similar to a SEM and uses a Ga^+ ion beam instead of an electron beam. Both a fine beam for high-resolution imaging on sensitive samples and a heavy beam for fast and rough milling can be obtained. The samples are mounted on a motorized five-axis stage inside a vacuum chamber. A variety of gases can be present in the chamber. The gases are used for faster and more selective etching as well as for the deposition of materials. Here we discuss only the etching mode of the system.

A high ion current beam is used for the etching. The result is physical sputtering of the substrate material, as illustrated in Fig. 15.11. By raster scanning the beam over the substrate, an arbitrary shape can be etched, including varying the etch depth from point to point. An etching gas (for example, Cl_2) introduced in the work chamber during milling will chemically assist the physical sputtering, i.e., will facilitate the removal of the reaction products, thus increasing the selectivity toward different materi-

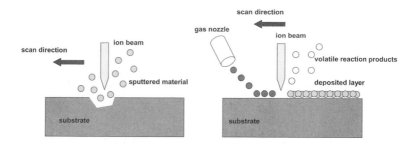

Figure 15.11 Focus ion beam: the principle of operation for the milling mode (left) and the deposition mode (right); adapted from Reyntjes and Puers.[79]

als and the etching rate. Typically, etch rates of 1 μm³/sec with a maximum aspect ratio of 15 and feature sizes of 100 nm to 50 μm with a fairly high geometrical flexibility are reported.

Depending on the application, FIB compares very favorably with other direct-write micromachining techniques in terms of resolution and accuracy. The main limitation is the processing time involved to machine large structures, since the etching rates are relatively low. Dimensions up to some tens of microns are easily feasible, but above 100 μm, the processing times become unacceptably high and other, less accurate techniques are preferred. Sometimes a two-step process is used to reduce the FIB milling time. First a high ion current is used to create a coarse trench; then a lower current ion beam, i.e., a finely focused beam, precisely defines the microstructure. Despite its shortcomings, the FIB technique is rather suitable for fast prototyping and small-scale post-processing.[81,82]

15.2.2.5 Powder Blasting

Powder blasting or abrasive jet machining is an interesting and promising directional etch technique for brittle materials like glass, silicon, and ceramics.[83-85] It is a fast process with etch rates of several microns per minute, depending on particle size, speed, and substrate material. Although able to etch silicon, it is more often used to micromachine glass wafers that are used in combination with silicon for microsystem applications.[83]

Powder blasting is based on physical etching only. The basic event is illustrated in Fig. 15.12. When an abrasive particle hits the target sample

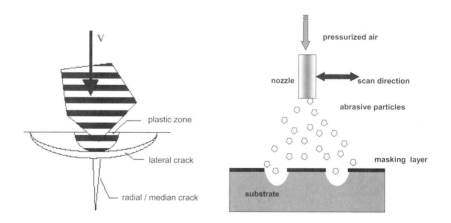

Figure 15.12 Powder blasting micromachining: schematic view of the etch mechanism of a sharp particle (left), from Slikkerveer et al.,[84] and of the process (right).

with a high velocity, high compressive stresses arise around the impact site, generating a plastically deformed area. This deformation leads to large tensile stresses that result in lateral (parallel to the substrate surface) and radial or median microcracks into the substrate. The lateral cracks lead to chipping of the substrate material, resulting in the actual etching. The radial cracks run into the substrate, decreasing its strength.[84,85]

Sharp alumina particles are available in a wide range of average sizes. For micromachining, application sizes ≤30 µm are generally used. The powder is ejected from a nozzle using a compressed-air jet and hits the substrate at high speed, between 80 to 200 m/sec. The nozzle is canned over the target several times to achieve a uniformly etched surface. Using a properly patterned mask layer, fine structures can be fabricated without the need to reduce the particle jet size. Metal masks, if sufficiently thick, are erosion-resistant enough to be used as masks. For fine structures, however, mask layers that can be patterned accurately using lithography are preferred. Thick layers of electroplated copper[85] or photosensitive elastomers have been used successfully.

The low cost of the equipment involved and the simple process make this technique attractive, especially for difficult-to-machine substrates such as glass wafers. Attention is being paid to improving both etch rates and profiles, the roughness of the etched surfaces, and the aspect ratio by using smaller size powder particles and more suitable masking materials.

15.3 Surface Micromachining

Surface micromachining is a quite different technology from the bulk micromachining processes described above. Surface micromachining basically involves depositing thin films on the wafer surface and selectively removing one or more of these layers to leave free-standing structures. In recent years, a number of new processes have been developed that use the epitaxial layer, or the upper few microns of the substrate, as a mechanical layer. These technologies are discussed in this section and have been given the collective title epi-micromachining.

15.3.1 Basic Process Sequence

Surface micromachining techniques can be traced back to the 1960s. The resonant gate structure of Nathanson *et al.* showed how free-standing structures could be fabricated with thin film (in this case metal).[5,86] In the 1970s, surface-micromachined devices were fabricated where the sacrificial layer was the epitaxial layer. An example of this was deflectable aluminum-coated oxide mirrors.[87]

In the 1980s came the first micromachining processes using layers produced entirely by chemical vapor deposition (CVD).[6,7,88] In this case, polysilicon and oxide were used as the mechanical and sacrificial layers, respectively. This early work showed the potential of this new process and were the first examples of moving mechanical parts.[7,88]

The basic principle of surface micromachining is shown in Fig. 15.13. Two types of layers can be seen: the sacrificial layer and the mechanical layer. The sacrificial layer is so called because it is removed during subsequent processing. In this example, first the sacrificial layer is deposited and defined, followed by the same process for the mechanical layer. At the end of the process, the sacrificial layer is removed to leave the free-standing mechanical structure. As shown in Fig. 15.13(b), the sacrificial layer is accessed from the side of the structure or through access holes.

15.3.2 Materials and Etching

As mentioned above, two types of materials are required for surface micromachining and there is a wide range of materials available. Some important characteristics are listed in Table 15.6.

A range of layer combinations can be used that meet these requirements. Table 15.7 gives some examples of suitable combinations; it is by no means a complete list.

Some examples of etchants and their properties are given in Table 15.8.

Selectivity is an important issue in determining the combination of layers. It is selectivity not only over etching of the mechanical layers but also of the other layers on the device, since aluminum is often present in integrated devices. A number of options for obtaining a high selectivity of oxide etching over that of aluminum are given in Table 15.9.[45]

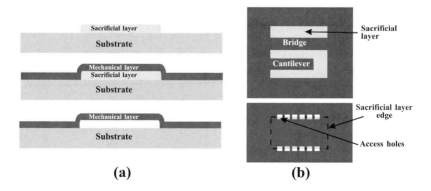

Figure 15.13 Basic surface micromachining process.

Table 15.6 Characteristics for Sacrificial and Mechanical Layers

Sacrificial Layer	Mechanical Layer
Ease of etching	Resistance to the sacrificial etchant
Stability	Low tensile stress and low stress profile

Table 15.7 Examples of Sacrificial and Mechanical Layers with the Appropriate Etchant

Sacrificial Layer	Mechanical Layer	Sacrificial Etchant
Oxide (PSG, LTO, etc.)	Polysilicon, silicon nitride, silicon carbide	HF
Oxide (PSG, LTO, etc.)	Aluminum	Pad etch
Polysilicon	Silicon nitride	KOH, TMAH
Polysilicon	Silicon dioxide	TMAH
Aluminum	Silicon dioxide/nitride	Al etch, Krumm etch
Resist	Aluminum	Acetone/oxygen plasma

Polysilicon, silicon carbide, silicon nitride, and oxide can be deposited using chemical vapor deposition. CVD can be divided into three categories: atmospheric-pressure CVD (APCVD), low-pressure CVD (LPCVD), and plasma-enhanced CVD (PECVD). LPCVD, which is widely used, operates at temperatures ranging from 400 to 1050°C. Examples of layers and deposition parameters are given in Table 15.10.

With all these processes, the gasses are brought into a sealed chamber at elevated temperatures. This results in a breaking down of the gasses, bringing the desired elements onto the surface of the substrate. The temperature of thin-film deposition can be significantly reduced with PECVD. In this case, a plasma is used to enhance the breakdown of the gasses so that thin films can be deposited at temperatures below 400°C, which means that they can be formed after aluminum deposition. An alternative technique for the formation of these layers is sputtering onto the substrate. With PECVD, the temperature is usually 400°C or less and a high deposition rate is maintained, often an order of magnitude higher than LPCVD. Table 15.11 gives some examples of processes. Note that there are many variations on these processes.

Table 15.8 Wet Etch Rates for Fresh Solutions, in Å/min.; from Williams and Muller[50]

						Material						
Etchant	Target	Si	Poly n+	Poly	Wet Ox	Dry Ox	LTO	PSG	PSG (*)	SiN 1	Sin 2	Al 2% Si
49% HF	SiO_2	—	0	—	23k	>10k	F	36k	140	52	42	
5:1 BHF	SiO_2	—	9	2	1000	1000	1200	6800	4400	9	4	1400
Phosphoric acid (85%, 160°C)	SiN	—	7	—	0.7	0.8	<1	37	24	28	19	9800
Si etch 126 HNO_3; 60 H_2O; 5NH_4F	SI	1500	3100	1000	87	W	110	4000	1700	2	3	4000
Aluminum etch 16H_3PO_4; HNO_3; 1Hac; 2H_2O (50°C)	Al	—	<10	<9	000—	<10	0	2	6600			

Table 15.9 Comparison of Oxide and Aluminum Etch Rates in Four Etch Mixtures

Etchant	Oxide rate (nm/min.)	Al rate (nm/min.)	Selectivity
73% HF	1500 (thermal)	2.2	680
73% HF:IPA 1:1	833 (thermal)	12	69
BHF/glycerol	95 (thermal)	0.55	170
Pad etch	200 (PSG)	5	40

Table 15.10 Commonly Used Gasses and Temperature Ranges for LPCVD Systems

Layer	Gasses	Temperature
Polysilicon	SiH_4	550 to 700°C
Silicon nitride	$SiH_2Cl_2 + NH_3$	750 to 900°C
	$SiH_4 + NH_3$	700 to 800°C
Silicon dioxide undoped	SiH_4+O_2	400 to 500°C
PSG (phosphorus-doped)	$SiH_4+O_2+PH_3$	400 to 500°C
BSG (boron-doped)	$SiH_4+O_2+BCl_3$	400 to 500°C
BPSG (phosphorus/boron-doped)	$SiH_4+O_2+PH_3+BCl_3$	400 to 500°C
Silicon carbide	SiH_4+CH_4	900 to 1050°C
Silicon germanium	GeH_4	300 to 400°C
Germanium SiH_4	SiH_4+GeH_4	400 to 550°C

Table 15.11 Commonly Used Gasses and Temperature Ranges for PECVD Systems

Layer	Gasses	Temperature
a-Si	SiH_4	400°C
Silicon nitride	$SiH_4 + NH_3 +N_2$	400°C
Silicon dioxide undoped	$SiH_4+ N_2 +N_2O$	400°C
Silicon dioxide (TEOS)	$TEOS+O_2$	350°C
Oxynitride	$SiH_4+ N_2 +N_2O +NH_3$	400°C
BPSG (phosphorus/boron doped)	$SiH_4+ N_2 +N_2O +PH_3 +B_2H_6$	400°C
Silicon carbide	$SiH_4+ CH_4$	400°C

The properties of CVD materials are all highly dependent on deposition temperature, gas flow, and chamber pressure. With PECVD systems, there is the additional factor of the low-frequency and high-frequency RF power.

15.4 Epi-Micromachining

Epi-micromachining is a group of micromachining technologies that use the epitaxial layer, or a similar thickness of the upper part of the substrate, as the mechanical layer and use a buried layer, or simply the substrate, as the sacrificial layer.

15.4.1 SIMPLE

The SIMPLE (silicon micromachining by plasma etching) process forms micromachined structures using a single etch step.[89] This process makes use of a Cl_2/BCl_3 chemistry that etches low-doped material anisotropically and n-type material above a threshold of about 8×10^{19} cm^{-2} isotropically. The basic process sequence is shown in Fig. 15.14. The first additional step to the bipolar process is a heavily doped buried layer since the bipolar buried layer has a doping level that is too low to be under-etched. This buried layer is formed by ion implantation of arsenic. The implantation required is thus 1×10^{16} cm^{-2} at 180 keV followed by a 1200°C, 4-hour anneal and drive-in. This is followed by the formation of the standard bipolar buried layer using implantation of antimony. The standard bipolar epitaxial layer is then grown [Fig. 15.14(a)], followed by an additional deep diffusion where the mechanical structure will be formed [Fig. 15.14(b)]. After this, a full standard bipolar process is performed [Fig. 15.14(c)]. A thick PECVD oxide is deposited, which serves

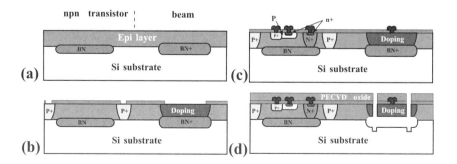

Figure 15.14 Main steps for the SIMPLE process.

as a mask for the plasma etching of the epilayer and the buried layer. The ratio of lateral to vertical etch rate and selectivity over mask etching is dependent on the gas ratio, power, and pressure. A limiting factor on under-etching is the drop in etch rate as a function of time. This is because the etching results from spontaneous chemical reactions related to the concentration of Cl atoms at the reaction surface and the rate of desorption of reaction products. Typically, for an epilayer thickness of 4 μm and mechanical structures of width 3 μm, a masking oxide layer of 1 μm would be required. In addition to limitations in lateral dimensions of the mechanical layers, narrow trenches will result in a reduction in the under-etch rate. For trench widths less than about 3 μm, the under-etch rate is reduced. This is due to the transportation limitation of the reactive agents as well as the reactive products. The final structure is given in Fig. 15.14(d), which clearly shows how the vertical etching continues in the trench during the lateral etching of the buried layer.

The SIMPLE technique has the advantage of simplicity; because it is performed after the aluminum deposition, it is therefore fully compatible with the electronics circuitry. The high-doped buried layer required for under-etching has some influence on the electronics due to autodoping. The increased epi-doping concentration results in a small increase in vertical npn transistor gain and a fall for lateral pnp devices.

15.4.2 SCREAM

The SCREAM (single-crystal reactive etching and metallization) process[90] uses a combination of anisotropic and isotropic plasma etching. The basic process sequence is shown in Fig. 15.15. After patterning the oxide, trenches are etched that define the sidewalls of the structure [Fig. 15.15(a)]. An oxide layer is deposited [Fig. 15.15(b)] and etched back using the plasma etching, resulting in the structure shown in

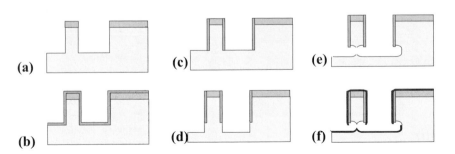

Figure 15.15 SCREAM process sequence.

Fig. 15.15(c). A further plasma etch deepens the trench beyond the sidewall protection [Fig. 15.15(d)]. The trenches now need to be underetched, for which an isotropic etch is required. This structure is shown in Fig. 15.15(e). Finally, aluminum is sputtered to form a contact to the mechanical structures, as shown in Fig. 15.15(f). High aspect ratios can be achieved, although they are limited by the thickness of the masking oxide, which must survive all the etch steps. Maximum reported beam height was 20 µm, with a width of total width (including metal) of 6 µm. If the devices are not to be integrated with electronics, the processing finishes at this point and the devices can be contacted with bond wires. The cross section shows how isolation is achieved through the poor step coverage of the aluminum. Integration with electronics can be achieved using thick resist layers to pattern the wafer after micromachining.[91] In this integrated version, the micromachining is performed after the processing of the electronics, including metallization. First an oxide layer is deposited to protect the electronic circuitry during subsequent processing. This oxide also forms the masking layer for the micromachining. After the formation of the free-standing structures [Fig. 15.15(e)], the oxide is etched back to reveal the metal pads of the electronics. A metal layer is then deposited as in the nonintegrated version. In this case, the metal serves three purposes: (1) capacitor electrodes, (2) interconnects, and (3) contacts to the IC metallization. Finally, a thick resist layer is used to pattern the metal. An OCG 895I 90cs resist was used for this purpose. Since the detailed metallization of the electronics was defined before the micromachining, only large structures need to be defined using the thick resist.

15.4.3 Black Silicon

An alternative process for SIMOX wafers is the black silicon method.[92] This uses dry etching in a similar manner to the SCREAM process. The important difference is that the sidewall passivation is achieved in the plasma etcher, and thus a multistep one-run process can be achieved. To fabricate the micromechanical structures, four etch steps are required. First the epilayer is etched anisotropically, using a liftoff metallic mask, until the buried oxide is reached [Fig. 15.16(a)]. A CHF_3 plasma etch is used to etch the underlying oxide and deposit a fluorocarbon (FC) on the sidewall [Fig. 15.16(b)], which protects the sidewall during further etching and also has a low surface tension to reduce sticking. This is followed by a trench floor etch using $SF_6/O_2/CHF_3$. Finally, an isotropic RIE etch removes the silicon from under the upper silicon, resulting in the structure shown in Fig. 15.16(c). This technique can also be used on bonded wafers, resulting in structures considerably thicker than normally

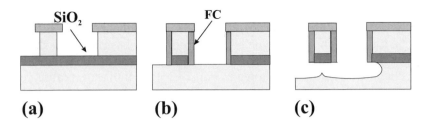

Figure 15.16 Basic process steps for black silicon.

achieved with epi structures. A further advantage of this technique is that the whole etch process can be achieved in a single plasma etcher.

15.4.4 MELO

The MELO (merged epitaxial lateral overgrowth) process is an extension of selective epitaxial growth (SEG). Selective epitaxial growth uses HCl added to the dichlorosilane; the HCl etches the silicon. However, if there is a pattern of bare silicon and silicon dioxide, the silicon deposited on the oxide will have a rough grain-like structure, with a large surface area, and will therefore be removed more quickly. Thus, although the growth rate will be lower than a normal deposition, a selective growth can be achieved. Once the silicon layer reaches the level of the oxide, both vertical and lateral growth will occur, yielding lateral overgrowth. This basic process is illustrated in Fig. 15.17.[93]

If two of these windows are close enough together, they will merge, giving the MELO process. The result is buried silicon dioxide islands. This process lends itself well to micromachining, as shown in Fig. 15.18.[94] It has the advantage of producing single-crystal structures, but the disadvantages are that beams must be orientated in the <100> direction, due to growth mechanisms of silicon,[95] and that the lateral dimensions are limited, since growth continues both laterally and vertically.

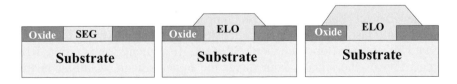

Figure 15.17 Basic SEG process extended to ELO.

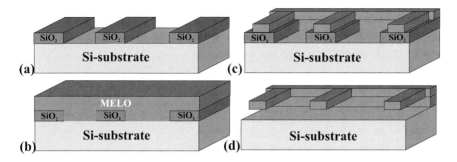

Figure 15.18 Basic MELO process.

15.4.5 Porous Silicon

The sacrificial porous silicon technique is a process that uses silicon as both the mechanical and sacrificial layers.[54,58,95-97] Regions of silicon are selectively made porous; these porous regions are used as the sacrificial material. The porous formation setup is very similar to the electrochemical KOH etching, although in this case the etchant is HF. A positive voltage is applied to a platinum electrode and a negative voltage to the back side of the wafer. If aluminum is used as the back contact, a special holder must be used to protect the back of the wafer from the HF etchant. The current flow between the platinum electrode and silicon substrate enhances the formation of holes on the surface, which result in pores. The high surface area of this material results in rapid etching, after porous formation, in KOH. This makes the material highly suitable as a sacrificial material. The porous silicon formation rate is highly dependent on the current density, HF concentration, lighting, and the doping in the substrate.[96] The process sequence is shown in Fig. 15.19. In this case, the selective etching of p-type material over the n-type epilayer is used. First

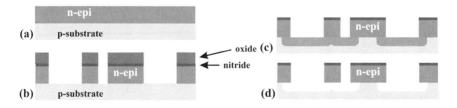

Figure 15.19 Basic process steps for sacrificial porous-silicon-based micromachining.

a plasma etch is used to etch through the epilayer to reveal the substrate [Fig. 15.19(b)]. The porous layer is then formed using the process described above as shown in Fig. 15.19(c). Finally, the porous layer is removed in KOH at room temperature [Fig. 15.19(d)].

Because the porous silicon technique is extremely simple and can be applied as a post-processing step, it is fully compatible with electronic circuitry. The only remaining problem is to protect the areas of electronics and metallization from the HF etchant; this can be achieved by using an alternative etchant.[46] One disadvantage of this technique is the added process complexity introduced by the requirement for a back-side electrical contact during etching.

15.4.6 SIMOX

An alternative technique is to start with a wafer with a buried sacrificial layer, such as a SIMOX (separation by implantation of oxide) wafer.[98] SIMOX wafers have a buried oxide layer above which is a single-crystal silicon layer. The substrates are prepared by implantation of oxygen with a typical dose of 1.8×10^{18} cm^{-2}. The implantation is performed at temperatures above 500°C to avoid amorphization of the silicon. A high-temperature anneal (>1300°C) is used to eliminate defects generated by the implantation. A typical resulting structure is a 2000Å upper silicon layer with a 4000Å buried oxide. The wafers may then be further processed with an additional epitaxial growth. A plasma etch through the epilayer is used to reveal the sacrificial oxide. This layer is then removed in the same manner as with surface micromachining. The basic process is shown in Fig. 15.20. This process has the following advantages:

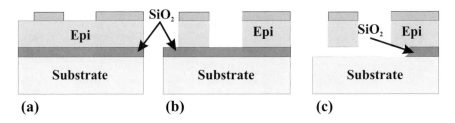

Figure 15.20 SIMOX-based micromachining.

- Substrate industrially available (SIMOX)
- CMOS compatibility
- Single-crystalline silicon surface layer
- Freedom on the surface structure thickness by standard epitaxial processing
- SiO_2 buried layer as a sacrificial and insulating layer

However, the higher costs of the start material may present problems for some applications. A further problem that may arise is unwanted under-etching at the epi-oxide interface, although in many applications this is not a significant problem. As mentioned above, the process is fully compatible with SOI electronic processing.

15.4.7 Epi-Poly

An interesting alternative to the processes described above is the epi-poly technique. In this case, polysilicon layers are grown in the epitaxial reactor. Although this technique departs from using single-crystal silicon as a mechanical material, it has greater flexibility in terms of lateral dimensions. It has been used as a thick-surface micromachining process.[99] Alternatively, the mechanical layers can be formed at the same time as the single-crystal epilayer required for the electronics.[100,101] The basic process is shown in Fig. 15.21. After the formation of the sacrificial oxide, a polysilicon seed is deposited [Fig. 15.21(a)]. This polysilicon seed ensures a uniform growth. A standard epilayer growth will then form epi-poly on the seed and single crystal where the substrate is bare [Fig. 15.21(b)]. The epilayer growth rate on the polysilicon seed is about 70% of that on the single-crystal silicon. Therefore the total thickness of the sacrificial layer and seed can be adjusted to ensure a planar surface after epilayer growth. The mechanical layer is then patterned

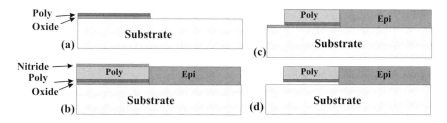

Figure 15.21 Basic epi-poly process.

[Fig. 15.21(c)] and released through sacrificial etching [Fig. 15.21(d)]. The process is extremely simple because the mechanical layer is formed at the same time as the epilayer for the electronics. One potential problem with epi-poly is the compressive stress that can be generated during oxidation. This problem can be eliminated by protecting the epi-poly with a thin silicon nitride throughout the bipolar processing. This technique has been successfully applied, resulting in zero or low tensile stress.[95,101]

The epi-poly process requires minimal additional processing before epitaxial deposition and has no detrimental effect on the characteristics of the electronics. The process is therefore fully compatible with electronics processing.

15.4.8 Release and Stiction

One of the last steps in surface micromachining, the sacrificial etch and release, is a critical stage. The sacrificial etch involves the removal of the sacrificial layer from underneath the mechanical layer. Release techniques include wet chemical etching, vapor etching, and dry etching. Wet etching can be highly selective and can be used for significant under-etching. For an oxide sacrificial layer, HF is a commonly used wet etchant. This presents the problem of stiction, however, as discussed below, since the wafers must go through a drying process. A possible option is to use vapor etching. Etchants such as HF give off sufficient vapor to etch the oxide, without wetting the surface. A further step away from wet etching is plasma etching. A number of processes have used this technique, including SIMPLE, SCREAM, and black silicon, which are described in detail below. The advantage of these techniques is that they are totally dry and therefore exhibit reduced stiction. However, they are generally limited in the lateral distances they can achieve.

Stiction remains a concern in achieving high yields with surface micromachining. Smooth microstructure surfaces at close proximity are ideal candidates for stiction. If drying is performed directly after a water rinse, van de Waal forces pull the free-standing structures toward the substrate, resulting in stiction. There are a number of techniques to reduce this problem. The simplest is to use a final rinse in a liquid with a lower surface tension, such as isopropyl alcohol (IPA). This will improve yield, but not to levels required for an industrial process. Super-critical drying can generate considerably higher yield.[102] The cycle of this process is illustrated in Fig. 15.22.

This technique yields good results, although the apparatus required is expensive. A cheaper alternative is freeze-drying,[103] which uses the sublimation of cyclohexane to avoid the liquid phase during drying. The wafers are held in a nitrogen flow to avoid the problem of water forming on the surface.

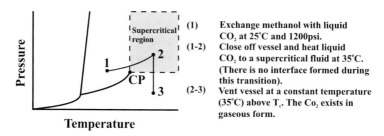

Figure 15.22 Sequence for supercritical drying.

15.5 IC Compatibility Issues

When the basic processes for micromachining, described above, are combined with electronics, a number of aspects must be considered. These aspects include the following:

- Compatibility with the cleanroom (contamination)
- Compatibility with the electronics (thermal budget, planarity)

With wet bulk micromachining, an important consideration is contamination of the cleanroom, in particular with KOH, although the temperature is not a problem. Etching is usually performed at below 100°C and thus does not adversely affect the electronic circuitry. Provided the wafers do not have to return to the process line, the processes are fully compatible. If a return to the clean line is required, TMAH is a better option. As described above, the front side of the wafer should be protected from the etchant or, alternatively, a metallization should be used that is not attacked. Plasma-etch processes are also at low temperature. In this case, a quite different problem can arise. If CMOS wafers are exposed to a plasma for long periods, a shift in the V_T can result.

15.5.1 Compatible Bulk Micromachining

Bulk micromachining is a post-processing, low-cost technique that has been applied to realize several microsensors and microstructures. The compatibility to both bipolar and CMOS processes has been proven. For bipolar processes, both front side and backside processes are possible, as indicated in Fig. 15.23. Generally, ECE on the epilayer/substrate junction is preferred as an etch-stop technique. If SOI wafers are the starting material, the buried oxide acts as the etch stop. For the front-side approach, time-stop and/or geometrical considerations are also used to define the

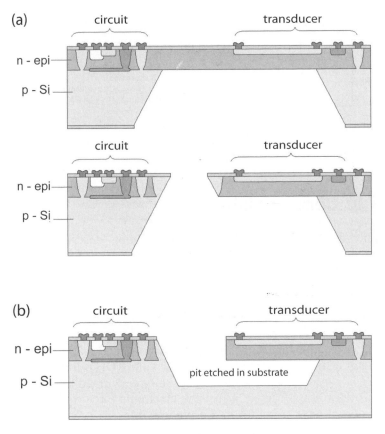

Figure 15.23 Compatibility to IC-bipolar process: (a) back-side etching (membrane and beam); (b) front-side approach.

cavities and the desired microstructure. Similar considerations apply for CMOS processes. The dielectric sandwich of a nitride and oxide layer is often used as the etch stop (see Fig. 15.24); ECE is also used for both front-side and back-side structuring (see Fig. 15.25). The pn junction is formed by the n well and p substrate.

The major drawbacks of conventional bulk micromachining in anisotropic etchants are the poor lateral dimensional control due the anisotropic nature of the etch process (i.e., the etching along the <111> planes) and the necessity for convex corner compensation when etching silicon mesas, resulting in large area devices. For the isotropic etchants, major problems are the limited choice in masking materials and the large lateral under-etch due to the isotropy of the etching. In cases where the

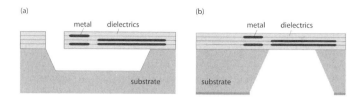

Figure 15.24 Compatibility with IC CMOS process: stop on dielectric (left) front-side approach; (right) back-side approach.

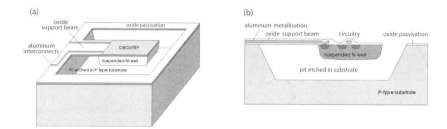

Figure 15.25 Compatibility with IC CMOS process using ECE etch-stop from the front side (adapted from Kovacs et al.[13]).

etch is done from the backside of the wafer, another disadvantage is the necessity of front-back alignment, together with the further increase in the lateral dimensions (and accuracy, due to wafer thickness variations) of the structures. However, the sensible improvement in front-back alignment equipment currently available has certainly eased this last concern.

New, upcoming bulk micromachining technologies address most of these drawbacks. For example, most of the dry-etching processes show less or no crystal-orientation dependence and enable high aspect ratio structures to be fabricated. The small lateral dimensions in combination with a large flexibility in vertical dimensions are essential for the realization of many microsensors and microstructures.

Although laser micromachining and focused ion beam micromachining are very attractive for prototyping or truly 3D microstructuring, the direct-write character of these processes makes them less suitable for mass production. DRIE is an IC-compatible process that is becoming the most promising alternative to conventional bulk micromachining due to the well-controlled and fast etching offered by the newly developed plasma etching equipment.

15.5.2 Compatible Surface Micromachining

With most surface micromachining techniques, contamination of the cleanroom is not a problem since standard semiconductor layers are used. The main issues are thermal budget and planarity. Planarity is an issue if small structures have to be defined afterwards, since spinning of thin layers of resist on highly nonplanar surfaces presents a number of problems. These can be solved either by ensuring that no small structures have to be defined at a later stage or by planarizing the surface.[104-108] The second issue is the thermal budget. If the micromachining is to be combined with electronics, it is important for the two thermal budgets to be compatible. Additional thermal steps can have a significant effect on the electronics.[109] Therefore, it is extremely important to choose carefully the position at which the micromachining is inserted into the standard process sequence. There are three options:

1. Preprocessing (additional layers formed before the start of standard processing)
2. Integrated processing (adding steps during standard processing)
3. Post-processing (using standard layers from the electronic devices, or using additional layers deposited after the completion of the electronics)

15.5.2.1 Preprocessing

Although described as preprocessing, there are usually some processing steps also required after the standard processing. In the case of preprocessing, the major consideration is to ensure that the mechanical properties are not detrimentally affected by the standard processing. Polysilicon has been used for preprocessing; this basic structure is given in Fig. 15.26.[110,111]

15.5.2.2 Integrated Processing

With the integrated processing option, the wafers are removed from the standard line and, after the addition of micromachining steps, return to the standard line. The position of the additional process steps is extremely important. In some cases, the additional depositions are added after the

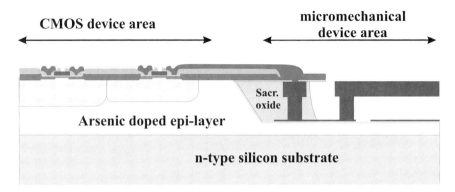

Figure 15.26 Structure fabricated using preprocessing combined with CMOS.

main thermal processing but before the aluminum. Depending on the sensitivity of the electronics devices to thermal budget, a maximum thermal budget for the micromachining can be tolerated. An example of an integrated process combining polysilicon-based surface micromachining with bipolar electronics is given in Fig. 15.27.[112]

A similar process has been developed by Fischer *et al.*, although it was used with an aluminum gate CMOS process.[113] The basic process is shown in Fig. 15.28. The interesting feature is that the deposition of the micromechanical structures is performed before the gate oxidation. As with the Delft process, this process uses photoresist to protect the electronics during sacrificial etching.

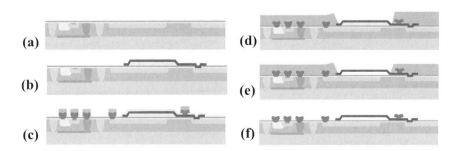

Figure 15.27 Polysilicon-based integrated surface micromachining process.

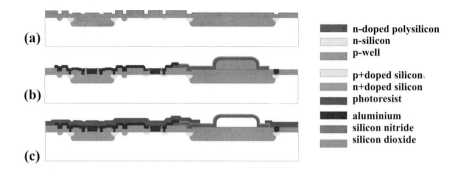

Figure 15.28 Compatible surface micromachining process: (a) after completion of the CMOS process (before gate oxidation) and capping silicon nitride; (b) formation of the sacrificial and mechanical layers followed by gate oxidation and aluminium deposition; (c) formation of the resist protection mask and sacrificial etching; from Fischer et al.[113]

15.5.2.3 Post-Processing

In the post-processing option, wafers go through the complete standard process. After standard processing, either existing layers can be used or additional layers can be added. Both approaches can be found in the literature. Examples of the first option include using the gate poly in a CMOS process[114] and a combination of oxide and metal used in standard processes.[115] In the first of these, the field oxide is used as the sacrificial layer; in the second, it is the substrate itself that is etched. These two options are shown in Fig. 15.29.

Since there are several aluminum layers in modern CMOS processes, aluminum can be used as both the sacrificial and mechanical layers. This is illustrated in Fig. 15.30.[116,117]

The potential problem with this technique is that the processes for the layers are in general not developed with the mechanical properties in mind and may not be suitable. This can, however, be overcome for many devices through careful design.[118] Alternatively, layers can be added after the completion of standard processing. This gives more freedom to optimize the layers, but the thermal budget is limited. If aluminum is used, the thermal budget is limited to about 400°C. Other metals can be used for the electronics to give more freedom for the micromachining. One such example of this approach is shown in Fig. 15.31. Tungsten, used as the metal in this case, can withstand the higher temperatures of the micromachining.[119]

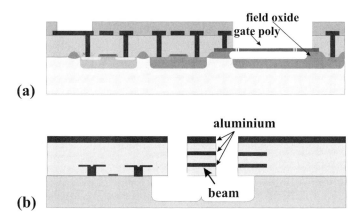

Figure 15.29 Post-processing micromachining with CMOS (a) using the gate poly; (b) using aluminium and oxide.

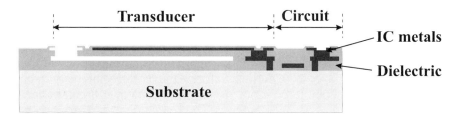

Figure 15.30 Micromachining integrated into a CMOS process.

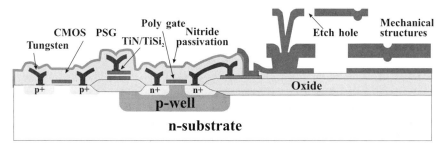

Figure 15.31 Post-processing micromachining using tungsten metallization for the electronics.

15.5.3 Compatible Epi-Micromachining

With epi-micromachining there are a number of compatibility issues specific to the technique. These are considered in turn.

The SIMPLE process has been implemented using standard bipolar technology. However, the buried layer for the micromachining had a much higher doping than that of the standard buried layer of bipolar processes. This special doping must then be added as an extra step before epilayer deposition. Although experiments showed that this resulted in a small rise in the doping level throughout the epilayer, this could be compensated for with adjustments in the processing or circuit design.

The SCREAM process uses a multietch step followed by an additional metal layer. Although in the early versions of the process the integration issue was not addressed, later publications showed this process to be fully compatible. Since the electronics metallization was already performed before micromachining, only large structures had to be defined; this could be done with thick photoresist.

Black silicon is also a dry-etching process. In terms of process temperature, it is fully compatible with electronics, although the electronics should then be fabricated in SOI wafers. This is also the case for the SIMOX process.

The MELO process requires more integration into the processing. It requires an oxide layer to be defined before epilayer deposition and a special epilayer technique to be used. However, in principle, it can be integrated with an epilayer-based process. The remaining problem is that a much thicker epilayer is required to achieve sufficient lateral coverage. A possible option is to grow a thicker layer and polish back.

Sacrificial porous silicon processes only require a process containing an n-type epilayer on a p-type substrate. Also, the electronics must be protected from the etchant during porous formation.

The epi-poly process also requires full access to the process line. As with the MELO process, patterned oxide is required on the substrate before epilayer formation and an additional nitride deposition is required to protect the polysilicon from subsequent oxidations.

All these processes can be integrated with electronics, although not always with full flexibility with regard to electronics technology. An additional consideration with some of these processes is the problem of non-pacified junctions that can be left open after etching.

15.6 Conclusions

This chapter has discussed the wealth of micromachining processes that are available to the device developer. These include both wet and dry etching techniques, and bulk and surface micromachining. If these are to be integrated with electronics on a single chip, there are a number of compatibility issues that need to be addressed.

References

1. K.E. Petersen, "Silicon as a mechanical material," *Proc. IEEE*, 70:420–457 (1982)
2. O.N. Tufte and G.D. Long, "Silicon diffused element piezoresistive diaphragm," *J. Appl. Phys.*, 33:3322 (1962)
3. K.E. Bean, "Anisotropic etching of silicon," *IEEE Trans Electron Devices*, ED-25: 1185–1193 (1978)
4. J.C. Greenwood, "Etched silicon vibrating sensor," *J. Phys. E. Sci. Instrum.*, 17: 650–652 (1984)
5. H.C. Nathanson and R.A. Wickstrom, "A resonant-gate silicon surface transistor with high-Q band pass properties," *Appl. Phys. Lett.*, 7:84 (1965)
6. R.T. Howe and R.S. Muller, "Polycrystalline and amorphous silicon micromechanical beams: annealing and mechanical properties," *Sens. Actuators*, 4: 447–454 (1983)
7. L-S. Fan, Y-C. Tai, and R.S. Muller, "Pin joints, gears, springs, cranks and other novel micromechanical structures," *Proc. Transducers 1987*, Tokyo, (1987), pp. 849–852
8. M. Elwenspoek and H. Jansen, *Silicon Micromachining*, Cambridge University Press (1998)
9. H. Seidel, H. Csepregi, A. Heuberger, and H. Baumgartner, "Anisotropic etching of crystalline silicon in alkaline solutions I & II," *J. Electrochem. Soc.*, 137:3612–3632 (1990)
10. D. Zielke and J. Frühauf, "Determination of rates for orientation-dependent etching," *Sens. Actuators*, A48:151–156 (1995)
11. K. Sato, M. Shikida, Y. Matsushima, T. Yamashiro, K. Asaumi, Y. Irie, and M. Yamamoto, "Characterization of orientation-dependent etching properties of single-crystal silicon: effects of KOH concentration," *Sens. Actuators*, A64: 87–93 (1998)
12. M. Shikida, K. Sato, K. Tokoro, and D. Uchikawa, "Differences in anisotropic etching properties of KOH and TMAH solutions," *Sens. Actuators*, A80:179–188 (2000)
13. G.T. Kovacs *et al.*, "Bulk micromachining of silicon," *Proc. IEEE*, 86(8):1536–1551 (1998)
14. T.A. Kwa, P.J. French, R.F. Wolffenbuttel, P.M. Sarro, L. Hellemans, and J. Snauwaert, "Anisotropically etched silicon mirrors for optical sensor applications," *J. Electrochem. Soc.*, 142:1226–1233 (1995)
15. R.M. Finne and D.L. Klein, "A water-amine-complexing agent system for etching silicon," *J. Electrochem. Soc.*, 114: 965–970 (1967)
16. A. Reisman, M. Berkenblit, S.A. Chan, F.B. Kaufman, and D.C. Green, *J. Electrochem. Soc.*, 126:1406–1410 (1979)
17. O. Tabata, "pH -controlled TMAH etchants for silicon micromachining," *Sens. Actuators*, A53: 335–339 (1996)

18. A. Merlos, M. Acero, M.H. Bao, J. Bauselles, and J. Esteve, "TMAH/IPA anisotropic etching characteristics," *Sens. Actuators*, A37–38:737–743 (1993)
19. P.M. Sarro, S. Brida, C.M.A. Ashruf, W.v.d. Vlist, and H.v. Zeijl, "Anisotropic etching of silicon in saturated TMAHW solutions for IC-compatible micromachining," *Sens. Mater.*, 10:201–212 (1998)
20. H. Baltes, "CMOS as sensor technology," *Sens. Actuators*, 37–38:51–56 (1993)
21. M.J. Dececlerq, L. Gerzberg, and J.D. Meindl, "Optimization of the hydrazine-water solution for anisotropic etching of silicon in integrated circuit technology," *J. Electrochem. Soc.*, 122(4):201–212 (1975)
22. M. Mehregany and S.D. Senturia, "Anisotropic etching of silicon in hydrazine," *Sens. Actuators*, 13:375–390 (1988)
23. M.A. Gajda, J.E.A. Shaw, A. Putnis, and H. Ahmed, "Anisotropic etching of silicon in hydrazine," *Sens. Actuators*, A40 :227–236 (1994)
24. U. Schnakenberg et al., "NH_4OH based etchants for silicon micromachining," *Sens. Actuators*, A21–23:1031–1035 (1990)
25. L.D. Clark, Jr., J.L. Lund, and D.J. Edell, "Cesium hydroxide (CsOH): a useful etchant for micromachining silicon," *Tech. Digest IEEE Solid State Sensor and Actuator Workshop*, Hilton Head Island, SC, June 6–9, 1988, pp. 5–8
26. E.D. Palik et al., "Study of the etch-stop mechanism in silicon," *J. Electrochem. Soc.*, 137:2051–2059 (1982)
27. Y. Gianchandani and K. Najafi, "A bulk dissolved wafer process for microelectromechanical systems," *IEDM Tech. Digest* (1991), pp. 757–760
28. A. Perez-Rodriguez, A. Romano-Rodriguez, J.R. Morante, M.C. Acero, J. Esteve, and J. Montserrat, "Etch-stop behaviour of buried layers formed by substoichiometric nitrogen ion implantation into silicon," *J. Electrochem. Soc.*, 143:1026–1033 (1996)
29. H.A.Waggener, "Electrochemically controlled thinning of silicon," *Bell Syst. Tech. J.*, 49:473–475 (1970)
30. B. Kloek, S.D. Collins, N.F.de Rooij, and R.L. Smith, "Study of electrochemical etch-stop for high precision thickness control of silicon membranes," *IEEE Electron Dev.*, 36:663–669 (1989)
31. P.M. Sarro and A.W.van Herwaarden, "Silicon cantilever beams fabricated by electrochemically controlled etching for sensor applications," *J. Electrochem. Soc.*, 133:1724–1729 (1986)
32. A.W. van Herwaarden, D.C. van Duyn, B.W. van Oudheusden, and P.M. Sarro, "Integrated thermal sensors," *Sens. Actuators*, A21–23:621–630 (1989)
33. htttp://www.ammt.de
34. E. Peeters, D. Lapadatu, R. Puers, and W. Sansen, "PHET, An electrodeless photovoltaic electrochemical etch-stop technique," *J. Microelectromech. Syst.*, 3:113–123 (1994)
35. D. Lapadatu, M. de Cooman, and R. Puers, "A double-sided capacitive miniaturized accelerometer based on photovoltaic etch-stop technique," *Sens. Actuators*, A53:261–266 (1996)

36. P.J. French, M. Nagao, and M. Esashi, "Electrochemical etch-stop in TMAH without externally applied bias," *Sens. Actuators*, A56:279–280 (1996)
37. C.M.A. Ashruf, P.J. French, P.M.M.C. Bressers, P.M. Sarro, and J.J. Kelly, "A new contactless electrochemical etch-stop based on gold/silicon/TMAH galvanic cell," *Sens. Actuators*, A66:284–291 (1998)
38. C.M.A. Ashruf, *Galvanic Etching of Silicon for Fabrication of Micromechanical Structures*, Delft University Press (2000), ISBN 90-407-2001-0
39. M.M. Abu-Zeid, "Corner undercutting in anisotropically etched isolation contours," *J. Electrochem. Soc.*, 131:2138–2142 (1984)
40. X. Wu and W. Ko, "Compensating corner undercutting in anisotropic etching of (100) silicon," *Sens. Actuators*, A18:207–215 (1989)
41. R. van Kampen and R.F. Wolffenbuttel, "Effects of <100>-oriented corner compensation structures on membrane quality and convex corner integrity in (100)-silicon using aqueous KOH," *J. Micromech. Microeng.*, 5:91–94 (1995)
42. H.L. Offereins, H. Sandmaier, K. Marusczyk, K. Kuhl, and A. Plettner, "Compensating corner undercutting of (100) silicon in KOH," *Sens. Mater.*, 3:127–144 (1992)
43. G.M. O'Hallaran, *Capacity Humidity Sensor Based on Porous Silicon*, Delft University Press (2000), ISBN 90-407-1919-5
44. G.M. O'Hallaran, M. Kuhl, P.J. Trimp, and P.J. French, "The effect of additives on the adsorption properties of porous silicon," *Sens. Actuators*, A61:415–420 (1997)
45. P.T.J. Gennissen and P.J. French, "Sacrificial oxide etching compatible with aluminium metallization," *Proc. Transducers 97*, Chicago, USA, June 1997, pp. 225–228
46. M. Kuhl, G.M. O'Halloran, P.T.J. Gennissen, and P.J. French, "Formation of porous silicon using an ammonium fluoride based electrolyte for application as a sacrificial layer," *J. Micromech. Microeng.*, 8:317–322 (1998)
47. H. Robbins and B. Schwartz, "Chemical etching of silicon," *J. Electrochem. Soc.*, 123:1903–1909 (1976)
48. S.D. Collins, "Etch stop techniques for micromachining," *J. Electrochem. Soc.*, 144:2242–2262 (1997)
49. N. Schweisinger and A. Albrecht, "Wet chemical isotropic etching procedure of silicon – a possibility for the production of deep structured microcomponents," *Proc. SPIE Micromachining and Microfabrication Process Technology III*, 3233:72–81 (1997)
50. K.R. Williams and R.S. Muller, "Etch rates for micromachining processes," *J. Microelectromech. Syst.*, 5:256–269 (1996)
51. K.C. Lee, "The fabrication of thin, freestanding, single-crystal, semiconductor membranes," *J. Electrochem. Soc.*, 137:2556–2574 (1990)
52. T. Bischoff, G. Muller, W. Welser, and F. Koch, "Front side micromachining using porous-silicon sacrificial-layer technology," *Sens. Actuators*, A60:228–234 (1997)
53. C. Ducso *et al.*, "Porous silicon bulk micromachining for thermally isolated membrane formation," *Sens. Actuators*, A60:235–239 (1997)

54. C.J.M. Eijkel, J. Branebjerg, M. Elwenspoek, and F.C.M. van de Pol, "A new technology for micromachining of silicon: dopant selective HF anodic etching for the realization of low-doped monocrystalline silicon structures," *IEEE Electron. Dev. Lett.*, 11:588–589 (1990)
55. M.J.J. Theunissen, "Etch channel formation during anodic dissolution of n-type silicon in aqueous hydrofluoric acid," *J. Electrochem. Soc.*, 119:351–360 (1972)
56. M. Esashi, H. Komatsu, T. Matsuo, M. Takahashi, T. Takishima, K. Imbayashi, and H. Ozawa, "Fabrication of catheter-tip and sidewall miniature pressure sensor," *IEEE Trans. Electron. Dev.*, 29:57–63 (1982)
57. G. Kaltsas and A.G. Nassiopoulou, "Front side bulk micromachining using porous-silicon technology," *Sens. Actuators*, A65:175–179 (1998)
58. W. Lang, P. Steiner, and H. Sandmaier, "Porous silicon: a novel material for microsystems," *Sens. Actuators*, A51:31–36 (1995)
59. P.T.J. Gennissen and P.J. French, "Development of silicon accelerometers using epi micromachining," *Proc. SPIE Micromachined Devices and Components*, Santa Clara, CA, USA, Sept. 1999, 3876:84–92
60. T.E.Bell and K.D.Wise, "A dissolved wafer process using porous silicon sacrificial layer and a lightly-doped bulk silicon etch-stop," *Proc. IEEE MEMS*, Heidelberg, Germany, Jan. 1997, pp. 251–256
61. T.Yoshida, T.Kudo, and K.Ikeda, "Photo-induced preferential anodization for micromachining," *Sensors Mater.*, 4/5:229–238 (1993)
62. C.M.A. Ashruf, P.J. French, P.M.M.C. Bressers, and J.J. Kelly, "Galvanic porous silicon formation without external contact," *Sens. Actuators*, A74:118–122 (1999)
63. H. Ohji, P.J. Trimp, and P.J. French, "Fabrication of free standing structures using a single step electrochemical etching in hydrofluoric acid," *Sens. Actuators*, A73:95–100 (1999)
64. H. Ohji, P.J. French, and K. Tsutsumi, "Fabrication of mechanical structures in p-type silicon using electrochemical etching," *Sens. Actuators*, A82(1–3):254–258 (2000)
65. H. Ohji, P.J. French, S. Izuo, and K. Tsutsumi, "Initial pits for electrochemical etching in hydrofluoric acid," *Sens. Actuators*, A85(1–3):390–394 (2000)
66. V. Lehmann, "Porous silicon- a new material for MEMS," *Proc. IEEE MEMS Workshop 1996*, San Diego, USA (1996), pp. 1–6
67. H. Ohji and P.J. French, "Single step electrochemical etching in ammonium fluoride," *Sens. Actuators*, A74:109–112 (1999)
68. J.W. Bartha, J. Greschner, M. Puech, and P. Maquin, "Low temperature etching of Si in high-density plasma using SF_6/O_2," *Microelectron. Eng.*, 27:453–456 (1995)
69. Alcatel, France, *http://www.alcatelvacuum.com*
70. Surface Technology Systems (STS), UK, http://www.stsystems.com
71. Plasmatherm/Unaxis, *http://www.plasmatherm.com*
72. F. Laemer, A. Schilp, K. Funk, and M. Offenberg, "Bosch deep silicon etching: improving uniformity and etch rate for advanced MEMS applications," *Proc. IEEE MEMS 1999 Conf.*, Orlando, FL, USA (1999)

73. G. Craciun, M.A. Blauw, E. van der Drift, and P.J. French, "Aspect ratio and crystallographic orientation dependence in deep dry silicon etching at cryogenic temperatures," *Tech. Digest Transducers 2001*, Munich, Germany, June 2001
74. http://www.el.utwente.nl/tt/projects/droog/study.htm
75. E. Klaassen et al., "Silicon fusion bonding and deep reactive ion etching; a new technology for microstructures," *Proc. Transducers 1995*, Stockholm, Sweden, June 1995, pp. 556–559
76. T.M. Bloomstein and D.J. Ehrlich, "Laser deposition and etching of three-dimensional microstructures," *Proc. Transducers 1991*, San Francisco, USA, June 1991, pp.507–511
77. M. Mullenborn, H. Dirac, J.W. Petersen, and S. Bouwstra, "Fast 3D laser micromachining of silicon for micromechanical and microfluidic applications," *Proc. Transducers 1995*, Stockholm, Sweden, June 1995, pp. 166–169
78. Resonetics, Nashua, NH, USA, http://www.resonetics.com
79. S. Reyntjes and R. Puers, "Focused ion beam applications in microsystem technology," *Micro-Mechanics Europe (MME) Workshop*, Uppsala, Sweden, October 1–3, 2000, pp. 87–96
80. R.J. Young, "Micro-machining using a focused ion beam," *Vacuum* 44:353–356 (1993)
81. G.J. Althas et al., "Focused ion beam system for automated MEMS prototyping and processing," *Proc. SPIE Micromachining and Microfabrication Process Technology III*, 3223:198–207 (1997)
82. J.H. Daniel and D.F. Moore, "A microaccelerometer structure fabricated in silicon-on insulator using a focused ion beam process," *Sens. Actuators*, A73:201–209 (1999)
83. E. Belloy, S. Thurre, E. Walckiers, A. Sayah, and M.A.M. Gijs, "The introduction of powder blasting for sensor and microsystem applications," *Sens. Actuators*, A84:330–337 (2000)
84. P.J. Slikkerveer, P.C. Bouten, and F.C.M. de Haas, "High quality mechanical etching of brittle materials by powder blasting," *Proc. XIII Eurosensors Conf.*, The Hague, The Netherlands, Sept. 12–15, 1999, pp.655–662; ISBN 90-76699-02-X
85. H. Wensink, J.W. Berenschot, H.V. Jansen, and M.C. Elwenspoek, "High resolution powder blast micromachining," *Proc. IEEE MEMS 2000 Conf.*, Miyazaki, Japan (2000), pp. 769–774
86. H.C. Nathanson, W.E. Newell, R.A. Wickstrom, and J.R. Davis, Jr., "The resonant gate transistor," *IEEE Trans. Electron. Dev.*, 14:117–133 (1967)
87. R.N. Thomas, J. Guldberg, H.C. Nathanson, and P.R. Malmberg, "The mirror matrix tube: a novel light valve for projection displays," *IEEE Electron. Dev.*, ED-22:765 (1975)
88. K.J. Gabriel, W.S.N. Trimmer, and M. Mehregany, "Micro gears and turbines etched from silicon," *Proc. Transducers 87*, Tokyo, 1987, pp. 853–856
89. Y.X. Li, P.J. French, P.M. Sarro, and R.F.Wolffenbuttel, "Fabrication of a single crystalline capacitive lateral accelerometer using micromachining based on single step plasma etching," *Proc. MEMS 95*, Amsterdam, Jan–Feb 1995, pp. 398–403

90. K.A. Shaw and N.C. MacDonald, "Integrating SCREAM micromachined devices with integrated circuits," *Proc. IEEE MEMS*, San Diego, USA, 11–15 February 1996, pp. 44–48
91. K.A. Shaw, Z.L. Zhang, and N.C. MacDonald, "SCREAM I: a single mask, single-crystal silicon, reactive ion etching process for microelectromechanical structures," *Sens. Actuators*, A40:63–70 (1994)
92. M. de Boer, H. Jansen, and M. Elwenspoek, "Black silicon V: a study of the fabricating of moveable structures for micro electromechanical systems," *Proc. Transducers 95*, Stockholm, Sweden (1995), pp. 565–568
93. M. Bartek, P.T.J. Gennissen, P.M. Sarro, P.J. French, and R.F. Wolffenbuttel, "An integrated silicon colour sensor using selective epitaxial growth," *Sens. Actuators*, A41–42:123–128 (1994)
94. A.E. Kabir, G.W. Neudeck, and J.A. Hancock, "Merged epitaxial lateral overgrowth (MELO) of silicon and its applications in fabricating single crystal silicon surface micromachining structures," *Proc. Techcon 93*, Atlanta, GA, USA (1993)
95. P.T.J. Gennissen, *Micromachining Techniques Using Layers Grown in an Epitaxial Reactor*, Delft University Press (1999), ISBN 90-407-1843-1
96. T.E. Bell, P.T.J. Gennissen, D. de Munter, and M. Kuhl, "Porous silicon as a sacrificial material", *J. Micromech. Microeng.*, 6:361–369 (1996)
97. P.T.J. Gennissen, P.J. French, D.P.A. De Munter, T.E. Bell, H. Kaneko, and P.M. Sarro, "Porous silicon micromachining techniques for acceleration fabrication," *Proc. ESSDERC 95*, Den Haag, The Netherlands, Sept. 1995, pp. 593–596
98. B. Diem, P. Rey, S. Renard, S. Viollet Bosson, H. Bono, F. Michel, M.T. Delaye, and G. Delapierre, "SOI 'SIMOX' from bulk to surface micromachining, a new age for silicon sensors and actuators," *Sens. Actuators*, A46–47:8–26 (1995)
99. T. Lisec, M. Kreutzer, and B. Wagner, "A surface micromachined piezoresistive pressure sensor with high sensitivity," *Proc. ESSDERC 95*, Den Haag, The Netherlands, Sept. 1995, pp. 339–342
100. P.T.J. Gennissen, P.J. French, M. Bartek, P.M. Sarro, A. van der Bogaard, and C. Visser, "Bipolar compatible epitaxial polysilicon for surface micromachined smart sensors," *Proc. SPIE Micromachining and Microfabrication Process Technology II*, Austin, Texas, USA, 14–15 Oct. 1996, pp. 135–142
101. P.T.J. Gennissen, M. Bartek, P.J. French, and P. M. Sarro, "Bipolar compatible epitaxial poly for smart sensor-stress minimization and applications," *Sens. Actuators*, A62:636–645 (1997)
102. G.T. Mulhern, D.S. Soane, and R.T Howe, "Supercritical carbon dioxide drying of microstructures," *Proc. Transducers 1993*, Yokohama, Japan, pp. 296–299
103. R. Legtenberg and H.A.C. Tilmans, "Electrostatically driven vacuum-encapsulation polysilicon resonators: Part 1. Design and fabrication," *Sens. Actuators*, A45:57–66 (1994)
104. P.J. French and R.F. Wolffenbuttel, "Reflow of BPSG for sensor applications," *J. Micromech. Microeng.*, 3:1–3 (1993)

105. A.C. Adams, "Plasma planarisation," *Solid State Technol.*, 24(1):178–181 (1981)
106. Y.X. Li, P.J. French, and R.F. Wolffenbuttel, "Plasma planarization for sensor applications," *J. Microelectromech. Syst.*, 4:132–138 (1995)
107. B. Roberts, "Chemical mechanical planarisation," *Proc. IEEE/SEMI Advanced Semiconductor Manufacturing Conference and Workshop 1992*, Cambridge, MA, USA, 30 Sept.–1 Oct. 1992, pp. 206–210
108. J.J. Sniegowski, "Chemical mechanical polishing: enhancing the manufacturability of MEMS," *Proc. SPIE Micromachining and Microfabrication Process Technology II*, Austin, Texas, USA, 14–15 Oct. 1996, pp. 104–115
109. K.M. Mahmoud and R.F. Wolffenbuttel, "Compatibility between bipolar read-out electronics and microstructures in silicon," *Sens. Actuators*, A31:188–199 (1992)
110. J.H. Smith, S. Montague, J.J. Sniegowski, J.R. Murray, R.P. Manginell, and P.J. McWhorter, "Characterisation of the embedded micromachined device approach to the monolithic integration of MEMS with CMOS," *Proc. SPIE Micromachining and Microfabrication Process Technology II*, Austin, Texas, USA, 14–15 Oct. 1996, 2879:306–314
111. Y.B. Gianchandani, M. Shinn, and K. Najafi, "Impact of long high temperature anneals on residual stress in polysilicon," *Proc. Transducers 97*, Chicago, USA, June 1997, pp. 623–624
112. B.P. van Drieënhuizen, J.F.L. Goosen, P.J. French, Y.X. Li, D. Poenar, and R.F. Wolffenbuttel, "Surface micromachined module compatible with BiFET electronic processing," *Proc. Eurosensors 94*, Toulouse, France, Sept. 1994, p. 108
113. M. Fischer, M. Nägele, D. Eichner, C. Schöllhorn, and R. Strobel, "Integration of surface micromachined polysilicon mirrors and a standard CMOS process," *Sens. Actuators*, A52:140–144 (1996)
114. C. Hierold, A. Hilderbrandt, U. Näher, T. Scheiter, B. Mensching, M. Steger, and R. Tielert, "A pure CMOS surface micromachined integrated accelerometer," *Proc. MEMS 96*, San Diego, USA, Feb. 1996, pp. 174–179
115. G.K. Fedder, S. Santhanan, M.L. Read, S.C. Eagle, D.F. Guillou, M.S.-C. Lu, and L.R. Carley, "Laminated high-aspect ratio microstructures in a conventional CMOS process," *Proc. MEMS 96*, San Diego, USA, Feb. 1996, pp. 13–18
116. D. Westberg, O. Paul, G.I. Andersson, and H. Baltes, "Surface micromachining by sacrificial aluminium etching," *J. Micromech. Microeng.*, 6:376–384 (1996)
117. O. Paul, D. Westberg, M. Hornung, V. Ziebart, and H. Baltes, "Sacrificial aluminum etching for CMOS microstructures", *Proc. MEMS 97*, Nagoya, Japan, Jan. 1997, pp. ...
118. H. Xie and G.K. Fedder, "A CMOS z-axis capacitive accelerometer with comb-finger sensing," *Proc. IEEE MEMS 2000*, Miyazaki, Japan, 23–27 January 2000, pp. 496–501
119. J.M. Bustillo, G.K. Fedder, C.T.-C. Nguyen, and R.T. Howe, "Process technology for the modular integration of CMOS and polysilicon microstructures," *Microsyst. Technol.*, 1:30–41 (1994)

16 LIGA Technology for R&D and Industrial Applications

Ulrike Wallrabe
Institute of Microsystem Technology IMTEK
University of Freiburg, Germany

Volker Saile
Research Centre Karlsruhe GmbH, Institute of Microstructure Technology,
and University of Karlsruhe, Germany

16.1 Introduction

The LIGA process[1] was first developed at the Forschungszentrum Karlsruhe GmbH. It is currently used worldwide by numerous other research institutes and by industry. LIGA allows for the manufacturing of microcomponents with almost arbitrary lateral geometry and resolution in the micron range, but with structure heights into the millimeter range. The typical materials used are polymers, metals, and ceramics, thus covering a wide range of "nonsilicon" candidates. The main process steps have given the technique its name, LIGA, a German acronym consisting of LI (Lithographie for lithography), G (Galvanik for electroplating), and A (Abformung for replication techniques, such as molding). In the classical understanding of LIGA, the lithography step is performed using the highly collimated and energetic X-rays of a synchrotron that can penetrate with little scattering into hundreds of microns of polymer resist and hence pattern it with extremely sharp, smooth, and vertical sidewalls. In addition, due to the short wavelengths of the X-rays, the spatial resolution of the process is very high, thus allowing for extremely high aspect ratios, i.e., the ratio of width over thickness of a structure. The overall process development requires establishing an entire process chain, e.g., a mask-making procedure, since X-ray masks need to be patterned with relatively high-aspect-ratio absorbing structures in a LIGA-like process, or an adopted sacrificial layer technique. Also, the deposition of thick resist films up to several millimeters is a typical LIGA task and is strikingly different from optical lithography.

With the advent of the UV-sensitive SU-8 resist family with its special optical properties, the LIGA concept has been extended to UV lithography,[2] with ultraviolet light replacing the X-rays, followed by subsequent electroplating and molding. This process is known by the names UV-LIGA or poor man's LIGA, the latter because it does not require access to a synchrotron.

This chapter briefly describes the classical LIGA technique, starting with the mask-making process and ending with the molding technique, with an emphasis on deep X-ray lithography. Then a variety of applications are presented, mostly from the field of micro-optics, but also from other industrial areas that take advantage of LIGA-based microsystem technology.

16.2 The LIGA Process

In deep X-ray lithography, synchrotron radiation is used to transfer the pattern of an X-ray mask into an X-ray-sensitive resist layer, typically polymethylmethacrylate (PMMA, Plexiglas), as shown in Fig. 16.1(a). Due to the scission of long molecular polymer chains upon exposure to X-rays, the irradiated areas can be dissolved with a suitable developer, resulting in a resist template on top of a normally electrically conductive substrate. Techniques involving microelectrodeposition can then be applied to grow a complementary structure of metal (e.g., copper, gold, nickel, and nickel alloys) by filling the template of the nonconducting resist [Fig. 16.1(b)]. In this process, either the substrate itself or a metal-plating base serves as a conductive baseplate for the electrochemical metal deposition. After resist removal, the electroplated metal structure itself may be used as the final product or, in view of mass production, as a mold insert. For the fabrication of a mold insert, the deposition of the metal is continued until a thick metal layer has developed above the resist structure, forming a continuous and solid metal baseplate for the fragile microstructures. With the mold insert, innumerable plastic copies can be mass-produced with high accuracy in detail and at a relatively low cost using injection molding, reaction injection molding, or the hot-embossing technique [Fig. 16.1(c)]. These replication techniques allow the use of a variety of other polymers, in addition to PMMA, such as POM (polyoxymethylene), PEEK (poly ether ether ketone), PVDF (poly vinylidene fluoride), PSU (polysulfone), and PC (polycarbonate). After demolding, these structures can be post-processed, for example, by dicing [Fig. 16.1(d)] or sputtering, or they can be filled with metal by electrodeposition, or serve as a lost template for the fabrication of, for example, ceramic structures.

The LIGA process is characterized by:

- Structure heights of up to 3 mm, depending on the design
- Freedom in lateral shape
- Smallest lateral dimension (structural detail) down to 200 nm (depending on design)
- Aspect ratios (ratio of structure height to smallest lateral dimension) of stand-alone structures of up to 50; supported structural details may show ratios of more than 500.
- Roughness of sidewalls <20 nm

Figure 16.1(a) X-ray lithography with a thick resist film using an X-ray mask.

Figure 16.1(b) Electrodeposition of metal in a resist template. Typical metals are Ni, Au, Cu, and FeNi.

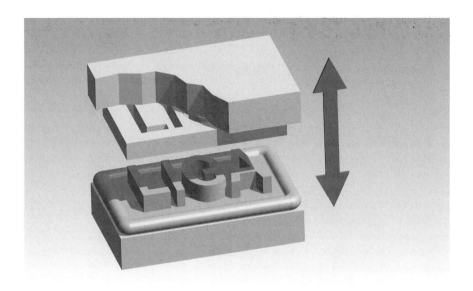

Figure 16.1(c) Molding (here, hot-embossing) of a polymer using a metallic molding tool.

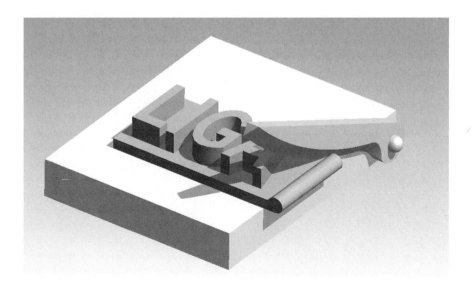

Figure 16.1(d) Post-processing of a molded component (here, dicing).

- Broad range of materials. Polymers: PMMA, POM, PSU, PEEK, PVDF, PC, LCP (liquid crystal polymer), PA (polyamide), and PE (polyethylene). Metals: nickel, copper, gold, NiFe, and NiP. Ceramics: PZT (lead zirconate titanate), PMNT (lead magnesium niobate), and Al_2O_3, ZrO_2.

LIGA involves a long process chain that, in many aspects, is comparable to semiconductor manufacturing. Hence, establishing LIGA as a true manufacturing technology requires stable individual process steps with high yield. Major efforts were therefore directed toward establishing well-controlled "standard" processes on the basis of the best-characterized PMMA resist. Furthermore, professional equipment such as X-ray scanners, megasonic development tanks, and electroplating facilities were developed. This section describes the standard X-ray lithography process and its characteristics, based on PMMA as the resist.

16.2.1 Mask Making

Classical chrome-quartz masks, the standard for UV lithography, cannot be used for X-ray lithography because they do not generate sufficient contrast for the X-rays. LIGA masks are shadow masks; i.e., they show the same lateral dimensions as the imaged product. The X-ray absorption in a material is proportional to the atomic number Z to the third power; that is, to assure a good contrast, an absorbing high-Z material with sufficient thickness needs to be patterned on a low X-ray absorption carrier, a thin plate, or a membrane of low-Z material. The combination that is typically used at the Forschungszentrum Karlsruhe is a titanium membrane ($Z = 22$) about 3 µm thick, above which are placed gold absorbers ($Z = 79$) up to 20 µm thick,[3] or a 500-µm-thick beryllium wafer ($Z = 4$) with up to 30 µm of gold absorber.[4]

Gold absorbers of such a thickness cannot be patterned with the required accuracy in one step. The masks are therefore made in two steps, first the so-called intermediate mask, then the working mask. For the intermediate mask, a 2.7-µm-thick PMMA layer is spun on a silicon substrate coated with 3-µm titanium. The PMMA is structured by a 100 keV electron beam writer, thus generating the master pattern, as indicated in Fig. 16.2(a). The developed PMMA is filled with 2.3 µm of gold by electroplating, and the PMMA is stripped. Then a carrier frame is glued to the substrate surrounding the area with the microstructures. Because of the purposely chosen bad adhesion of titanium to the silicon substrate, the frame with the membrane attached is easily pulled off the substrate [Fig. 16.2(b)].

This intermediate mask is a first X-ray mask; however, its contrast is only sufficient to structure a PMMA layer about 50 µm thick. There-

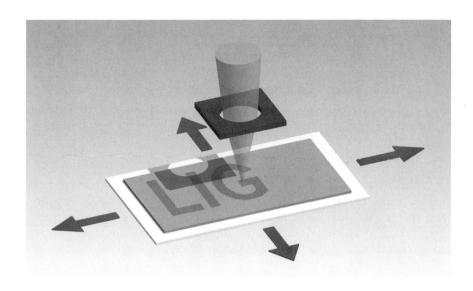

Figure 16.2(a) Writing of the master pattern with an electron-beam writer.

Figure 16.2(b) Separating the Ti membrane from the substrate.

fore, in a second step, the working mask is copied from the intermediate mask in a 50-μm-thick PMMA resist on another titanium membrane or on a Be wafer using X-ray lithography. The PMMA template can then be electroplated with 20-μm of gold or even more. The final result is a working mask with sufficient thickness of gold absorbers and thus with a high contrast for the X-rays, which lie in the range of several keV photon energy.

16.2.2 Deep X-Ray Lithography

In deep X-ray lithography, which is the first step of the LIGA process, an X-ray sensitive polymer layer several hundred micrometers or even several millimeters thick is applied to a metal carrier or an insulating carrier coated with a conductive layer that serves as a substrate. Normally, PMMA is used as the X-ray sensitive polymer layer (resist). The resist layer may be produced by direct polymerization of a resin on the substrate, by gluing a prefabricated PMMA sheet with polymerization glue to the substrate, or by bonding a prefabricated PMMA sheet to a spin-coated resist layer about 3 μm thick onto the substrate.

During exposure, synchrotron light is absorbed in the exposed PMMA areas, which results in a chemical modification; that is, the molecular weight is decreased due to chain scission and becomes soluble in an organic developer. To ensure complete dissolving, a minimum dose value of 4 kJ/cm^3 for PMMA has to be deposited at the bottom of the resist. The dose at the top of the resist layer should not exceed 14 to 20 kJ/cm^3, depending on the thickness of the resist. This is to avoid mechanical damage, so-called foaming, of the resist surface. With these minimum and maximum values for the absorbed dose, an exposure strategy has to be developed. At the interface between the resist and substrate, the lowest dose possible should be absorbed. Higher doses only lead to excessive production of secondary electrons generated at the interface by the X-rays. These secondary electrons break the chemical bonds between substrate and resist even in unexposed areas adjacent to the exposed areas, leading to a loss of adhesion between substrate and resist. On the other hand, a dose as high as possible is preferred for the resist surface to achieve acceptable exposure times. Dose values above a ratio of approximately 5:1 between surface and interface are optimal. This ratio depends on two parameters only: the thickness of the PMMA resist and, via the absorption spectrum of PMMA, the wavelength of the X-rays. In practice, a band of X-rays is filtered out of the "white" synchrotron radiation spectrum by using filters or an X-ray mirror, or by changing the parameters of the accelerator producing the X-rays. The central wavelength of this band is then tuned to achieve the optimal dose ratio for a given resist thickness. After irradiation, the exposed areas are placed in a suitable developer, the so-called GG developer, which is a mixture of 60% diethylenglykol-monobutylether, 20% morpholin, 5% ethanolamin, and 15% distilled aqua. To minimize defects, the developer must be free of residue, especially for the following electroplating process. The development process can be supported by application of megasound. For smooth developing of fragile microstructures, the megasound frequencies should be in the range of 1 MHz, and the intensity has to be adapted to the geometry of the microstructure.

As a consequence of the high energy and parallel nature of X-rays, very high structures can be produced with nearly vertical and extremely smooth sidewalls.[5] One measure indicating these characteristics is the aspect ratio, i.e., the height of the structure relative to the smallest lateral dimensions. Examples of aspect ratios achieved are given in Table 16.1, where a distinction is made between a component (A) that is relatively large (mm to cm) but contains very fine (µm to nm) details (e.g., the diffraction grating of the LIGA microspectrometer), and a stand-alone microstructural component (B) with lateral dimensions in the µm range (e.g., for actuators, inertial sensors, or optical components). Figures 16.3[(a) and (b)] and 16.4 show relevant microstructures as examples for each case.

Quite another approach for the fabrication of almost fully three-dimensional LIGA structures is described by Tabata *et al.*[6] During the X-ray exposure, the mask can be moved in the horizontal and vertical directions, the substrate can be rotated and tilted toward the synchrotron beam, and even a multimask process can be run. The X-ray scanner is schematically shown in Fig. 16.5(a). Combining the exposure features in

Table 16.1 Examples of Achieved Aspect Ratios by Deep X-Ray Lithography

Aspect Ratio	Minimum Lateral Dimensions (µm)	Structure Height (µm)	Type	Microstructure	Reference Product
650	0.2	130	A	Grating structure	Spectrometer for visible light [Fig. 16.3(a) and (b)]
250	2	500	A	Grating structure	IR spectrometer
40	Diameter = 26	1100	B	Columns, here in SU-8	Columns (Fig. 16.4)
30	100/100	3000	B	Lines and spaces	Lead frames

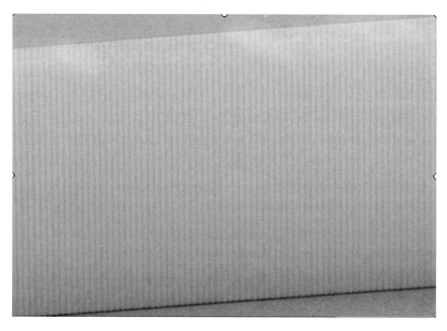

Figure 16.3(a) Microprecision component type A: The grating of a microspectrometer is a microscopic detail of a much larger overall device.

Figure 16.3(b) Close-up of the reflective grating.

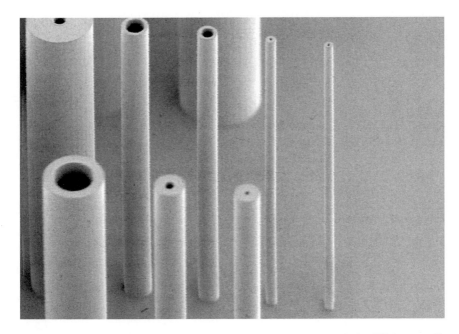

Figure 16.4 Microprecision component type B; columns made in SU-8 are self-standing microscopic components.

a clever way, gray-tone lithography can be performed, resulting in microstructures that have nonconstant lateral dimensions across their thickness. This becomes clear from an optical grating [Fig. 16.5(b)], where the line width at the top is smaller than at the bottom of the lines and the lines are tilted from the normal axis.

Due to an inhomogeneous X-ray dose deposition in the resist in such a procedure, special care must be taken with the dose profile and the development of the structures. Dedicated simulation tools allow an accurate prediction of the process parameters and the geometry of the features, providing a better understanding of the process.[7]

16.2.3 Electroplating and Micromolding

After X-ray exposure and development, the resist forms a nonconducting polymer template on a metal substrate. In an electroplating bath, deposition of metal will start at the bottom of the openings in the resist, where the bath is in contact with the metallic substrate surface. Micro-

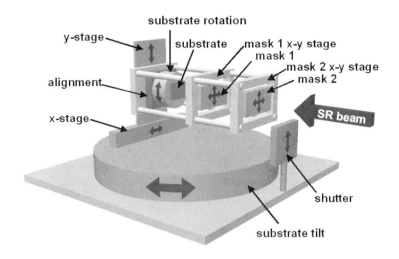

Figure 16.5(a) Multifunctional X-ray scanner for the fabrication of 3D LIGA structures.[6]

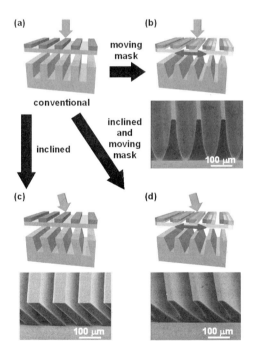

Figure 16.5(b) Grey-tone lithography for variable feature size across PMMA thickness and inclined sidewalls of PMMA structures.[6]

electroplating is essential in several LIGA process steps.[8] First, gold plating is the standard method for generating absorbers on X-ray masks. Here typical heights for plating are between 20 and 30 µm. For the production of metal components, the polymer template is filled with pure metals such as nickel and copper or, for specific magnetic properties, with NiFe and, when hardness needs to be high, NiCo. The height of such metal components ranges between 100 µm and several mm. Finally, when mold inserts are fabricated, the electroplating process is continued even after the open areas in the resist are completely filled. By such an overgrowth, a several-mm-thick metallic baseplate for the fragile metal structures is produced.

Electroplating of small features, especially ones with high aspect ratios, is challenging. Transport of material is limited by diffusion over larger distances and makes electroplating a slow process. Furthermore, the specifications on internal stress minimization and hardness optimization govern the choice of current density. These factors lead to an overall plating time of approximately 2 weeks for a typical mold insert. All relevant parameters, including chemical composition of the plating baths, must be kept within tight limits over such a long time. Monitoring the status of the bath continuously is therefore mandatory.

The final step in LIGA is the use of a LIGA mold as a high-precision replication tool. Polymer replication is a process of enormous economic importance for "macroscopic" products (most of our plastic articles are produced through injection molding) as well as in the microdimensions, with CD and DVD volume production through injection embossing providing a prominent example. For ultra-high resolution, various new techniques are currently being developed under the name "nanoimprinting." Microreplication appears to be a well-established technology; this is only partially true, however, and is limited to low-aspect-ratio structures. Replication with typical LIGA molds requires major modifications in process parameters and control as well as in equipment.[9] Special injection molding machines have been developed (e.g., by Battenfeld Corp.) that enable evacuation of the volume between the mold and the counterplate. Furthermore, the metal surfaces of the tool are heated in a variotherm process before the liquid polymer is injected. Pressure gradients within the polymer need to be minimized to prevent damage to the fragile high-aspect-ratio structures of the mold. Highly sophisticated simulation software is essential for modeling and optimizing the flow of the polymer.[10]

Whereas injection molding aims at high-volume production, hot embossing is highly appropriate for low-volume replication of fragile microstructures with high aspect ratios and small mold-form distortions. In the vacuum hot-embossing process, a microstructured mold insert is pressed into a thermoplastic film or sheet that is heated above its glass transition temperature. The polymer sheet, typically 1 or 2 mm thick, is placed between the upper and lower portions of the molding tool. The complete tool is then evacuated to ensure complete filling of the microstructured cavities, and the polymer is heated to above its softening temperature. The softened polymer is pressed into the microstructured cavities, then cooled down below the softening temperature, maintaining the applied force to avoid shrinkage and sinking marks. Finally, the machine is opened and the microstructured part is demolded. The low flow rates and molding speeds associated with this process ensure that even the smallest details, in the 100-nm range, are nearly perfectly replicated.

Hot embossing is a proper replication technique even for small series. Hot-embossing machines are now available commercially from the German company Jenoptik Mikrotechnik for wafers of up to 6-inch diameter. Machines for larger areas and shorter cycle time are under development. The current limits of the process are aspect ratios of up to 50 and structure heights of up to some hundreds of microns with lateral dimensions in the range of some tenths of microns. Even details of some hundred nanometers can be replicated, but the replication area is limited to approximately 30 x 70 mm for high-aspect-ratio applications. For medium-range aspect ratios (up to 5), the admissible embossing areas are in the range of 100 x 125 mm. For hot embossing, the field of process modeling is as essential as for injection molding. However, it is still in its infancy.

16.2.4 Sacrificial Layer Technique

An integrated fabrication technology, based on a sacrificial layer technique,[11] was developed to obtain partly or totally movable microstructures as shown in Fig. 16.6. In a first process step, a thin (<5 µm) sacrificial layer of Ti is deposited by sputtering onto the substrate and patterned by photolithography and etching [Fig. 16.6(a)]. If electrical connection of the microstructures is important, e.g., in case of actuators, chromium-gold tracks can be patterned below the titanium as shown by Kunz et al.[12] Then the standard LIGA process is used: polymerization of the thick resist onto the substrate, exposure to synchrotron radiation through a precisely adjusted mask [Fig. 16.6(b)], development of the resist, and electroforming. After stripping the resist, the sacrificial layer is etched selectively against all other materials [Fig. 16.6(c)]. Well-known examples of movable LIGA structures are accelerometers,[13] turbines,[14] electrostatic motors,[15] and comb drives.[16]

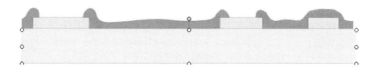

Figure 16.6(a) Substrate with patterned Cr/Au tracks for electrical connection and a patterned Ti layer as a sacrificial layer.

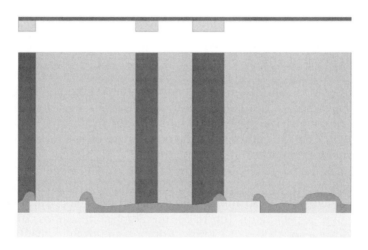

Figure 16.6(b) Substrate covered with PMMA; the X-ray exposure has to be performed aligned to a prepatterned substrate.

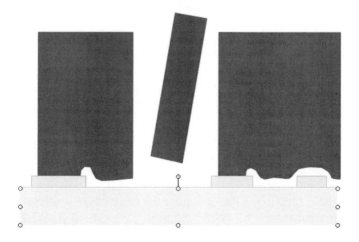

Figure 16.6(c) Filling of developed structures with metal and selective etching of Ti to release the structures from the substrate.

16.2.5 UV-LIGA Based on UV Lithography

Metallic spare parts, as well as mold inserts, can be fabricated by using conventional UV lithography and subsequent electroplating. The negative resist EPON® SU-8 is used for this process with the advantage for users of a standard chrome-quartz mask and a large functional area when used with a 6-inch wafer substrate. The maximum diameter for a mold insert (at the Forschungszentrum Karlsruhe) is 120 mm for circular shapes and 85 x 85mm side length for rectangular shapes. Application examples are capillary fluidic devices, lab-on-a-chip applications, and optical waveguides. The smallest structural resolution for this process is 5 µm, and the structure heights lie between 10 and 100 µm at the maximum aspect ratio of 3.5.[17] Figure 16.7(a) and (b) shows the SU-8 form of a microgear before electroplating. The mold shows rather smooth and steep sidewalls.

16.3 Application in Modular Micro-Optical Systems

Because a polymer, and very often the optically transparent PMMA, is typically microstructured using LIGA, the technology is well suited to applications in all fields of micro-optics. Striving for cost reduction through higher integration and on-going miniaturization, there are two basic technological approaches for micro-optical systems: a fully integrated fabrication sequence, and modular component ware. Deciding which approach to use depends on the number of pieces to be fabricated, the complexity of available processes and the respective yields, and the freedom in design and construction.[18] Each approach has its advantages. In the case of lower production numbers, and for the combination of several functional components (as, for example, the combination of micro-optical benches and actuators), modular design and fabrication is often the better choice. The processes are easier and safer, and working with pretested components will help to enhance the yield. This is a case of designing and manufacturing where LIGA perfectly matches the requirements.[19] Therefore, the applications described in this chapter concentrate on micro-optical systems for telecommunication and sensors that follow the modular approach. Pure optical benches, as well as complex systems based on MEMS or MOEMS techniques, are presented. At the end of the chapter, some non-optical LIGA applications are discussed that prove the flexibility of the process for diverse purposes.

16.3.1 Definition of a Modular Micro-Optical System

Generally, an "optical system" comprises a variety of functions in one unit: an optical function, and often an electrical and/or mechanical function.[18]

Figure 16.7(a) Nickel mold insert for a microgear made by UV-LIGA: 380 µm in diameter, 50 µm high; bearing shaft diameter 25 µm.

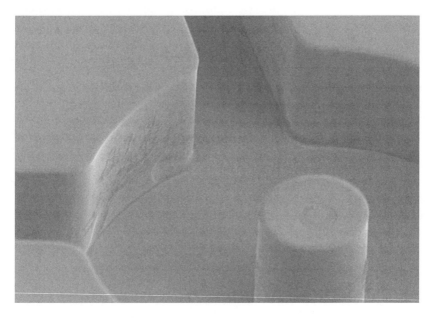

Figure 16.7(b) Close-up of the shaft detail in (a).

Examples for pure optical functions are light transfer, refraction and diffraction, imaging, filtering, beam splitting, and image superposition. Electrical or opto-electrical functions can be light generation, detection or analysis, electronic circuitry, and current/voltage supply to other functional components. Mechanical or electromechanical functions are provided by alignment structures or positioning aids, fixing and clamping structures, actuators, heaters, and coolers. An optical system is called "micro" when at least one of its components is fabricated by means of microtechnology. The overall size of the system or even of the diverse components need not necessarily be in the range of micrometers, but some structural detail may call for microtechnology (as described above as type A microstructure). Finally, a micro-optical system is defined as "modular" if its diverse functions are not fabricated in one common, fully integrated fabrication sequence and/or are not integrated on the same substrate. Here we consider an optical bench, e.g., fabricated with an adapted process on one substrate, and an actuator fabricated with another specific process on a second substrate, forming two modules that are finally put together.

16.3.2 Multifiber Connector from Polymer

As a first, relatively simple modular micro-optical system, we consider a multifiber connector that allows the parallel coupling of 16 optical fibers. It consists of two plastic pieces out of PMMA: one for the alignment of fibers and guide pins with five rows of highly precise alignment structures, the other for their fixation and protection. Elastic ripples in the sidewalls of the alignment structures, shown in a SEM photograph in Fig. 16.8(a), facilitate fiber and pin insertion. They also make the connector insensitive to variations in fiber diameter. The gap between the alignment structures decreases successively from the last row to the first row facing the opposite connector. This enables a very straightforward assembly and passive alignment of the fibers, without the need for micropositioning. Since the gap at the front face is 2 µm smaller than the fiber diameter, the fibers are clamped softly by the alignment structures, thus allowing easy handling during the on-going assembly. To bond both parts, UV-curing adhesive is filled into the device through a hole in the upper substrate. The glue spreads into the coupler by capillary forces alone. Finally, the front face is polished. The pins of both connectors to be coupled are inserted into a similarly precise coupling adapter, as indicated in Fig. 16.8(b). If the pins damage their respective holes in the adapter (for example, due to frequent reconnection), the adapter can be replaced easily, whereas the fibers and connectors can be maintained in their assembled form. The coupler is fabricated by means of plastic microinjection molding from PMMA. The molding tool was fabricated by LIGA technology and microprecision engineering as described in detail by

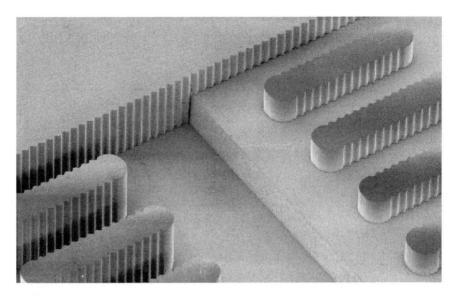

Figure 16.8(a) Rippled alignment structures for optical fibers; detail of the fiber connector.

Figure 16.8(b) Fully assembled multifiber connector with a common coupling adapter for protection of the connector pieces.

Wallrabe et al.[20] The connectors are commercialized under the name Rib-Con® by the Spinner Company in Munich, Germany.

Dunkel et al. describe a comparable connector concept as well as an injection molded coupler for which the tool was made in a similar manner;[21] however, the precision fiber grooves are obtained by V-grooves etched in silicon. They have extended this concept to an expanded beam coupler,[22] which has the advantage of decreased sensitivity to fiber misalignment and to impurities. In that case, the fibers are not coupled face to face; lens arrays are placed between them to expand the light beam for coupling and refocus it again into the output fibers, as sketched in Fig. 16.9(a). For this purpose, lens alignment structures are positioned in front of the V-grooves. Figure 16.9(b) shows a front face of such a micromolded coupler. The four smaller circular holes are the placeholders for the lenses; the two larger holes are for the guide pins. Finally, the coupler is equipped with fiber ribbon, a lens array of 500-µm-diameter lenses, and guide pins.

16.3.3 Heterodyne Receiver

The hybrid integration of light sources and detectors in optical benches underscores the idea of a micro-optical system. The totally different approaches of electro-optics and free-space optics call for a modular approach to avoid complex, difficult, and expensive processes. Figure 16.10(a) shows the setup of a heterodyne receiver, i.e., a wavelength filter for telecommunication. In this case, two incoming light beams need to be split and superimposed.[23] The signal beam and the beam from a local reference laser source are coupled into the system by means of monomode fibers. The light is collimated by ball lenses, then split into the two polarization states. Upon reaching the next optical surface, the beam of each polarization state is again split by 50% and is simultaneously superimposed with the component beams from the opposite light source. Each of the four final superimposed beams is detected by photodiodes. The system consists of a ceramic chip on which polymer alignment structures are patterned using LIGA technology. The fibers, ball lenses, prisms for the beam splitters, and diodes are separately manufactured components that are assembled in a fully passive manner on the chip. They are simply pushed toward the alignment structures and subsequently fixed by UV-curing glue. The accuracy is of the order of 1 µm. Since the altitude of the optical axis is defined by the radius of the ball lens (here, 450 µm), the fiber, which is only 125 µm in diameter, needs to be levered to the same height. This is done by a fiber mount, as shown in Fig. 16.10(b). The electrical connection of the diodes is achieved through gold tracks, which are prepatterned on the substrate by optical lithography and wet etching.

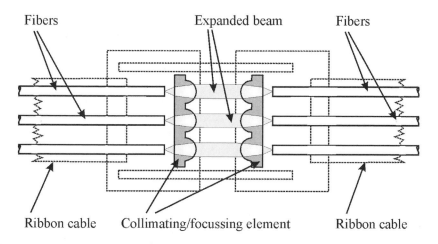

Figure 16.9(a) Schematic view of an expanded beam coupler in which the fibers are not coupled face to face but lens arrays improve coupling characteristics.

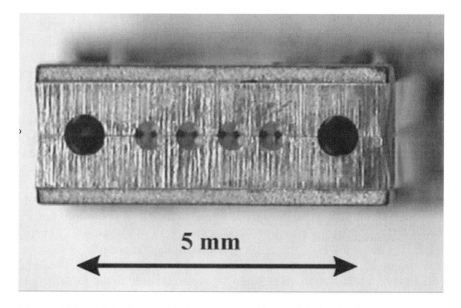

Figure 16.9(b) Injection-molded connector with small holes for lenses and large holes for guide pins.

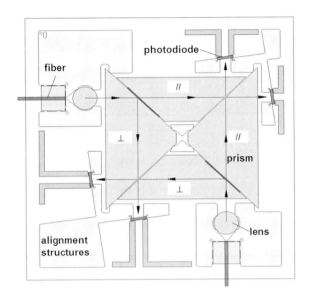

Figure 16.10(a) Schematic view of a heterodyne filter for infrared light with beam splitters and resuperposition on the diodes. The LIGA structures are used exclusively to position the optical components.

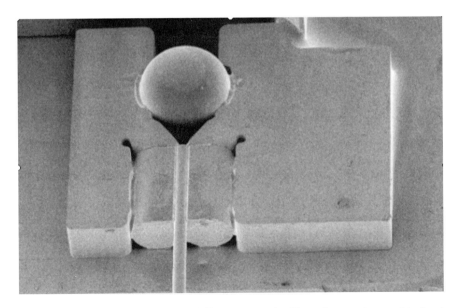

Figure 16.10(b) SEM of an optical fiber on top of a fiber alignment mount in front of a ball lens. The parts are assembled into the LIGA positioning structures.

16.3.4 Spectrometer

The LIGA-made microstructures of the heterodyne receiver have a purely mechanical function, i.e., they are used for alignment. However, they may also have a very distinct optical function, as is the case for microspectrometers,[24] one of which is illustrated in Fig. 16.11(a). The LIGA spectrometer was developed for visible and infrared light. It is produced by hot-embossing PMMA using a nickel molding tool. After the molding step, a gold layer is sputtered onto the PMMA components to make the optical surfaces reflective. The light is coupled into the device by an optical multimode fiber. The fiber is guided in the three-layer polymer waveguide toward a reflecting and self-focusing grating, which diffracts the light and thus separates it spatially into its constituent wavelengths. At the focal line, a 45-degree mirror is patterned that reflects the light onto a diode array that is aligned on top of the waveguide. The overall size of the device is 20.5 x 11.5 mm, which is fairly large. The thickness of the central microstructured layer is 125 µm, and the grating constant is 1.5 µm, with a depth of the teeth of only 0.2 µm. A photodiode array and some electronics are assembled upside-down on top of the spectrometer.

Spectrometers for use in the visible wavelength range are produced in volume by STEAG microParts. These modules, which are mainly applied to color measurement, form the key component of a portable measurement system for various applications. These measurement devices are developed, produced, and commercialized by companies that have many years of experience in their application areas. Three examples of color measurement systems are an instrument for the color determination of surfaces (Spectro-pen, Dr. Lange GmbH), an instrument for noninvasive measurement of the bilirubin concentration of newborn children [Fig. 16.11(b)] (Bili Check System, SpectRx Inc.), and a device to determine the color of diamonds (Diamond Colorimeter, Gran Computer Industries Ltd.).

16.3.5 Distance Sensor

Another application that relies on head-over assembly was used for a microdistance sensor that works on the basis of the triangulation principle;[25] i.e., the position of the spot on a sensor's detector is dependent on the distance to the object being measured. The distance sensor is divided into two functional units: a passive optical chip fabricated by LIGA technology and including curved mirrors, 45-degree mirrors, and alignment structures for cylindrical lenses; and an electro-optical chip with a laser, photo diode, and position-sensitive detector (PSD). The optical components on the LIGA chip are made from PMMA coated with an evaporated

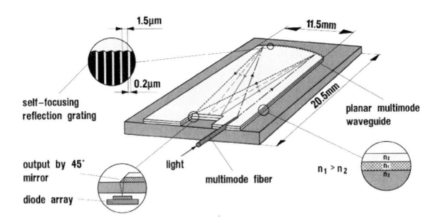

Figure 16.11(a) Schematic of a microspectrometer for visible light using PMMA as waveguiding material.

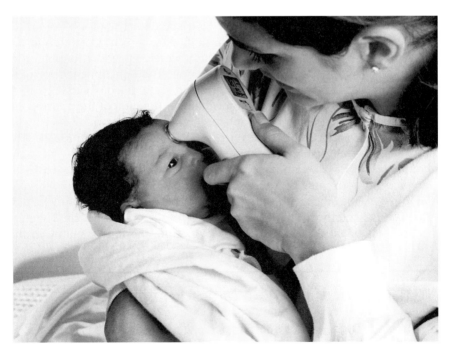

Figure 16.11(b) Measurement tool for the detection of bilirubin (a jaundice indicator) in newborn babies. (Manufactured by SpectRx Inc.)

gold layer. Both chips are again mounted head-over to form an optoelectromechanical sensor system. For easy assembly, two microspheres of glass are used as position aligners of the LIGA and the Si chip [Fig. 16.12(a)]. First, the microspheres are placed into pyramidal grooves on the Si base, thus centering them. Then the optical chip, with two cylindrical holes for the spheres, is passively aligned to the Si chip. The two chips are shown before the head-over assembly in Fig. 16.12(b). The Si chip is already mounted on a carrier in a TO8 housing.

16.3.6 Optical Cross-Connect with Rotating Mirrors

A fiber switch array combines a LIGA-fabricated micro-optical bench, as described above, with rotating mirrors arranged in a matrix scheme.[26] Figure 16.13(a) shows the concept for an $N \times N$ switch. The light from the fibers is collimated onto the mirror surfaces and refocused onto the output fibers by spherical glass lenses. Fibers and lenses are passively aligned with the help of fiber mounts and alignment stops in the optical bench. To achieve uniform insertion loses for all channels, the fibers are arranged so that all optical path lengths are the same. The mirrors are attached to the outer side of electrostatic micromotors that swivel the mirrors into and out of the beam path. Mechanical stops define the correct end position of the mirrors, which were designed as double mirrors to ensure a precise 90-degree reflection. The optical benches, static motor parts, and mirror stops are fabricated simultaneously using LIGA technology. Because of the electromechanical function of the motors, the structures are made from electroplated nickel on a sacrificial layer. The rotors with the attached double mirrors are manufactured similarly on a separate substrate and are manually placed above the stators. The fiber mounts out of PMMA are patterned on a third substrate. Figure 16.13(b) shows a close-up view of a fully assembled 2 x 2 switch matrix. Toward the left side, the fiber in the fiber mount and a lens can be distinguished. They are pushed toward the stop structures of the optical bench. The mirror in front of the lens is in the "on" position, the mirror below in the "off" position.

16.3.7 Oscillating Modulator for Infrared Light

Besides electrostatic actuation, LIGA technology also enables use of electromagnetic forces. In this case, the structures are electroplated in permalloy, which is a soft magnetic alloy of 80% nickel and 20% iron. A first application example is a modulator, i.e., a chopper for infrared

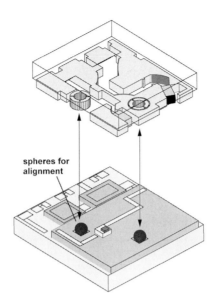

Figure 16.12(a) Two chips of a distance sensor are mounted head-over using glass balls as alignment aids.

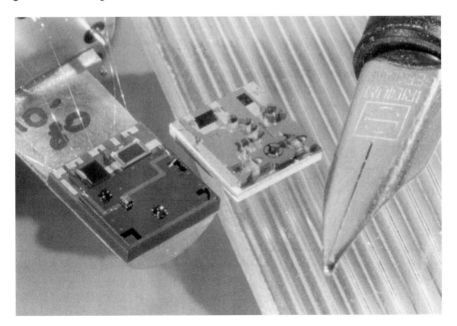

Figure 16.12(b) The LIGA-made optical bench and the electro-optical chip of a distance sensor before head-over assembly.

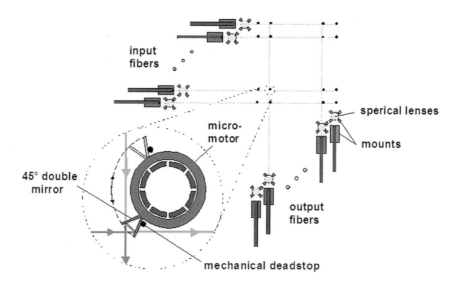

Figure 16.13(a) Schematic view of an optical cross-connect. Rotating mirrors at the beam junctions can be switched either to reflect the beam or to let it pass.

Figure 16.13(b) Optical bench with lens and fiber alignment structures and electrostatic micromotors made from nickel.

light, and is used for the suppression of noise in infrared spectrometers by using lock-in techniques.[27] The working principle becomes clear from the demonstrator, which is shown in Fig. 16.14(a). Alignment structures for input and output fibers are patterned onto a ceramic substrate in parallel with a movable shutter, which looks like a little hammer [Fig. 16.14(a)], and an electromagnetic actuator, as distinguished by the coil in the upper part of the picture. A current flowing through the coil generates a magnetic field, which is guided through the metallic yoke. Due to the decrease of the resistance of the magnetic circuit, the shutter is pulled backwards between the pole shoes of the circuit, thus enhancing the metallic cross section for the magnetic field. A sinusoidal current generates a periodic force field and hence an oscillation of the shutter. The yoke for the coil was manufactured on a separate substrate and released. The coil is wound manually and then assembled onto the substrate using alignment and clamping structures. For integration of the chopper in a spectrometer, the fiber alignment structures are skipped. The chopper is inserted head-over in the light input path of the spectrometer. This is facilitated by four stops on the edges of the yoke and the fixing block of the shutter, which were patterned simultaneously for this purpose. Figure 16.14(b) shows a close-up view of the very precise position of the shutter tip directly in front of the input fiber of a spectrometer in its alignment groove.

16.3.8 Laser Scanner for Barcode Reading Actuated by Electromagnetics

Here we consider a laser scanner for barcode reading[28] with an actuation principle similar to that for the chopper. Figure 16.15(a) shows a schematic drawing of the scanner. Again, a magnetic circuit with an assembled coil is fabricated from soft magnetic permalloy. A perpendicular mirror is attached to a freely suspended metal anchor that is part of the magnetic circuit. In this example, the magnetic resistance is minimized when the air gap in the circuit is minimized, which will happen when the anchor is deflected toward the ends of the pole shoes. As before, a periodic current will generate an oscillation of the mirror with twice the frequency of the current. The mirror is made from a 100-µm-thick Si wafer coated with a reflective Au layer and cut into 1.5 x 1.6 mm pieces by a wafer-dicing saw. It is placed into an alignment groove that was patterned in the anchor and is fixed by adhesive bonding. At a current of 20 mA, the scanner reaches a scan angle of 12 degrees in a nonresonant actuation mode. Figure 16.15(b) and (c) demonstrates the motion by means of stroboscopic photographs.

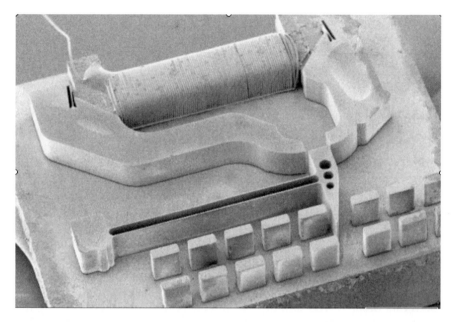

Figure 16.14(a) Electromagnetic actuator made from FeNi and operated as a chopper to modulate infrared light.

Figure 16.14(b) Close-up of the chopper shown in (a) with assembled fiber.

16.3.9 FTIR Spectrometer for Infrared Light

The third example is a Fourier transform infrared (FTIR) spectrometer. It is essentially a Michelson interferometer that consists of an optical bench for the passive alignment of the optical components, and an integrated actuator.[29] The optical bench and the actuator are made of permalloy with a height of 380 µm. The dimensions of the complete system are 11.5 x 9.4 mm². Figure 16.16(a) shows the developed FTIR spectrometer. The light that will be analyzed is delivered to the system via an optical fiber. It is collimated with a fused silica ball lens with a diameter of 650 µm. At the beam splitter, the light is separated into two rays: one reflects off the fixed mirror; the other travels toward the movable mirror. Both rays are reflected and travel back to the beam splitter, where they interfere with each other. The interference signal is recorded with an InGaAs PIN photodiode, which is sensitive in the wavelength range of 850 to 1700 nm. For the recording of the signal, the position of the movable mirror has to be determined very precisely. For this purpose, monochromatic laser light is used. The laser light is also delivered to the spectrometer through an optical fiber and propagates in a second optical channel parallel to the "white" light to be analyzed.

Here the chosen actuator is a variable reluctance motor. Figure 16.16(b) shows the movable plunger, which is attached to a set of four folded cantilever beams and is surrounded by fixed core structures, thus forming two magnetic circuits that are driven by two coils. Since variable reluctance motors only deliver pulling forces, the displacement length was

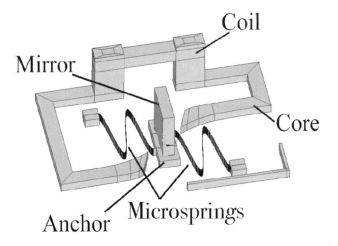

Figure 16.15(a) Schematic view of a microscanner with an electromagnetic actuator.

Figure 16.15(b) Tilting of a scanning mirror by the electromagnetic field.

Figure 16.15(c) Maximum displacement in the opposite direction.

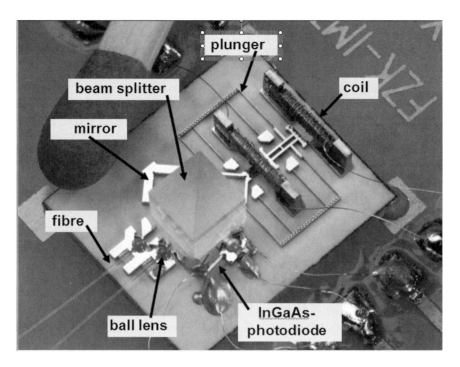

Figure 16.16(a) 3D view of a spectrometer including an optical bench (left side) and an actuator (right side).

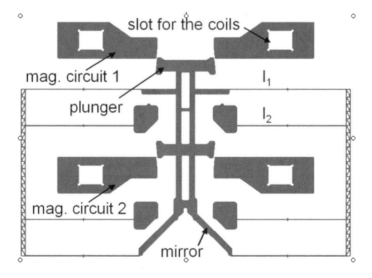

Figure 16.16(b) Schematic of the actuator and the magnetic circuits. The coils are placed into the respective slots.

increased by using two coils, one for pulling the plunger toward the beam splitter, the other to pull it in the opposite direction. The separate coils are assembled into slots prepared in the fixed magnetic circuit.

For spectral analysis, the optical bench was equipped with all optical components [Fig. 16.16(a)], and laser light at a wavelength of 1540 nm was delivered through a monomode fiber that ensures a good collimation of the beam. In this first experiment, only one coil was activated, for which a displacement of the actuator of 54 µm was achieved, leading to a spectral resolution of 25 nm.

16.3.10 Ultra-High X-Ray Lenses in SU8

All micro-optical systems described above are used in the visible and infrared wavelength range. However, a very new "optical" application for LIGA structures arises from the development and fabrication of X-ray lenses, described here as a final example of micro-optics. Research in X-ray optics is triggered by a large number of potential applications, ranging from space-based X-ray telescopes and focusing or imaging systems for synchrotron radiation analysis to medical imaging. Several concepts for focusing X-rays have been implemented, such as X-ray mirrors, hollow capillaries, crystal optics, and Fresnel lenses. A more recent approach, first proposed[30] and demonstrated[31] in the 1990s, is the application of lenses for X-rays. Such lenses function as their counterparts for the visible spectral range by refraction of electromagnetic radiation. For X-rays, however, materials have refractive indices of slightly less than 1, typically $(1-10^{-5})$. This small difference between air and materials is generally exploited for total reflection of X-rays at mirror surfaces when irradiated under small glancing angles. X-ray lenses are based on the fact that air (or vacuum) is the medium with higher refractive index than surrounding matter. A hole in a piece of metal will therefore act as a cylindrical lens.[31] However, the hole will show a very large focal length due to the small differences in refractive indices. For achieving useful focal lengths with X-ray lenses, many individual lenses are arranged along the optical axis. Various fabrication methods for X-ray lenses have been proposed, including mechanical machining of individual lenses and Si etching.

Recently, modified LIGA techniques were used to produce X-ray lenses with superior performance.[32] The lithographic process allows for producing, in parallel, several different lens systems, e.g., for different wavelength ranges and/or focal lengths. Furthermore, after making the X-ray mask, many lens systems can be printed with that mask. The lateral geometries can be freely chosen for optimization of optical properties and correction of aberrations.

Two materials have been used for the matrix in which the air lenses are embedded: SU-8 epoxy for lower photon energies 5 to 50 keV and electroplated Ni for higher energies >40 keV. A major breakthrough for X-ray lenses was the use of negative tone SU-8 resist that can be patterned to high accuracy with deep X-ray lithography. When exposed to X-rays, the polymer chains cross-link and the unexposed volume is removed in the development process. The cross-linked areas, however, are rather stable on further irradiation. They can be directly used for X-ray lenses, even when exposed to high intensity X-ray beams.

A large variety of different lens designs [Fig. 16.17(a)], for linear as well as double focusing, have been tested recently at various synchrotron radiation facilities, and the results have been compared to theoretical calculations. For the ESRF facility in Grenoble, dimensions of the focal spot are as small as 300 to 900 nm, limited only by the demagnified X-ray source size, by diffraction, and by fabrication inaccuracies [Fig. 16.17(b)]. The gain in intensity as compared to a pinhole with the same dimensions was found to be approximately 5000.[33]

16.4 Mechanical Applications

16.4.1 Cycloid Gear System

A simple shadow casting by deep X-ray lithography results in patterns with high-quality vertical walls, and hence in structures with high aspect ratios. However, the versatility requirements for current micropart applications include more complex 3D geometries and multilevel configurations. To meet such a diversity of design requirements, scores of efforts have been made to add further latitude to the simple mask pattern transfer by X-ray lithography by multilevel processes.

With regard to a sequential approach, additional processes are necessary between each conventional lithography cycle, such as replanarization after electroplating, as well as a mask-alignment procedure. These processes add another constraint to the maximum attainable fabrication precision, and it is generally considered to be in the range of ±5 µm with regard to the typical resist thickness of several tens to hundreds of microns.

One example of a 3D LIGA mechanical structure is a complete gearbox. A cycloid gear formation is advantageous for miniaturized systems, mainly because it does not need a planar arrangement of several gears. However, the layered formation of gear components requires an accurate planarization and height controllability of the microparts, for which sequential planar formation by the multiexposure LIGA technique can be

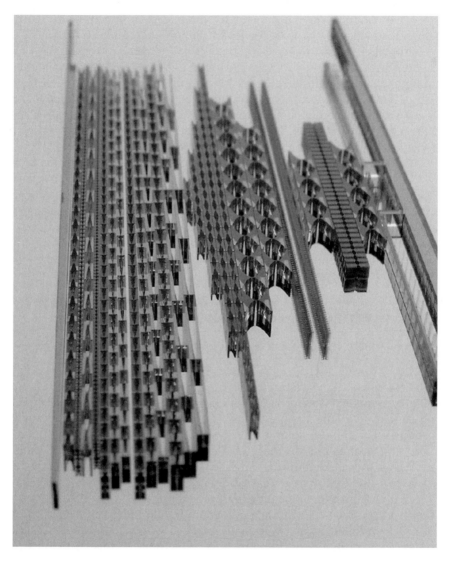

Figure 16.17(a) 20 X-ray lens systems printed onto a substrate. The optical axis is along the individual "rows." Lens details are displayed in the insert.

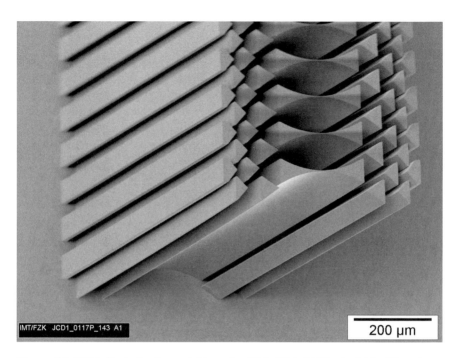

Figure 16.17(b) Part of a "mosaic" lens for 15 to 20 keV. The height of the structures is 1000 µm. The lenses are tilted with respect to the surface. A second lens system following this one is tilted to the other side. Using such an arrangement, a point focus can be achieved rather than a line focus.

quite fitting. Figure 16.18(a) and (b) shows a prototype of such a 2-mm-diameter microcycloid gear system.[34] The entire gear train consists of a casing and three vertically stacked disks and gears. Each part is composed of three different levels. The first level, 40 µm high, was fabricated by UV lithography; the second and third levels, 195 and 250 µm high, respectively, were processed by aligned deep X-ray lithography. The alignment error between two X-ray lithography processed layers was measured and found to be within the ±5 µm range. As a result of the height control process by mechanical surface machining, the deviation of structural height has been maintained to within the ±3 µm range for the UV lithography-processed structures, and to ±10 µm for the deep X-ray lithography (DXL)-processed structures.

For assembly of the gears, the three disks to be stacked were initially positioned just one on top of another using a vacuum gripper under a stereo microscope. The correct alignment was done afterward by introducing short pieces of optical fiber (125 µm in diameter) as alignment pins into the respective holes (135 µm in diameter) in the disks. Finally,

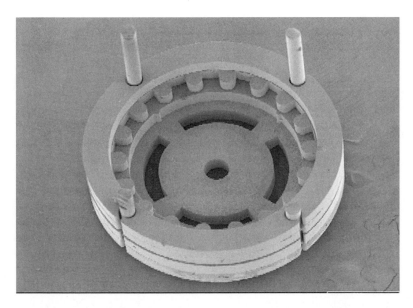

Figure 16.18(a) The casing of a microcycloid gear system fabricated by multiexposure X-ray lithography combined with UV lithography.

Figure 16.18(b) Micro-cycloid gear system with assembled central gear. The thin cylinders on the outer ring are pieces of optical fiber used as an alignment aid.

dynamic tests of the gear system were successfully conducted with a mechanical torque input by an electrical motor. A proper rotational speed reduction was observed in the operational input range of 3 to 1500 rpm with the designed gear ratio of 18.

16.4.2 LIGA Gyroscope

Figure 16.19(a) shows the schematic principle of the LIGA gyroscope.[35] It is based on a single vibrating inertial mass oscillating parallel to the substrate in the *x*-direction. The mass is driven by an electrostatic comb actuator, which is integrated into a frame surrounding the inertial mass detector. A rotation about the *z*-axis and hence perpendicular to the substrate generates a Coriolis force in the *y*-direction, leading to a *y*-deflection of the mass with respect to the actuator frame. The *y*-deflection is measured by a capacitive detector, which is integrated into the inertial mass.

An accelerometer is patterned next to the gyroscope for measuring and compensating the linear accelerations. It features almost the same configuration as the inertial mass of the gyroscope. Hence, both sensors have the same electromechanical properties and show the same dependence on inertial force, deflection, and temperature.

A detail of the inertial mass and the detector is shown in Fig. 16.19(a). Fingers of the comb electrode are attached to the frame of the inertial mass. They mesh with corresponding electrodes, which are fixed to the substrate. The finger pairs form a capacitor, whose gap width changes when the frame is deflected in the *y*-direction. With increasing temperature, the freely suspended frame expands with the thermal expansion coefficient of electroplated nickel; the fixed electrodes expand as well, although with the much lower coefficient of the Al_2O_3 substrate. This arrangement results in shrinkage of the gap width in capacitors C1 and C2 [see Fig. 16.19(a)]. To compensate for this effect, capacitors C3 and C4 are designed in such a way that the second gap will grow with increasing temperature.

The sensor is fabricated on an Al_2O_3 ceramic substrate by deep X-ray lithography and Ni electroplating. The overall size of the device is 9 x 4 mm² with heights of 100 or 250 µm and capacitance gap widths of 4 or 6 µm, respectively. An open framework architecture (trusslike) was chosen for the large frame structures; as such, the design reduces the total mass to be actuated and moved [Fig. 16.19(b)]. The gyroscope is characterized by an excellent linearity error of only 0.05% full-scale for a measurement range of ± 400 deg/sec. The resolution is 0.04 deg/sec at a bandwidth of 27 Hz. The variation in sensitivity is 0.7% for a temperature range of −40 to +80°C.

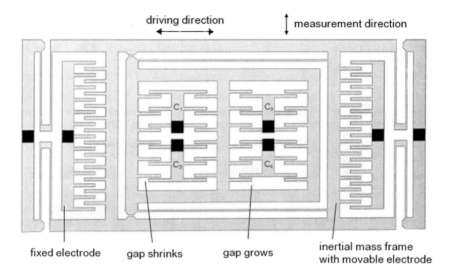

Figure 16.19(a) Detail of the electrostatic comb serving as a detector for the Coriolis force. The design of the comb structure is optimized for temperature compensation.

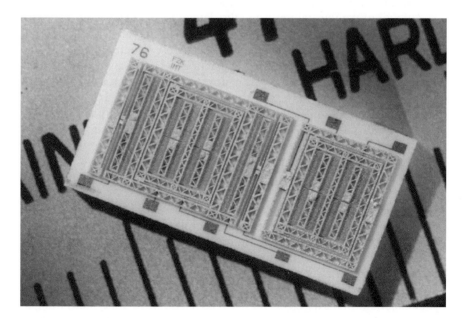

Figure 16.19(b) Gyroscope, on the right side of which an accelerometer is patterned to distinguish a linear acceleration from a Coriolis force due to a rotation.

16.4.3 Microturbines for Cardiac Catheters

A hydrodynamic rotating microdrive provides an alternative to electrostatic or electromagnetic micromotors when electricity has to be avoided either for medical reasons or due to risk of explosion hazards.[36] Therefore, a microturbine has been developed that is normally operated with water but may also run on gas. The turbine was designed for integration in a cardiac catheter to drive a cutter for arterial plaque removal.

The turbine [Fig. 16.20(a)] is assembled out of various microcomponents: an electroplated metallic rotor with radial blades, mounted on a shaft (distal catheter end); a nozzle plate made from PMMA by X-ray lithography surrounding the rotor; and a housing, which is also made from PMMA (using injection molding), which forms the distal tip of the cardiac catheter. The fluid is delivered from the back (proximal) side through the outer annular tube of the housing. It is injected into the rotor through the nozzle plate and is drained back through the inner annular ring. For medical reasons, the central tube cannot be used for fluid delivery. A so-called guide wire is threaded through this tube to place the catheter at the right position. The nozzle geometry (15-μm minimum width) is one of the parameters that define the power and the torque as well as the operation pressure.

The assembled turbine system is shown in Fig. 16.20(b). For better visibility of the nozzles, it is coated with gold. To ensure medical compatibility, the turbine is operated in a balanced salt solution at 37°C. At 88 ml/min., the pressure drop across the turbine is about 7 bar. The turbine achieves an idling speed of 1250 rev/sec and a maximum power of 45 mW, corresponding to a torque of 8 μNm at 900 rev/sec.

16.4.4 Watch Pieces Made by UV-LIGA

The Swiss company Mimotec SA uses a UV-LIGA based process for the fabrication of highly precise and customized watch pieces that cannot be manufactured using classical technologies such as wire electric discharge machining (WEDM). The total design freedom in the X-Y plane allows watch pieces to be improved with the addition of geometrical patterns of any shape. Numbers, letters, or notches directly integrated into the component help to identify different components with a similar macroscopic shape.[37] They may also provide a functional and aesthetic add-on value. For the Mimetal® process, thick layers of SU-8 resist are exposed to ultraviolet light, and nickel or nickel phosphorous is electroplated into

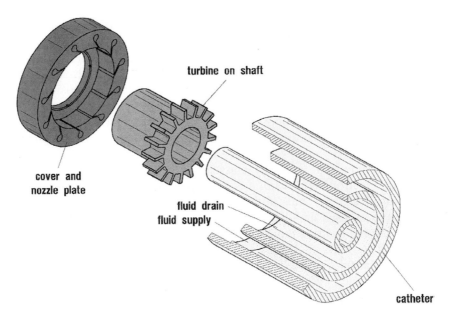

Figure 16.20(a) Schematic view of a turbine for cardiac catheters; the outer diameter is 2.5 mm.

Figure 16.20(b) Assembled turbine system. For visualization, the sealing cover plate is still missing.

the respective polymer form.[38] After stripping the resist, the watch pieces are released from the substrate by wet etching and are available for watch assembly.

Figure 16.21(a) is an image of a so-called cam of the days. Letters indicate the position of the cam as a function of the day of the week. Such indications help the positioning of the hand on the cam so that the hand and dial are synchronized. Figure 16.21(b) shows the toothed sector for actuating a pinion combined with a spring blade for reset actuation, and a cam reader in a single part. The teeth of the sector are carved to serve as a spring that compensates for play.[39] The smallest opening in the teeth is 25 µm wide. In this example, compensation is used to avoid having the second hand show an oscillation-like behavior in resetting the seconds back to zero. Play would induce a shift visible on the dial. Figure 16.21(c) shows a number of bridges for a mechanical movement; all have the same function but different designs.

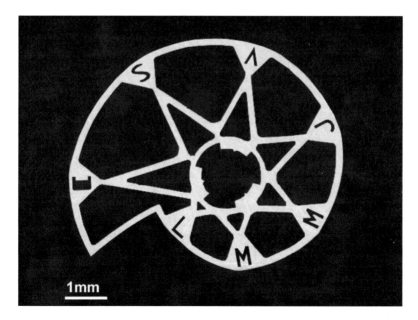

Figure 16.21(a) Customized cam of the days. Positions of the cam are indicated by the initial of the day (in French). Nickel phosphorus, 150 µm thick.

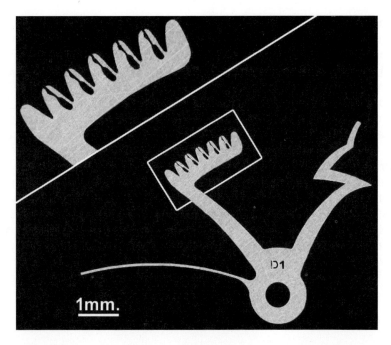

Figure 16.21(b) Toothed sector with identification number, integrated spring blade, cam reader, and compensation for play. (Picture courtesy of Agenhor SA.)

Figure 16.21(c) Bridges of watch mechanical movement. Nickel, 400 µm thick.

16.5 Outlook

Originally, LIGA was developed in Germany for a very specific purpose, namely, to manufacture nozzles for the separation of uranium isotopes in large enrichment facilities. The requirements on the technology were high aspect ratio and minimum feature sizes in the µm range. With the end of nuclear energy research in Germany, LIGA technology survived as a spinoff. Its potential became evident in the 1990s when numerous examples for structures and devices were presented. The lead in MEMS technologies, however, was assumed by Si micromachining and not by LIGA, in spite of its technical superiority for many applications. The main reasons for this were the required technical infrastructure, process know-how, and cost issues. Si-based technologies could exploit the vast technology base developed for chip making, with billions of dollars in existing investment. LIGA, on the other hand, was new and confined to research laboratories. Furthermore, as a key process step, LIGA required access to a synchrotron radiation facility, typically located at research laboratories, and was thus often unacceptable for industries establishing manufacturing plants. Nevertheless, several industrial LIGA products have been launched by industries, and synchrotron radiation facilities are currently widely used for microfabrication (see Table 16.2). This is a demonstration of the strength and superiority of LIGA.

Table 16.2 Synchrotron Radiation Centers with Facilities for Deep X-Ray Lithography and LIGA, and Commercial Providers of LIGA Services and Products

Synchrotron Radiation Facilities with LIGA Activities		Commercial LIGA Providers
Australia:	Australian synchrotron (under construction)	ANKA GmbH, Germany
Brazil:	LNLS	
Canada:	CLS (under construction)	Axsun Corp., USA
France:	SOLEIL (under construction)	
Germany:	ANKA, BESSY-II	MEZZO Corp., USA
Italy:	Elettra	
Japan:	NewSubaru, AURORA	Anwenderzentrum at
Korea:	PSL	BESSY, Berlin, Germany
Russia:	VEPP-III	
Singapore:	SSLS	
Sweden:	MAX-II	
Taiwan:	NSRRC	
USA:	ALS, APS, CAMD, NSLS-II, SPEAR-III, SRC	

As a consequence of the LIGA weaknesses—a long process sequence with highly specialized individual processes, the requirement for access to synchrotron radiation sources, and cost and throughput issues—the LIGA developers focused their efforts in past years on quality management, stabilization of processes, standards, and cost. As a consequence, LIGA laboratories like the IMT at Forschungszentrum Karlsruhe and IMM in Mainz introduced rigid quality management systems following ISO 9001:2000. Small-volume production was demonstrated and the reliability of processes was dramatically increased. The concept of delivering mold inserts on demand with quality as specified was verified, and at least small and medium-sized companies accept increasingly that they can now simply purchase a mold insert for production in their plants.

Furthermore, new concepts are currently tested where each component is produced directly by X-rays, similar to semiconductor manufacturing, at acceptable cost. This so-called direct LIGA approach uses only the first (LI) or the first two (LIG) steps of LIGA, thus reducing the complexity of mold fabrication and replication. The major concerns then are cost and throughput. Two highly encouraging approaches are being pursued: New resists such as SU-8 reduce the exposure times by one to two orders of magnitude. In addition, the ability to expose very large areas in one exposure step enables parallel processing of several wafers. The latter concept has already led an industrial consortium to propose a mostly automated fabrication line at the ANKA facility at Forschungszemtrun Karlsruhe.

Besides these efforts in making LIGA acceptable as a manufacturing technology for a large variety of industrial products, cutting-edge LIGA research remains a hot topic. Goals include research in new materials, new replication techniques, and new lithography approaches exploiting the short-wavelength nature of X-rays. Several questions are currently being addressed, such as "How small can we really get?" "Can we overlap our top-down technologies with the typical bottom-up approach in nanotechnology?" "At which dimensions?" Making devices much smaller than is possible today will open up entirely new fields in research and applications. Examples may include photonic crystals, X-ray microscopes, and devices for research with individual molecules, cells, and more.

Acknowledgments

This chapter summarizes the work of many individuals, most of whom are researchers at the Institute for Microstructure Technology of Forschungszentrum Karlsruhe. We would specifically like to thank Dr. Jürgen Mohr for his leadership in the LIGA activities at IMT. We further thank Prof. Jan Gerrit Korvink from the Institut für Mikrosystemtechnik,

IMTEK, at Freiburg University, for encouraging us to write this overview and for his support.

References

1. E.W. Becker, W. Ehrfeld, P. Hagmann, A. Maner, and D. Münchmeyer, "Fabrication of microstructures with high aspect ratios and great structural heights by synchrotron radiation lithography, galvanoformung and plastic molding (LIGA-process)," *Microelectr. Eng.*, 4:35–56 (1986)
2. H. Lorenz, M. Despont, N. Fahrni, N. LaBianca, and P. Renaud, "SU-8: a low-cost negative resist for MEMS," *J. Micromech. Microeng.*, 7:121–124 (1997)
3. L. Wang, Y. Desta, R. Fettig, H. Hein, P. Jakobs, and J. Schulz, "High resolution X-ray mask fabrication by a 100 keV electron-beam lithography system," *J. Micromech. Microeng.*, 14:722–726 (2004)
4. H. Hein, A. Janssen, B. Matthis, P. Jakobs, S. Achenbach, D. Maas, P. Henzi, and P. Meyer, "Arbeitsmasken für die Röntgentiefenlithographie, 5," *Statuskolloquium des Programms Mikrosystemtechnik*, Forschungszentrum Karlsruhe, Wissenschaftliche Berichte FZKA, Germany (2004), vol. 6990, pp. 183–184
5. F.J. Pantenburg, S. Achenbach, and J. Mohr, "Influence of developer temperature and resist material on the structure quality in deep X-ray lithography," *J. Vac. Sci. Technol.*, B16(6):3547–3551 (1998)
6. O. Tabata, N. Matsuzuka, T. Yamaji, S. Uemura, and K. Yamamoto, "3D fabrication by moving mask deep X-ray lithography (M^2DXL) with multiple stages," *IEEE MEMS 2002, 15th International Conference on Micro Electro Mechanical Systems, Las Vegas, Nevada* (2002), pp. 180–183
7. S. Hafizovic, Y. Hirai, O. Tabata, and J.G. Korvink, "X3D: 3D X-ray lithography and development simulation for MEMS," *Proc. Transducers '03, Boston* (2003), pp. 1570–1573
8. M Guttmann and K. Bade, "Electrochemistry in molecular and microscopic dimensions," *Galvanotechnik*, 11:2962–2968 (2002)
9. T. Hanemann, M. Heckele, and V. Piotter, "Current status of micromolding technology," *Polymer News*, 25(7):224–229 (2000)
10. D. Kauzlaric, A. Greiner, J.G. Korvink, "Modelling Micro-Rheological Effects in Micro Injection Moulding with Dissipative Particle Dynamics," *Nanotechnology Conference and Trade Show 2004*, vol. 2, Boston, Massachusetts (2004), pp. 454–457
11. J. Mohr, C. Burbaum, P. Bley, W. Menz, and U. Wallrabe, "Movable microstructures manufactured by the LIGA-process as basic elements for microsystems," *Micro System Technologies 90, 1st International Conference on Micro, Electro, Opto, Mechanical Systems and Components* (H. Reichl, ed.), Springer Verlag, Berlin, Germany (1990), pp. 529–537
12. T. Kunz, M. Kohl, A. Ruzzu, K. Skrobanek, and U. Wallrabe, "Adhesion of Ni-structures on Al2O3 ceramic substrates used for the sacrificial layer technique," *Microsyst. Technol.*, 6:121–125 (2000)

13. M. Strohrmann, F. Eberle, O. Fromheim, W. Keller, O. Krömer, T. Kühner, K. Lindemann, J. Mohr, and J. Schulz, "Smart acceleration sensor systems based on LIGA micromechanics," *Microsystem Technologies 94, 4th International Conference on Micro Electro, Opto, Mechanical Systems and Components* (H. Reichl, ed.), Springer Verlag, Berlin, Germany (1994), pp. 753–762
14. M. Himmelhaus, P. Bley, J. Mohr, and U. Wallrabe, "Integrated measuring system for the detection of the number of revolutions of LIGA microturbines in view of a volumetric flow sensor," *J. Micromech. Microeng.*, 2:196–198 (1992)
15. U. Wallrabe, P. Bley, B. Krevet, W. Menz, and J. Mohr, "Design rules and test of electrostatic micromotors made by the LIGA process," *J. Micromech. Microeng.*, 4:40–45 (1994)
16. M. Kohl, J. Göttert, and J. Mohr, "Verification of the micromechanical characteristics of electrostatic linear actuators," *Sensors and Actuators*, A53: 416–422 (1996)
17. S.J. Chung and H. Hein, "Nickel mould insert fabrication by using conventional optical lithography and electroforming up to 120 mm diameter, 4," *Statuskolloquium des Programms Mikrosystemtechnik*, Forschungszentrum Karlsruhe, Wissenschaftliche Berichte FZKA, Germany (2000), vol. 6423, pp. 195–196
18. U. Wallrabe and J. Mohr, "Modular Microoptical Systems for Sensors and Telecommunication," *Sensors Update*, vol. 12 (H. Baltes, J. Korvink, and G. Fedder, eds.), Wiley-VCH (2003), pp. 143–174
19. J. Mohr, A. Last, U. Hollenbach, T. Oka, and U. Wallrabe, "A modular fabrication concept for micro optical systems," *IEEE/LEOS International Conference on Optical MEMS 2001, Okinawa, Japan* (2001), pp. 77–78
20. U. Wallrabe, H. Dittrich, G. Friedsam, T. Hanemann, J. Mohr, K. Müller, V. Piotter, P. Ruther, T. Schaller, and W. Zißler, "Micromolded easy-assembly multi-fiber connector: RibCon.," *Microsyst. Technol.*, 8:83–87 (2002)
21. K. Dunkel, H.-D. Bauer, W. Ehrfeld, J. Hoßfeld, L. Weber, G. Hörcher, and G. Müller, "Injection-moulded fibre ribbon connectors for parallel optical links fabricated by the LIGA technique," *J. Micromech. Microeng.*, 8:301–306 (1998)
22. J. Schulze, W. Ehrfeld, J. Hoßfeld, M. Klaus, M. Kufner, S. Kufner, H. Müller, and A. Picard, "Parallel optical interconnection using self-adjusting microlenses on injection molded ferrules made by the LIGA technique," *EUROPTO Conference on Design and Engineering of Optical Systems, Berlin, Germany*, SPIE (1999), vol. 3737, pp. 562–571
23. P. Ziegler, J. Wengelink, and J. Mohr, "Passive alignment and hybrid integration of active and passive optical components on a microoptical LIGA bench," *3rd International Conference on Micro-Opto-Electro-Mechanical Systems, MOEMS'99, Mainz, Germany* (1999), pp. 186–189
24. P. Krippner, T. Kühner, J. Mohr, and V. Saile, " Microspectrometer system for the near infrared wavelength range based on the LIGA technology," *Micro- and Nanotechnology for Biomedical and Environmental Applications* (R.P. Mariella, Jr., ed.), SPIE (2000), vol. 3912, pp. 141–149

25. T. Oka, H. Nakajima, A. Shiratsuki, M. Tsugai, U. Wallrabe, U. Hollenbach, P. Krippner, and J. Mohr, "Development of a micro optical distance sensor with electric I/O interface," *The 11th International Conference on Solid-State Sensors and Actuators, TRANSDUCER'01, Munich, Germany* (2001), pp. 536–539
26. A. Ruzzu, D. Haller, J. Mohr, and U. Wallrabe, "Optical 2 x 2 switch matrix with electromechanical micromotors," *MOEMS and Miniaturized Systems* (M.E. Motamedi and R. Göhring, eds.), SPIE (2000), vol. 4178, pp. 67–76
27. P. Krippner, J. Mohr, and V. Saile, "Electromagnetically driven microchopper for integration into microspectrometers based on the LIGA technology," *Miniaturized Systems with Micro-Optics and MEMS* (M.E. Motamedi and R. Göhring, eds.), SPIE (1999), vol. 3878, pp. 144
28. B. Krevet, S. Hoffmann, M. Kohl, J. Mohr, and G. Oliva, "A miniaturized laser scanner in LIGA technology (MILS)," *ACTUATOR 2002 8th International Conference on New Actuators, Bremen, Germany* (2002), pp. 312–315
29. U. Wallrabe, J. Mohr, and C. Solf, "Miniaturized FTIR spectrometer with large travel electromagnetic linear actuator consuming low power," *ACTUATOR 2004, 9th International Conference on New Actuators, Bremen, Germany* (2004), pp. 321–324
30. T. Tomie, "X-ray lens," Japanese patent no. 1994000045288, priority 18.02.1994 (1996)
31. A. Snigirev, V. Kohn, I. Snigireva, and B. Lengeler, "A compound refractive lens for focusing high energy X-rays," *Nature*, 384: 49–51 (1996)
32. V. Nazmov, E. Reznikova, J. Mohr, V. Saile, A. Snigirev, I. Snigireva, M. DiMichiel, M. Drakopoulos, R. Simon, and M. Grigoriev, "Refractive lenses fabricated by deep SR lithography and LIGA technology for X-ray energies from 1 keV to 1 MeV," *AIP Conference Proceedings*, 705: 752–755 (2004)
33. V. Nazmov, E. Reznikova, A. Somogyi, J. Mohr, and V. Saile, "Planar sets of cross X-ray refractive lenses from SU-8 polymer," *SPIE Annual Meeting 2004, Denver, Colorado*, SPIE (2004), vol. 5539 (accepted for publication)
34. T. Hirata, S.J. Chung, H. Hein, T. Akashi, and J. Mohr, "Micro cycloid-gear system fabricated by multi-exposure LIGA technique," *SPIE International Symposium on Micromachining and Microfabrication 99, Santa Clara, California*, SPIE (1999), vol. 3875, pp. 164–171
35. K. Schumacher, O. Krömer, U. Wallrabe, J. Mohr, and V. Saile, "Micromechanical LIGA-gyroscope," *The 10th International Conference on Solid-State Sensors and Actuators, TRANSDUCERS'99, Sendai, Japan, June 99*, vol. 2, pp. 1574–157
36. U. Wallrabe, "Mikroturbinen als hydrodynamischer Kleinstantrieb," *F&M, Feinwerktechnik, Mikrotechnik, Mikroelektronik*, 9:646–649 (1998); published by Carl Hanser Verlag, München, Germany
37. European patent no. EP12254477A1
38. M. Despont, H. Lorenz, N. Fahrni, J. Brugger, P. Renaud, and P. Vettiger, "High aspect ratio, ultra thick, negative-tone near-UV photoresist for MEMS applications," *IEEE MEMS 1997, 10th International Conference on Micro Electro Mechanical Systems, Nagoya, Japan* (1997), pp. 518–522
39. European patent pending

17 Interface Circuitry and Microsystems

Piero Malcovati
Department of Electrical Engineering, University of Pavia, Pavia, Italy

Franco Maloberti
Department of Electronics, University of Pavia, Italy

17.1 Introduction

Sensing physical or chemical quantities is a fundamental task in information processing and control systems. A sensing element or transducer converts the quantity to be measured into an electrical signal, such as a voltage, a current, or a resistive or capacitive variation. The data obtained from the transducers then have to be translated into a form understandable by humans, computers, or measurement systems. An electronic circuit called a sensor interface usually performs this task. The functions implemented by a sensor interface can range from simple amplification or filtering to A/D conversion, calibration, digital signal processing, interfacing with other electronic devices or displays, and data transmission (through a bus or, recently, through a wireless connection, such as Bluetooth).

Very-large-scale integration (VLSI) technologies have been extensively used to make sensor interface circuits since the appearance of the first integrated circuit (IC) in the early 1960s. Most of the sensor systems realized so far consist of discrete sensors combined with one or more application-specific integrated circuits (ASICs) or commercial components on a printed circuit board (PCB) or hybrid board. Over the past few years, however, progress in silicon planar technologies has allowed miniaturized sensors (microsensors) to be realized by exploiting the sensing properties of IC materials (silicon, polysilicon, aluminum, silicon oxide, and nitride) or additional deposited materials (such as piezoelectric zinc oxide, sensitive polymers, or additional metallization layers).

When microsensors are realized using IC technologies and materials, it is possible to integrate the interface circuit and several sensors on the same chip or in the same package, leading to microsystems or micromodules.[1-6] The potential advantages of this approach are numerous: The cost of the measurement system is greatly reduced due to batch fabrication of both the sensors and the interface circuits; its size and interconnections are minimized; and its reliability is improved.

However, the choice of materials compatible with silicon IC technologies is limited and their properties are process-dependent. Therefore, integrated sensors often show worse performance than their discrete counterparts due to weak signals and to offset and nonlinear transfer characteristics; they increase demands on interface circuits.

Moreover, in several sensor applications (such as automotive, biomedical, environment monitoring, and industrial process control), the chip or the chips (in the same package) containing microsensors and interface circuits can be exposed to harsh environmental conditions, causing aging and degradation of on-chip electronic devices. This makes most circuit techniques, which rely on accurate component matching and complex analog functions, inconvenient for sensor applications.

Given these considerations, it is evident that microsensor interface features can be very diverse. They depend heavily on the quantity to be measured, the physical effect used, the system architecture, and the application. In any case, it is very important that the microsensor, the interface circuit, and often the package are designed together. Indeed, the optimum microsystem or micromodule is not necessarily obtained by interconnecting separately optimized sensors and interface circuits. Microsensor interface circuit design, therefore, requires specific and interdisciplinary knowledge as well as special techniques to achieve the reliability and performance demanded by the user.

Finally, profitable use of smart sensors in real products not only depends on good design; it is also related to a number of additional issues such as cost-effective production, packaging, and post-production testing. All these issues are considered in this chapter.

17.2 Microsensor Systems

There are two possible approaches for implementing microsensor systems: the microsystem approach and the micromodule approach. In the microsystem approach, the sensor and the interface circuitry are integrated on the same chip, as shown in Fig. 17.1. In this case, the whole system is realized using a fabrication process optimized for integrated circuits with a few compatible post-processing steps when necessary (typically, etching or deposition of materials). Therefore, the microsensor must be designed by taking into account the material characteristics (layer thickness, doping concentrations, and design rules) given by the standard IC process used (bipolar CMOS, BiCMOS) and any specific processing step has to be performed after the completion of the standard IC fabrication flow. Obviously, this situation reduces the degrees of freedom available for optimizing sensor performance, thus making the design more challenging. This approach also can raise cost and yield issues, especially when using modern technologies with small

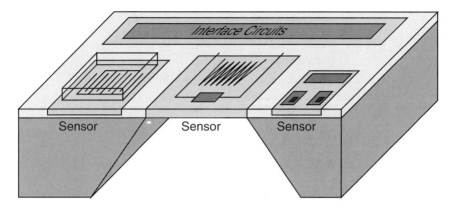

Figure 17.1 Microsensor system realized using the microsystem approach.

feature size (submicron technologies). In fact, while the silicon area occupied by the interface circuit is typically shrinking, together with the feature size of the technology, the sensor area in most cases remains constant. This is because it is determined by physical considerations, such as the mass of the structures or the angle of etched cavities, which are not changed by improvements in the technology. Therefore, while for integrated circuits the increasing cost per unit area is abundantly compensated by the reduction in the area, leading to an overall reduction of chip cost with the technology feature size, this might not be true for integrated microsystems. Moreover, a defect in the sensors may force us to discard the complete microsystem, even if the circuitry is working, thereby lowering the yield and again increasing the cost. (The yield for sensors is typically lower than for circuits.)

The microsystem approach also has considerable advantages. First, the parasitics due to the interconnections between the sensors and the interface circuitry are minimized and, more important, are well-defined and reproducible; this is very beneficial for system performance. In addition, the system assembly is simple, inexpensive, and independent of the number of connections needed, since all the interconnections are implemented during the IC fabrication process. Finally, when required, the use of the same technology allows us to achieve good matching between elements of the sensor and those of the interface circuitry, allowing accurate compensation of many parasitic effects.

In the micromodule approach, the sensors and the interface circuitry are integrated on different chips. They are included in the same package or mounted on the same substrate, as shown in Fig. 17.2. The interconnections between the sensor chip and the interface circuit chip can be realized with bonding wires or other techniques, such as flip-chip or wafer

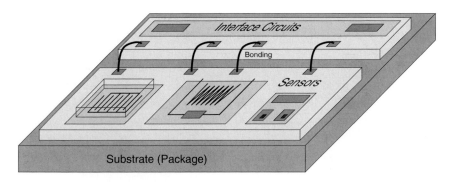

Figure 17.2 Microsensor system realized with the micromodule approach.

bonding. With this approach, the two chips can be fabricated with two different technologies, which are optimized for the sensors and the circuitry, respectively. Typically, expensive submicron technologies are adopted to realize interface circuitry, while low-cost technologies with large feature size and few masks are used for implementing sensors. Therefore, the sensor designer can adjust the material properties of the technology to optimize the performance of the devices, and the cost and yield issues mentioned for the microsystem approach are no longer a concern.

However, the micromodule approach also has a number of drawbacks. First, the assembling of the system can be quite expensive and a source of possible failures, and the number of interconnections allowed between the sensor and the circuitry is limited. Moreover, the parasitics due to the interconnections are some orders of magnitude larger, more unpredictable, and less repeatable than in the microsystem approach, thus destroying in many cases any improvements obtained in sensor performance by technology optimization. Finally, matching between elements of the sensor and elements of the interface circuitry cannot be guaranteed.

The advantages and disadvantages of both approaches are summarized in Table 17.1. From the considerations just discussed, it is evident that both approaches have merits. The choice of the approach to follow depends substantially on the application, the quantity to be measured, the kinds of sensors, the specifications of the interface circuits, and the available fabrication technologies, thus producing a number of tradeoffs that must be analyzed before a decision is made. For example, Fig. 17.3 illustrates the tradeoff between IC technology feature size and microsensor system cost. It can be clearly deduced from the indicative figures shown that if the sensor cost is constant, the microsystem approach is more convenient for technologies with large feature size, while the micromodule approach is more suitable for technologies with small feature size.

Table 17.1 Comparison Between the Microsystem and Micromodule Approaches

Microsystem Approach	Micromodule Approach
+ Reliability	+ Optimal yield
+ Minimal interconnection parasitics	+ Optimal processes for both sensors and circuitry
+ Simple and inexpensive assembling	+ Cost that scales with feature size
− Reduced yield	− Reliability
− Cost that does not scale with feature size	− Large interconnection parasitics
− Optimal process only for sensors	− Complex and expensive assembly

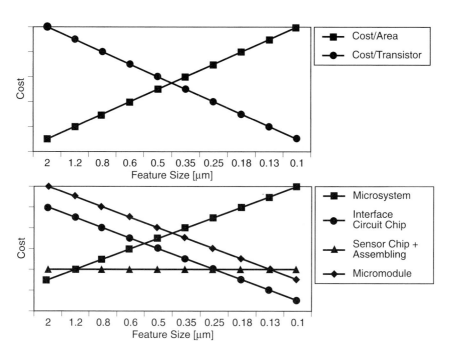

Figure 17.3 Tradeoff between IC technology feature size and integrated microsensor system cost.

17.3 Microsensor System Applications

As mentioned earlier, microsensor systems are currently replacing discrete sensors in a number of areas because of their inherent advantages; namely, batch fabrication, low cost, high reliability, and on-chip signal processing. In sensing applications, however, the environments where systems have to operate may be substantially different from the controlled and "mild" settings enjoyed by conventional integrated circuits. Sensors must be placed close to the quantity to be monitored; for example, in the human body, in an engine, or in an aggressive atmosphere. The operating environment of microsensor systems can consequently be both harsh and hostile. This introduces additional demands on the performance of integrated circuits, especially robustness. It is therefore useful to systematically consider the specific environmental conditions and requirements associated with the application areas where microsensors may be used (Table 17.2).

Table 17.2 Microsensors by Application Environment

Application	Quantity to Measure	
	Physical	**Chemical**
Automotive	• Acceleration • Flow • Magnetic field • Temperature • Pressure • Radiation • Images	• Exhaust gas composition • Combustion control • Air quality
Biomedical	• Pressure • Flow • Temperature • Acceleration • Magnetic field • Viscosity	• Biochemical species • Ionicity • Composition of body liquids
Household Appliances, Building Control, and Industrial Control	• Flow • Pressure • Acceleration • Magnetic field • Temperature • Radiation • Images • Viscosity	• Humidity • Gas composition • Liquid composition
Environmental	• Flow • Pressure • Temperature • Radiation • Images • Viscosity	• Gas composition • Liquid composition • Humidity • Biochemical compounds • Ionicity

17.3.1 Automotive Sensors

The number of sensors used in modern cars is growing because of government regulations and market expectations.[7-9] High fuel efficiency, clean emissions, improved safety, and comfort are the most important functions requiring on-board sensors. Moreover, because of the increasingly large number of parts per year involved, sensors for the automotive market must be extremely low priced. Automotive environmental conditions, summarized in Table 17.3, are especially difficult to address. Sensors can be placed inside the engine or outside the car, where they are subject to extreme temperature cycles, mechanical shocks, electromagnetic interference, and aggressive chemicals. Nevertheless, automotive sensors, like other car components, are required to be reliable and maintain their accuracy for 5 to 10 years.

It is evident from these considerations that automotive microsensor interface specifications are quite severe, making circuit design very challenging. An example of a microsensor system already introduced in the automotive market is the air-bag accelerometer.[10,11] In this case, special functions such as self-test, calibration, and bus interface have been included in the system.

Systems for engine emission and combustion control, based on microsensors, are also being developed.[12-17] They combine physical and chemical sensors that measure the state of the engine; for example, pressure in the cylinder, oxygen content in the exhaust gases, and engine rotation speed. These data are processed and used to regulate spark ignition and fuel injection, to optimize the performance of the engine in terms of fuel consumption and pollution.

Table 17.3 Microsensor Environment in Automotive Applications

Environmental Condition	Value
Temperature	• –40 to 150° C
Acceleration	• > 50 g
Vibrations	• > 15 g
Exposure to:	• Fuel
	• Brake fluid
	• Oil
	• Transmission fluid
	• Salt
	• Water
Electromagnetic Interference	• 200 V/m

17.3.2 Biomedical Sensors

Miniaturization and low power consumption are very important for biomedical and, in particular, implantable sensors. Microsensor technology is therefore very well suited for this type of application. Implantable microsystems or micromodules are indeed being developed for invasive monitoring of patients as well as for delivering electrical and chemical stimuli to the body.[18-23]

The specifications for implantable systems are very stringent. Low-voltage and low-power operation is imperative to ensure sufficiently long battery duration. The highest degrees of reliability and stability are required since substitution of a failed device requires surgical intervention and, in the worst case, a failure could be fatal. Packaging is crucial to protect the system from aggressive body fluids (such as blood) and vice versa. Finally, communication between the implanted system and the external world is limited, since only very few direct contacts are allowed.

Pacemakers are an important application of implantable microsensor systems. These devices are used to treat cardiac arrythmias such as bradycardia (slow heart rate) and tachycardia (rapid heart rate) by assisting the heart's natural pacemaking function with voltage pulses (5 to 10 V). Communication with the external world is provided by an inductive telemetry system operated at low frequencies or a transceiver, including an antenna, operated at radio frequencies. In some cases, the telemetry system also provides the power supply.[21,22]

17.3.3 Sensors for Household Appliances, Building Control, and Industrial Control

Electronic components, including sensors, are becoming more numerous and complex in modern white goods. Control systems for washing machines, ovens, air conditioning systems, alarm systems, vacuum cleaners, industrial process control, and so forth all require sensing elements. Sensor specifications in the industrial, household, and automotive markets are somewhat similar because of the large number of parts per year involved and the comparable environmental conditions (Table 17.3). In many cases, however, sensors for the household market must be particularly cheap, because the added value of the whole set of equipment is relatively low.

Recently, a number of microsensor systems for white goods have been proposed,[24-26] including flow sensors (for vacuum cleaners and air conditioning systems), temperature sensors (for ovens, washing machines, and air conditioning systems), magnetic sensors (for contactless switches), chemical sensors, and infrared sensors (for alarm systems).

17.3.4 Environmental Sensors

The release of various chemical pollutants into the atmosphere from industries, automobiles, and buildings, and into the hydrosphere causes environmental problems such as acid rain, greenhouse effect, ozone depletion, and polluted water. Microsensor systems for environmental monitoring are therefore being investigated.[27-29] The concentration of pollutants to be detected is usually quite small (in the parts-per-million range), and the selectivity of chemical microsensors is usually limited. This leads to tough specifications for on-chip interface circuits. Special nonlinear signal processing algorithms such as fuzzy logic and neural networks are widely used because by combining signals from several sensors, they allow significant information to be extracted, even in the presence of crosstalk and noise.

17.4 Interface Circuit Architecture

Figure 17.4 shows the block diagram of a generic microsensor interface circuit. This kind of architecture is valid for microsensor systems implemented with either the microsystem or the micromodule approach. In addition to the sensors, the system includes some analog front-end circuits (amplification and low-level processing), one or more analog-to-digital (A/D) converters, a digital signal processor, and an output interface.

It is well known that signal processing in the digital domain is more robust than in the analog domain thanks to the larger noise margin. Therefore, although processing is performed more efficiently with analog tech-

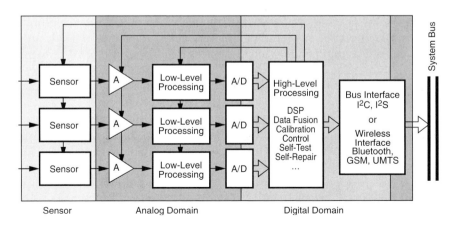

Figure 17.4 Block diagram of a generic microsensor interface circuit.

niques, in the presence of harsh environmental conditions, the trend is to minimize the analog section, moving the A/D converter toward the input and leaving complex processing to the digital section. This means that the overall performance of the system becomes strongly dependent on the quality of the A/D converter. Moreover, since less analog signal processing is performed, the bandwidth and dynamic range specifications for the A/D converter itself become more severe.

In the block diagram of Fig. 17.4, it is interesting to note the presence of feedback signals from the digital processor to the sensors and analog front-end, which allow the performance to be optimized by adjusting system parameters, depending on the output signal. For example, it is possible to correct the offset of the system by adjusting the reference voltages or to optimize the dynamic range by changing the gain of the analog front-end.

17.4.1 Requirements and Specifications

Sensor systems often emulate some kind of human sensing. Therefore, the properties of the electrical signals generated are related to the characteristics shown by natural perception: a relatively small bandwidth and quite a large dynamic range (often over a logarithmic span). Processing this kind of signal is usually not particularly difficult. However, when the dynamic range is large, noise contamination may become a problem. In this case, low-noise amplification and filtering, strictly limited to the band of interest, are necessary.

System specifications depend on both the function to be implemented and the application itself. Very often, however, a real-time response is required. In the case of human-like sensing, real-time means that a few milliseconds will always be available to process the signals; therefore, this is not a problem. By contrast, for control or recognition of fast-moving objects such as cars or planes, real-time can imply several megahertz of bandwidth.

Electronic equipment is becoming more and more portable, leading to battery-operated sensor systems with a small volume and weight. These features imply microsensor technologies, special packaging and assembling, low-voltage and low-power design methodologies, robustness, and shock resistance. For special cases (for example, in implanted devices, hearing-aids, or smart cards), it is also necessary to extract the power required for system operation from an electromagnetic flux irradiating the system itself. In this case, microactuators must be included in the system.

Before discussing the different sensor interface blocks in detail, it is useful to provide a sensor classification from the interface circuit point of view. The first parameter is the kind of output signal: current, voltage, or variation of resistance, capacitance, or inductance, as summarized in

Table 17.4. We can also consider different subclasses of sensors on the basis of signal level, bandwidth, and biasing requirements.

Table 17.5 provides critical parameters and tasks resulting from typical microsensor system specifications. It highlights challenging figures for signal level and bandwidth, and confers an immediate awareness of the difficulties to be faced in interface circuit design.

Table 17.4 Microsensors by Electrical Output Signal

Output Signal	Physical or Chemical Effect	Quantity to Measure	
		Physical	**Chemical**
Voltage	Pyroelectric	• Infrared radiation	• Reaction enthalpy
	Piezoelectric	• Strain • Acceleration • Pressure • Viscosity	• Gas composition • Liquid composition
	Thermoelectric	• Temperature • Infrared radiation • Flow • AC power	• Reaction enthalpy
	Hall	• Magnetic field	
Current	Lorentz	• Magnetic field	
	Photoelectric	• Radiation • Particles	• Gas composition • Liquid composition
	ISFET		• Ionicity • Liquid composition • Biochemical species
Capacitance	Dielectric permittivity variation		• Gas composition • Humidity • Biochemical species
	Dielectric thickness variation	• Strain • Acceleration • Pressure	• Humidity
Resistance	Piezoresistive	• Strain • Acceleration • Pressure	
	Magnetoresistive	• Magnetic field	
	Resistivity variation	• Temperature • Pressure • Flow • AC power	• Reaction enthalpy • Gas composition • Liquid composition

Table 17.5 Microsensor Interface Challenges

Classification	Parameter/Task	Challenges
Output Signal	• Analog voltage • Analog current • Change of resistance • Change of capacitance	• 10 µV • 10 nA • 1 mΩ • 10 aF
Operation	• Buffer • Amplification • Biasing • Filtering	• Low offset (100 µV) • Low noise (nV) • Low drift over time • 0.1% accuracy over temperature • Low cutoff frequency
Additional Functions	• Linearization • Self-test • Digital correction • Low power • Digital programming • Bus interface • Wireless interface	

17.5 Analog Front-End

The analog-front end of a microsensor interface circuit, directly connected to the sensing element, has to transform the raw sensor signal into something suitable for the subsequent A/D converter. The functions implemented in the analog front-end are typically limited to amplification and filtering, leaving more complex signal processing tasks to the digital section. Since the analog-front end is directly connected to the sensor, its features depend strongly on the kind of sensor considered. In this section, we consider in detail the characteristics of the analog front-end for sensors that provide voltage output, current output, or impedance variation.

17.5.1 Voltage Output

A wide variety of possible microsensors provide voltage output, typically including pyroelectric, thermoelectric, and piezoelectric devices; transistor-based temperature sensors; and magnetic sensors based on the Hall effect. Depending on specifications and applications, one of a variety of analog front-end circuits should be adopted. When the output signal of the sensor is sufficiently large (in the millivolt range) and noise is not a main concern, standard operational or instrumentation amplifiers, together with continuous-time or switched-capacitor filters, are sufficient. These solutions are quite conventional and require only customary design know-how. This task is made easier by libraries of standard analog cells, which are often provided by silicon foundries.

Situations in which the signal level approaches the limit given in Table 17.5 are much more challenging. Custom interface circuits must be designed, and special care must be taken to address noise and offset rejection. Bipolar transistors and high power supply voltages help to solve noise problems. Low-voltage supplies of 5 or 3.3 V (or even lower for portable equipment) in CMOS technology are an even more critical constraint.

Before discussing a specific interface implementation, we will recall the noise power spectral density of a simple MOS transistor. This is given by

$$S_{V_n} = \underbrace{\frac{8kT}{3g_m}}_{\text{Thermal}} + \underbrace{\frac{k_F}{2\mu f^\alpha C_{ox}^{k+1} WL}}_{\text{Flicker}}, \tag{1}$$

where g_m is the transconductance of the device, W and L are the gate dimensions, C_{ox} is the specific capacitance of the gate, k is the Boltzman constant, T is the absolute temperature, μ is the mobility of the channel, f is the frequency, and k_F is the flicker noise coefficient.

Equation (1) shows how to minimize noise contributions. The white thermal part is reduced by increasing the transconductance (at the expense of power consumption and DC gain, while improving speed), whereas flicker noise, which is higher than in bipolar devices because conduction takes place near the Si/SiO$_2$ interface, can be curtailed by increasing the gate area. When power and area consumption are critical parameters, however, noise should be reduced at the system level. When low-frequency noise (1/f) is the main concern, one of the following techniques is normally used:

- Auto-zero or correlated double sampling[30,31]
- Chopper stabilization[32, 33]

The auto-zero or correlated double sampling technique reduces the offset and low-frequency noise at the system level.[34,35] This technique requires sampled data operation. The low-frequency noise and the offset are, in fact, canceled in two steps, during two nonoverlapping clock phases. In the first step, the input signal (V_S) is disconnected from the circuit, and the spur signal (V_N) is sampled and stored. In the second step, V_S is connected to the circuit, and the previously stored spur component ($V_N\,z^{-1}$) is subtracted from the corrupted signal ($V_S + V_N$). If A is the gain of the circuit, the resulting transfer functions are ideally $H_S = A$ for the input signal, and

$$H_N = A\left(1 - z^{-1}\right) = 2A\sin\left(\frac{\pi f}{f_S}\right), \tag{2}$$

where f_S denotes the sampling frequency for the offset and the noise. Therefore, if $f_S \gg B$, and B denotes the bandwidth of the input signal, the in-band noise component is strongly attenuated. A practical implementation of this technique is reported by Malcovati et al.[36] The most important features of this circuit are summarized in Table 17.6.

The operating principle of the chopper stabilization technique is illustrated in Fig. 17.5.[37,38] A noisy operational amplifier with noise corner frequency f_{Corner} is preceded and followed by two identical modulators. The input signal, modulated with a square wave (Ck) having a frequency larger than f_{Corner}, is shifted into a region of the spectrum where the noise of the amplifier is dominated by the thermal component. After amplification, the signal is then modulated again and shifted back into the original band. The offset and the large low-frequency noise of the amplifier, superimposed on the signal by the amplification process, are also modulated and, therefore, pushed to a high frequency, where they can be removed by a subsequent low-pass filter.

This technique is often used in continuous-time systems, since it does not require sampled signals and allows quite good noise figures to be obtained. However, because of operational amplifier nonlinearities, the harmonics of the square wave may give rise to intermodulation products in the

Table 17.6 Features of the Low-Noise Operational Amplifier Based on the Auto-Zero Technique Reported by Malcovati et al.[38]

Parameter	Value
Application	Infrared thermoelectric sensor
Input referred noise	5 μV_{RMS}
Input referred offset voltage	1 μV
Bandwidth	10 Hz
Gain	Programmable

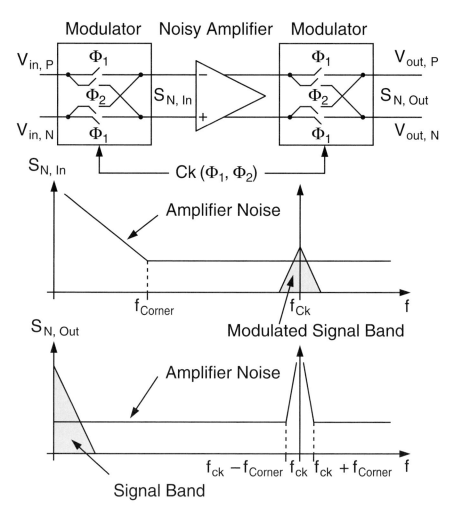

Figure 17.5 Operating principle of a low-noise operational amplifier based on the chopper stabilization technique.

signal band and degrade the noise performance of the circuit. Therefore, the operational amplifier has to be designed very carefully, possibly with a bandpass transfer function, to filter the offset and the high-order harmonics.

One of the best implementations of the chopper stabilization technique for microsensor applications is reported by Schaufelbühl et al.[39] and Menolfi and Huang.[40] The most important features of this circuit are summarized in Table 17.7.

Table 17.7 Features of the Low-Noise Operational Amplifier Based on the Chopper Stabilization Technique Reported by Schaufelbühl et al.[39] and by Menolfi and Huang[40]

Parameter	Value
Application	Infrared thermoelectric sensor array
Input referred noise	260 nV$_{RMS}$
Input referred offset voltage	600 nV
DC gain	77 dB
Bandwidth	600 Hz
Chopper frequency	11 kHz

17.5.2 Current or Charge Output

Sensors that provide a current output include particle detectors, optical sensors, ion-sensitive field effect transistors (ISFET), and magnetic sensors operated in the current (Lorentz) mode. The sensor currents can range from a few picoamperes to several microamperes. Charge preamplifiers are commonly used for very low currents (for example, in particle detectors).[41-48] Since a long integration time is required, the bandwidth of these circuits is quite small.

A typical structure of a charge preamplifier is shown in Fig. 17.6. Circuit operation is divided into two phases. First, the charge on the photodiode is integrated on capacitor C_F while the voltage across the

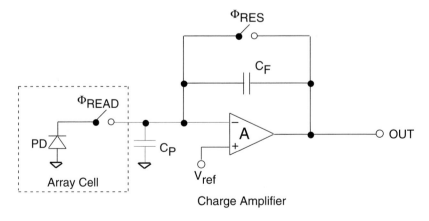

Charge Amplifier

Figure 17.6 Schematic of a typical charge amplifier.

photo-diode approaches V_{ref} (Φ_{READ} closed and Φ_{RES} open). Then, capacitor C_F is discharged (to be ready for a new sample), setting the photodiode voltage at V_{ref} (Φ_{READ} closed and Φ_{RES} closed). The error caused by the offset of the gain stage A is canceled because it is stored on C_P during Φ_{RES}.

In relatively wideband applications, nonintegrating transimpedance or current amplifiers should be used.[48-56] These circuits can achieve large amplification, but noise is a critical issue, especially when bipolar transistors are not available.

When the output current is sufficiently large, we can exploit the sensor to replace an active component (such as a current source) in an A/D converter or in a filter, thus minimizing the complexity of the system. In sampled data systems, the switched-current technique[57] can be used to implement auto-zeroed low-noise current amplifiers. In spite of some drawbacks that have not yet been fully solved (namely, charge injection and nonlinearity), this approach deserves further investigation, especially for low-voltage (battery-operated) applications.

As examples, we can consider the charge amplifier reported by Simoni et al.[58] and the current amplifier implemented in bipolar technology discussed by Bolliger et al.[59] The most important features of these two circuits are reported in Tables 17.8 and 17.9, respectively.

The schematic current amplifier designed for UV sensing applications[59] is shown in Fig. 17.7. The UV photodiode current and the DC bias

Table 17.8 Features of the Charge Amplifier Reported by Simoni et al.[58]

Parameter	Value
Application	CMOS digital camera
Dynamic range	82 dB
Input referred noise	780 electrons
Sensitivity (including photodiode)	4 lux

Table 17.9 Features of the Current Amplifier Reported by Bolliger et al.[59]

Parameter	Value
Application	Bipolar UV sensor
Input range	20 pA to 1 nA
Transresistance gain	0.95 GΩ
Bandwidth	20 Hz
Input referred noise	16.5 pA$_{RMS}$

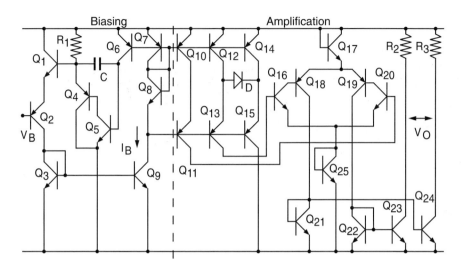

Figure 17.7 Schematic of the current amplifier reported by Bolliger et al.[59]

current (I_B) are amplified by the current gain (β) of transistors Q_{16} and Q_{18}. Moreover, a replica of I_B is also amplified by transistors Q_{19} and Q_{20}. The resulting currents, properly mirrored (Q_{21}, Q_{22}, Q_{23}, and Q_{24}), are transformed into voltage by R_3 and R_2. The left part of the circuit generates a suitable DC bias current by means of a feedback loop. Voltage V_B allows I_B to be controlled in the range of 5 to 15 nA. Assuming perfect element matching, the output voltage of the circuit can be expressed as

$$V_O = \beta_{16}\beta_{18}KR_3I_D. \tag{3}$$

With β around 200, $R_3 = 20$ kΩ, and a mirror factor K equal to 2, a transresistance gain of about 10^9 Ω can be achieved.

17.5.3 Impedance Variation

Silicon technologies allow a variety of resistive or capacitive structures whose value is controlled by a physical or a chemical quantity to be fabricated. Resistive sensors are usually based on piezoresistive effects (resistance variations due to stress in the material) or thermal effects (resistivity of conductors changes with temperature). Examples are strain gauges, piezoresistive pressure sensors, resistive Pirani gauges, and temperature sensors. Moreover, chemical sensors can be realized from materials whose conductivity changes according to the absorption of ambient gases. In contrast, capacitive microsensors are based on the variation in the permittivity or in the thickness of the dielectric layer of a capacitor induced by a physical or chemical quantity. Humidity, chemical compounds, pressure, and acceleration sensors can all be realized with this approach. The magnitude of the capacitive variation can range from hundreds of attofarads to a few picofarads.

DC Wheatstone bridges are frequently used to transform resistance variations into voltages and compensate for parasitic effects. Referring to Fig. 17.8, the output voltage of the bridge is given by

$$V_{out} = \left(\frac{Z_1}{Z_1 + Z_2} - \frac{Z_3}{Z_3 + Z_4} \right) V_{ref}. \qquad (4)$$

The amplitude of the bridge output signal depends on the magnitude of the resistive variation, which can range from a few milliohms to several hundreds of ohms. The whole bridge network, therefore, can be considered as a sensor providing voltage output.[60-62] When it is not possible or convenient to realize a bridge structure, a resistive sensor may replace any resistor in a circuit, which can therefore deliver an output signal sensitive to the quantity to be measured. For example, it is possible to design an RC oscillator whose output frequency is tuned by the resistive sensor.[63-65]

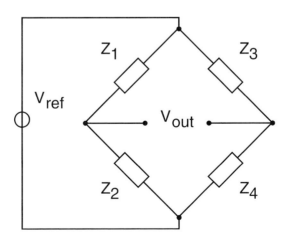

Figure 17.8 Schematic of the Wheatstone bridge.

Capacitive sensors can be read out with an AC bridge or used to build an oscillator. In CMOS technology, it is possible to place the sensor directly in a switched-capacitor circuit, as shown in Fig. 17.9.[66-70] In this circuit, the output voltage at the end of clock phase Φ_2 is given by

$$V_{out} = \frac{1}{C_2}\left(C_{1s}V_{sen} - C_{1r}V_{ref}\right), \tag{5}$$

where C_{1s} is the sensor and C_{1r} is a reference capacitor equal to the sensor but not sensitive to the quantity to be measured. By choosing suitable values for C_2, V_{sen}, and V_{ref}, it is possible to detect capacitance variations on the order of tens of attofarads (1 aF = 10^{-18} F).[71]

A practical implementation of this technique is reported by Gola et al.[72] The most important features of this circuit are summarized in Table 17.10.

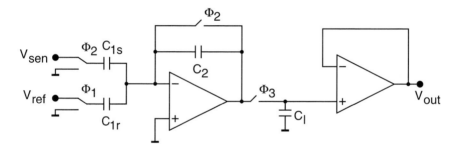

Figure 17.9 Switched-capacitor amplifier for capacitive sensors.

Table 17.10 Features of the Switched-Capacitor Interface for an Angular Accelerometer Reported by Gola et al.[72]

Parameter	Value
Application	Angular accelerometer
Resolution	2.5 rad/sec^2
Minimum capacitance variation	0.05 fF
Bandwidth	800 Hz

17.6 A/D Converter

A/D converters are becoming the most critical components of microsensor systems because the signal processing is reduced in the analog domain. For example, consider a sensor providing a maximum output signal of 10 mV on top of an offset voltage of ±100 mV (which is the case for Hall devices, for example). If we want to resolve 0.1% step by connecting an A/D converter directly to the sensor and performing the offset cancellation in the digital domain, we need 14-bit resolution. But if we implement some sort of offset cancellation in the analog domain in front of the A/D converter, the required resolution drops to 10 bits.

Before considering the most popular A/D converter architectures in detail, from the microsensor interface circuit point of view, it is useful to recall briefly a few concepts and definitions. The digitalization of an analog signal, or A/D conversion, involves discretization in both time and amplitude. For band-limited signals, sampling at the Nyquist rate (twice the signal bandwidth, or baseband) allows the original signal to be represented fully without distortion. However, discretization in amplitude, or quantization, always introduces an error (quantization error). During the quantization process, in fact, the input signal x is approximated with the closest quantized value x_n, giving rise to the quantization error $Q = x - x_n$. By considering Q as a stochastic variable, we can analyze the quantization effect with statistical methods, exploiting the same mathematical tools commonly used to handle noise. The quantization error is, therefore, generally identified as quantization noise. The power of the stochastic variable $Q(P_Q)$ is given by its variance, according to

$$P_Q = \sigma_Q^2 = \int_{-\Delta/2}^{\Delta/2} Q^2 \psi(Q) dQ, \tag{6}$$

where Δ denotes the quantization step amplitude and $\psi(Q)$ is the probability density of Q ($-\Delta/2 < Q < \Delta/2$). Assuming $\psi(Q)$ to be uniform ($\psi = 1/\Delta$), P_Q is given by

$$P_Q = \frac{\Delta^2}{12}. \tag{7}$$

Moreover, the spectrum of the quantization noise is generally considered to be white in the frequency range between DC and the sampling frequency f_S.[73]

A straightforward way to reduce the quantization error is to increase the resolution of the quantizer, thus making the step size (Δ) smaller (Nyquist rate A/D converters). Another way is to reduce the fraction of the quantization noise in the signal band B by increasing f_S above the Nyquist rate (oversampling), as shown in Fig. 17.10 (oversampled A/D converters). In this case, the resulting in-band quantization noise power is given by

$$P_Q = \int_0^B \frac{\Delta^2}{6f_S} df = \frac{\Delta^2 M}{12}, \qquad (8)$$

where $M = 2B/f_S$ is called the oversampling ratio. Naturally, a digital low-pass decimating filter must follow the oversampled A/D converter to eliminate the out-of-band quantization noise and resample the signal at the Nyquist rate.[74]

An overview of several different A/D converter architectures, both Nyquist rate and oversampled, with typical resolution and conversion time (expressed in number of clock cycles) is reported in Table 17.11.[75-115] The table also considers how well each architecture is suited to microsensor applications.

The use of Nyquist rate A/D converters is imperative in high-frequency applications, such as video processing or high-speed data transmission, because in these cases, oversampling would lead to an impractical speed of operation. Several architectures and algorithms are available to implement Nyquist rate A/D converters. The number of clock periods required to perform a complete conversion cycle and the corresponding hardware complexity are the most distinctive features of each architecture. For example, flash converters perform one conversion per clock period, but they require 2^N comparators and 2^N reference elements, N denoting the desired resolution. By contrast, successive approximation converters need N clock periods to complete one conversion, but they require only one comparator and N reference elements. Subranging, half-flash, and pipeline A/D converters need between two and N clock periods per conversion, with decreasing hardware complexity.

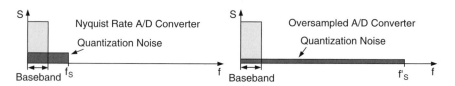

Figure 17.10 Signal and quantization noise spectra in Nyquist rate and oversampled A/D converters.

Table 17.11 Overview and Features of Different A/D Converter Architectures (N denotes the resolution, f_{ck} the clock frequency)

A/D Converter Architecture	Maximum Resolution	Conversion Time	Suitable for Microsensors
Nyquist Rate A/D Converters			
Flash A/D Converters[78-81]	6 bit	$1 / f_{ck}$	−
Subranging and Pipeline A/D Converters[82-86]	12 bit	$< N / f_{ck}$	=
Folding A/D Converters[87-88]	10 bit	$< N / f_{ck}$	−
Successive Approximation A/D Converters[89-92]	12 bit	N / f_{ck}	=
Algorithmic A/D Converters[93-95]	12 bit	N / f_{ck}	=
Oversampled A/D Converters			
Dual-Slope A/D Converters[96-97]	20 bit	$2^{N+1} / f_{ck}$	+
Incremental A/D Converters[98-100]	> 16 bit	$2^N / f_{ck}$	++
Sigma-Delta A/D Converters[101-118]	> 18 bit	$< 2^N / f_{ck}$	++

Figure 17.11 shows a schematic of a successive approximation A/D converter based on the charge redistribution principle. It consists of a binary weighted capacitive array, a comparator, and a successive approximation register (SAR).

At the beginning of each conversion cycle, switch S_1 is closed and the whole capacitive array is charged at the input voltage V_{in} (precharge and

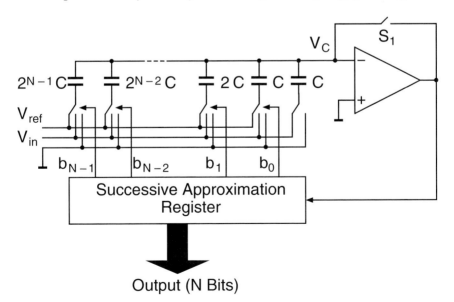

Figure 17.11 Schematic of a charge redistribution A/D converter based on the successive approximation algorithm.

auto-zero phase). Then S_1 is opened and capacitor $2^{N-1}C$, corresponding to the most significant bit (MSB), is connected to V_{ref} (b_{N-1} is set to one), while the rest of the array is connected to ground. Due to charge redistribution in the array, the voltage at the input of the comparator becomes

$$V_C = -V_{in} + \frac{V_{ref}}{2}. \tag{9}$$

If $V_C < 0$, the MSB (b_{N-1}) is confirmed to one and stored; otherwise, the MSB is set to zero. The same procedure is then repeated for the next bits. The capacitors that correspond to the already considered bits are connected to V_{ref} if the corresponding bit is one, or to ground if the corresponding bit is zero. At the end of the algorithm, V_C is therefore given by

$$V_C = -V_{in} + \sum_{i=0}^{N-1} b_i \frac{V_{ref}}{2^{N-i}}, \tag{10}$$

with $b_{N-1} \cdots b_0$ denoting the digital representation of the input signal.

Successive approximation A/D converters are widely used in sensor applications, especially for portable or battery-operated devices, in view of their low power consumption. However, because they rely on accurate analog component matching, their performance may degrade when the chip is exposed to aggressive post-processing steps or environmental conditions. By contrast, flash, half-flash, subranging, and pipeline A/D converters, although very useful for image sensing applications, are definitely impractical for low-frequency sensor applications, mainly because of their complexity.

Oversampling allows the resolution of an A/D converter to be improved by increasing the sampling frequency. This method alone is not very efficient, however, because we must quadruple the oversampling ratio M to gain one bit of resolution (or to attenuate the in-band quantization noise by 6 dB, according to Eq. (8). Better efficiency can be obtained by pushing part of the quantization noise outside the baseband while maintaining the oversampling ratio (M) constant, as shown in Fig. 17.12.

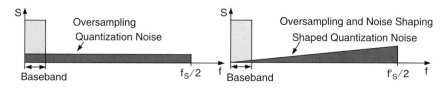

Figure 17.12 Noise shaping effect.

This effect, called noise shaping, can be achieved by introducing a suitable negative feedback around the quantizer, which modifies the transfer functions for the input signal (H_S) and for the quantization noise (H_Q).

In particular, when $H_S = 1$ and $H_Q = (1 - z^{-1})^L$, the oversampled A/D converter is called an Lth order sigma-delta ($\Sigma\Delta$) modulator. Thanks to the noise shaping effect, high-resolution $\Sigma\Delta$ modulators with a reasonable oversampling ratio can be realized using a single-bit quantizer (actually, a latched comparator). The quantization noise power spectral density in this case is given by[116-118]

$$S_Q = \frac{\Delta^2}{6 f_S} \left[2 \sin\left(\frac{\pi f}{f_S}\right) \right]^{2L}. \quad (11)$$

As usual, by integrating S_Q over the baseband, we can calculate the total in-band quantization noise power as

$$P_Q = \int_0^B S_Q df \cong \frac{\Delta^2 \pi^{2L}}{12(2L+1)M^{2L+1}}. \quad (12)$$

The maximum signal-to-noise ratio (the signal-to-noise ratio calculated using the maximum signal amplitude $\Delta/2$) is therefore given by

$$SNR_{max} = \frac{\Delta^2/8}{P_Q} = \frac{3(2L+1)M^{2L+1}}{2\pi^{2L}}. \quad (13)$$

The SNR is often used as figure of merit to quantify the accuracy of an A/D converter as an alternative to the number of significant bits (or resolution), N. From the SNR, we can easily calculate N as

$$N = \ln_2\left(\sqrt{\frac{2}{3}SNR} + 1\right) \cong \frac{1}{2}\ln_2\left(\frac{2}{3}SNR\right) = \frac{SNR|_{dB}}{6} - 0.292. \quad (14)$$

To remove the out-of-band shaped quantization noise efficiently, the low-pass decimating filter that follows an Lth order $\Sigma\Delta$ modulator must be at least of order $L + 1$. Usually, it consists of a "sinc" filter with a transfer function

$$H_D = \left[\frac{1 - z^{-D}}{D(1 - z^{-1})} \right]^{L+1}. \quad (15)$$

The most interesting oversampled A/D converters for sensor applications are first- and second-order $\Sigma\Delta$ modulators. Due to severe stability problems, which degrade the performance, reliability, and robustness of third-order or higher order modulators, their use in microsensor interfaces is generally impractical.

Figure 17.13 shows the block diagram and the linearized model of a first-order $\Sigma\Delta$ modulator. Due to the minimum number of analog components required (an integrator and a latched comparator), first-order $\Sigma\Delta$ modulators are the best candidates for sensor applications. However, because limit cycles (caused by the correlation between input signal and quantization noise) produce unpredictable noise tones in the baseband and degrade the SNR of the circuit, first-order modulators are not often used. Noise tones can be attenuated by introducing a high-frequency dither signal, which makes the input waveform sufficiently chaotic. However, this solution reduces the dynamic range of the modulator and complicates its design.

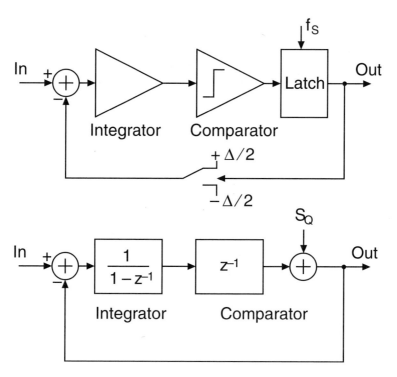

Figure 17.13 Block diagram and linearized model of a first-order sigma-delta modulator.

Another approach to solving the problem of limit cycles consists of resetting the integrator of the first-order ΣΔ modulator before each conversion cycle. The block diagram of such a circuit, called an incremental A/D converter, is shown in Fig. 17.14. Because of the periodic reset, the behavior of this circuit is deterministic rather than stochastic (i.e., for equal input signals, we obtain equal output bitstreams). Moreover, the decimating filter can be reduced to a simple up/down counter.

The incremental conversion algorithm is described by

$$\begin{cases} U_0 = In \\ U_{k+1} = U_k + \left[In - (-1)^{Q_k+1} \dfrac{\Delta}{2} \right], \end{cases} \quad (16)$$

where U and Q denote the output signals of the integrator and the comparator, respectively, and k denotes the current clock period. The N bit digital output signal obtained after 2^N clock periods is therefore given by

$$Out = 2^{N+1} \dfrac{In}{\Delta}. \quad (17)$$

Second-order ΣΔ modulators are much less sensitive to limit cycles than their first-order counterparts because the quantization noise is a more complex function of the design parameters. Consequently, they are less correlated to the input signal. Also, given the higher order noise shaping, they allow the same resolution to be achieved with a lower oversampling ratio, as shown in Fig. 17.15. The block diagram and the linearized model of a second-order ΣΔ modulator are shown in Fig. 17.16. The circuit con-

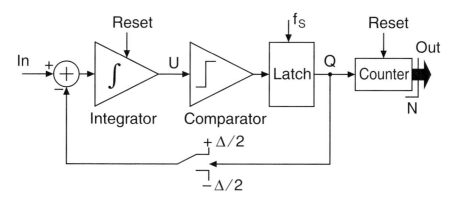

Figure 17.14 Block diagram of an incremental A/D converter.

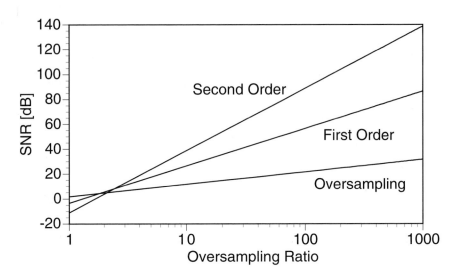

Figure 17.15 SNR versus oversampling ratio curves for different kinds of over-sampled A/D converters.

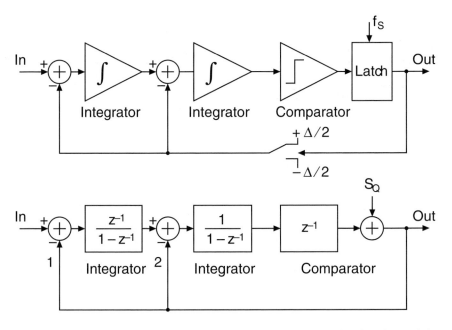

Figure 17.16 Block diagram and linearized model of a second-order sigma-delta modulator.

sists of two integrators and a comparator arranged in a closed-loop topology. Moreover, to ensure stability, a second feedback path connects the output of the comparator to the input of the second integrator.

Second-order $\Sigma\Delta$ modulators have been intensively used to realize microsensor interfaces because they can maintain a high level of robustness against aging and degradation by trading analog component accuracy with speed of operation. Moreover, they can be easily reconfigured to accept signals from different kinds of sensors.

As an example, consider the fourth-order, single-loop, single-bit sigma-delta modulator reported by Brigati et al.[115] The block diagram and the chip photograph of this circuit are shown in Figs. 17.17 and 17.18, respectively; the performance achieved is summarized in Table 17.12.

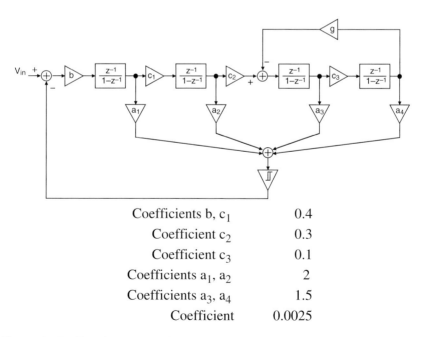

Coefficients b, c_1	0.4
Coefficient c_2	0.3
Coefficient c_3	0.1
Coefficients a_1, a_2	2
Coefficients a_3, a_4	1.5
Coefficient	0.0025

Figure 17.17 Fourth-order, single-loop, single-bit sigma-delta modulator reported by Brigati et al.[115]

Figure 17.18 Chip micrograph of the fourth-order sigma-delta modulator reported by Brigati *et al*.[118]

Table 17.12 Features of the Fourth-Order Sigma-Delta Modulator Reported by Brigati *et al*.[118]

Parameter	Value
Application	Sensor networks
Power consumption	50 mW
Input voltage range (peak-to-peak, differential)	2 V
Bandwidth	400 Hz
Sampling frequency	256 kHz
Noise power in band	−116.9 dB
Signal-to-noise ratio at full-scale signal	104.9 dB
Resolution	17.1 bits
Chip size (including pads)	3.2 x 3.8 mm

17.7 Digital Processing and Output Interface

In modern microsensor systems, most of the signal processing is performed in the digital domain. This section presents the most important functions typically included in microsensor systems, with particular emphasis on the wired and wireless output interfaces.

17.7.1 Digital Signal Processing

The most important signal processing functions required for sensor applications are filtering, calibration, and control. Filtering is obviously used to limit signal bandwidth and remove out-of-band spurs or to decimate the output signal of oversampled A/D converters. No particular solution needs to be adopted to implement filters for microsensor applications: Standard digital signal processing and design techniques can be used. Thus this topic will not be considered in detail in this chapter.

The response of integrated sensors is often nonlinear. In many cases, therefore, interface circuits have to include a calibration section to linearize the transfer characteristic of the sensor, avoiding the undesirable and unpredictable effects due to nonlinear terms. Moreover, since aging often modifies the response of the sensor during the lifetime of the device, the programmability of the calibration function is also important.

Linearization and calibration are typically implemented in the digital domain to exploit the flexibility of digital signal processing. The most common techniques for sensor calibration are based on lookup tables or polynomial correction.[119-120]

The last but not least important function typically implemented digitally in microsensor systems is the control of the system operation. This includes the timing generation, the selection of the mode of operation (for example, acquisition, calibration, transmission, and self-test), and the generation of the feedback signal for adjusting the sensor or analog front-end characteristics, mentioned in Section 17.4.

17.7.2 Wired Output Interfaces

A difficult and important task in large measurement and control systems is communication between a central computer and the sensors or the sensor subsystems, which are widely distributed throughout a plant, a building, or a car. The sensor output signals typically have different formats and may not be compatible with the input format of the central computer. Moreover, the number of wires involved can be very large, thus introducing cost and reliability problems.

Serial bus systems are the best candidates to solve these problems, since they require a minimum number of wires and allow simple transmission protocols to be implemented. Several serial bus standards have been proposed in recent years. Among them, the Philips I²C (Inter-IC) bus system has been specially developed to interconnect integrated circuits, including sensors.[121,122] This system allows relatively small distance data transmission through a serial connection using only four lines, namely two power supply lines, a clock line, and a serial data/address line (Fig. 17.19). The maximum transfer rate is 100 kbits/sec. Each device connected to the bus has its own unique 7-bit address and can operate as a transmitter or a receiver. A master starts the data transfer on the bus and generates the required clock signal. At the same time, any bus member, addressed by the master, is considered a slave. The I²C bus is a multimaster bus, since more than one device can initiate and terminate a data transmission. However, to avoid degradation of the message, only one device at a time can be the master.

A simplified version of the I²C bus system, called I²S[122], has been recently developed especially for sensor applications. In this case, the transmission is controlled by a single master (typically a microprocessor), which interrogates the various slaves (typically the sensors).

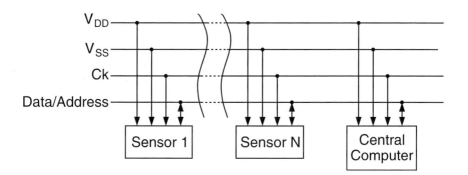

Figure 17.19 Block diagram of the I²C bus system.

Finally, depending on the application, specific bus interfaces can be used and possibly be compatible with standard computer systems. Among them, we can cite the SPI bus, the PCI bus, the VXI bus, and the Ethernet.[123-125]

17.7.3 Wireless Output Interfaces

In the past few years, new technologies based on infrared or radio transceivers have emerged to interconnect devices without using wires. These technologies are obviously very interesting for microsystem applications, since they allow great reductions in wiring costs.

The most promising approach for wireless interfaces, especially when short-range interconnections (less than 10 m) are required, is the Bluetooth standard.[126] Table 17.13 summarizes the most important features of the Bluetooth interface. Several fully integrated Bluetooth transceivers are available on the market, either as commercial parts or as IP blocks[127-130] to be included in custom integrated circuits, thereby allowing the wireless connection feature to be included in microsystems or micromodules without considerable design effort.

Other solutions for wireless interconnections, especially for applications operating over longer ranges, are based on cellular phone standards (GSM, UMTS, DECT)[131] or on wireless LANs (IEEE 802.11b or others).

Table 17.13 Features of the Bluetooth Standard

Parameter	Value
Frequency band	2.4 GHz
Operating range	10 m
Channel bandwidth	1 MHz
Number of channels	79
Modulation	GFSK
Point-to-point and point-to-multipoint connections	Yes
Fully integrated transceivers in standard technologies	Yes

17.8 Conclusions

This chapter has addressed the most important issues in the design of sensor interface circuitry and microsystems. The considerations presented show how essential interface circuits are in compensating for sensor shortages and increasing the functionality of microsensor systems. However, designing good interface circuits is not enough to realize microsensor systems with optimal performance and minimum risk. In fact, it is also very important to consider interface circuitry, packaging, and testing issues, right from the design of the very first sensor. All of these aspects have to be taken into account at the specification level, with the objective of creating an optimal system that is not necessarily simply the interconnection of optimum blocks. Although the issues related to sensors and circuitry have been considered in detail in this chapter, packaging and testing contribute significantly to the industrial success or failure of a specific microsensor system and are not less important.

We hope that the world of microsensor system designers will continue to grow in the future under the impetus provided by sales of an ever-increasing number of microsensors and micromodules.

References

1. H. Baltes, "CMOS as sensor technology," *Sensors and Actuators A*, 37–38:51–56 (1993).
2. S. Middelhock and S. A. Audet, *Silicon Sensors*, Academic Press, London (1989).
3. H. Baltes, D. Moser, E. Lenggenhager, O. Brand, and D. Jaeggi, "Thermomechanical microtransducers by CMOS and micromachining," *Micromechanical Sensors, Actuators and Systems*, vol. DSC-32, ASME, New York, NY (1991), pp. 61–75.
4. H. Baltes, O. Paul, J. G. Korvink, M. Schmeider, J. Bühler, N. Schneeberger, D. Jaeggi, P. Malcovati, M. Hornung, A. Häberli, M. von Arx, F. Mayer, and J. Funk, "IC MEMS microtransducers," *IEDM '96 Technical Digest* (1006), pp. 521–524.
5. H. Baltes and O. Brand, "CMOS-based microsensors and packaging," *Sensors and Actuators A*, 92(1–3):1–9 (2001).
6. C. Hagleitner, A. Hierlemann, D. Lange, A. Kummer, N. Kerness, O. Brand, and H. Baltes, "Smart single-chip gas sensor microsystem," *Nature*, 414(11):293–296 (2001).
7. P. Kleinschmidt and F. Schmidt, "How many sensors does a car need?," *Sensors and Actuators A*, 21:35–45 (1992).
8. B. Bertuol, "Sensors as key components for automotive systems," *Sensors and Actuators A*, 25–27:95–102 (1991).

9. I. Igarashi, "Special features of automotive on-board sensors," *Transducers '95—Eurosensors IX Digest of Technical Papers*, vol. 1, Stockholm, Sweden (1995), pp. 898–899
10. ADXL50, *Special Linear Reference Manual, Analog Devices*, Norwood, MA (1992)
11. M. Aikele, K. Bauer, W. Ficker, F. Neubauer, U. Prechtel, J. Schalk, and H. Seidel, "Resonant accelerometer with self-test," *Sensors and Actuators A*, 92(1–3):161–167 (2001)
12. W. Geiger, J. Merz, T. Fischer, B. Folkmer, H. Sandmaier, and W. Lang, "The silicon angular rate sensor system DAVED(r)," *Sensors and Actuators A*, 84(3):280–284 (2000)
13. J. Binder, "New generation of automotive sensors to fulfill the requirements of fuel economy and emission control," *Sensors and Actuators A*, 31:60–67 (1992)
14. L. Civardi, U. Gatti, F. Maloberti, and G. Torelli, "An integrated CMOS interface for lambda sensor," *IEEE Trans. Vehicular Technol.*, 43:40–46 (1994)
15. M. Komachiya, S. Suzuki, T. Fujita, M. Tsuruki, S. Ohuchi, and T. Nakazawa, "Limiting-current type air-fuel ratio sensor using porous zirconia layer without inner gas chambers: proposal for a quick-startup sensor," *Sensors and Actuators B*, 73(1):40–48 (2001)
16. H. Fritsch, R. Lucklum, T. Iwert, P. Hauptmann, D. Scholz, E. Peiner, and A. Schlachetzki, "A low-frequency micromechanical resonant vibration sensor for wear monitoring," *Sensors and Actuators A*, 62(1–3):616–620 (1997)
17. C. P. O. Treutler, "Magnetic sensors for automotive applications," *Sensors and Actuators A*, 91(1–2):2–6 (2001)
18. J. Ji and K. D. Wise, "An implantable CMOS circuit interface for multiplexed microelectrode recording arrays," *IEEE J. Solid-State Circuits*, 27:433–443 (1992)
19. S. Konishi, T. Kobayashi, H. Maeda, S. Asajima, and M. Makikawa, "Cuff actuator for adaptive holding condition around nerves," *Sensors and Actuators B*, 83(1–3):60–66 (2002)
20. H. Suzuki, T. Tokuda, and K. Kobayashi, "A disposable 'intelligent mosquito' with a reversible sampling mechanism using the volume-phase transition of a gel," *Sensors and Actuators B*, 83(1–3):53–59 (2002)
21. S. Takeuchi and I. Shimoyama, "Selective drive of electrostatic actuators using remote inductive powering," *Sensors and Actuators A*, 95(2–3):269–273 (2002)
22. R. Puers, M. Catrysse, G. Vandevoorde, R. J. Collier, E. Louridas, F. Burny, M. Donkerwolcke, and F. Moulart, "A telemetry system for the detection of hip prosthesis loosening by vibration analysis," *Sensors and Actuators A*, 85(1–3):42–47 (2000)
23. G. T. Kovacs, C. W. Storment, M. Halks-Miller, C. R. Belczynski, C. C. Santina, E. R. Lewis, and N. I. Maluf, "Silicon-substrate microelectrode arrays for parallel recording of neural activity in peripherals and cranial nerves," *IEEE Trans. Biomedical Eng.*, 41:567–576 (1994)

24. S. Wu, Q. Lin, Y. Yuen, and Y.-C. Tai, "MEMS flow sensors for nano-fluidic applications," *Sensors and Actuators A*, 89(1–2):152–158 (2001)
25. A. Glaninger, A. Jachimowicz, F. Kohl, R. Chabicovsky, and G. Urban, "Wide range semiconductor flow sensors," *Sensors and Actuators A*, 85(1–3):139–146 (2000)
26. E. Yoon and K. D. Wise, "An integrated mass flow sensor with on-chip CMOS interface circuit," *IEEE Trans. Electron Devices*, ED-39:1376–1386 (1992)
27. D. Harwood, "Something in the air electronic nose," *IEE Review*, 47(1):10–14 (2001)
28. P. I. Neaves and J. V. Hatfield, "A new generation of integrated electronic noses," *Sensors and Actuators B*, 27(1–3):223–231 (1995)
29. J. G. Ryan;L. Barry, C. Lyden, J. Alderman, B. Lane, L. Sciffner, J. Boldt, H. Thieme, "A CMOS chip-set for detecting 10ppb concentrations of heavy metals," *IEEE ISSCC '95 Digest of Technical Papers*, San Francisco, CA (1995), pp. 158–159
30. G. C. Temes and K. Haug, "Improved offset compensated schemes for switched-capacitor circuits," *Electronics Letters*, 20:508–509 (1984)
31. K. Haug, F. Maloberti, and G. C. Temes, "SC integrators with low finite gain sensitivity," *Electronics Letters*, 21:1156–1157 (1985)
32. R. Gregorian and G. C. Temes, *Analog MOS Integrated Circuits for Signal Processing*, Wiley, New York (1986)
33. K. C. Hsieh, P. R. Gray, D. Senderowicz, and D. G. Messerschmitt, "A low-noise chopper-stabilized differential switched-capacitor filtering technique," *IEEE J. Solid-State Circuits*, 16:708–715 (1981)
34. M. Tuthill, "A switched-current, switched-capacitor temperature sensor in 0.6 μm CMOS," *IEEE J. Solid-State Circuits*, 33(7):1117–1122 (1998)
35. M. Kayal and Z. Randjelovic, "Auto-zero differential difference amplifier," *Electronics Letters*, 36(8):695–696 (2000)
36. P. Malcovati, C. Azeredo Leme, R. Lenggenhager, F. Maloberti, and H. Baltes, "Low noise multirate SC read-out circuitry for thermoelectric integrated infrared sensors," *IEEE Trans. Instr. and Meas.*, 44:795–798 (1995)
37. A. Bakker, K. Thiele, and J. H. Huijsing, "A CMOS nested-chopper instrumentation amplifier with 100-nV offset," *IEEE J. Solid-State Circuits*, 35(12):1877–1883 (2000)
38. A. Bilotti and G. Monreal, "Chopper-stabilized amplifiers with a track-and-hold signal demodulator," *IEEE Trans. Circ. and Syst. I*, 46(4):490–495 (1999)
39. A. Schaufelbühl, N. Schneeberger, U. Münch, M. Waelti, O. Paul, O. Brand, H. Baltes, C. Menolfi, Q. Huang, E. Doering, and M. Loepfe, "Uncooled low-cost thermal imager based on micromachined CMOS integrated sensor array," *IEEE J. MEMS.*, 10:503–510 (2001)
40. C. Menolfi and Q. Huang., "A fully integrated, untrimmed CMOS instrumentation amplifier with submicrovolt offset," *IEEE J. Solid-State Circuits*, 34:415–420 (1999)

41. B. J. Pichler, W. Pimpl, W. Buttler, L. Kotoulas, G. Boning, M. Rafecas, E. Lorenz, and S. I. Ziegler, "Integrated low-noise low-power fast charge-sensitive preamplifier for avalanche photodiodes in JFET-CMOS technology," *IEEE Trans. Nucl. Sci.*, 48(6):2370–2374 (2001)
42. G. C. M. Meijer and V. P. Iordanov, "SC front-end with wide dynamic range," *Electronics Letters*, 37(23):1377–1378 (2001)
43. W. J. Marble and D. T. Comer, "Analysis of the dynamic behavior of a charge-transfer amplifier," *IEEE Trans. Circ. and Syst. I*, 48(7):793–804 (2001)
44. T. Kajita, G. C. Tentes, and U. K. Moon, "Correlated double sampling integrator insensitive to parasitic capacitance," *Electronics Letters*, 37(3):151–153 (2001)
45. Y. Hu, G. Deptuch, R. Turchetta, and C. Guo, "A low-noise, low-power CMOS SOI readout front-end for silicon detector leakage current compensation with capability," *IEEE Trans. Circ. and Syst. I*, 48(8):1022–1030 (2001)
46. A. Blanco, N. Carolino, P. Fonte, and A. Gobbi, "A new front-end electronics chain for timing RPCs," *IEEE Trans. Nuclear Science*, 48(4):1249–1253 (2001)
47. C. Guazzoni, M. Sampietro, A. Fazzi, and P. Lechner, "Embedded front-end for charge amplifier configuration with sub-threshold MOSFET continuous reset," *IEEE Trans. Nucl. Sci.*, 47(4):1442–1446 (2000)
48. C. Kapnistis, K. Misiakos, and N. Haralabidis, "A low noise small area self switched CMOS charge sensitive readout chain," *IEEE Trans. Nucl. Sci.*, 46(3):133–138 (1999)
49. H. H. Kim, S. Chandrasekhar, C. A. Burrus, Jr., and J. Bauman, "A Si BiCMOS transimpedance amplifier for 10-Gb/s SONET receiver," *IEEE J. Solid-State Circuits*, 36(5):776 (2001)
50. K. Chin-Wei, H. Chao-Chil, Y. Shih-Cheng, and C. Yi-Jen, "2 Gbit/s transimpedance amplifier fabricated by 0.35 $\propto$m CMOS technologies," *Electronics Letters*, 37(19):1160 (2001)
51. H. Barthelemy, I. Koudobine, and D. Van Landeghem, "Bipolar low-power operational transresistance amplifier based on first generation current conveyor," *IEEE Trans. Circ. and Syst. II*, 48(6):625 (2001)
52. G. Palmisano and S. Pennisi, "Dynamic biasing for true low-voltage CMOS class AB current-mode circuits," *IEEE Trans. Circ. and Syst. II*, 47(12):1575 (2000)
53. C. Feng-Tso and C. Yi-Jen, "Bandwidth enhancement of transimpedance amplifier by a capacitive-peaking design," *IEEE J. Solid-State Circuits*, 34(8):1170 (1999)
54. A. Pullia, C. Fiorini, E. Gatti, A. Longoni, and W. Buttler, "ROTOR: the VLSI switched current amplifier for high-rate high-resolution spectroscopy with asynchronous event occurrence," *IEEE Trans. Nucl. Sci.*, 45(6):3183 (1998)
55. C. Petri, S. Rocchi, and V. Vignoli, "High dynamic CMOS preamplifiers for QW diodes," *Electronics Letters*, vol. 34, n. 9, pp. 878 (1998)

56. G. Palmisano, G. Palumbo, and S. Pennisi, "High-drive CMOS current amplifier," *IEEE J. Solid-State Circuits*, 33(2):236 (1998)
57. C. Toumazou, J. B. Huges, and N. C. Battersby (eds.), *Switched Currents—An Analog Technique for Digital Technology*, Peter Peregrinus, London (1993)
58. A. Simoni, G. Torelli, F. Maloberti, A. Sartori, M. Gottardi, and L. Gonzo, "A 256 x 256-pixel CMOS digital camera for computer vision with 32 algorithmic ADC's on-board," *IEE Proceedings—Part G*, 146:184–190 (1999)
59. D. Bolliger, P. Malcovati, A. Häberli, P. Sarro, F. Maloberti, and H. Baltes, "Integrated ultraviolet sensor system with on-chip 1 Gtransimpedance amplifier," *IEEE ISSCC '96 Digest of Technical Papers*, San Francisco, USA (1996), pp. 328–329
60. S. Ekelof, "The genesis of the Wheatstone bridge," *Eng. Sci. and Education J.*, 10(1):40 (2001)
61. D. J. Yonce, P. P. Bey, Jr., and T. L. Fare, "A DC autonulling bridge for real-time resistance measurement," *IEEE Trans. Circ. and Syst. I*, 47(3):278 (2000)
62. D. W. Braudaway, "Precision resistors: a review of the techniques of measurement, advantages, disadvantages, and results," *IEEE Trans. Instr. and Meas.*, vol. 48, n. 5, pp. 888 (1999)
63. L. Xiujun and G. C. M. Meijer, "A smart and accurate interface for resistive sensors," *IEEE Trans. Instr. and Meas.*, 50(6):1651 (2001)
64. V. Ferrari, D. Marioli, and A. Taroni, "Oscillator-based interface for measur-and-plus-temperature readout from resistive bridge sensors," *IEEE Trans. Instr. and Meas.*, 49(3):590 (2000)
65. V. Ferrari, C. Ghidini, D. Marioli, and A. Taroni, "Oscillator-based signal conditioning with improved linearity for resistive sensors," *IEEE Trans. Instr. and Meas.*, 47(1):298 (1998)
66. S. Xiaojing, H. Matsumoto, and K. Murao, "A high-accuracy digital readout technique for humidity sensor," *IEEE Trans. Instr. and Meas.*, 50(5):1282 (2001)
67. S. Ogawa, Y. Oisugi, K. Mochizuki, and K. Watanabe, "A switched-capacitor interface for differential capacitance transducers," *IEEE Trans. Instr. and Meas.*, 50(5):1301 (2001)
68. G. C. M. Meijer and V. P. Iordanov, "SC front-end with wide dynamic range," *Electronics Letters*, 37(23):1378 (2001)
69. H. Morimura, S. Shigematsu, and K. Machida, "A novel sensor cell architecture and sensing circuit scheme for capacitive fingerprint sensors," *IEEE J. Solid-State Circuits*, 35(5):724 731 (2000)
70. W. Bo, K. Tetsuya, S. Tao, and G. Temes, "High-accuracy circuits for on-chip capacitive ratio testing and sensor readout," *IEEE Trans. Instr. and Meas.*, 47(1):20 (1998)
71. J. T. Kung and H. S. Lee, "An integrated air-gap-capacitor pressure sensor and digital readout with sub-100 attfarad resolution," *IEEE J. MEMS*, 1:121–129 (1992)

72. A. Gola, N. Bagnalasta, P. Bendiscioli, E. Chiesa, S. Delbò, E. Lasalandra, F. Pasolini, M. Tronconi, T. Ungaretti, and A.Baschirotto, "A MEMS-based rotational accelerometer for HDD applications with 2.5 rad Ú sec resolution and digital output," *Proc. ESSCIRC '01* (2001), pp. 336–339
73. J. C. Candy and O. Benjamin, "The structure of quantization noise from sigma-delta modulation," *IEEE Trans. Communication*, 29:1316–1323 (1981)
74. J. C. Candy, "Decimation for sigma-delta modulation," *IEEE Trans. Communication*, 34:249–258 (1986)
75. C. K. K. Yang, V. Stojanovic, S. Modjtahedi, M. A. Horowitz, and W. F. Ellersick, "A serial-link transceiver based on 8-GSamples/s A/D and D/A converters in 0.25-µm CMOS," *IEEE J. Solid-State Circuits*, 36(11):1684–1692 (2001)
76. V. Vujicic, "Generalized low-frequency stochastic true RMS instrument," *IEEE Trans. Instr. and Meas.*, 50(5):1089–1092 (2001)
77. M. Choi and A. A. Abidi, "A 6-b 1.3-Gsample/s A/D converter in 0.35-µm CMOS," *IEEE J. Solid-State Circuits*, 36(12):1847–1858 (2001)
78. K. Nagaraj, D. A. Martin, M. Wolfe, R. Chattopadhyay, S. Pavan, J. Cancio, and T. R. Viswanathan, "A dual-mode 700-Msamples/s 6-bit 200-Msamples/s 7-bit A/D converter in a 0.25-µm digital CMOS process," *IEEE J. Solid-State Circuits*, 35(12):1760–1768 (2000)
79. D. U. Thompson and B. A. Wooley, "A 15-b pipelined CMOS floating-point A/D converter," *IEEE J. Solid-State Circuits*, 36(2):299–303 (2001)
80. L. Sumanen, M. Waltari, and K. A. I. Halonen, "A 10-bit 200-MS/s CMOS parallel pipeline A/D converter," *IEEE J. Solid-State Circuits*, 36(7):1048–1055 (2001)
81. C. Hsin-Shu, S. Bang-Sup, and K. Bacrania, "A 14-b 20-Msamples/s CMOS pipelined ADC," *IEEE J. Solid-State Circuits*, 36(6):997–1001 (2001)
82. P. Hui, M. Segami, M. Choi, C. Ling, and A. A. Abidi, "A 3.3-V 12-b 50-MS/s A/D converter in 0.6-µm CMOS with over 80-dB SFDR," *IEEE J. Solid-State Circuits*, vol. 35, n. 12, pp. 1769–1780 (2000)
83. I. E. Opris, L. D. Lewicki, and B. C. Wong, "A single-ended 12-bit 20 Msample/s self-calibrating pipeline A/D converter," *IEEE J. Solid-State Circuits*, vol. 33, n. 12, pp. 1898–1903 (1998)
84. P. Hui, M. Segami, M. Choi, C. Ling, and A. A. Abidi, "A 3.3-V 12-b 50-MS/s A/D converter in 0.6-µm CMOS with over 80-dB SFDR," *IEEE J. Solid-State Circuits,* 35(12):1769–1780 (2000)
85. K. Nagaraj, F. Chen, and T. R. Viswanathan, "Efficient 6-bit A/D converter using a 1-bit folding front end," *IEEE J. Solid-State Circuits*, 34(8):1056–1062 (1999)
86. W. Claes, W. Sansen, and R. Puers, "A 40-µA/channel compensated 18-channel strain gauge measurement system for stress monitoring in dental implants," *IEEE J. Solid-State Circuits*, 37(3):293–301 (2002)
87. G. Promitzer, "12-bit low-power fully differential switched capacitor non-calibrating successive approximation ADC with 1 MS/s," *IEEE J. Solid-State Circuits*, 36(7):1138–1143 (2001)

88. S. Mortezapour and E. K. F. Lee, "A 1-V, 8-bit successive approximation ADC in standard CMOS process," *IEEE J. Solid-State Circuits*, 35(4):642–646 (2000).
89. J. M. Ingino and B. A. Wooley, "A continuously calibrated 12-b, 10-MS/s, 3.3-V A/D converter," *IEEE J. Solid-State Circuits*, 33(12):1920–1931 (1998).
90. P. Rombouts, W. De Wilde, and L. Weyten, "A 13.5-b 1.2-V micropower extended counting A/D converter," *IEEE J. Solid-State Circuits*, 36(2):176–183 (2001).
91. C. Dong-Young and L. Seung-Hoon, "Design techniques for a low-power low-cost CMOS A/D converter," *IEEE J. Solid-State Circuits*, 33(8):1244–1248 (1998).
92. C. Shu-Yuan and W. Chung-Yu, "A CMOS ratio-independent and gain-insensitive algorithmic analog-to-digital converter," *IEEE J. Solid-State Circuits*, 31(8):1201–1207 (1996).
93. R. V. Kochan, O. M. Berezky, A. F. Karachka, I. Maruschak, and O. V. Bojko, "Development of the integrating analog-to-digital converter for distributive data acquisition systems with improved noise immunity," *IEEE Trans. Instr. and Meas.*, 51(1):96–101 (2002).
94. E. Raisanen-Ruotsalainen, T. Rahkonen, and J. Kostamovaara, "An integrated time-to-digital converter with 30-ps single-shot precision," *IEEE J. Solid-State Circuits*, 35(10):1507–1510 (2000).
95. A. Yufera and A. Rueda, "S^{2}I first-order incremental A/D converter," *Circuits, Devices and Systems, IEE Proceedings*, 145(2):78–84 (1998).
96. A. Häberli, M. Schneider, P. Malcovati, R. Castagnetti, F. Maloberti, and H. Baltes, "2D magnetic microsensor with on-chip signal processing for contactless angle measurement," *IEEE J. Solid-State Circuits*, 31(12):1902–1907 (1996).
97. J. Robert, G. C. Temes, V. Valencic, R. Dessoulavy, and P. Deval, "A 16-bit low voltage A/D converter," *IEEE J. Solid-State Circuits*, 22(2):157–163 (1987).
98. K. Tai-Haur, C. Kuan-Dar, and Y. Horng-Ru, "A wideband CMOS sigma-delta modulator with incremental data weighted averaging," *IEEE J. Solid-State Circuits*, 37(1):11–17 (2002).
99. O. Oliaei, P. Clement, and P. Gorisse, "A 5-mW sigma-delta modulator with 84-dB dynamic range for GSM/EDGE," *IEEE J. Solid-State Circuits*, 37(1):2–10 (2002).
100. O. Bajdechi and J. H. Huijsing, "A 1.8-V delta-sigma modulator interface for an electret microphone with on-chip reference," *IEEE J. Solid-State Circuits*, 37(3):279–285 (2002).
101. C. B. Wang, "A 20-bit 25-kHz delta-sigma A/D converter utilizing a frequency-shaped chopper stabilization scheme," *IEEE J. Solid-State Circuits*, 36(3):566–569 (2001).
102. K. Vleugels, S. Rabii, and B. A. Wooley, "A 2.5-V sigma-delta modulator for broadband communications applications," *IEEE J. Solid-State Circuits*, 36(12):1887–1899 (2001).

103. K. Gulati and L. Hae-Seung, "A low-power reconfigurable analog-to-digital converter," *IEEE J. Solid-State Circuits*, 36(12):1900–1911 (2001)
104. E. Fogelman, J. Welz, and I. Galton, "An audio ADC Delta-Sigma modulator with 100-dB peak SINAD and 102-dB DR using a second-order mismatch-shaping DAC," *IEEE J. Solid-State Circuits*, 36(3):339–348 (2001)
105. K. Chien-Hung, C. Shr-Lung, H. Lee-An, and L. Shen-Iuan, "CMOS oversampling delta-sigma magnetic-to-digital converters," *IEEE J. Solid-State Circuits*, 36(10):1582–1586 (2001)
106. J. C. Morizio, I. M. Hoke, T. Kocak, C. Geddie, C. Hughes, J. Perry, S. Madhavapeddi, M. H. Hood, G. Lynch, H. Kondoh, T. Kumamoto, T. Okuda, H. Noda, M. Ishiwaki, T. Miki, and M. Nakaya, "14-bit 2.2-MS/s sigma-delta ADC's," *IEEE J. Solid-State Circuits*, 35(7):968–976 (2000)
107. P. C. Maulik, M. S. Chadha, W. L. Lee, and P. J. Crawley, "A 16-bit 250-kHz delta-sigma modulator and decimation filter," *IEEE J. Solid-State Circuits*, 35(4):458–467 (2000)
108. G. J. Gomez, "A 102-dB spurious-free DR sigma-delta ADC using a dynamic dither scheme," *IEEE Trans. Circ. and Syst. II*, 47(6):531–535 (2000)
109. Y. Geerts, M. S. J. Steyaert, and W. Sansen, "A high-performance multibit delta-sigma CMOS ADC," *IEEE J. Solid-State Circuits*, 35(12):1829–1840 (2000)
110. I. Fujimori, A. Nogi, and T. Sugimoto, "A multibit delta-sigma audio DAC with 120-dB dynamic range," *IEEE J. Solid-State Circuits*, 35(8):1066–1073 (2000)
111. I. Fujimori, L. Longo, A. Hairapetian, K. Seiyama, S. Kosic, C. Jun, and C. Shu-Lap, "A 90-dB SNR 2.5-MHz output-rate ADC using cascaded multibit delta-sigma modulation at 8x oversampling ratio," *IEEE J. Solid-State Circuits*, 35(12):1820–1828 (2000)
112. E. Fogelman, I. Galton, W. Huff, and H. Jensen, "A 3.3-V single-poly CMOS audio ADC delta-sigma modulator with 538-dB peak SINAD and 105-dB peak SFDR," *IEEE J. Solid-State Circuits*, 35(3):297–307 (2000)
113. F. Medeiro, B. Perez-Verdu, and A. Rodriguez-Vazquez, "A 13-bit, 2.2-MS/s, 55-mW multibit cascade sigma-delta modulator in CMOS 0.7-μm single-poly technology," *IEEE J. Solid-State Circuits*, 34(6):748–760 (1999)
114. S. Kawahito, C. Maier, M. Schneiher, M. Zimmermann, and H. Baltes, "A 2D CMOS microfluxgate sensor system for digital detection of weak magnetic fields," *IEEE J. Solid-State Circuits*, 34(12):1843–1851 (1999)
115. S. Brigati, F. Francesconi, P. Malcovati, and F. Maloberti, "A fourth-order single-bit switched-capacitor sigma-delta modulator for distributed sensor applications," *IEEE Trans. Instr. and Meas.*, 53(2):266–270 (2004)
116. R. M. Gray, "Oversampled sigma-delta modulation," *IEEE Trans. Communication*, 35:481–489 (1987)
117. J. C. Candy and G. C. Temes, "Oversampling methods for A/D and D/A conversion," in *Oversampling Delta-Sigma Data Converters* (J. C. Candy and G. C. Temes, eds.), IEEE Press, Piscataway, NJ (1991), pp. 1–29

118. S. Norsworthy, K. Schreier, and G. Temes, *Delta-Sigma Converters: Theory, Design and Simulation*, IEEE Press, Piscataway, USA (1997)
119. S. Xiaojing, H. Matsumoto, and K. Murao, "A high-accuracy digital readout technique for humidity sensor," *IEEE Trans. Instr. and Meas.*, 50(5):1277–1282 (2001)
120. P. Malcovati, C. Azeredo Leme, P. O'Leary, F. Maloberti, and H. Baltes, "Smart sensor Interface with A/D conversion and programmable calibration," *IEEE J. Solid-State Circuits*, 29(8):963–966 (1994)
121. M. J Rutka, *Integrated Sensor Bus*, Delft University Press, Delft, The Netherlands (1994)
122. F. Riedijk, *Integrated Smart Sensors with Digital Bus Interface*, Delft University Press, Delft, The Netherlands (1993)
123. K. B. Lee and R. D. Schneeman, "Distributed measurement and control based on the IEEE 1451 smart transducer interface standards," *IEEE Trans. Instr. and Meas.*, 49(3):627 (2000)
124. K. B. Lee and R. D. Schneeman, "Internet-based distributed measurement and control applications," *IEEE Instrumentation and Measurement Magazine*, 2(2):27 (1999)
125. M. Bertocco, F. Ferraris, C. Offelli, and M. Parvis, "A client-server architecture for distributed measurement systems," *IEEE Trans. Instr. and Meas.*, 47(5):1148 (1998)
126. The Bluetooth Special Interest Group, *Specification of the Bluetooth System, Version 1.1*, http://www.bluetooth.com (2001)
127. R. Schneiderman, "Bluetooth's slow dawn," *IEEE Spectrum*, 37(11):65 (2000)
128. J. C. Haartsen and S. Mattisson, "Bluetooth-a new low-power radio interface providing short-range connectivity," *Proc. IEEE*, 88(10):1661 (2000)
129. J. Grilo, I. Galton, K. Wang, and R. G. Montemayor, "A 12-mW ADC delta-sigma modulator with 80 dB of dynamic range integrated in a single-chip Bluetooth transceiver," *IEEE J. Solid-State Circuits*, 37(3):278 (2002)
130. H. Darabi, S. Khorram, C. Hung-Ming, P. Meng-An, S. Wu, S. Moloudi, J. C. Leete, J. J. Rael, M. Syed, R. Lee, B. Ibrahim, M. Rofougaran, and A. Rofougaran, "A 2.4-GHz CMOS transceiver for Bluetooth," *IEEE J. Solid-State Circuits*, 36(12):2024 (2001)
131. F. Torán, D. Ramírez, A. E. Navarro, S. Casans, J. Pelegrí, and J. M. Espí, "Design of a virtual instrument for water quality monitoring across the Internet," *Sensors and Actuators B*, 76(1–3):281–285 (2001)

Contributors

Henry Baltes, ETH, Zürich, Switzerland

Jens Ducrée, HSG-IMIT—Institute of Micro and Information Technology, Villingen-Schwenningen, Germany

Gary K. Fedder, Department of Electrical and Computer Engineering and The Robotics Institute, Carnegie Mellon University, Pittsburgh, Pennsylvania

Paddy J. French, Delft University of Technology, The Netherlands

Andreas Greiner, Institute of Microsystem Technology IMTEK, University of Freiburg, Germany

Andreas Hierlemann, ETH, Zürich, Switzerland

Jack W. Judy, University of California at Los Angeles, Los Angeles, California

Karim S. Karim, Simon Fraser University, Vancouver, Canada

Peter Koltay, Institute of Microsystem Technology IMTEK, University of Freiburg, Germany

Jan G. Korvink, Institute of Microsystem Technology IMTEK, University of Freiburg, Germany

Franz Laermer, Robert Bosch GmbH, Stuttgart, Germany

Zhenyu Liu, Institute of Microsystem Technology IMTEK, University of Freiburg, Germany

Whye-Kei Lye, Department of Electrical and Computer Engineering, University of Virginia, Charlottesville, Virginia

Piero Malcovati, Department of Electrical Engineering, University of Pavia, Italy

Franco Maloberti, Department of Electronics, University of Pavia, Italy

Arokia Nathan, University of Waterloo, Ontario, Canada

Pamela R. Patterson, HRL Laboratories, Malibu, California

Oliver Paul, Institute of Microsystem Technology IMTEK, University of Freiburg, Germany

Michael Reed, Department of Electrical and Computer Engineering, University of Virginia, Charlottesville, Virginia

Chavdar Roumenin, Bulgarian Academy of Sciences, Sofia, Bulgaria

Evgenii B. Rudnyi, Institute of Microsystem Technology IMTEK, University of Freiburg, Germany

Volker Saile, Research Centre Karlsruhe GmbH, Institute of Microstructure Technology, and University of Karlsruhe, Germany

Pasqualina M. Sarro, Delft University of Technology, The Netherlands

Osamu Tabata, Kyoto University, Japan

Toshiyuki Tsuchiya, Kyoto University, Japan

Friedemann Völklein, FH Wiesbaden, University of Applied Sciences, Rüsselsheim, Germany

Ulrike Wallrabe, Institute of Microsystem Technology IMTEK, University of Freiburg, Germany

Ming C. Wu, University of California, Berkeley, California

Hans Zappe, Institute of Microsystem Technology IMTEK, University of Freiburg, Germany

Roland Zengerle, Institute of Microsystem Technology IMTEK, University of Freiburg, Germany; HSG-IMIT—Institute of Micro and Information Technology, Villingen-Schwenningen, Germany

Index

1/f noise: 23, 42-43, 316, 462, 510, 515, 517
3ω method: 72

A

a-Se: 282-283, 316, 318-319, 321, 324
a-Si:H: 281-298, 303, 309-312, 315-316, 319, 332-335, 337, 339-340, 342-343
A/D conversion: 901, 921, 942
A/D converter: 269, 315, 910, 912, 918, 921-927, 939-941
above-threshold regime: 299
absolute sensitivity: 457-459
absolute temperature: 22, 36, 84, 231, 242-243, 247, 249, 913
absorptance: 6
absorption: 15, 26, 116, 236-237, 249, 268-269, 281-282, 286, 288, 293-294, 296, 311, 404, 436, 445, 499, 569-570, 574, 578, 581, 584, 596-597, 599-604, 607-608, 610, 630, 636, 646, 674, 731, 857, 859, 919
AC power: 258-259, 268, 911
AC-to-DC converter: 10
AC/DC converter: 38
acceleration: 4, 10, 17, 23, 25-27, 34, 53, 184, 195, 198, 221, 527-530, 533, 535, 537, 540-542, 547, 565, 679, 798, 851, 890, 898, 906-907, 911, 919
acceleration sensor: 537, 541, 564, 898
accelerometer: 211, 225, 529-536, 541-542, 544-547, 564-565, 798, 847, 850-851, 889-890, 907, 920, 935, 939
acetic acid: 814
acid rain: 909
acid/base neutralization: 598
acoustic device: 525
acoustic pressure: 4
acoustic resonator: 268
acoustic velocity: 792
acoustic wave velocity: 584
activation barrier: 576
active pixel sensor: 320, 324, 334, 342
activity: 16, 119, 403, 457, 526, 575-578, 637, 674, 707, 713, 743, 936
actuator: 8-11, 32-34, 38, 45, 48-49, 66, 68-69, 88, 106, 160, 200-201, 203-204, 207-209, 215, 218, 229-230, 276, 350-351, 354, 356, 367, 376, 384, 386, 388, 390, 395, 397, 400, 402, 453, 517, 532, 543, 564, 651, 653, 657, 666, 702, 708, 720-722, 734, 747-748, 753, 758, 773-774, 777-779, 782, 786-787, 799-803, 847, 869, 879-881, 883-884, 889, 899, 936
actuator element: 9-10
adaptive optical element: 10
adaptive remeshing: 143
add-drop multiplexer: 365, 370, 386
add-drop switch: 386
adhesion: 87, 534, 677, 769, 857, 859, 897
adhesive: 63, 694, 743, 869, 879
adhesive bonding: 879
aerosol: 671
AFM: 90, 269, 589
aging: 476, 556, 562, 902, 931
air conditioning: 268, 270, 908-909

air damping: 191, 194, 224, 564
air gap: 194, 202, 388, 390, 401, 594, 610, 634, 636, 757-759, 770, 879
air quality: 906
airbag: 523, 527, 535, 565
airbrush: 598
airflow sensor: 229
$Al_{0.3}Ga_{0.7}As$: 422
$Al_{0.8}Ga_{0.2}As$: 422
Al_2O_3: 231, 232, 486, 735, 857, 889, 897
albumin: 616, 731
alcohol: 588, 591, 837
aldehyde: 630
alignment structure: 869-871, 874, 878-879
alkaline fuel cell: 45
alkanethiol: 169
AlN: 528, 803
aluminum: 39, 82, 250-252, 257-258, 267, 269-271, 275-276, 297, 362, 383, 391, 400, 528, 581-582, 584, 586, 588, 596, 609-610, 641, 648, 735, 808-809, 814, 825-829, 831, 834, 842, 844, 851, 901
aluminum nitride: 528, 581-582, 735
ammonia: 571, 628, 636, 640
ammonium fluoride solution: 818
ammonium hydroxide: 807, 809
amorphization: 835
amorphous selenium: 282-283, 334, 339, 342
amorphous silicon: 75, 79, 87, 281-282, 285-286, 289, 316, 320-321, 324-325, 332, 337-343, 846
amorphous silicon multiplexer: 331
amorphous silicon nitride: 286, 335
amperometric: 498, 502, 516-517, 520, 620-622, 625, 647-648, 674, 707
amperometric mode: 516
amperometry: 617, 619
amplifier: 11, 35, 42-43, 49, 98, 252, 281, 314, 316-323, 330, 451, 475, 503, 505, 507-508, 518, 545, 550, 558, 585-586, 596, 619, 800, 914-918, 920, 936-938
amplitude gratings: 611
analog circuit design: 49, 221
analog front-end: 910, 912-913, 932
analog hardware definition language: 191
analog-to-digital: 9, 49, 51, 910, 940-941
analyte concentration: 568, 581, 597, 600, 619-620, 624, 633
analytical chemistry: 667, 688-689, 726
analytical system: 211, 567
angle multiplexing: 362
angular: 4, 22, 72, 214, 219, 359, 362, 376, 380, 395, 397, 399, 455, 527, 540-541, 543, 545, 547, 549, 565, 708, 755, 757, 783, 920, 935
angular resolution: 362
angular velocity: 4, 527, 543
anharmonicity: 541
anion: 628, 674, 688
anisotropic etching: 58, 67, 236, 245-246, 250, 255-256, 266-267, 273, 277, 349, 362, 364, 377, 383, 490, 551, 564, 616, 696, 805, 807, 845-847
anisotropic wet etching: 811
anisotropy: 520, 690, 807, 809, 819
anodic bonding: 256, 691

anodic dissolution: 813, 816, 848
anti-resonance: 437
antigen/antibody binding: 589
APCVD: 826
aperiodic grating: 392
aperture: 254, 346, 365, 392, 432-433, 435
application-specific integrated circuit: 901
array: 25, 29, 50, 125, 160-162, 164, 178, 203, 252-255, 281-282, 294, 320, 324, 331-334, 336, 342, 353, 355, 365-367, 369-370, 372, 386, 392, 395-397, 399-400, 451, 507, 521, 601, 609-611, 621, 644, 709-710, 712, 718, 777-778, 798-799, 871, 874, 876, 916, 924, 937
ASIC: 252-253
aspect ratio: 331, 357, 396-397, 748, 781, 818-819, 823-824, 840, 849, 851, 860, 867, 895, 899
assay: 709, 724
assembler: 123-124
asymmetric waveguide: 412, 433
atmospheric-pressure CVD: 826
atomic force microscope: 68
atomizer: 708
attenuator: 346, 365, 386, 391, 401
autocalibration: 512, 521
autodoping: 831
automotive: 17, 253-254, 449, 455, 523-524, 527, 532, 539, 565-566, 751, 902, 906-907, 909, 935
autonomous system: 47
autozeroing: 34-35
avalanche: 283, 761-764, 937

B

$(Bi_{0.5}Sb_{1.5})Te_3$: 264
back-etching: 584, 588, 590, 595, 598, 611, 622, 637, 640, 647
ball lens: 871, 873, 881
band bending: 291, 298, 634
band gap energy: 241
band-limited signal: 921
banded matrix: 131
bandpass: 189, 916
barcode: 879
base resistance: 491
base vibrating mode: 540
base-emitter contact: 241
base-emitter junction: 242
base-emitter voltage: 241
batch fabrication: 187, 361, 573, 690, 739, 901, 906
battery-operated sensor system: 910
BCB: 359, 362
beam divergence: 346-347
beam waist: 373-375, 431
beam-on-diaphragm: 559
bearing: 282, 648, 783-785, 868
behavioral model: 191, 216
bending: 63, 67-69, 75-76, 78, 80, 89-90, 192, 196-197, 206-207, 215, 335, 364, 392, 748, 752, 773, 776-777, 783, 790, 799
bending motion: 790
bending strength: 75
benzocyclobutene: 359, 362
Bernoulli equation: 196, 681, 704
beryllium wafer: 857
$Bi_{0.87}Sb_{0.13}$: 260
bias current: 8, 466-467, 918
bias signal: 7

biaxial scanner: 348
BiCMOS technology: 537
bidirectional: 705, 721, 796
bifurcation: 106
bimodal resonance: 364
bimorph actuation: 364, 388
bimorph actuator: 790
bimorph effect: 590-592, 753
biochemical compound: 906
biocompatibility: 695, 730-731, 733, 739, 747
bioluminescence: 601, 674
biomagnetometry: 454
biomedical system: 729, 731, 733, 735, 737, 739, 741, 743, 745, 747, 749
biomolecular chemistry: 667
biotin: 607
bipolar electronics: 738, 842
bipolar IC technology: 278, 476
bipolar magnetotransistor: 488, 520
bipolar transistor: 519, 520
BiSbTe: 250
black hole: 456
black silicon: 832, 837, 845, 850
blocking layer: 294
blood glucose: 743
bloodstream: 526, 709
bluetooth standard: 933-934
blur compensation: 525
bolometer: 25, 236-238, 277
Boltzmann constant: 84, 22, 232, 236, 264, 468, 497
Boltzmann distribution: 481
bond stretching: 117
bondgraph: 216, 226
bonding: 161, 174, 176, 186, 258, 274, 335, 343, 354, 358, 525, 534, 537-538, 550, 554-556, 562, 611, 615, 691, 694-696, 796, 802, 821, 849, 859, 879, 904
Boolean operation: 147
Born-Oppenheimer approximation: 112, 116
boron etch stop: 809
boundary element: 103, 133-136, 184
boundary integral: 128, 133, 136
boundary layer: 687
BPSG: 85, 828, 851
bradycardia: 908
Bragg reflector: 390, 450, 608
brain damage: 526
break down: 319
breakdown voltage: 761, 763-764
breast cancer: 588
brittle material: 62, 68, 823, 849
Brownian motion: 25, 675, 713
BSG: 85, 828
buckling: 57, 75, 78, 106, 267, 400
buffered HF solution: 814
bulge test: 67-68, 163
bulk absorption: 570
bulk acoustic wave sensor: 558
bulk material: 54, 58, 264, 638
bulk micromachining: 67, 69, 351, 354, 362, 530, 533, 551, 738, 740-741, 744, 805, 808, 818-819, 824, 838-840, 846, 848
bulk trap: 299
bulk-micromachined pressure sensor: 551-552, 554-555
bumping technology: 254

Index

buoyancy: 14, 679
buried layer: 829-831, 835, 844
buried oxide: 360, 552, 832, 835, 839
buried waveguide: 426
burning rate: 162, 164
bus: 326, 738, 901, 908, 912, 932-933, 942
butt coupling: 430-432

C

cadmium sulfide: 582
calibration: 19, 26-28, 32, 49, 258-259, 261-262, 279, 458-459, 461-462, 531-532, 543, 557, 561-562, 568, 571, 632, 705-706, 901, 908, 931-932, 942
calorimetric: 591-594, 597
cantilever beam: 57, 60, 67-68, 143, 197, 266-268, 273-276, 377, 379, 771
cantilever structure: 69-70
capacitance: 4, 23-24, 42, 136, 200-205, 207-209, 215, 218, 222, 224, 281, 283, 296, 314-321, 323-324, 326, 330-331, 333-334, 338, 350, 532-533, 535, 537-538, 548, 550, 558, 590, 632-633, 639, 646, 648, 682-683, 759-760, 770, 775, 778, 889, 911-913, 920-921, 937, 939
capacitance matrix: 200-201
capacitive detection: 529, 532, 540, 543, 550, 555, 780
capacitive element: 23, 528
capacitor: 22-23, 43, 202, 208, 218, 296, 315-316, 319, 397, 509, 532-535, 538, 547-549, 555, 558, 573, 624, 626, 633, 645, 647, 769-770, 792, 797, 831, 889, 913, 917, 919-921, 924, 936, 939-940, 942
capillary: 158, 168-169, 186, 681, 685-686, 688, 690, 694-695, 698, 700, 709-711, 716, 720, 722, 726, 746, 867, 869
capillary electrophoresis: 716, 720, 722, 726, 746
capillary pressure: 685-686, 698, 700
capping wafer: 531, 537
carbon monoxide: 571, 595, 610, 636, 643
carbon-black: 641
carcinogenesis: 731
carrier concentration: 232, 234, 244, 466, 468
cast molding: 692
catalytic chemical sensor: 576
catalytic combustion: 238-239
catalytic sensor: 592-593
catalyzer: 29
catheter: 738-739, 748, 799, 848, 891
cation: 695, 733
cavitation: 681, 703
cavity: 178-179, 186, 276, 349, 389-391, 401-402, 404, 437-438, 441-442, 445, 530, 551, 554, 608-610, 683, 744
CCD: 67, 434, 610-611
cell activity: 636
cell handling: 667
cellular phone: 526, 934
centrifugal force: 679, 703
ceramic substrate: 232, 879, 889
chalcogenide glass: 624
channel: 32, 46, 94, 277, 297, 299, 301-302, 306, 312, 326, 328, 338, 365, 370, 394, 401, 435, 437, 474-475, 489, 502, 606, 613, 627-628, 677, 683, 690-691, 697-698, 700, 710, 712-713, 716-717, 720, 725, 748, 848, 881, 913, 933, 940
channel length modulation: 328
characteristic equation: 411, 413, 420, 426
charge: 4, 17, 24, 45-48, 84, 134-135, 171, 199-200, 218, 230, 232, 244, 249, 263, 281, 286, 288-289, 294-297, 301-302, 304-305, 314-316, 318-324, 326, 331, 340-341, 434, 463, 466, 468, 496, 569-570, 574-575, 579, 581, 611, 616-617, 621, 624, 626-627, 630-631, 636, 639-641, 643, 674, 686, 688, 695, 703, 755-756, 759-760, 771, 786-787, 816, 916-918, 923-924, 937
charge accumulation: 627, 688
charge amplifier: 314, 316-318, 320-322, 916-917, 937
charge collection efficiency: 288
charge induction: 786
charge injection: 340, 917
charge preamplifier: 916
charge transfer: 312, 318, 569-570, 574, 579, 615, 639
charge-coupled device: 67, 434, 610-611
check valve: 697, 702, 720
chelating agent: 616
CHEMFET: 629, 631, 706-707
chemical microsensor: 649
chemical potential: 6, 577-580, 634
chemical process engineering: 667, 669, 713
chemical reaction: 15, 45, 239, 273, 569, 575-579, 591, 593, 596, 601, 614, 669, 714
chemical sensor: 567, 571, 573-574, 603, 611, 623, 663, 665, 706
chemical signal: 3, 6, 9
chemical vapor deposition: 60, 159, 285, 343, 631, 825-826
chemical warfare: 668
chemical-mechanical microactuation: 753
chemical-mechanical planarization: 367
chemically sensitive layer: 573, 607, 640
chemiluminescence: 601, 674
chemisorption: 574
chemodiodes: 617, 631
chemoelectrical transduction: 716
chemoluminescence: 15, 601
chemoresistor: 642, 648
chemotransistor: 617, 625
chip cost: 903
chipping: 824
chloride: 448, 623, 626, 636, 640, 648
chlorinated hydrocarbon: 585, 588, 591, 598, 646
chopper stabilization: 914-916, 941
chopping: 34-35, 372
chromatograph: 567
chrome-quartz mask: 867
circuit simulator: 327
circular membrane: 58
cladding layer: 409, 412, 422, 425, 606
clamped-free cantilever: 60
clamping voltage: 383
clark cell: 623
clark electrode: 649
cleaved facet: 430
clinical approval: 733
clipping loss: 374
clock jitter: 324
closed-loop control: 364, 743, 770
closed-loop operation: 533-534, 544
CMOS: 37, 42-43, 47, 49-50, 82-83, 85, 87, 89, 91, 96-99, 153, 159, 174-176, 183, 185, 227, 254, 266-272, 275-276, 278-279, 320, 322, 324, 342, 346, 391, 399, 453, 475-476, 480, 487-488, 490, 518-521, 531, 558, 566, 584, 589, 591, 596, 598, 609-611, 613, 620, 622, 625, 631, 637, 641, 644, 647, 738,

748, 797, 808, 810, 821, 835, 839, 841-844, 846, 851, 902, 913, 917, 921, 934-942
CMOS dielectric: 254, 276
CMOS MEMS: 89, 97, 153, 159
CMP: 367
co-energy: 200-201, 206
CO_2: 239, 534
coagulation: 731
coefficient of friction: 754
CoFe: 364
coherence: 5, 399
cold luminescence: 601
collimation: 450, 884
collimator lens: 374
colloid: 671
color measurement: 874
columnar-grained film: 80
comb drive: 204, 351, 356, 358, 396-397, 779-780, 783
comb structure: 367, 890
comb-drive actuator: 69, 367, 386, 390, 778-779
combustion control: 906, 908
common mode rejection ratio: 43
compact models: 95, 149, 158-159
complementary logic: 332
composite membrane: 59, 82
composite structure: 793
composition: 6, 29-30, 74, 84, 96, 145, 161, 163, 189, 191-192, 567, 576, 602, 615, 619, 815, 864, 906, 911
compressibility: 672, 703
compressive strain: 57
computational thermodynamics: 118
concentrated force: 196-197
concentration: 6, 10, 15, 67-69, 79-80, 230, 232-234, 238-239, 241, 249, 445, 476, 485, 499, 567, 578, 581, 591, 594-595, 617, 619, 621-622, 624-627, 639, 643, 646, 648-649, 673-676, 714, 742, 807-809, 812-814, 830-831, 834, 846, 874, 909
condensation heat: 597
conducting oxide: 571
conduction band: 232, 234, 298-299, 604, 640
conductivity modulation: 485
conductometric: 617, 638-639, 642, 707
conductometry: 617, 639
configuration interaction: 113
confinement factor: 418, 420, 434
confocal geometry: 373
conformal mapping: 203, 208, 477
connector: 869-872, 898
conservation of energy: 200, 210
conservative network: 216, 218, 220
constant-charge system: 755
constant-gap actuators: 765, 783
constant-voltage system: 756, 759, 778
constitutive equation: 788
consumer product: 524-525, 539
contact injection: 289
contact resistance: 265, 338, 639
contactless switch: 908
contactless temperature measurement: 235
continuous-time: 913, 916
control engineering: 205
control system: 185, 561
control theory: 150-152
control volume method: 125, 136

convection: 14, 137, 256, 264-265, 271, 619, 673, 678, 684
convergence problem: 218, 220
conversion: 2, 9-13, 16, 18, 46-47, 66,
conversion chain: 18
conversion efficiency: 18, 245, 282
convex corner: 267, 741, 812-813, 840, 847
cooler: 262-263, 279
coplanar-electrode structure: 297
copper: 231, 261-262, 269, 362, 634, 735, 798, 824, 854, 857, 864
Coriolis force: 539-541, 889-890
corner compensation structure: 812
corpuscular radiation: 6
correlated double sampling: 914, 937
correlation spectroscopy: 392-393, 611
corrosion: 97, 562, 704, 730
cost: 27-28, 31, 119, 157, 161, 181, 188, 192, 230, 240, 254, 276, 281, 320, 331-332, 335, 345-346, 368, 386, 401, 447-448, 523-525, 531, 533, 537-540, 544, 555, 560, 565, 571, 573, 668-669, 689-694, 697, 701, 719, 722, 729, 747, 806, 808, 824, 839, 854, 867, 895-897, 901-906, 932, 937, 940
Cottrell equation: 620
Couette damping: 194, 199, 214, 222
counter-electrode: 558, 619, 621, 813
coupler: 432, 437-444, 450, 602, 605, 869, 871-872
coupling angle: 432-433, 605
coupling coefficient: 788-789
coupling efficiency: 374, 431, 433
coupling technique: 404, 429-430, 432
covalent binding: 707
Cr layer: 289
crab-leg spring: 206-207
cracking: 55
creep: 458, 463
critical angle: 406-407, 413, 433
cross section: 46, 62, 197, 204, 233, 250, 260, 267-268, 294-295, 336, 383, 392-393, 409, 416, 418, 435, 474, 476, 494, 552-553, 593, 605, 609, 680, 682, 701, 705, 713, 806, 810, 813, 831, 879
cross-bar support: 354
cross-sensitivity: 19-20, 26, 35, 39, 458, 462, 476, 497, 499, 502, 512, 568, 571, 641
cross-sensitivity cancellation: 35
crosslinking: 707
crosstalk: 98, 162, 252, 355, 363, 367, 377, 909
cryogenic temperature: 516, 849
crystal optics: 884
crystal oscillators: 324
CsOH: 807, 809, 847
current filament: 478, 494
current modulation: 322, 771
current-related sensitivity: 458
curvature: 55, 81, 89, 196, 353-355, 373, 679, 793
curve fitting: 208
cutback method: 437
cutoff: 413, 426, 428, 912
CVD: 84, 230, 266-267, 269, 593, 631, 825-826, 829
cyanide: 626
cyclic loading: 69, 75, 81
cyclic voltammetry: 621
cyclohexane: 725, 837
cycloid gear: 885, 888

INDEX 949

cylindrical lens: 884
cytotoxicity: 731

D

damage-induced loss: 422
damping coefficient: 194
dangling bond: 341
dark current: 282-283, 289-291, 293, 311, 317, 320, 326
data reduction: 573
data transmission: 3, 901, 924, 933
decay time: 4
decimating filter: 922, 926-927
decoupling: 356
deep diffusion: 830
deep reactive ion etching: 354, 400, 525, 802, 818, 821, 849
defect: 75, 299-302, 304-305, 332, 338, 340-341, 754, 815, 903
defect-state creation: 301-302, 304
degenerately doped silicon: 39
degradation: 299, 302, 315, 320, 340, 461, 694, 730, 744, 902, 931, 933
degrees of freedom: 107, 162, 168, 170, 191, 201, 209, 221, 267, 738-739, 902
demultiplexer: 332-333
density: 2, 4-5, 14, 21-25, 42-43, 46, 60, 72, 74, 112, 114, 134, 144, 162, 194, 233, 250, 260, 262, 265, 286, 289, 297-301, 311-312, 320, 332, 338, 367, 447, 460, 466, 479-480, 482, 488, 514, 580-581, 603, 619, 624-625, 627, 639, 643, 670, 674-676, 678, 709-712, 725, 741, 754, 756-757, 759, 764-765, 789-790, 803, 814, 817-819, 834, 849, 864, 913, 922, 925
deoxyribonucleic acid: 601
depletion layer: 283, 286, 295, 476, 491
depletion zone: 234, 251
depth of focus: 431
design automation: 94-95, 98-99, 138
design rule: 159, 267
destructive interference: 410, 438
detection mode: 540-544, 548-550
detection of molecule: 567
detectivity: 18, 246-247, 252-253
detector saturation: 289
device design: 188, 282, 458, 499, 510, 520
device domain simulator: 94
dew-point sensing: 34, 265
diagnostic: 273, 275, 315-318, 320, 454, 729, 748
diagnostic microstructure: 275
diamond saw cutting: 695
diaphragm: 53, 444, 554, 566, 721, 845
dichlorosilane: 833
dielectric: 4, 49, 55, 62, 65, 84, 154, 180, 254, 257, 267, 269, 276, 282, 296, 309, 312, 333, 338, 341, 367-368, 389, 405-407, 422, 431-432, 435, 468, 510, 551-552, 556, 559, 589, 593, 596, 603, 608, 635, 645-646, 648, 674, 689, 757, 761, 773-777, 797, 814, 839, 911, 919
dielectric constant: 4, 65, 296, 332, 337, 468, 644-646
dielectric loss tangent: 4
dielectric material: 757
dielectric thin film: 55

dielectrometer: 643
dielectrophoresis: 689
diethylenglykolmonobutylether: 860
differential measurement: 66, 631
diffraction: 365, 372, 388, 390-391, 394, 402, 431, 435, 611, 860, 869, 885
diffraction efficiency: 394
diffraction grating: 365, 391, 402, 611, 860
diffractive optical: 346, 390-391, 393, 402, 611
diffuser: 46, 698-699, 720
diffuser-nozzle valve: 698
diffusion coefficient: 335, 620, 676
diffusion current: 240-241
diffusion length: 288, 471, 488, 498, 620
digital projection: 364
digital signal processing: 514, 613, 901, 931-932
dimensional control: 840
diode: 178, 240, 281, 284-287, 289-291, 293-295, 309, 311, 314, 319-320, 339, 390-391, 401, 404, 471, 485, 488, 491, 496-498, 518, 520, 611, 624, 626, 632-633, 874, 881, 917
diode capacitance: 314
diode Hall effect: 471, 491, 517
dipole moment: 112
direct etching: 821
direct-write micromachining: 823
direction of movement: 251
discharging time: 333
discretization: 42, 44, 106, 109, 125, 128, 131, 133, 135, 146, 148, 151, 159, 166, 181-182, 921
discretization noise: 42, 44
dispenser: 708, 724
dispensing: 46, 598, 667, 708, 724
dispersed phase: 671
dispersion force: 574
displacement: 4, 11, 25-26, 49, 57, 63, 66, 68-69, 122, 169, 171, 190, 194-197, 200-202, 204-209, 212, 215, 219, 222, 349, 360, 367, 376-377, 384, 388, 393, 442, 449-450, 452, 455, 514, 583, 588, 635, 683, 702, 772, 777-779, 788-789, 791, 794, 797, 802, 881-882, 884
displacement sensor: 449-450
display: 162, 335, 348, 365, 391, 394, 397-399, 402, 451, 684, 772, 801
dissipative particle dynamics: 120, 897
dissipative transport process: 21
distributed mass: 198, 215
distribution: 4, 43, 55, 57, 96, 110, 116, 122, 143, 145, 159, 172, 179-180, 208, 298-299, 374, 377-378, 417-420, 430, 434-436, 440, 444, 466, 491, 500, 603-604, 626, 641, 675, 686, 689, 691, 708
DMD: 95, 365, 525, 771-772
DNA: 601, 669, 707, 709, 712, 718, 720, 724, 729, 745-746, 749
domain simulation: 94
domain simulator: 138, 151
doping concentration: 248, 830
double layer: 688, 703
double-differential architecture: 43
double-resonant operation: 542, 544
doubly supported beam: 56-57, 60, 69
DRIE: 354, 357, 360, 362, 364, 377, 384-386, 402, 525, 530-531, 535, 546-547, 551-553, 781, 785, 819-820, 840

drift: 17, 37, 244, 270, 295, 368, 458, 463, 471, 478, 488, 490, 506-507, 509-510, 512, 516, 518, 556, 571, 628, 631, 688, 849, 912
drift velocity: 463, 477, 489
drive voltage: 352, 770, 790
drop-on-demand: 708, 711
drug delivery: 699, 701, 708-709, 724, 731, 743-744, 747, 749
drug discovery: 667-668, 724
dual-resonator bandpass filter: 189
dual-transistor structure: 241
dust: 51, 537, 554, 787
duty cycle: 4, 321, 325
duty-cycle modulation: 10
DWDM: 353
dynamic equilibrium: 575, 578, 624, 686
dynamic operation: 362
dynamic pressure: 680-681, 723
dynamic range: 282, 289, 320, 327-331, 370, 386-388, 399, 456, 528, 571, 910, 917, 927, 937, 939, 941-942
dynamical behavior: 262

E

e-grain: 17, 44
e-h pair: 284, 295
e-h pair: 295
E^2PROM: 557
ECE: 184-185, 614, 810-811, 813, 839
edge effect: 195, 289
EDP: 267, 588, 590, 595, 598, 644, 807-808
effective index: 415, 426-429, 432, 444
effective mass: 231
effective volume: 80
effectivity: 16-17, 32, 38
eigenvector: 107, 109
elastic compliance: 788
elastic measurement: 62
elastic modulus: 87, 754, 794
elastostatic: 207
electret: 51, 803, 941
electric field: 4, 109, 199, 208, 282, 284, 295, 312, 314, 407, 416-418, 426, 445, 463, 466, 481-482, 485, 490-491, 619, 627-628, 646, 688-689, 703, 756-758, 762, 764, 766, 787-791, 797
electrical breakdown strength: 65
electrical discharge machining: 735
electrical double layer: 688, 703
electrical insulation: 235, 251, 816-817
electrical potential: 24, 244, 579, 691, 758
electrical signal: 3-5, 9-10, 456, 497, 502, 535, 541, 559, 569-570, 583, 592, 901
electrical stiffness: 203
electrical thermometer: 241
electro-optical: 404, 446, 448, 607, 874, 877
electro-optical effect: 404, 446
electro-optics: 446, 452, 871
electrochemical discharge machining: 735, 748
electrochemical dissolution: 743
electrochemical energy: 44
electrochemical etching: 251, 258, 817-818, 849
electrochemical passivation: 810
electrochemical potential: 6, 579-580, 624, 706
electrochemical sensor: 580, 648
electrochemically controlled pn etch-stop: 810
electrohydrodynamic: 703, 722
electrokinetic: 667, 688
electrokinetic effect: 667
electrokinetic motion: 688
electroless dissolution: 814
electrolyte: 45-46, 617, 622-624, 628, 632-633, 636, 649, 674, 706, 735, 848
electrolytic conductivity: 674
electromagnet: 362, 364
electromagnetic: 2, 5-6, 11, 26, 42, 44, 46-48, 97, 131, 346, 349, 361-364, 372, 387, 398, 400-401, 407, 409, 416-417, 495, 543, 546-547, 563, 574, 598-599, 603, 702, 753, 797-798, 876, 879-882, 884, 891, 899, 907, 911
electromagnetic interference: 97, 574, 907
electromagnetic scanner: 362, 798
electromechanical memory: 774, 777
electromechanical transduction: 751-753
electromotive force: 464, 580
electron ejection: 286
electron trapping: 299
electron-hole pair: 47, 282, 322
electron-hole plasma: 485, 494, 499
electron-ion pair: 762
electronic stability program: 523
electroosmosis: 702, 716
electroosmotic flow: 687, 703
electrophoresis: 688, 694, 716, 719-720, 722, 724, 726, 746
electrophoretic mobility: 688
electrophoretic separation: 716
electroplated copper: 362, 824
electroplated Ni: 84, 885
electroplating: 15, 84, 364, 853, 857, 860, 862, 864, 867, 885, 889
electrostatic actuator: 66, 69, 106, 207, 218, 388, 532, 758, 782, 787, 802
electrostatic breakdown: 760-762, 764, 766
electrostatic drive: 540, 549-550
electrostatic energy: 755-757, 759-761, 764-765, 767, 778, 790
electrostatic excitation: 60
electrostatic field theory: 188
electrostatic force: 14, 34, 63-65, 68, 88, 91, 200-201, 204, 215, 224, 352, 383, 481, 756, 759-761, 766, 768, 770
electrostatic force grip: 63-64, 88, 91
electrostatic induction: 786
electrostatic scanner: 349, 351
electrostatic spring: 533
electrostriction: 797, 803
electrothermal: 97, 152, 163-164, 166, 185, 217, 532, 559, 797, 803, 798-799
elementary particle: 3, 6
embossing: 448, 692-693, 726, 854, 856, 864-865, 874
emissivity: 236, 264
emitter current: 241
emitter-injection modulation: 491, 518
emulsion: 671
end-fire coupling: 430-431
endoscopic: 345, 738, 740, 747-748
energy band structure: 297
energy density: 2, 5, 757, 759, 764-765, 789-790
energy field: 2
energy harvesting: 44, 47
energy minimization method: 58
energy storage: 18, 44

Index

energy stored in the spring: 767
enhancement-mode: 332
entropy: 3, 5, 24, 341, 578, 686
entropy flow: 24
environment domain simulator: 94
environmental condition: 907
enzymatic reaction: 592
enzyme immobilization: 598
epi-micromachining: 814, 825, 829, 844
epi-poly: 835-836, 845
epi-process: 805
epitaxial layer: 477, 489, 825, 829-830
epoxy glue: 562
equalizer: 370
equilibrium: 6, 17, 23-25, 47-48, 104, 109, 111, 118, 298, 328, 341, 466, 485, 494, 575-580, 591, 597, 617, 623-624, 630, 634, 673, 685, 758, 766, 768, 777
equilibrium constant: 575-576, 578-579, 597
equipartition theorem: 23
equipment domain simulator: 94
error estimation: 149
etch stop: 286, 530, 611, 815-816, 839, 848
etch-rate: 809, 815, 820
ethanolamin: 860
ethene: 628
ethernet: 933
ethylene oxide: 730
ethylene-diamine-pyrocatechol: 807
Euler-Bernoulli equation: 196
evanescent field: 438, 603-604
evaporation: 3, 231, 264, 384, 584, 588, 595, 610, 622, 637, 640, 644, 647, 670, 710
excess minority carrier: 498
exhaust gas: 906
expansion coefficient: 74-75, 145, 337, 672, 691, 694-695, 889
extinction ratio: 370
extrinsic stress: 55, 74

F

fabrication: 7, 26, 37, 58, 68, 158, 174, 184, 187-188, 191, 230, 238, 240-241, 254, 263, 265-267, 271, 273-275, 277-278, 283-285, 289, 299, 309, 335, 337-338, 340, 343, 346-347, 354, 364, 374, 380, 384, 386, 396-397, 402, 441, 445, 447, 471, 474, 476, 488, 510, 524, 530-531, 533, 540, 552, 555, 557, 562, 564, 566, 584, 587-588, 590-591, 595, 598, 606, 609-611, 614-615, 622, 625-626, 637, 640-641, 644, 647, 667, 689, 692-694, 701, 710-712, 719, 724, 730, 733-735, 738-739, 741, 744-748, 753, 780-781, 796, 801-802, 805, 807, 809-810, 847-851, 854, 860, 863, 865, 867, 869, 884-885, 891, 896-898, 902-904
fabrication technology: 187, 278, 340, 554-555, 692, 865
Fabry-Perot: 388, 401, 404, 437, 441, 608-609, 798
far-field: 178, 435-436
Faraday constant: 579, 620
fast prototyping: 816-817, 823-824
fast-etching plane: 812
fast-setting glue: 694
fatigue: 14, 55, 69-70, 75, 81, 84, 89, 97
fatigue fracture test: 81
feedback control: 356, 364, 368
FEM: 62, 103-104, 109, 122-123, 125, 128, 135-136, 141, 143-145, 166, 171, 173, 177, 226

Fermi level: 244, 299, 301-302, 634
ferromagnetic: 455-456, 483, 516, 753, 801
FIB: 822-824
fiber: 46, 63, 345, 353, 355, 365, 368, 372, 374-375, 384, 386-387, 389, 395-396, 399-400, 402, 431-433, 437, 449, 451, 603-604, 615-616, 655, 869-871, 873-874, 876, 878-881, 884, 887-888, 898
fiber diameter: 869
fibrinogen: 731
field-effect mobility: 338
field-effect transistor: 340, 626
filament magnetosensitivity: 491
fill factor: 252, 309, 320, 338, 353, 366-367, 373-374
film nucleation: 56
film shrinkage: 79
filter: 21, 34-35, 43, 189-190, 220, 227, 253-254, 388, 392, 399, 401, 441, 546, 682, 798-799, 871, 873, 916, 918, 927, 941
finesse: 389, 441
finite difference: 98, 103, 111, 128, 130-132, 175
finite difference method: 128, 132, 175
finite element: 67, 98, 100, 102-103, 111, 122-126, 128, 133, 144, 154, 163, 166, 173, 180, 182-184
finite element method: 103, 122-123, 125-126, 128, 133, 163, 173, 184
finite volume method: 103, 125, 129
first-order $\Sigma\Delta$ modulator 269
fixed-fixed beam: 195, 197, 207
flame safety lamp: 593
flame temperature: 162-163
flap valve: 225, 691
flash evaporation: 264
flat-band voltage: 298, 633
flexible plates: 193
flexural plate wave: 703
flexural-plate-wave device: 586-587
flicker noise: 23, 913
flicker noise coefficient: 913
flip-chip: 357, 396, 451, 799, 904
flip-up: 380
flow control: 667, 690, 696
flow rectifier: 697
flow sensor: 29, 49, 257-258, 269, 278, 592, 704, 723, 898, 936
flow velocity: 30, 229, 257, 680-681, 704-705, 738
flow-channel field: 46
flow-on-demand: 709
fluctuation-dissipation theorem: 21, 49
fluidic capacitance: 682-683
fluorescence: 601-603, 612, 674-675, 690, 694, 704, 716, 746
fluorescent dye: 709
fluorocarbon polymer: 819
fluoroscopy: 319-320, 324, 342
fluorpolymer: 734
fluxgate magnetic sensor: 36
foaming: 859
focused ion beam: 822, 840, 850
force: 3-4, 14-15, 34, 58, 62-63, 65-68,
force array: 777
force-feedback knife: 793
formaldehyde: 730
forward reaction: 575
forward voltage: 240-241, 496, 632-633
foundry process: 351, 367, 780-781
four-fold switching: 40-41

Fourier transform infrared (FTIR) spectrometer: 881
FPW: 587-588, 703
fracture force: 66
fracture strength: 55, 62, 68, 88, 800
fracture toughness: 55, 68, 81, 89-90, 92
frame rate: 282
free convection: 14, 256
free energy: 5, 574-576
free enthalpy: 5
free-space optical MEMS: 347, 349, 351, 353, 355, 357, 359, 361, 363, 365, 367, 369, 371, 373, 375, 377, 379, 381, 383, 385, 387, 389, 391, 393, 395, 397, 399, 401
free-space system: 403
freeze-drying: 837
frequency: 4, 22-24, 34, 38, 42-43, 46-47, 50, 60, 69, 72, 75, 107, 109, 145, 193-194, 200, 225, 259, 267, 329, 331, 340, 346, 349, 352, 354-356, 358-360, 367, 391, 397, 399, 441-442, 456-458, 461-462, 473, 487, 492, 494-495, 507, 512, 528-529, 533, 541-544, 546-547, 559, 562, 564-565, 569-570, 581-587, 589-590, 598-601, 603, 639, 642, 645-646, 651, 682, 689, 719, 723, 751, 771, 773-774, 781, 783, 792, 795, 798-799, 802, 829, 879, 912-916, 920, 922-925, 927, 930, 933, 935, 939, 941
frequency bandwidth: 22
frequency mixing: 267, 545
frequency shift: 69, 75, 570, 581, 586
frequency tuning: 542
Fresnel reflection: 431
friction coefficient: 65
friction force: 65
fringing field: 757, 764
front-end: 937, 939
FTIR: 881, 899
fuel cell: 45-46, 51
fugacity: 576
full-scale output: 16, 459, 461-462
full-width half-maximum: 441
functional block: 7
functional material: 53, 144, 188
fundamental mode: 418, 589
fusion: 354, 358, 554, 556-557, 611, 691, 802, 821, 849
fuzzy logic: 909

G

GaAlAs: 388, 401
GaAs: 388-389, 401, 414, 429, 434-435, 446-447, 449, 452, 467, 469, 473, 475-476, 482, 486-488, 491, 514, 518, 585, 604-605, 608
gallium oxide: 642
Galvani potential: 579, 634
galvanic cell: 811, 847
galvanic etch-stop: 810-811
galvanic isolation: 455
galvanomagnetic sensor: 467, 469
gap wavelength: 446-447
gap-closing actuators: 388, 765, 767, 777
gas bubble: 681
gas chromatography: 716
gas composition: 29-30, 906, 911
gas flow sensor: 29, 49, 257
gas interaction: 97
gate electrode: 296-297, 627, 771
gate-induced charge: 297

Gauss elimination: 103
Gaussian beam: 367, 373, 430
Gaussian spot: 430
generation of netlists: 220
geometrical: 40, 67, 98, 124, 139-140, 143-144, 161, 172, 238, 247, 264, 466, 471, 478, 482, 543, 556, 698, 823, 839, 891
giant magnetoresistance: 453, 483, 519
giant negative magnetoresistance: 483
Gibbs energy function: 118, 575
Gibbs enthalpy: 686
Gibbs free energy: 574-576
gimbal: 348
glass soldering: 696
glass substrate: 296, 310, 349, 354
glucose oxidase: 623, 631, 649
glucose sensor: 743
GMR: 483-484
Golay cell: 11
gold absorber: 857
gold bump: 361
gold plating: 864
gold/chromium: 816
graded index: 379
gradient-index (GRIN) lens: 450
grain boundary: 639
grain growth: 79
graphical user interface: 99, 101, 216
grating coupler: 432, 450, 602, 604
grating light modulator: 390, 801
grating light valve: 390, 402, 774, 776
gravimetric sensor: 581
gravitation: 3
gravitational wave: 3
gray-tone lithography: 862
Green function: 134-135
greenhouse effect: 909
GRIN: 379
grinding: 695
guided-end beam: 195, 197
gyration: 10
gyroscope: 211, 539-550, 564-565, 889-890, 899

H

Hagen-Poiseuille: 680, 682, 699
Hall angle: 466, 482
Hall cell: 464
Hall coefficient: 466, 468, 482
Hall effect: 8, 14, 39, 171-172, 453, 464-467, 471, 477, 491, 494, 500-502, 517-520, 913
Hall effect device: 453, 471, 491, 500, 516
Hall element: 464, 467, 489, 507, 517-518
Hall field: 466, 470, 478
Hall generator: 464
Hall plate: 33, 174, 186, 464, 466, 477, 518
Hall sensor: 33, 172, 174, 464-465, 467, 470, 510, 512, 515, 519, 521
Hall slab: 464
Hall transducer: 464, 503
halogenated compound: 571
Hankel norm approximation: 152
harsh environmental condition: 524, 902, 910
health care: 345, 729, 735-736, 745
heart valve: 732
heat capacitance: 24
heat capacity: 5, 14, 72, 89, 111, 118, 183, 275, 279,

INDEX 953

592, 678
heat diffusion: 675, 678
heat exchange: 17, 256, 713-714
heat exchanger: 684, 714, 719, 725
heat flux: 5, 109, 219, 259-262, 264-265, 268, 271, 279
heat flux sensor: 279
heat of chemical reaction: 239
heat of oxidation: 594
heat sink: 246, 251, 254, 262, 264, 271, 276
heat source: 683
heat transfer: 15, 154, 162-163, 165, 255, 259, 551, 683-684, 719, 725
heat transfer coefficient: 259, 683, 725
heat transmission coefficient: 260, 684
heavy-metal ion: 619
hematocrit: 649
hemolysis: 731
hetero-junction: 293
heterodyne receiver: 871, 874
heterogeneous catalysis: 715
heterojunction structures: 476
HF-based solution: 813
hierarchical approach: 189
high injection: 485
high-aspect-ratio micromachining: 817
high-density microarray: 710, 712
high-energy electron: 283
high-force actuation: 772
high-temperature superconducting cuprate: 628
high-throughput screening: 669
higher order mode: 417-418, 435
hologram: 362
horizontal comb drive: 783
horizontal force: 66
hostile: 732, 907
hot-embossing: 854, 856, 864-865, 874
hotplate: 594, 642, 644
humidity sensor: 269, 848, 939, 942
hybrid electrostatic microactuator: 785
hybrid integration: 451, 527, 537-539, 871, 898
hydraulic inductance: 682
hydrazine: 807-808, 846-847
hydrocarbon: 571, 585, 588, 591, 598, 606, 610, 640-641, 646
hydrodynamic focusing: 680-681
hydrodynamic resistance: 698
hydrofluoric acid: 89, 534, 554, 848-849
hydrogen peroxide: 623, 648-649
hydrogen sulfide: 623, 628
hydrogenation: 715
hydrophilic: 168, 686, 691, 700, 720
hydrophilicity: 686, 698
hydrophobic: 168-169, 632, 686, 698, 700, 720
hydrophobic break: 698
hydroxyl: 623, 630, 648
hysteresis: 16, 299, 458, 461, 507, 516, 568, 631, 775, 777

I

I2C (Inter-IC) bus: 932
I2S: 933
ideality factor: 282, 286
ignition temperature: 162-164
image processing: 67, 314-315
image sensor: 281, 287, 339, 342
imaging: 254, 279, 281-284, 286, 289, 295, 302, 309, 314-318, 320, 324, 326, 331-332, 334-335, 338-339, 342-343, 346, 348, 355, 370, 386, 394, 399, 434, 440, 452, 611, 823, 869, 884
immune and allergenic response: 731
immuno-isolation barrier: 744
immunoassay: 588
immunoreaction: 607
impedance: 4-5, 10, 21, 208, 218, 418, 458, 463, 531, 545, 557-558, 586, 617, 619, 638-639, 641, 682, 913, 919
impedance matching: 10
implant: 425, 731, 736
implantable: 709, 724, 743-744, 749, 797, 908, 935
impurity concentration: 233-234, 241, 485
impurity scattering: 234
in-band noise component: 914
in-plane strain: 61
InAs: 467, 469, 476, 482-483, 485, 518-519
incidence angle: 406-407, 409
incremental A/D converter: 927, 940
index profile: 179, 426
index-matching liquid: 386
indium oxide: 642
indium tin oxide: 289
inductance: 4, 136, 184, 912
inductively coupled: 752, 819
inertial device: 751
inertial force: 198-199, 889
inertial sensor: 527, 561, 565
inflammation: 731
information processing: 6, 901
information technology: 5, 48, 667
infrared detector: 49, 243
infrared sensor: 278
infrared window: 253
injection molding: 448, 693, 734, 854, 864-865, 891
ink-jet printing head: 525
inkjet technology: 667
InP: 389-390, 401, 446-447, 452, 604
input domain: 12
input impedance: 458, 463, 618
input signal: 7, 10, 12-13, 16, 18-20, 27, 31-32, 43, 72-73, 319-320, 322-324, 331, 914, 921, 924-927
input terminal: 17
input transducer: 9, 453
InSb: 467, 469, 473, 482-483, 485, 487, 517
InSb: 467, 469, 473, 482-483, 485, 486-487, 518
insect antennae: 630
insertion loss: 353, 355, 365, 369, 374, 376-377, 384-388
instrumentation amplifier: 49, 329, 936
insulin delivery: 722, 743
integrated circuit technology: 846
integrated laser: 429
integrated optic: 449
integrated processing: 841-842
integration mode: 321
intensity modulation: 404
interaction energy: 2, 574
interdigital transducer: 582-583, 586
interdigitated comb: 203, 534, 779, 782
interdigitated comb structure: 534
interdigitated electrode: 641
interface: 5, 50, 57, 98-99, 101, 110, 115, 135-137, 148, 153, 164, 169, 187, 190, 216, 220, 224, 282, 285-286, 289-291, 295, 297, 299, 301, 304, 311-

312, 314-315, 340-341, 406-407, 433, 437, 453, 486, 488, 502-504, 506-507, 509-511, 516, 517, 520, 521, 573-574, 603, 608, 609, 619, 621, 626, 627-628, 630, 631, 632-633, 634-635, 636-637, 738, 748, 751, 816, 835, 836, 859, 899, 901-905, 907-913, 915, 917, 919-921, 923, 925, 927, 929, 931, 933-942
interface circuitry: 5, 901-905, 907, 909, 911, 913, 915, 917, 919, 921, 923, 925, 927, 929, 931, 933-935, 937, 939, 941
interfacial surface tension: 667, 670, 685, 694, 707, 713
interference: 97, 390, 393, 405, 440-441, 443-444, 452, 456, 574, 602, 606, 608, 639, 716, 907
interference coupler: 440-441
interferometer: 441-444, 450, 452, 605-609
intermodulation: 916
internal stress: 55, 60, 67, 75, 80, 90, 335, 864
interoperability: 192, 219-220
intrinsic carrier concentration: 232
intrinsic semiconductor: 296
inversion layer: 474
inverted-staggered structure: 296-297
inverted-staggered structure: 297
ion beam milling: 741
ion implantation: 79-80, 90, 364, 810, 830, 847
ion-controlled diode: 625, 631
ion-selective electrode: 622-623, 648-649
ion-selective membrane: 623, 629, 707
ion-sensitive capacitor: 625
ion-sensitive field effect transistor: 916
ionic mobility: 689
ionicity: 906, 911
ionization factor: 241
ionization potential: 762
IPA: 828, 846
iridium: 627, 631, 643
iridium oxide: 628
irradiance: 236
irreversible characteristic: 567
ISFET: 625-626, 628-631, 633, 637, 706, 911, 917
isopropyl alcohol: 836
isotropic etching: 58, 274, 805, 813, 848
ITO: 289-295, 339-340, 662-663, 666

J
Jablonski diagram: 600
Jacobian matrix: 104-105
jet contraction: 680-681, 699
Johnson: 21, 50, 88, 91-92, 225, 246, 278, 659-660, 746
Joule heat: 70, 73

K
Kapton: 335, 337, 343
Kelvin probe: 634-635
kinetic energy: 210-211, 762
Kirchhoffian network: 191, 225
KOH: 45, 268, 273, 286, 377, 383, 530, 547, 551, 553, 584, 588, 590, 595, 598, 644, 690, 741, 807-809, 818, 826, 834, 838, 846, 848

L
lab-on-a-chip: 669, 715, 717, 867
Lagrange multiplier method: 138
Lamb wave: 587

Lambert-Beer relation: 600
laminar flow: 678-681, 704, 713
laminar nitrogen: 257
laminating: 690
lanthanium fluoride: 624
lapping: 695
large-area digital imaging: 334
large-area electronics: 335
large-volume product: 524
laser ablation: 692, 694, 748
laser drilling: 616
laser fluorescence: 704
laser light reflection: 589-590
laser micromachining: 719, 735, 821-822, 840, 850
laser tissue welding: 740
laser trimming: 28, 543
laser-scanning confocal microscope: 362
latched comparator: 925, 927
lateral comb drive: 204
lattice vibration scattering: 231, 234
lead magnesium niobate: 857
lead zirconate: 390, 528, 588, 789, 857
lead zirconate titanate: 390, 528, 588, 789, 857
leakage current: 234, 281-283, 286, 288, 291, 294, 297, 302, 311, 316, 320, 337-338, 937
length scale: 145, 346, 349
lens: 365, 369-370, 374, 386, 430-431, 434, 450, 614-616, 871-873, 876, 878, 881, 884-887, 899
lens array: 871-872
Levenberg-Marquardt algorithm: 208
levitation: 454, 787, 797
life science: 684, 717, 724
lifetime: 81, 335, 524, 561, 694, 783, 932
LIGA: 525, 690, 694, 734, 739, 749, 753, 781, 785, 797, 853-855, 857, 859-861, 863-865, 867-869, 871, 873-877, 879, 881, 883-885, 887, 889, 891, 893, 895-899
LIGA gyroscope: 889
light emission: 445-448
light intensity: 10, 28, 437, 495, 570, 599, 816-818
light-emitting diode: 401, 610
light-guiding: 404, 606
limit of detection: 569
limit stop: 769, 771, 775
$LiNbO_3$: 445, 582
linear response: 27, 286, 558
linearity: 16, 26, 252, 286, 322-323, 326, 328-331, 532, 889, 938
liquid chromatography: 716
liquid crystal polymer: 857
liquid density: 580
liquid-crystal display: 281, 340-341
liquid-solid interaction: 692
lithium niobate: 445, 582
lithium tantalate: 581-582, 592
lithium-ion battery: 45
Littman/Metcalf tunable resonator: 390
Littrow configuration: 370
load resistance: 245, 324, 332
load-deflection technique: 58-59, 61, 67, 82-83
lock-in technique: 879
locomotion: 10, 801
long-term stability: 88-89, 458, 476, 515-516, 556, 562, 571, 706-707, 713, 808
look-up table: 27, 557
Lord Kelvin: 243, 277

INDEX 955

Lorentz force: 349, 463, 466, 477, 481, 484-485, 489, 494, 546
loss parameter: 436-437
lost template: 854
Love wave: 582
low pressure CVD: 84
low-density microarray: 710
low-doped silicon resistor: 556
low-noise amplification: 43, 910
low-pass filter: 43, 545, 682, 914
low-pressure CVD: 826
Low-temperature fuel cell: 45
LPCVD: 60, 75, 77-78, 84, 86, 88, 90-91, 588, 590, 595, 606, 637, 644, 814, 828-829
LSI: 54, 75, 82
lubricant: 169-170
Lubrication: 225, 785
lumped-element: 682

M

μc-Si:H: 337, 338
μc-Si:H: 338
Mach-Zehnder: 404, 443, 452, 605-606, 608
macromodeling: 193, 205, 207-208
magnetic actuation: 361, 381
magnetic detection: 540
magnetic drive: 540
magnetic field: 2, 5, 8, 33-34, 50, 171-173, 182, 186, 361, 383, 407-408, 416, 453-454, 456-457, 460-466, 471, 476, 478, 481-482, 484-485, 488, 495, 497-500, 502-503, 507, 512, 514-515, 517, 519-520, 539, 543, 546, 587, 879, 906, 911
magnetic induction: 5, 40, 459-462, 484, 500
magnetic levitation: 454
magnetic moment: 5
magnetic permeability: 5
magnetic sensor: 34, 455-457, 484-485, 502, 514-515, 518
magnetic shield: 484
magnetic signal: 3, 5, 40, 495
magnetic torque: 361
magnetic transducer 454, 456, 514, 520,
magnetic vector: 489
magnetization: 5, 15, 460, 483
magnetoconcentration: 485, 491, 493, 498
magnetocoupler: 484
magnetodiode: 485-487, 498, 519-520
magnetodiode (MD) effect: 485
magnetogradiometer: 453
magnetohydrodynamic: 703
magnetomechanical microactuation: 753
magnetoresistance: 14, 481-483, 498
magnetoresistor: 483, 505
magnetosensitivity: 453, 457-458, 462-463, 470-471, 476, 490-491, 494, 498, 500, 503, 511-512, 514-516
magnetostatic force: 15
magnetostriction: 14-15, 752-753
magnetostrictive effect: 364
magnetotransistor: 14, 509, 512, 518, 520
majority carrier: 470,624,634
mammography: 319-320, 324, 326, 331
MARS: 370-37 1
masking material: 808
mass: 4, 6, 10, 24-25, 27, 29-31, 34, 63, 69, 75, 137, 193-195, 198, 209, 211-214, 217, 251-252, 268, 290, 349, 393, 455, 468, 524-525, 528-533, 535, 540, 542-543, 546-547, 549, 564, 569-570, 580-582, 584, 586-587, 589-591, 594, 619, 644, 646, 671, 674, 678, 682, 690, 692, 713-714, 719, 723, 729, 795, 814, 840, 854, 889, 903, 936
mass flow: 4, 10, 29-31, 268, 682, 723, 936
mass production: 251, 690, 729, 839, 854
mass-sensitive sensor: 580, 582
mass-sensitive sensor: 94,
master-slave system: 37
matching: 10, 42, 151-152, 205, 242, 379, 389, 430-434, 586, 616, 676, 902-904, 918, 925
material optimization: 138, 144-145
material property: 54, 60, 124, 157, 180, 192, 800
materials domain simulator: 94
maximum electrostatic-energy density: 764
Maxwell's equation: 190, 409
MBE: 264, 473
mean free path: 255, 483, 676, 763
meander: 232, 234, 236, 238-240, 255, 269, 271, 481, 587, 594
meander-shaped resistor:240
mechanical antireflection switch: 370
mechanical fluctuation: 25
mechanical grip: 63
mechanical microsensor: 524, 561
mechanical milling: 805
mechanical potential: 2, 213
mechanical property: 54, 87, 92
mechanical signal: 3-4
mechanical stability: 252, 355
mechanical wear: 769
medical: 46, 111, 254, 283, 295, 315, 317-318, 320, 326, 338-339, 342-343, 524, 526, 571-572, 655, 668, 717, 730, 733-736, 739, 743, 822, 884, 891
medical imaging: 283, 295, 317-318, 326, 339, 342, 884
medium-density arrays: 710
megasound: 860
Melloni: 243, 277
MELO: 833, 845, 851
mesh: 108-109, 111, 122-125, 127-128, 130-131, 133, 135-136, 138, 143, 145-149, 156, 159, 166, 181-182, 184, 889
mesh adaptivity: 138, 145-147, 149, 182
mesh generation: 138, 145-148, 159, 166, 184
mesh refinement: 130, 143
meshed model: 190
metabolism: 637
metal thin film: 560
metal-oxide semiconductor: 82, 626
methane: 239-240, 593-595, 610
metric: 188
Michelson: 404, 442, 444, 449-450, 452, 881
micro tensile test: 84
micro-optical element: 345
micro-opto-mechanical system: 403
microactuators: 10, 193, 207, 209, 229, 266, 276, 397, 541, 748, 751, 753-755, 757, 759-763, 765-767, 769-775, 777-781, 783-787, 789-793, 795-797, 799-803, 911
microarray: 669, 710-712, 718, 724
microbolometer: 236-239
microbridge: 594
microcalorimetric application: 591
microcamera: 738
microchannel: 716, 725-726

microcilia: 10, 753
microcoil matrix: 379
microcombustion engine: 753
microcontroller: 526
microcooler: 263-265
microcrack: 824
microcrystalline silicon: 294
microdrilling: 822
microdroplet: 673
microelectromechanical system: 1, 87, 89, 187, 189, 227, 229, 260, 245, 399, 401, 402, 367, 729, 747, 751, 846
microelectronic: 7, 21, 23-24, 42-43, 159, 243, 263, 453, 667-668, 729, 734, 747
microengine: 801
microfilament: 594
microflexure: 779
microfluidic assembly: 168, 182
microfluidic chip: 158, 715
microfluidics: 97, 153, 158, 224, 667-671, 673, 675, 677, 679-681, 683, 685-687, 689, 691-693, 695-699, 701-703, 705-707, 709, 711, 713, 715, 717-723, 725-727
microgear: 780, 867-868
microgripper: 799
microindenter: 69
microinstrument: 723, 739, 747
microlatches: 376
microluminometer: 611, 613
micromachining: 30, 47, 51, 53, 56, 91,
micromachining technology: 47, 53, 377, 539, 546, 551, 565, 573, 805, 807, 809, 811, 813, 815, 817, 819, 821, 823, 825, 827, 829, 831, 833, 835, 837, 839, 841, 843, 845, 847, 849, 851
micromechanical flexure: 786
micromirror: 189, 345-347, 349-350, 353, 355, 362, 364-370, 372, 374, 383-384, 388, 393-400, 615-616, 753, 771, 798, 803
micromodule: 902-905, 910
micromolding: 719, 862, 897
microphone: 803, 941
micropiercing structure: 745
micropositioner: 10
micropump: 225, 709, 719-722, 724, 749
microreactor:
microreactor: 713, 725-726
microreactors: 273, 669, 695, 713, 715, 718, 713, 725-726
microsphere: 876
microsystem: 1-3, 6-7, 10, 13, 16-17, 21, 25-26, 28, 32, 42, 44-47, 49, 93, 164, 182-183, 193, 199, 209, 215, 221, 225, 399, 403, 445, 450, 454-456, 484, 521, 597, 614-615, 667, 717, 719, 723-724, 730, 732-734, 747, 753, 799, 822, 824, 850, 853-854, 898, 902-905, 910, 934-935
microthermal analysis: 592
microthermopile: 250
microthruster: 152, 160-167, 182, 185
microtip: 273-274
microtiter plate: 668
microtransducer: 1, 3, 5, 7-9, 11, 13, 15, 17, 19, 21, 23, 25-27, 29, 31, 33, 35, 37, 39, 41-43, 45, 47, 49-51, 91, 182, 225, 278, 459, 461, 474
microtransducer operation: 1, 3, 5, 7, 9, 11, 13, 15, 17, 19, 21, 23, 25, 27, 29, 31, 33, 35, 37, 39, 41, 43, 45, 47, 49, 51

microtriangulation: 753
microturbine: 891
microturbomachinery: 753, 800
microvalve : 10, 721
microwave: 184, 225, 258, 651
milling: 741, 747, 822-823
miniaturization: 7, 46, 346, 361, 484, 488, 523, 526-527, 667, 669, 689, 692, 715, 724, 729, 759, 867, 908
minimal degree reordering algorithm: 104
minimally invasive surgery: 736-737
minimum breakdown voltage: 761, 763
minority carrier: 519
mirror curvature: 353-355
mirror optics: 251
mirror ringing: 377
MIS: 282
mixed-energy-domain system simulation: 215
mixing: 118, 123, 267, 545-546, 669, 679, 713-714, 725
Mo layer: 284-286, 288, 309, 311
mobility: 171, 230, 232, 234, 241, 294, 298, 316, 332, 338, 465-468, 470-471, 476-477, 482-483, 490, 497-498, 639-641, 913
mode matching: 389, 432-433
mode mismatch: 431
mode shape: 543
mode-matching: 431
model order reduction: 138, 149, 151, 163-164, 182, 192, 205
modeling: 26, 112, 73, 136, 154-155, 158, 161-164, 171, 178, 182-183, 185-186, 190, 192-193, 205, 207-209, 215, 217, 219, 224, 226-227, 246-247, 264, 299, 302, 340, 342-343, 682, 803, 864-865
modulating transduction effect: 11
modulation transfer function : 282
modulator: 269, 402, 440, 774, 876, 925-931, 941-942
Moeller-Plesset perturbation theory: 114
MOEMs: 183, 227, 394-398, 401, 403, 450, 604, 751, 867, 898-899
molar absorptivity: 600
mold insert: 692-693, 854, 864, 867-868, 896
molding tool: 856, 865, 869, 874
mole fraction: 576
molecular dynamics: 103, 109, 114-117, 120-121, 158, 183
molecular recognition: 571, 707
molecular simulation: 110-111, 145, 158, 183
molecular-based simulation: 96
molecular-beam epitaxial growth: 476
moment of inertia: 194, 197, 349
MOMS: 403
monochromatic radiation: 600
monocrystalline silicon: 247-248, 535, 556, 848
monolayer: 169
monolithic: 49, 168, 243, 277, 386, 388, 390, 399, 401, 449-450, 453, 527, 531, 537-539, 550, 557-558, 564-566, 586, 588, 591, 610, 720, 738, 745-746, 798, 851
monolithic integration: 168, 243, 386, 388, 531, 537-539, 549-550, 557-558, 588, 590, 745, 851
monolithic oscillator circuit: 586
Monte Carlo simulation: 116, 120
morpholin: 860
MOS: 48, 50, 342, 474, 518, 521, 613, 624, 626-627, 632-633, 770, 913, 936

INDEX 957

MOSFET: 298, 453, 471, 474, 491, 625-628, 632-633, 635-636, 706, 771, 937
most significant bit: 924
motion: 24-25, 44, 47, 81,107, 116, 170, 180, 194-195, 197, 200, 203, 207, 209-211, 215-218, 226, 281, 356, 360-361, 364, 388, 399-400, 495, 525-528, 539-544, 547, 549-550, 586, 590, 672, 675, 685-688, 698, 711, 738-739, 766, 769-770, 776-777, 780, 786, 790-791, 794, 801, 879
movable electrode: 766-770, 774, 776
moving front method: 148
multi-project-wafer (MPW): 99
multicollector: 489
multicomponent analysis: 571, 650
multiexposure: 885, 888
multifunctional integrated film: 738
multifunctional transducer: 229
multigrid method: 106
multilamination: 713-714
multimode: 440-441, 801, 874
multiple layer: 781, 787
multiple measurement: 41
multiple reflection: 405, 608
multiplexer: 252, 331-332, 334-335, 343
multipole expansion: 109, 135
multisensor chip: 268, 270, 520
multistacked poly-Si: 80
multiuser process: 80
MUMPS process: 69, 99
mylar: 338

N

n-i-p photodiode: 294
n-polysilicon: 47, 247, 252, 267-268, 271
Nafion: 572, 623
nanocrystalline: 643
nanoindenter: 63, 68-69
NaOH: 807
naphthalene: 614
native oxide: 224, 808
Navier-Stokes equation: 680
navigation system: 454, 526-527
negative temperature coefficient: 592
negative tone SU-8: 885
NEMS: 93, 111
NEP: 18, 247, 253
Nernst equation: 580, 617, 623-624, 629, 648-649
network model: 682
neural network: 909
neural probe array: 6
neuromagnetism: 454
Ni electroplating: 889
nickel: 69, 89, 91, 231, 238, 341, 525, 648, 738, 854, 857, 864, 868, 874, 876, 878, 889, 891, 893-894, 898
nickel resonator: 69
NiCo: 385, 864
NiCr: 258, 560
NiFe: 382, 857, 864
NiTi: 732, 735, 739, 743, 753
nitric acid: 814
nitrogen dioxide: 572, 585
nitrogen oxide: 571, 626, 640, 643-644, 648
node: 108, 122, 125, 127-128, 148, 166-167, 216, 221, 315-316, 318-319, 325, 327-328, 330, 334
noise: 2, 4, 16-18, 21-26, 42-44, 49, 98, 224, 246-247, 253, 260, 263, 288-289, 315-320, 322-324, 340, 342, 458, 462, 473, 475-476, 479-480, 484, 487, 492, 494, 498, 504, 515, 517-520, 532, 543-545, 550, 565, 569, 574, 613, 709, 879, 909-910, 912-918, 921-922, 924-927, 929-931, 936-937, 939-940
noise corner frequency: 914
noise equivalent power: 246
noise shaping: 924-925, 927
noncentrosymmetric: 581, 592, 787-788
nondestructive testing: 281
nonlinearity: 16, 33, 37, 42, 105, 136, 209, 243, 458-459, 473, 483, 492, 503, 507, 514, 558, 918
nonoverlapping clock: 914
nozzle: 162, 165, 698-699, 709, 712, 720, 822, 824, 891
nozzle plate: 891
NTC: 592
nuclear: 3, 6, 47-48, 456, 600, 895, 937
nuclear photovoltaics: 47
numerical aperture: 432-433
numerical instabilities: 142
Nusselt number: 684
Nyquist: 21-22, 921-924
Nyquist rate: 921-923

O

objective function: 138, 140, 144, 172-173
octree: 148
ocular pressure: 526
off-diagonal: 200, 203, 213-214, 218
offset: 16-20, 26-32, 34-35, 37-39, 42-43, 49-51, 171-174, 186, 201, 349, 360, 367, 385, 450, 458, 462, 473, 475-480, 492, 499, 504, 507-510, 512, 515-516, 518, 521, 532, 557-558, 712, 902, 910, 912-917, 921, 936-937
offset reduction: 50-51, 186, 507-509, 520-521
ohmic contact: 235, 286, 290
oligonucleotide: 718, 725
on/off current ratio: 337
Onsager relation: 40
op-amp: 330, 506, 512, 514
operation: 1, 3, 5-7, 9, 11, 13, 15-17, 19, 21, 23, 25-27, 29, 31, 33, 35, 37-39, 41, 43, 45, 47-49, 51, 69, 97-98, 161, 163-164, 168, 178-179, 185, 205, 229, 238, 257, 262, 264-265, 289, 296-299, 318-319, 322, 327-331, 339, 342, 356, 368, 372, 377, 388, 392, 438, 440, 450, 456, 463, 466, 470, 476, 488, 491, 494, 502, 507, 526, 532-534, 537, 540, 543-544, 550, 553-554, 561-563, 573, 589, 595, 628, 633, 642, 648, 670, 730, 751, 754-755, 764, 766, 769, 774, 781, 791, 822, 891, 908, 911-912, 914, 917, 924, 931-932
operational amplifier: 617, 914-916
optical absorption: 281, 445
optical amplifier: 450
optical bench: 798, 869, 876-878, 881, 883-884
optical crossconnect: 395-397
optical detection: 60, 529, 590, 674
optical fiber: 365, 395-396, 399-400, 402, 432-433, 614, 654, 873, 881, 887-888
optical function: 867, 874
optical intensity: 410, 418, 434-435, 443
optical MEMS: 345, 384, 393-404, 446, 448, 898
optical sensing: 388
optical switch: 353, 370, 377, 380, 399-401, 798
optical waveguide: 405, 430, 602, 607

optimization: 26, 94, 138-146, 148, 153, 162, 171-173, 176, 182, 184, 186, 229, 238, 246-247, 252, 260, 276-277, 309, 494, 846, 864, 884, 904
optimization problem: 138
opto-mechanical effect: 445
optode: 606
ordinary differential equation: 95, 327
organophosphorous compound: 630
orientation: 2, 4, 75, 180, 383, 461, 543, 551-552, 583, 791, 805-807, 821-822, 840, 846, 849
oscillator: 540, 559, 584, 586, 920-921, 938
osmosis: 753
out-of-plane force: 790
output domain: 12
output impedance: 458, 463
output signal: 7-9, 12-13, 16-17, 19-21, 32, 34, 38-40, 286, 320, 457, 459, 461-463, 470, 484, 488, 497, 503, 507, 510, 512, 515, 558, 752, 910-912, 919, 921, 927, 931
output transducer: 9, 18
oven: 562
overgrowth: 833, 851, 864
overload: 17, 531, 534, 562-563
overload and overrange protection: 562
overpressure load: 551
oversampling: 44, 922, 924-925, 928, 931, 941-942
OXC: 353, 355
oxy-nitride: 294
oxygen: 45, 75, 80, 239, 289-292, 534, 571-572, 594, 597, 608, 623, 627, 635, 643-645, 647-649, 738, 810-811, 816, 826, 835, 908
oxygen content: 75, 907
oxygen diffusion: 291
oxygen sensor: 643, 646-647
ozone depletion: 909

P

p-i-n a-Si:H photodiode: 282
p-polysilicon: 47, 267-269, 271
PA: 857
pacemaker: 731
Padé approximants: 150, 152
palladium: 627, 634, 636, 643, 648
palladium/polyaniline: 634
parallel-field Hall microsensor: 519
parallel-plate capacitor: 202, 208, 296
parameterized model: 209
parasitic: 7-8, 18-21, 27-29, 35-36, 38, 41-42, 283, 312, 316, 318, 338, 462-463, 537, 550, 558, 770, 903, 920, 937
parasitic back channel: 312
parasitic capacitance: 283, 337, 770, 937
parasitic sensitivity: 19
parasitic signal: 19, 28, 41
parsing: 101-102
partial differential equation: 96, 122
partial pressure: 576
particle image velocimetry: 704
particle trapping: 689
partition coefficient: 578, 581, 597
Paschen curve: 674, 761-762, 764
pass transistor: 332
pass-transistor logic: 331-332
passivation layer: 273-274, 294
passive instrument: 736
passive pixel: 314, 316, 343

path length difference: 410
path-following methods: 106
patient care: 729
pattern recognition: 571, 650
PC: 296, 7, 791, 854, 857
PCI bus: 933
PCR: 713, 746
PE: 857
PECVD: 85-86, 267, 285-286, 299, 335, 606, 609-610, 808, 828-830
PEEK: 854
pellistor: 49, 239-240, 592-593
pellistor chip: 240
Peltier: 14, 34, 243, 249, 262, 277
penicillin: 598, 630, 631
penicillinase: 631
peripheral: 1, 39
peristaltic sequence: 702
permalloy: 362, 379, 876, 879, 881
permittivity of free space: 296, 350
PET: 335, 735
petroleum industry: 524
PEUT: 581-582, 645
pH: 6, 51, 182, 225, 278-279, 340, 464, 564-565, 624, 626, 629-632, 636-637, 641-642, 647-648, 653, 686, 701, 730, 801-802, 846
pH sensor: 628
pharmacological therapy: 742
phase condition: 407, 411-412
phase gratings: 611
phase jitter: 545
phase noise: 544
phase shift: 4-5, 72, 407, 410-411, 417, 444, 607
phase velocity: 559
phase-selective demodulation: 545
phosphorescence: 601
phosphoric acid fuel cell: 46
phosphosilicate glass: 380
photocharge: 295
photochemistry: 15, 712
photocurrent: 282, 293, 316, 318, 636, 816
photodetector: 282, 390, 435, 437, 610
photodetector: 22, 339, 282, 390, 404, 435, 437, 444, 449-450, 452, 610
photodiode: 14, 282, 290-291, 293-296, 314-316, 320-321, 339-340, 342, 499, 608, 613, 615, 874, 917-918
photoelectron: 284
photonic circuit: 421
photonic crystal: 896
photonic network: 365, 397
photoresist hinge: 359
photosensitive elastomer: 824
photovoltaic cell: 46, 816
photovoltaic effect: 8, 15
photovoltaic etch-stop: 810, 816, 847
physical magnetoresistance effect: 481
physisorption: 574, 585, 589, 591, 598, 608, 648
pick-and-place: 168
piezoelectric: 14, 47-48, 69, 92, 97, 224, 364, 372, 528-529, 532, 539-540, 559-560, 564-565, 581-587, 590, 592, 694, 701-702, 719, 722, 738-739, 743, 748, 753-755, 787-796, 798, 802-803, 901, 911, 913
piezoelectric element: 48, 790
piezoelectric sensor: 528, 592
piezoresistive effect: 11, 53, 529, 555-556, 918

piezoresistive element: 704
piezoresistive pressure sensor: 851
piezoresistor: 12, 553
pin-hooking system: 63
pipeline A/D converters: 923-925
piston: 360
Pitot tube: 704
pixel integration: 309
plasma etching: 530, 805, 818, 830-831, 837, 840, 850
plasma-enhanced CVD: 826
plasma-etch: 838
plastic: 14, 75, 335, 337-338, 343, 429, 448, 516, 537, 692-694, 743, 854, 864, 869, 897
plastic deformation: 14, 75
plastic substrate: 335, 429
plate bending: 773
platinide metal: 626
platinum: 231-232, 236, 238-240, 592-593, 627, 643, 739, 810-811, 816, 834
platinum/chromium: 816
plunger: 881, 884
PMMA: 734-735, 854, 857, 859, 863, 866-867, 869, 874-876, 891
pn junction: 14, 444-445, 530, 551, 556, 810-811, 816, 840
pneumatic: 697, 702, 712, 721, 753
$POCl_3$: 77
point contact: 235, 376
point-of-care analysis: 668, 706
polarization: 4-5, 14, 178-180, 365, 377, 385-388, 394, 407, 412-413, 420, 460, 592, 598, 600, 603, 606, 617, 627-628, 632-633, 639, 787-788, 793, 797, 871
polarization mode dispersion: 377
polarization-dependent loss: 377, 387
Polarization-dependent loss polishing: 377, 387
polishing: 695, 818, 851
poly(ethylene terephthalate): 735
poly(methyl methacrylate): 734
poly-Si: 51, 55, 61-62, 65, 69-75, 77, 79-82
polyamide: 631, 857
polyaniline: 628, 636, 640
polycarbonate: 448, 735, 854
polychrometer: 392-393
polycrystalline ITO: 289-290, 338-339
polycrystalline SiGe: 272
polycrystalline silicon: 55, 88, 90-91, 785
polyethylene: 857
polyimide: 297, 343, 398, 448, 615, 648, 723, 738, 748, 777, 803
polymer electrolyte membrane fuel cell: 46
polymer hinge: 359
polymeric coatings: 610
polymethylmethacrylate: 448, 854
polynomial correction: 932
polyoxymethylene: 854
polypyrrole: 634
polypyrrole: 636
polysilicon seed: 835, 836
polysiloxane: 621
polystyrene: 448, 631, 735
polysulfone: 854
polytetrafluoroethylene: 734
polythiophene: 640
polytope: 123
polyurethane: 571, 631, 731
polyvinylidene fluoride: 738

pop-up: 361, 380
pore diameter: 818
pore tip: 818
porous silicon: 258, 554, 566, 799, 814, 816-818, 834, 845, 848-849, 851
porous thin metal film: 645
position: 4, 10, 25, 27, 43, 63, 107, 109, 162, 196, 219, 229, 251-252, 281, 298, 351, 354-355, 372, 376, 382-383, 388, 396, 435, 437, 455, 494, 501-502, 514, 521, 524-526, 532, 535, 542-543, 553, 561, 605, 613, 615, 681, 712, 743, 756, 766-768, 770, 776-778, 782, 792, 841-842, 873-874, 876, 879, 881, 891, 893
position control: 525
position-sensing photodiode: 614
position-sensitive detector: 874
positive temperature coefficient: 592
post-processing: 56, 101, 109-110, 159, 215, 266-268, 275, 821-822, 824, 834, 839, 841, 843-844, 856, 902, 925
potassium iodide: 636
potential energy: 25, 112, 117, 169, 210, 213, 681, 756
potential energy minimum: 169
potentiometric sensor: 604, 636
potentiometry: 616-617, 623, 625, 639, 648
potentiostat: 619
powder blasting: 823-824, 850
powder-injection molding: 693
power consumption: 17, 42-43, 361, 386, 527, 557-558, 644, 697, 699-701, 709, 908, 913, 925, 930
power content: 10, 38
power source: 44, 259, 456, 810-811, 816
power splitting ratio: 440
preamplifier: 917, 937
precision-machined device: 523
premature electrostatic breakdown: 764
preprocessing: 101-102, 573, 821, 841-842
pressure: 4, 11-12, 14-15, 18, 27-31, 50, 53, 58, 60, 62, 67, 79, 84, 91, 122, 193-194, 229, 255-256, 264-265, 270, 278, 286, 311, 444, 449, 452, 455, 457, 462-463, 495, 524-527, 530, 550-561, 566, 577-578, 588, 610, 670, 672-673, 677, 679-683, 685-688, 693, 695, 697-704, 709, 716, 721, 738, 751, 761, 763-765, 787, 791, 797, 802, 805, 807, 822, 826, 829-830, 848, 864, 891, 906, 908, 911, 919, 939
pressure drop: 679, 682, 698, 891
pressure sensor: 50, 452, 530, 551-553, 558, 565-566, 805, 848, 850, 938
priming: 703, 720, 722
principle of duality: 598
printed circuit board: 901
prism coupling: 432
process condition: 54
process control: 667, 754, 822, 902, 909
process design: 188
process domain simulator: 94
process flow: 188, 190
process monitoring: 55, 57, 571
process simulator: 97
projection display: 365, 399, 772
propagation constant: 414, 428
propane: 595
proportional loop regulator: 533
proportional-differential loop regulator: 533
prosthesis: 936
protocol: 693-694, 715

proximity: 4, 10, 106, 268, 431, 438-443, 455, 483, 495, 593, 837
proximity coupler: 438-440, 442-443
pseudomonas putida: 614
PSU: 854
Pt counterelectrode: 810
PTAT: 242
PTC: 592
PTFE: 734
pull-in: 351, 356, 365, 367, 377, 388, 394, 767-769, 771-772, 774-776, 782-783
pull-in voltage: 768, 771-772, 775-776
pull-up: 776
pull-up voltage: 776
push-pull: 796
PVDF: 738, 789-790, 854, 857
pyrex: 530-531
pyroelectric effect: 592
pyroelectric infrared detector: 35
PZT: 390, 528, 539-540, 560, 588, 789-794, 796, 803, 857

Q

quadrature discrimination: 544
quadtree: 148
quality factor: 69, 541, 543-544, 551
quality management: 896
quantization error: 921-922
quantization noise: 921-922, 924-927, 939
quantum chemistry 112-114, 117, 120-121, 181
quantum efficiency: 18, 282, 295, 604, 608
quantum mechanics: 3
quantum molecular dynamic algorithm: 116
quantum well: 450, 476, 518-519
quarter-wave plate: 365
quartz: 528, 539-540, 560, 564, 570, 581-582, 585, 611, 694-695, 787, 821, 857
quasi-Newton method: 106
quasi-stationary: 103

R

radiant flux: 498-499
radiant signal: 3, 5-6
radiation: 2, 5-6, 8, 10-11, 14-15, 18, 21, 25-26, 29-31, 34, 42, 46-48, 74, 98, 229, 235-239, 246, 250-252, 268-269, 281, 339, 457, 592, 599-603, 605, 609, 611, 674, 752, 854, 859, 865, 884-885, 895-897, 906, 911
radiation field: 2, 48
radiation power: 236-238
radio-frequency identification: 47
radio-frequency MEMS: 751
radioactivity: 674
radiography: 319, 324, 331
Raman scattering: 600
range: 74
range of motion: 356, 766, 769-770, 776, 786
rapid thermal process: 79
raster scanning: 348, 362, 823
rate constant: 111, 119, 575
rate of rotation: 527, 540
ray optics: 884
ray tracer: 98
Rayleigh dissipation function: 210
Rayleigh length: 373-374

Rayleigh scattering: 436, 600
Rayleigh surface acoustic wave: 582
reaction injection molding: 854
reaction rate: 6, 814
reactive ion beam etching: 696
reactive ion etching: 267, 744, 818, 850
read-write head: 525
readout mode: 321
recombination: 241, 295, 340, 444, 485, 488
rectangular membrane: 58-60, 62, 83, 87, 92
redox potential: 6, 625
reduced-order modeling: 185, 190, 205, 224
reentrant barb: 740, 742
reference beam: 450
reference electrode: 619, 621, 625-626, 628, 636, 647, 814
reference oscillator: 584
reference voltage: 910
reflectance: 6, 437, 442, 602-603
reflection: 371, 386-387, 389-390, 405-407, 409, 411-413, 415, 417, 433, 442, 599, 602-603, 613, 876, 884
refraction: 599, 602, 869, 884
refractive dielectric constant: 65, 296
refractive index 389, 405, 406-407, 412, 417, 420, 884
relative sensitivity: 457, 459
relaxation generator: 509
relaxation time: 231, 234
reliability: 55, 62, 89, 98, 230, 276, 299, 335, 376, 381, 456-458, 515, 523-524, 527, 531, 562-563, 571, 644, 703, 708, 739, 754, 772, 896, 902, 905-906, 908, 927, 932
repeatability: 16, 26, 362, 374, 376, 380, 458, 462, 800
replication: 452, 692-693, 719-720, 853-854, 864-865, 896
reset mode: 320
residual stress: 55, 69, 87-88, 91, 197, 224, 351, 353, 362, 851
resistance: 4-5, 22-24, 39, 72-74, 230-240, 247, 253, 257, 259-261, 265, 315-317, 321, 324-328, 330-332, 334, 471, 479, 482-483, 485, 494-495, 497-498, 503, 509, 529-530, 532, 556-557, 569-570, 592-595, 628, 638-643, 697-698, 704, 709, 754, 826, 879, 911-912, 919-920, 938
resistor: 21-24, 29, 31, 38-39, 43, 71-72, 161-164, 166, 234-240, 258-259, 269-271, 275, 326, 328-330, 332, 334, 364, 483, 510, 528, 556, 573, 595, 639, 688, 704, 920
resolution: 17, 21, 26, 44, 94, 98, 145, 238, 254-255, 279, 282, 320, 332, 338-339, 341, 346, 362, 364, 369, 399, 441, 443, 450, 458, 461, 496, 500-501, 515, 517, 525, 538, 545, 590, 637, 689, 705, 716, 738, 796, 801, 822-823, 850, 853, 864, 867, 884, 889, 897, 920-926, 930-931, 938-939
resonance frequency: 4, 34, 528-529, 533, 540-541, 558-559, 562, 570, 581, 589
resonant frequency: 60, 346, 349, 352, 354, 356, 358-360, 367, 581, 771, 781, 783, 795
resonant gate structure: 825
resonant magnetic sensor: 34
resonant-gate transistor: 771
response time: 17, 26, 236, 262, 282, 370, 388, 393, 458, 463, 633, 643
responsivity: 13, 26, 246, 260
restenotic response: 731

INDEX

restoring force: 170, 766, 768-770, 773, 776
restoring torque: 170, 350
return loss: 377
reverse bias: 242, 286, 290, 295-296, 315, 320, 340, 498
reverse reaction: 575
reversibility: 458, 461, 568, 571, 574
Reynolds number: 679
RF-ID: 47
rib waveguide: 422, 424
ridge waveguide: 422, 424
RIE: 80, 267, 606, 644, 818-819, 821, 832
rigid-body dynamic: 103
ring resonator: 543
ripple: 57, 62
ripple transition: 62
rollover protection: 527, 539
rotational damping: 214
rotor: 51, 203-205, 732, 747, 785, 891, 938

S

sacrificial layer: 269-270, 554, 595, 780, 814, 825-826, 829, 835-836, 844, 848-849, 853, 865-866, 876, 897
sacrificial layer etching: 595
sacrificial layer technique: 853, 865, 897
safety system: 527
sagging: 267
Sagnac: 404
sampling frequency: 914, 921, 924, 930
sampling time: 321, 324
saturation current: 241, 328, 496
SAW: 418, 560, 582-583, 585-587, 805, 879
scaling law: 353, 395
scan angle: 346-347, 351, 353, 356, 358-361, 367, 399, 879
scattering: 231, 234, 249, 284, 288, 436-437, 466-467, 482-483, 599-600, 602-603, 674, 853
scattering factor: 466-467
scattering of light: 600
Schottky: 282-287, 289-291, 293-294, 309, 311, 319, 339-340
Schottky contact: 290
Schottky photodiode: 282, 290, 338
Schrödinger equation: 112-113, 116
scintillation: 6
scratch drive: 380, 400
SCREAM: 830-832, 837, 845, 850
screen printing: 159, 616, 622, 637, 640, 644
second-order $\Sigma\Delta$ modulator 926-927, 929
secondary ion mass spectroscopy: 290
security system: 282
Seebeck coefficient: 39, 244, 247-250, 257, 260, 262, 264, 276, 596
Seebeck effect: 8, 14, 243, 245, 257, 277, 279, 595
seesaw structure: 388
segregation: 713
seismic mass: 34, 528-530, 532-533, 535, 547
selectivity: 16, 18-19, 392, 569, 571, 574, 576, 603, 625, 628, 641, 643, 808-809, 815-816, 818-820, 822-823, 826, 828, 830, 909
self-aligned process: 359
self-assembly: 168-169, 171, 186, 351, 354, 396
self-focusing: 874
self-generating transduction effect: 7, 12
self-oscillating reaction: 753

self-test:: 253, 529, 532, 534, 564, 798, 908, 912, 932, 935
semiconducting thermoresistor: 232, 234
semiconductor laser: 404, 431
semiconductor technology: 47, 235, 239, 263, 530, 553, 573, 592
sensactor: 10, 32-34
sensactor compensation: 32-34
sensactuator: 9-10
sense of touch: 736
sensing layer: 574, 578, 580, 584, 587, 590
sensitive area: 237, 245-246, 251-252
sensitivity: 8, 12-13, 16-20, 27-28, 31-32, 35-37, 49, 144, 172-173, 175-177, 184, 195, 230, 234, 236, 238-239, 243, 245-246, 252-253, 256-260, 262, 278, 318, 322, 342, 457-458, 470, 473, 475-477, 479-480, 483-484, 487-488, 491-495, 511-512, 515, 518-521, 528, 530-533, 535-536, 538, 542-543, 546, 557-558, 568, 571, 580, 586-587, 590, 605, 628, 630, 632-633, 641, 643, 646, 674, 716, 842, 851, 871, 889, 917, 936
sensitivity analysis: 184, 321
sensor bridge: 29
sensor cost: 523, 531, 904
sensor effect cube: 49
sensor element: 9, 36, 251, 314, 528, 545, 550, 554, 560-561, 594, 631
sensor microsystem: 10, 456, 934
sensor-actuator: 10
separation by implantation of oxide: 835
serial bus: 608, 932
Severinghaus electrode: 642
SGFET: 635, 637
shape factor: 346
shape function: 197
shape memory alloy: 748, 799
shape optimization: 143-144, 148
shear horizontal acoustic plate mode: 582
shear stress: 4, 196, 309
shear transverse wave: 582
shock: 4, 128, 182, 377, 398, 524, 911
shot noise: 24, 316
shrinkage: 562, 693, 865, 889
shutter: 34, 386-388, 879
side-airbag: 527
sidewall: 188, 380, 415, 506, 818-819, 831, 848
SiGe: 51, 247, 272
sigma-delta amplifier: 43, 51
signal conditioning: 4, 7, 9-10, 187, 268, 573, 938
signal difference: 30
signal domains: 2, 8, 12
signal flow graph: 216
signal processing: 227, 229, 252, 276, 453, 484, 496, 499, 502-503, 507, 509, 573, 613, 906, 909-910, 912, 921, 931, 936, 940
signal transduction: 6, 630, 707
Signal-to-noise: 17, 26, 246-247, 260, 315, 462, 543-544, 550, 569, 574, 709, 925, 930
silane: 75, 79
silanon group: 686
silica glass: 695
silicate glass: 695
silicon carbide: 628, 809, 826, 828
silicon crystal plane: 807
silicon foundry: 272

silicon fusion bonding: 354, 554, 556, 611, 691, 802, 819, 849
silicon membrane pressure sensor: 11
silicon microphone: 525
silicon nitride: 60-61, 84, 86, 88, 92, 236, 240, 246, 257, 260, 266-267, 309, 337, 340-341, 362, 370, 391, 586-587, 604, 606, 614, 622, 631, 637, 739, 774, 809, 814, 826, 828, 836, 842
silicon on insulator: 349
silicon oxide: 61, 84-85, 235-236, 246, 257, 260, 266-267, 269, 274, 534-535, 554, 556, 560, 608, 628, 631, 633, 636, 809, 819, 901
silicon scanner: 345
silicon technology: 573, 690-691, 848
silicon tip: 274
silicon-on-sapphire: 488
silicone gel: 554
silicone oil: 554
silicone rubber: 631
silver sulfate: 626
silver/silver-chloride: 621
SIMOX: 832, 834-835, 845, 851
SIMPLE: 829, 830, 836, 842
simulation: 26, 54, 89, 93-99, 101-103, 105, 107, 109, 111-123, 125, 127, 129, 131, 133, 135-139, 141, 143, 145-149, 151, 153-155, 157-165, 167-169, 171, 173-177, 179-187, 189-193, 195, 197, 199, 201, 203, 205, 207-209, 211, 213, 215-217, 219, 221-227, 247, 276, 327, 332, 334, 343, 682, 693, 862, 864, 897, 942
simulation strategy: 93, 153
simulation tool: 93, 101, 175, 180
sinc: 927
single-crystal silicon: 53, 69, 89, 349-350, 354, 741, 816, 835-836, 846, 850
single-mode: 379, 384, 400, 413, 419, 434-435, 441
single-mode fiber: 379, 384, 400
single-side-cracked specimen: 68
singular perturbation approximation: 152
sintered ceramic electrolytes: 626
sintered ceramic pill: 238
size optimization: 143
slab waveguide: 409, 421-422, 426
sliding prism coupler: 437
slow-etching (111) plane: 812
SMA: 738-739, 799
smart actuator: 10
smart mote: 17, 44
smart sensor: 10, 483, 573, 850, 942
SnO_2: 571-572
soft lithography: 692
SOI: 76, 349, -355, 359-360, 362, 364, 369, 384-386, 390, 396, 400, 518, 551-552, 554, 556-557, 560, 810, 836, 839, 845, 851, 937
sol-gel deposition: 796
solder sealing: 614-615
solid fuel: 160, 162, 164-165
solid model: 190-191
sorption constant: 578
source follower circuit: 320
source-drain current:: 298, 625-627
space charge region: 495, 628
spark: 908
sparse matrix: 103, 105

specific heat:: 55, 70, 74
specifications: 27, 165, 188-189, 191, 215, 219, 526, 730, 796, 864, 904, 907-910, 912-913
specificity: 569, 707
spectral energy density: 2
spectral equalization: 391
spectral noise power density: 21-22, 24
spectral response: 293-294, 367, 369
spectroscopy: 290, 340-341, 388, 600, 602-603, 938
SPI bus: 933
spinning-current offset reduction: 508
spinning-wheel gyroscope: 540
split beam: 812
splitter: 881, 884
spreading contact: 233, 235
spreading resistance: 233-234
spring constant: 25, 196-197, 203, 349-350, 528
spring-mass system: 528, 531, 540, 542
spur component: 914
sputtering: 82-83, 231, 251, 286, 290, 584, 595, 622, 637, 640, 644, 647, 796, 803, 823, 829, 854, 865
squeeze-film damping: 194, 214
stability: 17, 106, 142, 276, 283, 288-289, 330, 332, 339, 362, 367, 532, 556, 709, 768, 770, 783, 826, 908, 927, 931
stabilization: 36, 441, 507, 516, 637, 640, 647, 801, 896
stable solution: 768
staebler-wronski effect: 300
staggered-electrode structure: 296
state function: 568, 575
Statistical Mechanics: 49, 114-115, 117
statistical physics: 19, 23
stator: 204, 785
steer-by-wire: 561
steering angle: 524, 561
Stefan-Boltzmann law: 236
stent: 731-732, 735
stepper motor: 755, 773, 787, 792, 802
sterilization: 730, 739
Stern layer: 686
stiction: 376, 383, 391, 534, 565, 670, 836-837
stiffness: 66, 68, 74, 170, 193, 196-197, 203, 205-207, 210-214, 533, 766-767, 776-777, 779, 782
stiffness matrix: 205, 207, 212
stochastic signal: 16, 24
stoichiometric number: 576-577
stopper: 376, 380
storage capacitance: 319, 333
strain: 4, 11, 18, 56-59, 61-62, 66-67, 69, 86-87, 90, 175-176, 179-180, 194, 206, 337, 364, 444, 528, 530, 551, 555-556, 559, 561-562, 753, 787-792, 797, 800, 911, 919, 940
strain energy: 206, 790
strain gauge: 561, 800, 939
strain-stress curve: 59
streaming potential: 687-688
strength measurement: 67, 90
streptavidin: 607
stress: 4, 14, 31, 39-41, 50, 56, 58-62, 66-69, 75, 77-81, 87, 90-91, 96-98, 110, 143, 157, 175-176, 180, 196-197, 224-226, 255, 267, 275, 286, 299, 301-307, 309-312, 330, 335, 337-338, 340-341, 353-354, 377, 389, 391, 400, 428, 476, 510, 516, 521, 528-

INDEX

529, 537, 556, 558-559, 561, 581, 589-590, 592, 690, 693-694, 696, 730, 735, 754, 787-791, 793-794, 809, 826, 836, 851, 919, 940
stress concentration: 67-68
stress distribution: 55, 110, 143
stress gradient: 80, 389, 754
stress relaxation: 79
stress-concentration problem: 68-69
stripe waveguide: 421-422, 426-427
stripping voltammetry: 621
structural material: 53, 55, 58
structured design: 187, 191, 193, 215, 221, 224-225
SU-8: 725, 734, 747, 853, 860, 862, 867, 885, 891, 896-897, 899
subranging: 923-925
sucrose: 607
Suhl-Shockley effect: 485
sulphur dioxide: 626, 648
SUMMiT-V: 367
supercritical drying: 534, 837
superlattice structure: 476
superposition: 3, 134, 176, 485, 869
suppressed-side-wall injection: 509
surface acoustic wave: 558-559, 566, 583, 585, 803
surface leakage: 241
surface micromachining: 51, 56, 91, 99, 269, 271, 349, 362, 534, 538, 551, 553, 595, 609, 636, 738, 744, 805, 824-825, 834-836, 840-842, 845, 850-851
surface plasmon resonance: 603, 613, 615
surface quality: 346, 691
surface tension: 137, 158, 359, 397, 534, 669, 673, 685, 708-709, 814, 832, 837
surface treatment: 631
surface-drive actuator: 786
surface-micromachined pressure: 30, 68-69, 88, 187, 351-355, 377, 380, 382, 394-396, 451, 533, 535-537, 539, 541-542, 546, 548-550, 553-555, 557, 781, 785, 824
surface-micromachining: 53, 197, 353, 365, 367, 386, 388, 391, 539, 547, 564-565
surface-near-bulk-micromachining: 539
surface-to-volume ratio: 643, 671, 673, 754
surgical system: 729, 735-736, 749
surveillance: 252
suspended membrane: 236, 240, 252
suspended-gate FET: 634
suspension: 70, 72, 96, 367, 564, 782-783
SVGA: 364
swelling coefficient: 646
switch: 50, 164, 281, 309, 315, 319-321, 325, 336-338, 353, 355-356, 361, 366, 368-374, 376-381, 383, 385-386, 396, 399-400, 440, 444-445, 484, 495, 521, 668, 701, 818, 876, 899, 924
switched-capacitor: 43, 645, 913, 920, 936, 938, 941
switched-current: 917, 936
switching: 40-41, 43, 281, 315, 332, 339, 346, 355, 365, 367, 372-374, 376-377, 380-381, 383, 385, 387, 396-397, 399-400, 441, 507, 538, 597, 698, 701, 709, 795, 801-802
switching characteristic: 281, 332
switching time: 355, 365, 367, 377, 380, 385, 387, 400
symbolic tool: 94, 99
symmetric waveguide: 411, 413-414
symmetry: 40, 212-213, 215, 235, 246, 507, 521, 543

synchrotron: 14, 853-854, 859-860, 865, 884-885, 895-897
system design: 42, 188-190, 192, 218, 220, 527
system identification: 205, 225
system modeling: 224
system on a chip: 722
system simulator: 159
system-level simulation: 149, 155, 187, 189, 191, 193, 195, 197, 199, 201, 203, 205, 207-209, 211, 213, 215, 217, 219, 221, 223, 225, 227

T
tachometry: 455
tachycardia: 908
tactile sensor: 748
tandem transduction: 11, 752
tangential-drive: 785
tantalum oxide: 608, 631, 637
taper: 434
target molecule: 709, 712
TbFe: 364
TE: 47, 4-408, 411-413, 420, 428, 604-606
technology: CAD: 98, 152
teflon: 534, 735, 803
telecommunication: 346, 352, 394, 867, 898
telemetric sensing: 557
telemetry: 46, 908, 936
TEM: 407
temperature coefficient: 39, 74, 233-234, 237, 241-242, 260, 462, 491, 496, 502, 592
temperature coefficient of resistance: 39, 74
temperature coefficient of responsivity: 260
temperature compensation: 250, 509-511, 520, 531-532, 556-557, 559, 561-562, 571, 890
temperature dependence: 19, 27-28, 31, 36, 232-235, 238, 241, 341, 446-447, 483-484, 488, 494-495, 497, 509-510, 512, 515, 557, 571, 585
temperature drift: 270, 488, 509
temperature gradient: 8, 38, 40, 244, 249, 271, 276, 596, 714
temperature sensor: 28, 30, 38, 70-71, 231, 234, 239, 241, 258, 269, 276-277, 510, 705, 936
temperature stabilization: 36, 636, 639, 646
tensile strength: 65, 85, 88, 90
tensile stress: 56, 59-60, 67, 75, 79-80, 87, 197, 309, 354, 391, 555, 826, 836
tensile testing: 58, 61, 63, 66, 88-89, 91
tensor notation: 789
Tesla valve: 698
test structure: 55-57, 63, 71, 89, 275-276
tether beam: 547
tetra-methyl-ammonium hydroxide: 807
TFT: 281, 296-299, 301-304, 306-309, 312, 314-338, 342-343
therapeutic: 729, 742-744
thermal actuator: 66, 88, 708
thermal capacitance: 24
thermal capacity: 236
thermal conductivity: 5, 18, 30, 55, 70, 74, 82,
Thermal diffusivity: 72, 74
thermal equilibrium: 23
thermal expansion: 14, 91, 145, 184, 353, 388, 446-447, 560, 562, 672, 753
thermal expansion coefficient: 74-75, 145, 335, 691, 694-695, 889

thermal gas flow sensor: 29, 31
thermal generation: 289
thermal insulation: 236, 240, 245, 257-258, 267-268, 270
thermal mass: 24, 642
thermal modulation: 229
thermal noise: 21-22, 43, 316-317, 320
thermal Pressure Sensor: 50, 30
thermal property: 71
thermal radiation: 25, 47, 235-236, 246
thermal resistance: 5, 24, 236-240, 257, 260-261
thermal signal: 3, 5
thermal: SiO_2 85
thermal time constant: 5, 38
thermal-based microsensor: 226, 229-231, 233, 235, 237, 239, 241, 243, 245, 247, 249, 251, 253, 255, 257, 259, 261, 263, 265-267, 269, 271, 273, 275-277, 279
thermionic emission: 284
thermo-optic effect: 445
thermoconverter: 258-259, 268
thermocouple: 243-244, 262, 264, 274
thermocycling: 277, 684
thermodynamic degree of freedom: 23
thermodynamic equilibrium: 47, 577-578, 580, 591, 622, 673
thermoelectric: 18, 22, 30, 35, 38-41, 47, 51, 229, 236, 243-248, 250, 252-256, 258, 260, 262-265, 268-277, 279, 529, 592, 595-598, 911, 913-914, 916, 937
thermoelectric coefficient: 18
thermoelectric cooler: 262-263, 279
thermoelectric efficiency: 247, 264
thermoelectric generator: 271-272, 279
thermoelectric junction: 260
thermoelectric leg: 250
thermoelectric material: 18, 47, 252, 279
thermoelectricity: 243
thermometry: 267, 592
thermophotovoltaic: 47
thermophysical property: 185, 259, 592
thermopile: 22, 245-246, 250-255, 257-261, 264, 267-268, 279, 596
thermoplastic: 359, 692, 720, 865
thermoresistive: 229, 236
thermosensitivity: 497
thick-film conductor: 615
thickness shear mode resonator: 581
thin film: 54-60, 63, 65, 67, 70, 74, 87, 90, 194, 279, 338, 339-341, 343, 364, 529, 560, 602, 753, 790, 799, 803, 805, 824, 825
thin-film membrane: 29, 58, 60
thin-film resistor: 236-237, 271
thin-film transistor: 281, 296
Thomson: 14, 243, 249-250, 277, 719
Thomson coefficient: 250
Thomson heat: 249-250
3ω method: 72
three-phase contact point: 685
threshold voltage: 282, 299, 301-302, 316, 321-322, 325, 330, 332-333, 342, 627
throttle: 709
tilt: 3-4, 21, 170, 374-375
tilt sensor: 3
time stop: 809

time-of-flight method: 705
time-to-market: 538
tin dioxide: 571, 638, 642
TiO_2: 389, 572,388,
tip deflection: 793-794
tire pressure monitoring: 558
tissue glue: 740
tissue repair: 739
titanate: 390, 528, 588, 592, 789, 857
titanium membrane: 857, 859
TM: 407, 411-413, 420, 428, 604-606
TMAH: 807-808, 826, 838, 846-847
TO-5 housing: 252
toluene: 597, 613-614, 645-646
top-down design: 191
topography: 4, 159
topology optimization: 143-146, 171-173, 182, 184, 186
TopSpot principle: 711
torque: 169, 219, 221, 349-350, 356-357, 367, 524, 540, 561, 563, 755, 757, 785, 889, 891
torque sensor: 560
torsion bar: 356-357
total analysis system: 94, 717, 726
total internal reflection: 407, 417, 602
total reflection: 602-603, 884
toxicity: 743, 808
tradeoff: 143, 155, 170, 192, 318, 320, 323, 328, 335, 351, 362, 515, 766, 779, 783, 904-905
transconductance: 43, 322, 913
transconductance-capacitor filter: 43
transdermal delivery: 744, 746
transduction: 2, 6-14, 18-22, 38, 209, 284-285, 339, 457, 485, 488-489, 570, 583, 586, 589, 592, 595, 603, 608, 610-611, 613, 617, 622, 626-628, 631-632, 634, 637, 643, 706-707, 716, 751-755, 795, 797
transduction effect: 11-13
transfer function: 18, 20, 26, 152, 205, 282, 392, 915-916, 925, 927
transfer of momentum: 677
transimpedance: 917, 937-938
transistor: 48, 50, 94, 216, 218, 241-243, 253, 297, 335, 337, 341, 343, 471, 474, 488-489, 573, 624, 627, 706, 801, 831, 846, 850, 913
transition metal oxides: 592
transmission grating: 611
transmittance: 6, 289
transresistance gain: 917, 919
transverse electric: 407, 604
transverse electromagnetic: 407
transverse magnetic: 407, 604
transverse voltage: 51, 464
tree-based decomposition: 148
trench: 354, 380, 817, 819, 822-823, 830-832
triangulation: 874
trimming: 28, 232, 543-544, 566
truncated balanced approximation: 152
tunable filters: 388-390, 394
tunable laser: 390, 402
tungsten: 231, 843-844
tunnel effect: 11, 14
tunneling: 49, 282, 289, 291, 529
tunneling current sensing: 529

INDEX 965

turbine: 891-892
twisting: 364, 794
two-axis gyroscope: 547
two-dimensional array: 178, 252, 709, 777
two-dimensional electron gas: 476

U

ultrasonic cutting tool: 793
ultrasonic distance measurement: 526
ultrasound emitter: 10
underetching: 246, 267, 269, 273, 530, 830-831, 835-836
unijunction magnetotransistor: 503
unity-gain configuration: 330
urea: 592, 598, 648
uric acid: 592
UV lithography: 696, 853, 857, 867, 887-888
UV-LIGA: 853, 867-868, 891

V

v-groove: 451
vacuum cleaner: 526, 908
vacuum level: 633
vacuum sensor: 255-256, 269, 279
valence band: 234, 300, 604, 813-814
valve: 691, 697-698, 700-702, 709, 720, 753, 850
van der Pauw: 71-72, 89
van der Waals adhesion forces: 534
van der Waals interactions: 691
vapor etching: 836-837
vapor pressure: 673, 681
vapor sensor: 582
vapor-phase etching: 534
vaporization heat: 597
variable optical attenuator: 346, 365, 386, 391, 401
variable reluctance: 881
variable-capacitance: 778
variable-reactance actuator 778
vascular anastomosis: 740
VCSEL: 132, 178, 182, 390, 402, 608
vehicle dynamics control: 523, 527, 539
velocity: 4, 6, 137, 171, 194, 198, 200, 210, 455, 463-464, 466, 471, 478, 481, 490, 495, 502, 527, 541, 543, 584, 586, 588, 603, 676-677, 679, 687, 689, 697, 773, 824
velocity of sound: 584, 588
vertical cavity laser: 178
vertical comb: 356-360, 367, 369, 397, 399, 783
vertical comb drive: 351, 356, 397
vertical force: 66
vertical TFT structure: 338
vertical-cavity surface-emitting: 390, 402
vibrational energy: 47, 601
video processing: 924
virtual reality: 525, 539
viscosity: 111, 194, 580, 670, 675-677, 687, 689, 797, 906, 911
viscous damping: 199, 210, 225, 551
viscous force: 679
VLSI: 136, 224, 227, 653, 747, 901, 938
volatile organic: 29, 797
Volta potential: 634, 636
voltage divider: 325, 328, 769-770

voltage-related sensitivity: 458
voltammetry: 617-618, 639
volume-of-fluid (VOF) method: 136
Voronoi mesh: 127
VT shift: 283
VXI bus: 933

W

wafer scale: 354
wafer-level encapsulation: 537
water vapor: 7, 46
wave-front: 10
waveguide: 403-405, 407, 409-420, 422, 425-426, 428-445, 449-450, 452, 602-608, 874
waveguide core: 410, 413, 416, 418, 420, 422, 429, 433, 441, 443
waveguide loss: 437-438
waveguide thickness: 412, 417, 420, 434
wavelength: 4, 46, 236-237, 293, 346, 352, 365-370, 373, 377, 386-388, 390-391, 393-394, 397, 399, 402-403, 405, 410, 412-413, 415, 417, 420, 432, 434-435, 437, 441-443, 445-447, 449, 498, 583-584, 586, 599-600, 602-605, 608-610, 613-615, 859, 871, 874, 881, 884, 896, 898
WDM: 346, -368, 386, 388, 390, 394, 399
wetting: 685, 734, 837
Wheatstone bridge: 31, 237-239, 484, 555-557, 595, 639, 919, 938
white good: 908
white noise: 22-23
Wiedemann-Franz law: 83
wire bond sensor: 174
wireless energy transfer: 563
wireless LAN: 933
work function: 286, 617, 634-636
working electrode 619, 621
working fluid: 706, 722
working point: 18-19, 33-34, 36, 509

X

X-ray: 230, 281-284, 286-289, 311, 313-314, 316, 319, 324, 331, 339, 342, 734, 853-855, 857, 859-860, 862-864, 866, 884-889, 891, 895-897, 899
X-ray imaging: 281, 283, 314, 316, 339, 341
X-ray lithography: 230, 854-855, 857, 859-860, 885, 887-889, 891, 895, 897
X-ray microscope: 896
X-ray mirror: 859
X-ray sensitive polymer: 859
X-ray telescope: 884

Y

Y-coupler: 438-439, 444
yield strength: 55, 62
yoke: 879

Z

Zeeman laser: 606
zinc oxide: 528, 560, 581-582, 589, 592, 607, 643, 790, 901
zirconium dioxide: 571
ZnO: 84, 87, 92, 528, 590, 789-790, 792, 796, 803
ZrO_2: 571-572, 857